Physiological Tests for Elite Athletes

SECOND EDITION

Rebecca K. Tanner and Christopher J. Gore

EDITORS

Australian Institute of Sport

Australian Government

Australian Sports Commission

**AUSTRALIAN
INSTITUTE OF SPORT**

The Australian Institute of Sport is the High Performance Division of
the Australian Sports Commission.

Human Kinetics

Library of Congress Cataloging-in-Publication Data

Physiological tests for elite athletes / Rebecca K. Tanner and Christopher J. Gore, editors ; Australian Institute of Sport. -- 2nd ed.
    p. ; cm.
  Includes bibliographical references and index.
  ISBN 978-0-7360-9711-6 -- ISBN 0-7360-9711-2
  I. Tanner, Rebecca K. II. Gore, Christopher John, 1959- III. Australian Institute of Sport.
  [DNLM: 1. Physical Fitness--physiology. 2. Clinical Laboratory Techniques. 3. Exercise Test--methods. 4. Sports. QT 255]
  612'.044088796--dc23

                          2012009494

ISBN-10: 0-7360-9711-2 (print)
ISBN-13: 978-0-7360-9711-6 (print)

Acquisitions Editor: Amy N. Tocco; **Developmental Editor:** Judy Park; **Assistant Editors:** Brendan Shea, Katherine Maurer, Steven Calderwood, Susan Huls, and Anne Rumery; **Copyeditor:** Julie Anderson; **Indexer:** Bobbi J. Swanson; **Permissions Manager:** Dalene Reeder; **Graphic Designer:** Nancy Rasmus; **Graphic Artist:** Nancy Rasmus and Tara Welsch; **Cover Designer:** Keith Blomberg; **Photographer (cover):** Yuzuru Sunada; **Photographer (interior):** Chapter 11 photos courtesy of the International Society for the Advancement of Kinanthropometry (ISAK); chapter 15 photos courtesy of Damian Farrow; chapter 25 photos courtesy of the Australian Rugby Union; all other photos © Human Kinetics, except where otherwise noted; **Photo Asset Manager:** Laura Fitch; **Visual Production Assistant:** Joyce Brumfield; **Photo Production Manager:** Jason Allen; **Art Manager:** Kelly Hendren; **Associate Art Manager:** Alan L. Wilborn; **Illustrations:** © Human Kinetics; **Printer:** Edwards Brothers Malloy

Printed in the United States of America   10 9 8 7 6 5 4 3 2 1

The paper in this book is certified under a sustainable forestry program.

Human Kinetics
Website: www.HumanKinetics.com

United States: Human Kinetics
P.O. Box 5076
Champaign, IL 61825-5076
800-747-4457
e-mail: humank@hkusa.com

Canada: Human Kinetics
475 Devonshire Road Unit 100
Windsor, ON N8Y 2L5
800-465-7301 (in Canada only)
e-mail: info@hkcanada.com

Europe: Human Kinetics
107 Bradford Road
Stanningley
Leeds LS28 6AT, United Kingdom
+44 (0) 113 255 5665
e-mail: hk@hkeurope.com

Australia: Human Kinetics
57A Price Avenue
Lower Mitcham, South Australia 5062
08 8372 0999
e-mail: info@hkaustralia.com

New Zealand: Human Kinetics
P.O. Box 80
Torrens Park, South Australia 50620
800 222 062
e-mail: info@hknewzealand.com

E5232

CONTENTS

PART I   Laboratory and Athlete Preparation   1

1   Quality Assurance in Exercise Physiology Laboratories   3

2   Pretest Environment and Athlete Preparation   11

3   Data Collection and Analysis   35

PART II   Testing Concepts and Athlete Monitoring   43

4   Ergometer-Based Maximal Neuromuscular Power   45

5   Anaerobic Capacity   59

Contents

CONTRIBUTORS

Michael Blackburn; Yachting Australia

Darrell L. Bonetti; Australian Institute of Sport

Pitre Bourdon; South Australian Sports Institute; ASPIRE Academy of Sports Excellence

Matt B. Brearley; Northern Territory Institute of Sport; National Critical Care and Trauma Response Centre

Nicola Bullock; Australian Institute of Sport

Darren J. Burgess; Liverpool Football Club

Dale W. Chapman; Australian Institute of Sport

Sally A. Clark; Australian Institute of Sport

Stuart J. Cormack; Edith Cowan University; Faculty of Health Sciences, Australian Catholic University, Melbourne

Shaun D'Auria; Queensland Academy of Sport

Eric J. Drinkwater; Australian Institute of Sport; Charles Sturt University

Grant M. Duthie; Australian Institute of Sport; Newcastle Knights Rugby League Football Club

Tammie R. Ebert; Australian Institute of Sport, Cycling Australia

Damian Farrow; Australian Institute of Sport; Victoria University

Kate L. Fuller; Australian Institute of Sport

Tim J. Gabbett; Queensland Academy of Sport; School of Exercise Science, Australian Catholic University, Brisbane

Scott A. Gardner; Australian Institute of Sport; UK Sport

Laura A. Garvican; Australian Institute of Sport

Christopher J. Gore; Australian Institute of Sport

Daniel J. Green; Australian Institute of Sport

John Gregory; Tasmanian Institute of Sport

Allan Hahn; Australian Institute of Sport

Shona Halson; Australian Institute of Sport

Dean G. Higham; University of Canberra; Australian Institute of Sport; Australian Rugby Union

Stuart Karppinen; Cricket Australia

Aaron Kellett; Cricket Australia; Tennis Australia

Markus J. Klusemann; Charles Sturt University; Australian Institute of Sport; Basketball Australia

Hamilton Lee; Australian Institute of Sport

Michael J. Marfell-Jones; International Society for the Advancement of Kinanthropometry; Open Polytechnic Kuratini Tuwhera

David T. Martin; Australian Institute of Sport; Cycling Australia

Michael R. McGuigan; New Zealand Academy of Sport, North Island

Clare L. Minahan; Griffith University

John A. Mitchell; Australian Rugby Union

Paul G. Montgomery; Australian Institute of Sport; St Kilda Football Club

Mark A. Osborne; Queensland Academy of Sport

Peter Peeling; Western Australian Institute of Sport; University of Western Australia

Ted Polglaze; Western Australian Institute of Sport; Australian Institute of Sport

Marc Portus, Cricket Australia; Praxis Sport Science Pty Ltd

David B. Pyne; Australian Institute of Sport

Marc J. Quod; Australian Institute of Sport; GreenEDGE Professional Cycling Team

Claire Rechichi; Western Australian Institute of Sport; Australian Institute of Sport

Machar Reid; Tennis Australia

Anthony J. Rice; Australian Institute of Sport

Michael P. Riggs; South Australian Sports Institute; South Australian Cricket Association

Greg Rowsell; South Australian Sports Institute

Philo U. Saunders; Australian Institute of Sport

Bernard Savage; Swimming Australia

Jeremy M. Sheppard; Edith Cowan University; Australian Volleyball Federation

Narelle Sibte; Tennis Australia

Gary Slater; Australian Institute of Sport; University of the Sunshine Coast

Katie Slattery; New South Wales Institute of Sport

Matt Spencer; Western Australian Institute of Sport; Norwegian School of Sport Sciences

Tom Stanef; South Australian Sports Institute

Danielle Stefano; Victorian Institute of Sport

Frankie Tan; Western Australian Institute of Sport; Singapore Sports Council

Rebecca K. Tanner; Australian Institute of Sport

Kristie-Lee Taylor; Edith Cowan University; Australian Institute of Sport; ACT Academy of Sport

Stephen Timms; Cricket Australia; Praxis Sport Science Pty Ltd

Nicole E. Thomas; Australian Institute of Sport

Joanna Vaile; Australian Institute of Sport

Andrew Verdon; Yachting Australia

David Whiteside; University of Western Australia; Tennis Australia

Sarah M. Woolford; South Australian Sports Institute

PREFACE

At the Montreal Olympics (1976), Australia did not win a single gold medal, the first time since 1936. In an attempt to change this situation, the Australian government established a national system of state-based sport institutes. Just how successful these institutes have been in raising the level of athletic performance in the country is suggested by the fact that Australia won 9, 16, 17, and 17 gold medals, respectively, at the 1996 (Atlanta), 2000 (Sydney), 2004 (Athens), and 2008 Olympics (Beijing). Part of this success can be traced to the development of standardized test procedures and the collaboration of sport science staff from national sport institutes to allow comparison of results among laboratories.

Physiological Tests for Elite Athletes, Second Edition, contains the most current of these standardized physiological test procedures. Although other manuals of test protocols are in print, they generally address tests relevant to gymnasium and health club clients or cardiac rehabilitation patients. This volume is unique in its focus on testing the elite athlete. If you work with elite athletes or are a student who aspires to work with elite athletes, this manual will provide you with a comprehensive guide to the "how and why" of the principal physiological tests.

Physiological Tests for Elite Athletes, Second Edition, provides test protocols for the physiological assessment of elite athletes in 18 different sports and includes the rationales and normative data for these protocols. Although the protocols are used in Australia, they were developed from an understanding and scrutiny of international literature related to athlete assessment. As such, the extensive normative data for these tests provide excellent reference points for measuring elite athletes from any country. Readers will also find the reproducible forms for data collection and for preparticipation screening useful.

How This Book Is Organized

The book is divided into four sections. Part I, Laboratory and Athlete Preparation (chapters 1-3), deals with the often overlooked issue of quality assurance in the exercise laboratory, athlete preparation prior to testing, and approaches for data collection and analysis.

Part II, Testing Concepts and Athlete Monitoring (chapters 4-10), provides generalized test procedures for the determination of anaerobic capacity and ergometer-based neuromuscular power. It also discusses concepts for the measurement of maximal aerobic power and blood lactate thresholds, including practical applications. Practical and applied guidelines for the use of altitude and hypoxic exposure and the preparation of athletes for competition in hot and humid environments are presented. The relatively new area of physiological recovery is discussed.

Part III, Fundamental Assessment Principles and Protocols (chapters 11-15), presents principles and protocols for commonly used athlete assessment tools. Protocols for the assessment of athlete physique, agility, and strength and power are provided, as are protocols for field-based physiological assessment. Consideration is given to perceptual and decision-making capacities of performance. These procedures are extensively cross-referenced to the sport-specific chapters in part IV (chapters 16-33). The reader must keep this cross-referencing in mind, because many chapters are not complete unless read in conjunction with material elsewhere in the book. For example, the measurement techniques for skinfolds (assessment of physique), muscular strength tests (strength and power assessment protocols), and field-based tests such as 20 m sprint and multistage fitness test (field testing protocols) are referred to in the relevant sport-specific chapters.

Part IV, Physiological Protocols for the Assessment of Athletes in Specific Sports (chapters 16-33), provides specific test protocols for 18 sports. Many of these sports are those in which Australia has had international success—for example, basketball, cricket, cycling, hockey, netball, rowing, Rugby Union, sailing, swimming, triathlon, and water polo. Each chapter contains a rationale for the tests, lists of necessary equipment, and detailed test procedures. Where possible, normative data and reliability data for each test are tabulated.

What's New in This Edition

Key to the overall content of *Physiological Tests for Elite Athletes, Second Edition*, is the practical nature of all chapters. Content reflects key areas of expertise and strength in the Australian sport system and applied exercise physiology research. A number of new chapters have been included, and normative data and reference material have been extensively updated. Some new chapters presented in the book include these:

- Data Collection and Analysis—approaches for analyzing data from the physiological monitoring of individual athletes and groups of athletes in team sports

- Ergometer-Based Maximal Neuromuscular Power—practical definitions and test procedures for ergometer-based maximal neuromuscular power tests

- Altitude Training—practical methods for using altitude and hypoxic exposure and their potential effects on performance at altitude and at sea level

- Heat—an overview of physiological responses to hot environmental conditions and guidelines for preparing athletes for competition in the heat

- Physiological Recovery—scientific basis of various popular recovery strategies

- Perceptual–Cognitive and Perceptual–Motor Contributions to Elite Performance—discussion of perceptual–cognitive and perceptual–motor characteristics of sport performance and identification of attributes that reliably distinguish the elite performer

- Sport specific test protocols have also been included for the sports of Australian Football, Rugby League, sprint kayak, and volleyball (indoor and beach).

The purpose of this book is to provide working procedures for athletic testing that have a sound theoretical basis and known precision. When tests are carefully administered and have good precision, repeat tests are useful in tracking changes over time to determine the effectiveness of a training intervention.

The majority of chapters in this book were written by sport scientists who spend every working day with athletes as their sole focus. Often their work extends to 7 days a week, because that is the nature of elite sport. Most of the authors are not academics who theorize about what might help improve an athlete's performance; rather, they interact closely with coaches and athletes to optimize sporting success. This book includes their collective insight and experience.

eBook available at HumanKinetics.com

ACKNOWLEDGMENTS

This book is the collective effort of many generous scientists. We acknowledge the numerous contributors and thank them for taking the time to turn their expertise, academic and practical, into text. We also acknowledge and thank the authors and co-authors of chapters published in the first edition of *Physiological Tests for Elite Athletes*, which was the foundation for the current edition. The many athletes whose data are contained in this book must be thanked indirectly, as none of them can be identified by name for ethical reasons.

We remember our mentors who have passed away in recent years. In particular we acknowledge and recognize Douglas (Doug) Tumilty (August 29, 1941, to October 31, 2004) and Emeritus Professor Robert (Bob) Withers (August 26, 1938, to September 23, 2007). Doug and Bob made extraordinary contributions to sport science, research, and quality assurance in Australia and inspired many. We also acknowledge and recognize Dr. Frank Pyke (December 1, 1941, to November 22, 2011). Frank was a fantastic scientist, educator, manager, leader, and advocate for sport.

He combined exceptional knowledge and expertise with tremendous enthusiasm and optimism, and had an amazing ability to galvanize and inspire the people around him. We are proud to stand on their shoulders.

Thank you to Professor Christopher Gore, editor of the first edition, for affording me the opportunity to work with him and allowing me to steer this edition of the book. Thank you also to Kate Fuller for her support and assistance throughout the development of this book and for effectively running the National Sport Science Quality Assurance (NSSQA) program during this time.

Finally, the support of the Australian Sports Commission (ASC) and the Australian Institute of Sport (AIS) has been instrumental in the production of this book. Thank you for supporting the book and realizing the enormous value and intrinsic worth of capturing our collective physiology expertise in a printed text.

Rebecca K. Tanner

INTRODUCTION

Professor Allan Hahn, PhD
Professorial Research Fellow, University of Canberra
Research Leader—Coaching, Queensland Academy of Sport Centre of Excellence for Applied Sport Science Research
Honorary Emeritus Professor, Australian Institute of Sport

When the first edition of this book was published in 2000, the introductory section (written by Frank Pyke) focused on the value of physiological testing in identifying the strengths and weaknesses of particular athletes, monitoring progress, providing feedback, educating coaches and athletes, and predicting performance potential. The importance of test relevance, specificity, practicality, validity, and accuracy was emphasized, as was the need for standardization of test conditions. These points remain as salient today as they were 12 years ago.

During the past decade, sport science has continued to evolve at a rapid rate. It has become an even more integral component of high-performance sport. A major study published in 2008 identified "scientific research" as one of nine pillars of international sporting success and noted that it was an area in which investment could well provide a competitive advantage (De Bosscher et al. 2008). Recognition that effective use of sport science can yield performance benefits has led to increased employment of scientists in sport settings around the world. With more scientists engaged directly with high-level coaches and athletes, and with the collective experience of those scientists accumulating, there has been a powerful stimulus for the development of new test protocols. Although this is very positive, it has heightened the challenge associated with gaining broad acceptance for particular protocols, as all proposals are subjected to considerable scrutiny.

In recent years, there has been a gradually progressive shift of testing away from the laboratory and into the field, as evidenced by a number of the protocols presented in this book. The shift is being driven by a range of factors. Scientists themselves are spending more time in the field as a result of ever-increasing realization of the importance of constant, real-world interaction with sport programs. This is allowing greater appreciation of field testing possibilities. Scope for field testing has been enhanced by the advent of new and often miniaturized technologies that in some sports enable measurement of workloads and physiological responses during training and even competition. Although it might be difficult or impossible to standardize environmental conditions in the field, mathematical models can sometimes be developed to correct for their effects. Imperfect correction may be more than compensated by improved test specificity and the ability to collect data much more frequently. Obtaining measurements during the actual sporting activity, rather than during a laboratory simulation, is clearly the ultimate in specificity. Basing the process on standardized, very regularly repeated training sessions may allow monitoring of progress with high resolution, in contrast to the occasional snapshots provided by typical laboratory testing schedules. In many circumstances, the ideal might well be a combination of regular field monitoring and much less frequent laboratory testing, but we must always ask whether the two approaches can genuinely yield different and complementary information.

Another trend that is becoming apparent in the contemporary physiological testing of athletes is a quest to acquire targeted data with the maximum possible simplicity. Scientists are increasingly aware of the exact qualities that they want to measure and of the need for the measurement techniques to be fully incorporated into the overall programs of the athletes (rather than disrupting those programs). The use of relatively simple tests has several advantages. Less can go wrong, so the results tend to be more consistent. Test rationales, procedures, and outcomes are often better understood by coaches and athletes and accepted as having practical value. Also, simple tests can generally be administered to quite large numbers of athletes, facilitating development of comprehensive databases for subsequent analysis.

In practice, the test protocols that have emerged over recent years have been a product of continual interaction between scientists and coaches. This has ensured evolution of an ability to obtain high-quality data without placing excessive demands on the time or other resources of coaches and athletes. It has also fostered shared commitment to the protocols. True partnership between coaches and scientists, in which each party draws on the unique knowledge and expertise of the other, has become the driving force for effective applied sports science.

Physiological testing is now more commonly combined with testing in other disciplines, particularly in field situations. This reflects a rapidly emerging realization of the benefits of a multidisciplinary approach to sport science. Interpreting physiological and biomechanical data in concert can clearly produce greater insights than considering them separately.

For the testing of athletes to meet its objectives, considerable attention needs to be paid to the manner in which results are presented. Many testing processes involve measurement of multiple variables at high data rates, but scientists need to become skilled in summarizing the outcomes into just a few key points that are of practical relevance to coaches and athletes. Wherever possible, the key points should be associated with guidelines for action. The methods used to communicate the key points can be verbal, written, or visual and ideally should be customized according to the preference of the recipient. However, the iterative process inherent in good science demands that the key points should always be clearly recorded for future reference, along with the raw and summarized data on which they are based.

The value of maintaining specific test protocols over long periods of time is enormous, because this practice permits the detection of longitudinal trends, the establishment of standards for athletes at different points in their developmental pathways, and the identification of truly exceptional performances. Therefore, decisions to change protocols should not be taken lightly. At the same time, however, the principles of science dictate that the protocols should be continually evaluated to determine whether they are optimally serving their originally envisaged purposes. When shortcomings are evident, refinements aimed at addressing them should be introduced and assessed for effectiveness. It is only after this assessment has proven positive that the new protocol should replace the old. Even then, comprehensive studies should be conducted to carefully characterize the statistical relationship between the results of the two tests, with a view to enabling prediction of the results of one test from those of the other, thereby preserving continuity of information. In general, simplicity of test protocols may favor their longevity.

Although the development and refinement of test protocols are clearly a scientific process, they are not a sufficient end in themselves. Instead, the primary function of protocol development is to provide a tool that can be used to determine whether the effects of specific training activities or other interventions accord with the educated expectations of their designers. The results should lead to confirmation, refinement, or refutation of the conceptual models underpinning the interventions and therefore should affect future practice and plans. This is the essence of the scientific method and consequently the work of a sport scientist. It is never enough for a scientist to simply conduct a test protocol on several occasions throughout a year and report the results to coaches and athletes without understanding and taking account of the context in which it was programmed.

The protocols outlined in this book are by no means the only measurement tools that scientists working with the targeted sports will use. The protocols are intended to provide a core of consistent, long-term monitoring around which other, more transient procedures can be based. The latter may be implemented to answer specific questions arising at particular points in time. For example, measurements of total hemoglobin mass might be introduced in association with periods of altitude training. It would be logical to measure core body temperatures, skin temperatures, and sweating responses during interventions aimed at preparing athletes for competition in the heat. Specialized methods for assessing muscle activation might be added to the test battery when there is a major focus on the development of muscular strength or power. Consequently, adherence to core test protocols does not restrict scientific creativity and scope for innovation, characteristics that are essential to the pursuit of world-class science.

Standardizing test protocols is only one aspect of the task of ensuring comparability of test results between different scientific groups and institutions. The calibration of equipment used for the testing, and the exact ways in which measurements are performed, can obviously affect the results. Attention therefore must be given to the quality of the whole measurement environment, rather than just to conformity with test protocols. Multiorganizational participation in broad quality assurance programs involving regular checking of measurement accuracy is highly desirable in this regard. However, test outcomes can be affected by a range of factors apart from the purely technical. It is vital to create a situation that provides athletes with appropriate psychological encouragement. The professionalism and general demeanor of staff conducting the testing can have a critical influence and must take account of the fact that testing can often cause athletes to feel considerable pressure.

In any discussion of protocols for testing of high-level athletes, a few points deserve emphasis. It must always be remembered that the ultimate goal is to acquire information that can be used to help athletes enhance their performances in their competitive sporting domains. Improved test performance is not of central importance in its own right, and the creation of "laboratory champions" should be avoided lest it detract from the primary focus. Test results should be used only as indicators of potential sporting achievement and progress toward that achievement. When successive tests show a plateau in results, it should not be automatically assumed that no further progress is occurring. Instead, one should ask whether the test is sufficiently sensitive to detect small physiological changes that could be influential from a competitive performance perspective. The reliable identification of such changes is a very significant challenge, because it can require differentiating a relatively weak signal from substantial background noise. This necessarily entails a risk of obtaining false-negative results. Constant attempts to maximize test sensitivity are necessary, and perfection in this task should never be assumed. The sensitivity is likely to be greatest when tests are highly specific.

Congratulations are due to the National Sport Science Quality Assurance program (which operates under the auspices of the Australian Institute of Sport) for conceiving the idea for this book and seeing it through to publication. The first edition,

with inputs coordinated by Professor Christopher Gore, attained wide distribution and stimulated productive discussion among members of the international sport science community. This was partly because the test protocols were presented as guidelines rather than prescriptions and therefore were clearly open to debate and refinement. The same approach has been adopted for this edition, compiled by Rebecca Tanner, and will undoubtedly lead to further collaboration and progress.

Sport science is still a very young field of endeavor. It is at an exciting time in its history, with both the breadth and depth of its knowledge base undergoing exponential development. New technologies are producing unprecedented learning opportunities. Many of the trends identified here will continue, and others will emerge. The endeavors of scientists are likely to play an increasing role in aiding the performances of athletes and in augmenting the impetus of sport as a positive influence in the global community. The deliberate creation of high-performance sport environments that facilitate scientific enquiry and exploration could well accelerate these effects. However, as a dynamic future for applied sport science unfolds, the fundamental importance of documenting methods, subjecting concepts and models to peer scrutiny, and accommodating expert feedback will remain. This book reinforces those vital elements and will be a harbinger of a new phase of dialogue.

Reference

De Bosscher, V., Bingham, J., Shibli, S., van Bottenburg, M.D., and Knop, P. 2008. *The Global Sporting Arms Race: An International Comparative Study on Sports Policy Factors Leading to International Sporting Success.* Oxford, UK: Meyer & Meyer Sport (UK) Ltd.

PART I

Laboratory and Athlete Preparation

Quality Assurance in Exercise Physiology Laboratories

Rebecca K. Tanner and Christopher J. Gore

Quality assurance is common in many areas of manufacturing and service industries and can be described as "the overall measures that a laboratory uses to ensure the quality of its operations" (CITAC/Eurachem 2002, 7).

In the context of sport science and exercise physiology, *quality assurance* refers to systems used to give the scientist, coach, and athlete confidence that all results are accurate and reliable. These systems ensure that validated protocols are used, test equipment is appropriately calibrated, the level of uncertainty is quantified for specific tests, and test results are collected in a standardized format readily identifying all test details. A quality-assured service also means that test results are returned to the coach or athlete in a timely manner (e.g., within 1-2 days) and in a form that can be readily understood and interpreted. These systems mean that if the test has to be repeated at another location or several months (or even years) later, enough information has been recorded to allow replication.

Although seemingly straightforward, quality assurance in an exercise science laboratory is often perceived as difficult. This commonly relates to three issues faced by scientists conducting physiological testing:

- Test protocols evolve over time as techniques are refined and new approaches are implemented.
- Some testing may be experimental rather than routine.
- Measurement accuracy is compromised by variation in the athlete's presentation for testing (characteristics such as motivation, training on the day before testing, or even muscle glycogen levels).

Issues such as these highlight the importance of quality assurance in sport science and that quality assurance systems should be a fundamental aspect of operations for all sport science laboratories that test athletes. Adequate documentation provides easy identification of test protocols, allowing scientists to track whether tests were conducted with a current or historical protocol, and also helps to identify the reason for an aberrant test score—for example, because an athlete was tested after an unusually hard training session on the previous day.

Because sport science is a dynamic field, and is constantly evolving, the need to develop and trial new protocols will always exist. The aim of quality assurance is certainly not to stifle creativity or to make all laboratories clones of each other. Rather, well-implemented quality assurance systems enable data from different locations to be pooled so users can compare results between laboratories and have confidence in the test results. For example, if a national sporting association stipulates that all athletes from its sport be tested with specific protocols, quality assurance procedures enable sport scientists to conduct the testing and report the results in a standardized and uniform format, regardless of the location where the athletes are tested.

Identifying the uncertainty of athletic testing is an important component of quality assurance in exercise physiology laboratories. Repeat trials can be conducted on a representative subpopulation of athletes to identify the imprecision or uncertainty of a specific test protocol. These trials can be used to calculate a statistic called the *typical error of measurement* (TE), which incorporates error due to both the equipment and biological variation of athletes (Hopkins 2000). In addition, this statistic can be used to generate 68% or 95% confidence intervals of a real change. No longer does an exercise physiologist have to guess whether a change from 62 to 63 ml · kg^{-1} · min^{-1} is meaningful or simply a measurement artifact. The TE provides a statistical basis for the interpretation of test results. In addition, TE data provide an objective measurement of the competency of a sport science laboratory. For a laboratory to provide a quality service to coaches and athletes, it is not sufficient to simply possess a set of skinfold calipers, a gas analysis system, and a treadmill. Sport scientists must prove, for example, that they can reliably use skinfold calipers with a TE of better than 2% and that they can measure maximal oxygen consumption of athletes to better than 3% (see table 1.1).

Evaluating Quality Assurance

Worldwide, the effectiveness of quality assurance systems in manufacturing and service industries is measured against the International Organization for Standardization document, ISO 9001 (ISO 2008). Companies actively pursue this certification because it improves the quality and marketability of their product or service and the competitiveness of their organization. A company seeking ISO 9001 certification must prepare extensive documentation addressing each of the eight quality management principles of the ISO standard and must submit to an external assessment of how well the company implements the documented procedures. The company must also complete and maintain records of regular internal audits of its quality assurance procedures and submit to a full reassessment about every 3 years. The ISO 9001 certification has prestige and credibility because it is a truly international standard.

Quality assurance in laboratories is well established in areas such as pathology, hematology, materials testing, and metrology. For example, if your doctor orders a cholesterol test for you, a laboratory technician uses documented procedures to analyze

TABLE 1.1 Target Typical Error of Measurement (TE) Data

Measurement	Units of raw data	Target TE
Anthropometry		
Σ7 skinfold sites	mm	<5%
Field testing		
Multistage shuttle run test	laps	<4 laps
Vertical jump	cm	<2 cm
Sprint tests (20 m and 30 m)	s	<0.03 s
Sprint tests (40 m)	s	<0.04 s
5-0-5 agility test	s	<0.05 s
Oxygen consumption (at $\dot{V}O_2$max)		
$\dot{V}O_2$	L · min^{-1}	<0.15 L · min^{-1}
Ventilation	L · min^{-1}	<5.0 L · min^{-1}
Heart rate	beats · min^{-1}	<2 beats · min^{-1}
Respiratory exchange ratio	n/a	<0.05
Blood lactate threshold (and associated measures)		
Lactate Threshold 1:		
Blood lactate concentration	mmol · L^{-1}	<0.2 mmol · L^{-1}
Power output	watts	<10 W
$\dot{V}O_2$	L · min^{-1}	<0.15 L · min^{-1}
Heart rate	beats · min^{-1}	<5 beats · min^{-1}
Lactate Threshold 2:		
Blood lactate concentration	mmol · L^{-1}	<0.5 mmol · L^{-1}
Power output	watts	<10 W
$\dot{V}O_2$	L · min^{-1}	<0.15 L · min^{-1}
Heart rate	beats · min^{-1}	<4 beats · min^{-1}

These targets have been developed from data acquired from a number of exercise physiology laboratories certified by the National Sport Science Quality Assurance program of the Australian Sports Commission.

your blood sample with a calibrated analyzer. The result printed on the report that your doctor receives pertains to the sample submitted for analysis. Many government laboratories must, by law, have quality assurance certification.

Because exercise physiology laboratories often have fewer than six staff and cannot afford to have a full-time quality assurance officer, the ISO 9001 standard is considered to be excessive for them. An appropriate and functional certification scheme that is within the constraints of time and resources of such laboratories has been developed by the National Sport Science Quality Assurance (NSSQA) program (formerly known as the Laboratory Standards Assistance Scheme, LSAS) of the Australian Sports Commission.

Although the essence of the ISO 9001 standard was retained, only the sections relevant to exercise physiology laboratories are included in the NSSQA scheme. Although some have argued that a watered-down version of the ISO 9001 standard is worthless because the standard is already minimalist, those exercise physiologists who have implemented quality assurance are adamant that the NSSQA scheme has resulted in quantifiable improvements in the assessment of athletes in Australia. These scientists are more confident about the accuracy and interpretation of results that they return to coaches and athletes.

The following information outlines some of the key elements for implementing successful quality assurance systems in Australian exercise physiology laboratories.

Australian Certification Model

This section summarizes the approach and certification model that is implemented by the NSSQA program in Australia.

National Sport Science Quality Assurance

The NSSQA program is an initiative of the Australian Sports Commission and Australian Institute of Sport. NSSQA takes a national leadership role in overseeing quality assurance in the delivery of sport science services to athletes and coaches through the network of national and state institutes and academies of sport and other laboratories and facilities (e.g., universities, private laboratories) involved in the assessment of Olympic-, national-, and state-level athletes in Australia.

The NSSQA program is focused on quality assurance in sport science laboratories that test elite Australian athletes—although quality assured testing can readily occur in the field with instrumentation such as light gates or with the combination of a transportable ergometer, heart rate monitor, and lactate analyzer. Quality assurance is a catch-cry of business and industry today, and there is good reason why it should also be the basic tenet of sport science servicing. By spending their sport science budgets only at NSSQA-accredited laboratories and facilities, Australian sports organizations can be assured of high-quality measures and services that are beneficial to the preparation of Australian athletes for international competition.

The NSSQA program was initially established in response to demands from personnel at network physiology laboratories who were concerned that different laboratories around Australia obtained different results when testing the same athlete. The primary aim was to standardize physiology laboratory testing of elite Australian athletes, such that an athlete tested in one location (e.g., Perth) will achieve the same test scores in another (e.g., Canberra, or Adelaide, or any other accredited laboratory). Given the landscape of the Australian high-performance network, where national team athletes are often required to train and receive sport science services in different states and are often required to move from state to state when selected for various teams, comparability of results is paramount. Furthermore, testing data are often pooled in central databases for access by national team coaches and are commonly used for selection purposes. Accordingly, athletes, coaches, and scientists need to have confidence that all test results are accurate, reliable, and ultimately comparable. This is achieved through the implementation of quality assurance systems in participant laboratories.

Goals of the Accreditation Process

The main aim of the NSSQA program and the exercise physiology accreditation program is to promote continuous improvement in exercise physiology testing standards in Australia and assist laboratories involved with the assessment of athletes to establish and maintain a testing environment of national standard. The program works with laboratory staff to monitor quality assurance issues and works with sport scientists to critically evaluate all aspects of laboratory function likely to affect the reliability and accuracy of test results.

The primary aim of the accreditation process is to achieve comparability of results of testing between different laboratories.

Secondary aims of the accreditation scheme are these:

- Establish greater confidence in exercise physiology services among coaches and athletes.

- Encourage sport scientists to exchange information.

- Develop a national database of accredited laboratories and encourage national sporting organizations and coaches to use it.

- Develop a database of appropriate measurement error tolerances for commonly conducted tests.

- Develop a database of test results on elite athletes by pooling results from all accredited laboratories.

The process continues to be refined over time because quality assurance is, by definition, a process of ongoing improvement. It should also be pointed out that quality assurance in Australian exercise physiology laboratories has been initiated and implemented by the sport scientists themselves. Laboratory accreditation is not merely an examination of a laboratory but a collaborative effort to improve standards of athlete testing. It is the responsibility of NSSQA to provide advice to applicant laboratories on how to rectify any perceived deficiencies observed during assessment of a laboratory.

The Australian Sports Commission has introduced a clause to funding agreements with national sporting organizations (NSOs) stating that where testing is deemed necessary by the NSO, testing of National program athletes must be undertaken using laboratories accredited under the National Sport Science Quality Assurance program (Australian Sports Commission, Sport Collaboration Agreement, 8). This essentially means that all NSOs in Australia are required to undertake testing of national program athletes in NSSQA accredited laboratories.

Two phases of accreditation have been implemented.

- **Phase 1: Precision and reliability**: The first phase of NSSQA accreditation, implemented between 1995 and 1997, required laboratories to demonstrate the precision (or reliability) of athletes' test results and to establish the typical measurement error associated with these test results. Equipment limitations combined with biological, day-to-day changes in the performance of athletes mean that all measures have an "uncertainty" or error associated with them. Thus, instead of reporting a sum of skinfolds as 65 mm, the laboratory reports this value to an athlete and coach as 65 ± 3 mm. The expression "±3 mm" indicates the

level of uncertainty or TE associated with the measure. Similarly, maximal oxygen consumption ($\dot{V}O_2max$) would be reported as 70 ± 2 ml · kg^{-1} · min^{-1} rather than 70.1 ml · kg^{-1} · min^{-1}. This is the same kind of measurement error that is associated with any scientific measurement; for instance, cholesterol results are usually reported as a value with a measurement error range of ±0.1 mmol · L^{-1}.

- **Phase 2: Accuracy:** In the second and current phase of the accreditation scheme, laboratories are required to demonstrate both reliability and the accuracy of measures, which means that they must show not only that the measures have a small level of uncertainty but also that they are accurate when compared with a criterion (or "first principles") measure. For example, an athlete in a maintenance period of training might have his $\dot{V}O_2max$ measured as 73 ± 2 ml · kg^{-1} · min^{-1}, but 2 weeks later, when he is tested at another laboratory as part of a training camp, his $\dot{V}O_2max$ could be measured as 68 ± 2 ml · kg^{-1} · min^{-1}. Establishing the accuracy of measures helps coaches and athletes to interpret any difference in measurements between laboratories. To this end NSSQA has collaborated with others to develop devices such as a dynamic calibration rig for skinfold calipers (Carlyon et al. 1998; Gore et al. 2000), a gas analysis calibrator that can simulate the ventilation pattern and expired gas fractions of an elite athlete (Gore et al. 1997), and a torque meter for calibration of cycle ergometers (Stanef 1988; Woods et al. 1994).

Accreditation Streams

Accreditation is granted for demonstrated competence in five streams, which group the most common physiology tests conducted on athletes. The streams, with examples of tests within each, are as follows:

- **Anthropometry**: skinfolds, height, body mass, girths, bone lengths and breadths

- **Field testing**: multistage shuttle run test, straight line sprint, vertical jump, agility run

- **Oxygen consumption**: $\dot{V}O_2max$ and maximal accumulated oxygen deficit tests

- **Ergometry**: performance tests on calibrated cycle ergometers, rowing ergometers, and treadmills

- **Blood analysis**: measurement of blood lactate and blood gas

Laboratories can apply for accreditation in all five streams or in selected streams. A laboratory that is accredited in all five streams offers a greater range, not necessarily higher quality, than a laboratory that is accredited in only one or two streams. The quality of measurements from any certified laboratory meets similar criteria based on calibration records and the reliability of duplicate measures on typical error of measurement.

For each stream, requirements for accreditation are divided into the following areas (followed by examples):

- **Staffing**: listing of staff including job descriptions, qualifications, CPR/first aid qualifications; staff training and induction processes
- **Documentation**: policy statement, equipment inventory, reliability calculation and interpretation procedures, informed consent forms, document control procedures, thermal environment information
- **Measurement and equipment calibration**: comprehensive logs of calibration and maintenance of equipment in line with NSSQA tolerance and frequency requirements
- **Test methods**: documented test protocols
- **Athlete reports**: detailing required fields such as identification of laboratory, sport scientist's name and contact details, test protocol, equipment
- **Reliability data**: TE data for all routine tests

Accreditation Model

The basis of accreditation is a collaborative peer-based assessment process of the areas listed previously. It entails an on-site assessment by three expert peers to critically evaluate all aspects of the laboratory management, staff, facilities, and operations likely to affect the reliability and accuracy of the laboratory's test results. Accreditation is granted by the National Accreditation Committee (NAC), convened by the Australian Institute of Sport and the National Sport Science Quality Assurance program. Thus, the process of accreditation is a partnership between the NAC/NSSQA and the individual laboratories.

Accreditation is granted for 4 years and is reviewed every 4 years thereafter. Some Australian laboratories have now been re-accredited four times. Reaccreditation requires a laboratory to collect new reliability data to demonstrate competence in all techniques used to test athletes and entails another on-site assessment of the laboratory quality systems. An annual report process is implemented to further facilitate quality control procedures in laboratories and encourage annual data collection so that the quadrennial accreditation visit is merely a summary of the preceding 4 years combined with a thorough check that the documented quality procedures are being used effectively. This is an attempt to reduce

(or spread) the workload of both the applicant laboratories and NSSQA. In summary, the annual review process is intended to keep quality assurance in focus for each laboratory by using this report as an internal audit to verify that processes and procedures are working appropriately.

Implementing Quality Assurance

The successful implementation of quality assurance in an exercise physiology laboratory relies on the establishment of a functional and practical quality system that is carefully integrated into the everyday operation of the laboratory. Such a system must clearly and thoroughly document and record information relating to key areas mentioned previously (e.g., laboratory staffing, laboratory quality policies and practices, measurement and equipment calibration requirements, test protocols and procedures, test results and test reports, reliability data).

These are the elements from the ISO 9001 standard that NSSQA has determined to be most pertinent to small-scale sport science laboratories.

The quality system must be clearly documented (hereafter referred to as *laboratory documentation*) and can be stored electronically as a compilation of files and documents or can be printed in hard copy. All laboratory documentation must be easily accessible to all laboratory staff and readily updated.

One important aspect of laboratory documentation is that it must be truly relevant, practical, and useful to the sport science laboratory. There is little value in merely copying laboratory documentation from another institution, because the documented procedures from one laboratory may not be appropriate or relevant in another.

Laboratory Staffing

The influence of staff training, staff proficiencies, and competencies on the collection of accurate and reliable data is significant. Aside from possessing educational qualifications, laboratory staff must have adequate training and high levels of competence to perform specific tasks and tests. Thorough staff induction and training procedures should be implemented to ensure that all staff are familiar with the specific operational requirements of the laboratory (e.g., code of conduct, emergency procedures, laboratory policies and procedures, documentation of control systems). Annual staff competency checks are recommended to ensure ongoing maintenance of staff competence and the continuous development of skills. A staff competency register should be implemented to record and monitor the specific competencies for each laboratory employee (e.g., anthropometry, blood sampling, use of metabolic cart, operation of treadmill, use of electronic light gates), along with pertinent information relating to aspects such as vaccination status (e.g., hepatitis B), first aid qualifications, and job-specific certification (e.g., anthropometry accreditation).

It is useful to include general job descriptions of all staff in this section of the laboratory documentation. A diagram of the organizational structure should clearly indicate who has overall and day-to-day responsibility for quality assurance within the laboratory. Key among the roles of this person are maintaining document control, keeping calibration records, writing test procedures, and ensuring that staff are adequately trained in quality issues. This person is also responsible for identifying deficiencies in the quality systems and developing procedures for review and continuous improvement.

Laboratory Quality Policies and Practices

There is no one "right" way to develop and store laboratory documentation; however, the methods chosen must be useful and effective. The aim of the quality assurance system is to make the operation of the laboratory more efficient and to provide traceability of results. For example, if a sport scientist changes jobs, adequate documentation must be available so that another sport scientist could replicate the protocols that were being used. Ideally, the records of previous tests should contain sufficient information to identify exactly which test protocol was used, what equipment was used and how it was calibrated, and the error associated with that test. It should be possible to longitudinally track test results for an individual athlete across many years and identify whether different protocols were used during that time.

Policy Statement for Quality Assurance

This policy statement should specify the objectives and commitment of the laboratory, staff, and management to the quality assurance system and practices. Full support of quality assurance is essential from the highest level of management because such a system requires resources to implement and maintain.

Equipment Inventory

A comprehensive inventory of laboratory equipment assists with the management of laboratory assets as well as the ordering, reordering, and repair of equipment. The inventory should include equipment name, make and model, serial number, quantity, storage location in the laboratory, supplier details (including address, contact numbers and names, reorder numbers), date purchased, and approved service agent (and his or her contact details).

Equipment operating manuals should be kept in an identified (documented) central location for ease of reference to operating and troubleshooting procedures. Operating manuals can also be scanned for electronic storage.

Determination of Test Precision and Reliability

Laboratories should establish and use a standard statistical procedure (e.g., typical error of measurement) to quantify the uncertainty associated with specific test procedures. The principles of these calculations along with how the data are applied in the day-to-day operation of the laboratory should be documented for staff to review as required. The raw error data should be kept in laboratory documentation.

Schedules for the collection of error data at regular intervals should be prepared and implemented to ensure that the precision estimates are relevant to the current staff and equipment.

As a guide to the uncertainty of common tests used for athletes, table 1.1 presents typical error of measurement (TE) data that have been acquired from a number of NSSQA-certified exercise physiology laboratories. For detailed information on TE, refer to Hopkins (2000) and chapter 3, Data Collection and Analysis.

Consent Forms

Obtaining informed consent is an important aspect of laboratory practice. Informed consent is one of the guiding principles of the Declaration of Helsinki for Research Involving Human Beings (*Journal of Applied Physiology* 1996). Routine monitoring of athletes also falls under the intent of this declaration, and therefore consent must be obtained from all athletes prior to testing. Laboratory documentation should include consent form templates for all testing categories undertaken by the laboratory (e.g., scholarship athletes, nonscholarship athletes, research). See chapter 2, Pretest Environment and Athlete Preparation, for more information regarding recommended content of informed consent forms and a sample form.

Document Control

Although it is important to document aspects of laboratory function such as calibration procedures, test protocols, and data record sheets, this process will quickly deteriorate without a system of document control. For these documents to be effective, a master control document must be implemented to enable the ready identification of current calibration procedures and current test protocols for any sport. The master control document should detail the following information for all procedures and protocols: document name and file path, author, date when procedure or protocol was implemented, who authorized its use, what procedure or protocol it supersedes, and where superseded procedures and protocols are located. Obsolete documents should be removed from use but kept on record. This may require that protocols be numbered uniquely to provide easy reference to them in case old protocols need to be reused. Ideally, the process for developing, authorizing, issuing, reviewing, and updating all new documents should itself be documented. Traceability tables at the front of all documents are one way to facilitate the tracking of changes to procedures and protocols.

Thermal Environment

A controlled and stable testing environment is vitally important not only for the comfort of athletes but for reliable and accurate testing results. The thermal environment of exercise physiology laboratories should be kept within 18 to 23 °C and a relative humidity less than 70% during testing. It is recommended that laboratories use a system where the laboratory thermal environment can be logged and tracked longitudinally and recalled as required.

Measurement and Equipment Calibration Requirements

A prerequisite to accuracy is that equipment is calibrated, ideally against the relevant first principles of time, distance, or mass. Most equipment can be calibrated against external standards. In some instances, calibration procedures may be completed in-house, whereas in other cases equipment may need to be calibrated by an outside organization. Outside organizations that calibrate laboratory equipment should have ISO certification and issue calibration documents showing the traceability of the equipment that was used for calibration. Equipment used within the laboratory should be calibrated or verified according to documented procedures and assessed against strict tolerances.

Calibration Schedule

Laboratory equipment should be calibrated or verified in a timely manner. Laboratories should prepare a complete schedule, on a chart, calendar, or computer, that alerts staff when weekly, monthly, or yearly calibration events are due. Quality assurance software packages for this purpose are commercially available.

Calibration and Maintenance Procedures

Calibration and maintenance procedures help to ensure that equipment is maintained in optimal working condition. Procedures for the calibration of all laboratory equipment should be carefully documented. Instructions should include the acceptable calibration tolerances as well as the corrective action to be taken if calibrations are outside the tolerances. Where calibration services are provided by an outside organization, the name and contact details for this organization should be recorded.

Calibration Logs

Calibration logs allow one to readily identify departures from calibration over time or even a sudden loss of calibration as a consequence of mechanical or electronic failure. Calibration events should be recorded longitudinally in calibration logs for each item of equipment. It is recommended that calibration logs include equipment name, model, and serial number; calibration frequency and tolerance information; name of staff conducting the calibration; reference to calibration protocol; date calibration was performed; raw data for calibration event; identification of calibration pass or fail; and comments relating to any action taken or required after the calibration event. Copies of calibration reports provided by outside organizations should be incorporated into laboratory documentation. Some laboratories attach a sticker to equipment to ensure that staff are acutely aware of its calibration status and date of next calibration prior to use.

Test Protocols and Procedures

The methods used for athlete assessment should be adequately documented to allow for replication and to allow traceability of altered protocols. If the test procedure departs from the standard protocol, this should be noted on the results sheet so that possible reasons for unusual results can be traced and understood. If the protocol is taken from a published article, a copy of the article should be included in laboratory documentation. Even if the protocol is a modification of a published article, it is relevant to include a copy of the article to allow easy reference to the source material.

Although some laboratories may have very extensive test protocols, others may write procedures that contain only the essential information that could be followed by a competent scientist. If particular methods of data smoothing or analysis are used, explanations of these must be included with the test protocol because a different interpretation of the data is likely if different methods of analysis are used.

To facilitate day-to-day use of the documented test protocols and calibration procedures, it is helpful to have summary work instructions in close proximity to the operator. For example, a plastic-laminated summary of the operating and calibration procedure for an oxygen consumption system can be posted on the side of the machine; wash-up instructions for respiratory mouthpieces can be attached to the wall next to the sink; and a summary of the workloads and sampling times for a progressive lactate profile test can be made available in a folder next to the ergometer. A master copy of all work summary instructions should be maintained with laboratory documentation.

Athletes' Test Results and Reports

While allowing for flexibility according to coach requirements, standardized templates for results sheets and reports should ensure that the following primary information is identified:

- Institution or laboratory name
- Name and contact details for the sport scientist responsible for testing
- Athlete or group name and sport
- Test date, time, location, surface, and environmental conditions
- Test protocol or method and equipment identifier
- Error data for relevant tests

References

Australian Sports Commission. Sport Collaboration Agreement. Canberra, Australia: Australian Sports Commission.

Carlyon, R.G., Bryant, R.W., Gore, C.J., and Walker, R.E. 1998. Apparatus for precision calibration of skinfold calipers. *American Journal of Human Biology* 10(6):689-697.

CITAC/Eurachem. 2002. *Guide to Quality in Analytical Chemistry*. Prague: Cooperation of International Traceability in Analytical Chemistry/Eurachem. http://www.eurachem.org/guides/pdf/CITAC%20EURACHEM%20GUIDE.pdf.

Gore, C.J., Catcheside, P.G., French, S.N., Bennett, J.M., and Laforgia, J. 1997. Automated $\dot{V}O_{2max}$ calibrator for open-circuit indirect calorimetry systems. *Medicine and Science in Sports and Exercise* 29(8):1095-1103.

Gore, C.J., Carlyon, R.G., Franks, S.W., and Woolford, S.M. 2000. Skinfold thickness varies directly with spring coefficient and inversely with jaw pressure. *Medicine and Science in Sports and Exercise* 32(2):540-546.

Hopkins, W.G. 2000. Measures of reliability in sports medicine and science. *Sports Medicine* 30(1):1-15.

ISO 9001. 2008. *Quality Management Systems—Requirements*. Geneva: International Organization for Standardization.

Stanef, T. 1988. Preliminary report on dynamic calibration rig for ascertaining accuracy of ergometer in the health and exercise sciences. *Motion Technology* October:2-3.

Woods, G.F., Day, L., Withers, R.T., Ilsley, A.H., and Maxwell, B.F. 1994. The dynamic calibration of cycle ergometers. *International Journal of Sports Medicine* 15(4):168-171.

World Medical Association. 2012. Declaration of Helsinki—ethical principles for medical research involving human subjects. http://www.wma.net/en/30publications/10policies/b3/.

Pretest Environment and Athlete Preparation

Kate L. Fuller and Nicole E. Thomas

The physiological testing of athletes should be performed in a controlled scientific environment, and the preparation for testing involves numerous considerations. Most physiological testing, whether maximal or submaximal, carries inherent risks to the participant, and it is important that each laboratory has thorough pretest preparation processes to minimize risk while ensuring that full duty of care toward athlete and laboratory staff is maintained. In addition to managing the risks associated with physiological testing, laboratory staff must monitor the many variables that can influence the accuracy and reliability of test results, such as quality assurance issues, pretest diet, fatigue, medications, illness, injury, and environmental conditions.

When preparing to test the physiological function of athletes, laboratory staff must consider both the testing environment and preparation of the athlete. This chapter provides a framework for laboratories to develop and use comprehensive pretest procedures for staff involved in the physiological assessment of athletes.

The following is covered in this chapter:

- Preparation of the testing environment
  - Risk management processes
  - Quality assurance considerations
- Athlete preparation
  - Medical screening
  - Musculoskeletal screening
  - Informed consent
  - Pretest athlete questionnaire
  - Pretest diet considerations
- Pretest checklists for testing in the laboratory and field

The forms and sample documents provided as appendixes to this chapter are examples only, and each laboratory should modify the material presented here to suit its needs. Furthermore, attention should be paid to the special considerations recommended for the sport-specific protocols presented in this manual.

Risk Management Processes

When preparing the testing environment, staff must take all necessary steps to manage hazards that have the potential to place the health and safety of athletes, testing staff, and observers at risk. This process is commonly referred to as *risk management*. Staff who coordinate physiological testing sessions have a moral and legal responsibility for the health and safety of those involved. The Australian Occupational Health and Safety Act (1991) states, "An employer must take all reasonably practical steps to protect the health and safety at work of the employer's employees." This document, along with the Occupational Health and Safety Code of Practice (2008), is a valuable resource for new laboratories and should be consulted to establish workplace policies and procedures.

The Occupational Health and Safety Code of Practice (2008) outlines risk management as a four-step process:

1. **Hazard identification**: systematic surveying of the testing environment for hazards. A hazard identification checklist is useful in guiding this process, and a template for this has been provided in form 2.1. It is recommended that this template be customized to meet the requirements of each testing environment.

2. **Risk assessment**: evaluation of risk through assessment of probability and consequence of injury or illness arising from exposure to an identified hazard. A risk assessment matrix is a commonly used tool for qualitatively assessing risks associated with a particular hazard. The matrix allows you to assess the likelihood of an undesirable incident and associated consequences of exposure to the hazard. A sample risk assessment matrix is provided in table 2.1.

3. **Implementation of controls**: actions taken to reduce identified risks to an acceptable level. Actions should be prioritized in the following way:

- Eliminate the hazard.
- Substitute or modify the hazard.
- Isolate the hazard.
- Use engineering controls to control the hazard at its source.
- Use administrative controls.
- Use personal protective equipment.

4. **Review of the process**: ongoing monitoring and review of hazards and any implemented controls.

Risk management strategies should be formalized and implemented when the laboratory is established, including processes to ensure that all strategies are reviewed periodically. Each laboratory varies in design and layout, but common structural considerations include floor surface, temperature control, electricity points, and lighting. Laboratory procedures for emergency evacuation, medical emergencies, dangerous or hazardous goods handling, and general safety are important and should be communicated to all testers and athletes involved in the testing session. All laboratory equipment must be well maintained and in good operating condition. When developing the testing schedule the layout of the laboratory should be considered to ensure that staff and athletes execute test procedures in a controlled and efficient manner. The immediate testing area should be clear of trip hazards and should provide adequate segregation between the testing area and any areas of general public access.

Conducting physiological testing in the field presents its own unique set of risk management considerations. In many situations, testing in the field involves establishing a temporary testing area in a location where such work would not typically occur. It is the responsibility of the tester to ensure that all safety considerations noted when testing in a laboratory are applied to testing in the field and that any potential hazards are identified and controlled prior to testing. The location should be selected to suit the type of testing session to be conducted, and all testing surfaces must be carefully inspected for hazards. Consideration needs to be given to the transport of equipment and staff, managing exposure to the weather, and suitability of electrical power supply setup.

TABLE 2.1 Example of a Simple Qualitative Risk Assessment Matrix

	LIKELIHOOD			
Consequences	Very likely	Likely	Unlikely	Highly unlikely
Fatality	High risk	High risk	High risk	Medium risk
Major injuries	High risk	High risk	Medium risk	Medium risk
Minor injuries	High risk	Medium risk	Medium risk	Low risk
Negligible injuries	Medium risk	Medium risk	Low risk	Low risk

Reprinted from Australian Government 2008. Available: http://www.comlaw.gov.au/Details/F2008L02054.

Collecting Biological Samples

Athlete blood, sweat, and urine samples are frequently collected and analyzed to investigate physiological responses to training. The collection and handling processes for these samples present a number of biological and health hazards to both the athlete and the tester. Cross-contamination of bodily fluids through needle-stick injury, splashes to the face, or through exposure to damaged skin can result in major injury and, in the instance of hepatitis or HIV transmission, may be fatal. Potential hazards must be minimized by thoroughly training staff on correct sampling and handling techniques; implementation of first aid and emergency response procedures; and the use of appropriate equipment, including personal protective equipment.

Quality Assurance Considerations

When preparing the testing environment, staff should identify elements that may affect the reliability or validity of the data collected. The following section highlights some examples of quality assurance considerations when preparing the testing environment. Because many quality assurance considerations also have an impact on health and safety, it is recommended that both elements be considered concurrently. Quality assurance practices are covered in further detail in chapter 1, Quality Assurance in Exercise Physiology Laboratories.

- **Environmental conditions**: Controlling environmental conditions during physiological testing ensures that factors such as temperature and humidity have minimal impact on the athlete's performance and that conditions are repeatable for proceeding tests. Where the environmental conditions cannot be controlled, such as in the field, an environmental sensor should be used to monitor and record the conditions. This information is valuable to assist with the interpretation of results and for the comparison of future test results.

- **Instruments and equipment**: The instruments and equipment used during each testing session must be periodically calibrated and maintained in good operating condition. This not only provides for the safety of those working in the testing environment but also ensures that the data collected are valid and reliable. Equipment registers, calibration and maintenance schedules, and documented procedures are useful tools to manage equipment appropriately.

- **Testers**: It is the responsibility of the supervising tester to ensure that the physical testing environment, equipment, and protocols used during the testing meet appropriate safety and quality assurance standards. Appropriate training in the operation of equipment and the use of standardized test protocols are important to ensure that the tests are conducted in a valid and reliable manner.

Medical Screening

Although the overall population of athletes is generally at low risk for sudden death, a number of largely congenital but clinically unsuspected cardiovascular diseases have been causally linked to sudden death in trained athletes, usually in association with physical exertion (Maron 2003). As such, medical screening of athletes prior to the commencement of a training program or fitness testing is an important consideration. Although systematic medical screening cannot completely prevent sudden death in athletes, it minimizes risk by identifying at-risk athletes who require further investigation for cardiac abnormalities.

For athletes entering an ongoing sports program with a sports institute or academy, detailed systematic medical screenings performed by physicians are recommended and standard practice in Australia. Brukner and colleagues (2004), in "Screening of Athletes: Australian Experience," identify the aims of elite athlete medical screenings: prevent sudden death, ensure optimal medical and musculoskeletal health, optimize performance, prevent injury, review medications and vaccinations, collect baseline data (such as blood test results), develop a professional relationship with the athlete, and educate the athlete.

Medical screenings should be performed by an appropriately qualified physician prior to any fitness testing. Screening should include a physical examination and collection of the athlete's medical history. The physician should assess the athlete for cardiovascular disease risk factors, general health, and nutritional issues; obtain the athlete's injury and surgery history; and review vaccinations and medications. It is also suggested that a preparticipation electrocardiograph be obtained for all athletes.

Many laboratories perform fitness testing for "one-off" clients and as such require an expedient way of screening athletes prior to testing without the cost and logistical difficulties associated with a full medical screening and physician examination. Pretest medical screening questionnaires such as the American College of Sports Medicine (ACSM) Physical Activity Readiness Questionnaire (PAR-Q; ACSM 2010), American Heart Association/ACSM Health/Fitness Facility Preparticipation Screening Questionnaire (ACSM 2010), and Sports Medicine Australia (SMA) Pre-exercise Screening System (2005) have been designed to screen for high-risk individuals who require medical clearance prior to fitness testing or engagement in physical activity. Although developed for use with the general public, these self-administered screening tools provide a useful way to identify athletes at high risk of exercise-related complications during fitness testing due to underlying cardiovascular, cerebrovascular, respiratory, or metabolic diseases (SMA 2005). However, no screening questionnaire can serve as a fail-safe tool in the prevention of exercise-related injuries and illnesses. An example of a self-administered preparticipation medical screening questionnaire is included in form 2.2.

Musculoskeletal Screening

For elite athletes, preparticipation or preseason musculoskeletal screenings are a common component of the annual training plan. Performed by sports physiotherapists, musculoskeletal screening helps prevent injury and maximize performance by identifying potential weaknesses and limitations that may predispose an athlete to both acute and overuse injuries. Screenings should be tailored to include accurate and reliable tests that are specific to the demands of the sport. The information collected should be used to enhance training programs through addressing identified deficits in stability, strength, mobility, or flexibility.

In general, musculoskeletal screening tests can be grouped as shown in table 2.2.

An example of a commonly used musculoskeletal screening protocol for the Australian Institute of Sport Basketball program is shown in form 2.3.

Informed Consent

Obtaining genuine informed consent from athletes (or, where required, their parent or legal guardian) prior to testing is an important aspect of laboratory procedure, as it has both legal and ethical implications. The documentation of such consent is generally good practice and can protect the interests of both the testing organization and athletes. The content of informed consent forms used by laboratories will necessarily vary depending on the testing organization and the nature of the testing. Following are some things to consider when developing or reviewing current informed consent forms used within a laboratory. The list is provided for general guidance only and is not intended to be exhaustive or to constitute

legal advice. Laboratories should seek their own legal or other professional advice, as appropriate, when dealing with issues of informed consent or preparing informed consent forms for their particular purposes.

- Clearly identify the testing organization and the departments or individuals involved in conducting the testing.

- Provide (as an attachment if convenient) in clear, plain language an explanation of the nature of any sport science test to be performed. This should include the risks and discomforts associated with completing each test. For example, where relevant it may be appropriate to advise athletes that they will be
  - undertaking physical exercise at or near their capacity and there is possible risk in the physical exercise at this level, including episodes of light headedness, fainting, abnormal blood pressure, chest discomfort, and nausea; and
  - undertaking activities that involve risks and dangers of serious bodily injury, such as disability, paralysis, and death.

- Include, where relevant, an acknowledgement by the athlete that he or she
  - has been provided with a copy of and read the written information about the sport science tests and their associated risks;
  - has been given an opportunity to ask questions and receive a satisfactory explanation about the sport science tests to be conducted;
  - understands that he or she can withdraw consent, freely and without prejudice, at any time before, during, or after testing;

TABLE 2.2 Musculoskeletal Screening Tests

Type of screening test	Example
Posture tests	• Spinal curves • Hip, knee, and foot alignment
Mobility tests	• Shoulder flexion active range of movement • Shoulder prone internal and external rotation range in 90°
Flexibility tests	• Levator scapulae muscle length • Modified Thomas test
Strength tests	• Calf heel raises • Shoulder internal and external rotation strength using handheld dynamometer
Stability tests—joint and functional	• Ankle medial ligament stress test • Shoulder apprehension test for anterior glenohumeral joint instability
Special tests	• Shoulder combined elevation test • Ankle posterior impingement test

o understands any special risks involved in the testing;

o consents to the specific uses of the information obtained from the testing sessions (such uses should be clearly specified, e.g., for statistical and scientific analysis, performance analysis, publication in scientific journals or papers without identification of the individual athlete); and

o is otherwise satisfied that the confidentiality of information obtained during testing will be protected.

• Include a statement to the effect that the athlete understands the nature of the sport science tests and their risks and that the athlete consents, on that basis, to participate in the tests for the purpose of constructing a valid and reliable profile of his or her condition or performance.

• Require that athletes advise the tester if they have any injury, illness, or physical defect at the time of testing; if they believe that they cannot complete the test safely; or if there is any other reason that may impair their ability and or capacity to complete the test.

• To the extent determined appropriate by the testing organization, require that the athlete release the testing organization and its employees from any liability associated with the testing.

• Obtain the athlete's signature. When the athlete is a minor (<18 years of age) and the consent of a parent or guardian is required, the consent documentation should provide a statement for the parent or guardian to sign.

Form 2.4 is a sample informed consent form, and form 2.5 is a sample of the information that can be given to the athlete to explain the assessment procedures.

Pretest Athlete Questionnaire

A pretest questionnaire completed by the athlete immediately prior to testing is a valuable way of documenting information regarding the many variables that can potentially influence testing results. This information can be particularly useful when data are retrospectively evaluated and unique results examined. Form 2.6 is an example of a pretest questionnaire designed to gather relevant information quickly from the athletes immediately prior to testing.

Pretest Diet Considerations

The athlete's pretest diet can influence the accuracy and reliability of physiological testing results. Because an athlete's dietary intake can influence metabolism and exercise performance, consideration should be given to standardizing the diet 24 to 72 hours before physiological testing.

Jeacocke and Burke (2010), in a review of methods of standardizing dietary intake prior to performance testing, summarized the effects of different nutrients and dietary components on exercise performance and outlined common approaches to standardizing nutrients prior to performance testing. The authors identify key nutrients that may require standardization prior to testing due to their potential effect on performance, including total energy intake, carbohydrate intake, protein intake, and fluid, caffeine, and alcohol consumption. Their review found that in some instances athletes are prescribed specific diets to follow whereas in other cases athletes are asked to maintain habitual intakes in the lead-up to a performance test, each approach having advantages and disadvantages. Jeacocke and Burke (2010) reported that no studies are available quantifying the effect of standardizing pretest diet on the reliability of different exercise test protocols and, as such, research into this topic could provide valuable insights and potentially change the way athletes prepare for testing.

Table 2.3 is a modified excerpt from Jeacocke and Burke (2010) that summarizes key nutrient and dietary components that may require standardization prior to exercise testing and outlines common approaches used by scientists to control these components.

Pretest Checklists

Each facility should develop pretest checklists for testing of athletes so that preparation is standardized and all safety considerations are covered. Forms 2.7 and 2.8 are sample pretest checklists for testing in the laboratory and the field. These checklists summarize important tasks involved in the preparation of athletes and their environment prior to testing, in both the laboratory and the field. These checklists should be modified to suit each facility so that they reflect the individual operating requirements of the laboratory and incorporate any additional instructions regarding equipment specific to the facility. It is recommended that the checklists be posted in the laboratory and kept in the field testing kit to ensure that they are easily accessible for staff.

TABLE 2.3 Nutrients and Dietary Components Requiring Standardization or Control

Nutrient or component	Background	Common approaches to dietary standardization or control
Energy	Acute or chronic periods of energy deficit or low energy availability (<30 kcal · kg⁻¹ of fat-free mass per day) alter metabolic rate, nitrogen balance, glycogen storage, and hormonal responses. Such alterations could be expected to alter metabolism and exercise performance. It is desirable that athletes be in a state of reasonable energy balance and adequate energy availability.	Several strategies can be used to promote reasonable energy balance in the days leading to performance testing: Impose a standard energy intake for the athlete by setting requirements from predictions of RMR and the energy cost of daily activity or from a standard energy availability of 45 kcal · kg⁻¹ fat-free mass. A disadvantage is that the true energy requirements of individuals may be different than predicted values and thus energy balance may not be achieved. Ask athletes who are weight stable to follow their usual diet before testing. Ask athletes to keep a record of their usual diet for a period before the study to provide an estimate of their usual energy intake. This may then be used to set targets for self-chosen dietary control methods or preparation of prepackaged foods.
CHO	Consuming CHO in the hours before a test restores liver glycogen after an overnight fast and tops up suboptimal muscle glycogen stores, but it also changes substrate use (reduced fatty acid oxidation, increased CHO use) during subsequent exercise at submaximal intensities. The net effect on CHO availability during exercise depends on the amount of CHO consumed preexercise vs. the increase in CHO utilization. In some athletes, intake of small amounts of CHO in the hour before exercise is associated with hypoglycemia and reduced endurance.	As in the case of energy, scientists may choose to provide a standardized approach to CHO intake before and during a performance test or allow athletes to follow their usual dietary practices. The advantages and disadvantages identified above also apply to CHO intake. The guideline for CHO in the 1-4 h before exercise is 1-4 g · kg⁻¹.
Protein	Habitual protein intake will affect protein metabolism and requirements, and a pretrial standardization period is needed for certain interventions such as measurements of nitrogen balance or whole-body protein utilization. Although intake of protein before or during a trial, particularly in comparison with habitual intake, may alter metabolism or protein kinetics, it is unlikely that these issues will directly affect performance of a single exercise bout.	Researchers can choose between setting a standardized protein intake and allowing participants to follow habitual intakes (see comments for energy). For convention, protein intakes are often standardized in the 24 h pretrial diet to provide a certain protein amount (g · kg⁻¹ body mass) or a certain % of total energy intake. Habituation to different levels of protein intake is usually carried out over longer periods (e.g., 5-7 days).
Fluid	Dehydration is likely to impair performance of prolonged or strenuous exercise undertaken in hot environments, especially when additional fluid loss incurred during a performance trial is added to a preexisting fluid deficit. In most situations, it is desired or assumed that athletes begin a test in a euhydrated state.	To ensure euhydration, dietary standardization protocols may require athletes to consume a certain volume of fluid on the evening before or hours before a performance test, leaving sufficient time for excess fluid to be excreted as urine. Urine characteristics such as specific gravity or osmolality can be measured before a test to monitor pretest hydration.

Nutrient or component	Background	Common approaches to dietary standardization or control
Caffeine	Caffeine has various profound effects on metabolism and can enhance the performance of a range of exercise protocols. It is consumed as part of a normal diet for most adults as well as for specific ergogenic effects on performance. Withdrawal from habitual caffeine can cause side effects such as lethargy and headaches.	It is common practice to require participants to avoid the acute effects of caffeine by abstaining from caffeine-containing products in the pretest period of up to 24 h. This includes caffeine-containing products such as drinks, foods, and specialized "energy" or sports products. This is often achieved by providing athletes with a region-specific list of caffeine products that they should avoid. A disadvantage of this approach is that the athlete can suffer withdrawal symptoms. An alternative approach is to allow athletes to follow and document habitual caffeine practices.
Alcohol	Alcohol intake, particularly in excess, affects CHO metabolism, protein synthesis, and hydration status. It can also distract athletes from following their normal dietary practices or complying with instructions. Acute intake of alcohol and an alcohol "hangover" (period of up to 24 h after excessive alcohol intake) can impair performance of various exercise protocols.	It is common practice to require participants to completely avoid intake of alcohol for 24 h before a performance test. In some situations, strict avoidance may not be required, and alcohol intake might be limited to small amounts (<20 g per day) in line with population guidelines for healthy use of alcohol.

CHO = carbohydrate; RMR = resting metabolic rate.

Adapted, by permission, from N. Jeacocke and L. Burke, 2010, "Methods to standardize dietary intake before performance testing," *International Journal of Sports Nutrition and Exercise Metabolism* 20: 87-103.

References

American College of Sports Medicine. 2010. *ACSM's Guidelines for Exercise Testing and Prescription*, 8th ed. Philadelphia, PA: Lippincott Williams & Wilkins.

Brukner, P., White, S., Shawdon, A., and Holzer, K. 2004. Screening of athletes: Australian experience. *Clinical Journal of Sport Medicine* 14(3):169-177.

Jeacocke, N., and Burke, L. 2010. Methods to standardize dietary intake before performance testing. *International Journal of Sport Nutrition and Exercise Metabolism* 20(2):87-103.

Maron, B. 2003. Sudden death in young athletes. *New England Journal of Medicine*. 349:1064-1075.

Occupational Health and Safety Code of Practice. 2008. Australian Federal Register of Legislative Instruments F2008L02054.

Occupational Health and Safety Act. 1991. Australian Federal Register of Legislative Instruments F2008L02054.

Sports Medicine Australia (SMA). 2005. *Pre-Exercise Screening System*. Dickson, ACT: Australian Government Department of Health and Ageing.

FORM 2.1 Hazard Identification Checklist

Assessment team name(s)_____

Description of location/testing session: _____ Date: _____

Potential source of hazard	Description of hazards	Risk assessment (circle one)	Description of controls	Risk assessment with controls in place (circle one)
TESTING ENVIRONMENT				
Emergency response		High/medium/low		High/medium/low
Electrical hazards		High/medium/low		High/medium/low
Thermal hazards		High/medium/low		High/medium/low
Mechanical hazards		High/medium/low		High/medium/low
Slip and trip hazards		High/medium/low		High/medium/low
EQUIPMENT AND INSTRUMENTS				
Exposure to moving parts		High/medium/low		High/medium/low
Oversize or heavy		High/medium/low		High/medium/low
Ergonomics		High/medium/low		High/medium/low
Operation, transport, storage		High/medium/low		High/medium/low
PROTOCOLS AND PROCEDURES				
Test methods		High/medium/low		High/medium/low
Equipment operations		High/medium/low		High/medium/low
STAFF TRAINING AND SKILLS				
Staff training and qualifications		High/medium/low		High/medium/low
Bystanders and general public		High/medium/low		High/medium/low
DANGEROUS AND HAZARDOUS GOODS				
Dangerous goods		High/medium/low		High/medium/low
Hazardous goods		High/medium/low		High/medium/low

Other comments:

From Australian Institute of Sport, 2013, *Physiological tests for elite athletes*, 2nd ed. (Champaign, IL: Human Kinetics).

FORM 2.2 AHA/ACSM Health/Fitness Facility Preparticipation Screening Questionnaire

Name: _____ ☐ M ☐ F

DOB: _____ Date: _____

Home address: _____ Home phone: _____ Mobile: _____

Email: _____ Sport: _____

Assess your health needs by marking all **true** statements.

History

You have had:

_____ A heart attack

_____ Heart surgery

_____ Cardiac catheterization

_____ Coronary angioplasty (PTCA)

_____ Pacemaker / implantable cardiac defibrillator / rhythm disturbance

_____ Heart valve disease

_____ Heart failure

_____ Heart transplantation

_____ Congenital heart disease

Symptoms

_____ You experience chest discomfort with exertion

_____ You experience unreasonable breathlessness

_____ You experience dizziness, fainting, or blackouts

_____ You take heart medications

Other Health Issues

_____ You have diabetes

_____ You have asthma or other lung disease

(continued)

_____ You have burning or cramping sensations in your lower legs when walking short distances

_____ You have musculoskeletal problems that limit your physical activity

_____ You have concerns about the safety of exercise

_____ You take prescription medications

_____ You are pregnant

If you marked any of these statements in this section, consult your physician or other appropriate health care provider before engaging in exercise. You may need to use a facility with a medically qualified staff.

Cardiovascular Risk Factors

_____ You are a man older than 45 years

_____ You are a woman older than 55 years, have had a hysterectomy, or are postmenopausal

_____ You smoke, or quit smoking within the previous 6 months

_____ Your blood pressure is >140/90 mmHg

_____ You do not know your blood pressure

_____ You take blood pressure medication

_____ Your blood cholesterol level is >200 mg · dL^{-1}

_____ You do not know your cholesterol level

_____ You have a close blood relative who had a heart attack or heart surgery before age 55 (father or brother) or age 65 (mother or sister)

_____ You are physically inactive (i.e., you get <30 minutes of physical activity on at least 3 days per week)

_____ You are >20 pounds overweight

If you marked two or more of the statements in this section you should consult your physician or other appropriate health care provider before engaging in exercise. You might benefit from using a facility with a professionally qualified exercise staff * to guide your exercise program.

_____ None of the above

You should be able to exercise safely without consulting your physician or other appropriate health care provider in a self-guided program or almost any facility that meets your exercise program needs.

*Professionally qualified exercise staff refers to appropriately trained individuals who possess academic training, practical and clinical knowledge, skills, and abilities commensurate with the credentials defined in ACSM 2010.

Source: American Heart Association /ACSM Health/Fitness Facility Preparticipation Screening Questionnaire (American College of Sports Medicine 2010). From Australian Institute of Sport, 2013, *Physiological tests for elite athletes*, 2nd ed. (Champaign, IL: Human Kinetics).

FORM 2.3   Basketball Musculoskeletal Screening Form

Tester: _____ Date: _____ Squad: _____

Testing location: _____ Testing surface: _____ Temperature: _____ Humidity: _____

Testing time: _____ Name of athlete: _____

Date of birth: _____/_____/_____ Age: _____

Gender: ❒ Male   ❒ Female

Height (cm): _____   Mass (kg): _____

Home address: _____ Home telephone: _____

E-mail address: _____ Mobile telephone: _____

Position: _____

Shoots: R   L

Jumps off: R   L

Lands on: R   L

Coach: _____ Telephone: _____

Time in sport (years): _____ Current level: _____

Training volume per week: _____ Games per week: _____

Shoes (brand/style): _____

Do You Wear?

Foot orthotics	Yes	No	Type _____
Any braces or tape	Yes	No	Type _____
Contact lenses	Yes	No	Type _____
Mouth guard	Yes	No	Type _____

Previous Injuries

Date	Injury	Code
____/____/____	_____	_____
____/____/____	_____	_____

(continued)

21

___/___/___ _____ _____

___/___/___ _____ _____

___/___/___ _____ _____

___/___/___ _____ _____

Current Injuries

Date	Injury	Code
___/___/___	_____	_____
___/___/___	_____	_____

SCREENING TESTS

Standing	Left	Right	Comments
Posture: spinal curves			
Posture: hip–knee–foot alignment			
Lumbar spine FF			
Lumbar spine extension			
Lumbar spine quadrant			
GHJ flexion			
GHJ abduction			
GHJ IR: hand behind back			
GHJ ER: hand behind head			
GHJ empty can			
One-leg squat: quality			
Hop: quality			
Decline squat: pain			
Decline squat: range			
Decline squat: quality			
Dorsiflexion lunge (knee to wall)			

Standing	Left	Right	Comments
Gastrocnemius length			
Calf heel raises: endurance	Reps.....	Reps.....	
Proprioception EC 30 s			
Proprioception mat EC 30 s			

Sitting	Left	Right	Comments
Slump			
Thomas test: ITB			
Thomas test: iliopsoas			
Thomas test: RF			
Tx–Lx rotation			
GHJ IR in abduction			
GHJ impingement tests			
Elbow flexion			
Elbow extension			
Forearm supination			
Forearm pronation			
Wrist ROMs			
TFCC			
Thumb MCL			
Scaphoid palpation (if history of injury)			

Supine	Left	Right	Comments
GHJ apprehension			
GHJ anterior drawer			
Abdominal musculature			
Leg lengths: lying			
Leg lengths: sitting			
Squish			

(continued)

Supine	Left	Right	Comments
Pelvic position			
Hip flexion ROM			
Hip IR in 90/90			
Hip ER in 90/90			
Hamstrings 90/90			
Straight-leg raise			
Long adductor length			
Squeeze test 0°			
Squeeze test 45°			
Squeeze test 90°			
Knee flexion			
Knee hyperextension			
Tibial IR at 90°			
Tibial ER at 90°			
PFJ medial glide			
PFJ medial tilt			
Lachman test			
Pivot shift			
Posterior drawer			
Anterior drawer			
MCL stress			
LCL stress			
McMurray's test			
VMO bulk	Atrophy or normal	Atrophy or normal	
Palpation: patellar tendon			
Palpation: joint line			
Palpation: tibial tubercle			

Supine	Left	Right	Comments
Anterior lower-leg compartment			
Posterior lower-leg compartment			
Palpation: medial tibial border			
Subtalar joint mobility			
Talar tilt			
TCJ anterior drawer			
Resisted eversion (peroneus group)			

If indicated:

Ankle posterior impingement			
Ankle anterior impingement			
Palpation navicular bone			
Palpation talar dome			

Prone	Left	Right	Comments
Hip extension ROM			
Hamstring bulk	Atrophy or normal	Atrophy or normal	
Calf bulk	Atrophy or normal	Atrophy or normal	
Palpation: Achilles tendon			

EC = eyes closed; ER = external rotation; FF = forward flexion; IR = internal rotation; ITB = iliotibial band; GHJ = glenohumeral joint; MCL = medial collateral ligament; LCL = lateral collateral ligament; PFJ = patellofemoral joint; RF = rectus femoris; ROM = range of movement; TCJ = triangular fibrocartilage complex; Tx-Lx = thoracic-lumbar; VMO = vastus medialis oblique.

FORM 2.4   Sample Informed Consent Form

I (print name) _____ consent to participation in this physiological assessment on the following terms:

1. I have read the Explanation of Physiological Assessment Procedures attached and understand what I will be required to do. I have had the opportunity to ask questions and have received satisfactory explanations about the tests to be conducted.

2. I understand that I will be undertaking physical exercise at or near the extent of my capacity and there is possible risk in the physical exercise at that level, such as episodes of transient light-headedness, fainting, abnormal blood pressure, chest discomfort, and nausea.

3. I understand that this may occur although the staff in this laboratory will take all proper care in the conduct of the assessment, and I fully assume that risk.

4. I understand that I can withdraw my consent, freely and without prejudice, at any time before, during, or after testing.

5. I have told the person conducting the assessment of any illness or physical defect I have that may contribute to the level of that risk.

6. I understand that the information obtained from the test will be treated confidentially with my right to privacy assured. However, the information may be used for statistical or scientific reasons with privacy retained. (Note: Members of sport teams should have made special arrangements about treatment of individual data with team coach or manager.)

7. I release this laboratory and its employees from any liability for any injury or illness that I may experience during the assessment as well as any subsequent injury or illness that is connected to or to any extent influenced by the assessment.

8. I will indemnify this laboratory in respect to any liability it may incur in relation to any other person in connection with the assessment.

9. I hereby agree that I will present myself for testing in a suitable condition having abided by the requirements for diet and activity prescribed to me by laboratory staff.

Participant signature: _____   Date: _____

_____   _____   _____
Parent/guardian name (required if age less than 16)   Signature   Date

_____   _____   _____
Witness name   Signature   Date

From Australian Institute of Sport, 2013, *Physiological tests for elite athletes*, 2nd ed. (Champaign, IL: Human Kinetics).

FORM 2.5 Explanation of Physiological and Anthropometric Assessment Procedures

Information for Participating Athletes

Capillary Blood Test

Capillary blood samples (typically 5-75 µL or 1-15 small droplets) are taken from either the earlobe or fingertip before, during, or after exercise testing and are conventionally used to assess pH, lactate, and bicarbonate values. Additional parameters can be measured at the discretion of the physiologist. A capillary blood sample is obtained by using a lancet device, which makes a small puncture into the skin. Gloves and lancets are single use only and are discarded after every sample. This is a simple procedure and causes only minor discomfort.

Venous Blood Test

Venous blood samples (typically 2-10 ml) are taken by a trained phlebotomist to analyze routine blood measures including red and white blood cells and proteins. This test involves a standard forearm vein sampling technique whereby a small needle is inserted into the forearm vein while the athlete is lying down. There may be some temporary minor discomfort associated with this procedure, and there is potential for some local bruising to occur. Measures are taken to limit discomfort and the risk of bruising.

Anthropometry

Anthropometric assessment involves simple measurements of stature (height), body mass, skinfolds, girths, limb lengths, and bone breadths. In addition, skinfold thickness is measured using handheld calipers across several sites depending on the level of assessment and needs of the athlete. The athlete undertaking the anthropometry measures is typically required to be dressed in underclothing. There is minimal physical discomfort associated with these measurements.

Maximal Oxygen Consumption Test

Maximal oxygen consumption ($\dot{V}O_2max$) can be measured directly using a gas analysis system or can be extrapolated from the multistage shuttle run. Both tests require the athlete to exercise to a maximal level of exertion, and the test ceases when the athlete can no longer maintain the required pace. As with any exercise to maximal exertion there are potential associated risks, including temporary heavy breathing, muscular fatigue, episodes of light-headedness, fainting, abnormal blood pressure, nausea, and chest discomfort.

- **Multistage shuttle run**: The multistage shuttle run (and similar tests) is used to estimate an athlete's maximal aerobic capacity in the field or on a court. The test involves athletes running between two lines that are placed 20 m apart. The aim is to reach the 20 m line and turn as an audible beep is emitted from an audio device. The time between beeps progressively shortens, thus increasing the workload on the athlete. The test starts at a moderate intensity and progressively increases until the athlete becomes fatigued and cannot maintain the required pace. The athlete finishes the test when he or she fails to maintain the required pace for two successive beeps.

- **Gas analysis**: The athlete undertakes a progressive incremental test (i.e., the initial workload is light and then increases gradually to such a level that the athlete cannot sustain the work rate) that is performed on a sport-specific ergometer (i.e., cycling ergometer, rowing ergometer, or treadmill). Gas analysis measures ventilation, which is the athlete's maximal ability to transport and use oxygen for energy production. The athlete is required to wear a mouthpiece and headwear to collect expired ventilation. Heart rate is usually monitored with a transmitter belt attached to the chest.

- **Work capacity test**: A 7 min work capacity test is used to determine the maximal amount of work an athlete is able to complete in 7 min. This test is most commonly used with athletes who compete in a wheelchair. The test requires athletes to push around a 20 m square as often as possible in the allocated time.

(continued)

Submaximal Aerobic Power Test

Submaximal aerobic power tests assess cardiovascular fitness or aerobic power and involve exercising on an ergometer at low to moderate intensity that progressively increases throughout the test. For example, in a cycling protocol, an athlete pedals on a cycle ergometer at a light rate (25 W). This load is increased in small increments (25 W) every minute until the heart rate reaches 75% of the age predicted maximum, at which time the test will stop.

Heart Rate Monitoring

Heart rate is monitored with a specific device. A transmitter belt is worn around the chest, and the measured heart rate is displayed on a watch receiver attached to the wrist or ergometer or is recorded by the transmitter belt itself.

Anaerobic Capacity Tests

Anaerobic capacity tests involve short-duration maximal exercise efforts and can be conducted in the laboratory or in the field using different modalities depending on the parameter to be measured and the protocol to be used. As with any exercise to maximal exertion there are potential associated risks, including temporary heavy breathing, muscular fatigue, episodes of light-headedness, fainting, abnormal blood pressure, nausea, and chest discomfort. Some examples of anaerobic laboratory tests include these:

- **Cycling**: This test, conducted on the cycle ergometer, assesses anaerobic power and capacity (i.e., outright speed and sustained speed). Anaerobic power is assessed using a 10 s all-out effort and the anaerobic capacity by an all-out 30 s effort. The 30 s test is typically very exhausting.

- **Running**: For runners, sustained running speed is measured using the treadmill. Athletes are required to run to exhaustion (approximately 1-2 min) on a quickly moving treadmill on a moderate gradient. This test requires a high level of skill and fitness and is only used with trained runners.

- **Lower-body strength tests** (force plate measurements): The force plate enables the measurement of specific strength and power variables such as starting strength, explosive strength, and isoinertial stretch–shortening cycle strength (ISSC). Most of these strength measures are assessed with simple one- and two-leg jump tests.

- **Speed and acceleration**: Speed over a given distance (typically 20 or 40 m) is measured through the use of light gates that are activated when the athlete passes through them.

- **Vertical jump**: Explosive leg power is assessed through vertical jump height, which is measured using a vertical jump testing device (e.g., yardstick).

Agility Tests

Agility tests are sport-specific; however, the purpose of all agility tests is to assess the ability of the athlete to rapidly change direction and position. Athletes are expected to complete the tests with maximal speed, and the risks associated with maximal exertion are inherent.

Strength Tests

Different strength testing protocols are used to determine muscular strength for a given activity. A range of tests can be used that assess either the single or repeat maximal capacity of different muscles and muscle groups. Following the directions of the tester, the athlete performs different exercises using either free weights or his or her own body weight using a set number of repetitions or repetitions to volitional exhaustion. Athletes are expected to perform tests with maximal effort and may experience delayed-onset muscle soreness after testing.

From Australian Institute of Sport, 2013, *Physiological tests for elite athletes*, 2nd ed. (Champaign, IL: Human Kinetics).

FORM 2.6   Athlete Pretest Questionnaire

Tester: _____   Date: _____

Sport: _____   Squad: _____

Testing location: _____   Testing surface: _____

Temperature: _____ Humidity: _____   Testing time: _____

Name of athlete: _____   Date of birth: ____/____/____

Age: _____ Gender: ❏ Male ❏ Female   Height (cm): _____ Mass (kg): _____

Home address: _____   Home telephone: _____

E-mail address: _____   Mobile telephone: _____

Diet

How would you rate your diet over the last 2 days?

❏ Poor   ❏ OK   ❏ Good   ❏ Excellent

How many hours ago did you eat your last meal?   _____

Record foods eaten over the last 24 h:

Meal	Foods including drinks	Portion size (cups, grams)
Breakfast		
Snacks		
Lunch		
Snacks		
Dinner		

(continued)

Environment

Have you been training in hot conditions over the last 2 weeks?   Yes ❑   No ❑

If yes, provide details: _____

Have you been training or sleeping at altitude over the last 2 weeks?   Yes ❑   No ❑

If yes, provide details: _____

Illness

Are you currently suffering from any type of illness?   Yes ❑   No ❑

If yes, provide details: _____

Have you had any type of illness or health problem over the last 2 weeks?   Yes ❑   No ❑

If yes, provide details: _____

Injury

Do you currently have any injuries?   Yes ❑   No ❑

If yes, provide details: _____

Have you had any injuries in the last 2 weeks?   Yes ❑   No ❑

If yes, provide details: _____

Medications and Supplements

Are you currently taking any medication?   Yes ❑   No ❑

If yes, provide details: _____

Have you taken any medication over the last 2 weeks?   Yes ❑   No ❑

If yes, provide details: _____

Have you taken any supplements over the last 2 weeks?   Yes ❑   No ❑

If yes, provide details: _____

Motivation

Evaluate your motivation for training today.

❏ Poor ❏ OK ❏ Good ❏ Excellent

Evaluate your motivation for testing today.

❏ Poor ❏ OK ❏ Good ❏ Excellent

Training

Evaluate your last week of physical training.

❏ Easy ❏ Moderate ❏ Hard ❏ Very hard

How fatigued are you today? (0 = not at all; 5 = extremely)

❏ 0 ❏ 1 ❏ 2 ❏ 3 ❏ 4 ❏ 5

How many hours ago did you last exercise? _____

Describe your last three training sessions:

Time	Training session	Difficulty (easy, moderate, hard)
Today		
Yesterday		
2 days ago		

(continued)

FORM 2.6 *(continued)*
Travel

Have you had to travel over the last 7 days?   Yes ❐   No ❐

If yes, provide details: _____

_____

_____

_____

_____

Miscellaneous

Please provide any additional information that you believe may influence your testing results:

_____

_____

_____

_____

_____

_____

_____

_____

_____

_____

_____

_____

_____

_____

_____

FORM 2.7 Pretest Checklist for Testing in the Laboratory

Testing Environment

1. The risk minimization process has been completed for the test protocols you are conducting.

2. The laboratory is clean and free of hazards. The testing area has been defined and cordoned off if required.

3. The equipment to be used is calibrated and operating correctly.

4. You have prepared all equipment for the testing session (i.e., make up mouthpieces, warm-up metabolic cart, stock and prepare blood trolley for use).

5. Laboratory conditions are appropriate for testing. The temperature is maintained between 18 and 23 °C, relative humidity is less than 70%, and environmental conditions have been noted on the testing recording sheet.

6. The time of day is recorded on the testing recording sheet.

7. The use of fans during the testing session is noted on the testing recording sheet.

8. Testers and athletes have appropriate access to first aid facilities and medical assistance.

Procedural Considerations

1. All pretest conditions, as specified in the protocol, have been implemented.

2. You have established the test order (as per the test protocol) if multiple tests are being conducted.

3. A copy of the test protocol is printed and readily available for staff conducting the testing.

4. Testing recording forms, rating of perceived exertion (RPE) chart, and all other relevant documentation and forms are available.

5. Staff members are appropriately trained and competent to conduct the tests.

Athlete Preparation

1. Athletes have been medically screened prior to testing.

2. Athletes have signed the informed consent form.

3. Athletes have completed the pretest questionnaire.

4. Athletes are in good health and injury free.

5. Athletes are dressed appropriately (light and nonrestrictive clothing).

6. Athletes are familiar with the testing protocols.

7. All athletes are encouraged to give their best efforts.

FORM 2.8 Pretest Checklist for Testing in the Field

Testing Environment

1. The risk minimization process has been completed for the test protocols you are conducting.

2. The equipment to be used is calibrated and operating correctly. Spare batteries and consumables, where appropriate, are packed.

3. All equipment is packed safely before leaving for the field testing location.

4. The testing surface is safe and appropriate. If tests are to be done on grass, the grass is not too long and there are no potholes where athletes will be running. If tests are to be conducted on courts, there are no hazards such as water or foreign objects and there is enough room for athletes to run through after sprints or the multistage shuttle run.

5. There is sufficient lighting at the testing venue.

6. The environmental conditions are measured and recorded on testing recording sheet.

7. The time of day is recorded on testing recording sheet.

8. Testers and athletes have appropriate access to first aid facilities and medical assistance.

Procedural Considerations

1. All testing recording sheets and documentation are packed.

2. All pretest conditions, as specified in the protocol, have been implemented.

3. Testers arrive at the venue with sufficient time to set up equipment prior to testing.

4. A copy of the test protocol is printed and readily available for staff conducting the testing.

5. The test order (as per the test protocol) has been established, if multiple tests are conducted.

6. Staff members are appropriately trained and competent to conduct the tests.

Athlete Preparation

1. All athletes have been medically screened prior to testing.

2. All athletes have signed the informed consent form.

3. All athletes have completed the pretest questionnaire.

4. Athletes are in good health and injury free.

5. Athletes are dressed appropriately (light and nonrestrictive clothing).

6. Athletes are familiar with the testing protocols.

7. The athletes' warm-up is appropriate and thorough.

8. All athletes are encouraged to give their best efforts.

From Australian Institute of Sport, 2013, *Physiological tests for elite athletes*, 2nd ed. (Champaign, IL: Human Kinetics).

Data Collection and Analysis

David B. Pyne

Physiological testing of athletes is conducted to provide evidence-based advice for coaches, athletes, and scientists. The credibility and value of the results of physiological testing depends largely on the confidence in the underlying data. A systematic approach to collection, processing, analysis, and reporting of testing results is needed. It is often said that the value of a scientific enterprise, in this case the physiological testing of athletes, is only as good as the underlying data. Effective analysis of data involves much more than applying the appropriate statistical test. Various analytical approaches have evolved for analyzing data collected from the physiological monitoring of individual athletes and for groups of athletes in team sports or research studies.

Data Capture and Storage

Results of physiological testing are captured by either manual or automated methods. Despite advances in technology and communication, a substantial amount of data are still collected manually and handwritten on a recording sheet. Data captured electronically should be stored and filed for subsequent processing, analysis, and retrieval. All computer systems should have an automatic or manual backup in case files are accidentally erased, hardware malfunctions, or data are corrupted in some other way. Portable devices such as a laptop personal computer, tablet computer, smartphone, portable hard drive, or USB memory drive can be misplaced or lost. Manual copies of data files should be made and stored in a different location if automatic backup is not available. This is particularly important for field testing away from the main laboratory: All test results should be secured immediately upon completion of the test-

ing. This responsibility usually falls on the supervising physiologist, but junior or technical staff assigned to individual testing stations should also ensure the security of data sets at all times. Where appropriate, the confidentiality of data should be maintained to avoid the prying eyes of other athletes, officials, and the media. Raw data must be accessible so that the veracity of test results can be double-checked in response to a query by a coach, athlete, or scientist.

Data Cleansing

Data processing is prone to the occasional error if there is a large volume of data to transfer, the format of the data is complex, or time is short. The accuracy of all data manually entered into a database or spreadsheet should be double-checked. This is best achieved by cross-referencing of data between the original source and the final destination. With larger data sets, two people can quickly check material: one person with the original data and one with the uploaded data. Descriptive statistics including the mean, standard deviation, minimum value, maximum value, count, and any other metrics of choice should be prepared on the uploaded data. Visual inspection of these statistics can reveal the presence of miscoded data—knowledge of expected sport-specific values assists this process. Statistical packages are useful in outputting outliers and identifying missing values. Miscoded or outlier values can be dealt with by reference to predetermined exclusion criteria related to the number of standard deviations away from the mean value. Suspect values that might be erroneous should be checked against the raw data and, if necessary, with the staff member responsible for that particular test. When missing values are imputed by interpolation (either linear or nonlinear), a systematic approach should be used and documented. Other measures of viewing the data include a frequency distribution (table or histogram), stem-and-leaf plot, and plots of individual athlete data as part of a team or group analysis.

Data Transformation

Coaches and athletes prefer to see raw data of the actual test scores in reports of physiological testing. However, data transformation is often undertaken to improve the rigor of the statistical analysis and make the data more accessible and comparable for the scientist and coach. Application of parametric tests is based on the assumption that the sampling distribution of outcome statistic (not the raw data) is normal. When the distribution of data is partly or grossly nonnormal, some scientists erroneously believe that the only solution is to use a nonparametric test. However, mathematical transformation of data prior to analysis is preferred over nonparametric testing (Atkinson et al. 2010; Hopkins et al. 2009). The choice of data transformation depends on the format of the experimental data and the type of statistical analysis being performed. The preferred method for continuous or interval data is log-transformation using natural logs. Log-transformation (and then back-transformation) yield effects as percents that are useful in generalizing the outcome of an intervention to other athletes or groups. Counts and proportion also present some challenges in the analysis of data. Other options for data transformation include percentile rankings, square-root transformation, and arcsine-root transformation. Consulting a statistician is a good idea at the planning stage of a research project or preceding analysis of complex data for an individual athlete.

Analytical Approaches to Testing Athletes

Physiological testing can be conducted to monitor effects in groups (teams or squads of athletes or subjects in a research study) and in individual athletes. Sport scientists and coaches have for decades made inferences on test scores using basic descriptive statistics and hard-earned practical experience. The mean value and standard deviation are used to characterize a data set. The mean value and confidence limits are used to interpret a within-subject change (from test to test for an individual athlete) or a between-subjects difference in test scores between groups. A more thorough approach involves comparison of

the magnitude of the signal (change in test score) to noise (the typical error [TE] of measurement). Is the observed change in fitness or performance a real change or simply error or noise in the test? Another level of interpretation involves determining whether the magnitude of the (unknown) true change is substantially greater than the smallest sport-specific worthwhile or practically important change. The concept of the smallest worthwhile change in a performance or physiological measure in sport or exercise sciences (Batterham and Hopkins 2006) is analogous to the meaningful clinically important difference (Barrett et al. 2005, 2007) and patient-based effect size approach (Lehmann 2005) used in clinical medicine.

The traditional approach to making an inference on clinical or research data has centered on whether the change of difference in performance or physiological measure is statistically significant. Most sport scientists and exercise physiologists are familiar with interpretation of p values, typically with a threshold value of $p < .05$. However there are several shortcomings in using a traditional analytical approach involving statistical significance, hypothesis testing, and p values (Stapleton et al. 2009). The null hypothesis is always false, there are no null effects in nature, and with a large enough sample size all effects are statistically significant. There is also frequent misinterpretation of nonsignificant effects as null or trivial when in fact they may have a small or moderate effect on an important dependent measure (Shrier and Batterham 2002). To address these shortcomings, more contemporary statistical approaches are emerging that provide options for analysis of individual and group data in the exercise sciences. These approaches are used to characterize the practical and clinical significance of effects, ensure precision of estimation using confidence limits, and make magnitude-based inferences about population effects of training and lifestyle interventions (Hopkins et al. 2009).

Individual Athletes

The most common form of testing for athletes involves repeated measurements of fitness and performance parameters within a season and from season to season during the athlete's career. Repeated measurements yield a change score from test to test and over specified periods of time. The key data required to interpret test results at the level of an individual athlete are as follows (Hopkins 2004):

- Raw data for the original and subsequent tests and the resulting (observed) change score

- The smallest worthwhile change or reference value for the particular test

- The typical error (noise or uncertainty) in the measurement derived from a reliability study

For example, a team sport athlete improves her 20 m sprint test time from 3.52 to 3.42 s with a change score (improvement) of 0.10 s. The smallest worthwhile change has been established as 0.05 s and the typical error for the 20 m sprint test as 0.04 s. The question is whether the change (0.10 s) is greater than the reference change (0.05 s) after accounting for the noise or error in the test (0.04 s). The computations (data not shown) reveal an 81% likelihood that the improvement is worthwhile in practical terms. Systematic rules, using a probabilistic approach and likelihoods, have been devised to determine whether the true change in a fitness or performance measure is beneficial, trivial, harmful, or unclear (Batterham and Hopkins 2006). The noise in many common laboratory and field physiological and performance tests is greater than the smallest worthwhile change. The difficulties in identifying small but worthwhile changes in fitness and performance in the face of noisy tests present a challenge for the physiologist. In some cases, the physiologist may be confident of identifying only moderate or large changes. Using this approach, the physiologist needs access to the typical error values for all tests used in routine testing of athletes, an estimate of the smallest worthwhile change, and confidence that the raw data are collected under conditions that satisfy the quality assurance program for the laboratory.

Team Athletes

The general approach to characterizing sample or group effects involves confidence limits and reference or threshold values for important effects (Hopkins et al. 2009). An effect is rated as unclear if the confidence interval, which represents uncertainty about the true (unknown) value of the population, overlaps values that are substantial in a positive and negative sense. Otherwise the effect is characterized with a statement on the likelihood that it is trivial, positive, or negative. The effect is considered to be unclear if its chance of benefit is at least promising but its risk of harm is unacceptable; alternatively, the effect is rated as trivial, beneficial, or harmful. Default levels of confidence for both mechanistic (90%) and clinical (benefit 25% and harm 0.5%) inferences have been proposed (Hopkins et al. 2009). Using 90% rather than 95% confidence intervals discourages the scientific community from reinterpreting the outcome as statistically significant or nonsignificant at the level of $p < .05$ (Sterne and Smith 2001). Assessment of clinical benefit and harm can involve raw effects; percentages and factor differences in means; ratios of rates, risks, and odds; standard deviations; and correlations.

Methods have been established for estimating the magnitude of the individual response to a treatment with a group (sample) research trial or randomized controlled trial. The preferred approach is to quantify the true rather than observed individual response to a particular training or lifestyle intervention (Pyne et al. 2005). The true individual response in a randomized trial is computed as the square root of the differences in the variances in the change scores between the treatment and control groups. For example, the mean (group) improvement in a repeat sprint test to a 6-week interval training program might be 5% ± 3% (mean ± 90% confidence interval). In other words the observed mean effect for the group is 5%, but the unknown true improvement is likely to range from 2% to 8%. What about individuals within the group? Assume that the true individual response to the interval training program, which is the magnitude of the improvement in the repeat sprint test score after accounting for the typical error of the test and any other training or maturation effects (control group), is 2.5%. We would advise the coach that the mean effect for the group is about 5% but the magnitude for an individual will typically vary by about 2.5% from athlete to athlete.

Measurement Studies

Every test needs to be assessed for reliability and validity to give the coach and physiologist confidence in the results of testing (Atkinson and Nevill 2007). Suitability of a test is also established by evaluating the signal to noise ratio using the following criteria:

- Signal < noise: test is marginal—find ways to reduce the noise or a better test.
- Signal ≈ noise: test is acceptable—use with caution.
- Signal > noise: test is good—use with confidence.

When the test is rated either marginal or acceptable, its performance can be improved by using repeat trials and attention to standardization of test protocols and procedures. Other factors influencing the choice of a physiological or performance test include the cost, staffing, and practicality of implementing the test in the laboratory or field. Athletes are less willing to undertake physiological tests that are invasive (e.g., muscle biopsy, repeated venipuncture) or exhaustive (prolonged high-intensity time trials). Time and money are always a consideration for the laboratory and scientist. A test must be substantially better in one or more of these factors to warrant its inclusion in a routine battery of tests.

Reliability

Reliability, or the consistency (precision) of a test or measurement, is frequently quantified in the physiology laboratory (Weir 2005). Reliability is important for sport scientists because a high level of reproducibility of test scores is needed to confidently monitor and interpret changes in test scores. A measurement study is needed to establish the reliability of each performance and physiological test used in the laboratory or field. The physiologist needs to determine the number and background of subjects and the number of trials. When possible, subjects should be recruited from the same sport and level of performance as the proposed routine testing. At least 20 to 40 subjects should be recruited to undertake two or more trials in a test–retest framework (Hopkins 2000). Two trials may well be the limit in exhaustive tests, such as a progressive incremental maximal oxygen consumption test, but several trials would be appropriate for tests that are easy to administer (e.g., anthropometric testing, jumps, or short sprints). The time between trials should account for practice, learning, potentiation and fatigue effects, and expected time between repeated measurements in the field or laboratory. A matrix of subjects and trials will allow the scientist to calculate both within-subject and between-subjects standard deviations. Subjects should be given a familiarization trial prior to formal data collection.

The primary effect statistics for reliability are the change in the mean between measurements or trials, the standard error of measurement, and the intraclass (ICC) or Pearson r correlation coefficient (Atkinson and Nevill 1998). The most common measure of reliability, the typical error (TE) of measurement, is computed as the standard deviation of the change scores between repeated measurements divided by $\sqrt{2}$. Table 3.1 details an example of calculating the typical error of measurement in a reliability study. A comprehensive inspection of reliability involves both systematic (bias) and random (imprecision) error. Systematic error can be further divided into constant error (affecting all scores equally) or bias (certain scores are affected differently than others). Limits of agreement (Bland and Altman 1986) have been widely used, although questions about the suitability of this measurement as an index of reliability have been raised. Several different terms are used for measures of reliability, including *accuracy*, *precision*, *repeatability*, *consistency*, and *agreement*—not all of these terms are interchangeable, and some are operationalized differently (Weir 2005). For example, the standard error of measurement should not be confused with the standard error of the mean.

Validity

Validity is defined as the extent to which a measurement of a new physiological or performance measure (i.e., the practical measure) is well-founded and corresponds accurately to the real world (criterion or "gold standard" measure). Many validity studies in the literature have examined new and modified test protocols and testing equipment, devices, and instruments. The first decision in a validity study is choice of the criterion measure. In most cases the criterion measure will be a well-established method used widely in practice and the sport science literature. The measurement error of the criterion measure should be stated. The analysis of validity study typically involves linear or nonlinear regression of the relationship between the practical and criterion measures. A calibration equation, the standard error of the estimate, and a validity correlation coefficient are then computed.

Error

Test–retest error in a physiological test includes biological and technological components. A portion of the error in tests can be attributed to the inherent natural biological variation in human physiology. Noise or error associated with technological or measurement aspects of a test and test equipment can be reduced in several ways. The physiologist should be alert to the emergence or availability of a better test or a new, more reliable ergometer, instrument, device, or item of testing equipment. Possible sources of noise can be reduced by standardizing the test conditions: the athlete's familiarity with the protocol, instructions for testing, warm-up before the test, clothing worn during the test, and pretest exercise and diet. Repeat tests or trials reduce the noise by a factor of $1/\sqrt{(n)}$. Thus, duplicate ($n = 2$) trials reduce the noise by a factor of $1/1.41 = 0.70$ (that is, a ~30% reduction), triplicate ($1/1.73$) trials = 0.58 (~40% reduction), and quadruplicate trials ($1/2.0$) = 0.50 (50% reduction), so repeated tests are certainly worthwhile where available. The practicality of undertaking repeat trials depends on the nature of the test being conducted. Repeat trials are relatively straightforward in anthropometric and brief explosive tests but are more difficult in sustained or prolonged maximal effort tests to exhaustion. Cost and time involved with testing are other considerations in determining the preferred number of duplicates.

Table 3.1 Calculation of Typical Error of Measurement

Subject	Trial 1	Trial 2	ΔT2-T1	100-ln (T1)	100-ln (T2)	ΔT2-T1
1	9.9	9.8	–0.1	229.3	228.2	–1.0
2	5.5	5.3	–0.2	170.5	166.8	–3.7
3	20.2	19.2	–1.0	300.6	295.5	–5.1
4	11.1	11.5	0.4	240.7	244.2	3.5
5	29.2	29.2	0.0	337.4	337.4	0.0
6	20	20.3	0.3	299.6	301.1	1.5
7	20.2	19.2	–1.0	300.6	295.5	–5.1
8	27.2	26.8	–0.4	330.3	328.8	–1.5
9	14.1	14.5	0.4	264.6	267.4	2.8
10	8.1	7.9	–0.2	209.2	206.7	–2.5
11	11.1	11.5	0.4	240.7	244.2	3.5
12	13.7	13.1	–0.6	261.7	257.3	–4.5
13	31.8	32.7	0.9	345.9	348.7	2.8
14	16.6	16.3	–0.3	280.9	279.1	–1.8
15	5.2	5.3	0.1	164.9	166.8	1.9
16	12.5	12.1	–0.4	252.6	249.3	–3.3
17	28.1	27.5	–0.6	333.6	331.4	–2.2
18	11.4	12.2	0.8	243.4	250.1	6.8
19	8.6	8.4	–0.2	215.2	212.8	–2.4
20	5.2	5.2	0.0	164.9	164.9	0
Mean	15.5	15.4	–0.1	259.3	258.8	–0.5
SD	8.4	8.3	0.5	57.0	56.9	3.3

Reliability in raw values			Reliability as a percentage via log-transformation	
Change in mean		–0.1	Change in mean (%)	–0.5
Lower 90% confidence limit		–0.3	Lower 90% confidence limit	–1.8
Upper 90% confidence limit		0.1	Upper 90% confidence limit	0.8
Confidence limits as ± value		0.2	Confidence limits as approximate ± value	1.3
Typical error		**0.4**	**Typical error as a CV (%)**	**2.4**
Lower 90% confidence limit		0.3	Lower 90% confidence limit	1.9
Upper 90% confidence limit		0.5	Upper 90% confidence limit	3.2

The typical error is calculated as the standard deviation of the change scores divided by √2. In this example the typical error = 0.5/√2 = 0.5/1.42 = 0.4. CV = coefficient of variation; ln = natural logarithm; SD = standard deviation.

Smallest Worthwhile Change

The smallest worthwhile change represents a sport-specific reference value or threshold beyond which an effect is likely to be important and meaningful in practical (performance) terms. There are two approaches for estimating the smallest worthwhile change either directly in time-based individual sports or indirectly in other circumstances. The smallest worthwhile change should be estimated for each group of athletes tested depending on the sport, the athletes' age and gender, and the level of competition. Data from a large enough set of routine test results can be used to generate reliable estimates. Relevant studies in the literature are also a useful source of data provided the demographics of the sample group are comparable with those athletes under the physiologist's care.

For individual, time-based sports such as running, cycling, swimming, and canoeing, the smallest worthwhile change is computed as about half of the typical within-subject coefficient of variation in performance time from race to race or competition to competition (Hopkins et al. 1999). The coefficient of variation in performance for a range of individual sports is typically 0.8% to 3%, and therefore the smallest worthwhile change ranges from 0.4% to 1.5%. For example, in competitive swimming the smallest worthwhile change in performance time has been established as 0.4% for international-level swimmers (Pyne et al. 2004). This reference value of 0.4% can be used to evaluate the relative effectiveness of training interventions in elite swimmers using time trials and to assess the likely impact on competitive performance. Both the progression and variability of performance need to be considered in the preparation of these athletes for international competitions.

For team and other sports in which there is no clear relationship between the fitness or test performance measure and (actual) competitive performance, the smallest worthwhile change is computed as 0.2 of the between-subjects standard deviation. The effect of a training intervention is calculated as the difference or change in the mean score as a proportion of the between-subjects standard deviation (Δmean/standard deviation [SD]). Corresponding reference values for a moderate (0.60), large (1.2), and very large (2.0) standardized changes or differences might also be useful in both routine testing and research projects. For example, a sample of 50 vertical jump test scores for young male athletes with a mean value of 60 cm and a standard deviation of 10 cm yields standardized effects of 2 cm (small), 6 cm (moderate), 12 cm (large), and 20 cm (very large). In terms of rankings, a change of 1 standard deviation is associated with an athlete moving from the 50th to 84th percentile.

Presentation of Results

There are probably as many different ways of presenting test results as there are sport scientists. In most circumstances, one of the testing staff (often a junior staff member or a student) will be assigned the task of processing the data for the final report to send to the coaches and athletes. In this case, it is the responsibility of the supervising physiologist to proofread the report for completeness and accuracy. This task can also be delegated to another member of the testing team involved in the testing but not the initial preparation of reports. Like a manuscript submitted to a peer-reviewed journal, a testing report must be scrutinized closely. The same level of scrutiny should be applied to all reports emanating from a laboratory or department irrespective of whether they are submitted for formal examination (academic thesis) or peer-review publication.

There are no formal requirements for the format of a testing report. Although the format should adhere to institutional, statutory, and conventional guidelines, there is always scope for creative and elegant presentation of numeric data and results. A report typically contains the following elements: title page, group report, individual reports, brief description or explanation of the tests conducted, and a metadata sheet detailing key information on protocols, equipment, staffing, and name and contact detail of the supervising sport scientist (see chapter 1, Quality Assurance in Exercise Physiology Laboratories). Copies of previous test results for a team, reference values, and other comparative information can be included. Rank ordering and percentile rankings are useful when comparing test scores of large numbers of athletes. Exact duplication of numeric data in text, tables, and figures within a report should be avoided. Plain language statements and summaries should be used rather than expansive lists of numeric values with or without statistical effects. Text overloaded with numeric results is often too dense and inaccessible for most readers.

Numeric Values

A scientific report will be greatly enhanced by concise and systematic presentation of numeric values (Hopkins et al. 2009). Avoid excessive use of decimal places, and round up numbers to improve clarity and reduce clutter. The mean and SD (and confidence limits for effect statistics) should have the same number of significant figures: for example, 5% ± 3%, 1.2 ± 1.1 kg; 7.43 ± 0.20 pH units. Providing two significant digits is the most common numeric presentation, although there are occasions in which a single digit or three digits are appropriate. Raw data are often more meaningful for coaches and athletes, but standardized effects (change and difference scores with confidence limits) are useful in making magnitude-based inferences and in comparing large number of tests (with different raw units). A confidence interval can be presented using the word *to*: for example, 5.6 (3.8 to 7.4 units). Confidence limits should be presented using the ± symbol, for example, 0.7% ± 0.5%. Qualitative descriptors of magnitude (e.g., trivial, small, moderate, large, or very large) or substantiveness (e.g., trivial or stable, lower, higher) are useful for reporting magnitude of effects in plain language.

Graphical Presentation

Figures are a useful way to display and summarize data (Pyne 2007). Line charts should be used for serial or repeated measurements of an individual athlete from test to test. Bar graphs are preferred for single observations of means of different tests or different groups of players, teams, or some other grouping variable. An example of bar graph featuring Z scores of different fitness tests for a team sport player is shown in figure 3.1. Error bars should be shown for both group means

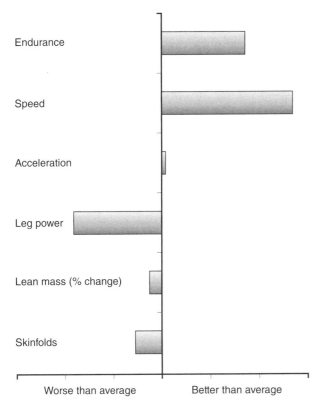

FIGURE 3.1 A Z-score bar chart for comparison of standardized test scores across several different measures of fitness for an individual player in comparison with the team averages. Each positive and negative tick mark on the horizontal axis represents an increment of 0.6 of a between-subjects standard deviation (a standardized moderate difference).

(standard deviation) and effect statistics (confidence limits). The range or region of a trivial effect can be indicated by a line or light shading. Symbols should be legible and abbreviations and supplementary information detailed in the key, legend, or footnote. Horizontal or vertical dotted or dashed lines that assist data interpretation can be useful. Ensure that font sizes are sufficiently large and axes labels are aligned horizontally for the reader.

Statistical Analysis Section

There are several occasions when a physiologist is required to outline the analytical or statistical approach to scientific data. Most people are familiar with the statistical analysis section or paragraph, which is usually located at the end of the methods section of a manuscript. Detailing the statistical analysis and tests used is also necessary in preparation of grant proposal and applications and occasionally in unpublished reports prepared for coaches, teams, institutions, or sporting organizations. The following checklist of points should be a useful guide:

- An opening statement outlining the overall analytical approach
- Details of the data collection and screening processes for miscodings (errors) and missing or imputed values
- Details on data transformation (e.g., log transformation) and descriptive statistics (e.g., mean, standard deviation)
- The model used, including dependent, predictor, moderator, and mediator variables as required
- Justification of the sample size using traditional power analysis or emerging approaches for estimating the smallest (clinically or practically) important effects
- The confidence limits (typically 90% or 95%) used in making inferences about the true (unknown) value of the effect in the population
- Criteria for interpretation of magnitude-based inferences for standardized effects

Distribution of Results

The final step in physiological testing is the distribution of the report. The supervising physiologist should determine the distribution list for the final report prior to the camp or testing session. The report might simply be for the coach and athlete in an individual sport or for the head coach, assistant coaches, and athletes in a team sport. Other support staff such as the team manager, sport physician, athletic trainer, strength and conditioning coach, therapist, dietitian, or psychologist might receive the testing report. On occasions, a report might be distributed more widely to a national sporting organization, parents of junior athletes, and even the media. Confidentiality of individual results must be respected at all times, and results must not be released to a third party without authority or permission of the coach or institution, and the individual athletes, as required. Many coaches and athletes are comfortable with print copies of a testing report, but these days most reports are distributed in electronic form via e-mail, the Internet, or some other form of digital media or social networking.

References

Atkinson, G., and Nevill, A.M. 1998. Statistical methods in assessing measurement error (reliability) in variables relevant to sports medicine. *Sports Medicine* 26(4):217-238.

Atkinson, G., and Nevill, A.M. 2007. Method agreement and measurement error in the physiology of exercise. In: E.M. Winter, A.M. Jones, R.C.R. Davison, P.D. Bromley, and T.H. Mercer (eds), *Sport and Exercise Physiology Testing*. London, UK: Routledge, pp 41-48.

Atkinson, G., Pugh, C., and Scott, M. A. 2010. Exploring data distribution prior to analysis—benefits and pitfalls. *International Journal of Sports Medicine* 31(12):841-842.

Barrett, B., Brown, D., Mundt, M., and Brown, R. 2005. Sufficiently important difference—expanding the framework of clinical significance. *Medical Decision Making* 25(3):250-261.

Barrett, B., Harahan, B., Brown, D., Zhang, Z., and Brown, R. 2007. Sufficiently important difference for common cold—severity reduction. *Annals of Family Medicine* 5(3):216-223.

Batterham, A.M., and Hopkins, W.G. 2006. Making meaningful inferences about magnitudes. *International Journal of Sports Physiology and Performance* 1(1):50-57.

Bland, J.M., and Altman, D.G. 1986. Statistical methods for assessing agreement between two methods of clinical measurement. *Lancet* 1(8416):307-310.

Hopkins, W.G. 2000. Measures of reliability in sports medicine and science. *Sports Medicine* 30(1):1-15.

Hopkins, W.G. 2004. How to interpret changes in an athletic performance test. *Sportscience* 8:1-7.

Hopkins, W.G., Hawley, J.A., and Burke, L.M. 1999. Design and analysis of research on sport performance enhancement. *Medicine and Science in Sports and Exercise* 31(3):472-485.

Hopkins, W.G., Marshall, S.W., Batterham, A.M., and Hani, J. 2009. Progressive statistics for studies in sports medicine and exercise science. *Medicine and Science in Sports and Exercise* 41(1):3-13.

Lehmann, H.P. 2005. Are we ready for patient-based effect sizes in clinical-trials research? *Medical Decision Making* 25(3):248-249.

Pyne, D.B. 2007. The art and practice of presenting results. *International Journal of Sports Physiology and Performance* 2(4):345-346.

Pyne, D.B., Hopkins, W.G., Batterham, A.M., Gleeson, M., and Fricker, P.A. 2005. Characterising the individual performance responses to mild illness in international swimmers. *Clinical Journal of Sports Medicine* 39(10):752-756.

Pyne, D.B., Trewin, C.B., and Hopkins, W.G. 2004. Progression and variability of competitive performance of Olympic swimmers. *Journal of Sports Sciences* 22(7):613-620.

Shrier, I. and Batterham, A.M. 2002. Clinically relevant? *Clinical Journal of Sports Medicine* 12(6):328-330.

Stapleton, C., Scott, M.A., and Atkinson, G. 2009. The "so what" factor: Statistical versus clinical significance. *International Journal of Sports Medicine* 30(11):773-774.

Sterne, J.A.C., and Smith, G.D. 2001. Sifting the evidence—What's wrong with significance tests. *British Medical Journal* 322(7295):226-231.

Weir, J.P. 2005. Quantifying test-retest reliability using the intraclass correlation coefficient and the SEM. *Journal of Strength and Conditioning Research* 19(1):231-240.

PART II

Testing Concepts and Athlete Monitoring

Ergometer-Based Maximal Neuromuscular Power

Mark A. Osborne, Dale W. Chapman, and Scott A. Gardner

Numerous methods have been used to assess maximal power output in humans. Early methods to measure maximal power output included the maximal stair climb tests described by Margaria and colleagues (1966) and the vertical jump test described by Davies and Rennie (1968). Although useful, the data from these tests did not provide any information regarding the force and velocity contributing to maximal muscle power and were only relevant to a non–sport-specific activity (i.e., they only provided a measurement of peak power output; Finn et. al. 2000). To quantify maximal neuromuscular power output as it relates to sport performance, the sport scientist must measure an optimal combination of load or resistance and the speed of movement applied to that load.

Cycle ergometry has traditionally been the easiest exercise modality for testing maximal power output when repeated cyclic muscle contraction is required. Unlike other modes of exercise, such as swimming, where power production of arm pulling and leg kicking is so complex that its accurate calculation is almost impossible, cycling is a sport in which force and velocity can be measured with relative ease (Maud and Shultz 1989). In addition, many aspects believed to affect power output production such as changes in muscle length and additional mass can be controlled for during seated cycling. In the scientific literature, the terms *maximal power output* and *peak power output* have often been used interchangeably, yet they can describe two different physiological variables.

Muscular Power Output

Muscular power output (measured in watts) is defined as the rate of doing work and can be understood as work done (in joules) divided by time taken (in seconds). Because the work done is the product of force and distance, power output can be calculated as the product of force and velocity. Therefore, during cycling, instantaneous power output can be calculated as a function of torque (in Newton meters, Nm) and angular velocity (rads per second, rad $\cdot$ s^{-1}) (Enoka 1988). Put simply into cycling terms, power output is the function of the rotational force applied (torque) to the pedals multiplied by the cadence:

$$\text{Power} = \text{Torque (Nm)} \times (\text{Cadence} \times 2\pi/60).$$

Maximal Power Output

Maximal power output is generally the power output displayed on ergometers and is the external mechanical work required to keep a fan or wheel moving at a constant angular velocity. The total external power is a function of the power required to overcome air resistance, the power required to overcome any rolling resistance of the wheels and bearings, and the power required to impart kinetic energy (KE) to accelerate a flywheel. Although the effects of air and rolling resistance during ergometer work must be acknowledged, the measurement of such variables presents inherent difficulties. For example, given that the speed of movement on a rowing ergometer is relatively slow, aerodynamics has minimal impact on rowing ergometer performance. For all published studies, these variables have generally been assumed to be negligible.

In rowing, the work performed to repeatedly accelerate and decelerate the rower's mass along the ergometer seat rail does not contribute to the maintenance of average flywheel velocity (Gordon 2003). Because the external power output is affected by the ability to accelerate and decelerate an athlete's body mass (such as when on a rowing ergometer) yet does not take into account the additional work involved, the acceleration and deceleration of the body (or ergometer mass when using a sliding ergometer) must be considered as a separate contribution and energetic cost.

Peak Power Output

Peak power output (PPO) refers to the highest instantaneous power output achieved during any activity. In jumping activities, PPO usually refers to the highest instantaneous power achieved prior to takeoff. In cycling, PPO usually refers to the highest power output achieved during an all-out maximal sprint test. PPO is usually reported as a peak 1 s average or as the highest average of one entire crank revolution or stroke throughout a test. Research has shown that PPO and its corresponding cadence are strongly influenced by the magnitude of load applied to the ergometer (Barnett et al. 1996). This means that peak power tests need to be conducted on exactly the same equipment under exactly the same conditions to be interpreted on a test–retest basis.

Measuring peak power on rowing and kayak ergometers involves the same principles as cycle ergometry; however, there are additional interpretation issues. On the majority of cycle ergometers, a change in cadence has a proportional increase in power output, and there is no change in the length of each pedal revolution as the crank length remains constant. In contrast, with rowing and kayaking ergometry the stroke length is not constrained, so the athlete is able to manipulate both flywheel speed and stroke length (or ergometer handle displacement). This has implications for tests determining power output because once the flywheel has started spinning, the athlete is able to exaggerate the flywheel acceleration by significantly reducing the drive phase component of the stroke; this is done by eliminating the leg drive and using only the arms to spin the flywheel faster and faster. Therefore, the technique associated with the PPO may not necessarily represent any relevant rowing or kayaking technique. Any power testing, therefore, has to be constrained to ensure that the technique is relevant to the appropriate sport.

Maximal Neuromuscular Power

Maximal power output is the optimal combination of muscle force and contraction velocity at a specific junction in time and movement. Maximal human power output is primarily limited by physiological, biomechanical, and neurological factors. These factors include muscle size and morphology, metabolic characteristics, force–velocity properties, force–length characteristics, and muscle activation properties (Jones et al. 1989). Furthermore, the central nervous system plays a large role in the ability to generate maximal muscular contraction, whereas the tendon and collagen structures are integral for transmission (Huijing 2003). The total internal power required must be derived from anaerobic and aerobic sources, or both, depending on the duration and intensity of exercise and is also a function of metabolic efficiency ($g\eta$).

During cycling tests, maximal power output refers to the apex of the parabolic power output–cadence relationship (figure 4.1). Maximal power output is derived from a range of maximal fatigue-free torques and corresponding cadence combinations via linear regression (Martin et al. 1997a). Therefore, maximal power output can be derived even if the optimal cadence has not been achieved during a given bout of maximal cycling. The benefit of using maximal power output as a major variable of interest is that it can be measured using many different types of ergometers. These primarily include inertial load, friction braked, and isokinetic ergometers. However, each variety of ergometer has a number of limitations regarding interpretation and application. Maximal power output is usually reported as the maximal average recorded for one entire crank revolution (or stroke cycle). However, some testing methods also report maximal power output as an instantaneous value within a crank revolution or over 1 s duration (1 Hz). This instantaneous power output is usually about 40% higher than the average per pedal revolution (Martin et al. 1997b).

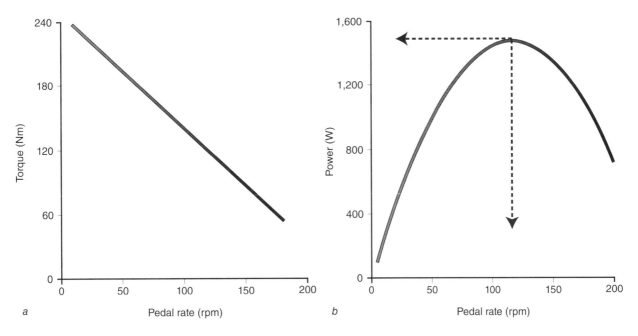

FIGURE 4.1 A schematic representation of the classic inverse linear force–velocity relationship *(a)* and the parabolic power output–velocity relationship observed during maximal bicycle tests *(b)*. Maximal power output is considered velocity specific and is located at the apex of the curve as depicted by the arrow. This figure is constructed from unpublished data during a maximal cycling trial of a world-class Australian track sprint cycling athlete.

Equipment Checklist

All equipment should be calibrated using appropriate procedures prior to the testing session, and the results of these calibration tests must be recorded and stored appropriately. The authors use a system for data management in accordance with recommendations of the Australian National Sport Science Quality Assurance (NSSQA) program. Determination of the maximal power relies on accurately measuring the force and velocity of the activity. The accuracy of any ergometer needs to be verified throughout the range of measurement using either a dynamic calibration rig (Gardner et al. 2004; Woods et al. 1994) or a static calibration rig (Wooles et al. 2005). Data collection and processing software is available for various ergometers that enables a rate of 2 samples per second or better. The following equipment is required:

- Environmental sensor (temperature, relative humidity, barometric pressure)
- Weighing scale (accurate to 0.1 kg)
- Cycle ergometer (examples of custom-built ergometers include the Wombat, Australian Institute of Sport, Canberra, Australia, and the AXIS cycle ergometer, Swift Performance Equipment, Brisbane, Australia)
- Cadence and work monitor, preferably interfaced with a computer

Test Protocols

Only inertial load tests can provide a measure of maximal power in a single bout. All other measures require multiple bouts in order to build a torque–cadence profile. More measures will increase the sensitivity and accuracy in determining the correct value. A regression should be determined with a minimum of four data points (e.g., bouts of maximal cycling). See the tables in this chapter for an indication of how many trials researchers have used to attempt to identify maximal power.

INERTIAL LOAD TESTS

These tests can be performed as laboratory-based tests or field tests using appropriate and reliable power meters (Gardner et al. 2007). During inertial load tests, resistance is provided solely via flywheel inertia. These tests are analogous to cycling in that the resistance varies as the cadence increases (Martin et al. 1997a). To avoid fatigue, the load of the ergometer should be such that the subject can accelerate through the full cadence range within 4 to 6 s.

Recent advances in technology in the areas of portable crank hubs, bottom brackets, and pedals provide sport scientists with an opportunity to better understand the relationship between force and velocity both in the laboratory and in the field. Although not routinely used by elite rowers and national teams, inertial load tests have also been used to assess power–velocity profiles during rowing (Sprague et al. 2007). These advances allow for greater examination of how supplemental training modalities such as resistance training affect sport-specific power generation. For a full description of the rowing ergometer modifications and the method used, refer to Sprague and colleagues (2007).

TEST PROCEDURE

Pretest Procedures

- Measure and record temperature, relative humidity, and barometric pressure.

- Measure and record the athlete's body mass along with other pertinent information (e.g., ergometer settings and positions).

- Instruct the athlete to perform a warm-up that is specific to their event and level of competition.

- Inform the athlete of the number of efforts (e.g., 2-4) and test duration (e.g., 6-10 s maximum).

- Instruct the athlete to complete as much work as possible during the test period.

- For cycling tests, instruct cyclists to start with the preferred pedal 30° above horizontal and remain seated throughout the test.

- Begin the test as soon as possible after completion of the warm-up.

- Record power output at high frequencies (up to 100 Hz) if the ergometer is interfaced with a computer.

DATA ANALYSIS

Although the SRM (Schoberer Rad Meβtechnik) power meter measures torque from strain gauges mounted inside the crank, downloaded data from the power meter are provided as power output and cadence. Torque is calculated from the power and cadence data. Use the following formula if you are using an SRM or similar power measurement hub:

Torque = Power Output / [(Cadence/60) $\times$ (2π)],

where power is expressed in units of watts, torque in Newton meters, and cadence in revolutions per minute (rpm).

Linear regression is used to determine the torque–cadence relationships (e.g., figure 4.1). This regression allows fatigue-free torque to be predicted from any cadence. The torque intercept of these regression equations represents maximal torque (Tmax) and the cadence intercept represents maximal cadence (PRmax). Similarly, the fatigue-free torque–cadence regression equation is used to predict fatigue-free power output using the following equation:

Fatigue-Free Power Output (W) = Fatigue-Free Torque (Nm) $\times$ [(Cadence (rpm)/60) $\times$ (2π)].

The apex of the power–cadence relationship occurs at half the maximal cadence, so optimal cadence (PR$_{opt}$) is PRmax/2 and maximal power (Pmax) is Tmax $\times$ PR$_{opt}$/2 $\times$ π/30.

Peak power refers to the highest power measured as a single per-revolution data point during an all-out maximal sprint test and usually also corresponds to the apex of the power output–cadence curve. The area under the time–power output curve reflects the work done.

ASSUMPTIONS

The athlete has no residual fatigue prior to testing and none accumulates during the session. This can be ensured by only using linear regression data with an r^2 greater than .95.

VALIDITY AND RELIABILITY

The validity and reliability of various power meters can vary significantly, because many of the devices use different principles at various points through the drive chain process. As a result, scores achieved with one system may vary significantly from those obtained with another system simply because one may measure power in the crank itself and another may measure power in the rear hub with various losses encountered through chains and chain rings. Thus, it is important to develop an accurate and reliable calibration method for each type of power meter.

LIMITATIONS

- The inertial load ergometers such as the SRM are accurate only if they account for the KE required to accelerate the flywheel.

- The validity of all maximal testing procedures depends on the motivation of the subject. Strong verbal encouragement may be warranted.

AIR-BRAKED ERGOMETERS

Air-braked ergometers are commonly used in Australia to test track cyclists, rowers, and sprint kayak athletes because these types of ergometers are the most commonly used during training. The limitation with tests carried out using air-braked ergometers is that during non–steady-state efforts such as an all-out sprint, they do not measure the kinetic energy required to accelerate the flywheel and therefore they underestimate true maximal power output.

TEST PROCEDURE

- Follow the pretest procedures given in the section Inertial Load Tests (page 48). To achieve a valid torque–cadence regression, depending on the inertial load of your ergometer, multiple bouts across a number of gear settings are required to obtain enough data points across the cadence range.

- Select the appropriate gear ratio for the athlete.

- Inform the athlete of the test duration: cycling, 6, 10, 30, or 60 s; rowing and kayaking, ~6 to 10 strokes at each stroke rate (e.g., 28, 32, 36, and 40 strokes per minute for rowing tests).

- Tell the athlete to complete as much work as possible during the test period.

- For cycling tests, have riders start with the preferred pedal 30° above horizontal and remain seated throughout the test.

- For rowing tests, have athletes start in their normal start position at approximately half slide with the knees bent at approximately 90°.

- For kayak testing, have paddlers start with tension on one blade replicating the normal starting position on the water.

- Begin the test as soon as possible after completion of the warm-up.

- Record power output at high frequencies (up to 100 Hz) if the ergometer is interfaced with a computer.

DATA ANALYSIS

When air-braked cycle ergometers are used, the following procedures are involved.

- If a Repco work monitor unit is used, this will display power (in watts) and record total work performed (in kilojoules). If the unit is not interfaced with a computer, the maximal deflection of the needle on the power scale needs to be noted as the peak power. The total work performed is displayed on the digital readout.

- When using an Exertech work monitor unit, press the button marked "Hold" immediately after the test ceases. This terminates the accumulation of work displayed on the digital readout meter. Once this is recorded, release this button and press the "Watts" key to display peak power. The total work performed is converted to mean power by the following equation:

$$\text{Mean Power (W)} = [\text{Work (kJ)} \times 1{,}000] / \text{Time (s)} \ (1 \text{ W} = 1 \text{ J} \cdot \text{s}^{-1}).$$

- Work monitor units cannot be relied upon to accurately measure the power output of an ergometer. This can be corrected if the difference between the actual and displayed power outputs is determined with a dynamic or static calibration rig. Accurate measurement of power output is critical to reporting peak power. If dynamic or at least static calibration is not possible, the accuracy of test results is limited. After dynamic calibration, peak power outputs can be corrected. Determining the relationship between actual and displayed power outputs involves a polynomial equation. A fourth-order polynomial expression generally provides the line of best fit. A fourth-order polynomial equation takes the following form:

$$y = k_1 + k_2 x - k_3 x^2 + k_4 x^3 - k_5 x^4,$$

where
y = corrected power output,
x = uncorrected power output, and

k_1, k_2, k_3, k_4 = constants from polynomial regression.

- The following equation is used to correct the peak power outputs from the air-braked cycle ergometer for barometric pressure, temperature, and humidity:

$$\frac{P_2 - (0.38 \times RH_2 \times WV_2)}{P_1 - (0.38 \times RH_1 \times WV_1)} \times \frac{T_1}{T_2}$$

where
P_2 = barometric pressure during test (mmHg),
P_1 = barometric pressure during ergometer calibration (mmHg),
RH_2 = relative humidity during test (decimal),
RH_1 = relative humidity during ergometer calibration (decimal),
$ppWV_2$ = partial pressure of water vapor during test (mmHg),
$ppWV_1$ = partial pressure of water vapor during ergometer calibration (mmHg),
T_2 = temperature during test (K), and
T_1 = temperature during ergometer calibration (K).

- When an air-braked ergometer is interfaced directly with a computer, the relationship between flywheel revolution and power output for each gear ratio is recorded and fit with a fourth-order polynomial to determine the various constants and coefficients. These are then used along with environmental conditions to determine peak power output over 1 s duration.

- The athlete's peak power is expressed in watts (W) and watts per kilogram (W · kg[-1]). It is equal to the highest power output recorded in 1 s and is corrected for both the work monitor calibration and ambient conditions. The corrected power outputs should be recorded on any data sheet along with the environmental conditions and the mass of the athlete.

- Rowing and kayaking tests of peak power are often recorded from the display monitor as the highest power output number over a single stroke period. This is actually the average power output over the entire stroke period and is a function of the power applied over the duration of the drive phase of the stroke and the vent setting, which determines the associated drag factor setting of the ergometer. This in turn is a function of how much the flywheel slows down between each drive phase. The ergometer can be interfaced with a computer to determine an actual peak power over a 1 s period; however, because the KE is still not considered, the result will underestimate the true power output. To gain a more accurate assessment of the peak force applied to the rowing ergometer handle, a load cell or other force-measuring device must be attached to the rowing handle and interfaced with a computer to measure the instantaneous peak force within each stroke.

ASSUMPTIONS

The athlete has no residual fatigue prior to testing and none accumulates during the session.

VALIDITY AND RELIABILITY

The validity of all maximal testing procedures depends on the motivation of the athlete. Strong verbal encouragement may be warranted.

LIMITATIONS

- The inertial load of the system must be accounted for to generate an accurate measurement of power output; this is, the calculation must account for the KE required to accelerate the flywheel (Martin et al. 1997a).

- Peak power during rowing and kayak ergometry should only be recorded when stroke rates are delimited within normal ranges where stroke length (handle displacement distance) and technique are the same as used in normal rowing or kayaking. Scores should be discounted when the stroke length decreases from the athlete's normal range.

- Changes in kayak ergometer setup, in particular, seat height, foot rest length, and simulated paddle length for each athlete, can greatly influence stroke length and stroke rate; these settings must be clearly recorded and maintained over time. When working with novice or young athletes for whom skill development and athletic maturation occur, you may need to repeat testing at the old and new preferred settings to ensure consistency of the data set.

FRICTION BRAKED ERGOMETER TESTING

Tests using friction-braked ergometers include the Wingate (30 s) anaerobic test and two shorter-duration (~6 s) force–velocity tests. When calculating power output, some but not all researchers include the extra torque required to overcome flywheel inertia (tables 4.1 and table 4.2). These tests usually require a number of different resistance settings to enable researchers to calculate absolute maximal power output

TABLE 4.1 Maximal Power Output From Force–Velocity Tests That Have Accounted for the Forces Needed to Overcome Flywheel Inertia in the Calculation of Power Output

Investigators	Subjects	Training status	Body mass, kg	Dura-tion, s	No. of efforts	Recovery time	MPO + iner-tial calcula-tion, W	MPO − inertial calcula-tion, W	Optimal PR, rpm
Lakomy (1986)	5 M 5 F	Students	64.5 ± 11.9				1,004 ± 304	774 ± 138	
Hirano et al. (1993)	8 M	Healthy	67 ± 4.8		5		1,434		
Gaitanos et al. (1993)	8 M	Physical education	71.8 ± 11.4	6	10	30 s	1,253 ± 335	17.5	
Seck et al. (1995)	7	Unknown	Not reported	7	4	5 min	1,119 ± 213	Unknown	
Linossier et al. (1996)	15 M	Untrained	67.7 ± 8.1	5-7		5 min	979 ± 115	879 ± 103	126 ± 9 (with*) 145 ± 9 (without*)
Arsac et al. (1996)	15 M	Trained	75.3 ± 7.5	8	6	5 min	899 ± 148		127 ± 10
Hibi et al. (1996)	22 M	Healthy college	71.1 ± 7.3	3		<5 min	Graphical estimation** (~1,100 ± 200)	Graphical estima-tion** (780 ± 100)	
Hautier et al. (1996)	8 M 2 F	Trained	65.5 ± 8.6	5			941 ± 157		120 ± 8
Linossier et al. (1997)	7 M	Trained	67.8 ± 8.5	5-7		5 min	1,246 ± 79 (post-training values)		
MacIntosh and MacEachern (1997)	9 M	Trained	74.6 ± 2.1	7	5	5 min	1,369 ± 50		12.1 ± 0.4 m · s⁻¹
MacIntosh et al. (2000)	8 M	Active	83.4 ± 10	~5			1,280 ± 83		
Hautier et al. (2000)	8 M 2 F	Trained	65.5 ± 8.6	5	15	25 s	957 ± 217		125 ± 2
Dorel et al. (2003b)	13 M	Cyclists	68.0 ± 7.1	5	3	4 min	948 ± 133		118 ± 5
Dorel et al. (2003a)	12 M	Sprint cyclists	82.5 ± 5.2	5	3	4 min	1,586 ± 105		129.5 ± 5

Note: Wherever possible, the difference when inertial forces are removed from the calculation is also included. All data are presented mean ± SE. MPO = maximal power output; PR = cadence; rpm = revolutions per minute. *With and without refers to the optimal cadence calculated with or without inertial load calculations. **Graphical estimation indicates that data is estimated from published figures.

TABLE 4.2   Maximal Power Output Using Force–Velocity Tests on Friction-Braked Ergometers

Investigators	Subjects	Training status	Body mass, kg	Duration, s	No. of efforts	Recovery, min	MPO, W	MPO, W · kg⁻¹	Optimal PR, rpm
Nakamura et al. (1985)	26 M	Active	69.4	<10	3-8	<2	930 ± 111	13.4 ± 1.6	115
Vandewalle et al. (1985)	7 M 7 F	Students	61.0-72.0 55.0-63.0	~6	7-8	5	813 ± 137 594 ± 70	12.4 ± 1.3 10.2 ± 1.2	
Vandewalle et al. (1987)	116 M 7 M	Athletic Track cyclists	64.0-91.0 68.0 ± 6.0	~6	7-8	5	1,221-758 1,150 ± 127	11.7-17.1 16.8 ± 1.23	
Anselme et al. (1992)	10 M; 4 F	Recreation sport	67.2 ± 2.5			5	964 ± 66	14.4 ± 1.0	137.5 ± 9.5
Linossier et al. (1993)	8 M; 2 F	Trained	63.7 ± 8.5	4-8	7-8	5	968 ± 134 (post-training values)	15.2 ± 2.1 (post-training values)	145.1 ± 20.1
Jaskolska et al. (1999)	32 M	Healthy	82.2 ± 14.5	5	6	5	877 ± 177	10.7 ± 2.2	
Baker and Davies (2002)	8	Students	75.3 ± 11.0	~6-8	8	5	1,020 ± 134	13.5 ± 1.8	141 ± 7

All data are presented mean ± SE. MPO = maximal power output; PR = cadence; rpm = revolutions per minute.

TEST PROCEDURE

- Follow the pretest procedures listed in the section Inertial Load Tests (page 48).
- Select the appropriate mass or resistance setting for each trial.
- Inform the athlete of the number of efforts (e.g., 4 as a minimum) and the test duration.
- Tell the athlete to complete as much work as possible during the test period.
- For cycling tests, have riders start with the preferred pedal 30° above horizontal and remain seated throughout the test.
- Begin the test as soon as possible after completion of the warm-up.
- Record power output every second if the ergometer is interfaced with a computer.
- Conduct multiple trials to develop a profile of the athlete and to calculate the maximal power output.

DATA ANALYSIS

The power profile is limited by the number of trials the athlete can complete within a session with-out experiencing residual fatigue that can affect subsequent efforts. The resolution is correlated to the number of trials performed. The peak powers achieved are a function of the gearing selected and the flywheel resistance.

ASSUMPTIONS

The athlete has no residual fatigue prior to testing and none accumulates during the session. This can be ensured by only using linear regression data with an r^2 greater than .95.

VALIDITY AND RELIABILITY

The validity of all maximal testing procedures depends on the motivation of the athlete. Strong verbal encouragement may be warranted.

LIMITATIONS

The inertial load ergometers such as the Monarch are accurate only if they measure the KE required to accelerate the flywheel.

ISOKINETIC TESTING

Isokinetic testing requires several short bouts of maximal effort cycling in which the contraction velocity (cadence) is held constant throughout each but changed for the subsequent bout. Like force–velocity testing, isokinetic testing (to calculate maximal power output) usually requires the subject to complete multiple maximal cycling bouts to develop a rider profile.

TEST PROCEDURE

- Follow the pretest procedures listed in the section Inertial Load Tests (page 48).
- Measure and record the athlete's body mass and mass of the flywheel along with other pertinent information (e.g., ergometer settings and positions).
- Select the appropriate gear ratio and flywheel mass for the athlete.
- Inform the athlete of the likely number of maximal efforts required (e.g., 4 as a minimum) and the approximate test duration.
- Instruct the athlete to complete as much work as possible during the test period.
- Tell the athlete to start with the preferred pedal 30° above horizontal and remain seated throughout the test.
- Begin the test as soon as possible after completion of the warm-up.
- Record power output at the highest possible sampling frequency if the ergometer is interfaced with a computer.
- Conduct multiple trials to develop a cadence-specific profile of the athlete.
- Provide adequate rest between trials to ensure that fatigue does not influence results.

DATA ANALYSIS

The peak power output for each effort is derived from each cadence-limited trial. These are then plotted against cadence to develop a peak power profile over a specific cadence range.

ASSUMPTIONS

The athlete has no residual fatigue prior to testing and none accumulates during the session.

VALIDITY AND RELIABILITY

If you are using an SRM ergometer with very powerful athletes, it is very difficult to maintain a true isokinetic mode. A motor can be coupled to the SRM ergometer to accelerate the flywheel to the desired cadence, hence removing any potential fatigue from the test and keeping the cadence as constant as possible during the test.

LIMITATIONS

The power profile developed is limited to the number of trials that the athlete can complete within a session without experiencing residual fatigue that can affect subsequent efforts. The resolution of the power profile is a function of the number of trials performed. The peak powers achieved depend on the gear selected and the inertial mass of the flywheel.

Expected Test Scores and Data Interpretation

When comparing values reported in the literature for peak power, you must consider the method of calculation and ergometer used and understand that there is substantial intersubject variation in the values reported for peak power. Unfortunately, it is difficult to produce a summary of pooled data for peak power because investigators report standard deviations and because athletes are classified according to event and competitive level. However, table 4.3 outlines some of the studies that have used adults to perform inertial load tests. The table shows reported values for optimal cadence and maximal power output (MPO) per revolution (P_{rev}) relative to body mass and in absolute terms. Where possible, maximal instantaneous power output ($P_{instant}$) is also displayed. It includes the duration and number of exercise efforts required to calculate maximal power output and the recovery given between bouts. Table 4.2 shows the peak power values associated with force–velocity tests using friction-braked ergometers; this table indicates a number of studies that have not accounted for overcoming flywheel inertia, whereas the previously presented table 4.1 provides a summary of studies that have accounted for the forces required to overcome this flywheel inertia.

Developing a cadence-based peak power profile is a useful process to identify an athlete's strengths and weaknesses across a range of cadences as well as to identify his optimal pedal rate. A summary of cross-sectional studies that have used adults performing maximal isokinetic tests at different cadences is presented in table 4.4. The table shows reported values for optimal cadence and maximal power output (MPO) per revolution (P_{rev}) relative to body mass and in absolute terms, maximal instantaneous power output ($P_{instant}$) is displayed in brackets. The table includes the duration and number of exercise

TABLE 4.3 Reported Inertial Load Tests Using Adult Subjects on Cycle Ergometers

Investigators	Subjects	Training status	Body mass, kg	Duration	No. of efforts	Recovery, min	MPO, W P_{rev} ($P_{instant}$)	MPO, $W \cdot kg^{-1}$ P_{rev} ($P_{instant}$)	Optimal PR, rpm
Martin and Coyle (1995)	23 M 13 M	Non-cyclists Cyclists	79.3 ± 17.7 71.6 ± 8.5	6.5 cranks 3-4 s	4	2	1,143 ± 176 (2,292 ± 238) 1,217 ± 158 (2,012 ± 236)	18.2 ± 2.2 (28.9 ± 3.0) 17.0 ± 2.2 (28.1 ± 3.3)	
Martin et al. (1997a)	13 M	Athletic	80.6 ± 9.2	6.5 cranks 3-4 s	4	2	1,317 ± 66 (2,137 ± 101)	16.4 ± 0.8 (26.6 ± 1.2)	122 ± 2 (P_{rev}) 131 ± 2 ($P_{instant}$)
Martin et al. (1997b)	5 M 25 M	Elite U19 Masters	83.3 ± 8.3 79.2 ± 9.2	6.5 cranks 3-4 s	4	2	1,658 ± 208 (2,632 ± 392) 1,362 ± 198 (2,131 ± 269)	19.9 ± 2.5 (31.6 ± 4.7) 17.2 ± 2.5 (26.9 ± 3.4)	
Martin et al. (2000a)	13 M 35 M	Cycle trained Active	72.0 ± 9.0 78.0 ± 17.0	6.5 cranks 3-4 s	4	2	~1,188 ~1,310	~16.5 ~16.8	~122 ~127
Martin et al. (2000b)	38 M	Active	76 ± 12	6.5 cranks 3-4 s	4	2	1,322 ± 38	17.4 ± 0.4	124 ± 8
Martin and Spirduso (2001)	16 M	Trained cyclists	73.0 ± 7.0	6.5 cranks 3-4 s	4	2	1,194 ± 47	16.4 ± 0.6	~117 ± 3
Gardner et al. (2007)	7 M	Elite track	86.2 ± 6.3	6 s	2	~3	Lab 1,791 ± 169 Field 1,792 ± 156	~20.8	Lab 128 ± 7 Field 129 ± 9

All data are presented mean ± SE. MPO = maximal power output; $P_{instant}$ = maximal instantaneous power output; PR = cadence; P_{rev} = maximal power output per revolution; rpm = revolutions per minute; U19 = under 19 years of age.

efforts required to calculate maximal power output and the recovery given between bouts. Track sprint cyclists and BMX riders have significantly higher peaks at higher cadences than do MTB riders, who in turn tend to have higher peaks than do road cyclists. This form of profiling can be used to identify an athlete's talent in other cycling disciplines. This form of velocity cadence–based profile can also be used to confirm velocity-based deficiencies identified from force plate–derived lower-body power profiling (see chapter 13 on strength and power assessment), which can then further inform the type of strength training required.

For an indication of cycling-specific peak power outputs using air-braked ergometers, refer to chapter 19, which describes protocols for the physiological assessment of high-performance cyclists.

Peak power on a rowing ergometer is presented in table 4.5. External power output was recorded as the highest average power over a complete stroke cycle across a range of race-relevant stroke rates. This form of testing requires significant skill to maintain stroke rate within the desired range while pulling maximally on the ergometer handle. Peak power data in rowers and kayakers may be useful indicators as to the relative strengths and weaknesses of the athletes. This form of profiling can also be used to assess the transfer of other forms of training (such as strength training) to the specific actions of either rowing or kayaking. As mentioned before, this form of stroke rate–based profile can be used to confirm stroke rate–based deficiencies identified from force plate–derived lower-body power profiling (see chapter 13), which can then further inform the type of strength

TABLE 4.4 Cross-Sectional Studies Using Adults Performing Maximal Isokinetic Tests on Cycling Ergometers at Different Cadences

Investigators	Subjects	Training status	Body mass, kg	Duration, s	No. of efforts	Recovery, min	MPO, W, P_{rev} ($P_{instant}$)	MPO, $W \cdot kg^{-1}$, P_{rev} ($P_{instant}$)	Optimal PR, rpm
Sargeant et al. (1981)	4 M 1 F	Active	71.8 ± 9	20			840 ± 153 (1,387 ± 222)	11.8 ± 2 (18.5 ± 3)	110
McCartney et al. (1983a)	12 M	Students	76 ± 8.6	45, 10, and 30	1, 6, and 2	2	Not reported	Not reported	
McCartney et al. (1983b)	13 M	Students	80.8 ± 10.2	10 and 30	6 and 3	2 and 1day*	1,050 ± 175 (1,826 ± 287)	13.0 ± 2.2 (22.6 ± 3.6)	<120
McCartney et al. (1983c)	6 M	Healthy	85.2 ± 14	30	3		972 ± 92 (1,558 ± 185)	11.4 ± 1.1 (18.3 ± 2.2)	
McCartney et al. (1985)	7 F	University employees	54.9 ± 3.9	<10	10	2	767-1187	12.3-21.6	<120
Sargeant (1987)	3 M 1 F	Active	68.7 ± 6.6				790 ± 158 (1,326 ± 199)	11.5 ± 2.3 (19.3 ± 2.9)	
Davies and Sandstrom (1989)	12 M 22 M 10 M	Sprint cyclists Pursuit cyclists Students	72.5 ± 9.8 68.0 ± 9.9 74.8 ± 10.1		8-10		1,241 ± 264 962 ± 206 1019 ± 183	17.1± 3.6 14.1 ± 3.0 13.6 ± 2.4	132 ± 3 122 ± 6 118 ± 8
Beelen and Sargeant (1991)	6 M	Healthy	77.1 ± 8.9	25	7		1,542 ± 221	20.0 ± 2.8	130 ± 21

All data are presented mean ± SE. MPO = maximal power output; $P_{instant}$ = maximal instantaneous power output; PR = cadence; P_{rev} = per revolution; rpm = revolutions per minute.

* McCartney et al. (1983b): two testing sessions separated by 1 day.

TABLE 4.5 Peak Power Output From Elite National- and International-Level Rowers

Category	Subjects, n	Status	Height, cm	Body mass, kg	~28 Stroke rate, W	~32 Stroke rate, W	~36 Stroke rate, W	~40 Stroke rate, W
Lightweight women	2	1x Senior A 1 x U23	173.8 ± 1.2	63.0 ± 3.0	335.5 ± 4.5	368.0 ± 9.0	404.5 ± 4.5	427.5 ± 9.5
Heavyweight women	6	3x Senior A 3 x U23	176.4 ± 2.8	79.3 ± 3.2	412.7 ± 28.5	452.7 ± 18.7	494.2 ± 20.1	536.7 ± 23.1
Lightweight Men	2	1x Senior A 1 x U23	182.8 ± 0.8	75.5 ± 0.5	505.0 ± 4.0	569.5 ± 1.5	639.0 ± 6.0	689.5 ± 1.5
Heavyweight Men	13	9x Senior A 4 x U23	196.2 ± 2.3 193.0 ± 4.0	97.8 ± 1.6 95.4 ± 3.6	668.0 ± 20.0 582.0 ± 26.0	736.0 ± 18.1 698.3 ± 25.9	814.3 ± 20.1 775.3 ± 20.8	881.8 ± 24.9 828.8 ± 28.5

Values recorded as the highest average power over a complete stroke cycle on an air-braked rowing ergometer (Concept II) at different stroke rates. Status describes the participants age. Senior A = athletes over 23 years of age; U23 = athletes under 23 years of age. Acceptable scores were recorded if the stroke rate displayed on the monitor was ±1 of the target stroke rate. All data are presented as mean ± SE.

training required. Peak instantaneous force from an instrumented handle during 2,000 m ergometer rowing is generally around 1,200 N for elite heavyweight males and between 800 and 900 N for elite heavyweight women. In stroke rate–limited peak power trials lasting only a few strokes, these values would be expected to increase significantly.

References

Anselme, F., Collomp, K., Mercier, B., Ahmaidi, S., and Prefaut, C. 1992. Caffeine increases maximal anaerobic power and blood lactate concentration. *European Journal of Applied Physiology* 65(2):188-191.

Arsac, L.M., Belli, A., and Lacour, J.R. 1996. Muscle function during brief maximal exercise: Accurate measurements on a friction-loaded cycle ergometer. *European Journal of Applied Physiology* 74(1-2):100-106.

Baker, J.S., and Davies, B. 2002. High intensity exercise assessment: Relationships between laboratory and field measures of performance. *Journal of Science and Medicine in Sport* 5(4):341-347.

Barnett, C., Jenkins, D.G., and Mackinnon, L. T. 1996. Relationship between gear ratio and 10-s sprint cycling on an air-braked ergometer. *European Journal of Applied Physiology and Occupational Physiology* 72(5-6):509-514.

Beelen, A., and Sargeant, A.J. 1991. Effect of fatigue on maximal power output at different contraction velocities in humans. *Journal of Applied Physiology* 71(6):2332-2337.

Davies, C.T., and Rennie R. 1968. Human power output. *Nature* 217(5130):770-771.

Davies, C.T., and Sandstrom, E.R. 1989. Maximal mechanical power output and capacity of cyclists and young adults. *European Journal of Applied Physiology and Occupational Physiology* 58(8):838-844.

Dorel, S., Hautier, C.A., and Lacour, J.R. 2003a. *Mechanical factors of sprint cycling measured by torque-velocity test in top-level cyclists.* Paper presented at the European College of Sports Science, Salzburg, Austria, July 9-12, 2003.

Dorel, S., Bourdin, M., Van Praagh, E., Lacour, J.R., and Hautier, C. A. 2003b. Influence of two pedalling rate conditions on mechanical output and physiological responses during all-out intermittent exercise. *European Journal of Applied Physiology* 89(2):157-165.

Enoka, R.M. 1988. *Neuromechanical Basis of Kinesiology,* 2nd ed. Champaign, IL: Human Kinetics.

Finn, J.P., Maxwell, B.F., and Withers, R.T. 2000. Air-braked cycle ergometers: Validity of the correction factor for barometric pressure. *International Journal of Sports Medicine* 21(7):488-491.

Gaitanos, G.C., Williams, C., Boobis, L.H., and Brooks, S. 1993. Human muscle metabolism during intermittent maximal exercise. *Journal of Applied Physiology* 75(2):712-719.

Gardner, A.S., Stephens, S., Martin, D.T., Lawton, E., Lee, H., and Jenkins, D. 2004. Accuracy SRM and power tap power monitoring systems for bicycling. *Medicine and Science in Sports and Exercise* 36(7):1252-1258.

Gardner, A.S., Martin, J.C., Martin, D.T., Barras, M., and Jenkins, D.G. 2007. Maximal torque- and power-pedaling rate relationships for elite sprint cyclists in laboratory and field tests. *European Journal of Applied Physiology* 101(3):287-292.

Gordon S. 2003. A mathematical model for power output in rowing on an ergometer. *Sport Engineering* 6(4):221-234.

Hautier, C.A., Linossier, M.T., Belli, A., Lacour, J.R., and Arsac, L.M. 1996. Optimal velocity for maximal power production in non-isokinetic cycling is related to muscle fibre type composition. *European Journal of Applied Physiology* 74(1-2):114-118.

Hautier, C.A., Arsac, L.M., Deghdegh, K., Souquet, J., Belli, A., and Lacour, J.R. 2000. Influence of fatigue on EMG/force ratio and cocontraction in cycling. *Medicine and Science in Sports and Exercise* 32(4):839-843.

Hibi, N., Fujinaga, H., and Ishii, K. 1996. Work and power outputs determined from pedalling and flywheel friction forces during brief maximal exertion on a cycle ergometer. *European Journal of Applied Physiology* 74(5):435-442.

Hirano, Y., Tagawa, T., and Miyadhita, M. 1993. Effect of braking load on maximal anaerobic muscle power output during short-term cycling exercise. *Journal of Biomechanics* 26(Suppl 1):1-157.

Huijing, P.A. 2003. Muscular force transmission necessitates a multilevel integrative approach to the analysis of function of skeletal muscle. *Exercise and Sport Sciences Reviews* 31(4):167-175.

Jaskolska, A., Goossens, P., Veenstra, B., Jaskolski, A., and Skinner, J.S. 1999. Comparison of treadmill and cycle ergometer measurements of force-velocity relationships and power output. *International Journal of Sports Medicine* 20(3):192-197.

Jones, D.A., Rutherford, O.M., and Parker, D.F. 1989. Physiological changes in skeletal muscle as a result of strength training. *Quarterly Journal of Experimental Physiology* 74:233-256.

Lakomy, H.K. 1986. Measurement of work and power output using friction-loaded cycle ergometers. *Ergonomics* 29(4):509-517.

Linossier, M.T., Denis, C., Dormois, D., Geyssant, A., and Lacour, J.R. 1993. Ergometric and metabolic adaptation to a 5-s sprint training program. *European Journal of Applied Physiology and Occupational Physiology* 67(5):408-414.

Linossier, M.T., Dormois, D., Fouquet, R., Geyssant, A., and Denis, C. 1996. Use of the force-velocity tests to determine the optimal braking force for a sprint exercise on a friction-loaded ergometer. *European Journal of Applied Physiology* 74(5):420-427.

Linossier, M.T., Dormois, D., Geyssant, A., and Denis, C. 1997. Performance and fibre characteristics of human skeletal muscle during short sprint training and detraining on a cycle ergometer. *European Journal of Applied Physiology and Occupational Physiology* 75(6):491-498.

MacIntosh, B.R., and MacEachern, P. 1997. Paced effort and all-out 30-second power tests. *International Journal of Sports Medicine* 18(8):594-599.

MacIntosh, B.R., Neptune, R.R., and Horton, J.F. 2000. Cadence, power, and muscle activation in cycle ergometry. *Medicine and Science in Sports and Exercise* 32(7):1281-1287.

Margaria, R., Aghemo, P., and Rovelli, E. 1966. Measurement of muscular power (anaerobic) in man. *Journal of Applied Physiology* 21(5):1662-1664.

Martin, J.C., and Coyle, E.F. 1995. Inertial loading: Design and validation of a method for measuring peak cycling power. *Medicine and Science in Sports and Exercise* 27(5):s58.

Martin, J.C., Wagner, B.M., and Cyle, E.F. 1997a. Inertial load method determines maximal cycling power in a single exercise bout. *Medicine and Science in Sports and Exercise* 29(11):1505-1512.

Martin, J.C., Martin, S.M., and Spirduso, W.W. 1997b. Maximum power and optimal pedaling rate of masters track cyclists. *Medicine and Science in Sports and Exercise* 29(5):S155.

Martin, J.C., Diedrich, D., and Coyle, E.F. 2000a. Time course of learning to produce maximum cycling power. *International Journal of Sports Medicine* 21(7):485-487.

Martin, J.C., Farrar, R.P., Wagner, B.M., and Spirduso, W.W. 2000b. Maximal power across the lifespan. *Journal of Gerontology: Medical Sciences* 55(6):M311-M316.

Martin, J.C., and Spirduso, W.W. 2001. Determinants of maximal cycling power: Crank length, pedaling rate and pedal speed. *European Journal of Applied Physiology* 84(5):413-418.

Maud, P.J., and Shultz, B.B. 1989. Norms for the Wingate anaerobic test with comparison to another similar test. *Research Quarterly for Exercise and Sport* 60(2):144-151.

McCartney, N., Heigenhauser, G. J., Sargeant, A.J., and Jones, N. L. 1983a. A constant-velocity cycle ergometer for the study of dynamic muscle function. *Journal of Applied Physiology* 55(1 Pt 1):212-217.

McCartney, N., Heigenhauser, G.J., and Jones, N.L. 1983b. Power output and fatigue of human muscle in maximal cycling exercise. *Journal of Applied Physiology* 55(1 Pt 1), 218-224.

McCartney, N., Heigenhauser, G.J., and Jones, N.L. 1983c. Effects of pH on maximal power output and fatigue during short-term dynamic exercise. *Journal of Applied Physiology* 55(1 Pt 1):225-229.

McCartney, N., Obminski, G., and Heigenhauser, G.J. 1985. Torque-velocity relationship in isokinetic cycling exercise. *Journal of Applied Physiology* 58(5):1459-1462.

Nakamura, Y., Mutoh, Y., and Miyashita, M. 1985. Determination of the peak power output during maximal brief pedalling bouts. *Journal of Sports Sciences* 3(3):181-187.

Sargeant, A.J., Hoinville, E., and Young, A. 1981. Maximum leg force and power output during short-term dynamic exercise. *Journal of Applied Physiology* 51(5):1175-1182.

Sargeant, A.J. 1987. Effect of muscle temperature on leg extension force and short-term power output in humans. *European Journal of Applied Physiology and Occupational Physiology* 56(6):693-698.

Seck, D., Vandewalle, H., Decrops, N., and Monod, H. 1995. Maximal power and torque-velocity relationship on a cycle ergometer during the acceleration phase of a single all-out exercise. *European Journal of Applied Physiology* 70(2):161-168.

Sprague, R.C., Martin, J.C., Davidson, C.J., and Farrar, R.P. 2007. Force-velocity and power-velocity relationships during maximal short-term rowing ergometry. *Medicine and Science in Sports and Exercise* 39(2):358-364.

Vandewalle, H., Pérès, G., Heller, J., and Monod, H. 1985. All out anaerobic capacity tests on cycle ergometers: A comparative study on men and women. *European Journal of Applied Physiology and Occupational Physiology* 54(2):222-229.

Vandewalle, H., Peres, G., Heller, J., Panel, J., and Monod, H. 1987. Force-velocity relationship and maximal power on a cycle ergometer: Correlation with the height of a vertical jump. *European Journal of Applied Physiology and Occupational Physiology* 56(6):650-656.

Woods, G.F., Day, L., Withers, R.T., Ilsley, A.H., and Maxwell, B.F. 1994. The dynamic calibration of cycle ergometers. *International Journal of Sports Medicine* 15(4):168-171.

Wooles, A.L., Robinson, A.J., and Keen, P.S. 2005. A static method for obtaining a calibration factor for SRM bicycle power cranks. *Sports Engineering* 8(3):137-144.

Anaerobic Capacity

Mark A. Osborne and Clare L. Minahan

There is an increasing requirement for anaerobic capacity to be defined and measured independently of anaerobic power. The peak rate of energy produced via anaerobic metabolism (anaerobic power) is difficult to measure directly and is therefore deduced from the peak power output measured during brief (<30 s) sprint-type exercise bouts. Nevertheless, coaches and scientists are interested not only in the peak power output that can be instantaneously generated but also in the total amount of energy that can be produced via the anaerobic energy systems (i.e., anaerobic capacity). The energy derived from intramuscular phosphagen stores is limited, and it has been suggested that the rate and total energy released from anaerobic glycolysis are also limited (Medbø et al. 1988). This suggests that the anaerobic energy systems have a maximal capacity. A large anaerobic capacity may be an important characteristic of athletes who compete in events that stress the anaerobic energy systems maximally, such as 400 to 1,500 m running, 200 to 400 m freestyle swimming, 1 to 4 km track cycling, 2,000 m rowing, and 500 to 1,000 m kayaking. Interestingly, few researchers have measured the anaerobic capacity of team sport athletes (Moore and Murphy 2003: rugby union; Ogura et al. 2006: football and soccer; Wadley and Le Rossignol 1998: Australian rules football).

In this chapter we

- describe the metabolic processes that contribute to the total amount of energy derived anaerobically during exercise,
- describe the theory of the maximal accumulated oxygen deficit and discuss the underlying

assumptions and limitations associated with this method, and
- explain how to measure anaerobic capacity during various modes of exercise.

The following definitions of anaerobic capacity and supramaximal exercise are used here to ensure consistency in the description of anaerobic capacity tests.

Anaerobic capacity has been defined as "the maximal amount of adenosine triphosphate resynthesised via anaerobic metabolism (by the whole organism) during a specific mode of short-duration maximal exercise" (Green and Dawson 1993, 312).

Supramaximal exercise is defined as exercise performed at a power output that is higher than that achieved at peak oxygen consumption ($\dot{V}O_2$peak). Supramaximal exercise may be performed as a sprint, that is, all-out exercise, typically less than 60 s in duration, or at a constant power output, typically 2 to 4 min in duration. Supramaximal exercise at constant power output is typically expressed as a percentage of $\dot{V}O_2$peak (e.g., 120% peak oxygen consumption), as a percentage of "personal best" time (PB) (e.g., 90% PB for 800 m running). When anaerobic capacity is determined, athletes exercise at supramaximal intensities during a single exercise bout termed the *performance test*.

Energy Systems

Whether the physical activity is one of power or endurance, energy demands for muscle contractions are met through the degradation of adenosine triphosphate (ATP). However, stores in the muscle are small and only sufficient to maintain intense muscle contraction for a few seconds. For continued work, ATP must be regenerated by aerobic or anaerobic metabolic pathways.

The phosphagen system produces ATP during intense exercise when energy demands exceed the ATP that can be provided from anaerobic glycolysis or aerobic metabolism. During the breakdown of phosphocreatine, the energy released is used to regenerate ATP. Quantitatively, ATP and phosphocreatine are the most important immediate energy sources. However, the amount of ATP stored, as well as the resynthesis of ATP via the phosphagen system, is only sufficient to sustain activities for 3 to 15 s during an all-out sprint (Brooks et al. 1996). Formation of ATP via the degradation of glycogen or glucose is called *glycolysis*, and the term *anaerobic glycolysis* specifically refers to the biochemical pathway that results in a net production of lactic acid and ATP. Anaerobic glycolysis can be termed the *glycolytic energy system* and is also capable of producing ATP rapidly without the involvement of oxygen. The glycolytic energy system is the predominant contributor to supramaximal exercise of between 6 and 60 s, after which the aerobic energy system becomes the major contributor for the resynthesis of ATP (Medbø and Tabata 1989; Withers et al. 1991). No single energy system exclusively provides the energy production required for exercise; rather, the energy systems operate with considerable overlap, and the relative contributions vary depending on the intensity and duration of the exercise bout.

Lactate and Fatigue

During short-duration supramaximal exercise, insufficient oxygen is available to accept the hydrogen produced during glycolysis, so lactate is formed. Blood lactate concentration is representative of lactate production in the muscle, the transport of lactate from the muscle to the blood, and the removal of lactate from the blood (Gollnick and Hermansen 1973; Jacobs 1986). Blood lactate concentration is often measured in recovery from supramaximal exercise to reflect the level of acidosis endured (Chwalbinska-Moneta et al. 1989). Although it was previously suggested that the accumulation of lactate in the muscle and blood was responsible for the decline in muscle force or power output that impairs exercise performance, it is now generally accepted that lactate accumulation inside the muscle is not responsible for decreased performance and muscle fatigue (Bangsbo and Juel 2006; Cairns 2006; Lamb and Stephenson

2006). Nevertheless, prolonged anaerobic glycolysis required for supramaximal exercise is associated with adverse consequences that may establish the upper limits of performance by inhibiting both contractile and energy-producing processes. For example, intense muscle contraction induces cellular potassium efflux as well as sodium and chloride influx, all of which cause pronounced perturbations in extracellular and intracellular potassium and sodium concentrations. These ionic perturbations are critical to membrane excitability and skeletal muscle contractility and therefore serve as potential causes of muscle fatigue (McKenna et al. 2008). Thus, an individual with an increased ability to produce energy via anaerobic glycolysis and at same time resist the deleterious effects of the associated ionic perturbations will demonstrate a large anaerobic capacity.

Anaerobic ATP Production

The direct measurement of ATP production is difficult (e.g., muscle biopsy). Thus, indirect methods of measuring ATP turnover have been developed. Oxygen consumption ($\dot{V}O_2$) measured during exercise using open-circuit spirometry reflects ATP production via aerobic metabolism. The accurate measurement of $\dot{V}O_2$ during a graded exercise test to exhaustion is simple and can provide information about the peak power of the aerobic energy system. However, information about the peak capacity of the anaerobic energy systems is more difficult to obtain in athletes. Several methods have been used to determine anaerobic ATP production during exercise, including the following:

- Changes in muscle metabolites measured from muscle biopsy
- Blood lactate concentration following supramaximal exercise
- Oxygen debt following supramaximal exercise
- Total amount of work or mean power output measured during short-duration maximal exercise
- Accumulated oxygen deficit

Muscle biopsies are highly invasive, and although blood lactate concentration can be used to indicate the extent of anaerobic glycolysis, blood lactate concentration cannot give an accurate estimate of the total ATP produced anaerobically (Gastin 1994). Moreover, the concept of paying back a debt that is incurred during exercise by restoring anaerobic stores has been discredited (Powers and Howley 1997).

Total Work and Mean Power Output During Short-Duration Sprinting

In the assessment of anaerobic power, peak power output determined during the 30-s Wingate Anaerobic Test (WAnT) has gained widespread acceptance as the test of choice (Inbar et al. 1996; Weber et al. 2006; Woolf et al. 2008). The calculation of the total work performed and the mean power output achieved during the WAnT has been and is still used to predict anaerobic capacity (Klasnja et al. 2010; Zupan et al. 2009). Nevertheless, it is now well documented that short-duration exercise tests (i.e., <60 s) are inadequate to exhaust the anaerobic energy systems and thus cannot be used to determine anaerobic capacity (Bar-Or 1987; Hill and Smith 1993; Medbø et al. 1988; Serresse et al. 1988). A significant percentage (9%-40%) of the energy provided in the WAnT is aerobically derived and is not accounted for when quantifying anaerobic capacity as the total work performed (Kavanagh and Jacobs 1988; Medbø and Tabata 1989). In addition, there is evidence to suggest that mean power output and anaerobic capacity expressed relative to body mass are not significantly related, and, as such, mean power output cannot be used to quantify anaerobic capacity (Minahan et al. 2007).

Accumulated Oxygen Deficit

Over the past 40 years the oxygen deficit has been used to quantify the anaerobic ATP production during exhaustive supramaximal exercise (Karlsson and Saltin 1970; Medbø and Tabata 1993; Zouhal et al. 2010). The method used to determine the accumulated oxygen deficit takes advantage of the ease and precision by which $\dot{V}O_2$ can be measured in the laboratory. For exercise performed at a power output below the ventilatory threshold, steady-state $\dot{V}O_2$ reflects the total rate of energy release (oxygen demand) during exercise. The relationship between power output and $\dot{V}O_2$ is called the $\dot{V}O_2$–power output relationship and appears to remain linear within a range of intensities (Medbø et al. 1988). Once the $\dot{V}O_2$–power output relationship has been established, the oxygen demand can be estimated for supramaximal exercise intensities by extrapolating the regression model. The accumulated oxygen demand represents the total energy required to fuel the supramaximal exercise bout and is the product of the oxygen demand and exercise duration. The accumulated oxygen deficit can then be determined as the difference between the accumulated oxygen demand and the total $\dot{V}O_2$ (i.e., accumulated oxygen uptake)

that is measured during the exercise bout. The accumulated oxygen deficit is expressed in oxygen equivalents (e.g., 40 ml O_2 eq $\cdot$ kg^{-1}) and represents energy that is derived anaerobically from intramuscular phosphagens (ATP and phosphocreatine) as well as energy derived via anaerobic glycolysis.

Previous literature demonstrates that the accumulated oxygen deficit reaches a maximal value during exhaustive supramaximal exercise (Medbø 1991). This maximal value has been defined as the maximal accumulated oxygen deficit (MAOD) and has been described as a valid and reliable method of quantifying an individual's anaerobic capacity (Medbø et al. 1988, Weber and Scheider 2001). In the last 40 years, the MAOD test has become the most used and accepted method for the quantification of anaerobic capacity (for review, see Noordhof et al. 2010).

Maximal Accumulated Oxygen Deficit

Although the MAOD test has been accepted as the gold standard to determine anaerobic capacity (Noordhof et al. 2010), there remains a strong opposition to its validity (Bangsbo 1992, 1996, 1998). Nevertheless, the MAOD test remains the best practical measure of anaerobic capacity, so we have addressed the assumptions and methodological issues that should be carefully considered when measuring MAOD.

Regardless of the mode of exercise used to determine MAOD, two main assumptions remain:

- The mechanical efficiency of supramaximal work is identical to that for submaximal work.
- The oxygen or energy demand for supramaximal exercise can be determined by extrapolating the $\dot{V}O_2$–power output relationship determined from $\dot{V}O_2$ measured during a series of submaximal exercise bouts.

There is no universal protocol for the MAOD test. Therefore, several methodological considerations arise:

- The number, duration, and intensity of the submaximal exercise bouts used to determine the $\dot{V}O_2$–power output relationship
- The duration of the performance test used in the MAOD test
- The type of exercise bout performed during the performance test (i.e., all-out or a constant power output)

$\dot{V}O_2$–Power Output Relationship

The $\dot{V}O_2$–power output relationship is lineally extrapolated to estimate the supramaximal power output for a given oxygen demand (e.g., 120% peak oxygen consumption) or vice versa (i.e., to estimate the oxygen demand for a given supramaximal power output). Medbø and colleagues (1988, 1989, 1993) used ten 10 min submaximal treadmill bouts, performed at an intensity ranging between 35% and 100% $\dot{V}O_2$peak, to determine the $\dot{V}O_2$–power output relationship. However, since these publications, several researchers have modified the method used to determine the $\dot{V}O_2$–power output relationship (Green and Dawson 1996; Minahan and Wood 2008; Roberts et al. 2003; Simmonds et al. 2010). There are two primary ways that researchers have modified these methods:

- Decreased the duration of the submaximal exercise bouts (Ogura et al. 2006; Russell et al. 2002; Simmonds et al. 2010)
- Decreased the number of submaximal exercise bouts (Doherty et al. 2000; Feriche et al. 2007; Simmonds et al. 2010)

There are two primary reasons for modifying these methods:

- To decrease the time required to perform the test and determine the $\dot{V}O_2$–power output relationship
- To minimize the continued increase in $\dot{V}O_2$ (i.e., slow component) during submaximal exercise bouts performed above the ventilatory threshold

Although decreasing the duration of the submaximal exercise bouts is a more efficient use of time, shorter submaximal exercise bouts used to determine the $\dot{V}O_2$–power output relationship will also decrease the calculated oxygen deficit. Using 4 min submaximal exercise bouts to determine the $\dot{V}O_2$–power output relationship results in a 20% to 25% lower MAOD than when ~10 min submaximal exercise bouts are used (Bangsbo 1992; Buck and McNaughton 1999a). Although some researchers have argued that the use of 10 min submaximal exercise bouts is necessary to prevent the underestimation of MAOD, we and others have previously used 4 min submaximal exercise bouts (Bickam et al. 2002; Minahan and Wood 2008; Simmonds et al. 2010) to minimize the slow component of $\dot{V}O_2$ during exercise bouts above the ventilatory threshold, thus preventing an overestimation of the supramaximal oxygen demand (Bangsbo 1992). Although not all researchers agree with the concept of a delayed onset (McDonald et al. 1997; Yano et al. 2004), it is generally accepted that the slow component begins 90 to 150 s after the onset of heavy-intensity constant power output exercise. Therefore, the oxygen demand predicted for a 2 to 3 min, constant power output, performance test may not include a large $\dot{V}O_2$ slow component. Thus, the oxygen demand of the maximal performance test predicted from 10 min submaximal exercise bouts that include a $\dot{V}O_2$ slow component may be overestimated.

Decreasing the number of submaximal exercise bouts reduces test time. However, the number of submaximal exercise bouts may also influence the predicted oxygen demand for a given supramaximal power output. Buck and McNaughton (1999b) systematically studied the effect of different numbers of submaximal exercise bouts on the $\dot{V}O_2$–power output relationship. These researchers suggested that the effect of reducing the number of submaximal exercise bouts in the $\dot{V}O_2$–power output relationship was dependent, in part, on the intensity of the exercise bouts removed (i.e., the effect of removing the lowest or the highest submaximal exercise bouts was the greatest). By using fewer submaximal exercise bouts with power outputs that range from 40% to 70% of $\dot{V}O_2$, testers can obtain regression characteristics (strength of the regression, $\dot{V}O_2$ gain, standard error of the individual $\dot{V}O_2$–power output regressions, and the standard error of the individual $\dot{V}O_2$ gain) similar to those obtained when 10 submaximal exercise bouts are used to establish the $\dot{V}O_2$–power output relationship.

Duration and Mode of the Performance Test

The duration of the performance test is reported to influence the magnitude of the calculated MAOD (Medbø et al. 1988, Medbø and Tabata 1989). The performance test must be long enough to allow maximal anaerobic energy release while brief enough to minimize the aerobic contribution to energy production. Medbø and Tabata (1989) reported an increase in the MAOD when the constant power output exercise bout was increased from 1 min to 2-3 min. They reported that 2 min or more was required for the MAOD to reach its highest value during exhaustive exercise at constant power output. Constant power output exercise bouts performed at an intensity equivalent to 115% to 130% of peak oxygen consumption will result in the athlete's exhaustion in about 2 to 4 min (Maughan and Poole 1981; Rupp et al. 1983). Alternatively, MAOD has been determined from sprint-type exercise tests (all-out) in which athletes are required either to exert their maximal effort for a specified duration or distance of a test (Pripstein et al. 1999) or to "go out hard" at the start of the test and then settle into a pace that can be maintained for the remainder of the test (Foley et al. 1991). There is no difference in MAOD determined during exhaustive constant power output performance tests compared with results obtained with either all-out (Gastin et al. 1995) or competition-specific time trial test protocols (paced) using well-trained cyclists (Foley et al. 1991).

Oxygen Stores of the Body

In the transition from rest to exercise, the $\dot{V}O_2$ measured at the mouth can underestimate oxygen that is consumed by the tissues. This is due to changes in the oxygen stores in the body, which consist of oxygen bound to hemoglobin and myoglobin, oxygen dissolved in the body fluids, and oxygen present in the lungs. The contribution of oxygen stores is measured as part of the oxygen deficit (estimated to contribute approximately 9%), and a reduction in the oxygen stores of the body contributes significantly to the production of ATP (Medbø et al. 1988). Consequently, this value should be subtracted from the absolute MAOD to give a more accurate estimation of the true whole-body anaerobic energy release. It has previously been argued that this stored oxygen should not be subtracted from the calculated oxygen deficit simply because a significant amount of early published literature did not account for these oxygen stores. It has also been argued that any overestimation of the MAOD that results when the contribution of stored oxygen is ignored may be countered by a potential underestimation of the MAOD that results when the reputed lower mechanical efficiency at maximal power outputs is ignored (Withers et al. 1993). However, because technology has advanced and efficiency can be estimated, the MAOD should be reported in both its adjusted and unadjusted formats for stored oxygen.

Validity of the MAOD

Both the theoretical and the biochemical measures of anaerobic energy yield have been favorably compared with MAOD (Bangsbo et al. 1990; Medbø et al. 1988; Withers et al. 1991). MAOD is unaffected by changes in inspired oxygen concentration, indicating that MAOD is independent of aerobic energy release. Linnarsson (1974) and Medbø and colleagues (1988) validated this concept by demonstrating no change in MAOD under hypoxic conditions despite a reduction in exercise duration and the accumulated oxygen uptake. To further validate the method of MAOD as a measure of anaerobic energy release, several studies have compared MAOD values between groups of untrained, endurance-trained, and sprint-trained athletes. Significantly larger MAOD values have been measured repeatedly in sprint-trained athletes when compared with both endurance-trained and untrained subjects (Gastin and Lawson 1994; Medbø et al. 1988; Saltin 1990; Scott et al. 1991). Differences of 15% to 36% have been reported, suggesting that sprint-trained athletes are capable of achieving a higher MAOD because of their greater capacity to produce ATP anaerobically. The comparison of anaerobic capacity with biochemical measures is regarded as the best method to validate the MAOD (Gastin 1994). Bangsbo and colleagues (1990) compared biochemical measures of MAOD calculated during intense one-leg, dynamic knee extensor exercise. This provided a precise quantitative measure of the changes in ATP and phosphocreatine concentration and lactate production in the muscle, which offered a model to study the anaerobic energy yield. In addition, the accumulations of alanine and glycolytic intermediates in the muscle were estimated. Finally, the investigators calculated the loss of pyruvate from the muscle and the uptake of lactate in the leg flexors. The total anaerobic energy release was 91.2 mmol $\cdot$ kg^{-1}. The measured oxygen deficit expressed in the same units was 91.6 mmol $\cdot$ kg^{-1} following adjustment for the decrease in oxygen stores.

Equipment Checklist

The following equipment is required:

- Environmental sensor (temperature, relative humidity, barometric pressure)
- Weighing scale
- Metabolic cart and associated equipment
- Ergometer
- Stroke rate or pedaling rate and work monitor, preferably interfaced with a computer

GENERAL MAOD MEASUREMENT PROCEDURES

Treadmill running was the mode of exercise used to determine MAOD in the original studies by Medbø and colleagues (1988). Using these original protocols as the basic template, others have since measured MAOD during cycling (Craig et al. 1995; Gardner et al. 2004), rowing (Bangsbo et al. 1993; Pripstein et al. 1999), swimming (Faina et al. 1997; Ogita et al. 1996, 1999; Reis et al. 2010a, 2010b), and kayaking (Bishop et al. 2003; Faina et al. 1997). MAOD has been measured regularly in the assessment of elite athletes within the Australian high-performance system.

The following procedures should be followed before the commencement of any MAOD test protocol:

- Measure and record temperature, relative humidity, and barometric pressure.
- When the athlete arrives, determine whether the guidelines for athlete preparation have been met and ask the athlete to complete the pretest procedures.
- Measure and record the athlete's body mass along with other pertinent information (e.g., ergometer settings and positions).
- Have athletes perform their own warm-up, specific to their sport or event and level of competition.

Several considerations should be noted regardless of the protocol or mode of exercise to ensure a true MAOD.

$\dot{V}O_2$–POWER OUTPUT RELATIONSHIP

- $\dot{V}O_2$ measured at low power outputs and speeds should be checked for linearity, because $\dot{V}O_2$ can be elevated by factors other than the exercise intensity (e.g., an inefficient running technique at low speeds can artificially increase oxygen consumption).
- $\dot{V}O_2$ determined during the first few stages of a mode-specific incremental exercise test

(for the determination of $\dot{V}O_2$peak) can be used to calculate the $\dot{V}O_2$–power output relationship. The 4 to 5 min exercise stages performed below the ventilatory threshold can be performed consecutively with short rest periods (e.g., 1 min) to obtain a blood sample for the subsequent determination of blood lactate concentration. It is recommended that higher-intensity (65%-80% $\dot{V}O_2$peak) submaximal exercise bouts be performed separately with longer recovery between (e.g., 30 min). Athletes should be encouraged to drink fluids *ad libitum* during rest periods and to perform low-intensity exercise to assist recovery between exercise bouts.

TEST PROCEDURE

Ideally the test is performed after at least 12 h of rest from any high-intensity exercise. Athletes should be hydrated and should have abstained from caffeine for at least 6 h before the test. It is important that athletes are not glycogen depleted. For test–retest comparison purposes, athletes should be tested at the same time of day each time they attend the laboratory.

For the majority of exercise modes, the power or velocity displayed on the ergometer is limited to a measure of the kinetic energy required to move the flywheel (or treadmill) and does not account for the internal mechanical cost of moving the limbs (when cycling, kayaking, or running) or moving the body mass along the ergometer slide (rowing). This internal mechanical cost increases nonlinearly with stroke rate or cadence and although accounted for in the measured $\dot{V}O_2$ is not accounted for in the measure of ergometer power or velocity that forms the basis of the $\dot{V}O_2$–power output relationship. The assumption that this nonlinear increase in internal energy cost does not change efficiency is a significant limitation of the measurement of accumulated oxygen deficit and the calculation of anaerobic capacity.

CYCLING MAOD

An electromagnetically braked cycle ergometer is preferred for this test. Replicate the setup of the athlete's own bike on the test ergometer (seat height, head stem, crank length, pedals) and record for future reference.

$\dot{V}O_2$–POWER OUTPUT RELATIONSHIP

The first four to five submaximal $\dot{V}O_2$ values included in the $\dot{V}O_2$–power output relationship can be determined during a continuous step cycling protocol

or during the first stages of an incremental cycling test for the determination of peak $\dot{V}O_2$ (see chapter 19 for physiological protocols for the assessment of high-performance cyclists). If higher submaximal exercise bouts are to be included in the $\dot{V}O_2$–power output relationship (i.e., above ventilatory threshold or a blood lactate concentration >5 mmol $\cdot$ L^{-1}), these should be performed with adequate recovery between each bout.

TEST PROCEDURE

- The athlete performs a mild warm-up. Because this test begins at 150 W for elite male cyclists and 125 W for elite female cyclists, stretching and easy locomotor activity at a very low intensity are sufficient for a warm-up. During the warm-up, outline the test procedures to the athlete (i.e., four to six cycling bouts performed at power outputs of up to 80%-90% of peak $\dot{V}O_2$; these consist of 5 min at 150 W + 5 min at 50 W for men and 3 min at 125 W + 3 min at 25 W for women).

- Place a piece of tape across the bridge of the athlete's nose. This helps prevent the nose clip from slipping during the test.

- Position the headset and mouthpiece on the athlete to ensure maximal comfort. Instruct the athlete to position the nose clip comfortably to block nasal air flow.

- Start the athlete pedaling while simultaneously starting the metabolic cart and timing device. During each power output, indicate the time remaining and check with the athlete about their ability to continue.

- The average $\dot{V}O_2$ value for the last 2 min of each work stage is recorded for inclusion in the regression equation of the $\dot{V}O_2$–power output relationship. A blood sample can be drawn during the final minute of each work stage from the athlete's hyperemic earlobe to determine blood lactate concentration ($[La^-]$). If blood $[La^-]$ is above 5 mmol $\cdot$ L^{-1}, then it is unlikely that the athlete is achieving steady-state $\dot{V}O_2$, and any further submaximal $\dot{V}O_2$ values should be determined during separate tests.

- In contrast to the continuous step protocol, after each 5 min work stage, the athlete is allowed to dismount the cycle ergometer after a mild cool-down (100 W) and rest for 30 min.

CONSIDERATIONS

- Athletes should complete the incremental test to exhaustion before MAOD is determined; this will provide the peak $\dot{V}O_2$ and establish the power output achieved at peak $\dot{V}O_2$.

- $\dot{V}O_2$ at low power outputs should be checked for linearity, because it is often elevated by factors other than the exercise intensity.

- Controlling environmental conditions is important to prevent oxygen drift associated with elevated body temperatures.

- If cadence differs significantly between the submaximal stages and the supramaximal effort, then the mechanical efficiency may also change because there is a greater internal mechanical cost at faster cadences. Results can be affected as a result of the increased energy cost of moving the limbs at faster pedal cadences, which is not accounted for in the measurement of cycling power output.

PERFORMANCE TEST PROCEDURE

The MAOD should be determined during a separate test performed at a power output corresponding to 120% of the power output achieved at peak $\dot{V}O_2$.

- Have the athlete perform their own sport- or event-specific warm-up and allow them adequate time to stretch, hydrate, and become comfortable with the impending performance test.

- Prepare the athlete for gas exchange (see submaximal protocol).

- Start the athlete pedaling against no resistance while you simultaneously start the metabolic cart and timing device.

- Have the athlete increase the cadence to about 85 to 90 revolutions per minute (rpm) and aim to maintain a constant cadence for the majority of the test.

- Apply the predetermined power output for the performance test at the beginning of a full 30 s gas collection bin. Measure $\dot{V}O_2$ continuously through the entire test.

- During the performance test, give the athlete strong verbal encouragement but do not reveal the time elapsed.

- Consider the athlete to have reached exhaustion when the cadence drops below 40 rpm.

- Record the exact time to exhaustion and stop the metabolic measurement system and remove the mouthpiece and headset from the athlete.

DATA ANALYSIS

Below is a worked example of the calculation of MAOD.

1. Power output achieved at peak $\dot{V}O_2$ = 450 W

2. Determination of $\dot{V}O_2$–power output relationship (table 5.1)

The $\dot{V}O_2$ for each submaximal stage is regressed against the respective power output so that an oxygen demand (expressed in oxygen equivalents) for the performance test (supramaximal power output)

TABLE 5.1 Values of the $\dot{V}O_2$–Power Output Relationship (Cycling)

Power output, W	Time, min	Stage	$\dot{V}O_2$, L · min^{-1}
100	0-5	1	1.66
150	5-10	2	2.21
200	10-15	3	2.82
250	15-20	4	3.41
300	20-25	5	3.99
30 min rest period			
350	0-5	6	4.56

can be predicted according to the following linear relationship:

$$y = a + bx$$

where

y = oxygen demand in liters,
x = power output in watts,
b = slope of regression line (0.0117), and
a = y-intercept of regression line (0.4816).

Here, $y = 0.4816 + 0.0117x$.

Thus, the oxygen demand for a power output equal to 120% of the power output achieved at peak $\dot{V}O_2$ (450 W × 1.2 = 540 W) is 6.55 L · min^{-1}.

3. Determination of MAOD: The total oxygen demand (accumulated oxygen demand) of the constant load performance test can be calculated by multiplying the oxygen demand by the duration of the performance test:

Oxygen Demand = 6.55 L · min^{-1}
Duration = 2.5 min.

Thus,

Accumulated Oxygen Demand = 16.375 L.

The total oxygen consumption (accumulated oxygen uptake) measured via gas exchange during the performance test can be calculated by converting each $\dot{V}O_2$ value from L · min^{-1} to L · 30 s^{-1} (i.e., divide each value by 2) (table 5.2) and then adding together:

Accumulated Oxygen Uptake = 11.05 L.

The accumulated oxygen deficit is calculated simply by subtracting the accumulated oxygen uptake from the accumulated oxygen demand:

16.37 – 11.05 = 5.32 (L O$_2$ eq · kg^{-1}).

Then, subtract 9% of the accumulated oxygen deficit to account for oxygen stores:

5.32 × 0.91 = 4.84 (L O$_2$ eq · kg^{-1}).

If a percentage contribution of the energy systems is to be reported, the 9% value subtracted from the accumulated oxygen deficit should be added to the accumulated oxygen uptake value.

Here, aerobic contribution to the performance test

= Accumulated Oxygen Uptake / Accumulated Oxygen Demand × 100
= (11.05 × 1.09) / 16.375 L O$_2$ eq · kg^{-1}
= 12.04 / 16.37 L O$_2$ eq · kg^{-1}
= 74%.

CONSIDERATIONS

When the performance test does not finish exactly on a 30 s interval, the $\dot{V}O_2$ value for the outstanding duration can be either portioned accordingly or removed

TABLE 5.2 Oxygen Uptake Values Obtained During Cycling Performance Test

Power output, W	Time, min:s	$\dot{V}O_2$, L · min^{-1}	$\dot{V}O_2$, L · 30 s^{-1}
540	0:00-0:30	2.13	1.11
540	0:30-1:00	4.55	2.28
540	1:00-1:30	4.90	2.45
540	1:30-2:00	5.16	2.58
540	2:00-2:30	5.35	2.68

from the analysis. The inability of some metabolic cart systems to determine oxygen consumption to the exact end of the test will often cause a degree of error in measurement of the actual amount of oxygen consumed during the performance test. For breath-by-breath metabolic carts, any remaining oxygen consumption can be portioned out relatively simply by stopping the data collection immediately after the test ceases and portioning out the average. For metabolic carts using a mixing chamber, the averaging interval can often be changed to a 5 or 10 s period as the smallest time-based increment depending on the software. In cases where the averaging interval can be changed, the oxygen equivalent can be determined by averaging over these smaller increments and can be added to the 30 s oxygen equivalent from the initial minute or minutes. If the time to exhaustion is only 1 or 2 s on either side of the smallest time-based increment, the portion can be ignored during analysis.

RUNNING MAOD

Accurate measurement of power output is critical to calculating oxygen demand and therefore calculating MAOD. However, power output is difficult to determine on a treadmill. This may be overcome by performing the submaximal exercise bouts at the same treadmill grade as the performance test, therefore negating the need to quantify power. Constant power output performance tests are generally performed on motorized treadmills. Lakomy (1984, 1987) and Cheetham and colleagues (1985) offer some insights into all-out performance tests performed on a non-motorized treadmill. A motorized treadmill that is used for testing should be capable of operating at greater than 25 km $\cdot$ h^{-1}.

$\dot{V}O_2$–POWER OUTPUT RELATIONSHIP

- In addition to measuring the athlete's body mass, measure and record the running shoe type, model, and mass.

- Place a piece of tape across the bridge of the athlete's nose. This helps prevent the nose clip from slipping during the test.

- Instruct the athlete to take a position at the front of the treadmill, and when they are ready, allow the athlete a brief warm-up and familiarization period on the treadmill. During this period, outline the test procedures to the athlete (e.g., the athlete will spend 4 min at each work stage with 1 min of rest, each stage will increase in velocity by 2 km $\cdot$ h^{-1}, and capillary blood samples will be taken after each work stage). For elite males, the initial stage is 12 km $\cdot$ h^{-1}; for elite females, the initial stage is 10 km $\cdot$ h^{-1}. The initial speeds may have to be altered to match the caliber of the athlete. Gradient is to be maintained at 1% throughout the entire test.

- Position the headset and mouthpiece on the athlete, ensuring maximal comfort. Instruct the athlete to position the nose clip comfortably to block nasal air flow.

- Prior to commencement of the test, have the athlete hold the hand rails. Start the treadmill while simultaneously starting the metabolic cart and timing device. During each work stage, indicate the time remaining and check with the athlete about their ability to continue. Stop the treadmill at the end of the 4 min, giving the athlete prior warning with a countdown over the last 5 s.

- Immediately after the athlete stops exercising, have a blood sample drawn from the hyperemic earlobe and analyzed for blood lactate concentration.

- During the rest interval the athlete may remove the nose clip and mouthpiece, however, the headset should remain in place. With approximately 30 s before the start of the next work stage, instruct the athlete to replace the mouthpiece and nose clip, if removed, and assume the starting position.

- Proceed as outlined previously, with the athlete completing four to six different power outputs at 80% to 90% of the subject's $\dot{V}O_2$max.

- To assess the appropriateness of a further stage, observe the athlete's lactate value and keep going until you record a value greater than 4 mmol $\cdot$ L^{-1}. If you are not sure whether the value is going to be greater than 4 mmol $\cdot$ L^{-1}, wait an extra 30 s (90 s total) to see what the value is and whether another stage is necessary.

- Encourage athletes to complete the incremental test to exhaustion, because this will determine the running velocity at $\dot{V}O_2$max (v $\dot{V}O_2$) and enable you to determine the $\dot{V}O_2$peak.

- Record the average running velocity throughout each stage.

- Record the $\dot{V}O_2$ for the final 2 min of each 4 min stage for inclusion in the regression equation. A minimum of 2 min of exercising before sampling at each running velocity should ensure steady-state oxygen consumption. A blood sample may be drawn during the 1 min interval to determine blood [La⁻]. If blood [La⁻] is above 5 mmol · L⁻¹, it is unlikely that the athlete is achieving steady-state oxygen consumption.

- Ensure that the athlete achieves steady-state oxygen consumption at all intensities used for the $\dot{V}O_2$–power output regression.

CONSIDERATIONS

- Check oxygen consumption at low velocities for linearity, because it is often elevated due to factors other than the exercise intensity. During treadmill running, there may be inefficiency at particularly slow running velocities.

- If rest periods are scheduled between stages, encourage subjects to make appropriate use of them. Hydration is important and subjects should be encouraged to drink fluids *ad libitum*. Continued exercising at low intensity after high-intensity exercise will assist recovery between stages.

- Control environmental conditions to prevent oxygen drift associated with elevated body temperatures.

TEST PROCEDURE

The MAOD can be determined during a separate test to exhaustion at a percentage of the velocity at $\dot{V}O_2$max. The velocity corresponding to 120% of the v $\dot{V}O_2$ is programmed into the treadmill. Because this is a maximal test to exhaustion at a predetermined velocity, MAOD is determined as opposed to accumulated oxygen deficit (AOD), where the subject may or may not be completely exhausted at the cessation of the test period. The same slope must be used for the performance test as the submaximal power outputs to ensure that efficiency is similar. The maximal capacity of the treadmill may dictate the potential speed of the performance test. If this is likely to be exceeded by the athlete at a particular treadmill grade, then a higher grade may be required during the entire test. If the test is conducted at zero gradient, then the speed during the performance test is likely to be high, so there is an increased safety risk. Determining the appropriate protocol is a compromise between grade and potential final speed.

- Prepare the athlete for gas exchange (see submaximal protocol).
- Instruct the athlete to take a position at the front of the treadmill with their feet straddling

the treadmill belt, and when the athlete is ready, allow a brief warm-up and familiarization period on the treadmill. During this period, outline the test procedures to the athlete.

- Tether the athlete with a safety harness to ensure safety when being transferred to and from the treadmill belt at speed.

- Prior to commencement of the test, have the athlete hold the hand rails and place their feet straddling the treadmill belt. When ready, start the treadmill to ensure it is at the appropriate speed before the athlete begins running. The athlete starts with one foot matching the treadmill belt speed; when they begin with both feet running on the treadmill belt, start the metabolic cart and timing device at that point.

- Give athletes verbal encouragement and have them continue exercising until exhaustion. With 10 s remaining in each 30 s period, ask the athlete if they are capable of running for at least 30 s more. The athlete should use hand signals to indicate their decision (i.e., thumbs up is yes; thumbs down is no).

- Stop the test when the athlete voluntarily presses the emergency stop button or indicates that they cannot continue.

- Consider the athlete to have reached exhaustion when they are unable to continue at the required velocity.

CONSIDERATIONS

- Other treadmill speeds or percentages of v $\dot{V}O_2$ can be used based on the capacity of an athlete (e.g., 400 m runners vs. 1,500 m runners); however, the speed should be selected so exhaustion occurs between 60 s and 4 min.

- Oxygen consumption is recorded continuously through the entire exercise bout.

- Other gradients up to 4% can be used to lower the maximal speed required for the test; this is particularly important from a safety perspective or if the treadmill does not have the capacity to reach appropriate speeds.

- If stride rate differs significantly between the submaximal stages and the supramaximal effort, then the mechanical efficiency may also change because there is a greater mechanical cost at faster stride frequencies. Results will be affected if the athlete moves their limbs at faster stride rates that are not accounted for in the measurement of running velocity.

DATA ANALYSIS

- The oxygen consumption for each submaximal stage is regressed against the respective velocity so that an oxygen equivalent for the maximal performance test can be predicted according to the linear relationship:

$$y = a + bx$$

where

y = oxygen equivalent,
x = velocity for each 30 s period,
b = slope of regression line, and
a = intercept of regression line.

- The oxygen equivalent (converted from $L \cdot min^{-1}$ to $L \cdot 30 s^{-1}$) is estimated for each 30 s effort. These can be totaled to determine an oxygen demand. The measured oxygen consumption (also converted from $L \cdot min^{-1}$ to $L \cdot 30 s^{-1}$; this can be totaled to determine an oxygen cost) is then subtracted from the estimated oxygen equivalent in order to determine an oxygen deficit for each 30 s period. These are then totaled for the entire maximal performance test to calculate an accumulated oxygen deficit. When the test to exhaustion does not finish exactly on the minute or 30 s mark, the oxygen consumption for the outstanding duration can be either portioned accordingly or removed from the analysis. The inability of some metabolic carts to determine oxygen consumption to the exact end of the test can cause a degree of error in determining the amount of oxygen consumed during the maximal effort. For breath-by-breath metabolic carts, portioning out any remaining oxygen consumption can be done relatively simply by stopping the data collection immediately after the test ceases and portioning out the average. For metabolic carts using a mixing chamber, the averaging interval can be changed to a 5 or 10 s period as the smallest time-based increment depending on the software. In cases in which the averaging interval can be changed, the oxygen equivalent can be determined by averaging over these smaller increments and added to 30 s oxygen equivalent from the initial minute or minutes. If the time to exhaustion is only 1 or 2 s on either side of the smallest time-based increment, the portion can be ignored from the analysis. For example, if the test lasted 116 s, then the $\dot{V}O_2$ over the initial 90 s can be converted from $L \cdot min^{-1}$ to $L \cdot 30 s^{-1}$ from the 30 s averaged data. Most metabolic carts are flexible enough to provide a 20 s sampling option, which will need to be converted from $L \cdot min^{-1}$ to $L \cdot 20 s^{-1}$, and a 5 s sampling option, which will need to be converted from $L \cdot min^{-1}$ to $L \cdot 5 s^{-1}$. The final second of data can be discounted from the $\dot{V}O_2$. If there are concerns about the stability of the metabolic cart, particularly if a constant load or constant velocity test is used, then the $\dot{V}O_2$ from the previous 30 s period can be extrapolated to the 26 s period and converted from $L \cdot min^{-1}$ to $L \cdot 20 s^{-1}$. These are then totaled over the duration to determine the oxygen demand.

- 9% should be subtracted from the total MAOD to account for stored oxygen.

- The athlete's anaerobic capacity or MAOD is expressed as $L\ O_2\ eq \cdot kg^{-1}$.

- The oxygen cost can be divided by the oxygen demand to determine a percentage contribution of aerobic energy sources. The percentage contribution of anaerobic energy sources can be determined by subtracting the percentage aerobic demand from 100%.

ROWING AOD

An accumulated oxygen deficit can be determined either during the final stage of an incremental test or during a stand-alone 2,000 m maximal performance ergometer test. In both cases, the preliminary stages of an incremental test can be used to establish the submaximal $\dot{V}O_2$–power output relationship. An accumulated oxygen deficit can be inferred from both trials; however, because neither maximal stage is a constant load test to exhaustion, it cannot be assumed that anaerobic metabolism is completely exhausted as a result of potential pacing limitations. Therefore, it is incorrect to use the term *maximal accumulated oxygen deficit* in tests of set distance or set duration.

In addition to the standard equipment listed previously, a Concept IID or IIE rowing ergometer and Concept II ergometer slides are required. Considerations for this test match those provided for the running MAOD.

$\dot{V}O_2$–POWER OUTPUT RELATIONSHIP

A common mechanical efficiency cannot be assumed because of the large interindividual variability.

Furthermore, power as displayed on the ergometer is a function of both stroke rate and stroke length, both of which are variable; this is unlike the situation in cycling, where only the pedal cadence varies because the cyclists are constrained by the length of the pedal crank. In addition, the power displayed on the rowing ergometer is only a measure of the kinetic energy of the flywheel and does not take into account the additional work of moving the body mass along the rowing ergometer slide, or the movement of the ergometer mass and limbs when using the Concept II slides. This internal mechanical cost increases with stroke rate and although accounted for in the measured $\dot{V}O_2$ is not accounted for in the measure of ergometer power that forms the basis of the regression equation. The assumption that this internal mechanical cost does not affect the calculation of anaerobic capacity is a major limitation of measuring accumulated oxygen deficit.

- Instruct the athlete to sit on the ergometer and take a few strokes to set the appropriate drag factor. During this period, outline the test procedures to the athlete (i.e., 4 min at each work stage with 1 min of rest between stages; see chapter 23 for protocols for the physiological assessment of rowers).

- Have the athlete perform a mild warm-up. Because this test begins with low power outputs, stretching and easy locomotor activity at a low intensity are sufficient for the warm-up.

- Place a piece of tape across the bridge of the athlete's nose. This helps prevent the nose clip slipping during the test.

- Position the headset and mouthpiece on the athlete, ensuring maximal comfort. Instruct the athlete to position the nose clip comfortably to block nasal air flow.

- Ensure that the ergometer display is set up with appropriate work to rest ratios. When the rower pulls on the handle for the first stage, the ergometer display begins the countdown automatically; the metabolic cart also needs to be started at this time.

- Four to six power outputs up to approximately 90% of the subject's $\dot{V}O_2$max are sufficient to develop an accurate submaximal $\dot{V}O_2$–power output relationship. Steady-state oxygen consumption must be achieved at all power outputs. This is generally the first five or six stages of a rowing incremental test (see chapter 23). The maximal intensity that can be used depends on the caliber and fitness of the athlete.

- Record the average power output over each stage.

- Immediately on ceasing exercise for each stage, have a blood sample drawn from the earlobe and analyzed for blood [La$^-$].

- During the rest interval the athlete may remove the nose clip and mouthpiece; however, the headset should remain in place. With approximately 30 s before the start of the next work stage, instruct the athlete to replace the mouthpiece and nose clip, if removed, and assume the starting position.

- Proceed as outlined previously, with the athlete completing all stages of the incremental test.

- Record the $\dot{V}O_2$ for the final 2 min of each 4 min stage (until the athlete reaches a blood [La$^-$] >4 mmol $\cdot$ L^{-1}); if the athlete is in a steady state, this value can be used in the regression equation. A minimum of 2 min of exercising before sampling at each power output should ensure steady-state oxygen consumption.

TEST PROCEDURE: 2,000 M

An accumulated oxygen deficit calculated during the final 4 min stage of an incremental test, with the first five or six stages used to develop the $\dot{V}O_2$–power output relationship, will invariably underestimate any AOD determined during a stand-alone 2,000-m trial given the elevated $\dot{V}O_2$ prior to the onset of the final stage. Thus, the oxygen deficit is an accumulated oxygen deficit (AOD) and not a maximal accumulated oxygen deficit (MAOD).

- Have the athlete perform a suitable warm-up. Because this test requires a maximal effort, athletes are recommended to begin with low power outputs, stretching, and easy locomotor activity at a low intensity prior to a number of intense efforts in preparation for an all-out effort.

- Set the appropriate drag factor on the ergometer flywheel (see chapter 23).

- Following a suitable warm-up, set the distance on the rowing ergometer display to 2,000 m with intervals of 100 m, or record the distance covered as shown on the display screen every 30 s.

- Record oxygen consumption continuously throughout the entire bout of exercise.

- Have the athlete complete the 2,000 m to the best of their ability.

- Provide verbal encouragement.

- Retrieve the average power output or duration of each 100 m interval from the ergometer.

DATA ANALYSIS

- The duration of each 100 m or distance covered in each 30 s interval of the 2,000 m performance test can be retrieved from the Concept II display monitor or recorded manually throughout the test. The average velocity is converted to an average power (in watts) using the following equation: Power (P) = 2.8 v^3, where v is the average velocity. This is measured as speed in meters per second. Following is a worked example of the calculation of AOD.

1. Determination of $\dot{V}O_2$–power output relationship (table 5.3)

The $\dot{V}O_2$ for each submaximal stage is regressed against the respective power output so that an oxygen demand (expressed in oxygen equivalents) for the performance test can be predicted according to the linear relationship:

$$y = a + bx$$

where

y = oxygen demand in liters,
x = power output in watts,
b = slope of regression line (0.0123), and
a = y-intercept of regression line (0.6922).

Here,

$$y = 0.6922 + 0.0123x.$$

TABLE 5.3 Values of the $\dot{V}O_2$–Power Output Relationship (Rowing)

Power output, W	Time, min	Stage	$\dot{V}O_2$, L · min^{-1}
190	0-4	1	2.96
230	5-9	2	3.60
270	10-14	3	4.00
310	15-19	4	4.53
350	20-24	5	5.07
390	25-29	6	5.42

2. Determination of AOD
The distance travelled in every 30 s effort of the 2,000 m test is converted to a power output (watts).

The total oxygen demand (accumulated oxygen demand) of the 2,000 m performance test can be calculated by converting each 30 s power output to an oxygen cost using the $\dot{V}O_2$–power output relationship and then summing the oxygen demand of each segment for the duration of the performance test (table 5.4).

Accumulated Oxygen Demand = 38.35 L.

The total oxygen consumption (accumulated oxygen uptake) measured via gas exchange during the performance test can be calculated by converting each $\dot{V}O_2$ value from L · min^{-1} to L · 30 s^{-1} (i.e., divide each value by 2) or liters per segment when the duration differs and it has to be portioned out accordingly and then added together. See subsequent note. Here,

Accumulated Oxygen Uptake = 35.15 L.

The accumulated oxygen deficit is calculated simply by subtracting the accumulated oxygen uptake from the accumulated oxygen demand:

$$38.31 - 35.15 = 3.20 \ (L \ O_2 \ eq \cdot kg^{-1}).$$

Then, subtract 9% of the accumulated oxygen deficit to account for oxygen stores:

$$3.20 \times 0.91 = 2.91 \ (L \ O_2 \ eq \cdot kg^{-1}).$$

If a percentage contribution of the energy systems is to be reported, the 9% value subtracted from the accumulated oxygen deficit should be added to the accumulated oxygen uptake value.

Here, aerobic contribution to the performance test

= Accumulated Oxygen Uptake / Accumulated Oxygen Demand $\times$ 100
= (35.15 $\times$ 1.09) / 38.35 L O$_2$ eq · kg^{-1}
= 38.31 / 38.35 L O$_2$ eq · kg^{-1}
= 99.92%.

Note: When the maximal performance test does not finish exactly on the minute or 30 s, the oxygen consumption for the outstanding duration can be either portioned accordingly or removed from the analysis. The inability of some metabolic carts to determine oxygen consumption to the exact end of the test can cause a degree of error in determining the actual amount of oxygen consumed during the performance test. For breath-by-breath metabolic carts, portioning out any remaining oxygen consumption can be done relatively simply by stopping the data collection immediately after the test ceases and portioning out the average. For metabolic carts using a mixing chamber, the averaging interval can be changed to a 5 or 10 s period as the smallest time-based increment depending on the software. When the software

TABLE 5.4   Oxygen Uptake Values Obtained During the 2,000 m Rowing Performance Test

Time, min:s	Distance remaining, m	Interval distance, m	Interval power, W	Actual O_2 uptake, $L \cdot min^{-1}$	O_2 uptake per segment, $L \cdot segment^{-1}$	O_2 demand, $L \cdot min^{-1}$	O_2 demand per segment, $L \cdot segment^{-1}$
0:00.0	2,000		489	1.430*			
0:30.0	1,832	168	489	2.660	1.33	6.71	3.35
1:00.0	1,668	164	459	5.075	2.54	6.34	3.17
1:30.0	1,504	164	458	5.758	2.88	6.33	3.16
2:00.0	1,340	164	461	5.992	3.00	6.36	3.18
2:30.0	1,176	163	452	6.002	3.00	6.25	3.12
3:00.0	1,013	163	448	6.040	3.02	6.21	3.10
3:30.0	851	162	443	6.126	3.06	6.15	3.07
4:00.0	690	161	431	6.173	3.09	6.00	3.00
4:30.0	529	161	436	6.257	3.13	6.06	3.03
5:00.0	368	161	433	6.239	3.12	6.02	3.01
5:30.0	206	162	444	6.227	3.11	6.16	3.08
6:00.0	40	165	468	6.258	3.13	6.45	3.23
6:07.2	0	40	487	6.232	0.75	6.69	0.80

*Baseline $\dot{V}O_2$ immediately upon starting.

allows, the oxygen equivalent can be determined by averaging over these smaller increments and added to 30 s oxygen equivalent over the initial minutes. If the time to exhaustion is only 1 or 2 s on either side of the smallest time-based increment, the portion can be ignored from the analysis. For example, if the test lasted 5 min and 56 s, then the $\dot{V}O_2$ over the initial 5 min and 30 s can be converted from $L \cdot min^{-1}$ to $L \cdot 30 \, s^{-1}$ from the 30 s averaged data. Most metabolic carts are flexible enough to provide a 20 s sampling option, which will need to be converted from $L \cdot min^{-1}$ to $L \cdot 20 \, s^{-1}$, and a 5 s sampling option, which will need to be converted from $L \cdot min^{-1}$ to $L \cdot 5 \, s^{-1}$. The final second of data can be discounted from the $\dot{V}O_2$. If there are concerns about the stability of the metabolic cart, particularly if a constant load or constant velocity test is used, then the $\dot{V}O_2$ from the previous 30 s period could be extrapolated to the 26 s period and converted from $L \cdot min^{-1}$ to $L \cdot 20 \, s^{-1}$. These subsections are then totaled over the duration to determine the oxygen demand.

CONSIDERATIONS

- Familiarity trials may be necessary if the athlete is unaccustomed to the 2,000 m maximal performance test, because poor pacing can affect results. If athletes are unable to adequately pace themselves during the test, they may start the test too hard and fatigue prematurely, potentially failing to complete the test. In contrast, if they start too conservatively, they will have too much left toward the latter stages of the test and will not completely exhaust all energy stores.

- Controlling environmental conditions is important to prevent oxygen drift associated with elevated body temperatures.

TEST PROCEDURE: 4 MIN PERFORMANCE TEST

- The athlete completes all stages of the standard rowing incremental test as per the submaximal $\dot{V}O_2$–power output relationship above (see chapter 23 for protocols used to assess rowers).

- The average power output or distance covered for each 30 s duration is subsequently retrieved from the ergometer memory.

DATA ANALYSIS

- The average velocity over each 30 s period is converted to an average power (in watts) using the following equation: Power (P) = $2.8 \, v^3$, where v is the average velocity. This is measured as speed in meters per second.

- The oxygen consumption for each submaximal stage is regressed against the respective average power output so that an oxygen equivalent for each 30 s interval of the seventh stage can be predicted according to the linear relationship:

$$y = a + bx$$

where

y = oxygen equivalent,
x = power output for each 30 s period,
b = slope of regression line, and
a = y-intercept of regression line.

- The oxygen equivalent (converted from $L \cdot min^{-1}$ to $L \cdot 30 s^{-1}$) is estimated for each 30 s effort. These can be totaled to determine an oxygen demand. The measured oxygen consumption (also converted from $L \cdot min^{-1}$ to $L \cdot 30 s^{-1}$ and totaled to determine an oxygen cost) is then subtracted from the estimated oxygen equivalent to yield an oxygen deficit for each 30 s period. These are then totaled for the entire 4 min performance test to calculate an accumulated oxygen deficit.

- Subtract 9% from the total AOD to account for stored oxygen.

- Express the subject's anaerobic capacity or AOD as $L \, O_2 \, eq \cdot kg^{-1}$.

- Divide the oxygen cost by the oxygen demand to determine a percentage contribution of aerobic energy sources. The percentage contribution of anaerobic energy sources can be determined by subtracting the percentage aerobic demand from 100%.

CANOEING AND KAYAKING AOD

As with the rowing example, accumulated oxygen deficit can be determined during a maximal 4 min effort (refer to chapter 28 for protocols used to assess sprint kayak athletes). The preliminary stages of an incremental test can be used to establish the submaximal $\dot{V}O_2$–power output relationship. An accumulated oxygen deficit can be inferred; however, because the maximal 4 min effort is not a constant load test to exhaustion, it cannot be assumed that anaerobic metabolism is completely exhausted. Therefore, it is incorrect to use the term *maximal accumulated oxygen deficit* in tests of set distance or set duration. In addition to the standard equipment listed previously, a kayak ergometer is required (e.g., Dansprint kayak ergometer).

$\dot{V}O_2$–POWER OUTPUT RELATIONSHIP

A common mechanical efficiency cannot be assumed, because there is large interindividual variability and because power as displayed on the ergometer display is a function of both stroke rate and stroke length. Both of these are variable, unlike the case in cycling, where only the pedal cadence varies as the cyclists are constrained by the length of the pedal crank.

- Instruct the athlete to sit on the ergometer, and ensure that the tension on the elastic shock cords is appropriately set. During this period, outline the test procedures to the athlete (i.e., 4 min at each work stage with 1 min of rest between stages) (refer to chapter 28).

- Have the athlete perform a mild warm-up. As this test begins with low power outputs, stretching and easy locomotor activity at a low intensity are sufficient for warm-up.

- Place a piece of tape across the bridge of the athlete's nose. This helps prevent the nose clip from slipping during the test.

- Position the headset and mouthpiece on the athlete, ensuring maximal comfort. Instruct the athlete to position the nose-clip comfortably to block nasal air flow. Set up the ergometer display with appropriate work to rest ratios. When the athlete pulls on the handle for the first stage, the ergometer display begins the countdown automatically; the metabolic cart also needs to be started at this time.

Four to six power outputs up to approximately 90% of the subject's $\dot{V}O_2$max are sufficient to develop an accurate submaximal $\dot{V}O_2$–power output relationship. Steady-state oxygen consumption must be achieved at all power outputs. The submaximal stages of the AOD test are the same as the 7×4 min step test. However, athletes are required to complete only as many 4 min submaximal work stages as needed to elicit a blood lactate concentration of 4 mmol $\cdot L^{-1}$ or greater. Once this has been achieved, the submaximal component of the test is stopped in order to avoid fatiguing the athlete prior to the ensuing 4 min performance test (time-trial effort) (see chapter 28). The maximal intensity that can be used depends on the caliber and fitness of the athlete.

- The average power output over each stage is recorded.

- Immediately after the athlete ceases exercise for each stage, a blood sample should be drawn from the earlobe and analyzed for blood [La⁻].

The 20 min rest interval:

- On completion of the last submaximal power output, 20 min of active recovery is allowed to prepare the athlete for the maximal performance test (for full details, see chapter 28).

TEST PROCEDURE

This final component of the AOD testing protocol is treated the same as the 7 × 4 min maximal component. At the 18 min mark of the active recovery period, ask the athlete to be seated on the kayak ergometer. Position the metabolic cart appropriately and ensure that the athlete's feet are secured in position on the footrest. At the 20 min mark of the recovery period, provide a 10 s count-in for the athlete to commence the 4 min performance test. The average power output or distance covered for each 30 s duration is recorded in real time or subsequently retrieved from the ergometer memory. Considerations and data analysis procedures for this test match those given previously for the rowing MAOD.

References

Bangsbo, J., Gollnick, P.D., Graham, T.E., Juel, C., Kien, S.B., Mizuno, M., and Saltin, B. 1990. Anaerobic energy production and $\dot{V}O_2$ deficit-debt relationship during exhaustive exercise in humans. *Journal of Physiology* 422:539-559.

Bangsbo, J., and Juel, C. 2006. Counterpoint: Lactic acid accumulation is an advantage/disadvantage during muscle activity. *Journal of Applied Physiology* 100:1412-1413.

Bangsbo, J., Michalsik, L., and Petersen, A. 1993. Accumulated O_2 deficit during intense exercise and muscle characteristics of elite athletes. *International Journal of Sports Medicine* 14(4):207-213.

Bangsbo, J. 1992. Is the O_2 deficit an accurate quantitative measure of the anaerobic energy production during intense exercise? *Journal of Applied Physiology* 73(3):1207-1209.

Bangsbo, J. 1996. Oxygen deficit: a measure of the anaerobic energy production during intense exercise? *Canadian Journal of Applied Physiology* 21(5):350-363.

Bangsbo, J. 1998. Quantification of anaerobic energy production during intense exercise. *Medicine and Science in Sports and Exercise*. 30(1):47-52.

Bar-Or, O. 1987. The Wingate anaerobic test: An update on methodology, reliability and validity. *Sports Medicine* 4(6):381-394.

Bickham, D., Le Rossignol, P., Gibbons, C., and Russell, A. 2002. Re-assessing accumulated oxygen deficit in middle-distance runners. *Journal of Science and Medicine in Sport* 5(4):372-382.

Bishop, D., Bonetti, D., and Spencer, M. 2003. The effect of an intermittent, high-intensity warm-up on supra-maximal kayak ergometer performance. *Journal of Sports Sciences* 21(1):13-20.

Brooks, G.A., Fahey, T.D., and White, T.P. 1996. *Exercise Physiology*, 2nd ed. Mountain View, CA: Mayfield.

Buck, D., and McNaughton, L. 1999a. Maximal accumulated oxygen deficit must be calculated using 10-min time periods. *Medicine and Science in Sports and Exercise* 31(9):1346-1349.

Buck, D., and McNaughton, L. 1999b. Changing the number of submaximal exercise bouts effects calculation of MAOD. *International Journal of Sports Medicine* 20(1):28-33.

Cairns, S.P. 2006. Lactic acid and exercise performance: Culprit or friend? *Sports Medicine* 36(4):279-291.

Cheetham, M.E., Williams, C., and Lakomy, H.K. 1985. A laboratory running test: metabolic responses of sprint and endurance trained athletes *British Journal of Sports Medicine* 19(2):81-84.

Chwalbinska-Moneta, J., Robergs, R.A., Costill, D.L., and Fink, W. 1989. Threshold for muscle lactate accumulation during progressive exercise. *Journal of Applied Physiology* 66(6):2710-2716.

Craig, N.P., Norton, K.I., Conyers, R.A., Woolford, S.M., Bourdon, P., Stanef, T., and Walsh, C.M. 1995. Influence of test duration and event specificity on maximal accumulated oxygen deficit of high performance track cyclists. *International Journal of Sports Medicine* 16(8):534-540.

Doherty, M., Smith, P.M., and Schroder, K. 2000. Reproducibility of the maximum accumulated oxygen deficit and run time to exhaustion during short-distance running. *Journal of Sports Sciences* 18(5):331-338.

Faina, M., Billat, V., Squadrone, R., De Angelis, M., Koralsztein, J.P., and Dal Monte, A. 1997. Anaerobic contribution to the time to exhaustion at the minimal exercise intensity at which maximal oxygen uptake occurs in elite cyclists, kayakists and swimmers. *European Journal of Applied Physiology and Occupational Physiology* 76(1):13-20.

Feriche, B., Delgado, M., Calderón, C., Lisbona, O., Chirosa, I.J., Miranda, M.T., Fernández, J.M., and Alvarez, J. 2007. The effect of acute moderate hypoxia on accumulated oxygen deficit during intermittent exercise in nonacclimatized men. *Journal of Strength and Conditioning Research* 21(2):413-418.

Foley, M.J., McDonald, K.S., Green, M.A., et al. 1991. Comparison of methods for estimation of anaerobic capacity. *Medicine and Science in Sports and Exercise* 23:202.

Gardner, A., Osborne, M., D'Auria, S., and Jenkins, D. 2004. A comparison of two methods for the calculation of accumulated oxygen deficit. *Journal of Sports Sciences* 21(3):155-162.

Gastin, P., and Lawson, D. 1994. Variable resistance all-out test to generate accumulated oxygen deficit and predict anaerobic capacity. *European Journal of Applied Physiology* 69(4):331-336.

Gastin, P.B., Costill, D.L., Lawson, D.L., Krzeminski, L., and McConell, G.K. 1995. Accumulated oxygen deficit during supramaximal all-out and constant intensity exercise. *Medicine and Science in Sports and Exercise* 27(2):255-263.

Gastin, P. 1994. Quantification of anaerobic capacity. *Scandinavian Journal of Medicine & Science in Sports* 4(2):91-112.

Gollnick, P.D., and Hermansen, L. 1973. Biochemical adaptation to exercise: Anaerobic metabolism. *Exercise and Sport Sciences Reviews* 1:1-43.

Green, S., and Dawson, B. 1993. Measurement of anaerobic capacities in humans: Definitions, limitations and unsolved problems. *Sports Medicine* 15(5):312-327.

Green, S., and Dawson, B.T. 1996. Methodological effects on the VO$_2$-power regression and the accumulated O2 deficit. *Medicine and Science in Sports and Exercise* 28(3):392-397.

Hill, D.W., and Smith, J.C. 1993. Gender difference in anaerobic capacity: Role of aerobic contribution. *British Journal of Sports Medicine* 27(1):45-48.

Inbar, O., Bar-Or, O., and Skinner, J.S. 1996. *The Wingate Anaerobic Test*. Champaign, IL: Human Kinetics.

Jacobs, I. 1986. Blood lactate: Implications for training and sports performance. *Sports Medicine* 3(1):10-25.

Karlsson, L., and Saltin, B. 1970. Lactate, ATP and CP in working muscles during exhaustive exercise in man. *Journal of Applied Physiology* 29(5):598-602.

Kavanagh, M.F., and Jacobs, I. 1988. Breath-by-breath oxygen consumption during performance of the Wingate Test. *Canadian Journal of Sport Sciences* 13(1):91-93.

Klasnja, A., Drapsin, M., Lukac, D., Drid, P., Obadev, S., and Grujic, N. 2010. Comparative analysis of two different methods of anaerobic capacity assessment in sedentary young men. *Vojnosanit Pregl* 67(3):220-224.

Lakomy, H.K.A. 1984. An ergometer for measuring the power generated during sprinting. *Journal of Physiology* 354:33P.

Lakomy, H.K.A. 1987. The use of a non-motorized treadmill for analysing sprint performance. *Ergonomics* 30(4):627-637.

Lamb, G.D., and Stephenson, D.G. 2006. Point:Counterpoint: Lactic acid accumulation is an advantage/disadvantage during muscle activity. *Journal of Applied Physiology* 100(4):1410-1412.

Linnarsson, D. 1974. Dynamics of pulmonary gas exchange and heart rate changes at start and end of exercise. *Acta Physiologica Scandinavica Supplement* 415:1-68.

MacDonald, M., Pedersen, P.K., and Hughson, R.L. 1997. Acceleration of VO$_2$ kinetics in heavy submaximal exercise by hypoxia and prior high-intensity exercise. *Journal of Applied Physiology* 83(4):1318-1325.

Maughan, R.J., and Poole, D.C. 1981. The effects of a glycogen loading regime on the capacity to perform anaerobic exercise. *European Journal of Applied Physiology* 46(3):211-219.

McKenna, M.J., Bangsbo, J., and Renaud, J-M. 2008. Muscle K+, Na+, and Cl- disturbances and Na+-K+ pump inactivation: Implications for fatigue. *Journal of Applied Physiology* 104(1):288-295.

Medbø, J.I., and Tabata, I. 1993. Anaerobic energy release in working muscle during 30 s to 3 min of exhausting bicycling. *Journal of Applied Physiology* 75(4):1654-1660.

Medbø, J.I., and Tabata, I. 1989. Relative importance of aerobic and anaerobic energy release during short-lasting exhausting bicycle exercise. *Journal of Applied Physiology* 67(5):1881-1886.

Medbø, J.I., Mohn, A.C., Tabata, I., Bahr, R., Vaage, O., and Sejersted O.M. 1988. Anaerobic capacity determined by maximal accumulated O$_2$ deficit. *Journal of Applied Physiology* 64(1):50-60.

Medbø, J.I. *Quantification of the Anaerobic Energy Release During Exercise in Man* (PhD thesis). Oslo: National Institute of Occupational Health.

Minahan, C., Chia, M., and Inbar, O. 2007. Does power indicate capacity? 30-s Wingate anaerobic test vs. maximal accumulated O$_2$ deficit. *International Journal of Sports Medicine* 28(10):836-843.

Minahan, C., and Wood, C. 2008. Strength training improves supramaximal cycling but not anaerobic capacity. *European Journal of Applied Physiology* 102(6):659-666.

Moore, A., and Murphy, A. 2003. Development of an anaerobic capacity test for field sport athletes. *Journal of Science and Medicine in Sport* 6(3):275-284.

Noordhof, D.A., De Koning, J.J., and Foster, C. 2010. The maximal accumulated oxygen deficit method. *Sports Medicine* 40(4):285-302.

Ogita, F., Hara, M., and Tabata, I. 1996. Anaerobic capacity and maximal oxygen uptake during arm stroke, leg kicking and whole body swimming. *Acta Physiologica Scandinavica* 157(4):435-441.

Ogita, F., Onodera, T., and Tabata, I. 1999. Effect of hand paddles on anaerobic energy release during supramaximal swimming. *Medicine and Science in Sports and Exercise* 31(5):729-735.

Ogura, Y., Katamoto, S., Uchimaru, J., Takahashi, K., and Naito, H. 2006. Effects of low and high levels of moderate hypoxia on anaerobic energy release during supramaximal cycle exercise. *European Journal of Applied Physiology* 98(1):41-47.

Powers, S.K., and Howley, E.T. 1997. *Exercise Physiology: Theory and Application to Fitness and Performance*, 3rd ed. Dubuque, IA: Times Mirror.

Pripstein, L.P., Rhodes, E.C., McKenzie, D.C., and Coutts, K.D. 1999. Aerobic and anaerobic energy during a 2-km race simulation in female rowers. *European Journal of Applied Physiology and Occupational Physiology* 79(6):491-494.

Reis, V.M., Marinho, D.A., Barbosa, F.P., Reis, A.M., Guidetti, L., and Silva, A.J. 2010a. Examining the accumulated oxygen deficit method in breaststroke swimming. *European Journal of Applied Physiology* 109(6):1129-1135.

Reis, J.F., Millet, G.P., Malatesta, D., Roels, B., Borrani, F., Vleck, V.E., and Alves, F.B. 2010b. Are oxygen uptake kinetics modified when using a respiratory snorkel? *International Journal of Sports Physiology and Performance* 5(3):292-300.

Roberts, A.D., Clark, S.A., Townsend, N.E., Anderson, M.E., Gore, C.J., and Hahn, A.G. 2003. Changes in performance, maximal oxygen uptake and maximal accumulated oxygen deficit after 5, 10 and 15 days of live high/train low altitude exposure. *European Journal of Applied Physiology* 88(4-5):390-395.

Rupp, J.C., Bartels, R.L., Zuelzer, W., Fox, E.L., and Clarke, R. 1983. Effect of sodium bicarbonate ingestion on blood and muscle pH exercise performance. *Medicine and Science in Sports and Exercise* 15(2):115.

Russell, A.P., Wadley, G., Snow, R.J., Giacobino, J.P., Muzzin, P., Garnham, A. and Cameron-Smith, D. 2002. Slow component of $\dot{V}O_2$ kinetics: The effect of training status, fibre type, UCP3 mRNA and citrate synthase activity. *International Journal of Obesity and Related Metabolic Disorders* 26(2):157-164.

Salitn, B. 1990. Anaerobic capacity: Past, present and prospective. In: A.W. Taylor, P.D. Gollnick, H.J. Green HJ, et al (eds), *Biochemistry of Exercise VII, International Series on Sport Sciences*, Vol 21. Champaign, IL: Human Kinetics, pp 387-412.

Scott, C.B., Roby, F.B., Lohman, T.G., and Bunt, J.C. 1991. The maximally accumulated oxygen deficit as an indicator of anaerobic capacity. *Medicine and Science in Sports and Exercise* 23(5):618-624.

Serresse, O., Lortie, G., Bouchard, C., and Boulay, M.R. 1988. Estimation of the contribution of the various energy systems during maximal work of short duration. *International Journal of Sports Medicine* 9(6):456-460.

Simmonds, M., Minahan, C., and Sabapathy, S. 2010. Caffeine improves supramaximal cycling but not the rate of anaerobic energy release. *European Journal of Applied Physiology* 109(2):287-295.

Wadley, G., and Le Rossignol, P. 1998. The relationship between repeated sprint ability and the aerobic and anaerobic energy systems. *Journal of Science and Medicine in Sport* 1(2):100-110.

Weber, C., Chia, M., and Inbar, O. 2006. Gender differences in anaerobic power of the arms and legs - a scaling issue. *Medicine and Science in Sports and Exercise* 38(1):129-137.

Weber, C.L., and Schneider, D.A. 2001. Reliability of MAOD measured at 110% and 120% of peak oxygen uptake for cycling. *Medicine and Science in Sports and Exercise* 33(6):1056-1059.

Withers, R., Sherman, W., Clark, D., Esselbach, P.C., Nolan, S.R., Mackay, M.H., and Brinkman, M. 1991. Muscle metabolism during 30, 60, and 90s of maximal cycling on an air-braked ergometer. *European Journal of Applied Physiology* 63(5):354-362.

Withers, R.T., Van Der Ploeg, G., and Finn, J.P. 1993. Oxygen deficits incurred during 45, 60, 75 and 90-s maximal cycling on an air-braked ergometer. *European Journal of Applied Physiology* 67(2):185-191.

Woods, G.F., Day, L., Withers, R.T., Ilsley, A.H., and Maxwell, B.F. 1994. The dynamic calibration of cycle ergometers. *International Journal of Sports Medicine* 15(4):168-171.

Wooles, A.L., Robinson, A.J., and Keen, P.S. 2005. A static method for obtaining a calibration factor for SRM bicycle power cranks. *Sports Engineering* 8(3):137-144.

Woolf, K., Bidwell, W.K., and Carlson, A.G. 2008. The effect of caffeine as an ergogenic aid in anaerobic exercise. *International Journal of Sport Nutrition and Exercise Metabolism* 18(4):412-429.

Yano, T., Okuyama, T., Reihan, A., and Ogata, H. 2004. Effect of blood lactate level on oxygen uptake at the offset of middle-intensity exercise. *Biology of Sport* 21(3):231-239.

Zouhal, H., Jabbour, G., Jacob, C., Duvigneau, D., Botcazou, M., Ben Abderrahaman, A., Prioux, J., and Moussa, E. 2010. Anaerobic and aerobic energy system contribution to 400-m flat and 400-m hurdles track running. *Journal of Strength and Conditioning Research* 24(9):2309-2315.

Zupan, M.F., Arata, A.W., Dawson, L.H., Wile, A.L., Payn, T.L., and Hannon, M.E. 2009. Wingate anaerobic test peak power and anaerobic capacity classifications for men and women intercollegiate athletes. *Journal of Strength and Conditioning Research* 23(9):2598-2604.

Blood Lactate Thresholds: Concepts and Applications

Pitre Bourdon

Blood lactate accumulation during incremental exercise tests has commonly been used to evaluate the effects of training, to set training intensities, and to predict performance. Typically this is done through the determination of deflection points or thresholds on the blood lactate versus exercise intensity curve. Although the concept of blood lactate thresholds has been developing for more than 70 years, there is still much controversy both about the explanation of these phenomena and about the methods that should be used to identify them. In fact, points of contention probably outnumber points of agreement, which can frustrate those searching for practical applications of these concepts (Thoden 1991). In their review, Jones and Ehrsour (1982) suggested that many studies on the blood lactate response to exercise offer valid information that has applications to sport; however, use of these concepts requires precise description of increment intensity and duration as well as carefully defined calculation techniques.

The popularity of the use of blood lactate–related thresholds as performance indicators has increased dramatically over the past 20 years, with many exercise science laboratories around the world now routinely measuring various blood lactate thresholds as an integral component of the physiological assessment of endurance athletes. Probable reasons for this burgeoning popularity are the following:

- The predictive and evaluative power associated with the lactate response to exercise
- The development of automated lactate analyzers that offer ease of sampling and improved accuracy
- The reliability of such measures under standardized conditions

- Increased levels of coaching education and understanding of such modern training methods

Although lactate testing has proven useful for evaluating endurance performance, prescribing exercise intensities, monitoring training adaptations, and subsequently enhancing performance, there are many different approaches to test methods, data analysis, and interpretation. This chapter presents an overview of the major concepts and controversies relating to the practical or applied implications of the blood lactate response to exercise as they relate to endurance exercise. A review of the more theoretical and mechanistic considerations pertinent to this issue is beyond the scope of the chapter.

Blood Lactate Testing Rationale

Measurement of the blood lactate response to exercise in conjunction with heart rate, oxygen consumption ($\dot{V}O_2$), and intensity is often a part of the routine physiological assessment of the high performance athlete. There are three main reasons for these measures:

- They serve as indicators of training adaptation.
- They correlate with endurance performance.
- They may indicate optimal training stimuli.

As outlined in the sections that follow, numerous studies strongly support the evaluative and predictive power of the blood lactate response to exercise, suggesting that it can serve well as a monitoring test for endurance performers.

Indicator of Training Adaptation

In the past, endurance training and detraining studies commonly used changes in maximal oxygen consumption $\dot{V}O_2max$) to indicate alterations in the capacity to perform endurance exercise (Coyle et al. 1986; Daniels et al. 1978). However, more recently, research has suggested that the blood lactate response to training adapts to a greater degree than does $\dot{V}O_2max$ (Denis et al. 1984; Hurley et al. 1984; Katch et al. 1978; Keith et al. 1992; Sjodin et al. 1982). In particular, blood lactate thresholds have been shown to be more sensitive indicators of training adaptations (Bishop et al. 1998; Carter et al. 1999b; Denis et al. 1982, 1984; Faude et al. 2009; Gaesser and Poole 1988; Gollnick et al. 1986; Hurley et al. 1984; Sjodin et al. 1982; Weltman et al. 1992; Yoshida et al. 1982a). This is especially true in highly trained athletes, in whom there may be little or no change in $\dot{V}O_2max$ but significant changes in endurance performance (Coyle et al. 1986; Daniels et al. 1978; Foster et al. 1982). Figure 6.1 and table 6.1 provide evidence supporting this fact by following changes in various physiological parameters in a national-level, middle-distance runner across 2 years of his training history, during which time running performance and blood lactate threshold values improved but $\dot{V}O_2max$ did not. Furthermore, numerous studies have demonstrated that blood lactate related thresholds can increase with training beyond the point where $\dot{V}O_2max$ fails to increase (Davis et al. 1979; Denis et al. 1982; Foster et al. 1995; Hurley et al. 1984; Jacobs 1986; Karlsson et al. 1972; MacRae et al. 1992; Poole and Gaesser 1985; Poole et al. 1990; Sjodin et al. 1982; Svedenhag and Sjodin 1985; Tanaka et al. 1986; Weltman 1995; Weltman et al. 1992). This would suggest that $\dot{V}O_2max$ and the ability to perform at prescribed submaximal intensities are limited by different mechanisms (Gollnick and Saltin 1982; Karlsson and Jacobs 1982; Saltin and Rowell 1980; Weltman 1995). A further extension of this concept suggests the need for specific training programs for the body's central (heart and lungs) and peripheral (skeletal muscles) components of endurance performance (Gollnick 1982; Saltin et al. 1976).

Correlation With Endurance Performance

Blood lactate–associated thresholds are highly related to performance in various types of endurance activities. In fact, many researchers have suggested that these parameters are a better indicator of endurance performance than the traditional gold standard: $\dot{V}O_2max$ (Allen et al. 1985; Bishop et al. 1998; Conconi et al. 1982; Craig et al. 1993; Farrell et al. 1979; Fohrenbach et al. 1987; Foster et al. 1995; Hagberg and Coyle 1983; Ivy et al. 1981; Jacobs 1986; Kumagai et al. 1982; LaFontaine et al. 1981; Londeree and Ames 1975; Olbrecht et al. 1985; Sjodin and Svendenhag 1985; Weltman 1995; Weltman et al. 1987). Numerous studies also testify to blood lactate

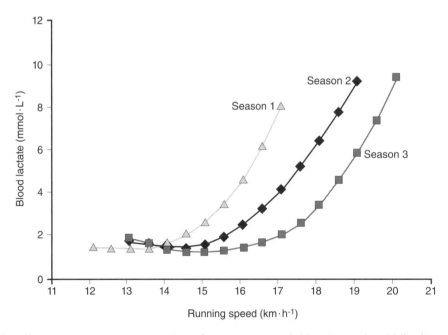

FIGURE 6.1 Blood lactate response curves monitored over two years. Subject is a male middle-distance runner. (Data were collected during routine physiological assessments in the physiology laboratory of the South Australian Sports Institute [SASI].)

TABLE 6.1   Blood Lactate Threshold Changes Monitored Over Two Seasons on a Male Middle-Distance Runner

Date	Threshold name	Blood lactate, mmol · L⁻¹	Running speed, km · h⁻¹	Heart rate, beats · min⁻¹	Percent-age maxi-mal heart rate,%	V̇O₂max, L · min⁻¹	Percentage V̇O₂max, %	800 m time, min:s
Autumn: season 1	LT1	1.41	13.0	166	83.4	3.61	64.9	n/a
	LT2	3.02	15.2	186	93.5	4.04	77.7	n/a
	Max	8.18	17.0	199	100.0	5.21	100.0	1:51.9
Autumn: season 2	LT1	1.40	14.7	163	81.5	3.67	66.2	n/a
	LT2	3.84	16.8	185	92.5	4.39	79.2	n/a
	Max	8.34	19.0	200	100.0	5.54	100.0	1:50.1
Autumn: season 3	LT1	1.36	15.8	162	85.3	3.99	73.2	n/a
	LT2	3.79	18.2	179	94.2	4.61	84.6	n/a
	Max	9.25	20.0	190	100.0	5.45	100.0	1:48.1

Data were collected during routine physiological assessments in the South Australian Sports Institute (SASI) exercise physiology laboratory. LT1 = lactate threshold 1; LT2 = lactate threshold 2; Max = maximal effort as assessed in the final stage of a 5 min progressive incremental test; n/a = not applicable.

thresholds as being powerful predictors of endurance performance (Amann et al. 2006; Bishop et al. 1998; Coen et al. 1991; Farrell et al. 1979; Faude et al. 2009; Jones and Carter 2000; Jones and Doust 1998; Nicholson and Sleivert 2001; Simoes et al. 2005; Stallknecht et al. 1998; Tanaka and Matsuura 1984).

Optimal Training Stimulus

Accumulated data suggest that the various blood lactate–related thresholds provide strong indices of exercise intensity by which to prescribe guidelines for training (Coen et al. 1991; Craig 1987; Fohrenbach et al. 1987; Fukuba et al. 1999; Heck et al. 1985; Jacobs 1986; Kindermann et al. 1979; Nicholson and Sleivert 2001; Olbrecht et al. 1985; Schnabel et al. 1982; Sjodin et al. 1982; Stegmann and Kindermann 1982; Weltman 1995; Weltman et al. 1992; Yoshida et al. 1982a). This is of particular interest to coaches, because these parameters can potentially provide them with a means to optimize training intensity and help prevent overreaching and overtraining (Bosquet et al. 2001).

Concepts and Controversies

Although the blood lactate response to exercise is used widely to control and monitor training programs, many factors in addition to training adaptation can affect the blood lactate response. Therefore, one needs to consider the following points when collecting, analyzing, and interpreting blood lactate measurements.

Terminology

One problem in understanding and interpreting the available literature about discontinuity in the blood lactate response to exercise is the plethora of terms used to describe similar phenomena. These terms include *lactate threshold* (Allen et al. 1985; Beaver et al. 1985; Craig 1987; Ivy et al. 1981; Tanaka et al. 1985; Weltman et al. 1990; Yoshida et al. 1987), *aerobic threshold* (Aunola and Rusko 1986; Skinner and McLellan 1980), *anaerobic threshold* (Aunola and Rusko 1986; Heck et al. 1985; Skinner and McLellan 1980; Wasserman and McIlroy 1964), *individual anaerobic threshold* (McLellan et al. 1991; Stegmann and Kindermann 1982; Stegmann et al. 1981), *aerobic–anaerobic threshold* (Kindermann et al. 1979), *onset of blood lactate accumulation* (Bentley et al. 2001; Karlsson and Jacobs 1982; Sjodin and Jacobs 1981; Sjodin et al. 1981), *onset of plasma lactate accumulation* (Farrell et al. 1979), *lactate turnpoint* (Davis et al. 1983), *maximal lactate steady state* (Beneke and Petelin von Duvillard 1996; Heck et al. 1985), *lactate minimum* (Jones and Doust 1998; MacIntosh et al. 2002; Tegtbur et al. 1993), and *Dmax* (Cheng et al. 1992). Without doubt there are other relevant terms, but clearly this is an extensive and complicated topic of discussion.

The situation is even more complicated in that some researchers have used the same term that another investigator used but to refer to a different phenomenon. For example, the term *lactate threshold* has been defined as the highest V̇O₂ (intensity) attained during an incremental work task not associated with an increase in blood lactate concentration above the resting level (Beaver et al. 1985; Ivy et al. 1980; Tanaka et al. 1985; Weltman et al. 1990; Yoshida et al. 1987), as the exercise intensity corresponding to a lactate concentration that is 1.0 mmol · L⁻¹ above the baseline (Coyle et al. 1983), or is the highest exercise intensity that elicits a blood

lactate concentration of 2.5 mmol · L⁻¹ after 10 min of steady-state exercise (Allen et al. 1985). The sidebar presents a number of commonly used terms for the blood lactate response and their corresponding definitions. Irrespective of the name assigned to an assessment technique, the user must have a clear

Sample Terms Used to Classify Changes in the Blood Lactate Response to Progressive Exercise

Lactate Threshold 1 (LT1)

The lowest intensity at which there is a sustained increase in blood lactate concentration above resting values.

aerobic threshold (Kindermann et al. 1979)—Fixed 2.0 mmol · L⁻¹ value

lactate threshold (ADAPT; software for the automated data analysis of progressive tests, Australian Institute of Sport 1995)—Intensity preceding a 0.4 mmol · L⁻¹ increase in blood lactate above the baseline

lactate threshold (Beaver et al. 1985)—Point of deflection in the log [blood lactate] versus log V̇O₂ transformation

lactate threshold (Ivy et al. 1980)—Intensity preceding nonlinear increase in blood lactate during progressive work

lactate threshold (Dickhuth et al. 1999)—The lowest value of the performance–lactate ratio

maximal steady state (LaFontaine et al. 1981)—Fixed 2.2 mmol · L⁻¹ value

onset of plasma lactate accumulation (Farrell et al. 1979)—Exercise intensity that elicited a blood lactate concentration 1.0 mmol · L⁻¹ greater than baseline

Lactate Threshold 2 (LT2)

The intensity that causes a rapid increase in blood lactate indicating the upper limit of equilibrium between lactate production and clearance.

aerobic–anaerobic threshold (Mader et al. 1976), **onset of blood lactate accumulation (OBLA)** (Sjodin and Jacobs 1981), and **4 mmol · L⁻¹ threshold** (Heck et al. 1985)—Fixed 4 mmol · L⁻¹ value

anaerobic threshold (ADAPT) (Australian Institute of Sport 1995)—Modified Dmax using LT instead of first exercise intensity as starting point

anaerobic threshold (Kindermann et al. 1979)—Steep part of exponential increase in lactate concentration, approximately 4 mmol · L⁻¹

Dmax (Cheng et al. 1992)—Point on blood lactate–exercise intensity curve at maximal distance from a line connecting starting and finishing intensities

individual anaerobic threshold (Keul et al. 1979)—A fixed slope point on the lactate power curve whose tangent is equal to 51°

individual anaerobic threshold (Stegmann et al. 1981)—Value based on a tangential model to define the intensity at maximal lactate steady state

individual anaerobic threshold (Dickhuth et al. 1999)—Value 1.5 mmol · L⁻¹ above LT

lactate minimum (Australian Institute of Sport 1995; Cheng et al. 1992; Coen et al. 2001; Keul et al. 1979; Stegmann et al. 1981; Tegtbur et al. 1993)—Lactate steady state determined through a series of decreasing exercise intensities following an initial maximal effort

maximal lactate steady state (MLSS) (Beneke et al. 2001)—Highest intensity at which the blood lactate concentration increases by no more than 1.0 mmol · L⁻¹ during the final 20 min of a 30 min constant intensity test

maximal steadystate workload (MSSW) (Borch et al. 1993)—Fixed 3 mmol · L⁻¹ value

understanding of the protocol required to detect the blood lactate related threshold.

Definitions

A number of researchers have independently suggested that there are at least two apparent discontinuities or thresholds in the blood lactate response to incremental exercise that may serve as general concepts for many of the terms proposed by other researchers (Beaver et al. 1986; Faude et al. 2009; Heck et al. 1985; Kindermann et al. 1979; Skinner and McLellan 1980; Wasserman 1984). The first of these is associated with the first exercise intensity at which there is a sustained increase in blood lactate above resting levels (Australian Institute of Sport 1995; Beaver et al. 1985; Coyle et al. 1984; Dickhuth et al. 1999; Ivy et al. 1980). This point is generally consistent with blood lactate concentrations of less than 2.0 mmol · L^{-1}. The second of these discontinuities is marked by a very rapid increase in blood lactate concentration. This point is representative of a shift from oxidative to partly anaerobic energy metabolism during incremental intensity exercise, and it refers to the upper limit of blood lactate concentration indicating an equilibrium between lactate production and lactate elimination (i.e., maximal lactate steady state) (Beneke 1995, 2003; Beneke et al. 1996, 2001; Harnish et al. 2001; Heck et al. 1985; Palmer et al. 1999). This second point is generally associated with blood lactate concentrations between 2.5 and 5.5 mmol · L^{-1}. Numerous researchers, however, have contested the suggestion these thresholds exist on the basis that there are no break points and no

thresholds in the ventilatory and metabolic responses to incremental exercise (Cheng et al. 1992; Hagan and Smith 1984; Hughson et al. 1987).

Although we acknowledge arguments concerning the correct nomenclature for these two discontinuities or thresholds, as well as the problems of whether thresholds actually exist, this chapter uses the terms *lactate threshold 1 (LT1)* and *lactate threshold 2 (LT2)* to describe the first and second thresholds, respectively. These terms are recommended for use in Australian sport science by the National Sport Science Quality Assurance Program (NSSQA). Figure 6.2 is a visual representation of LT1 and LT2 in relation to their respective positions on the blood lactate–exercise response curve as interpreted in the context of this chapter.

Categories of Blood Lactate Thresholds

Among the many terms and definitions used for blood lactate thresholds, most can be categorized into one of three broad classifications: (1) fixed blood lactate concentrations, (2) individualized lactate and anaerobic thresholds, and (3) maximal lactate steady-state assessments.

Fixed Blood Lactate Concentrations

As a strategy for minimizing the problems of biological noise associated with detecting inflections in the blood lactate response curve, work or physiological variables at fixed blood lactate concentrations of 2.0 mmol · L^{-1} (Kindermann et al. 1979), 2.2 mmol · L^{-1} (LaFontaine et al. 1981), 2.5 mmol · L^{-1} (Allen

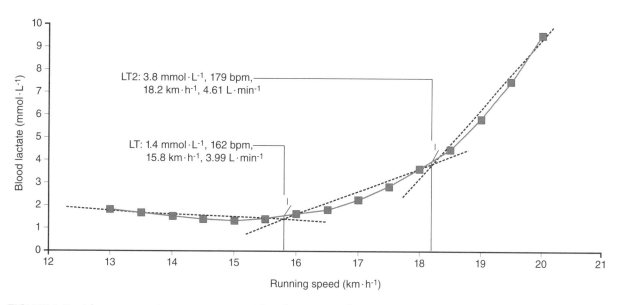

FIGURE 6.2 A lactate-exercise response curve identifying LT1 and LT2. Subject is a male middle-distance runner. (Data were collected during routine physiological assessment in the SASI exercise physiology laboratory.)

et al. 1985; Foster et al. 1995; Hagberg 1984), 3.0 mmol · L⁻¹ (Borch et al. 1993), and 4.0 mmol · L⁻¹ (Bentley et al. 2001; Bishop et al. 1998; Foster et al. 1993; Foxdal et al. 1994, 1996; Heck et al. 1985; Kindermann et al. 1979; Mader and Heck 1986; Mader et al. 1976; Oyono-Enguelle et al. 1990; Sjodin and Jacobs 1981; Weltman et al. 1989) have been used to assess the response to incremental exercise. The actual intensity associated with fixed blood lactate concentrations is determined by interpolation from visual plots of exercise intensity versus blood lactate, as illustrated in figure 6.3. Fixed blood lactate concentrations are, however, strongly influenced by an athlete's nutritional, training, and recovery state (Bosquet et al. 2001; Carter et al. 1999a; Dotan et al. 1989; Hughes et al. 1982; Ivy et al. 1981; Jacobs 1981; Maassen and Busse 1989; Yoshida 1984a) and care must be taken to control for such factors when testing an athlete.

Individualized Lactate and Anaerobic Thresholds

Stegmann and colleagues (1981) reported that steady-state blood lactate concentrations can vary greatly among individuals. On the basis of this fact, and in combination with arguments founded on the diffusion of lactate from active muscle to blood, they proposed the concept of individualizing blood lactate threshold determinations. Numerous others have since proposed methods such as log transformations, rates of metabolite accumulation, tangential methods, and even subjective assessments to determine individualized LT1 (Australian Institute of Sport 1995; Beaver et al. 1985; Bourdon et al. 2009; Coyle et al. 1984; Newell et al. 2007) and LT2 intensities (Anderson et al. 1995; Australian Institute of Sport 1995; Bourdon et al. 2009; Cheng et al. 1992; Coen et al. 2001; Keul et al. 1979; Stegmann et al. 1981; Tegtbur et al. 1993). Figure 6.4 provides a schematic presentation of eight methods commonly used to determine blood lactate thresholds (Australian Institute of Sport 1995; Beaver et al. 1986; Cheng et al. 1992; Coyle et al. 1984; Heck et al. 1985; Keul et al. 1979; Stegmann and Kindermann 1982).

Maximal Lactate Steady State (MLSS) Assessments

The MLSS defines the highest exercise intensity that can be maintained over time without continual blood lactate accumulation (Beneke 1995, 2003; Faude et al. 2009; Heck et al. 1985; Smekal et al. 2002). Thus, MLSS appears to indicate an exercise intensity above which the rate of anaerobic glycolysis exceeds the rate of mitochondrial pyruvate utilization, causing an accumulation of lactate (Beneke 2003; Billat et al. 2003; Heck et al. 1985; Mader and Heck 1986). Exercise above MLSS is also associated with a constant increase of pulmonary ventilation and $\dot{V}O_2$ (Gaesser and Poole 1996; Poole et al. 1988) and is poorly tolerated by athletes for extended periods of time (Billat et al. 2003; Gaesser and Poole 1996). Consequently, it is considered that the MLSS can discriminate qualitatively between sustainable exercise intensities, in which continuous work is limited by

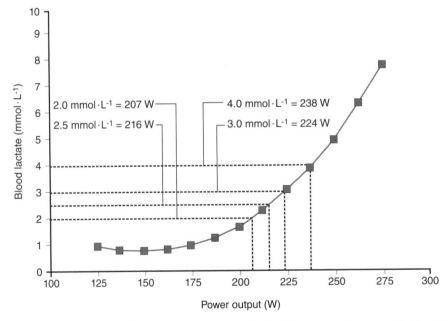

FIGURE 6.3 Comparison of fixed blood lactate concentrations during incremental exercise. Subject is a female heavyweight rower. (Data were collected during routine physiological assessment in the SASI exercise physiology laboratory.)

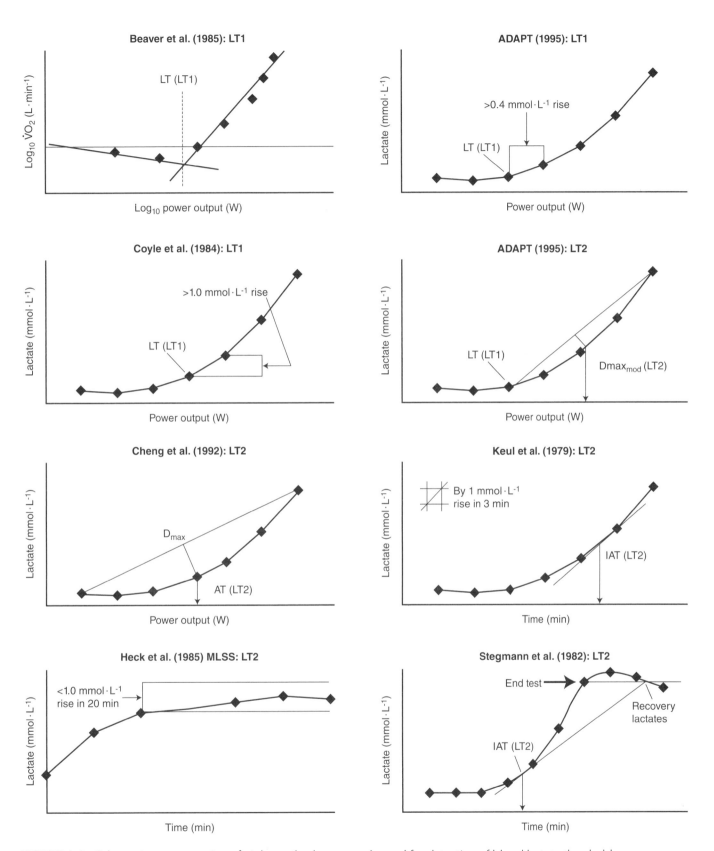

FIGURE 6.4 Schematic representation of eight methods commonly used for detection of blood lactate thresholds.

Adapted, by permission, from D. Bishop, D.G. Jenkins and L.T. MacKinnon, 1998, "The relationship between plasma lactate parameters, W peak and 1-h cycling performance in women," *Medicine and Science in Sport and Exercise* 30(8): 1270-1275.

stored energy, and exercise intensities that have to be ended because of a disturbance of cellular homeostasis through the accumulation of fatiguing metabolic by products (Beneke 2003). Furthermore, MLSS represents a quantitative measure of the exercise-related behavior of the blood lactate concentration (Beneke et al. 2001) and has been considered the best index of the capacity for endurance exercise (Jones and Carter 2000).

Because the terms and their corresponding definitions affect the interpretation of the blood lactate response to exercise, before selecting a method the user must understand the protocol required to assess and detect the response. One must determine the appropriateness of the method for evaluating or prescribing training or performance. Most laboratories in Australia dealing with high-performance athletes are using the ADAPT program to determine LT1 and LT2 (Australian Institute of Sport 1995). Typical error (TE) (refer to chapter 3, Data Collection and Analysis) measurements for LT1 and LT2 using ADAPT have been shown to have good precision for athletes tested at NSSQA-accredited laboratories in the endurance sports of cycling, running, and rowing.

Test Protocols

Protocol-related factors such as the sampling site, increment duration and intensities, continuous versus discontinuous exercise bouts, mode of exercise, environmental factors, and subject attributes can all affect the measurement of the blood lactate response to incremental exercise. Therefore, one must consider such factors carefully when establishing a test protocol.

Blood Collection, Handling, and Analysis

The blood vessel sampling site has been shown to affect the measurement of the relationship between exercise intensity and blood lactate concentration (El-Sayed et al. 1993; Foxdal et al. 1990; Welsman 1992; Yoshida et al. 1982b). Different samples sites include arterial blood, venous blood, capillary blood, and arterialized venous blood. The blood lactate thresholds determined from different sample sites are not comparable unless one has applied a correction factor. Samples taken from the same vessel type but from different locations on the body—for example, capillary samples drawn from the toe, fingertip, and ear lobe—have shown fingertip capillary blood lactate levels to be higher than earlobe capillary levels during incremental cycling, running (Dassonville et al. 1998; Feliu et al. 1999), and rowing (Feliu et

al. 1999) tests. Forsyth and Farrally (2000), however, found no difference in simultaneous capillary samples drawn from the ear, fingertip, and toe during rowing.

Blood handling practices prior to analysis for lactate concentration can also affect the measurement of the relationship between exercise intensity and blood lactate concentration. Differences can arise depending on whether measurements are taken using untreated whole blood, plasma, serum, deproteinized blood, or lysed whole blood. Whole-blood lactate concentrations, for example, have been reported to be between 10% and 30% lower than plasma values (Foxdal et al. 1994; Telford 1984; Welsman 1992).

Once the blood has been sampled and treated, the way in which it is analyzed for lactate concentration can also have a significant effect on the result obtained. Various brands of automated blood lactate analyzers use different assaying techniques to determine blood lactate concentration (Bishop 2001; Buckley et al. 2003; Medbo et al. 2000; Pyne et al. 2000; Shimojo et al. 1993; Thin et al. 1999). Figure 6.5 and table 6.2 demonstrate the difference in blood lactate concentration values for duplicate whole-blood samples between three automated lactate analyzers. As these data show, the results from these three analyzers should not be compared unless appropriate correction factors have been applied. These data also highlight a potential problem in using fixed blood lactate thresholds (e.g., 4 mmol · L^{-1}), because one can see that the threshold intensities derived from different analyzers differ markedly for the fixed 4 mmol · L^{-1} values calculated from the same samples.

Within a laboratory it is therefore important to standardize blood sampling site, treatment procedures, and assay method for lactate determinations in order to allow direct and meaningful comparisons between results collected under test–retest conditions.

Increment Duration

Another factor that affects the blood lactate–exercise response curve is the incremental stage duration. Numerous studies have demonstrated that the longer the increment duration, the lower the LT2 values (Bentley et al. 2001; Bentley et al. 2007; Carter et al. 1999a; Coen et al. 2001; Foxdal et al. 1996; Freund et al. 1989; Heck et al. 1985; McLellan 1987; Rusko et al. 1986; Stockhausen et al. 1997; Yoshida 1984b). Most exercise physiology laboratories around the world use incremental intensity exercise tests with work durations between 3 and 5 min. These durations are considered adequate for measurements of $\dot{V}O_2$

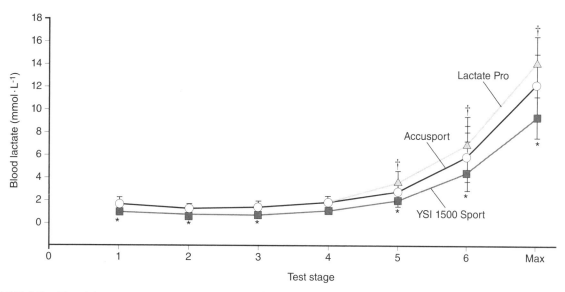

FIGURE 6.5 Blood lactate concentrations measured using the Lactate Pro, Accusport, and YSI 1500 Sport blood lactate analyzers during incremental exercise. *YSI significantly different from Lactate Pro ($p < .0001$) and Accusport ($p < .001$). †Lactate Pro significantly different from Accusport ($p < .0001$).

Adapted from *Journal of Science and Medicine in Sport*, Vol. 6(4), J.D. Buckley, P.C. Bourdon and S.M. Woolford, "Effect of measuring blood lactate concentrations using different automated lactate analysers on blood lactate transition thresholds," pgs. 408-421, copyright 2003, with permission from Elsevier.

and heart rate, because these variables are usually in steady state within these time frames for trained individuals. However, these durations seem less adequate when the aim of the test is to determine the highest exercise intensity corresponding to a maximal lactate steady-state concentration or LT2. It is thus possible that the use of exercise durations shorter than 5 min may overpredict the LT2 exercise intensity. A number of studies have reported that exercise periods of at least 5 to 8 min may be required to attain steady-state blood lactate concentrations and therefore allow an accurate determination of LT2 (Foxdal et al. 1994, 1996; Heck et al. 1985; LaFontaine et al. 1981; Oyono-Enguelle et al. 1990; Rieu et al. 1989; Stegmann and Kindermann 1982; Stockhausen et al. 1997).

Figure 6.6 and table 6.3 clearly demonstrate that varying increment duration leads to different blood lactate response curves, which in turn affects the calculation of the exercise intensity corresponding to a specific blood lactate threshold. The selection of increment duration should depend largely on the reason for conducting the test. If one is testing to prescribe exercise intensities for endurance training, it is necessary to remember that such training generally takes place under steady-state conditions. Thus, a test protocol using a number of exercise bouts of at least 5 min should provide more valid data for prescription of training intensities (Foxdal et al. 1996; Oyono-Enguelle et al. 1990).

Rest Interval Duration

The intrastage rest interval can affect the determination of blood lactate thresholds. With longer breaks between work bouts in discontinuous protocols, such as interruptions in the rowing and kayaking protocols to allow blood sampling, blood lactate thresholds tend to move to higher work intensities (Foster et al. 1995; Heck et al. 1985). Another factor associated with the intrastage rest interval is the timing of blood collection. There is no consensus as to how long the sampling interval should be after the work bout; however, one study indicated that there is no difference in blood lactate values collected across a range of 15 to 45 s after exercise in male rowers during ergometer testing (Kass and Carpenter 2009). Therefore, any intervals between work stages should be kept as brief and standardized as possible, with 1 min being a possible maximum.

Number of Increments

With many of the methods outlined earlier, mathematical manipulations are performed to best fit the data curve and generate the blood lactate thresholds (Australian Institute of Sport 1995; Beaver et al. 1985; Cheng et al. 1992; Foster et al. 1995; Keul et al. 1979; Stegmann et al. 1981; Tegtbur et al. 1993). Such manipulations require a minimum number of increments; five to seven increments are generally sufficient.

TABLE 6.2   Blood Lactate Concentration, Heart Rate, and Power Output at Various Blood Lactate Transition Thresholds Measured Using the YSI 1500 Sport, Accusport, and Lactate Pro Lactate Analyzers

			YSI 1500 Sport	Accusport	Lactate Pro
LT1 measures	ADAPT LT (n = 23)	HLa, mmol · L⁻¹	1.0 ± 0.2*	1.5 ± 0.3	1.5 ± 0.3
		Power, W	158.0 ± 24.9	158.1 ± 25.6	158.1 ± 25.6
		HR, beats · min⁻¹	149 ± 8	149 ± 9	149 ± 9
	Log–Log LT (n = 23)	HLa, mmol · L⁻¹	1.0 ± 0.3*	1.8 ± 0.5*	1.6 ± 0.4
		Power, W	164.8 ± 24.9	175.5 ± 30.8	166.8 ± 27.6
		HR, beats · min⁻¹	151 ± 9	156 ± 12	152 ± 10
LT2 measures	Dmax (n = 23)	HLa, mmol · L⁻¹	2.2 ± 0.5*	3.0 ± 0.7*	3.3 ± 0.6*
		Power, W	205.8 ± 30.5	207.4 ± 31.5	204.4 ± 31.1
		HR, beats · min⁻¹	168 ± 8	169 ± 8	168 ± 8
	ADAPT AT (n = 23)	HLa, mmol · L⁻¹	3.1 ± 0.5*	4.1 ± 0.8*	4.7 ± 0.7*
		Power, W	220.8 ± 31.7	221.2 ± 32.3	219.4 ± 32.4
		HR, beats · min⁻¹	174 ± 8	175 ± 8	174 ± 8
	4 mmol · L⁻¹ (n = 23)	HLa, mmol · L⁻¹	4.0 ± 0	4.0 ± 0	4.0 ± 0
		Power, W	231.9 ± 32.0*	219.7 ± 30.7*	212.5 ± 30.7*
		HR, beats · min⁻¹	179 ± 8*	174 ± 8*	171 ± 8*

ADAPT LT = the real data point preceding point on lactate curve where lactate reached 0.4 mmol · L⁻¹ above minimum recorded lactate reading; Log–Log LT = point at which blood lactate began to increase when log–log lactate concentration plotted against log of exercise parameter of interest; Dmax = point on third-order polynomial regression of blood lactate yielding maximal perpendicular distance to straight line formed by two end data points; ADAPT AT = point on a third-order polynomial regression curve yielding maximal perpendicular distance to straight line formed by Dmax threshold and final lactate point; 4 mmol · L⁻¹ = fixed blood lactate concentration of 4 mmol · L⁻¹; HLa = blood lactate concentration; HR = heart rate.

*Significantly different from other two analyzers ($p < .0001$).

Adapted from Journal of Science and Medicine in Sport, Vol. 6(4), J.D. Buckley, P.C. Bourdon and S.M. Woolford, "Effect of measuring blood lactate concentrations using different automated lactate analysers on blood lactate transition thresholds," pgs. 408-421, copyright 2003, with permission from Elsevier.

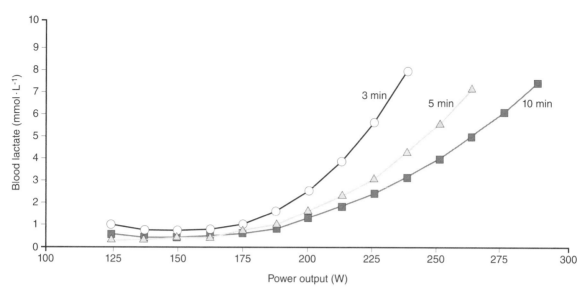

FIGURE 6.6 Effect of increment duration on the blood lactate response curve. Subject is a heavyweight female rower. (Data were collected during routine physiological assessments in the SASI exercise physiology laboratory).

TABLE 6.3 Comparison of Select Blood Lactate Thresholds Determined From Incremental Blood Lactate Tests With Varying Work Duration; Subject Is a Female Heavyweight Rower

Blood lactate threshold	Increment duration, min	Blood lactate, mmol · L⁻¹	Power output, W	$\dot{V}O_2$max, L · min⁻¹	Percentage $\dot{V}O_2$max, %	HR, beats · min⁻¹	Percentage HRmax, %
LT1-ADAPT	3	0.72	172.4	2.83	70.8	166	81.8
	5	0.75	170.6	2.84	71.2	168	82.8
	10	0.96	169.9	2.85	71.4	170	83.7
LT2-ADAPT	3	2.26	221.8	3.41	85.4	185	91.1
	5	2.79	220.7	3.52	88.1	188	92.6
	10	2.55	200.6	3.31	82.9	185	91.1
2.0 mmol · L⁻¹	3	2.0	216.4	3.35	83.8	183	90.1
	5	2.0	207.8	3.34	83.6	183	90.1
	10	2.0	193.7	3.21	80.3	181	89.2
4.0 mmol · L⁻¹	3	4.0	250.5	3.75	93.9	197	97.0
	5	4.0	235.9	3.72	93.1	195	96.1
	10	4.0	214.2	3.52	88.1	191	94.1
MLSS		2.4	199.9	3.29	82.4	183	90.1

Data were collected during routine physiological assessments in the South Australian Sports Institute (SASI) exercise physiology laboratory. LT1-ADAPT = the real data point preceding point on lactate curve where lactate reached 0.4 mmol · L⁻¹ above minimum recorded lactate reading; LT2-ADAPT = point on a third-order polynomial regression curve yielding maximal perpendicular distance to straight line formed by Dmax threshold and final lactate point; 4 mmol · L⁻¹ = fixed blood lactate concentration of 4 mmol · L⁻¹; 2 mmol · L⁻¹ = fixed blood lactate concentration of 2 mmol · L⁻¹; MLSS = maximal lactate steady state; HR = heart rate; HRmax = maximal heart rate; $\dot{V}O_2$max = maximal oxygen consumption.

Intensity Increments

Most investigators agree that the work intensity increments should be relatively small to allow more precise determination of the thresholds (Coen et al. 2001; Foxdal et al. 1994; Stockhausen et al. 1997). Stockhausen and colleagues (1997) reported that the time constant of lactate kinetics and the time to elicit 95% MLSS both increased significantly with the magnitude of the increment intensity while subjects were cycling on an ergometer. Similarly, Coen and colleagues (2001) reported that a reduction in increment running speed from 2 to 1 km · h^{-1} significantly increased the LT2 velocity by 6%.

Within sport science laboratories, decisions regarding increment duration, intensity increase, interval between increments, and number of increments have generally been a matter of convenience and logistics (Foster et al. 1995; Foxdal et al. 1994; Weltman et al. 1990). However, the evidence presented here suggests that longer work stages (>5 min) are definitely preferred over shorter stages, that breaks between stages should be as short as possible (<1 min), and that there should be at least five stages with relatively small increases in intensity.

Ergometer Type

The choice of test ergometer will influence the intensity at which a blood lactate threshold will occur. In assessment of high-performance athletes, the choice of ergometers should be dictated by the principle of training specificity (Beneke et al. 2001; Craig 1987; Jacobs 1986; Withers et al. 1981). For example, cycle ergometers should be used to test cyclists, treadmills for runners, and rowing ergometers for rowers. Beneke (2003) reported that irrespective of training status, LT2 (determined as MLSS) occurs at different blood lactate concentrations for different exercise modes. This suggests that LT2 depends on the motor pattern of exercise and may be a function of the relationship between intensity per unit muscle mass and the mass and training status of the muscle primarily engaged in the activity (Beneke 2003; Beneke and Petelin von Duvillard 1996). This factor has major implications for triathletes, because here the need is to assess the blood lactate response to exercise for each sporting discipline. Table 6.4 supports this concept by presenting blood lactate threshold data for each discipline generated from three male triathletes during routine physiological assessments in the SASI exercise physiology laboratory.

One may also need to consider event specificity within a sport. For example, varying cadence can affect the work output at various blood lactate thresholds during cycling ergometry (Beneke 1994; Beneke and Petelin von Duvillard 1996; Craig et al. 1993, 1995; Denadai et al. 2006; Woolford et al. 1999). This point is relevant in the assessment of road cyclists, whose

TABLE 6.4 Mean Blood Lactate Threshold Values From Three Male Triathletes

Blood lactate threshold	Event discipline	Blood lactate, mmol · L^{-1}	Work intensity*	$\dot{V}O_2$max, L · min^{-1}	Percentage $\dot{V}O_2$max, %	HR, beats · min^{-1}	Percentage HRmax, %
ADAPT–LT1	Swim	1.59	1.136	n/a	n/a	143	80.5
	Ride	1.01	213.2	3.08	59.2	133	72.3
	Run	1.41	14.0	3.69	68.1	156	81.0
ADAPT–LT2	Swim	3.60	1.270	n/a	n/a	163	91.6
	Ride	3.09	315.7	4.24	81.5	161	87.6
	Run	3.17	16.7	4.51	83.2	178	92.7
2.0 mmol · L^{-1}	Swim	2.00	1.179	n/a	n/a	149	83.7
	Ride	2.00	280.2	3.84	73.8	151	82.3
	Run	2.00	15.3	4.08	75.3	167	86.5
4.0 mmol · L^{-1}	Swim	4.00	1.286	n/a	n/a	165	92.6
	Ride	4.00	337.9	4.49	86.4	167	90.9
	Run	4.00	17.5	4.75	87.7	184	95.8

Data were collected during routine physiological assessments in the South Australian Sports Institute (SASI) exercise physiology laboratory. Swim intensity, m · s^{-1}; ride intensity, W; run intensity, km · h^{-1}. ADAPT-LT1 = the real data point preceding point on lactate curve where lactate reached 0.4 mmol · L^{-1} above minimum recorded lactate reading; ADAPT-LT2 = point on a third order polynomial regression curve yielding maximal perpendicular distance to straight line formed by Dmax threshold and final lactate point; 2 mmol · L^{-1} = fixed blood lactate concentration of 2 mmol · L^{-1}; 4 mmol · L^{-1} = fixed blood lactate concentration of 4 mmol · L^{-1}; HR = heart rate; HRmax = maximal heart rate; n/a = not available; $\dot{V}O_2$max = maximal oxygen consumption. The calculated threshold values are reported for each discipline.

preferred cadence tends to be 90 to 100 revolutions per minute (rpm). In contrast, track endurance cyclists tend to pedal at cadences around 110 to 120 rpm, suggesting that their protocol needs to be modified from that of road cyclists (Craig et al. 1993, 1995). Pool length in swimming has been shown to affect the blood lactate–velocity relationship with lower blood lactates and heart rates and higher swimming velocities being reported for submaximal efforts conducted in short course (25 m) compared with long course (50 m) pools (Keskinen et al. 2007).

As with all testing equipment, ergometers should be regularly calibrated to minimize potential errors from one test to another. Heck and colleagues (1985) support this point by reporting a distinct difference in the lactate curves with the use of identical loading procedures but different treadmills.

Environmental and Athlete Conditions

The blood lactate response to exercise intensity and derived thresholds have been shown to be reproducible under standardized conditions (Celik et al. 2005; Coen et al. 2001; Dickhuth et al. 1999; Heitcamp et al. 1991; Jacobs 1986; Karlsson and Jacobs 1982; McLellan and Jacobs 1993; Pfitzinger and Freedson 1998; Zhou and Weston 1997). Table 6.5 offers further support by showing TE values for blood lactate threshold values generated in the SASI exercise physiology laboratory as well as target TE values recommended by NSSQA. However, a number of factors have been identified that affect the blood lactate response to exercise; these factors generally fall into one of the following classifications: temperature, altitude, circadian rhythms, nutritional status, or other subject attributes.

Temperature and Altitude

High ambient temperatures have been shown to increase blood lactate concentrations both at rest and during exercise (Fink et al. 1975; MacDougall et al. 1974). In contrast, during exercise in the cold, LT1 has been shown to occur at higher exercise intensities than in normothermic environments (Blomstrand et al. 1984; Flore et al. 1992; Therminarias et al. 1989).

Acute exposure to altitude causes enhanced lactate production (Friedmann et al. 2005; Robergs et al.

TABLE 6.5 Sample and Target Typical Error Values for Blood Lactate Thresholds

Test measure	TE	TE%	95% CI	Range	NSSQA target TE
Peak values					
Blood lactate, mmol · L^{-1}	1.0	7.5	2.8	11.8-17.8	<1.0
Power output, W	5.4	1.7	15.0	254.8-376.7	n/a
Heart rate, beats · min^{-1}	2	1.1	5.5	178-196	<2
$\dot{V}O_2$, L · min^{-1}	0.11	2.5	0.30	3.73-5.64	<0.15
LT1					
Blood lactate, mmol · L^{-1}	0.3	15.2	0.7	1.1-2.7	<0.2
Power output, W	1.5	0.9	4.2	138-194	<10
Heart rate, beats · min^{-1}	2	1.6	6.4	140-160	<5
% HRmax	1.6	2.1	4.4	74.5-84.4	n/a
$\dot{V}O_2$, L · min^{-1}	0.04	1.3	0.1	2.39-3.47	<0.15
% $\dot{V}O_2$max	2.2	3.3	6.2	56.0-79.0	n/a
LT2					
Blood lactate, mmol · L^{-1}	0.5	10.9	1.3	3.6-6.0	<0.5
Power output, W	5.2	2.2	14.4	183-292	<10
Heart rate, beats · min^{-1}	2	1.0	4.6	156-190	<4
% HRmax	1.1	1.2	3.0	85.7-97.4	n/a
$\dot{V}O_2$, L · min^{-1}	0.11	2.8	0.3	3.14-4.92	<0.15
% $\dot{V}O_2$max	2.4	2.9	6.7	75.0-92.0	n/a

Calculated from rowing 2 in 1 test data (Bourdon et al. 2009). $n = 10$ subjects: 5 males and 5 females. Data were collected during routine physiological assessments in the South Australian Sports Institute (SASI) exercise physiology laboratory. LT1 and LT2 calculated using ADAPT (Australian Institute of Sport 1995). n/a = not available; TE = typical error of measurement; TE% = typical error as a percentage of the actual measure; 95% CI = 95% confidence interval; NSSQA target TE = Target TE values based on the data collected from seven NSSQA-accredited laboratories.

1998; Weltman 1995) with an associated reduction of exercise intensity corresponding to the blood lactate thresholds (Clark et al. 2007; Friedmann et al. 2004, 2005; Koistinen et al. 1995; Robergs et al. 1998; Weltman 1995). Care must be taken when interpreting lactate values achieved under the hypoxic conditions of altitude: It has been observed that after acclimatization to high altitudes, lactate values begin to approach sea level values. This phenomenon is known in the literature as the *lactate paradox*, the underlying mechanisms of which are yet to be fully understood (Kayser 1996). Such findings are potentially of major importance when taking blood lactate measures in the field.

Circadian Rhythms

Circadian rhythms were reported for $\dot{V}O_2$ and heart rate at LT2 in male rowers due to circadian effects (Forsyth and Reilly 2004). Madsen and Lohberg (1987) also reported different threshold velocities when assessing swimmers in the morning and afternoon. The recommendations from these data are that blood lactate–intensity profiles should ideally be conducted at the same time of the day that athletes normally train to ensure precision and relevance of blood lactate threshold determination.

Nutritional Status

Blood lactate concentration is decreased at any given exercise intensity when muscle glycogen stores are depleted (Dotan et al. 1989; Hughes et al. 1982; Ivy et al. 1981; Jacobs 1981; Maassen and Busse 1989; Yoshida 1984a). Muscle glycogen depletion can result from either dietary (Greenhaff et al. 1987; Jacobs 1981; Yoshida 1984a, 1986) or exercise (Genovely and Stanford 1982; Jacobs 1981) manipulations. Although a glycogen-depleted state leads to decreased lactate production and a resultant shift to the right of the blood lactate–performance curve, neither the shape nor the slope of the curve is changed (Frohlich et al. 1989). Therefore, when a subject is in a glycogen-depleted state, endurance capacity is overestimated in the determination of fixed blood lactate thresholds, whereas individually based thresholds calculated from the individual shape of the curve are only affected minimally (Frohlich et al. 1989; Urhausen and Kindermann 1992). Obviously, one must consider the implications of depleted muscle glycogen stores when evaluating the blood lactate response to exercise, particularly if fixed blood lactate thresholds are to be calculated. Additionally, dietary manipulations such as creatine (Chwalbinska-Monta 2003), caffeine (Gaesser and Rich 1985), and bicarbonate ingestion (Davies et al. 1986; Kowalchuk et al. 1984) have been shown to affect blood lactate

thresholds. Standardization of pretest diet and exercise patterns is therefore recommended.

Subject Attributes

A number of factors specific to female athletes need to be considered when assessing the blood lactate responses to exercise. The reproductive cycle phase may influence physiological assessments, especially when using fixed blood lactate concentrations. Forsyth and Reilly (2005) reported that the 4 mmol · L^{-1} work intensity was significantly higher during the midluteal phase compared with the midfollicular phase, whereas Dmax LT2 blood lactate concentrations were significantly lower. Oosthuyse and colleagues (2005) also reported a trend for faster time trials during the late follicular phase. Contrary to these findings, a review by Janse de Jonge (2003) found that most research reported no changes to the lactate response to exercise over the menstrual cycle. It has been suggested that pregnancy may influence the determination of blood lactate thresholds, with one study reporting a reduced buffering capacity of lactic acid during pregnancy (Lotgering et al. 1995).

Other personal factors such as smoking (Gorecka and Czernicka-Cierpisz 1995; Hirsch et al. 1985), motivation, and anemia (Celsing and Ekblom 1986) have been shown to affect the blood lactate response to exercise. Therefore, it is advisable to take a detailed record of an athlete's pretest condition and, where possible, to standardize these conditions for future test sessions.

Relationship of Lactate Concentrations in Muscle and Blood

A strong correlation between blood and muscle lactate concentrations exists during exercise (Foster et al. 1995; Jacobs 1986; Karlsson and Jacobs 1982). However, it is erroneous to interpret blood lactate accumulation as wholly reflective of muscle lactate production. Blood lactate concentration depends on the existence of a net positive gradient for lactate between muscle and blood and is affected by dilution in the body water; by removal by organs such as the liver, heart, and inactive skeletal muscle; and by the temporal lag before lactate produced in the muscle appears in the blood (Foster et al. 1995; Jacobs 1986; Stallknecht et al. 1998; Weltman 1995). Jorfeldt and colleagues (1978) reported that the rate of lactate release from the muscle to blood is partially dependent on concentration, with a maximal release occurring in the range of 4 mmol · kg^{-1} wet weight of muscle. Therefore, at high muscle lactate concentrations there may be a significant time lag before lactate equilibrates with the blood. This is supported by a number of studies that have simultaneously measured lactate concentrations in the muscle and blood. Rusko and colleagues (1986) used work bouts of about 5 min duration and demonstrated that blood lactate was consistently lower than muscle lactate. Jacobs (1981) reported that the blood lactate concentration progressively underestimated the muscle lactate concentration with use of shorter work durations (<4 min). This is supported by the data from Green and colleagues (1983), who reported that in a rapidly incremented exercise test (1 min stages), the difference between muscle and blood lactate concentrations was considerable.

Thus, although blood lactate concentrations reflect changes in lactate metabolism in skeletal muscle (Foster et al. 1995), the relationship between blood and muscle lactate concentrations is influenced by the duration of the exercise stages, with longer work durations (at least 5 min) allowing more time for equilibration between muscle and blood such that changes in blood lactate more accurately reflect changes in muscle lactate concentrations (Krustrup et al. 2006).

Laboratory Versus Field Testing

The use of laboratory-derived values to assess and regulate exercise intensity in the field is open to question. For example, does rowing an ergometer in the laboratory require the same technique as rowing a racing shell on water? How do you account for the lack of environmental factors (e.g., wind resistance in cyclists, water resistance and drag in swimming and kayaking) in the laboratory? Does the difference in running surface (i.e., treadmill belt vs. grass) affect the blood lactate–exercise response curve? Further research is needed to answer such applied questions. However, numerous studies have investigated the heart rate–blood lactate relationship in both the laboratory and field environments. Nonsignificant differences between these test environments have been reported for running (Brettoni et al. 1989; Coen et al. 1991), rowing (Bourdin et al. 2004; Coen et al. 2003; Lormes et al. 1987; Urhausen et al. 1993), kayaking (van Someren and Oliver 2002), and cycling (Brettoni et al. 1989; Foster et al. 1993). Unpublished applied research conducted at the South Australian Sports Institute comparing the blood lactate–heart rate response of kayakers on water and in the laboratory supported these findings. This latter study showed no significant difference between the heart rates corresponding to the calculated LT1 and LT2 (Australian Institute of Sport 1995) for either environment. Table 6.6 presents the results of this research (reported with the permission of SASI). These findings are not unequivocal, as Schmidt and

colleagues (1984) reported that heart rates for cross-country skiers were 16 beats · min^{-1} higher in the laboratory than in the field.

Compared with laboratory investigations, field-based tests monitoring the blood lactate response to exercise have the following potential advantages:

- They are extremely specific to the actual athletic performance.
- They have more practical significance to the athlete and coach.
- They potentially enable the investigator to test more than one subject at a time.

Possible disadvantages of field-based tests are as follows:

- Environmental conditions (e.g., ambient temperature, wind speed and direction, surface conditions) cannot be controlled.
- The number of physiological parameters that can be monitored is limited.

Conversely, laboratory-based tests have following advantages:

- They are performed in a controlled, stable, and reproducible environment.
- Physiological data acquisition is easier.
- There is greater accuracy in the control of exercise intensities.

The potential disadvantages of laboratory-based tests include the following:

- Specific technique may be questionable because of ergometer design.
- Generally, only one athlete can be tested at a time given equipment limitations.

In deciding which environment is most suitable for athlete assessment, the coach and sport scientist need to collaborate and form a clear understanding of the test's purpose and desired outcomes. Furthermore, if the aim is to make direct comparisons between laboratory- and field-based test protocols to determine the blood lactate response to exercise, it is advisable that the protocols mimic each other as closely as possible.

Practical Applications

The way blood lactate levels change with increased exercise intensity can provide valuable information about how the athlete is adapting to training. The following sections describe how this information

TABLE 6.6   Comparative Heart Rates (in beats · min⁻¹) at the Lactate Thresholds for 16 Kayakers

Athlete	Lactate threshold 1		Lactate threshold 2		Maximal test	
	HEART RATES		HEART RATES		HEART RATES	
	Field	Laboratory	Field	Laboratory	Field	Laboratory
Females						
DA	149	148	165	164	180	179
RJ	156	157	176	180	189	197
JP	153	151	177	178	190	194
DB	144	141	171	168	187	189
LL1	149	147	175	175	189	190
LL2	148	147	176	174	194	191
LL3	145	145	176	176	189	189
Males						
BF	155	161	178	176	191	195
BM	156	153	173	169	189	190
BS	144	143	167	168	185	189
JB	139	140	166	163	181	178
SA	142	138	165	165	186	186
MD	148	145	169	167	185	184
DF	154	154	185	182	194	193
DG	157	154	178	179	196	197
TD	141	143	170	172	200	201
SD	152	153	184	183	206	207
DS	154	150	180	180	198	197
Mean	149.2	148.3	173.9	173.3	190.5	191.4
SD	5.7	6.3	6.1	6.5	6.6	7.2

Values were determined from blood lactate response curves conducted both in the laboratory and in the field. No significant difference was found between any of the three heart rate groups.

Optimal Blood Lactate Concentration for Training

Several groups have proposed that blood lactate thresholds be monitored regularly during training in order to provide information relative to training intensity (Antonutto and Di Prampero 1995; Coen et al. 1991; Fohrenbach et al. 1987; Keith et al. 1992; Kindermann et al. 1979; Madsen and Lohberg 1987). However, there is a lack of scientific consensus about the best way to use blood lactate concentrations in the design and modification of training programs. In the case of endurance training, researchers have reported optimal blood lactate levels for training between 1.3

helps in the evaluation of the athlete's endurance adaptations and training status.

and 10.0 mmol · L⁻¹ (Billat 1996; Borch et al. 1993; Fohrenbach et al. 1987; Hagberg and Coyle 1983; Heck et al. 1985; Katch et al. 1978; Mader et al. 1976; Madsen and Lohberg 1987; Sjodin and Jacobs 1981; Stegmann and Kindermann 1982). Many studies, however, favor the use of individually derived LT1 and LT2 values over fixed blood lactate concentrations (Beneke 1995; Beneke et al. 2001; Bishop et al. 1998; Buckley et al. 2003; Carter et al. 1999a; Cheng et al. 1992; Coen et al. 2001; Dickhuth et al. 1999; Harnish et al. 2001; Keith et al. 1992; MacIntosh et al. 2002; Palmer et al. 1999; Urhausen and Kindermann 1992). The question of optimal training intensities will remain unanswered until additional controlled training studies with athletes examine differences in performance that result from training at different blood lactate thresholds.

Prescribing Endurance Training Intensities

A myriad of terms exist to describe similar endurance training intensities and have been used across sports as well as across institutes and academies. This is highlighted in table 6.7, where the nomenclature used to describe the same basic training classifications varied widely among sports despite having been published in the same reference text (Gore 2000). This can cause confusion among coaches as well as sport scientists who are trying to draw reference across sports.

Although different terms, descriptions, and definitions are used within sports, there is evidence of some commonality between these training zones (Gore 2000). It was on these grounds that in June 2004, NSSQA hosted a workshop for affiliated exercise physiology laboratories and subsequently

established a national work group to review and discuss the standardization of nomenclature and terms of reference of commonly used physiological and training parameters. A synthesis of the outcomes of the NSSQA workshop deliberations and the recommendations of the work group form the basis of the recommended terms used in this chapter.

Endurance work intensities are generally divided into five training zones (T1-T5) based on their relationship to LT1 and LT2 on the lactate–exercise intensity curve. Figure 6.7 and table 6.8 present these five endurance training zones and their associated physiological descriptions and definitions as recommended for implementation on a national level by NSSQA. Training performed below LT1 is regarded as *light aerobic* (T1); that performed between LT1 and LT2 is divided into two halves, with the lower half termed *moderate aerobic* (T2) and the upper half *heavy aerobic* (T3). Anaerobic threshold work (i.e.,

TABLE 6.7   Sample Endurance Training Zone Classifications Used in Australian Sports

Sport	Training zone descriptor	Endurance training zones
Cycling	Endurance	E1, E2, E3, E4
Kayaking	Aerobic	A1, A2, A3, A4, A5, A6
Rowing	Utilization	U3, U2, U1, AT, Transport
Running	Aerobic	A1, A2, A3, A4, A5, $\dot{V}O_2$max
Triathlon*	Training Zone	T1, T2, T3, T4, T5, T6

*Note different definition for training zone determination applied.

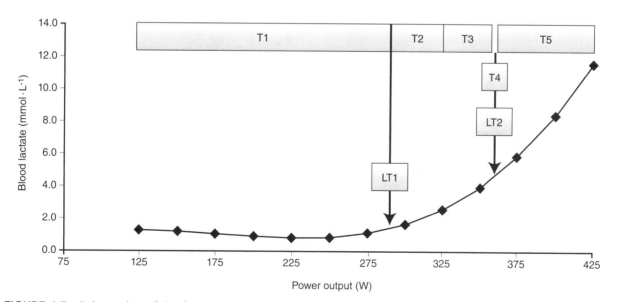

FIGURE 6.7   Relationship of the five endurance training zones to the blood lactate–incremental exercise response curve. Subject is a male cyclist. (Data were collected during routine physiological assessments in the SASI exercise physiology laboratory.)

TABLE 6.8 Classification of Training Zones as a Function of LT1 and LT2

Training zone	Prescriptive description	Blood lactate threshold relationship	Blood lactate, mmol · L⁻¹	Percentage HRmax, %	Percent V̇O₂max, %	RPE scale	Critical duration, h:min:s
T1	Light aerobic	Below LT1	<2.0	60-75	<60	Very light	>3 h
T2	Moderate aerobic	Lower half between LT1 and LT2	1.0-3.0	75-84	60-72	Light	1-3 h
T3	Heavy aerobic	Upper half between LT1 and LT2	2.0-4.0	82-89	70-82	Somewhat hard	30-90 min
T4	Threshold	LT2	3.0-6.0	88-94	80-85	Hard	20-60 min
T5	Maximal aerobic	Above LT2	>5.0	92-100	85-100	Very hard	2-12 min

Data from thousands of incremental endurance tests conducted by the various Sports Institutes/Academies within Australia were used to compile these guidelines. Prescriptive description = general classification of the five training zones according to aerobic intensity; blood lactate threshold relationship = position of training zones on "standard" blood lactate curve; blood lactate = blood lactate concentrations typically associated with intensity of exercise in each training zone; percentage HRmax = percentage of maximal heart rate typically associated with intensity of exercise in each training zone; percentage V̇O₂max = percentage of maximum oxygen uptake typically associated with intensity of exercise in each training zone; RPE scale = ratings of perceived exertion according to Borg's 6-20 rating (15 point) of perceived exertion table (Borg 1970); critical duration = time to exhaustion typically associated with intensity of exercise in each training zone.

work performed at LT2) constitutes a category in itself (T4). Training performed above LT2 is classified as V̇O₂max or *maximal aerobic* work (T5). Note that there are areas of overlap within table 6.8; its purpose is to provide general, not individual, guidelines to define the five commonly used endurance training zones.

Shifts in Blood Lactate Response Curves

The relationship between blood lactate and increasing exercise intensity can provide useful information for the evaluation of adaptations to endurance training (Denis et al. 1984; Hurley et al. 1984; Katch et al. 1978; Sjodin et al. 1982). Any analysis or interpretation of the blood lactate response curve is best performed at the individual level, because values can be affected by the various factors already discussed in connection with experimental conditions. Even after sport scientists have controlled for as many of these confounding factors as possible, each athlete's results should be analyzed and interpreted separately, although one should not discount the value of intersubject or group comparisons in some situations.

There are two commonly used methods of assessing the blood lactate–intensity curves for improvement, maintenance, or degradation of fitness. These involve (1) subjective evaluation of the graphical and raw data and (2) objective or statistical evaluation of the data.

Graphical overlays of individual athletes' test–retest blood lactate profiles, with a subjective assessment of any curve shifts, are the most commonly used method of determining the extent of any adaptation. Some of the most common changes in the blood lactate–intensity curve and their interpretations (Madsen and Lohberg 1987; Pyne 1989; Weltman 1995) are as follows:

- A shift in the curve down, to the right, or both indicates increases in the event-specific aerobic capacity of the athlete. This is expressed by the athlete's ability to exercise at a greater intensity for a given blood lactate level or to express lower blood lactate levels for the same intensity (see figure 6.8, *a* and *b*).

- A shift in the curve up, to the left, or both indicates a deterioration in the event-specific aerobic capacity of the athlete. This is expressed by the athlete's exercising at a lesser intensity for a given blood lactate level or expressing a higher blood lactate level for the same intensity (see figure 6.8, *c* and *d*).

Changes in specific blood lactate threshold values (i.e., LT1 and LT2) can also serve as distinct indicators of change in an athlete's training status. Increases in the intensity at LT1 reflect an improvement in base aerobic condition. This adaptation presents itself as a shift down, to the right, or both in the initial

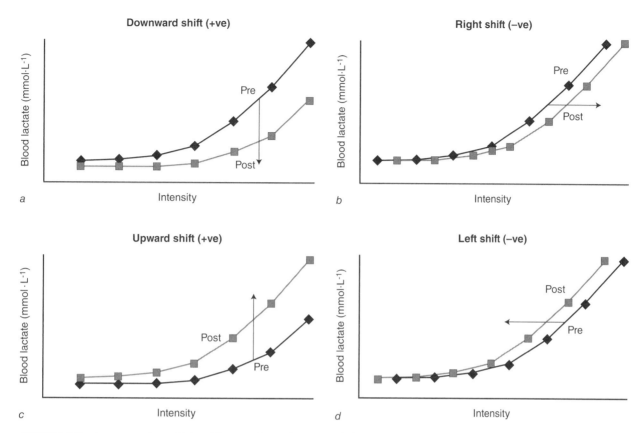

FIGURE 6.8 Interpretation of blood lactate–intensity curve shifts.

three to four stages. This is thought to be the result of delayed blood lactate production due to increased fat oxidation and enhanced aerobic mechanisms. Similarly, increases in exercise intensity at LT2, represented by graphical shifts down, to the right, or both, may indicate an improvement in higher-level aerobic endurance. Possible causes are decreased lactate production, improved lactate clearance, or improved acid buffering.

Determining TE (or precision) values for the various blood lactate thresholds allows for a more meaningful interpretation of these points of reference. These values represent a quantitative means of assessing significant changes in the threshold related variables (see table 6.5). They also give the scientist the ability to account for errors that are influenced by method and tester (Norton et al. 2000).

More objective means of interpreting the significance of curve shifts also exist. These usually involve the use of least squares linear regression analysis (i.e., ADAPT [Australian Institute of Sport 1995]). Such analytical methods, however, should not override the need for subjective interpretations—they simply serve to present the data in a way that aids in interpretation. Objective methods tend to be very rigid and as such do not fully account for the individual nature of the data. The tester must always take into account the raw data together with information relating to the athlete's state at the time of the test. This should involve asking the coach and athlete questions about current training and medical and psychosocial status. Such a process allows the application of the "art of the science" and often leads to more meaningful interpretation.

Final Thoughts

Discontinuities—or *thresholds*, as they are commonly termed—in the blood lactate response to exercise can be used as indexes of endurance capacity and markers for training prescription. Irrespective of methodological variations, there are two main areas of discontinuity in the blood lactate–intensity curve. The first of these discontinuities, LT1, is associated with the first exercise intensity at which there is a sustained increase in blood lactate above resting levels. The second, LT2, is marked by a very rapid increase in blood lactate concentration. This point represents a shift from oxidative to partly anaerobic energy metabolism during constant intensity exercise, and it refers to the upper steady-state limit of blood lactate concentration indicating equilibrium between lactate production and lactate elimination.

A number of factors (e.g., increment duration; calculation methods; techniques associated with blood collection, handling, and analysis; ergometer type; ambient environment; pretest exercise and diet) can all modify the blood lactate–intensity relationship and its dependent thresholds. Therefore, it is necessary to control for these factors if meaningful data are to be collected. As identified throughout this chapter, blood lactate thresholds based on fixed blood lactate concentrations are generally affected more by such factors. The recommendation, therefore, is to use individually based methods of determining LT1 and LT2.

The aim of this chapter has been to provide a brief overview of the major concepts and controversies associated with the use of the blood lactate accumulation to evaluate responses to exercise. The discussion is by no means exhaustive: The frequency with which research related to this area is appearing in the sport science literature makes comprehensiveness extremely difficult yet attests to the growing acceptance of blood lactate and its associated thresholds as valuable diagnostic, interpretive, and predictive tools.

References

Allen, W.K., Seals, DR., Hurley, B.F., Ehsani, A.A., and Hagberg, J.M. 1985. Lactate threshold and distance running performance in young and older endurance athletes. *Journal of Applied Physiology* 58:1281-1284.

Amann, M., Subudhi, A.W., and Foster, C. 2006. Predictive validity of ventilatory and lactate thresholds for cycling time trial performance. *Scandinavian Journal of Medicine & Science in Sports* 16:27-34.

Anderson, J., Johnstone, B., and Cook-Newell, M. 1995. Meta-analysis of the effects of soy protein intake on serum lipids. *New England Journal of Medicine* 333:276-282.

Antonutto, G., and Di Prampero, P.E. 1995. The concept of anaerobic threshold. *Journal of Sports Medicine and Physical Fitness* 35:6-12.

Aunola, S., and Rusko, H. 1986. Aerobic and anaerobic thresholds determined from venous lactate or from ventilation and gas exchange in relation to muscle fiber composition. *International Journal of Sports Medicine* 7:161-166.

Australian Institute of Sport. 1995. ADAPT Software Program Sport Science Division. Belconnen, ACT, Australia: Australian Institute of Sport.

Beaver, W.L., Wasserman, K., and Whipp, B.J. 1985. Improved detection of lactate threshold during exercise using a log-log transformation. *Journal of Applied Physiology* 59:1936-1940.

Beaver, W.L., Wasserman, K., and Whipp, B.J. 1986. A new method for detecting anaerobic threshold by gas exchange. *Journal of Applied Physiology* 60:2020-2027.

Beneke R. 1994. Maximal lactate steady state and individual concepts. *Perfusion* 10:387-388.

Beneke R. 1995. Anaerobic threshold, individual anaerobic threshold, and maximal lactate steady state in rowing. *Medicine and Science in Sports and Exercise* 27:863-867.

Beneke R. 2003. Maximal lactate steady state concentration (MLSS): Experimental and modelling approaches. *European Journal of Applied Physiology* 88:361-369.

Beneke, R., Heck, H., Schwarz, V., and Leithauser, R. 1996. Maximal lactate steady state during the second decade of age. *Medicine and Science in Sports and Exercise* 28:1474-1478.

Beneke, R., Leithauser, R.M., and Hutler, M. 2001. Dependence on the maximal lactate steady state on the motor pattern of exercise. *British Journal of Sports Medicine* 35:192-196.

Beneke, R., and Petelin von Duvillard, S. 1996. Determination of maximal lactate steady state response in selected sports events. *Medicine and Science in Sports and Exercise* 28:241-246.

Bentley, D., McNaughton, L., and Batterham, A. 2001. Prolonged stage duration during incremental cycle exercise: Effects on the lactate threshold and onset of blood lactate accumulation. *European Journal of Applied Physiology* 85:351-357.

Bentley, D.J., Newell, J., and Bishop, D. 2007. Incremental exercise test design and analysis: Implications for performance diagnostics in endurance athletes. *Sports Medicine* 37:575-586.

Billat, L.V. 1996. Use of blood lactate measurements for prediction of exercise performance and control of training. *Sports Medicine* 22:157-175.

Billat, V.L., Sirvent, P., Py, G., Koralsztein, J-P., and Mercier, J. 2003. The concept of maximal lactate steady state. *Sports Medicine* 33:407-426.

Bishop, D. 2001. Evaluation of the Accusport blood lactate analyser. *International Journal of Sports Medicine* 22:525-530.

Bishop, D., Jenkins, D.G., and MacKinnon, L.T. 1998. The relationship between plasma lactate parameters, W_{peak} and 1-h cycling performance in women. *Medicine and Science in Sports and Exercise* 30:1270-1275.

Blomstrand, E., Bergh, U., Essen-Gustavsson, B., and Ekblom, B. 1984. Influence of low muscle temperature on muscle metabolism during intense dynamic exercise. *Acta Physiologica Scandinavica* 120:229-236.

Borch, K.W., Ingjer, F., Larsen, S., and Tomten, S.E. 1993. Rate of accumulation of blood lactate during graded exercise as a predictor of anaerobic threshold. *Journal of Sports Sciences* 11:49-55.

Borg, G. 1970. Perceived exertion as an indicator of somatic stress. *Scandinavian Journal of Rehabilitation Medicine,* 23:92-98.

Bosquet, L., Leger, L., and Legros, P. 2001. Blood lactate response to overtraining in male endurance athletes. *European Journal of Applied Physiology* 84:107-114.

Bourdin, M., Messonier, L., and Lacour, J.R. 2004. Laboratory blood lactate profile is suited to on water training monitoring in highly trained rowers. *Journal of Sports Medicine and Physical Fitness* 44:337-341.

Bourdon, P.C., David, A.Z., and Buckley, J.D. 2009. A single exercise test for assessing physiological and performance parameters in elite rowers: The 2-in-1 test. *Journal of Science and Medicine in Sport* 12:205-211.

Brettoni, M., Alessandri, F., Cupelli, V., Bonifazi, M., and Martelli, G. 1989. Anaerobic threshold in runners and cyclists. *Journal of Sports Medicine and Physical Fitness* 29:230-233.

Buckley, J., Bourdon, P., and Woolford, S. 2003. Effect of measuring blood lactate concentrations using different automated lactate analysers on blood lactate transition thresholds. *Journal of Science and Medicine in Sport* 6:408-421.

Carter, H., Jones, A.M., and Doust, J.H. 1999a. Effect of incremental test protocol on the lactate minimum speed. *Medicine and Science in Sports and Exercise* 31:837-845.

Carter, H., Jones, A.M., and Doust, J.H. 1999b. Effect of six weeks of endurance training on the lactate minimum speed. *Journal of Sports Sciences* 17:957-967.

Celik, O., Kosar, S.N., Korkusuz, F., and Bozkurt, M. 2005. Reliability and validity of the modified Conconi test on Concept 11 rowing ergometers. *Journal of Strength and Conditioning Research* 19:871-877.

Celsing, F., and Ekblom, B. 1986. Anaemia causes a relative decrease in blood lactate concentration during exercise. *European Journal of Applied Physiology* 55:74-78.

Cheng, B., Kuipers, H., Snyder, A.C., Keizer, H.A., Jeukendrup, A., and Hesselink, M. 1992. A new approach for the determination of ventilatory and lactate thresholds. *International Journal of Sports Medicine* 13:518-522.

Chwalbinska-Monta, J. 2003. Effect of creatine supplementation on aerobic performance and anaerobic capacity in elite rowers in the course of endurance training. *International Journal of Sports Medicine* 13:173-183.

Clark, S.A., Bourdon, P.C., Schmidt, W., Singh, B., Cable, G., Onus, K.J., Woolford, S.M., Stanef, T., Gore, C.J., and Aughey, R.J. 2007. The effect of acute simulated moderate altitude on power, performance and pacing strategies in well-trained cyclists. *European Journal of Applied Physiology* 102:45-55.

Coen, B., Schwartz, L., Urhausen, A., and Kindermann, W. 1991. Control of training in middle and long distance running by means of the individual anaerobic threshold. *International Journal of Sports Medicine* 12:519-524.

Coen, B., Urhausen, A., and Kindermann, W. 2001. Individual anaerobic threshold: Methodological aspects of its assessment in running. *International Journal of Sports Medicine* 22:8-16.

Coen, B., Urhausen, A., and Kindermann, W. 2003. Sport specific performance diagnosis in rowing: An incremental graded exercise test in coxless pairs. *International Journal of Sports Medicine* 24:428-432.

Conconi, F., Ferrari, M., Ziglio, P.G., Droghetti, P., and Codeca, L. 1982. Determination of the anaerobic threshold by a noninvasive field test in runners. *Journal of Applied Physiology* 52:869-873.

Coyle, E.F., Coggan, A.R., Hemmert, M.K., and Walters, T.J. 1984. Glycogen usage and performance relative to lactate threshold. *Medicine and Science in Sports and Exercise* 16:120-121.

Coyle, E.F., Hemmert, M.K., and Coggan, A.R. 1986. Effects of detraining on cardiovascular responses to exercise: role of blood volume. *Journal of Applied Physiology* 60:95-99.

Coyle, E.F., Martin, W.H., Ehsani, A.A., Hagberg, J.M., Bloomfield, S.A., Sinacore, D.R., and Holloszy, J.O. 1983. Blood lactate threshold in some well-trained ischemic heart disease patients. *Journal of Applied Physiology* 54:18-23.

Craig, N.P. 1987. The measurement of blood lactate to monitor and prescribe aerobic and anaerobic training: Concepts and controversies. In proceedings of the Australian Sports Medicine Federation National Scientific Conference, Adelaide, pp 118-135.

Craig, N.P., Norton, K.I., Bourdon, P.C., Woolford, S.M., Stanef, T., Squires, B., Olds, T.S., Conyers, R.A.J., and Walsh, C.B.V. 1993. Aerobic and anaerobic indices contributing to track endurance cycling performance. *European Journal of Applied Physiology* 67:150-158.

Craig, N.P., Norton, K.I., Conyers, R.A.J., Woolford, S.M., Bourdon, P.C., Stanef, T., and Walsh, C.B.V. 1995. Influence of test duration and event specificity on maximal accumulated oxygen deficit of high performance track cyclists. *International Journal of Sports Medicine* 16:534-540.

Daniels, J.T., Yarborough, R.A., and Foster, C. 1978. Changes in VO_2max and running performance with training. *European Journal of Applied Physiology* 39:249-254.

Dassonville, J., Beillot, J., Lessard, Y., Jan, J., Andre, A.M., Le Pourcelet, C., Rochcongar, P., and Carre, F. 1998. Blood lactate concentrations during exercise: effects of sampling site and exercise mode. *Journal of Sports Medicine and Physical Fitness* 38:39-46.

Davies, S.F., Iber, C., Keene, S.A., McArthur, C.D., and Path, M.J. 1986. Effect of respiratory alkalosis during exercise on blood lactate. *Journal of Applied Physiology* 61:948-952.

Davis, H.A., Bassett, J., Hughes, P., and Gass, G.C. 1983. Anaerobic threshold and lactate turnpoint. *European Journal of Applied Physiology* 50:383-392.

Davis, J.A., Frank, M.H., Whipp, B.J., and Wasserman, K. 1979. Anaerobic threshold alterations caused by endurance training in middle aged men. *Journal of Applied Physiology* 46:1039-1046.

Denadai, B.S., Ruas, V.D., and Figueira, T.R. 2006. Maximal lactate steady state concentration independent of pedal cadence in active individuals. *EuropeanJournal of Applied Physiology* 96:477-480.

Denis, C., Dormis, D., and Lacour, J.R. 1984. Endurance training, VO2max and OBLA: A longitudinal study of two different age groups. *International Journal of Sports Medicine* 5:167-173.

Denis, C., Fouquet, R., Poty, P., Geyssant, A., and Lacour, J.R. 1982. Effect of 40 weeks of endurance training on the anaerobic threshold. *International Journal of Sports Medicine* 3:208-214.

Dickhuth, H-H., Yin, L., A. N., Rocker, K., Mayer, F., Heitkamp, H-C., and Horstmann, T. 1999. Ventilatory, lactate-derived and catecholamine thresholds during incremental treadmill running: Relationship and reliability. *International Journal of Sports Medicine* 20:122-127.

Dotan, R., Rotstein, A., and Grodjinovsky, A. 1989. Effect of training load on OBLA determination. *International Journal of Sports Medicine* 10:346-351.

El-Sayed, M.S., George, K.P., and Dyson, K. 1993. The influence of blood sampling site on lactate concentration during submaximal exercise at 4 mmol/L lactate level. *European Journal of Applied Physiology* 67:518-522.

Farrell, P.A., Wilmore, J.H., Coyle, E.F., Billing, J.E., and Costill, D.L. 1979. Plasma lactate accumulation and distance running performance. *Medicine and Science in Sports and Exercise* 11:338-344.

Faude, O., Kindermann, W., and Meyer, T. 2009. Lactate threshold concepts: How valid are they? *Sports Medicine* 39:469-490.

Feliu, J., Ventura, J.L., Sequra, R., Rodas, G., Riera, J., Estruch, A., Zamora, A., and Capdevila, L. 1999. Differences between lactate concentration of samples from ear lobe and the finger tip. *Journal of Physiology and Biochemistry* 55:333-339.

Fink, W.J., Costill, D.L., and Van Handel, P.J. 1975. Leg muscle metabolism during exercise in the heat and cold. *European Journal of Applied Physiology* 34:183-190.

Flore, P., Therminarias, A., Oddou-Chirpaz, M.F., and Quirion, A. 1992. Influence of moderate cold exposure on blood lactate during incremental exercise. *European Journal of Applied Physiology* 64:213-217.

Fohrenbach, R., Mader, A., and Hollmann, W. 1987. Determination of endurance capacity and prediction of exercise intensities for training and competition in marathon runners. *International Journal of Sports Medicine* 8:11-18.

Forsyth, J.J., and Farrally, M.R. 2000. A comparison of lactate concentration in plasma collected from the toe, ear, and fingertip after a simulated rowing exercise. *British Journal of Sports Medicine* 34:35-38.

Forsyth, J.J., Reilly, T. 2004. Circadian rhythms in blood lactate concentration during incremental ergometer rowing. *European Journal of Applied Physiology* 92:69-74.

Forsyth, J.J., and Reilly, T. 2005. The combined effect of time of day and menstrual cycle on lactate threshold. *Medicine and Science in Sports and Exercise* 37:2046-2053.

Foster, C., Cohen, J., Donovan, K., Gastrau, P., Killian, P.J., Schrager, A., and Snyder, A.C. 1993. Fixed time versus fixed distance protocols for the blood lactate profile in athletes. *International Journal of Sports Medicine* 14:264-268.

Foster, C., Pollock, M.L., Farrell, P.A., Maksud, M.G., Anholm, J.D., and Hare, J. 1982. Training responses of speed skaters during a competitive season. *Research Quarterly for Exercise and Sport* 53:243-246.

Foster, C., Schrager, M., and Snyder, A.C. 1995. Blood lactate and respiratory measurement of the capacity for sustained exercise. In: P.G. Maud and C. Foster (eds), *Physiological Assessment of Human Fitness*. Champaign, IL: Human Kinetics, pp 57-62.

Foxdal, P., Sjodin, A., and Sjodin, B. 1996. Comparison of blood lactate concentrations obtained during incremental and constant intensity exercise. *International Journal of Sports Medicine* 17:360-365.

Foxdal, P., Sjodin, B., Rudstam, H., Ostman, C., Ostman, B., and Hedenstierna, G. 1990. Lactate concentration differences in plasma, whole blood, capillary finger blood and erythrocytes during submaximal graded exercise in humans. *EuropeanJournal of Applied Physiology* 61:218-222.

Foxdal, P., Sjodin, B., Sjodin, A., and Ostman, B. 1994. The validity and accuracy of blood lactate measurements for prediction of maximal endurance running capacity: Dependency of analysed blood media in combination with different designs of the exercise test. *International Journal of Sports Medicine* 15:89-95.

Freund, H., Oyono-Enguelle, S., Heitz, A., Marbach, J., Ott, C., and Gartner, M. 1989. Effect of exercise duration on lactate kinetics after short muscular exercise. *European Journal of Applied Physiology* 58:534-542.

Friedmann, B., Bauer, T., Menold, E., and Bartsch, P. 2004. Exercise with the intensity of the individual anaerobic threshold in acute hypoxia. *Medicine and Science in Sports and Exercise* 36:1737-1742.

Friedmann, B., Frese, F., Menold, E., and Bartsch, P. 2005. Individual variation in the reduction of heart rate and performance at lactate thresholds in acute normobaric hypoxia. *International Journal of Sports Medicine* 26:531-536.

Frohlich, L., Urhausen, A., Seul, U., and Kindermann, W. 1989. The influence of low-carbohydrate and high-carbohydrate diets on the individual anaerobic threshold. *Leistungssport* 19:18-20.

Fukuba, Y., Walsh, M.L., Morton, R.H., Cameron, B.J., Kenny, C.T.C., and Banister, E.W. 1999. Effect of endurance training on blood lactate clearance after maximal exercise. *Journal of Sports Sciences* 17:239-248.

Gaesser, G.A., and Poole, D.C. 1988. Blood lactate during exercise: time course of training adaptation in humans. *International Journal of Sports Medicine* 9:284-288.

Gaesser, G.A., and Poole, D.C. 1996. The slow component of oxygen uptake kinetics in humans. *Exercise and Sport Sciences Reviews* 24:35-71.

Gaesser, G.A., and Rich, R.G. 1985. Influence of caffeine on blood lactate response during incremental exercise. *International Journal of Sports Medicine* 6:207-211.

Genovely, H., and Stanford, B.A. 1982. Effects of prolonged warm-up exercise above and below anaerobic threshold on maximal performance. *European Journal of Applied Physiology* 48:323-330.

Gollnick, P. 1982. Peripheral factors as limitations to exercise capacity. *Canadian Journal of Applied Sport Sciences* 7:14-21.

Gollnick, P.D., Bayley, W.M., and Hodgson, D.R. 1986. Exercise intensity, training, diet, and lactate concentration in muscle and blood. *Medicine and Science in Sports and Exercise* 18:334-340.

Gollnick, P.D., and Saltin, B. 1982. Significance of skeletal muscle oxidative enzyme enhancement with endurance training. *Clinical Physiology* 2:1-12.

Gore, C.J. (ed). 2000. *Physiological Tests For Elite Athletes*. Champaign, IL: Human Kinetics.

Gorecka, D., and Czernicka-Cierpisz, E. 1995. Effects of smoking on exercise tolerance in healthy subjects. *Pneumonologia i Alerglogia Polska* 63:632-638.

Green, H.J., Hughson, R.L., Orr, G.W., and Ranney, D.A. 1983. Anaerobic threshold, blood lactate, and muscle metabolites in progressive exercise. *Journal of Applied Physiology* 54:1032-1038.

Greenhaff, P.L., Gleeson, M., and Maughan, R.J. 1987. The effects of dietary manipulation on blood acid-base status and the performance of high intensity exercise. *European Journal of Applied Physiology* 56:331-337.

Hagan, R., and Smith, M. 1984. Pulmonary ventilation in relation to oxygen uptake and carbon dioxide production during incremental load work. *International Journal of Sports Medicine* 5:193-197.

Hagberg, J.M. 1984. Physiological implications of the lactate threshold. *International Journal of Sports Medicine* 5:106-109.

Hagberg, J.M., and Coyle, E.F. 1983. Physiological determinants of endurance performance such as studied in competitive racewalkers. *Medicine and Science in Sports and Exercise* 15:287-289.

Harnish, C.R., Swensen, T.C., and Pate, R.R. 2001. Methods for estimating the maximal lactate steady state in trained cyclists. *Medicine and Science in Sports and Exercise* 33:1052-1055.

Heck, H., Mader, A., Hess, G., Mucke, S., Muller, R., and Hollman, W. 1985. Justification of the 4-mmol/l lactate threshold. *International Journal of Sports Medicine* 6:117-130.

Heitcamp, H., Holdt, M., and Scheib, K. 1991. The reproducibility of the 4 mmol/l lactate threshold in trained and untrained women. *International Journal of Sports Medicine* 12:363-368.

Hirsch, G.L., Sue, D.Y., Wasserman, K., Robinson, T.E., and Hansen, J.E. 1985. Immediate effects of cigarette smoking on cardiorespiratory responses to exercise. *Journal of Applied Physiology* 58:1975-1981.

Hughes, E.F., Turner, S.C., and Brooks, G.A. 1982. Effects of glycogen depletion and pedaling speed on "anaerobic threshold." *Journal of Applied Physiology* 52:1598-1607.

Hughson, R.L., Weisiger, K.H., and Swanson, G.D. 1987. Blood lactate concentration increases as a continuous function in progressive exercise. *Journal of Applied Physiology* 62:1975-1981.

Hurley, B.F., Hagberg, J.M., Allen, W.K., Seals, D.R., Young, J.C., Cuddihee, R.W., and Holloszy, J.O. 1984. Effect of training on blood lactate levels during submaximal exercise. *Journal of Applied Physiology* 56:1260-1264.

Ivy, J.L., Costill, D.L., Van Handel, P.J., Essig, D.A., and Lower, R.W. 1981. Alteration in lactate threshold with changes in substrate availability. *International Journal of Sports Medicine* 2:139-142.

Ivy, J.L., Withers, R.T., Van Handel, P.J., Elger, D.H., and Costill, D.L. 1980. Muscle respiratory capacity and fibre type as determinants of the lactate threshold. *Journal of Applied Physiology* 48:523-527.

Jacobs, I. 1981. Lactate, muscle glycogen and exercise performance in man. *Acta Physiologica Scandinavica Supplementum* 495:1-35.

Jacobs, I. 1986. Blood lactate: implications for training and sports performance. *Sports Medicine* 3:10-25.

Janse de Jonge, X.A. 2003. Effects of the menstrual cycle on exercise performance. *Sports Medicine* 33:833-851.

Jones, A.M., and Carter, H. 2000. The effect of endurance training on parameters of aerobic fitness. *Sports Medicine* 29:373-386.

Jones, A.M., and Doust, J.H. 1998. The validity of the lactate minimum test for determination of the maximal lactate steady state. *Medicine and Science in Sports and Exercise* 30:1304-1313. Jones NL, and Ehrsour RE. 1982 The anaerobic threshold. *Exercise and Sport Science Reviews* 10: 49-83

Jorfeldt, L., Juhlin-Dannfelt, A., and Karlsson, J. 1978. Lactate release in relation to tissue lactate in human skeletal muscle during exercise. *Journal of Applied Physiology* 44:350-352.

Karlsson, J., and Jacobs, I. 1982. Onset of blood lactate accumulation during muscular exercise as a threshold concept. *International Journal of Sports Medicine* 3:190-201.

Karlsson, J., Nordesjo, L-O., Jorfeldt, L., and Saltin, B. 1972. Muscle lactate, ATP, and CP levels during exercise after physical training in man. *Journal of Applied Physiology* 33:199-203.

Kass, L., and Carpenter, R. 2009. The effect of sampling time on blood lactate concentration ([Bla]) in trained rowers. *International Journal of Sports Physiology and Performance* 4:218-228.

Katch, V., Weltman, A., Sady, S., and Freedson, P. 1978. Validity of the relative percent concept for equating training intensity. *EuropeanJournal of Applied Physiology* 39:219-227.

Kayser, B. 1996. Lactate during exercise at high altitude. *European Journal of Applied Physiology* 74:195-205.

Keith, S.P., Jacobs, I., and McLellan, T.M. 1992. Adaptations to training at the individual anaerobic threshold. *European Journal of Applied Physiology* 65:316-323.

Keskinen, O.P., Keskinen, K.L., and Mero, A.A. 2007. Effect of pool length on blood lactate, heart rate, and velocity in swimming. *International Journal of Sports Medicine* 28:407-413.

Keul, J., Suison, G., Berg, A., Dickhuth, H.H., Goesttler, I., and Kuebel, R. 1979. Determination of the individual anaerobic threshold in the assessment of efficiency and in the designing of training. *Deutsche Zeitschrift fuer Sportmedizin* 7:212-218.

Kindermann, W., Simon, G., and Keul, J. 1979. The significance of the aerobic–anaerobic transition for the determination of work load intensities during endurance training. *European Journal of Applied Physiology* 42:25-34.

Koistinen, P., Takala, T., Martikkala, V., and Leppaluoto, J. 1995. Aerobic fitness influences the response of maximal oxygen uptake and lactate threshold in acute hypobaric hypoxia. *International Journal of Sports Medicine* 16:78-81.

Kowalchuk, J.M., Heigenhauser, G.J.F., and Jones, N.L. 1984. Effect of pH on metabolic and cardiorespiratory responses during progressive exercise. *Journal of Applied Physiology* 57:1558-1563.

Krustrup, P., Mohr, M., Nybo, L., Jensen, J.M., Neilsen, J.J., and Bangsbo, J. 2006. The Yo-Yo IR2 test: Physiological response, reliability, and application to elite soccer. *Medicine and Science in Sports and Exercise* 38:1666-1673.

Kumagai, S., Tanaka, K., Matsuura, Y., Matsuzaka, A., Hirakoba, K., and Asano, K. 1982. Relationships of the anaerobic threshold with 5km, 10km and 10 mile races. *European Journal of Applied Physiology* 49:13-23.

LaFontaine, T.P., Londeree, B.R., and Spath, W.K. 1981. The maximal steady state versus selected running events. *Medicine and Science in Sports and Exercise* 13:190-192.

Londeree, B.R., and Ames, S.A. 1975. Maximal steady state versus state of conditioning. *European Journal of Applied Physiology* 34:269-278.

Lormes, W., Steinmacker, J.M., Michalsky, R., Grumert-Fuchs, M., and Wodick, R.E. 1987. Comparison of the multi-stage test on a rowing ergometer in a racing shell. *International Journal of Sports Medicine* 8:165.

Lotgering, F.K., Struijk, P.C., van Doorn, M.B., Spinnewijn, W.E., and Wallenburg, H.C. 1995. Anaerobic threshold and respiratory compensation in pregnant women. *Journal of Applied Physiology* 78:1772-1777.

Maassen, N., and Busse, M.W. 1989. A measure of endurance capacity or an indicator of carbohydrate deficiency? *European Journal of Applied Physiology* 58:728-737.

MacDougall, J.D., Reddan, W.G., Layton, C.R., and Dempsey, J.A. 1974. Effects of metabolic hyperthermia on performance during heavy prolonged exercise. *Journal of Applied Physiology* 36:538-544.

MacIntosh, B.R., Esau, S., and Svendahl, K. 2002. The lactate minimum test for cycling: Estimation of the maximal lactate steady state. *Canadian Journal of Applied Physiology* 27:232-249.

MacRae, H.S.H., Dennis, S.C., Bosch, A.N., and Noakes, T.D. 1992. Effects of training on lactate production and removal during progressive exercise in humans. *Journal of Applied Physiology* 72:1649-1656.

Mader, A., and Heck, H. 1986. A theory of the metabolic origin of "anaerobic threshold." *International Journal of Sports Medicine* 7:45-65.

Mader, A., Heck, H., and Hollmann, W. 1976. Evaluation of lactic acid anaerobic energy contribution by determination of post-exercise lactic acid concentration in ear capillary blood in middle distance runners and swimmers. In: F. Landing and W. Orban (eds), *Exercise Physiology*. North Miami, FL: Symposia Specialists.

Madsen, O., and Lohberg, M. 1987. The lowdown on lactates. *Swimming Technique* May-July Volume 4 Issue 1:21-28.

McLellan, T.M. 1987. The anaerobic threshold: Concept and controversy. *Australian Journal of Science and Medicine in Sport* 19:3-8.

McLellan, T.M., Cheung, K.S.Y., and Jacobs, I. 1991. Incremental test protocol, recovery mode and the individual anaerobic threshold. *International Journal of Sports Medicine* 12:190-195.

McLellan, T.M., and Jacobs, I. 1993. Reliability, reproducibility and validity of the individual anaerobic threshold. *European Journal of Applied Physiology* 67:125-131.

Medbo, J., Mamen, A., Holt Olsen, O., and Eversten, F. 2000. Examination of four different instruments for measuring blood lactate concentration. *Scandinavian Journal of Clinical and Laboratory Investigation* 60:367-380.

Newell, J., Higgins, D., Madden, N., Cruickshank, J., Einbeck, J., McMillan, K., and McDonald, R. 2007. Software for calculating blood lactate endurance markers. *Journal of Sports Sciences* 25:1403-1409.

Nicholson, R., and Sleivert, G. 2001. Indices of lactate threshold and their relationship with 10-km running velocity. *Medicine and Science in Sports and Exercise* 33:339-342.

Norton, K., Marfell-Jones, M., Whittingham, N., Kerr, D., Carter, L., Saddington, K., and Gore, C. 2000. Anthropometric assessment protocols. In: Gore CJ (ed), *Physiological Tests for Elite Athletes*. Champaign, IL: Human Kinetics, pp 66-85.

Olbrecht, J., Madsen, O., Mader, A., Liesen, A., and Hollmann, W. 1985. Relationship between swimming velocity and lactate concentration during continuous and intermittent training exercises. *International Journal of Sports Medicine* 6:74-77.

Oosthuyse, T., Bosch, A.N., and Jackson, S. 2005. Cycling time trial performance during different phases of the menstrual cycle. *European Journal of Applied Physiology* 94:268-276.

Oyono-Enguelle, S., Heitz, A., Marbach, J., Ott, C., Gartner, M., Pape, A., Vollmer, J.C., and Freund, H. 1990. Blood lactate during constant-load exercise at aerobic and anaerobic thresholds. *European Journal of Applied Physiology* 60:321-330.

Palmer, A.S., Potteiger, J.A., Nau, K.L., and Tong, R.J. 1999. A 1-day maximal lactate steady-state assessment protocol for trained runners. *Medicine and Science in Sports and Exercise* 31:1336-1341.

Pfitzinger, P., and Freedson, P. 1998. The reliability of lactate measurements during exercise. *International Journal of Sports Medicine* 19:349-357.

Poole, D.C., and Gaesser, G.A. 1985. Response of ventilatory and lactate thresholds to continuous and interval training. *Journal of Applied Physiology* 58:1115-1121.

Poole, D.C., Ward, S.A., Gardner, G.W., and Whipp, B.J. 1988. Metabolic and respiratory profile of the upper limit for prolonged exercise in man. *Ergonomics* 31:1265-1279.

Poole, D.C., Ward, S.A., and Whipp, B.J. 1990. The effects of training on the metabolic and respiratory profile of high-intensity cycle ergometer exercise. *European Journal of Applied Physiology* 59:421-429.

Pyne, D., Boston, T., Martin, D., and Logan, A. 2000. Evaluation of the Lactate Pro blood lactate analyser. *European Journal of Applied Physiology* 82:112-116.

Pyne, D.B. 1989. The use and interpretation of blood lactate testing in swimming. *Excel* 5:23-26.

Rieu, M., Miladi, J., Ferry, A., and Duvallet, A. 1989. Blood lactate during submaximal exercises. Comparison between intermittent incremental exercises and isolated exercises. *European Journal of Applied Physiology* 59:73-79.

Robergs, R.A., Quintana, R., Parker, D.L., and Frankel, C.C. 1998. Multiple variables explain the variability in the decrement in VO_2max during acute hypobaric hypoxia. *Medicine and Science in Sports and Exercise* 30:869-879.

Rusko, K., Luhdtanen, P., Rahkila, P., Viitasalo, J., Rehunen, S., and Harkonen, M. 1986. Muscle metabolism, blood lactate and oxygen uptake in steady state exercise at aerobic and anaerobic thresholds. *European Journal of Applied Physiology* 55:181-186.

Saltin, B., Nazar, K., Costill, D.L., Stein, E., Jansson, E., Essen, B., and Gollnick, P.D. 1976. The nature of the training response; peripheral and central adaptations to one-legged exercise. *Acta Physiologica Scandinavica* 96:289-305.

Saltin, B., and Rowell, L.B. 1980. Functional adaptations to physical activity and inactivity. *Federation Proceedings* 39:1506-1513.

Schmidt, P., Jacob, E., Huber, G., Lehmann, M., and Keul, J. 1984. The behaviour of heart rate and blood lactate levels

in laboratory tests and field tests at cross-country skiing (Abstract). *International Journal of Sports Medicine* 5:297.

Schnabel, A., Kindermann, W., Schmitt, W.M., Biro, G., and Stegmann, H. 1982. Hormonal and metabolic consequences of prolonged running at the individual anaerobic threshold. *International Journal of Sports Medicine* 3:163-168.

Shimojo, N., Naka, K., Uenoyama, H., Hamamoto, K., Yoshioka, K., and Okuda, K. 1993. Electrochemical assay system with single-use electrode strip for measuring lactate in whole blood. *Clinical Chemistry* 39:2312-2314.

Simoes, H.G., Denadai, B.S., Baldissera, V., Campbell, C.S., Hill, and D.W. 2005. Relationships and significance of lactate minimum, critical velocity, heart rate deflection and 3000m track-tests for running. *Journal of Sports Medicine and Physical Fitness* 45:441-451.

Sjodin, B., and Jacobs, I. 1981. Onset of blood lactate accumulation and marathon running performance. *International Journal of Sports Medicine* 2:23-26.

Sjodin, B., Jacobs, I., and Karlsson, J. 1981. Onset of blood lactate accumulation and enzyme activities in m vastus lateralis in man. *International Journal of Sports Medicine* 2:166-170.

Sjodin, B., Jacobs, I., and Svedenhag, J. 1982. Changes in onset of blood lactate accumulation (OBLA) and muscle enzymes after training at OBLA. *European Journal of Applied Physiology* 49:45-57.

Sjodin, B., and Svendenhag, J. 1985. Applied physiology of marathon running. *Sports Medicine* 2:83-89.

Skinner, J.S., and McLellan, T.H. 1980. The transition from aerobic to anaerobic metabolism. *Research Quarterly for Exercise and Sport* 51:234-248.

Smekal, G., Scharl, A., von Duvillard, S.P., Pokan, R., Baca, A., Baron, R., Tschan, H., Hofmann, P., and Bachl, N. 2002. Accuracy of neuro-fuzzy logic and regression calculations in determining maximal lactate steady-state power output from incremental tests in humans. *European Journal of Applied Physiology* 88:264-274.

Stallknecht, B., Vissing, J., and Galbo, H. 1998. Lactate production and clearance in exercise. Effects of training. A mini-review. *Scandinavian Journal of Medicine & Science in Sports* 8:127-131.

Stegmann, H., and Kindermann, W. 1982. Comparison of prolonged exercise tests at the individual anaerobic threshold and the fixed anaerobic threshold of 4 mmol.l{+-1} lactate. *International Journal of Sports Medicine* 3:105-110.

Stegmann, H., Kindermann, W., and Schnabel, A. 1981. Lactate kinetics and individual anaerobic threshold. *International Journal of Sports Medicine* 2:160-165.

Stockhausen, W., Grathwohl, D., Burklin, C., Spranz, P., and Kuel, J. 1997. Stage duration and increase of work load in incremental testing on a cycle ergometer. *European Journal of Applied Physiology* 76:295-301.

Svedenhag, J., and Sjodin, B. 1985. Physiological characteristics of elite male runners in and off season. *Canadian Journal of Applied Sport Sciences* 10:127-133.

Tanaka, K., and Matsuura, Y. 1984. Marathon performance, anaerobic threshold, and onset of blood lactate accumulation. *Journal of Applied Physiology* 57:640-643.

Tanaka, K., Nakagawa, T., Hazana, T., Matsuura, Y., Asano, K., and Iseki, T. 1985. A prediction equation for direct assessment of anaerobic threshold in male distance runners. *European Journal of Applied Physiology* 54:386-390.

Tanaka, K., Watanabe, H., Konishi, Y., Mitsuzono, R., Sumida, S., Tanaka, S., Fukuda, T., and Makadomo, F. 1986. Longitudinal associations between anaerobic threshold and distance running performance. *European Journal of Applied Physiology* 55:248-252.

Tegtbur, U., Busse, M.W., and Braumann, K.M. 1993. Estimation of an individual equilibrium between lactate production and catabolism during exercise. *Medicine and Science in Sports and Exercise* 25:620-627.

Telford, R. 1984. Lactic acid measurements—are they useful? *Sports Science and Medicine Quarterly* 1:2-7.

Therminarias, A., Flore, P., Oddou-Chirpaz, M.F., Pellerei, E., and Quirion, A. 1989. Influence of cold exposure on blood lactate response during incremental exercise. *European Journal of Applied Physiology* 58:411-418.

Thin, A., Hamzah, Z., FitzGerald, M., McLoughlin, P., and Freaney, R. 1999. Lactate determination in exercise testing using an electrochemical analyser: With or without blood lysis? *European Journal of Applied Physiology* 79:155-159.

Thoden, J.S. 1991. Testing aerobic power. In: J.D. Mac-Dougall, H.A. Wenger, and H.J. Green (eds). *Physiological Testing of the High-Performance Athlete*. Champaign, IL: Human Kinetics, pp 107-174.

Urhausen, A., and Kindermann, W. 1992. Exercise physiology: Performance diagnostics and training control. In: H. Haag, O. Grupe, and A. Kirsch (eds). *Sports Science in Germany*. Berlin, Germany: Springer-Verlag, pp 69-103.

Urhausen, A., Weiler, B., and Kindermann, W. 1993. Heart rate, blood lactate, and catecholamines during ergometer and on water rowing. *International Journal of Sports Medicine* 14(Suppl 1):S20-S23.

van Someren, K.A., Oliver, J.E. 2002. The efficacy of ergometry determined heart rates for flatwater kayak training. *International Journal of Sports Medicine* 23:28-32.

Wasserman, K. 1984. The anaerobic threshold measurement to evaluate exercise performance. *American Review of Respiratory Disease* 129:S35-S40.

Wasserman, K., McIlroy, M.B. 1964. Detecting the threshold of anaerobic metabolism in cardiac patients during exercise. *American Journal of Cardiology* 14:844-852.

Welsman, A. 1992. Methodological problems of lactate testing. *Coaching Focus* 21:14-15.

Weltman, A. 1995. *The Blood Lactate Response to Exercise*. Champaign, IL: Human Kinetics.

Weltman, A., Seip, R., Levine, S., Snead, D., Rogol, A., and Weltman, A. 1989. Prediction of lactate threshold and fixed blood lactate concentrations from 3200 meter time trial running performance in untrained females. *International Journal of Sports Medicine* 10:207-211.

Weltman, A., Seip, R.L., Snead, D., Weltman, J.Y., Haskvitz, E.M., Evans, W.S., Veldhuis, J.D., and Rogal, A.D. 1992. Exercise training at and above the lactate threshold in previously un-trained women. *International Journal of Sports Medicine* 13:257-263.

Weltman, A., Snead, D., Seip, R., Schurrer, R., Levine, S., Rutt, R., Reilly, T., Weltman, J., and Rogol, A. 1987. Prediction of lactate threshold and fixed blood lactate concentrations from 3200-m running performance in male runners. *International Journal of Sports Medicine* 8:401-406.

Weltman, A., Snead, D., Stein, P., Seip, R., Schurrer, R., Rutt, R., and Weltman, J. 1990. Reliability and validity of a continuous incremental treadmill protocol for the determination of lactate threshold, fixed blood lactate concentrations, and VO$_2$max. *International Journal of Sports Medicine* 11:26-32.

Withers, R.T., Sherman, W.M., Miller, J.M., and Costill, D.L. 1981. Specificity and anaerobic threshold in endurance trained cyclists and runners. *European Journal of Applied Physiology* 47:93-104.

Woolford, S., Withers, R., Craig, N., Bourdon, P., Stanef, T., and McKenzie, I. 1999. Effect of pedal cadence on the accumulated oxygen deficit, maximal aerobic power and blood lactate transition thresholds of high-performance junior endurance cyclists. *European Journal of Applied Physiology* 80:285-291.

Yoshida, T. 1984a. Effect of dietary modification on lactate threshold and onset of blood lactate accumulation during incremental exercise. *European Journal of Applied Physiology* 53:200-205.

Yoshida, T. 1984b. Effect of exercise duration during incremental exercise on the determination of anaerobic threshold and the onset of blood lactate accumulation. *European Journal of Applied Physiology* 53:196-199.

Yoshida, T. 1986. Effect of dietary manipulation on anaerobic threshold. *Sports Medicine* 3:4-9.

Yoshida, T., Chida, M., Ichioka, M., and Suda, Y. 1987. Blood lactate parameters related to aerobic capacity and endurance performance. *European Journal of Applied Physiology* 56:7-11.

Yoshida, T., Suda, Y., and Takeuchi, N. 1982a. Endurance training regimen based upon arterial blood lactate: effects on anaerobic threshold. *European Journal of Applied Physiology* 49:223-230.

Yoshida, T., Takeuchi, N., and Suda, T. 1982b. Arterial versus venous blood lactate increase in the forearm during incremental bicycle exercise. *European Journal of Applied Physiology* 50:87-93.

Zhou, S., and Weston, S.B. 1997. Reliability of using the D-max method to define physiological responses to incremental exercise testing. *Physiological Measurement* 18:145-154.

Determination of Maximal Oxygen Consumption ($\dot{V}O_2$max)

Christopher J. Gore, Rebecca K. Tanner, Kate L. Fuller, and Tom Stanef

This chapter is a technical note for the sport scientist and addresses quite selected aspects of measuring maximal oxygen consumption ($\dot{V}O_2$max). The chapter does not attempt to provide a rationale for the value of assessing or interpreting maximal oxygen consumption or enter the debate about whether the cardiovascular system or brain limits $\dot{V}O_2$max (Brink-Elfegoun et al. 2007; Levine 2008; Noakes 2008; Beltrami et al. 2012). Instead, the chapter focuses on calibration issues for ergometers and $\dot{V}O_2$ systems, with particular emphasis on volume-monitoring devices and gas analyzers, which are potentially the sources of the largest error. Detail is also provided on the calculations used for data reduction when either expired or inspired volume is measured.

Theoretical Rationale

Oxygen consumption ($\dot{V}O_2$) is the product of the cardiac output ($\dot{Q}$) and arteriovenous oxygen difference ($CaO_2 - C\bar{v}O_2$):

$$\dot{V}O_2 = \dot{Q} \times (CaO_2 - C\bar{v}O_2).$$

Oxygen consumption is therefore reflected by both central and peripheral physiological variables and can be calculated via the direct Fick method:

$$\dot{V}O_2 = \dot{Q} \times (CaO_2 - C\bar{v}O_2),$$

where

$$\dot{Q} = \text{Heart Rate (HR)} \times \text{Stroke Volume (SV)}$$

and therefore

$$\dot{V}O_2 = (HR \times SV) \times (CaO_2 - C\bar{v}O_2).$$

The direct measurement of stroke volume and arteriovenous oxygen difference is highly invasive and ethically difficult to justify for the routine determination of $\dot{V}O_2$max in athletes. However, oxygen consumption and the related variables can be determined by open circuit indirect calorimetry. This involves measuring the pulmonary ventilation and comparing inspired and expired carbon dioxide and oxygen concentrations. Although this is theoretically simple, the practical aspects related to technology are surprisingly difficult to implement with precision.

During an incremental exercise test, the increase in $\dot{V}O_2$ is essentially linearly related to increasing power output (figure 7.1). Eventually a point is reached where the oxygen consumption will not increase despite an increase in power output, thereby indicating that $\dot{V}O_2$max has been attained (Hawkins et al. 2007). Oxygen consumption may then plateau or even decline. Over the period of a few weeks, in a healthy athlete, $\dot{V}O_2$max is stable and reproducible with test–retest reliability of approximately 2% (Saunders et al. 2009).

$\dot{V}O_2$max can be expressed absolutely as liters per minute ($L \cdot min^{-1}$) or relative to body mass per minute ($ml \cdot kg^{-1} \cdot min^{-1}$). The latter is a better indicator of average running speed during a 5 km race than a scaled predictor ($ml \cdot kg^{-2/3} \cdot min^{-1}$), which is designed to remove the confounding effect of body mass (Heil 1997; Nevill et al. 2005). The foregoing emphasizes the distinction between measures of physical performance and physiological function. $\dot{V}O_2$max data are presented in $L \cdot min^{-1}$ when total power output is important, as in rowing, whereas $ml \cdot kg^{-1} \cdot min^{-1}$ is generally used when reporting $\dot{V}O_2$max for activities in which the subject's weight

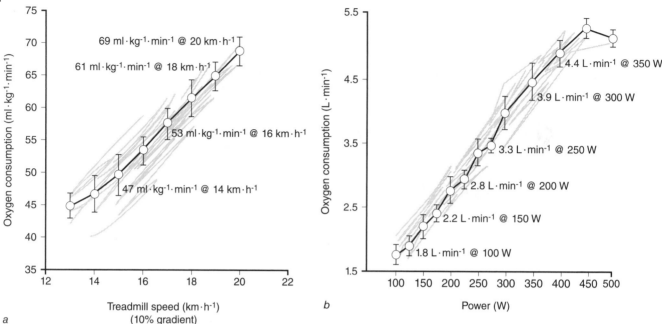

FIGURE 7.1 Steady-state oxygen consumption to work ratio on a treadmill for elite male and female middle-distance runners *(a)* and on a cycle ergometer for elite male and female cyclists *(b)*. (Mean ± SD in black line; individual data in gray lines; values are indicative of oxygen consumption at typical speeds or power outputs.)

is unsupported (e.g., running). A wide variation in $\dot{V}O_2$max can be expected: It is not unusual to find a range of 30 to 80 ml · kg⁻¹ · min⁻¹ for 20- to 30-year-old men and 25 to 65 ml · kg⁻¹ · min⁻¹ for similarly aged women.

$\dot{V}O_2$max decreases with age and physical inactivity, and men generally have higher $\dot{V}O_2$max values than women. The high $\dot{V}O_2$max values of elite middle- and long-distance athletes are attributable to a combination of genetic endowment and training. Bouchard and colleagues (1998) attributed 30% to 50% of the variation in $\dot{V}O_2$max to genetic factors. Direct measures of total hemoglobin mass also confirm a strong relationship with $\dot{V}O_2$max that is independent of sex and indicate that an increase in hemoglobin by 1 g is associated with an increase in $\dot{V}O_2$max of approximately 4 ml · min⁻¹ (Schmidt and Prommer 2010). The literature indicates that physical training can increase $\dot{V}O_2$max by 5% to 30%, but greater improvements have been observed in cardiac patients, persons of very low initial fitness, and those who achieve substantial weight loss (ACSM Position Stand 1998).

Given that increases in $\dot{V}O_2$max are associated with adaptations in both central and peripheral physiological variables, it is important that athletes simulate the movement pattern of competition by being tested on a sport-specific ergometer (e.g., treadmill for runners, cycle ergometer for cyclists, and rowing ergometer for rowers). However, with highly trained athletes, the $\dot{V}O_2$max is a relatively weak predictor of race performance (Robertson et al. 2010) given the importance of other factors such as mechanical efficiency, lactate threshold, anaerobic capacity, and motivation.

Ergometers

Comprehensive calibration of testing equipment is critical when sport scientists follow athletes over a period of time, be it weeks or even years. The same applies to any single test of an athlete, because sport scientists have an obligation to provide accurate data for every test. If the accuracy of the ergometer is unknown, the test should not proceed. Indeed, calibration was one part of a debate about the mechanical efficiency of Lance Armstrong (Martin et al. 2005) before and after chemotherapy. Calibration of both the ergometer, as described in this section, and the indirect calorimetry system (see later discussion) is a critical prerequisite for confident interpretation of test results.

Treadmills

Both treadmill speed and gradient should be checked using first-principles techniques (in this instance that means calculating from time and distance) or using calibrated digital tools (e.g., digital inclinometers and tachometers). Zero grade on the treadmill should be checked with a spirit level and shims placed under the treadmill feet as required. Once the treadmill bed has been checked for level, a digital inclinometer can be used for gradient calibration with sufficient accuracy. For instance, the resolution of digital inclinometers ranges from approximately 0.02% to 0.2% with accuracy of ±0.2% up to 17% gradient. The various elevations can then be verified using a digital inclinometer or by first-principles "rise over run" method:

$$\% \text{ Grade} = \frac{\text{vertical height}}{\text{horizontal distance}} \times 100$$

Treadmill speed can be verified by marking a line across the belt, measuring the length of the belt, and counting the number of belt revolutions per minute with timing from a handheld stopwatch:

$$\text{Speed (km/h)} = \frac{\text{rpm} \times 60 \times \text{belt length (m)}}{1,000}$$

Digital contact tachometers provide an expedient calibration alternative but they must be calibrated regularly using traceable standards to ascertain that they provide better than 0.5% accuracy as a minimum in the speed range of 6 to 24 km · h⁻¹. A treadmill calibration protocol should include four speed settings between 0 and 24 km · h⁻¹ (e.g., 6, 12, 18, 24 km · h⁻¹) and four gradient settings between 0% and 24% (e.g., 0%, 8%, 16%, 24%) unless limited by the treadmill (e.g., maximal gradient of 20%). Minimally, the calibration should span the likely physiological range (e.g., 6 and 24 km · h⁻¹ or 0% and 24%) as well as two equidistant intermediary points. The accuracy for all elevations and speeds should be ±0.2% for elevation and ±0.2 km · h⁻¹ for speed.

A further consideration of treadmill calibration is that calibrations are usually conducted when the treadmill is "unloaded" (i.e., no one is running on the treadmill). However, when an athlete runs on a treadmill, the belt speed can "brake" and the gradient can "bounce" due to foot strike and vertical loading, respectively. Both are as problematic, particularly given that the extent of perturbation may vary for athletes of different masses. Notwithstanding these errors, it is recommended that all calibrations be conducted when the treadmill is unloaded. Our experience over 10 years indicates that calibration of treadmill gradient and speed every 12 months is sufficient.

Cycle Ergometers

Mechanically braked ergometers, which use a balance (e.g., Monark), may be statically calibrated by sus-

pending a known mass from the balance at the point of belt attachment after checking the zero. However, dynamic calibration of mechanically braked, electromagnetically braked, and air-braked cycle ergometers is unequivocally the method of choice, because this method calibrates the complete system including the frictional resistance of the transmission from the pedal shaft (Cumming and Alexander 1968; Russell and Dale 1986; Telford et al. 1980). A dynamic calibration rig, which is depicted in figure 7.2, can be either purchased from VacuMed or fabricated in accordance with the recommendations of Woods and colleagues (1994).

Pedal cadence must be measured accurately when using a mechanically braked cycle ergometer. This can be accomplished by using a revolution counter in conjunction with either a metronome or a cadence indicator. The frictional load in the transmission of a mechanically braked cycle ergometer can usually be kept to a minimum by regular servicing. This involves lubricating the chain, adjusting the tension so that the chain moves vertically about 1 cm, checking all sets of ball bearings for wear, and greasing them or replacing them together with worn bearing cups.

The advantages of electrically braked ergometers are that the power output, which should range from 0 to 750 W, is relatively independent of pedal cadence, and small increases in power output can be made more accurately. The former advantage can also be achieved by gearing air-braked and mechanical ergometers. However, gearing changes can affect the physiological responses to a specific power output (Woolford et al. 1999).

An alternative method of ensuring accuracy of cycle ergometer data is to use instrumented cranks

FIGURE 7.2 The dynamic calibration of an air-braked cycle ergometer.

Photo courtesy of the South Australian Sports Institute.

to bypass other methods of determining power. The SRM (Schoberer Rad Messtechnik, Julich-Weldorf, Germany) Science version crank with eight strain gauges has a power accuracy of ±0.5% for the range of 0 to 4,800 W, according to the manufacturer. Our experience indicates that verification of the SRM calibration is warranted, which requires a dynamic calibration rig. Longitudinal data from the South Australian Sports Institute (SASI) indicate good stability of the eight strain gauge model SRM crank for periods of 12 months.

All cycle ergometers (e.g., electromagnetically braked, air braked, and friction braked) should be calibrated dynamically with a first-principles torque meter calibration rig at cadences from 60 to 130 revolutions per minute (rpm) at power outputs of approximately 100 to 600 W. Higher power loads (up to 2,000 W) and higher cadence (up to 180 rpm) should be assessed for ergometers used to measure track sprint cyclists. Care must be taken to ensure that the ergometer components (e.g., flywheel vanes) are not damaged or distorted before and after calibration at high power outputs. A further consideration is that accuracy of power measurement on an ergometer in stochastic tests may be different from the accuracy during constant power tests (Abbiss et al. 2009). This finding reinforces the need for dynamic calibration of any ergometer.

Calibration frequency should match usage patterns and the importance of testing. Ideally, dynamic calibration should occur immediately before each block of athlete tests or not less than every 12 months.

Kayak Ergometers

The Dansprint Pro kayak ergometer (Dansprint ApS, Hvidovre, Denmark) is currently the ergometer of choice. A first-principles torque meter that is used for the dynamic calibration of cycle ergometers can also be used to verify the Dansprint kayak ergometer (see figure 7.3). SASI has developed a modification of the torque meter that enables determination of the total load presented to the paddler. The calibration, therefore, includes the flywheel load plus that presented by the retractable (bungee cord) drive mechanism of the kayak ergometer. This development allows calibration over a range of stroke rates so that a more accurate relationship of power and flywheel velocity can be established. As in calibration of air-braked cycle ergometers, a third-order polynomial can be derived to estimate power from the frequency of the flywheel or, alternatively, a correction chart can be created to determine actual power from the displayed power on the Dansprint display unit.

The calibration rig provides a method of mechanical actuation of the ropes on the kayak ergometer to

FIGURE 7.3 The dynamic calibration of a Dansprint air-braked kayak ergometer.

Photo courtesy of the South Australian Sports Institute.

simulate paddling at various stroke rates from 60 to 120 strokes per minute. The length of the rotating arms can be adjusted to provide stroke lengths from 131 cm to 175 cm. The calibration rig measures the angular velocity of the rotating drive arms and simultaneously measures the reaction torque created in driving the ergometer to determine true mechanical power delivery.

$$\text{Mechanical Power (P)} = \text{Torque (T)} \times \text{Angular Velocity } (\omega)$$

where P is measured in watts, T in Nm, and ω in rad · s^{-1}.

To standardize testing of athletes, three load factors are used. A load factor of 30 is used for junior and senior female athletes, 35 for junior males, and 40 for senior male paddlers. To standardize the calibration procedure, the ergometer is clamped to the calibration rig in a standard marked position to provide good repeatability results. To standardize bungee tension (rope return cord), the ropes are alternately extended to a distance of 210 cm, and the tension is measured and adjusted to 1.5 kg ± 20 g. Data are collected at specified stroke rates to produce power levels in the typical load range of 50 to 350 W. A table of actual versus indicated power is generated, and a correction chart for each load factor is established (see form 7.1).

The modified SASI calibration rig was used to evaluate the accuracy of 12 Dansprint ergometers. Collectively, they underestimated true mean power by 25 to 95 W (21%-23%) across the range 50 to 350 W. When a correction was made for the tension associated with the bungee cord, the errors were reduced to 1% to 3%. These data indicate that it is necessary to calibrate the kayak ergometers against a first-principles torque meter to establish a calibration

FORM 7.1 Sample Dansprint Ergometer Calibration Report (Actual vs. Indicated Power)

CALIBRATION REPORT

Dansprint Kayak Ergometer	Australian Institute of Sport–September 2010
Display ID: 2007807774	Combined CF30 and CF40 data

Calibration regression generated to correct the approximate 22% error of the Dansprint LCD panel

To correct for the error, use the following correction factor:

True power = LCD display power $\times$ 1.271971 + 6.164912 (R^2 = 0.9974)

Note: Set the rubber bungee tensions (L&R) for 1.5 kg load when paddle is extended to 60 cm mark of ergometer rail.

This calibration data were established on September 1, 2010, using the SASI kayak calibration rig.

Conversely, if you want to set the actual load via the LCD display power:

LCD display power = 0.7842 $\times$ desired power – 4.4695 (R^2 = 0.9974).

This look-up table is for setting a desired power for testing athletes (e.g., to work at 150 W, get the athlete to maintain 113.2 W on LCD display).

Desired actual power, W	LCD display power, W	Desired actual power, W	LCD display power, W	Desired actual power, W	LCD display power, W	Desired actual power, W	LCD display power, W
50	34.7	150	113.2	250	191.6	350	270.0
55	38.7	155	117.1	255	195.5	355	273.9
60	42.6	160	121.0	260	199.4	360	277.8
65	46.5	165	124.9	265	203.3	365	281.8
70	50.4	170	128.8	270	207.3	370	285.7
75	54.3	175	132.8	275	211.1	375	289.6
80	58.3	180	136.7	280	215.1	380	293.5
85	62.2	185	140.6	285	219.0	385	297.4
90	66.1	190	144.5	290	222.9	390	301.4
95	70.0	195	148.4	295	226.9	395	305.3
100	74.0	200	152.4	300	230.8	400	309.2
105	77.9	205	156.3	305	234.7	405	313.1
110	81.8	210	160.2	310	238.6	410	317.1
115	85.7	215	164.1	315	242.6	415	321.0
120	89.6	220	168.1	320	246.5	420	324.9
125	93.6	225	172.0	325	250.4	425	328.8
130	97.5	230	175.9	330	254.3	430	332.7
135	101.4	235	179.8	335	258.2	435	336.7
140	105.3	240	183.7	340	262.2	440	340.6
145	109.2	245	187.7	345	266.1	445	344.5
150	113.2	250	191.6	350	270.0	450	348.4

Calibrated by:

Date:

From Australian Institute of Sport, 2013, *Physiological tests for elite athletes*, 2nd ed. (Champaign, IL: Human Kinetics).

regression or correction that takes into consideration the total true load exerted by the paddler. At a first approximation, a correction of +22% to the power displayed by the Dansprint will provide a more accurate indicate of the true power.

Dynamic calibration should occur at least every 12 months or, ideally, immediately before each block of tests.

Rowing Ergometers

The protocols for testing Australian rowers (refer to chapter 23) recommend the Concept2 Model D or E rowing ergometer. The manufacturer claims that, provided appropriate maintenance procedures are undertaken regularly, the measured power output is within ±2% of the true value. The calibration reproducibility data from the Dansprint kayak ergometer (approximately ±0.8% or ±2 W, 50-350 W), which uses a similar flywheel to the Concept2 Model D/E, support the concept of excellent repeatability for strokes completed within a few minutes of each other. However, the accuracy of the Concept2 D/E data for power is more challenging to verify independent of the manufacturer's claims.

There is no commercial system that can calibrate the Concept2 Model D/E. One method to approximately verify power output is to calculate power from measurements of force applied to the chain connecting the handle to the ergometer flywheel, the duration of the force, and the horizontal linear displacement of the handle (Macfarlane et al. 1997):

$$Power = \frac{Force \times Distance}{Time}$$

If you are using a load cell and handle position or velocity (assessed with a rotary potentiometer), the load cell should be calibrated with masses at four points (0-60 kg) at least every 12 months or, preferably, immediately before any large group of tests. A rod of known length (e.g., 1.2 m) should be used to verify the accuracy of the potentiometer immediately before each test. A further consideration is that power is also exerted by the rower through the foot plate of the ergometer and that the total force generated should take account of both aspects including the three-dimensional orthogonal components of the forces (Macfarlane et al. 1997) and the forces exerted by the rower vertically through the seat (Colloud et al. 2006; Murphy et al. 2010).

If you do not have access to calibration equipment for Concept2 rowing ergometers, at least ensure that correct "drag force" settings are implemented for testing (refer to chapter 23) and that ergometers undergo regular maintenance procedures (e.g., oil chain, check runner under seat for deterioration, change batteries in display).

$\dot{V}O_2$ Test Systems

In the past decade, several reviews have comprehensively critiqued the different approaches to assessing $\dot{V}O_2$ (Hodges et al. 2005; Macfarlane 2001). In addition, Roecker and colleagues (2005) summarized the advantages and disadvantages of the three common approaches: Douglas bag method, mixing chamber systems, and breath-by-breath methods.

The traditional approach to assessing $\dot{V}O_2$ is the Douglas bag method (Douglas 1911), and despite being laborious, it is still the method of choice in some laboratories (Rosdahl et al. 2010; Truijens et al. 2008; Withers et al. 2006). Tubing connects the athlete via a headpiece and respiratory valve to a Douglas bag or meteorological balloon in which the expired gas is collected. This causes extra weight for the athlete, but the limitation of gas permeability of the Douglas bags can be overcome by using aluminized Mylar bags instead of PVC; VacuMed is one supplier (www.vacumed.com). All Douglas bags should be periodically checked for leaks by filling with expired gas and serially monitoring the percentages of carbon dioxide and oxygen; Douglas bags also can be evacuated under heavy negative pressure to verify that they and their associated valves are air-tight. Because of the robustness of the Douglas bag method, the indirect calorimetry system used by the Australian Institute of Sport (AIS) is an in-house, custom-built, fully automated Douglas bag system (see figure 7.4) that has been described previously (Russell et al. 2002).

Mixing chamber systems measure ventilation and gas fractions at slightly different times, because they need to accumulate volume over a set period, for instance, 30 seconds, and sample the gases that correspond to the ventilation period. Even in the best-case scenario it takes time for the gases to be transported to the analyzers and for the analyzers to respond fully. This is not an issue during steady-state conditions, but it becomes critical during the latter stages of a $\dot{V}O_2$max test when conditions are not in a steady state (Wasserman et al. 1994). Indeed, the time constant appropriate to the mixing chamber can change as a function of ventilation (Hughson et al. 1980; Roecker et al. 2005), which is difficult to circumvent unless the mixing chamber itself can change volume.

Breath-by-breath systems have the major advantage of being able to characterize oxygen consumption kinetics (Hughson 2009), but the alignment issue associated with mixing chambers becomes even more critical. The challenge is to identify

FIGURE 7.4 The AIS automated Douglas bag system. All expirate is collected into aluminized Mylar bags, the volume of which is measured in a large piston (with real-time assessment of position, temperature, and pressure using calibrated transducers). Gas fractions are measured on commercial analyzers calibrated with three primary gravimetric-standard gases (±0.02% absolute accuracy) before each test and checked for drift after each test.

Photo courtesy of the Australian Institute of Sport.

As commented by Hodges and colleagues (2005), it is challenging to quantify and mitigate errors when manufacturers do not release algorithms. Those authors noted that assessment of $\dot{V}O_2$ is often a compromise between the expediency of fast response breath-by-breath data and the imperative that data are of sufficient rigor to stand up to close scrutiny. When sport scientists assess elite athletes, accuracy is imperative, given that relatively small changes in their $\dot{V}O_2$max may have a significant effect on performance.

The AIS indirect calorimetry system (figure 7.4) is based on the Douglas bag principle and has been in use for approximately 12 years. At times it is in operation for 14 to 16 h per day for 5 to 7 days in a row, sometimes with test protocols that run for more than 60 min. These pragmatic requirements of high usage led us to develop a system that could withstand such extended use and still achieve high accuracy. We developed a system in-house for two main reasons: first to have complete control of the algorithms used to calculate $\dot{V}O_2$ and second because nothing was commercially available that we thought could survive such abuse.

Volumetric and Gas Analysis Equipment

The following sections deal with equipment and measurement issues related to the airways system, gas volume, and gas analysis. Understanding potential sources of error and their magnitude in each of these three areas can help mitigate these issues.

Airways System

The overall pressure due to inspiratory and expiratory resistance of the system must be checked with a flow meter, manometer, and exhaust pump (figure 7.5), and these pressures should each be less than 6 cm H_2O at flows up to 300 L · min^{-1} (Jones 1988).

Respiratory valves should have a low resistance and a dead space less than 100 ml excluding the mouthpiece, and they should not leak (e.g., Hans Rudolph 2700 T valve and 2730 Y valve; www.rudolphkc.com). The dead space, which is the volume that is common to both inspiration and expiration, is inversely proportional to the resistance. A balance must therefore be struck between these two variables, and this will largely depend on the flow rate.

Resistance is proportional to length and inversely proportional to the radius raised to the fourth power. Hence, tubing should be greater than 30 mm in internal diameter and, ideally, not longer than 1.5 m on either the inspiratory or expiratory sides. Consider using 51 mm internal diameter tubing if the tubing

single breaths and align the signals corresponding to volume and gas fractions integrated over time. As commented by Beck and Gore (2004, 555), "Its appeal is in ease of use and the high time resolution of measurements, but users should be aware that, of the three major techniques available for gas exchange measurement, it is technically the most difficult to implement and may be the most prone to technical failure."

A further consideration is that assessment of $\dot{V}O_2$ in the field is even more technically challenging than in the laboratory (Meyer et al. 2005), given the obstacles of miniaturizing equipment and the need to remove water vapor from the respiratory gases prior to analysis or at least standardize the amount of water vapor in both calibration gases and expirate. It is very worthwhile to quantify $\dot{V}O_2$ in the field, but it appears that errors may be somewhat greater for portable systems than for laboratory-based indirect calorimetry systems (Vogler et al. 2010).

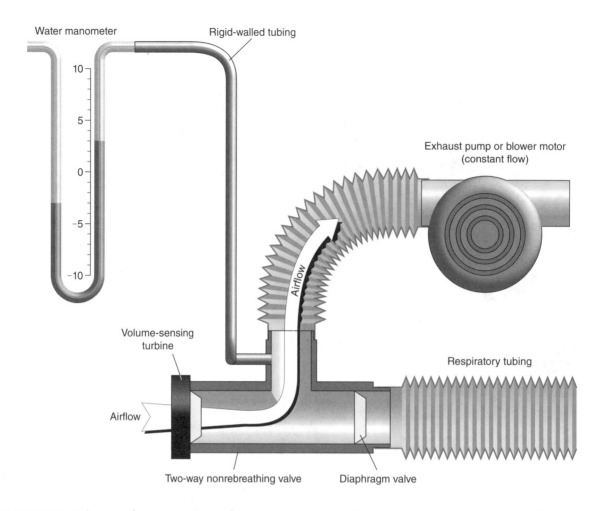

FIGURE 7.5 Schematic for measurement of pressure due to inspiratory resistance of a respiratory circuit.

length has to be 2 m, such as when testing rowers, and there is substantial travel of the athlete using a rowing ergometer. Sudden angulations of the tubing should be avoided. VacuMed (www.vacumed.com) manufactures corrugated tubing (Clean-bor) that minimizes kinking; the smooth inside surface also minimizes turbulence. Tubing of this type is strongly recommended.

Regularly check for leaks with a manometer by placing gas in the circuit under pressure. Leaky junctions can be located by using a liquid leak detector (e.g., Snoop, available from Swagelok, www.swagelok.com).

Gas Volume

Ideally, the instrument that measures gas volume should have accuracies of ±2.0% for pulsatile flows and ±1% for continuous flows (Gardner 1979). In practice, we have used a standard of ±3% given the challenges associated with calibration at high flow rates (Gore et al. 1997; Tanner and Gore 2006).

A water-sealed spirometer (e.g., Tissot or Stead-Wells) is the primary calibration standard against which other secondary calibration standards such as syringes and sinusoidal artificial lungs should be initially checked. However, the constancy of the cross-sectional volume of the spirometer's bell should still be verified throughout its elevation by withdrawing saturated gas with a calibration syringe. Saturation of gas in the spirometer can be attained when the bell has been left raised overnight. It is also advisable to use a water manometer to confirm that the gas pressure within the bell corresponds with that of the atmosphere throughout its elevation.

The calibration procedure must be specific to the method of measuring the volume (Hart and Withers 1996). Thus, if the measuring device is on the inspiratory side of the circuit, calibrations should be conducted with a Tissot spirometer, syringe, or sinusoidal artificial lung using pulsatile flows of 50, 100, 150, 200, and 250 L · min⁻¹. In practice, delivering flow rates much above approximately 150 L · min⁻¹ is problematic with a 3 L syringe, because rapid impact of the syringe plunger against the caps at the top and bottom of the stroke can cause artifacts. Therefore, a motor-driven syringe is the more reliable and accurate method of calibration, although motor-driven pumps also have limitations for determining peak flows (Miller et al. 2000). If the Douglas bag method is used in conjunction with an exhaust pump to draw the expirate through a gas meter, then volume calibrations should be conducted at a constant flow rate.

The latest generation of metabolic carts use pneumotachographs to determine volume; for instance, the Parvomedics TrueOne 2400 system (www.parvo.com) uses a Hans Rudolph screen pneumotachograph with a claimed accuracy of $\pm2\%$ for the range of 0 to 800 L · min^{-1}. Pneumotachographs measure the decrease in pressure of airflow across a screen or through a tube. Bernoulli's law (flow is proportional to the square root of the pressure difference) then enables a pressure change to be converted to a volume via electronics and firmware. These instruments have been demonstrated to have good precision and accuracy up to flow rates of 250 L · min^{-1} at ambient temperature and pressure (ATP) (Tang et al. 2003).

Gas collection should be made for a minimum of 30 s at each workload. Gas temperature and barometric pressure should be measured to an accuracy of ± 0.5 °C and ± 1.0 mmHg, respectively. Mercury barometers are rarely used, and digital barometers can be calibrated to this tolerance. Our annual records over the past decade indicate that calibration of digital barometers every 2 years is adequate. The timing devices should be accurate to ±0.1 s.

Gas Analysis

Electronic oxygen and carbon dioxide analyzers should have an absolute accuracy of at least $\pm0.05\%$. It is important that they are warmed up for at least an hour to eliminate electrical drift. The analyzers must be calibrated prior to any testing and at regular intervals over the expected physiological range of measurement (oxygen: 18%-15%; carbon dioxide: 3%-5%) with gases of either primary gravimetric-standard grade (alpha grade; $\pm0.02\%$ tolerance) that have been prepared gravimetrically or of beta grade ($\pm0.2\%$ tolerance) that have been verified on analyzers that have been calibrated successfully with three primary gravimetric-standard grade gases. For an example, see table 7.1.

Wasserman and colleagues (1994, p 448) advocate a two-point calibration check; they recommend atmospheric air and a previously verified calibration gas whose carbon dioxide and oxygen concentrations are in the middle of the anticipated range of the mixed expirate. If this method is adopted, the

TABLE 7.1 Example Calibration Gases

	CO_2, %	O_2, %
Atmospheric air	0.03	20.93
Tank 1 (α grade)	5.58	14.55
Tank 2 (α grade)	4.25	16.50
Tank 3(α grade)	3.05	18.05
Tank 1 (β grade)	5.07	14.89
Tank 2 (β grade)	3.65	17.88

accuracy of the analyzers should still be checked regularly using multiple gases.

Because most analyzers are pressure dependent, and therefore flow dependent, it is essential that the resistance to flow be similar for both calibration and measurement and should remain constant throughout a test. It is conceivable that condensation or a globule of saliva on the internal lumen of the gas sample tubing could alter flow and would be very difficult to adjust post hoc because the time at which the partial blockage occurred could not be identified. A posttest calibration check of the analyzers is therefore critically important.

The gas sample is usually passed through a desiccant column of anhydrous calcium sulfate or calcium chloride. The calculations in the final section of this chapter therefore assume that the carbon dioxide and oxygen percentages are for dry expirate. If the water vapor is not removed, then dilution of the carbon dioxide and oxygen reduces their concentrations as follows (Wasserman et al. 1994, pp 458-460):

$$\% \text{ of true value} = \left(\frac{P - pH_2O}{P}\right)\times100$$

where P = ambient barometric pressure and pH_2O = partial pressure of water vapor (at a given temperature).

Hence, if after drying, the humidity of the mixed expirate is 30% (pH_2O = 5.26 mmHg at 20 °C; assume P = 750.0 mmHg), then true carbon dioxide and oxygen values of 4.0% and 17.0% will be reduced to 3.97% and 16.88%, respectively. The $\dot{V}O_2$ in this example will therefore be spuriously elevated by approximately 3% if no correction is made. It should also be noted that Nafion tubing, which only removes water vapor such that an equilibrium is achieved with ambient humidity, does not fully dry the gas unless a countercurrent of dry air is drawn continuously past its exterior. The data in figure 7.6 provide reason for some concern about using Nafion within a metabolic cart for a prolonged test such as a 1 h cycling time trial. More data are required about the drying efficiency of Nafion and whether longer lengths are more effective than shorter lengths.

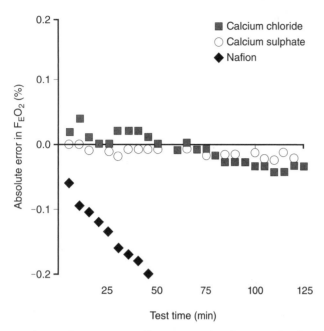

FIGURE 7.6 Drying effectiveness for three methods indicated by stability of fraction of oxygen. Data were collected on one calibrated AEI S-3A/I oxygen analyzer (AEI Technologies Inc., Naperville, Illinois). Room air was directed over a 70 °C water bath to generate saturated expirate of approximately 37 °C. Anhydrous calcium sulfate and calcium chloride are both effective drying media for approximately 1 h. New Nafion tubing (Perma Pure LLC, Toms River, New Jersey) shows rapid effects of water vapor.

To verify room air, a number of laboratories use software that allows the operator to enter values for fractional concentrations of oxygen and carbon dioxide in inspired gas (F_IO_2 and F_ICO_2, respectively) that are not the standard ones of 20.93% and 0.03%, respectively. This procedure should be reviewed carefully because a 0.035% change in F_IO_2 will affect the calculated $\dot{V}O_2$ by 1.0%. The following data illustrate this point. Samples of laboratory and outside air were collected at the one institution and analyzed at two laboratories. At laboratory 1 there appeared to be a 0.07% reduction in the ambient F_IO_2 (table 7.2). When the same foil bags of laboratory and outside air were analyzed at laboratory 2 (table 7.2), the results indicated that the air in laboratory 1 was not contaminated. Rather, the results from laboratory 2 indicated a potential alinearity problem with the oxygen analyzer at laboratory 1 that caused this analyzer to read room oxygen as 20.85%/20.86% rather than 20.93%. Unless chemical gas analysis can be used to establish unequivocal contamination of room air, it is recommended that any oxygen software default values for F_IO_2 and F_ICO_2 be set at 20.93% oxygen and 0.03% carbon dioxide, respectively.

Recent trends in global warming suggest that a value of 0.04% (>350 ppm) might be more relevant for room air carbon dioxide (www.esrl.noaa.gov/gmd/ccgg/trends/).

TABLE 7.2　Comparison of Laboratory Air and Outdoor Air at Two Laboratories

	NOMINAL VALUES		LABORATORY 1, MEASURED VALUES		LABORATORY 2, MEASURED VALUES	
	O_2	CO_2	O_2	CO_2	O_2	CO_2
†Laboratory 1 calibration gas 1: BOC α gas	15.05*	2.51	15.05	2.49	n/a	n/a
†Laboratory 1 calibration gas 2: BOC α gas	16.47	3.75	16.49	3.69	n/a	n/a
†Laboratory 1 calibration gas 3: BOC α gas	17.94	5.00*	17.94	5.00	n/a	n/a
†Laboratory 1 calibration gas 4: special BOC α gas	20.93	0.03	20.89	0.01	20.92	0.02
Laboratory 1 air from sample drying tube of $CaCl_2$	20.93	0.03*	20.86	0.03	n/a	n/a
Foil bag of laboratory 1 air	20.93	0.03	20.85	0.02	20.92	0.04
Foil bag of laboratory 1 outdoor air	20.93	0.03	20.86	0.01	20.93	0.03
‡Laboratory 2 calibration gas	15.10*	5.08*	n/a	n/a	15.10	5.08
Laboratory 2 air from sample drying tube of $CaSO_4$	20.93	0.03	n/a	n/a	20.93	0.03

Laboratory 1 calibrated analyzers with four α standard gases† and used software to fit a linear regression before the foil bags were measured. The analyzers at laboratory 2 were calibrated on one chemically analysed gas mixture‡ and room air that had been previously chemically analysed at 20.93% v/v O_2 and 0.03% v/v CO_2.

BOC = British Oxygen Company; n/a = not applicable.

*Set or zeroed/spanned analyzers on this gas.

General Procedures

In addition to requiring calibration of individual items of equipment, accurate $\dot{V}O_2$max data depend on controlled preparation of the laboratory and subject. This section also covers the criteria for attaining $\dot{V}O_2$max, the biological variability of $\dot{V}O_2$max, measurement error, and the calibration of indirect calorimetry systems.

Environmental and Laboratory Conditions

It is recommended that the laboratory be kept within 18 and 23 °C with a relative humidity less than 70% during testing (figure 7.7). Both these variables should be recorded on the test data sheet.

Athlete Preparation

Athletes must be given instructions regarding how to prepare for the test and what will happen during the test.

The athlete should refrain from eating at least 2 h before the $\dot{V}O_2$max test. Although many protocols recommend no exercise on the day of the test, and preferably none on the previous day, this is imprac-

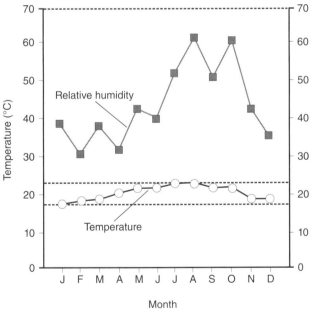

FIGURE 7.7 Plot of yearly laboratory environmental conditions. Data were collected using a verified Vaisala PTU Thermometer/Environmental Tracker across a period of 12 months. The recommended limits for laboratory conditions can be a challenge to attain in very hot or cold weather. Testing should be scheduled during hours when acceptable limits are attainable and air conditioning is operational (e.g., 8 a.m. to 6 p.m.).

tical when testing high-performance athletes. These athletes should undertake no unaccustomed exercise on the previous day, and the lead-up to testing should be individually standardized. For example, a cyclist could have a 50 km recovery ride on the morning of an afternoon $\dot{V}O_2$max test, but a similar ride should be completed on subsequent testing occasions. Probably more important than the single preceding day is an appreciation for the volume of unaccustomed training and racing completed over 7 days prior to testing, as this is a more integrated indicator of accumulated fatigue. Practically, it is very difficult for athletes to present themselves in the same state of test readiness several weeks or months apart.

Before the test begins, each subject must be familiarized with the breathing apparatus, ergometer, and, if a treadmill is used, the emergency stop procedure. For a novice, this introduction should take place the day prior to the test.

A piece of 3M millipore paper tape or similar material placed on the nose will ensure that sweating does not cause the noseclip to slip off during the test. The patency of the noseclip should be checked by asking the subject to try to expire through the nose. Generally, we find that the VacuMed reusable nose clip (R9015) is satisfactory without tape (www.vacumed.com) provided that skin preparation is good, which can be attained by using an alcohol swab to remove moisturizer or makeup.

A heart rate monitor, such as Polar or Suunto, should be fitted and worn throughout the test to provide cardiac frequency. An electrocardiogram is generally not used to test athletes except to screen for cardiovascular disorders (Panhuyzen-Goedkoop 2009).

A communication system must be devised between the subject and the investigator so that the subject can indicate the termination of the test.

The athlete should have a cool-down period after the $\dot{V}O_2$max test. This will minimize the venous pooling of blood, assist with the metabolism of lactate, and provide a transition from vigorous exercise to rest. The subject could therefore be asked to exercise at a low workload until the heart rate recovers to below approximately 120 beats/min.

Criteria for $\dot{V}O_2$max

Several criteria have been proposed to indicate that $\dot{V}O_2$max has been attained. The main criterion, although debated, is that $\dot{V}O_2$ reaches a plateau despite increases in workload. Other commonly used criterion are a respiratory exchange ratio (R) value greater than 1.10 and a 5 min postexercise blood lactate of greater than 8.0 mmol $\cdot$ L^{-1}. These latter two points are merely supplementary to the main criterion (plateau in $\dot{V}O_2$) and do not in themselves indicate that $\dot{V}O_2$max has or has not been attained. In our experience, most high-level athletes can readily achieve all three criteria.

Which criterion should be used to identify the plateau in $\dot{V}O_2$ is debated (Howley et al. 1995; Beltrami et al. 2012). The most frequently quoted leveling criterion is that of Taylor and colleagues (1955), who advocated that an increment of less than 2.1 ml · kg^{-1} · min^{-1} for an increase in treadmill elevation of 2.5% at 11.3 km · h^{-1} should indicate the plateau. However, this criterion was based on the mean ± standard deviation of 4.2 ± 1.1 ml · kg^{-1} · min^{-1} for the increase in $\dot{V}O_2$ that was associated with a step increment in their protocol. Taylor and colleagues stated that there was only a small chance of erroneously deciding that $\dot{V}O_2$max had been attained if the increase in $\dot{V}O_2$ was two standard deviations below the expected mean rise if $\dot{V}O_2$max had not been reached. The absolute leveling criterion therefore depends on the magnitude of the workload increment. However, Hawkins and colleagues (2007) demonstrated that a supramaximal workload approximately 30% greater than that attained in a 2 min incremental protocol did not elicit a meaningful change in $\dot{V}O_2$max because both means were within approximately 0.5% of each other. This result suggests that the 2 min protocol was effective for eliciting a plateau in well-trained athletes.

Increases in work rate should be selected such that the incremental part of the protocol is completed within 8 to 12 min (Buchfuhrer et al. 1983). Such a procedure has been demonstrated to yield the highest $\dot{V}O_2$max, but differences with durations outside this range are small (Buchfuhrer et al. 1983). Unpublished data from the AIS and SASI have demonstrated that both an incremental protocol of seven 4 min rowing stages and a 5 min all-out cycling performance test produce $\dot{V}O_2$max values that are equal to those attained during a continuous 1 min incremental protocol of approximately 12 min duration. In highly motivated athletes, relatively short (8-12 min) or long (~30 min) protocols do not affect attainment of $\dot{V}O_2$max.

If the main criterion for $\dot{V}O_2$max is not reached, then the subject should be asked to return 2 or 3 days later for an additional incremental exercise test. After a standard warm-up, commence this test three levels below the terminating point of the preceding one and use increments in workloads that are 50% of those in the original test. Unfortunately, for elite athletes, training schedules and the goals of their coach often preclude a repeat test.

All volumes expired and inspired on which $\dot{V}O_2$max is calculated should be collected for a duration of exactly 60 s. If data are collected for shorter durations, then adjacent sampling periods (e.g., 2 × 30 s) should be averaged. Similarly, 60 s of breath-by-breath data should be considered to calculate the $\dot{V}O_2$max.

Biological Variability of $\dot{V}O_2$max

The measurement of $\dot{V}O_2$max includes both technical error related to equipment calibration and biological variability because the athlete may not reproduce exactly the same effort when assessed on two or more occasions. Although there is little information in the literature about the biological variation in $\dot{V}O_2$max, Katch and colleagues (1982) reported that the combined technical error and biological variability for $\dot{V}O_2$max was 5.6%. They furthermore concluded that biological variability accounted for 90% of the total error.

The National Sport Science Quality Assurance program (NSSQA) of the Australian Institute of Sport/Australian Sports Commission now requires laboratories that seek certification to submit data for duplicate measures of $\dot{V}O_2$max on a group of athletes. The intraindividual variability of these data is a combination of technical or equipment error and biological variability. Calculation of the typical error of measurement (TE), which is the within-subject standard deviation of a single score (Hopkins 2000), enables quantification of the reliability with which a laboratory can measure $\dot{V}O_2$max.

In one case, TE data for $\dot{V}O_2$max from five certificated Australian laboratories yielded a mean of 2.2%, which is less than half the value reported by Katch and colleagues (1982). Data from the AIS have yielded values of approximately 2% for both maximal and submaximal $\dot{V}O_2$ (Clark et al. 2007; Robertson et al. 2010; Saunders et al. 2009). This suggests that the combined biological and analytical variation in $\dot{V}O_2$max is likely to be approximately 2% with well-calibrated equipment and well-habituated subjects.

TE data can also be used to determine the probability that a change in $\dot{V}O_2$max is a true change as a consequence of training or detraining rather than results that are associated with measurement error and/or biological variation:

- The 68% confidence limits of the change between two successive measurements = ±TE × $\sqrt{2}$ because both the first and second measures have error or uncertainty associated with them.

- The 95% confidence limits for a change = ±1.96 × TE × $\sqrt{2}$.

- A further factor for interpretation is assessing the likelihood of change relative to a "worthwhile" change, which for an elite athlete may be as small as 1%. Hopkins (2001) has a comprehensive description of both concepts (TE and worthwhile change) (http://sportsci.org/resource/stats/relyappl.html#assess), as well as a spreadsheet for calculating the likelihood of a change (http://sportsci.org/resource/stats/xprecisionsubject.xls).

Measurement Error Analysis

Even if one could eliminate the biological and analytical variability in the $\dot{V}O_2$max of an athlete, the need for precision of measurement is illustrated clearly by a mathematical analysis of the effects of changes in ventilation, gas fractions, temperature, barometric pressure, and relative humidity on the calculated $\dot{V}O_2$max. Such an analysis permits all but one variable to be held constant so that its effect on the calculated $\dot{V}O_2$max can be determined. This analysis (see table 7.3) indicates that ventilation ($\dot{V}_I$) and the fraction of oxygen in the dry expirate (F_EO_2) are the two variables in which small changes will have the greatest impact on the calculated oxygen. A 5% error in the measured ventilation translates directly to a 5% error in $\dot{V}O_2$, whereas a 1% error in F_EO_2 (e.g., 0.18%) translates to a 6.5% error in the $\dot{V}O_2$. For this reason, any error caused by inadequate drying of gases can also result in a larger error in the $\dot{V}O_2$. In contrast, a 1% error in barometric pressure (7.55 mmHg), which is extremely large compared with the likely errors obtained from a suitably calibrated digital barometer, translates to only a 1% error in $\dot{V}O_2$. Furthermore, the errors in ventilation and in F_EO_2 can be cumulative; for example, if ventilation is 5% high and F_EO_2 is 1% low, then the calculated $\dot{V}O_2$ will be 11.7% in error. This analysis illustrates that careful calibration of both the ventilation device and the oxygen analyzer is critical if errors are to be minimized.

Calibration of Indirect Calorimetry Systems

The individual components of an indirect calorimetry system should be calibrated before and after each test, and it is also important to check the function of the overall system on a regular basis.

Component Calibration

The calibration of indirect calorimetry systems often uses the technique of component calibration. Calibration of the ventilation device is conducted with either pulsatile or constant flows to span the physiological flow rate of 10 to 250 L · min⁻¹ BTPS (Body Temperature, ambient Pressure, and Saturated with

water vapor). Separate calibration of the gas analyzers at 3 points establishes their linearity over the range for dry expirate of approximately 15% to 18% for oxygen and approximately 3.0% to 5.0% for carbon dioxide. The gold standard for ventilation calibration is a water-sealed spirometer such as a Tissot gasometer (Warren E. Collins, Inc., Braintree, Massachusetts) or Stead-Wells spirometer. Both of these instruments have a precision cylinder with a known and constant cross-sectional area; hence, the volume and bell factor can be calculated from first principles:

$$\text{Volume} = \pi r^2 \times h,$$
$$\text{where}$$
$$r = \text{cylinder radius and}$$
$$h = \text{vertical displacement of cylinder.}$$
$$\text{Bell factor (ml} \cdot \text{mm}^{-1}) = \text{volume (ml) / h (mm).}$$

Tissot gasometers are no longer fabricated, but Rosdahl and colleagues (2010) described a 160 L custom-made copy based on the Collins Tissot.

The gold standard for gas analysis is the manometric chemical method, such as the Lloyd Haldane or micro-Scholander analyzer, but commercial companies with precision scales provide high-quality gas mixtures that are determined gravimetrically (from the molecular masses of the component gases) with a claimed absolute accuracy of ±0.02%. In practical terms, these high-precision, primary gravimetric-standard mixtures are usually within tolerance, but very occasionally an aberrant tank will be provided by a supplier. The magnitude of error is usually so great that it is readily apparent compared with the two other gases used for calibration. Accordingly, it is good practice to verify a new set of three high-precision gases using analyzers that have been calibrated with three standard gas tanks that have been in use previously. Such an approach is not possible if you are setting up a gas analysis system from scratch. The value of having three tanks is that you can use room air and the other three gases to verify linearity, by using one of these four gases as a reference point and checking the other three.

Biological Calibration

Calibration of the critical components before each test and checking for drift after each test are crucial steps in the calibration of an indirect calorimetry system, but it is also important to conduct an integrated system calibration at least annually or when a new system is purchased. Biological approaches to calibration are a useful and expedient first step toward integrated calibration. Tables such as those published by Åstrand and Rodahl (1986) indicate approximate figures for steady-state oxygen consumption at a variety of cycling work rates; for example, the $\dot{V}O_2$ =

TABLE 7.3 Effect of Likely Variation in Ventilation, Gas Fractions, Barometric Pressure, Temperature, and Relative Humidity on Calculated Indirect Calorimetry Values

Likely error	Reference values	+5% $\dot{V}O_I$	+1% F_EO_2	+1% F_ECO_2	+1% P_B
F_EO_2, % v/v	**17.51**	17.51	*17.69*	17.51	17.51
F_ECO_2, % v/v	**3.79**	3.79	3.79	*3.83*	3.79
$\dot{V}O_I$ ATP, L · min^{-1}	**150.00**	*157.50*	150.00	150.00	150.00
$\dot{V}O_I$ STPD, L · min^{-1}	**136.10**	142.90	136.10	136.10	137.48
$\dot{V}O_E$ BTPS, L · min^{-1}	**166.63**	174.96	167.00	166.71	166.54
$\dot{V}O_E$ STPD, L · min^{-1}	**136.70**	143.53	137.00	136.76	138.08
$\dot{V}O_2$, L · min^{-1}	**4.542**	4.770	4.249	4.532	4.589
Error in $\dot{V}O_2$ vs. reference values, %	**0.00**	+5.00	−6.46	−0.23	+1.01
$\dot{V}CO_2$, L · min^{-1}	**5.146**	5.403	5.158	5.195	5.198
RER	**1.133**	1.133	1.214	1.146	1.133
Temperature, C	**22.0**	22.0	22.0	22.0	22.0
P_B, mmHg	**755.0**	755.0	755.0	755.0	*762.55*
Relative humidity, %	**50.0**	50.0	50.0	50.0	50.0

1. Reference values are shown in bold, and the variable that has been altered in subsequent analyses is shown in bold italics. F_EO_2 = fractional concentration of expired oxygen; F_ECO_2 = fractional concentration of expired carbon dioxide; $\dot{V}O_I$ ATP = inspired volume, ambient temperature, and pressure; $\dot{V}O_I$ STPD = inspired volume, standard temperature, and pressure dry; $\dot{V}O_E$ BTPS = expired volume, body temperature, and pressure saturated with water vapor; $\dot{V}O_E$ STPD = expired volume, standard temperature, and pressure dry; $\dot{V}O_2$ = volume oxygen; $\dot{V}CO_2$ = volume carbon dioxide; RER = respiratory exchange ratio; P_B = barometric pressure.

2. *Depression of gas fraction attributable to 30% water vapor in analyzers.

1.5 L · min^{-1} at 100 W, 2.8 L · min^{-1} at 200 W, and 4.2 L · min^{-1} at 300 W. Although these figures ignore interindividual variation in the mechanical efficiency of cycling and assume calibrated cycle ergometers, they can provide a basic guide to the overall function of an indirect calorimetry system. Data from our laboratory for elite middle-distance runners and cyclists are presented in figure 7.1; for elite cyclists, the $\dot{V}O_2$ = 1.8 L · min^{-1} at 100 W, 2.8 L · min^{-1} at 200 W, and 3.9 L · min^{-1} at 300 W.

Calibration Machines

A number of attempts have been made to manufacture a mechanical device that can deliver to an indirect calorimetry system precise gas fractions and ventilation that mimic those of an athlete. Huszczuk and colleagues (1990) describe a calibrator that simulates $\dot{V}O_2$ up to 5.0 L · min^{-1} by mixing room air and 21% carbon dioxide (balance N_2). There are two commercially available versions of this calibrator (VacuMed, www.vacumed.com/zcom/product/Product.do?compid=27&prodid=294) that are targeted to hospital and athletic markets based on the range of ventilations and simulated $\dot{V}O_2$.

The Australian Institute of Sport has developed two $\dot{V}O_2$max calibrators, the first of which has been described (Gore et al. 1997); it has the simulated capacity of an athlete and can calibrate both $\dot{V}_E$ and $\dot{V}_I$ systems. The second-generation calibrator (see figure 7.8) has much greater capacity than the first and can easily generate ventilation of 10 to 300 L · min^{-1} and a simulated $\dot{V}O_2$ of 0.3 to 10 L · min^{-1}. Compared with values measured by criterion indirect calorimetry systems, those predicted by the first-generation calibrator demonstrated an accuracy of approximately ±2% and a precision of approximately ±1% for predicted $\dot{V}O_2$ ranging from 2.9 to 7.9 L · min^{-1} and ventilation ranging from 89 to 246 L · min^{-1}. The second-generation device can generate gas fractions with an absolute accuracy of ±0.05% for both oxygen and carbon dioxide and ventilation within ±3%. Both $\dot{V}O_2$max calibrators provide a method to interrogate any $\dot{V}O_2$ system in terms of the accuracy of the component oxygen and carbon dioxide analyzers as well as the ventilation device and software used for data reduction. Unfortunately, the AIS calibrator and the associated software are not available commercially.

+1% Temperature	+1% RH/pH$_2$	Gas drying*	CUMULATIVE ERRORS			
			+5% $\dot{V}O_I$ & −1% F_EO_2	+5% $\dot{V}O_I$ & +1% F_EO_2	−5% $\dot{V}O_I$ & −1% F_EO_2	−5% $\dot{V}O_I$ & +1% F_EO_2
17.51	17.51	*17.37*	*17.34*	*17.69*	*17.34*	*17.69*
3.79	3.79	*3.76*	3.79	3.79	3.79	3.79
150.00	150.00	150.00	*157.50*	*157.50*	*142.50*	*142.50*
135.99	136.08	136.10	142.90	142.91	129.29	129.29
166.51	166.61	166.25	174.58	175.35	157.95	158.65
136.60	136.68	136.39	143.22	143.86	129.58	130.15
4.539	4.542	4.794	5.075	4.461	4.592	4.036
−0.07	−0.02	+5.54	+11.73	−1.79	+1.09	−11.14
5.142	5.145	5.087	5.391	5.415	4.878	4.900
1.133	1.133	1.061	1.062	1.214	1.062	1.214
22.22	22.0	22.0	22.0	22.0	22.0	22.0
755.0	755.0	755.0	755.0	755.0	755.0	755.0
50.0	*50.5*	50.0	50.0	50.0	50.0	50.0

FIGURE 7.8 The AIS Max calibrator. The system uses two pistons to mix known volumes of room air with 21% carbon dioxide (balance N$_2$) to generate a range of physiological expirates of known composition.

Calculations

The fundamental calculations underlying most commercially available systems are, understandably, not released by the manufacturers, but the following sections underpin every system of which we are aware. Not shown here are the additional correction factors that manufacturers may use to adjust for gas transit times between the athlete's mouth and the analyzers, the time for a full response of the analyzers, the compensation for mixing chambers, volumetric calibration factors, changes in the temperature of expirate at very high compared with low rates of ventilation, how they deal with changing gas viscosity, how they verify the efficacy of the drying method, or the temporal integration of the volume and gas fraction signals. Hodges and colleagues (2005) have also lamented the fact that algorithms are not made available by the manufacturers, but the devil is in the detail of these correction factors, which can be responsible for errors of up to approximately 10% in a $\dot{V}O_2$ (Beck and Gore 2004). When one is searching for changes in athletes of a few percentage points, an uncertainty of a $\dot{V}O_2$max by this extent means that the data would be worthless.

When Expired Volume Is Measured

The volume of expired gas is measured under ambient temperature and pressure, saturated with water vapor (ATPS) conditions. The volume of expired gas measured over 1 min is known as either the pulmonary ventilation or minute ventilation ($\dot{V}_E$) and is usually expressed in L · min^{-1} BTPS. It may be computed by either multiplying the $\dot{V}_E$ L · min^{-1} at ATPS by the appropriate conversion factor (Diem and Lentner 1974, p 259) or using the combined gas laws formula. Note that in this latter method, the temperature in degrees Celsius (°C) is converted to absolute temperature in Kelvin (K) by adding 273.16:

$$\frac{P_1 \times V_1}{T_1} = \frac{P_2 \times V_2}{T_2}$$

$$\therefore \dot{V}_E \text{ L} \cdot \text{min}^{-1} \text{ BTPS} = V_1 \times \frac{P_1 \times T_2}{P_2 \times T_1}$$

where

$V_1 = \dot{V}_E$ L · min^{-1} ATPS,

P_1 = barometric pressure of the ambient air minus the partial pressure of water vapor at the temperature of the expired gas (see table 7.4),

T_2 = 310.16 K (273.16 K + 37 °C for body temperature),

P_2 = barometric pressure of the ambient air in mmHg minus 47.07 mmHg for the partial pressure of water vapor at body temperature, and

T_1 = 273.16 K plus the temperature of the expired gas in °C.

The volume of the expired air needs to be adjusted to standard temperature of 0 °C and standard pressure of one atmosphere or 760 mmHg and must be dry, indicating the absence of water vapor (standard conditions for temperature, pressure and dry [STPD]), to determine the $\dot{V}O_2$. This is accomplished by multiplying the $\dot{V}_E$ ATPS L · min^{-1} by the appropriate conversion factor (Diem and Lentner 1974, pp 260-269). The combined gas laws formula can also be used:

$$\frac{P_1 \times V_1}{T_1} = \frac{P_2 \times V_2}{T_2}$$

$$\therefore \dot{V}_E \text{ L} \cdot \text{min}^{-1} \text{ STPD} = V_1 \times \frac{P_1 \times T_2}{P_2 \times T_1}$$

where

$V_1 = \dot{V}_E$ L · min^{-1} ATPS,

P_1 = barometric pressure of the ambient air minus the partial pressure of water vapor at the temperature of the expired gas (see table 7.4), which is assumed to be completely saturated with water vapor,

T_2 = 273.16 K,

P_2 = 760 mmHg, and

T_1 = 273.16 K plus the temperature of the expired gas in °C.

The concentrations of oxygen and carbon dioxide in dry atmospheric air are constant at 20.93% and 0.03%, respectively, but these may be altered if the testing environment is not properly ventilated. It is also known that the remaining gas (79.04%), which contains primarily N_2, does not participate in physiological reactions. The volume of inspired gas ($\dot{V}_I$) can therefore be calculated using what has frequently been called the Haldane transformation (Haldane 1912) but should, according to Poole and Whipp (1988), be more correctly attributed to Geppert and Zuntz (1888):

$$(\dot{V}_I \text{ L} \cdot \text{min}^{-1} \text{ STPD}) \times (79.04) = (\dot{V}_E \text{ L} \cdot \text{min}^{-1} \text{ STPD}) \times$$
$$(\%N_2 \text{ in dry expired gas})$$

$$\therefore \dot{V}_I \text{ L} \cdot \text{min}^{-1} \text{ STPD} =$$

$$\frac{\dot{V}_E \text{ L} \cdot \text{min}^{-1} \text{ STPD} \times (\%N_2 \text{ in dry expired gas})}{79.04}$$

where the terms are defined as follows:

% N_2 in Dry Expired Gas = 100 – (% O_2 in Dry Expired Gas + % CO_2 in Dry Expired Gas).

$$\text{Volume of } O_2 \text{ Inspired} = \dot{V}_I \text{ L} \cdot \text{min}^{-1} \text{ STPD} \times \frac{20.93}{100}$$

$$\text{Volume of } O_2 \text{ Expired} = \dot{V}_E \text{ L} \cdot \text{min}^{-1} \text{ STPD} \times$$
$$\frac{(\%O_2 \text{ in dry expired gas})}{100}$$

$\therefore \dot{V}O_2$ L · min^{-1} STPD = Volume O_2 Inspired L · min^{-1} STPD – Volume O_2 Expired L · min^{-1} STPD.

$$\dot{V}O_2 \text{ ml} \cdot \text{kg}^{-1} \cdot \text{min}^{-1} = \frac{(\dot{V}O_2 \text{ L} \cdot \text{min}^{-1} \text{ STPD})}{\text{body mass in kg}} \times 1,000$$

The respiratory exchange ratio (R) may then be calculated:

$$R = \frac{CO_2 \text{ produced}}{O_2 \text{ uptake}}$$

$$R = \frac{\dot{V}CO_2 \text{ L} \cdot \text{min}^{-1} \text{ STPD}}{\dot{V}O_2 \text{ L} \cdot \text{min}^{-1} \text{ STPD}}$$

where

$$CO_2 \text{ produced } (\dot{V}CO_2 \text{ L} \cdot \text{min}^{-1}) =$$

$$\left(\dot{V}_E \text{ L} \cdot \text{min}^{-1} \text{STPD} \times \frac{(\%CO_2 \text{ in dry expired gas})}{100} \right) -$$

$$\left(\dot{V}_I \text{ L} \cdot \text{min}^{-1} \text{STPD} \times \frac{0.03}{100} \right)$$

TABLE 7.4 Partial Pressure (mmHg) of Water Vapor in Saturated Gas

°C	0	0.1	0.2	0.3	0.4	0.5	0.6	0.7	0.8	0.9
12	10.51	10.58	10.65	10.72	10.79	10.87	10.94	11.01	11.08	11.15
13	11.23	11.30	11.38	11.45	11.52	11.60	11.68	11.75	11.83	11.91
14	11.98	12.06	12.14	12.22	12.30	12.38	12.46	12.54	12.62	12.70
15	12.78	12.87	12.95	13.03	13.12	13.20	13.29	13.37	13.46	13.54
16	13.63	13.72	13.81	13.89	13.98	14.07	14.16	14.25	14.34	14.43
17	14.53	14.62	14.71	14.81	14.90	14.99	15.09	15.18	15.28	15.38
18	15.47	15.57	15.67	15.77	15.87	15.97	16.07	16.17	16.27	16.37
19	16.47	16.58	16.68	16.79	16.89	17.00	17.10	17.21	17.32	17.42
20	17.53	17.64	17.75	17.86	17.97	18.08	18.19	18.31	18.42	18.53
21	18.65	18.76	18.88	18.99	19.11	19.23	19.35	19.46	19.58	19.70
22	19.82	19.95	20.07	20.19	20.31	20.44	20.56	20.69	20.81	20.94
23	21.07	21.19	21.32	21.45	21.58	21.71	21.84	21.98	22.11	22.24
24	22.38	22.51	22.65	22.78	22.92	23.06	23.19	23.33	23.47	23.61
25	23.76	23.90	24.04	24.18	24.33	24.47	24.62	24.76	24.91	25.06
26	25.21	25.36	25.51	25.66	25.81	25.96	26.12	26.27	26.43	26.58
27	26.74	26.90	27.05	27.21	27.37	27.53	27.70	27.86	28.02	28.18
28	28.35	28.52	28.68	28.85	29.02	29.19	29.36	29.53	29.70	29.87
29	30.04	30.22	30.39	30.57	30.75	30.92	31.10	31.28	31.46	31.64
30	31.83	32.01	32.19	32.38	32.56	32.75	32.94	33.13	33.32	33.51
31	33.70	33.89	34.08	34.28	34.47	34.67	34.87	35.07	35.26	35.47
32	35.67	35.87	36.07	36.28	36.48	36.69	36.89	37.10	37.31	37.52
33	37.73	37.95	38.16	38.37	38.59	38.81	39.02	39.24	39.46	39.68
34	39.90	40.13	40.35	40.58	40.80	41.03	41.26	41.49	41.72	41.95
35	42.18	42.41	42.65	42.89	43.12	43.36	43.60	43.84	44.08	44.33
36	44.57	44.82	45.06	45.31	45.56	45.81	46.06	46.31	46.56	46.82
37	47.08	47.33	47.59	47.85	48.11	48.37	48.64	48.90	49.17	49.53

Reprinted from K. Diem and C. Lentner, 1974, *Scientific tables* (Basel, Germany: Novartis International), 257. With courtesy of Novartis International.

The ventilatory equivalents for oxygen and carbon dioxide can also be calculated. By convention, these are expressed as

$$\frac{\dot{V}_E}{\dot{V}O_2} = \frac{\dot{V}_E \, L \cdot min^{-1} \, \text{BTPS}}{\dot{V}O_2 \, L \cdot min^{-1} \, \text{STPD}}$$

and

$$\frac{\dot{V}_E}{\dot{V}CO_2} = \frac{\dot{V}_E \, L \cdot min^{-1} \, \text{BTPS}}{\dot{V}CO_2 \, L \cdot min^{-1} \, \text{STPD}}$$

When Inspired Volume Is Measured

If the volume measuring device is on the inspiratory side of the circuit, it is necessary to measure the relative humidity with a hygrometer. The computations are as follows:

$$\therefore \dot{V}_I \, L \cdot min^{-1} \, \text{STPD} = V_1 \times \frac{P_1 \times T_2}{P_2 \times T_1}$$

where

V_1 = inspired Volume ($\dot{V}_I$) L · min^{-1} ATPH (ambient temperature, pressure, and humidity);

P_1 = ambient pressure in mmHg minus pH$_2$O, which depends on the temperature and relative humidity. Thus, the pH$_2$O at 24 °C and 40% relative humidity is 8.95 mmHg, that is, 40% of 22.38 mmHg, which is the pH$_2$O when gas is completely saturated with water vapor at 24 °C (see table 7.4);

T_2 = 273.16 K;

P_2 = 760 mmHg; and

T_1 = 273.16 K plus the temperature of the inspired gas in °C.

$$\dot{V}_E \, L \cdot min^{-1} \, \text{STPD} = \frac{\dot{V}_I \times L \cdot min^{-1} \, \text{STPD} \times (79.04)}{\% \, N_2 \, \text{in dry expired gas}}$$

where

% N_2 in Dry Expired Gas = 100 − (% O_2 in Dry Expired Gas + % CO_2 in Dry Expired Gas).

$$\dot{V}_E \, L \cdot min^{-1} \, \text{BTPS} = \dot{V}_I \times \frac{P_1 \times T_2}{P_2 \times T_1}$$

where

$V_1 = \dot{V}_E \, L \cdot min^{-1}$ STPD,

P_1 = 760 mmHg,

T_2 = 310.16 K (273.16 K + 37 °C for body temperature),

P_2 = ambient pressure in mmHg minus 47.07 mmHg for the partial pressure of water vapor at body temperature, and

T_1 = 273.16 K.

All the subsequent calculations are done in the same way as outlined for measurement of expired volume.

Many laboratories now have facilities for online acquisition, reduction, and display of the preceding variables. This involves interfacing the electrical outputs from the physiological analyzers with a microcomputer via an analogue to digital converter (12 bit) or the same calculations can be performed in Microsoft Excel.

Equipment Suppliers

This guide is included because similar information is not readily available elsewhere. No attempt was made to include or exclude listings based on product quality. Possible omissions were inadvertent. Inclusion in this guide does not represent endorsement by the Australian Sports Commission or the authors.

Calibration Gases

Air Liquide America Specialty Gases LLC
  www.airliquide.com
Air Liquide Australia
  www.airliquide.com.au
BOC Gases Australia Ltd.
  www.boc.com.au
Coregas (includes Linde gas)
  www.coregas.com
Specialty Gases (branches in Singapore, Africa, Europe)
  www.specialty-gases.com
VacuMed
  www.vacumed.com

Calibration Syringes

A-M Systems Inc. (distributors worldwide)
  www.a-msystems.com
Hans Rudolph Inc.
  www.rudolphkc.com
VacuMed
  www.vacumed.com

Sinusoidal and Motor-Driven Calibration Syringes

JV Precision Engineering
Australian Capital Territory 2607
Australia

Dry Gas Meters

American Meter Company (eight distributors throughout the United States)
Elster American Meter
  www.elster-americanmeter.com
Apex Instruments
  www.apexinst.com
Harvard Apparatus
  www.harvardapparatus.com

Gas Analyzers

AEI Technologies Inc. (Formerly Ametek, Applied Electrochemistry)
  www.aeitecno.com
VacuMed
  www.vacumed.com

Kayak Ergometers

Dansprint Pro
  www.dansprint.com
Kayak Pro
  www.kayakpro.com

Leak Detector–Snoop (and Gas-Tight Pressure Tubing and Fittings)

Swagelok
  www.swagelok.com
Swagelok (Australia)
Adelaide, SA 5087
Australia

Manometers–U tube

Beacon Engineering Products (supply Dwyer manometers)
New South Wales 2116
Australia
Dwyer Instruments Pty. Ltd.
  www.dwyer-inst.com.au

Pneumotach

Hans Rudolph Inc.
  www.rudolphkc.com

Respiratory Valves, Tubing, and Hand-held Volumetric Calibration Syringes

Hans Rudolph Inc.
  www.rudolphkc.com
VacuMed
  www.vacumed.com

Rowing Ergometers

Concept 2 Inc. (suppliers worldwide)
  www.concept2.com
Concept 2 Australia
Jeff Sykes and Associates
Geelong, Victoria 3220
Australia

Treadmills

AusTredEx
  www.ausmanufacturers.com.au/austredexh/p/cosmos
  www.h-p-cosmos.com
Cardiac Science Corporation (Quinton systems)
  www.cardiacscience.com

$\dot{V}O_2$ Systems and Metabolic Carts

AEI Technologies Inc. (Formerly Ametek, Applied Electrochemistry)
  www.aeitecno.com
CareFusion (Jaeger & SensorMedics products; distributors worldwide)
  www.carefusion.com
Cortex Biophysik GmbH (distributors worldwide)
  www.cortex-medical.de
Cosmed Pulmonary Function Equipment (distributors worldwide)
  www.cosmed.com
Medical Graphics Corporation (distributors worldwide)
  www.medgraphics.com
ParvoMedics
  www.parvo.com
VacuMed
  www.vacumed.com

Weighing Scales

A & D Mercury Pty. Ltd.
  www.andmercury.com.au
American Weigh Scales
  www.americanweigh.com
Inscale Ltd
  www.inscale-scales.co.uk
ScaleSmart
  www.scalesmart.co.uk/
Wedderburn Scales (Branches Australia-wide)
  www.wedderburn.com.au

References

Abbiss, C.R., Quod, M.J., Levin, G., Martin, D.T., and Laursen, P.B. 2009. Accuracy of the velotron ergometer and SRM power meter. *International Journal of Sports Medicine* 30(2):107.

American College of Sports Medicine Position Stand. 1998. The recommended quantity and quality of exercise for developing and maintaining cardiorespiratory and muscular fitness, and flexibility in healthy adults. *Medicine and Science in Sports and Exercise* 30(6):975-991.

Åstrand, P.O., and Rodahl, K. 1986. *Textbook of Work Physiology*, 3rd ed. New York: McGraw-Hill.

Beck, K.C., and Gore, C.J. 2004. Optimizing breath-by-breath VO2 measurements. *Medicine and Science in Sports and Exercise* 36(3):554-555.

Beltrami, F.G., Froyd, C., Mauger, A.R., Metcalfe, A.J., Marino, F., and Noakes, T.D. 2012. Conventional testing methods produce submaximal values of maximum oxygen consumption. *British Journal of Sports Medicine* 46(1):23-29.

Bouchard, C., Daw, E.W., Rice, T., Pérusse, L., Gagnon, J., Province, M.A., Leon, A.S., Rao, D.C., Skinner, J.S., and Wilmore, J.H. 1998. Familial resemblance for VO2max in the sedentary state: the HERITAGE family study. *Medicine and Science in Sports and Exercise* 30(2):252-258.

Brink-Elfegoun, T., Kaijser, L., Gustafsson, T., and Ekblom, B. 2007. Maximal oxygen uptake is not limited by a central nervous system governor. *Journal of Applied Physiology* 102(2):781-786.

Buchfuhrer, M.J., Hansen, J.E., Robinson, T.E., Sue, D.Y., and Wasserman, K. 1983. Optimizing the exercise protocol for cardiopulmonary assessment. *Journal of Applied Physiology: Respiratory, Environmental and Exercise Physiology* 55(5):1558-1564.

Clark, S.A., Bourdon, P.C., Schmidt, W., Singh, B., Cable, G., Onus, K.J., Woolford, S.M., Stanef, T., Gore, C.J., and Aughey, R.J. 2007. The effect of acute simulated moderate altitude on power, performance and pacing strategies in well-trained cyclists. *European Journal of Applied Physiology* 102(1):45-55.

Colloud, F., Bahuaud, P., Doriot, N., Champely, S., and Chèze, L. 2006. Fixed versus free-floating stretcher mechanism in rowing ergometers: mechanical aspects. *Journal of Sports Sciences* 24(5):479-93.

Cumming, G.R., and Alexander, W.D. 1968. The calibration of bicycle ergometers. *Canadian Journal of Physiology and Pharmacology* 46(6):917-919.

Diem, K., and Lentner, C. (eds). 1974. *Scientific Tables*. Basle, Switzerland: Ciba-Geigy.

Douglas, C.G. 1911. A method for determining the total respiratory exchange in man. *Journal of Physiology (London)* 42:xvii-xviii.

Gardner, R.M. 1979. American Thoracic Society Statement: Snowbird workshop on standardization of spirometry. *American Review of Respiratory Disease* 199:831-838.

Geppert, J., and Zuntz. N. 1888. Ueber die Regulation der Athmung. *Pflüegers Archives* 42:189-245.

Gore, C.J., Catcheside, P.G., French, S.N., Bennett, J.M., and Laforgia, J. 1997. Automated $\dot{V}O_2$max calibrator for open-circuit indirect calorimetry systems. *Medicine and Science in Sports and Exercise* 29(8):1095-1103.

Haldane, J.S. 1912. *Methods of Air Analysis*. London: Charles Griffin.

Hart, J.D., and Withers, R.T. 1996. The calibration of gas volume measuring devices at continuous and pulsatile flows. *Australian Journal of Science and Medicine in Sport* 28(2):61-65.

Hawkins, M.N., Raven, P.B., Snell, P.G., Stray-Gundersen, J., and Levine, B.D. 2007. Maximal oxygen uptake as a parametric measure of cardiorespiratory capacity. *Medicine and Science in Sports and Exercise* 39(1):103-107.

Heil, D.P. 1997. Body mass scaling of peak oxygen uptake in 20- to 79-yr-old adults. *Medicine and Science in Sports and Exercise* 29(12):1602-1608.

Hodges, L.D., Brodie, D.A., and Bromley, P.D. 2005. Validity and reliability of selected commercially available metabolic analyzer systems. *Scandinavian Journal of Medicine and Science in Sports* 15(5):271-279.

Hopkins, W.G. 2000. Measures of reliability in sports medicine and science. *Sports Medicine.* 30(1):1-15.

Hopkins, W.G. 2001. A new view of statistics. *Sportscience.* http://sportsci.org/resource/stats/contents.html.

Howley, E.T., Bassett, D.R., and Welch, H.G. 1995. Criteria for maximal oxygen uptake: review and commentary. *Medicine and Science in Sports and Exercise* 27(9):1292-1301.

Hughson, R.L. 2009. Oxygen uptake kinetics: Historical perspective and future directions. *Applied Physiology, Nutrition and Metabolism* 34(5):840-850.

Hughson, R.L., Kowalchuk, J.M., Prime, W.M., and Green, H.J. 1980. Open-circuit gas exchange analysis in the non-steady-state. *Canadian Journal of Applied Sport Sciences* 5(1):15-18.

Huszczuk, A., Whipp, B.J., and Wasserman, K. 1990. A respiratory gas exchange simulator for routine calibration in metabolic studies. *European Respiratory Journal* 3(4):465-468.

Jones, N.L. 1988. *Clinical Exercise Testing*, 3rd ed. Sydney: Saunders.

Katch, V.L., Sady, S.S., and Freedson, P. 1982. Biological variability in maximum aerobic power. *Medicine and Science in Sports and Exercise* 14(1):21-25.

Levine, B.D. 2008. $\dot{V}O_2$max: what do we know, and what do we still need to know? *Journal of Physiology* 586(1):25-34.

Macfarlane, D.J. 2001. Automated metabolic gas analysis systems: a review. *Sports Medicine* 31(12):841-861.

Macfarlane, D.J., Edmond, I.M., and Walmsley, A. 1997. Instrumentation of an ergometer to monitor the reliability of rowing performance. *Journal of Sports Sciences* 15(2):167-173.

Martin, D.T., Quod, M.J., Gore, C.J., and Coyle, E.F. 2005. Lance Armstrong's cycling efficiency: Are conclusions supported by the data? (Letter to the Editor). *Journal of Applied Physiology.* 99(4):1628-1629.

Meyer, T., Davison, R.C., and Kindermann, W. 2005. Ambulatory gas exchange measurements—Current status and future options. *International Journal of Sports Medicine* 26(Suppl 1): S19-S27.

Miller, M.R., Jones, B., Xu, Y., Pedersen, O.F., and Quanjer, P.H. 2000. Peak expiratory flow profiles delivered by pump systems. Limitations due to wave action. *American Journal of Respiratory and Critical Care Medicine* 161(6):1887-1896.

Murphy, A.J., Chee, S.T.H., Bull, A.M.J., and McGregor, A.H. 2010. The calibration and application of a force-measuring apparatus on the seat of a rowing ergometer. *Proceedings of the Institution of Mechanical Engineers, Part P: Journal of Sports Engineering and Technology* 224(1):101-107.

Nevill, A.M., Bate, S., and Holder, R.L. 2005. Modeling physiological and anthropometric variables known to vary with body size and other confounding variables. *American Journal of Physical Anthropology* Suppl 41:141-153.

Noakes, T.D. 2008. Testing for maximum oxygen consumption has produced a brainless model of human exercise performance. *British Journal of Sports Medicine* 42:551-555.

Panhuyzen-Goedkoop, N.M. 2009. Preparticipation cardiovascular screening in young athletes. *British Journal of Sports Medicine* 43(9):629-630.

Poole, D.C., and Whipp, B.J. 1988. Letter to the editor. *Medicine and Science in Sports and Exercise* 20:420-421.

Roecker, K., Prettin, S., and Sorichter, S. 2005. Gas exchange measurements with high temporal resolution: the breath-by-breath approach. *International Journal of Sports Medicine* 26(Suppl 1):S11-S18.

Robertson, E.Y., Saunders, P.U., Pyne, D.B., Aughey, R.J., Anson, J.M., and Gore, C.J. 2010. Reproducibility of performance changes to simulated live high/train low altitude. *Medicine and Science in Sports and Exercise* 42(2):394-401.

Rosdahl, H., Gullstrand, L., Salier-Eriksson, J., Johansson, P., and Schantz, P. 2010. Evaluation of the Oxycon Mobile metabolic system against the Douglas bag method. *European Journal of Applied Physiology* 109(2):159-171.

Russell, J.C., and Dale, J.D. 1986. Dynamic torquemeter calibration of bicycle ergometers. *Journal of Applied Physiology* 61(3):1217-1220.

Russell, G., Gore, C.J., Ashenden, M.J., Parisotto, R., and Hahn, A.G. 2002. Effects of prolonged low doses of recombinant human erythropoietin during submaximal and maximal exercise. *European Journal of Applied Physiology* 86(5):442-449.

Saunders, P.U., Telford, R.D., Pyne, D.B., Hahn, A.G., and Gore, C.J. 2009. Improved running economy and increased hemoglobin mass in elite runners after extended moderate altitude exposure. *Journal of Science and Medicine in Sport* 12(1):67-72.

Schmidt, W., and Prommer, N. 2010. Impact of alterations in total hemoglobin mass on VO$_2$max. *Exercise and Sport Sciences Reviews* 38(2):68-75.

Tang, Y., Turner, M.J., Yem, J.S., and Baker, A.B. 2003. Calibration of pneumotachographs using a calibrated syringe. *Journal of Applied Physiology* 95(2):571-576.

Tanner, R.K., and Gore, C.J. 2006. Quality assurance in Australian exercise physiology laboratories in quest of excellence. *International Journal of Sports Physiology and Performance* 1(1):58-60.

Taylor, H.L., Buskirk, E., and Henschel, A. 1955. Maximal oxygen intake as an objective measure of cardio-respiratory performance. *Journal of Applied Physiology* 8:73-80.

Telford, R.D., Hooper, L.A., and Chennells, M.H.D. 1980. Calibration and comparison of air-braked and mechanically-braked bicycle ergometers. *Australian Journal of Sports Medicine* 12:40-46.

Truijens, M.J., Rodríguez, F.A., Townsend, N.E., Stray-Gundersen, J., Gore, C.J., and Levine, B.D. 2008. The effect of intermittent hypobaric hypoxic exposure and sea level training on submaximal economy in well-trained swimmers and runners. *Journal of Applied Physiology* 104(2):328-337.

Vogler, A.J., Rice, A.J., and Gore, C.J. 2010. Validity and reliability of the Cortex MetaMax3B portable metabolic system. *Journal of Sports Sciences* 28(7):733-742.

Wasserman, K., Hansen, J.E., Sue, D.Y., Whipp, B.J., and Casaburi, R. 1994. *Principles of Exercise Testing and Interpretation,* 2nd ed. Philadelphia: Lea & Febiger.

Withers, R.T., Brooks, A.G., Gunn, S.M., Plummer, J.L., Gore, C.J., and Cormack, J. 2006. Self-selected exercise intensity during household/garden activities and walking in 55 to 65-year-old females. *European Journal of Applied Physiology* 97(4):494-504.

Woods, G.F., Day, L., Withers, R.T., Ilsley, A.H., and Maxwell, B.F. 1994. The dynamic calibration of cycle ergometers. *International Journal of Sports Medicine* 15(4):168-171.

Woolford, S.M., Withers, R.T., Craig, N.P., Bourdon, P.C., Stanef, T., and McKenzie, I. 1999. Effect of pedal cadence on the accumulated oxygen deficit, maximal aerobic power and blood lactate transition thresholds of high-performance junior endurance cyclists. *European Journal Applied Physiology* 80(4):285-291.

Altitude Training

Sally A. Clark, Philo U. Saunders, and Christopher J. Gore

Since the first confirmed ascent to the summit of Mount Everest in 1953 by Sir Edmund Hillary and Sherpa Tensing Norgay, scientists have been intrigued by the physiological responses to altitude and the limitations induced by chronically low levels of oxygen. The impact of hypoxia on athletic performance was first evident at the 1968 Olympic Games held in Mexico City at an altitude of approximately 2,200 m. At these Games, performances were markedly compromised in those events requiring a predominately aerobic component (i.e., >2 min duration). Furthermore, the sustained success of long-distance runners from several African countries who live and train at altitude has highlighted the potential benefit to elite athletes of using altitude to prepare for competition.

This chapter describes the various methods for using altitude and hypoxic exposure and their potential effect on improving performance at altitude and at sea level. The potential mechanisms responsible for these performance gains are not described in this chapter, because others have done so comprehensively (Gore et al. 2007; Millet et al. 2010). There have been several debates about the classification of levels of altitude. In 2007, a group of leading researchers in altitude training agreed on a new classification (Bärtsch et al. 2008b). The resultant classification system, outlined as followed, is referred to in this chapter:

- Near sea level (0-500 m)
- Low altitude (>500 to 2,000 m)
- Moderate altitude (>2,000 to 3,000 m)
- High altitude (>3,000 to 5,500 m)
- Extreme altitude (>5,500 m)

Classical Altitude Training

Classical altitude training refers to the process of athletes living and training at natural altitude ranging from 1,600 to 3,000 m for a period of 2 to 4 weeks. The potential benefits of classic altitude training over other modalities of altitude exposure are that altitude acclimatization provides the stimulus for both central and peripheral adaptations as well as an additional training stimulus at the same absolute submaximal workload compared with sea level (Bärtsch and Saltin 2008). In addition to anecdotal evidence, published reports describe elite athletes who produce world-class performances subsequent to classic altitude training (Bärtsch and Saltin 2008; Daniels and Oldridge 1970). Many factors can contribute to improved performance outcomes after classic altitude training, and some of these cannot be measured, for example, spending 4 weeks away from home in a novel, exciting environment, where quality food is provided and the opportunity to rest and recover between training sessions exceeds that which normally occurs during an athlete's regular training regime consequent to work, study, and family commitments. Several weeks spent training at moderate altitude has the potential to provide a high-quality training program, making the effect of altitude alone difficult to measure. This is further complicated by the fact that even in well-controlled studies, the control group is usually aware that they are not at altitude. Overall, we cannot discount the potential placebo effect of classic altitude training given the widespread belief in the athletic community that it is an effective supplement to training.

One of the potential shortfalls of classic altitude training is the reduced absolute intensity of training that athletes can undertake (Levine and Stray-Gundersen 1997), which may dampen any potential performance improvements. With this in mind, scientists have developed other methods that athletes can adopt to prepare for competition using altitude, which are discussed subsequently. However, appropriate periodization of training while at altitude can negate the potential loss of training intensity with classic altitude. Methods such as reducing duration of intervals and increasing recovery times are ways that intensity can be maintained; these are described later.

Live High, Train Low

An alternative approach to classic altitude training is the combination of "living high" (e.g., 2,500 m) to obtain beneficial physiological changes associated with altitude acclimatization and "training low" (e.g., <1,250 m), which allows maintenance of high-intensity training. There is growing scientific and popular support that the "live-high, train-low" (LHTL) model may be the most advantageous hypoxic strategy to improve subsequent sea-level performance in well-trained athletes (Bonetti and Hopkins 2009; Fulco et al. 2000; Levine and Stray-Gundersen 1992; Rusko et al. 1995). Because the geography of many countries does not readily permit LHTL, a further refinement involves athletes' living at a simulated altitude under normobaric conditions and training at or close to sea level (Rusko 1996). In recent years, endurance athletes have used several new devices and modalities to complement the LHTL approach. These modalities include normobaric hypoxia via nitrogen enrichment generated with molecular sieves, which allows athletes to undertake LHTL, as well as supplemental oxygen to simulate normoxic or hyperoxic conditions during exercise and sleep at natural altitude. In the past 15 years many studies that used LHTL modalities have been published, and the mechanisms that support a change in performance have been strongly debated (Levine et al. 2005).

Indeed, more recently, a double-blind placebo study has failed to find any performance (Siebennman et al. 2012) or mechanistic (Nordsborg et al. 2012) benefit of LHTL.

Intermittent Hypoxic Exposure and Training

Intermittent hypoxic exposure (IHE) in the strict sense consists of repeated switching between breathing hypoxic and normoxic air at rest, usually totaling 60 to 90 min. Each hypoxic interval only lasts a few minutes, allowing a more extreme hypoxia between the fraction of inspired oxygen (F_IO_2) of 12% (equivalent to the inspired partial pressure of oxygen at ~4,500 m) to 10% (~ 6,000 m), a level that is not usually reached with other modalities (Bärtsch et al. 2008a). When IHE is used as a single interval without any intervening periods of normoxia, this modality can be referred to as prolonged hypoxic exposure (PHE), and because the time is too short (~1-4 h) to develop symptoms of acute mountain sickness, the hypoxia can be high (e.g., equivalent to 4,000 to 5,500 m) (Bärtsch et al. 2008b).

Intermittent hypoxic training (IHT) refers to training under hypoxic conditions and spending the rest of the time at or near sea level (Dufour et al. 2006). This method is usually performed using simulated hypoxia in a laboratory on some form of ergometry. Protocols for this method vary but typically consist of training under hypoxic conditions for 1 to 2 h a day for 2 to 5 days per week for several weeks.

Using Hypoxia to Prepare for Competition at Altitude

Most international sporting events are often limited to less than 3,500 m. Examples of such events include stages in the Tour de France, Vuelta Espana, and Giro d'Italia cycling races. Some exceptions are football matches in La Paz, Bolivia (3,600 m), and the cycling tour of Qinghai Lake, China, which has multiple peaks higher than 3,500 m. It is well known that performance is compromised at altitude as low as 1,000 m (Clark et al. 2007), and therefore it is in the athlete's best interest to prepare for competition at altitude by performing some method of altitude acclimatization and training.

Modality

Traditionally, classic altitude training has been the preferred method to prepare an athlete for competition at altitude. Classic altitude training allows for central and peripheral adaptations that result in the restoration of $\dot{V}O_2$ toward sea-level values via changes in exercise arterial oxygen saturation, which is considered crucial for maintaining performance. The hypoxic conditions provide an additional training stimulus, at the same submaximal absolute workload, that may enhance performance if the athlete can adapt appropriately to new training load.

The decision about whether to use classic altitude training methods to prepare the athlete will depend on the altitude at which the competition is taking place. For altitudes between approximately 1,000 and 3,000 m, preparing at the competition level is probably optimal. If the competition occurs at altitudes above 3,000 m, then spending time at those altitudes prior to competition may be detrimental to performance (Levine and Stray-Gundersen 1997). Altitudes higher than 3,000 m have greater potential to cause overtraining and to compromise the ability of an athlete to absorb and respond to the hypoxic and training stimuli. In addition, prolonged exposure to high altitude can lead to a significant loss of muscle fiber cross-sectional area and a decrease of mitochondrial volume density, which combined reduce mitochondrial volume by up to 30% (Hoppeler et al. 2003). This loss of muscle mass is likely to be detrimental in terms of performance. In this instance, other approaches to acclimatization need to be considered. An alternative that has not been extensively investigated in the scientific literature is to live at high altitude and include some intensity training with hyperoxia, equivalent to sea level F_IO_2 (Wilber et al. 2004). In sports such as cycling or running, the "hyperoxic" training may be possible during ergometer training sets but would be logistically cumbersome. Hyperoxic training would comprise only a portion of the total training conducted at high altitude, and the total training load would need to be carefully managed.

Length of Exposure

The time spent at altitude is an important factor; there is no point making the effort to travel and train at altitude if the length of exposure is insufficient to stimulate worthwhile adaptations. A minimum period of 2 weeks is recommended for altitude training designed to improve competition performance at altitude (Schuler et al. 2007). Anecdotally, many elite endurance athletes spend longer than 4 weeks training at moderate altitude, which may achieve better responses; however, such claims require further systematic investigation. The additional benefits of extended altitude training, if any, need to be quantified and evaluated against the cost and time away from home. Classic altitude studies that have measured competitive performance show that 2 to 4 weeks appears to elicit the best results in performance, and longer durations at altitude do not necessarily lead to further performance improvements and may even cause deterioration in performance gains from the initial 2 to 4 weeks (Saunders et al. 2009a). However, one potential benefit of prolonged residence at moderate altitude has recently been demonstrated in army recruits, whose hemoglobin mass (Hb_{mass}) took more than 7 months to reach a new, higher plateau at 2,210 m (Brothers et al. 2007).

Training Intensity

When using classic altitude training to prepare for competition at altitude, the intensity at which the athlete trains needs to be modified because of the reduction in oxygen transport. Athletes will not able to maintain the same training velocities or recoveries in interval sessions as they would at sea level. Performance in shorter events (~1-2 min) are relatively unaffected at moderate altitude (Peronnet et al. 1991), and this fact can be used when training athletes at altitude. To avoid a reduction in race-specific fitness, athletes should undertake a series of shorter race-pace efforts in which velocity is not compromised (or possibly is enhanced because of the reduced air density; Peronnet et al. 1991) and have longer recoveries than at sea level to maintain speed during the entire training session. With acclimatization and partial restoration of $\dot{V}O_2max$ at altitude (Adams et al. 1975), the duration of the interval efforts can be increased or the recovery times decreased. This approach fits with the general model that is often used when training athletes at altitude, that is, taking the first few days to a week to acclimatize to the altitude. Lower-intensity, higher-volume training is accompanied by shorter interval work to maintain competition velocities. As acclimatization occurs, the intensity of longer training intervals can be increased. It seems that the more often an athlete trains at altitude, the easier it becomes to adapt to the stress and to undertake normal training loads. This is one reason why multiple altitude training camps are undertaken during a season and over an athlete's career.

Training Location

Using the classic altitude training model to prepare for competition at altitude is not always practical, and factors such as the cost, access to good training facilities, and time away from the home training environment can make it undesirable. Furthermore, the reduction in absolute training intensity when at altitude has resulted in an increased use of the LHTL approach to altitude training. The LHTL model has received a lot of scientific attention in the past 2 decades, and the majority of the work has focused on improving sea-level performance. However, there is some evidence that this approach may enhance performance at altitude. Schuler and colleagues (2007) measured $\dot{V}O_2$ and performance at 2,340 m among eight cyclists after 1, 7, 14, and 21 days living at 2,340 m (19 h/d) and training at or below 1,100 m. In this study, $\dot{V}O_2max$ and time to exhaustion at 2,340 m improved over time, and this improvement occurred within the first 14 days of altitude exposure (Schuler et al. 2007). There are few places in the world where athletes can live at moderate altitude but are able to come down to low altitude to do race specific work prior to competition. For example, in Australia there are no mountain ranges high enough to enable athletes to perform either classic altitude training or the LHTL approach effectively. At the Australian Institute of Sport (AIS), we have built a nitrogen-enriched altitude facility where athletes can live and sleep to prepare for competition. All of the studies from the AIS have focused on sea-level performance, given that it is quite rare for our athletes to compete at venues of moderate altitude or higher. Another use of the LHTL facility is to reduce the time taken to acclimatize to natural altitude. Athletes may spend 1 to 2 weeks of living and sleeping in the altitude facility prior to departing for an overseas classic altitude training camp. Although this concept needs further scientific examination, it may be a useful model to establish partial acclimation so that the time spent acclimatizing at natural altitude is reduced and training is maximized.

In some instances, time and money may be important considerations when one is deciding what type of altitude or hypoxic exposure to use. Some professional sports teams may not have the luxury of spending 2 weeks at the competition venue or have access to an altitude facility to prepare for one event held at moderate to high altitude. In this situation, the use of IHE or IHT may be a viable option because such training can be done with minimal cost, travel, and inconvenience. The empirical evidence regarding the efficacy of IHE and IHT on erythropoietin response and athletic performance is not compelling (Hamlin et al. 2010; Tadibi et al. 2007; Wilber 2007); however, there is some evidence that using IHT or IHE may be beneficial for preparing athletes for competition at altitude (Beidleman et al. 2008; Gore et al. 2008; Muza 2007; Roels et al. 2007; Vogt and Hoppeler 2010). For example, Roels and colleagues (2007) reported that 3 weeks of training (5 × 1-1.5 h training sessions per week) in hypoxia (equivalent to 3,000 m) improved cycling peak power performance by 11.3% under hypoxic conditions. The extensive studies of Vogt and Hoppeler (2010) provide strong evidence of beneficial adaptations to muscle at a structural, biochemical, and molecular level after training in hypoxia, and the investigators conclude that moderate evidence supports the use of IHT to prepare for competition at altitude. However, performances changes from IHT are unclear. Insightfully, Vogt and Hoppeler (2010) suggest that relatively subtle changes in performance might still be worthwhile to elite athletes.

Using Hypoxia to Prepare for Competition at Sea Level

Elite athletes are always seeking a competitive edge. Given the well-known success of East African runners who live and train at moderate altitude, many endurance athletes now include some form of altitude training to prepare for sea-level competition.

Modality

Endurance athletes have been using classic altitude training for nearly half a century in pursuit of improving sea-level performance (Dick 1992; Levine et al. 2005), and such training is still used today by many athletes (Wilber 2007). The efficacy of this model for improving sea-level performance remains unclear; some studies report significant improvements in sea-level endurance performance (Daniels and Oldridge 1970), whereas others have failed to do so (Buskirk et al. 1967; Jensen et al. 1993). Given that the relative improvement in performance required by an elite athlete to increase his chance of winning medals at international competition is about 0.5% (Hopkins and Hewson 2001), it is not surprising that with sample sizes less than 20, many studies have been underpowered to detect a change of this magnitude (Gore et al. 2007). A meta-analysis (Bonetti and Hopkins 2009) that consolidated the existent literature concluded that for elite athletes, classic altitude is the only one of all the hypoxic modalities that is clearly beneficial (by a few percentage points) for sea-level performance. This may explain why classic altitude training remains popular among the world's elite endurance athletes; the scientific literature is struggling to catch up with the case-study approach of coaches for the last 60 years. It matters not to a coach whether a group effect is statistically significant; rather, coaches want to know whether classic altitude training can help a specific athlete deliver a personal best performance at competition.

With the introduction of hypoxic modalities such as LHTL, IHE, and IHT, there have been considerable changes in the way athletes use altitude training to prepare for competition at sea level. Access and convenience of altitude and hypoxic training have increased with the introduction of simulated hypoxic devices, which athletes can use on a more regular basis throughout the training and competition year. When determining how to use altitude training in a yearly plan, coaches and athletes must consider a number of factors. The severity of altitude, time spent training at altitude, history of altitude training, timing of training leading into competition, whether there is a lower altitude training option, and the type of altitude that is used are all important factors that may maximize the benefits of altitude training.

Access to different modalities is the first factor to consider; not everyone can readily access simulated altitude devices and others do not have close access to natural altitude venues of sufficient altitude. Therefore, availability, cost, travel time, and time away from home are important considerations. If all options are available, natural moderate altitude can be used in a period of high-volume training when athletes are able to use the hypoxic training stress and when high-intensity training is not as important. During a competition phase, natural altitude venues with lower altitude training options are perfect because athletes spend most of the time at moderate altitude and perform only key sessions at the lower altitude. In this scenario, athletes can complete high-quality training while still receiving the physiological adaptations that result from spending most of the day at moderate altitude and performing most of their training at moderate altitude. If using simulated altitude, athletes need to be diligent in spending more than 12 h at altitude (Clark et al. 2009); training on an ergometer at simulated altitude, if possible, may be of some additional benefit (Robertson et al. 2010a). Simulated LHTL can be an attractive option over classic altitude training if the athlete is able to stay in their home environment and have continual access to their normal support systems. Classic altitude camps can be used when travel is necessary for competitions and an appropriate altitude training venue is close by for preparation.

Length of Exposure

The duration of classic altitude training prior to competition at sea level should be at least 2 weeks, and closer to 4 weeks is preferable. Four weeks is ideal because it allows the athletes time to acclimatize to the hypoxia by keeping training intensity and volume lower in the first week before increasing the intensity and duration of efforts over the following 3 weeks. When the LHTL approach is used, a duration of at least 3 weeks of more than 12 h exposure per day is recommended (Clark et al. 2009). When using this approach at the AIS, we have consistently reported increases in Hb_{mass} and improved performance (Robertson et al. 2010a; Saunders et al. 2010). However, in our earlier work conducted at the AIS when athletes were only spending approximately 21 days of 8 to 9 h per night at 3,000 m simulated altitude, we reported no change Hb_{mass} (Ashenden et al. 1999) but we did find increases in muscle buffering capacity and improved submaximal exercise efficiency (Gore et al. 2001), each of which may be beneficial to performance (Levine et al. 2005).

Training Location

The severity of altitude chosen to prepare for sea-level competition may depend on whether classic altitude or the LHTL method is used; however, a range of about 2,000 to 3,000 m is recommended. In a natural altitude environment, altitudes of approximately 2,000 m to 2,500 m are ideal, whereas an altitude of closer to 3,000 m is recommended when one is using LHTL; this is so because the number of hours per day of exposure is less and it appears that the overall response is dose related (Levine and Stray-Gundersen 2006). Altitudes greater than 3,000 m are not recommended for the same reasons that have previously been identified. Poor sleep quality (Kinsman et al. 2002) and overall recovery may be compromised at high altitude, which is not ideal leading into competition. Altitudes greater than 3,000 m may be used when one is performing IHE or PHE of up to 4 to 5 h per day. But as described in a previous section, the evidence that these modes of hypoxia improve sea-level performance remains inconclusive, particularly for elite athletes (Bonetti and Hopkins 2009).

Multiple Exposures

With regard to multiple-altitude training camps, recent work with athletes at the AIS indicates that longer than 8 weeks is required between altitude training stints to maximize the training that follows (Robertson et al. 2010b). A longer exposure also ensures that athletes are not excessively fatigued going into a subsequent altitude training camp, training phase, or competition. The periodization of the training year and the training phase prior to an altitude camp should be considered carefully. It is advisable for endurance athletes to use altitude training several times throughout a competition year. The emphasis of training can be tailored to meet demands of the training phase of the athletes. For example, in the early build-up period, when athletes are trying to increase the volume of training and when high-quality training is not as critical, a longer period of altitude training can be undertaken that focuses on accumulating a high volume of training using the hypoxic stimulus rather than accumulating any high-intensity training. It is recommended that no more than 2 months be spent at altitude at any one time and that athletes undertake short blocks (2-4 weeks) more frequently throughout the year, with two or three stints optimal.

Athletes should spend adequate periods at sea level between these multiple-altitude exposures (>8 weeks) to capitalize on increased capacities gained from altitude and to ensure that athletes are fresh and motivated for training at altitude for each camp. However, we have used low to moderate altitude training during a competitive season in elite middle-distance runners (Saunders et al. 2009b). In this particular study, seven elite middle-distance runners lived at approximately 1,800 m and did all their low- to moderate-intensity running at 1,700 to 2,200 m. However, because the athletes were in their competitive season, they completed all high-quality sessions at 900 m altitude to maintain 800/1,500 m race-pace interval training required to stay race fit. This protocol resulted in improved competitive performance by 1.9% (90% confidence limits, 1.3%-2.5%).

Competition Timing

An area of great contention is the time that athletes should spend back at sea level after altitude training to maximize their performance. Most studies have measured performance immediately after altitude training; however, coaches have often reported performance gains in the 2 to 3 weeks and even up to 8 weeks after athletes return to sea level. Improved performance in the first 2 to 4 days on return to sea level may be related to hemodilution and persistence of ventilatory adaptations to altitude, whereby improvements observed 2 to 3 weeks after return to sea level may be a result of increased oxygen transport (Millet et al. 2010). There is little scientific evidence concerning the time course of performance after either natural or simulated LHTL methods of altitude training. Furthermore, there appears to be large individual variation, and therefore it is recommended that athletes and coaches experiment with the timing to determine, if possible, individual guidelines.

Accounting for individual response is critical in the effective use of hypoxic training. Levine and Stray-Gundersen (1997) identified responders and nonresponders to LHTL in terms of their simulated race performance and found wide variation between individuals. But Roberston and colleagues (2010b) were the first to attempt to quantify the magnitude of true individual response to hypoxic training. For a LHTL intervention, the individual variation in performance was of similar magnitude to the mean response, approximately 1%. If this estimate is correct for other modalities of hypoxia, it will remain challenging for single scientific studies to identify the potential benefit or otherwise of hypoxic training to elite athletes who require performance improvements of just 0.5% to improve their medal-winning chances (Hopkins and Hewson 2001). But for the individual athlete and coach, this finding means that there is moderate likelihood that altitude training will be beneficial for sea-level performance in endurance events. The challenge for an athlete and coach is to use hypoxia in a prudent, systematic manner that does not induce injury or illness and to fit hypoxic training within the annual and quadrennial periodization plan. Altitude training is not a panacea for a poorly performing athlete, but does appear to be one method that can be used to induce overload, adaptation, and rebound for even the most elite athletes.

Altitude and Illness

A further consideration of undertaking altitude training is the possibility of an increased risk of illness. Exercise at altitude can lead to immune suppression through sympathoadrenal pathways that increase the release of epinephrine and impairment of T-cell activation and proliferation, which increases the risk of infection during initial exposure to altitude (Mazzeo 2005). Anecdotally, we have reported some instances of illness during training at moderate altitude (Gore et al. 1998), but whether the incidence of illness is elevated beyond what would occur during a similar training camp at sea level remains unclear. Avoiding illness is not always possible, but athletes can reduce the risk of illness by easing into training initially at altitude so that the immune system is not placed under excessive stress from both hypoxia and hard training. With acclimatization, the immune suppression is lessened over time, epinephrine release returns to sea-level values, and T-cell function returns to near sea-level values (Mazzeo 2005).

Hypoxia is a stressor and has been shown to negatively affect many markers of immune function. There is modest evidence to conclude that a 3-week camp at moderate altitude would increase the likelihood of illness in a previously healthy and injury-free athlete. The environmental influences of local sanitation and exposure to novel pathogens might be just as important as the hypoxic stress; for instance, a training camp in the Ethiopian highlands might present a greater risk of illness than a similar altitude camp held in the United States. The same increased risk might be associated with long-haul travel, for instance, from Australia to train in the Italian Alps. Further studies are required to provide more information about the propensity to illness during moderate altitude training. Compared with using control-group and treatment-group interventions with small sample sizes, an efficient approach is to keep records over many years of the frequency, severity, and duration of illness associated with altitude camps compared with equivalent sea-level camps. Such data may exist in many countries but have not been data-mined with this specific question in mind.

References

Adams, W.C., Bernauer, E.M., Dill, D.B., and Bomar, J.B., Jr. 1975. Effects of equivalent sea-level and altitude training on VO₂max and running performance. *Journal of Applied Physiology* 39(2):262-266.

Ashenden, M.J., Gore, C.J., Dobson, G.P., and Hahn, A.G. 1999. "Live high, train low" does not change the total

Summary

Effective altitude training requires a foundation of at least several years of training at a high level. We do not recommend this method of training for developing athletes who lack fundamental experiences, such as adequate competition experience, and who can readily gain more than a 1% performance benefit from conventional training at sea level. Used prudently, hypoxic training appears to improve the endurance performance of the most elite athletes at sea level as well as at moderate altitude. The challenge for coaches, scientists, and athletes is to learn how to consistently gain more from an altitude training camp than can commonly be gained from a well-conducted training camp near sea level (Saunders et al. 2010). For the elite athlete, well-planned altitude exposures appear to be worthwhile given the underlying physiological adaptation (Millet et al. 2010; Saunders et al. 2009a), even if the performance benefit is more nebulous, because the magnitude of benefit (~0.5%-1%) is similar to the extent to which an athlete's performance varies from race to race (Pyne et al. 2004). Athlete preparation as well as careful monitoring of training and well-being while at altitude seem crucial to allow the body to adapt and respond to the stress of altitude.

haemoglobin mass of male endurance athletes sleeping at simulated altitude of 3000 m for 23 nights. *European Journal of Applied Physiology* 80(5):479-484.

Bärtsch, P., and Saltin, B. 2008. General introduction to altitude adaptation and mountain sickness. *Scandinavian Journal of Medicine & Science in Sports* 18(Suppl 1):1-10.

Bärtsch, P., Dehnert, C., Friedmann-Bette, B., and Tadibi, V. 2008a. Intermittent hypoxia at rest for improvement of athletic performance. *Scandinavian Journal of Medicine & Science in Sports* 18(Suppl):50-56.

Bärtsch, P., Saltin, B., and Dvorak, J. 2008b. Consensus statement on playing football at different altitudes. *Scandinavian Journal of Medicine & Science in Sports* 18(Suppl 1):96-98.

Beidleman, B.A., Muza, S.R., Fulco, C.S., Cymeman, A., Sawka, M.N. Lewis, S.F., and Skrinar, G.S. 2008. Seven intermittent exposures to altitude improves exercise performance at 4300 m. *Medicine and Science in Sports and Exercise* 40(1):141-148.

Bonetti, D.L., and Hopkins, W.G. 2009. Sea-level exercise performance following adaptation to hypoxia: a meta analysis. *Sports Medicine* 39(2):107-127.

Brothers, M.D., Wilber, R.L., and Byrnes, W.C. 2007. Physical fitness and haematological changes during acclimatization to moderate altitude: A retrospective study. *High Altitude Medicine & Biology* 8(3):213-224.

Buskirk, E.R., Kollias, J., Akers, R.F., Prokop, E.K., and Reategui, E.P. 1967. Maximal performance at altitude and on return from altitude in conditioned runners. *Journal of Applied Physiology* 23(2):259-266.

Clark, S.A., Bourdon, P.C., Schmidt, W., Singh, B., Cable, G., Onus, K.J., Woolford, S.M., Stanef, T., Gore, C.J., and Aughey, R.J. 2007. The effect of acute simulated moderate altitude on power, performance and pacing strategies in well-trained cyclists. *European Journal of Applied Physiology* 102(1):45-55.

Clark, S.A., Quod, M.J., Clark, M.A., Martin, D.T., Saunders, P.U., and Gore, C.J. 2009. Time course of haemoglobin mass during 21 days live high: train low simulated altitude. *European Journal of Applied Physiology* 106(3):399-406.

Daniels, J., and Oldridge, N. 1970. The effects of alternative exposure to altitude and sea level on world class distance runners. *Medicine and Science in Sports* 2:107-112.

Dick, F.W. 1992. Training at altitude in practice. *International Journal of Sports Medicine*. 13:S203-S206.

Dufour, S.P., Ponsot, E., Zoll, J., Doutreleau, S., Lonsdorfer-Wolf, E., Geny, B., Lampert, E., Fluck, M., Hoppeler, H., Billat, V., Mettauer, B., Richard, R., and Lonsdorfer, J. 2006. Exercise training in normobaric hypoxia in endurance runners. 1: Improvement in aerobic performance capacity. *Journal of Applied Physiology* 100(4):1238-1248.

Fulco, C.S., Rock, P.B., and Cymerman, A. 2000. Improving athletic performance: Is altitude residence or altitude training helpful? *Aviation, Space, and Environmental Medicine* 71(2):162-171.

Gore, C.J., Clark, S.A., and Saunders, P.U. 2007. Nonhematological mechanisms of improved sea-level performance after hypoxic exposure. *Medicine and Science in Sports and Exercise* 39(9):1600-1609.

Gore, C.J., Hahn, A.G., Aughey, R.J., Martin, D.T., Ashenden, M.J., Clark, S.A., Garnham, A.P., Roberts, A.D., Slater, G.J., and McKenna, M.J. 2001. Live high:train low increases muscle buffer capacity and submaximal cycling efficiency. *Acta Physiologica Scandinavica* 173:275-286.

Gore, C.J., Hahn, A.G., Rice, A., Bourdon, P., Lawrence, S., Walsh, C., Stanef, T., Barnes, P., Parisotto, R., Martin, D., and Pyne, D. 1998. Altitude training at 2690 m does not increase total haemoglobin mass or sea level VO_{2max} in world champion track cyclists. *Journal of Science and Medicine in Sport* 1(3):156-170.

Gore, C.J., McSharry, P.E., Hewitt, A.J., and Saunders, P.U. 2008. Preparation for football competition at moderate to high altitude. *Scandinavian Journal of Medicine & Science in Sports* 18(Suppl 1):85-95.

Hamlin, M.J., Marshall, H.C., Hellemans, J., and Ainslie, P.N. 2010. Effect of intermittent hypoxia on muscle and cerebral oxygenation during a 20-km time trial in elite athletes: a preliminary report. *Applied Physiology, Nutrition, and Metabolism* 35(4):548-559.

Hopkins, W.G., and Hewson, D.J. 2001. Variability of competitive performance of distance runners. *Medicine and Science in Sports and Exercise* 33(9):1588-1592.

Hoppeler, H., Vogt, M., Weibel, E.R., and Fluck, M. 2003. Response of skeletal muscle mitochondria to hypoxia. *Experimental Physiology* 88(1):109-119.

Jensen, K., Nielsen, T.S., and Fiskastrand, A. 1993. High-altitude training does not increase maximal oxygen uptake or work capacity at sea level in rowers. *Scandinavian Journal of Medicine & Science in Sports* 3(4):256-262.

Kinsman, T.A., Hahn, A.G., Gore, C.J., Wilsmore, B.R., Martin, D.T., and Chow, C.M. 2002. Respiratory events and periodic breathing in cyclists sleeping at 2650 m simulated altitude. *Journal of Applied Physiology* 92(5):2114-2118.

Levine, B.D., and Stray-Gundersen, J. 1992. A practical approach to altitude training: where to live and train for optimal performance enhancement. *International Journal of Sports Medicine* 13(Suppl 1):S209-S212.

Levine, B.D., and Stray-Gundersen, J. 1997. "Living high-training low": effect of moderate-altitude acclimatization with low-altitude training on performance. *Journal of Applied Physiology* 83(1):102-112.

Levine, B.D., and Stray-Gundersen, J. 2006. Dose-response of altitude training: How much altitude is enough? *Advances in Experimental Medicine and Biology* 588(Part 8):233-247.

Levine, B.D., Stray-Gundersen, J., Gore, C.J, and Hopkins, W.G. 2005. Point: Counterpoint: Positive effects of intermittent hypoxia (live high:train low) on exercise are/are not mediated primarily by augmented red cell volume. *Journal of Applied Physiology* 99:2053-2055, discussion 2055-2058.

Mazzeo, R.S. 2005. Altitude, exercise and immune function. *Exercise and Immunology Review* 11:6-16.

Millet GP, Roels B, Schmitt L, Woorons X, and Richalet, J.P. 2010. Combining hypoxic methods for peak performance. *Sports Medicine* 40(1):1-25.

Muza, S.R. 2007. Military applications of hypoxic training for high-altitude operations. *Medicine and Science in Sports and Exercise* 39(9):1625-1631.

Norsborg, N.B., Siebenmann, C., Jacobs, R.A., Rasmussen, P., Diaz, V., Robach, P., and Lundby, C. 2012. Four weeks of normobaric "Live high-train low" does not alter muscular or systemic capacity for maintaining pH and K+ homeostasis during intense exercise. *Journal of Applied Physiology* March 29 [Epub ahead of print].

Peronnet, F., Thibault, G., and Cousineau, D.L. 1991. A theoretical analysis of the effect of altitude on running performance. *Journal of Applied Physiology* 70(1):399-404.

Pyne, D., Trewin, C., and Hopkins, W. 2004. Progression and variability of competitive performance of Olympic swimmers. *Journal of Sports Sciences* 22(7):613-620.

Robertson, E.Y., Saunders, P.U., Pyne, D.B., Gore, C.J., and Anson, J.M. 2010a. Effectiveness of intermittent training in hypoxia combined with live high/train low. *European Journal of Applied Physiology* 110(2):379-387.

Robertson, E.Y., Saunders, P.U., Pyne, D.B., Aughey, R.J., Anson, J.M., and Gore, C.J. 2010b. Reproducibility of performance changes to simulated live high/train low altitude. *Medicine and Science in Sports and Exercise* 42(2): 394-401.

Roels, B., Bentley, D.J., Coste, O., Mercier, J., and Millet G.P. 2007. Effects of intermittent hypoxic training on cycling performance in well-trained athletes. *European Journal of Applied Physiology* 101(3):359-368.

Rusko, H.K. 1996. New aspects of altitude training. *American Journal of Sports Medicine.* 24:S48-S52.

Rusko, H.K., Penttinen, J.T.T., Koistinen, P.O., Vahasoyrinki, P.I., and Leppaluoto, J.O. 1995. A new solution to simulate altitude and stimulate erythropoiesis at sea level in athletes. In: J. Viitasalo and U. Kujala (eds), *The Way to Win.* Helsinki: International Congress on Applied Research in Sports, pp 287-289.

Saunders, P.U., Pyne, D.B., and Gore, C.J. 2009a. Endurance training at altitude. *High Altitude Medicine & Biology* 10:135-148.

Saunders, P.U., Telford, R.D., Pyne, D.B., Gore, C.J., and Hahn, A.G. 2009b. Improved race performance in elite middle-distance runners after cumulative altitude exposure. *International Journal of Sports Physiology and Performance* 4(1):134-138.

Saunders, P.U., Ahlgrim, C., Vallance, B., Green, D.J., Robertson, E.Y., Clark, S.A., Schumacher, Y.O., and Gore, C.J. 2010. An attempt to quantify the placebo effect from a three week simulated altitude training camp in elite race walkers. *International Journal of Sports Physiology and Performance* 5(4):521-534.

Schuler, B., Thomsen, J.J., Gassmann, M., and Lundby, C. 2007. Timing the arrival at 2340 m altitude for aerobic performance. *Scandinavian Journal of Medicine & Science in Sports* 17(5):588-594.

Siebenmann, C., Robach, P., Jacobs, R.A., Nordsborg, N., Diaz, V., Christ, A., Olsen, N.V., Maggiorini, M., and Lundby, C. 2012. "Live high-train low" using normobaric hypoxia: A double-blinded, placebo-controlled study. *Journal of Applied Physiology* 112(1):106-117.

Tadibi, V., Dehnert, C., Menold, E., and Bärtsch, P. 2007. Unchanged anaerobic and aerobic performance after short-term intermittent hypoxia. *Medicine and Science in Sports and Exercise* 39(5):858-864.

Vogt, M., and Hoppeler, H. 2010. Is hypoxia training good for muscles and exercise performance? *Progress in Cardiovascular Diseases* 52(6):525-533.

Wilber, R.L., Holm, P.L., Morris, D.M., Dallam, G.M., Subudhi, A.W., Murray, D.M., and Callan, S.D. 2004. Effect of FIO2 on oxidative stress during interval training at moderate altitude. *Medicine and Science in Sports and Exercise* 36(11):1888-1894.

Wilber, R.L. 2007. Application of altitude/hypoxic training by elite athletes. *Medicine and Science in Sports and Exercise* 39(9):1610-1624.

Heat

Matt B. Brearley and Philo U. Saunders

Within the context of thermal physiology, the human body consists of a core that is maintained within a narrow temperature band and a periphery that varies in temperature depending on environmental influence. Body core temperature stability is sought to provide a constant physiological state by matching heat loss to metabolic heat production. Such thermoregulation requires a sophisticated set of homeostatic mechanisms and reflexes to enable physical activity in diverse environments, particularly for elite endurance athletes. The preoptic anterior hypothalamus houses the thermoregulatory control center that receives afferent impulses from thermoreceptors located in the core and periphery. Afferent input is referenced against a set point temperature that is subject to circadian rhythm with a nadir of early morning and peak in the late afternoon (Kräuchi and Wirz-Justice 1994). If the integrated body temperature is higher than the set point, effector heat loss mechanisms, such as sweating and increased blood flow to the cutaneous circuit, are initiated in an attempt to maintain thermal homeostasis (figure 9.1). Cutaneous blood flow enhances the potential for dry heat exchange, whereas increased sweating allows for heat loss through evaporation.

Physiological Responses to Hot Environmental Conditions

Endurance activities in the heat challenge human thermoregulation. For the purposes of this chapter, the terms *hot* and *humid* are used to describe environments that limit heat dissipation. Although it is acknowledged that the combination of heat and elevated environmental water vapor pressure pose a more severe thermoregulatory challenge than a hot but dry environment, the responses to physical activity and strategies to maximize performance contained within this chapter do not differentiate between environmental conditions. Rather, this chapter provides an overview of the physiological responses to exercise in the heat and provides strategies that can be tailored to meet the needs of the individual athlete.

Cardiovascular Responses

Skin temperature and cutaneous blood flow are higher for a given workload in hot environments (Patterson et al. 1994; Lee et al. 1995) because blood pools in the compliant cutaneous venous plexus at the expense of smooth muscle (Rowell, 1986) and possibly skeletal muscle (Gonzalez-Alonso et al. 1998). These responses displace central blood volume (Johnson and Rowell, 1975) and cause mean arterial pressure and venous return to decrease (Nybo and Nielsen, 2001a). A compensatory increase in heart rate eventuates in an effort to defend cardiac output (Gonzalez-Alonso et al. 1999) and meet demand from the contracting musculature and cutaneous circuit. A higher cardiac frequency for a given workload is therefore observed for physical activity in hot conditions.

High ambient temperatures limit endogenous heat transfer by convection, conduction, and radiation compared with a cool environment. A greater reliance is consequently placed on evaporative heat loss to achieve thermal equilibrium. The higher sweat rates observed in the heat (Galloway and Maughan 1997; Maxwell et al. 1996) contribute to cardiovascular

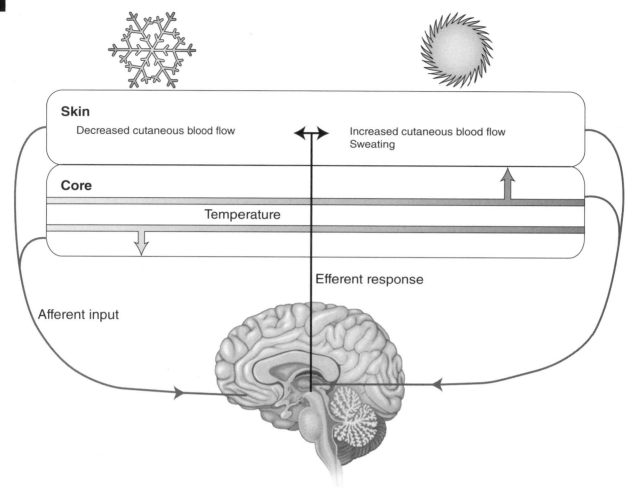

FIGURE 9.1 Schematic of effector responses to increased body temperature.

strain by lowering plasma volume (Singh et al. 1993). The concomitant loss of plasma volume and increase in plasma osmolality alter the body temperature to sweat production relationship, eventually causing sweat rate to plateau or decrease (Fortney et al. 1981; Sawka et al. 1989).

Thermal Responses

The combination of physical activity and restricted dry and evaporative heat exchange in a hot and humid environment elevates peripheral and core tissue temperatures compared with cooler conditions. For example, 25 min of moderate cycling in 40 °C resulted in a core body temperature of 38.9 °C, whereas the same exercise bout in a thermoneutral 20 °C produced approximately 37.9 °C (Parkin et al. 1999). A similar experimental design extended to 40 min of cycling resulted in core body temperatures of 38.7 and 39.7 °C for temperate and hot conditions, respectively (Febbraio et al. 1994). Similar findings are reported for less intense exercise. Jentjens and

colleagues (2002) reported a core body temperature of 39.1 °C following 90 min of 55% $\dot{V}O_2$max cycling in 35 °C compared with a core body temperature of 38.3 °C cycling in 16 °C. Time trial performances permit self-selection of pace (power output) based on physiological and perceptual cues. Such performances also result in higher (Tucker et al. 2004) or similar core body temperature (Tatterson et al. 2000) responses despite lower power outputs and heat production in hot conditions. Peripheral temperatures are also higher during exercise in the heat as demonstrated by the 3 to 5 °C skin temperature discrepancy between hot and thermoneutral environments (Adams et al. 1975; Jentjens et al. 2002).

Higher rates of body heat storage in hot conditions coupled with the maintenance of moderate metabolic rates ensure the onset of hyperthermia, a condition observed to occur at an esophageal temperature of approximately 40 °C during fixed workload cycling for well-trained subjects (Gonzalez-Alonso et al. 1998, 1999; Nybo and Nielsen 2001b). These findings purport high internal temperatures (~40 °C) as an independent cause of exhaustion during prolonged exercise in the heat. Although evidence of a set point temperature for fatigue is persuasive, observations of substantially cooler internal temperatures at exhaustion following fixed workload exercise (Cheung and McLellan 1998; Mitchell et al. 2003; Saboisky et al. 2003) signify that other factors contribute to fatigue in the heat. Furthermore, when athletes are free to select their workload, exercise intensity appears to be regulated prior to the failure of the thermoregulatory system that manifests as heat stroke. The integration of afferent feedback from the body's physiological systems combined with motivation, experience, and expected distance or duration therefore determines pace selection (Noakes 2011; Noakes et al. 2005). The aforementioned observations of fatigue that coincides with core body temperatures of ~40 °C (Gonzalez-Alonso et al. 1998, 1999; Nybo and Nielsen 2001b) precedes the development of heat stroke support such a protective mechanism (Noakes 2011). Hence the physiological responses observed during athletic competition in the heat contribute to the selection of a pace that allows physical activity to be completed within the physiological limits of the body, ultimately serving the preservation of life.

Metabolic Responses

Evidence exists for an alteration in metabolism with physical activity in a hot environment. Some (Dimri et al. 1980; Febbraio et al. 1994; Young et al. 1985) but not all investigators (Maxwell et al. 1996; Snow et al. 1993) have concluded that exercise in a hot environment relies more on anaerobic processes than does exercise in a cool environment based on observations of elevated blood and plasma lactate concentrations during submaximal exercise in the heat. Because blood lactate concentration is the net product of lactate production and removal, blunted lactate clearance may account for the observed differences, rather than a greater reliance on anaerobic processes per se. Exercise-induced heat stress results in redistribution of blood flow to the periphery to permit heat exchange, which causes vasoconstriction of renal and splanchnic vascular beds (Rowell 1986), thereby mobilizing vascular reserves to defend central blood volume such that splanchnic vasoconstriction may reduce the rate of hepatic lactate removal. However, observations of slower rates of adenosine diphosphate rephosphorylation (Willis and Jackman 1994), elevated respiratory exchange ratio, and reduced muscle blood flow in the heat support the concept of anaerobic processes (Febbraio et al. 1994; Gonzalez-Alonso and Calbet 2003), particularly during intense exercise.

Diversion of blood flow to the cutaneous circuit lowers cardiac output (Gonzalez-Alonso et al. 1999; Nadel et al. 1979), thus decreasing maximal aerobic power in the heat (Gonzalez-Alonso and Calbet 2003), whereas hot conditions increase the aerobic energy cost of submaximal physical activity (MacDougall et al. 1974) that might be related to additional cost of sweating, circulatory, and respiratory mechanisms. Exercise in the heat also increases glycogen usage (Febbraio et al. 1994) and accelerates adenosine triphosphate degradation and muscle glycolysis (Kozlowski et al. 1985). Overall, the metabolic response to exercise in the heat is characterized by increased aerobic and anaerobic energy cost during submaximal exercise and reduced maximal aerobic power.

Hematological Responses

Expansion of plasma volume has been commonly reported after heat adaptation and leads to an increase in stroke volume (Nielsen et al. 1993; Patterson et al. 2004). The increase in plasma volume associated with training in the heat may be a result of an influx of protein from the cutaneous interstitial space to the vascular compartments or perhaps from sodium and water retention that results in an isosmotic plasma volume expansion (Nielsen et al. 1993). Patterson and colleagues (2004) demonstrated that exercise in the heat induced a volume expansion that was present across the entire extracellular compartment and not restricted to just plasma volume. The expansion of extracellular fluid and plasma volume is a factor of water movement between the body fluid compartments and is driven by the balance of osmotic and hydrostatic forces acting across the capillary beds. This can result from reduced capillary hydrostatic pressure, elevated interstitial hydrostatic pressure, or an elevated intravascular osmotic pressure due to an increased plasma protein content (Patterson et al. 2004).

Performance in the Heat

In addressing physical performance in hot conditions, most investigators have used continuous exercise modes to show that sustained exercise performance is compromised in the heat. Time to exhaustion at a constant workload is lower when conducted in hot conditions across a variety of research designs (Febbraio et al. 1994; Galloway and Maughan 1997; Parkin et al. 1999). Figure 9.2 demonstrates the relationship between time to exhaustion and environmental temperature.

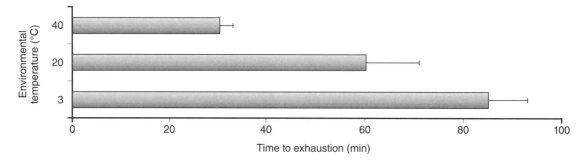

FIGURE 9.2 Time to exhaustion at 70% $\dot{V}O_2$max in 3, 20, and 40 °C (<50% relative humidity). Data given as mean with error bars representing standard deviation.

Adapted from J.M. Parkin et al., 1999, "Effect of ambient temperature on human skeletal muscle metabolism during fatiguing submaximal exercise," *Journal of Applied Physiology* 86(3):902-908. Used with permission.

Time trial performance is also detrimentally influenced by high environmental temperatures. Thirty minutes of high-intensity cycling in the heat (32 °C) reduces self-selected power output by 6.5% compared with cycling in temperate (23 °C) conditions (Tatterson et al. 2000). Similar findings have been reported for 20 km time trial performances in 15 and 35 °C respectively, where power output significantly decreased toward the end of the cycling bout (Tucker et al. 2004). Hence, continuous exercise performance in the heat consistently demonstrates an adverse influence for high environmental temperatures.

Although the magnitudes of effect comparisons between continuous and intermittent tests are problematic given the diversity of experimental designs, prolonged intermittent exercise also demonstrates an adverse effect for hot conditions. This point has been supported by studies showing shorter time to exhaustion during (Kraning and Gonzalez 1991; Maxwell et al. 1996; Morris et al. 1998) and following intermittent activity (Finn et al. 2001). The detrimental effect of heat on prolonged intermittent exercise should be delineated from the performance benefits of brief supramaximal exercise in the heat (Ball et al. 1999; Falk et al. 1998). Although high muscle temperatures are conducive to increasing power output (Sargeant 1987), prolonging the duration of activity induces rapid heat storage that becomes detrimental for endurance.

Hot conditions exert a profound effect on physiological, perceptual, and performance responses to continuous and intermittent exercise. Higher core and peripheral temperatures, cardiac frequencies, sweat rates, cutaneous blood flow, and thermal sensation eventuate during physical activity in the heat. While all physiological responses may contribute to the curtailment of endurance performance, core body temperature is recognized as the key physiological variable for athletes competing in hot conditions (Sawka et al. 1992). Core body temperature measure-

ment and analysis are therefore priorities in efforts to maximize performance.

Thermal Assessment of Athletes

Responses to physical activity in the heat are highly individual, the product of heat generation and dissipation. Heat production is dependent upon muscle mass and work rate, whereas heat dissipation relies on sweat rate, skin blood flow, morphology, and the environment. Although work rate, sweat rate, and skin blood flow can be routinely measured, the impact of athlete morphology is difficult to model yet extremely important, with in excess of half an individual's body tissue located within 2.5 cm of the heat dissipation site, the skin (DuBois 1951).

Advancements in technology for measuring core body temperature have permitted the monitoring of athletes in the field via an ingestible thermometer (Mittal et al. 1991). Assessing an individual's core body temperature response to training sessions or, preferably, competitions allows sport scientists to analyze the net result of the interactions among the environment, heat production, and heat loss, negating the need for complex modeling. Ideally, the thermal assessment takes place in a hot environment with enough lead time to allow athletes to adjust to and become accustomed to heat management strategies prior to their key competition. Such analysis can predict which athletes require individualized heat management strategies, a process deemed critical to the performance of team sports in hot conditions given that it may not be practical to provide cooling for the entire team.

The data need to be analyzed in conjunction with additional physiological observations such as cardiac frequency and sweat rate; perceptual data including rate of perceived exertion, thermal sensation, and thermal discomfort; and performance data inclusive of movement (global positioning system) and coaches' feedback.

Checklist for Thermal Assessment Using Core Body Temperature Pills

1. Administer core body temperature pill with meal a minimum of 4 h prior to assessment.
2. Immediately prior to the assessment session, measure core body temperature before and following consumption of cool fluid. A rapid reduction in temperature indicates that the pill is located within the stomach and is susceptible to underestimating core body temperature following fluid consumption.
3. Assess core body temperature throughout the session or during breaks. Collect as many data points as possible.
4. Measure core body temperature at end of session.

Such analysis will identify athletes who consistently achieve high core body temperatures (>39.5 °C) in conjunction with the aforementioned factors. Figure 9.3 summarizes the core body temperature response of three international field hockey athletes to a test match played in hot and humid conditions. The analysis and the coaches' feedback indicated that athlete 3 had a high workload that manifested in the core body temperature response displayed. This athlete was prioritized as the most likely to benefit from heat management strategies that include acclimatization or acclimation, hydration, and cooling.

Acclimatization

Heat acclimatization is the process of adaptation to sustained thermal stress in the natural environment. Although acute heat storage causes increased cutaneous blood flow and sweating, repeated bouts of thermal stress lead to physiological and perceptual adaptation thereby improving tolerance of hot and humid conditions. Heat acclimation refers to the same process achieved through simulated conditions and is addressed in a subsequent section.

Heat acclimatization can be achieved passively or actively. Substantial heat storage in the absence of physical activity is rare, requiring extremely hot conditions. Heat acclimatization can be supplemented by sustained passive heat exposure; however, this is not recommended because of limited advantages observed during physical activity, this method's poor time efficiency, high levels of perceptual strain, and impact on recovery from training. Active heat acclimatization uses physical training as an endogenous heat source in warm to hot conditions to limit heat dissipation and promote heat storage.

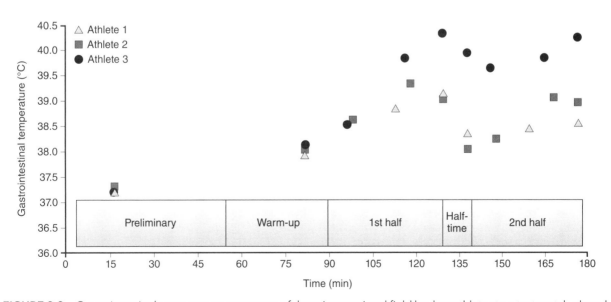

FIGURE 9.3 Gastrointestinal temperature responses of three international field hockey athletes to a test match played in hot and humid conditions.

For acclimatization to be beneficial in maintaining thermal homeostasis in hot conditions, metabolic heat production or heat dissipation pathways must be modulated. Body heat loss pathways at rest and during physical activity are augmented by acclimatization, such that a lower resting metabolic rate and improved capacity for evaporative heat exchange are observed. Heat acclimatization promotes a series of adaptations as summarized in the sidebar, including lower resting core body temperature (Kampmann et al. 2008), an earlier onset of sweating (Shido et al. 1999), and an increased capacity to sweat (Armstrong and Kenney 1993).

Augmentation of cutaneous blood flow following heat acclimatization allows for greater dry heat exchange, and although cutaneous blood flow has been observed to increase following acclimatization (Armstrong and Kenney 1993), this may be principally a result of training (Aoyagi et al. 1994). Collectively, these adaptations provide a greater window for heat storage prior to attainment of an abnormally high core body temperature.

The classic view of heat acclimatization dictates that individuals require up to 14 days of daily heat storage to confer full acclimatization (Armstrong and Maresh 1991). Such observations have come from a small field of research examining trained but not elite athletes. Adaptations to chronic endurance training mimic those of heat acclimatization; therefore, elite endurance athletes are considered to be partially heat acclimatized (Pandolf et al. 1977), especially those who train and compete in temperate climates. Elite athletes who predominantly train and compete in warm to hot environments can be classified as heat acclimatized. For these athletes, the primary goal of undertaking sessions in hot conditions in the lead-up to an event or competition in the heat is to practice pacing and heat management strategies rather than to promote physiological adaptation per se.

The time required to maximize performance in hot conditions requires individual consideration of the following factors:

- Training status
- History of training and competing in hot conditions
- Climate of key event or competition
- Familiarity with pacing strategies in the heat

Elite endurance athletes are unlikely to require 2 weeks to induce heat acclimatization. They may, however, require a similar amount of time to practice pacing and to become accustomed to pacing and heat management strategies. Training should be specific and should induce elevations of core body temperature beyond 38.5 °C with a concomitant thermal sensation of warm to hot, some thermal discomfort, and a high sweat rate. Self-pacing requires a delicate balance between enduring a level of physiological stress to promote adaptation and preventing the development of excessive heat storage, particularly during the early stages of a heat acclimatization program. Therefore, the initial sessions upon visiting a hot climate should account for the additional effect of the new climate.

If the heat acclimatization or pacing period occurs immediately prior to key competition, planning is required to ensure that tapering doesn't limit opportunities to adapt and sustain efforts similar to those of the competition. This potentially results in the earlier commencement of heat acclimatization and maintenance throughout the latter stages of the taper. Again, this will be an individual consideration.

Other factors to take into account for heat acclimatization planning are seasonal influence, age, and gender. It's intuitive to expect that daily activity during the summer months would contribute to heat acclimatization. Behaviors such as avoiding the hottest part of the day and using air conditioning limit physiological strain and the associated stimulus for adaptation. Unless specific sessions that induce substantial heat storage are undertaken during the summer months, minimal benefit is anticipated (Bain and Jay 2011). Heat acclimatization appears to

Adaptations Observed During Heat Acclimatization

Initial

- Expanded plasma volume
- Increased cutaneous blood flow
- Decreased heart rate
- Decreased perception of effort

Subsequent

- Decreased resting core body temperature
- Decreased exercise core body temperature
- Increased sweat secretion

be equally effective for individuals irrespective of age or gender when matched for fitness and morphological variables (Avellini et al. 1980).

Acclimation

As previously mentioned, the negative effects of exercising in heat and humidity can be attenuated to a large degree by a period of heat acclimatization, a process that involves training under similar conditions to allow the body to adapt and to perform better. Heat acclimation is another method of preparing for competition in the heat, making use of heat rooms or climate chambers during winter months or when in a location where conditions are much cooler than the competition. Acclimation can be used for longer periods, because there is constant respite from the heat and athletes are not consequently exposed to oppressive conditions. It is recommended that heat acclimation be supplemented with a shorter session (1-2 weeks) of acclimatization in a climate similar to that of competition to allow the body full adaptation for competing in the heat. The type of training done should be specific to the sport as long as the particular ergometer (treadmill, cycle, rowing) can be sourced. Nonspecific exercise can be used (e.g., cycling for a runner) if this is the only option and will still drive adaptations from exercising in the heat. Initially, low-intensity exercise should be performed, with duration and intensity added during the acclimation process.

Work done at the Australian Institute of Sport (AIS) found good performance benefits for race walkers who were preparing for competitions in the heat at the Osaka World Championships 2007 and Beijing Olympic Games 2008. The athletes undertook 7 to 8 weeks of 60 to 90 min of exercise in a heat tent once per week prior to departure and a further 10 days of acclimatization in the race venue prior to competition. This strategy provided good preparation for the athletes by exposing them to the heat once per week but allowed adequate respite and recovery between these sessions. In this particular heat acclimation, the 60 to 90 min exercise sessions were conducted at easy intensity at the start of the 8 weeks and then increased in intensity, with the last 2 weeks including intervals at race pace. This strategy can be used in preparation for summer competition; during the spring, athletes undertake 1 or 2 sessions per week outdoors in the middle of the day, or in a heated gym or a hot room on a treadmill or stationary bike to start the acclimatization process.

Hydration

In the absence of fluid replenishment during prolonged physical activity in the heat, the production of sweat results in dehydration. Sweating can reduce plasma volume (Maw et al. 1998) and diminish extracellular fluid reserves during high-intensity exercise of prolonged duration. This decreases the volume of blood that can be distributed between the active muscles and the cutaneous circuit. Dehydration-induced decreases in blood volume have the primary effects of decreasing venous return, decreasing cardiac output (Nadel et al. 1980), and increasing plasma osmolality (Kamijo et al. 2005). Such responses have led sport scientists to use hydration strategies to maximize performance in the heat. In this regard, hydration may have been overemphasized as a quick-fix for athletes competing in hot conditions. In terms of heat management strategies, hydration appears to be more familiar to athletes than the alternative or complementary strategies of heat acclimatization and cooling. Furthermore, this awareness exists despite the lack of consensus in the published literature regarding optimal athlete hydration practices. Athletes and coaches should understand that hydration is one factor that can contribute to performance in the heat and that consideration of all strategies is a prerequisite for best results. So how do athletes ensure that their hydration practices are helping them maximize their performance in the heat?

One viewpoint is to ensure that fluid losses are limited to less than 2% of body mass to prevent performance decrement (Cheuvront et al. 2003). To satisfy this criterion, athletes need to assess sweat rates and dehydration levels across a variety of sessions and climatic conditions. Hence, the relatively simple procedures of monitoring athlete fluid balance are now common in elite sport. Dehydration can be assessed by monitoring pre- and postcompetition seminude body mass on calibrated scales. Postcompetition body mass is measured following the removal of sweat from the skin surface and hair by toweling down, and dehydration is estimated by the following equation:

Dehydration (% body mass) = (Precompetition Body Mass – Postcompetition Body Mass) / (Postcompetition Body Mass × 100).

In addition to body mass, fluid consumption, urine or fecal output, and analysis time can be monitored to permit estimation of sweat rate (see following equation). Athletes can be allocated individual drink bottles from which they exclusively consume fluids. Fluid consumption is calculated by measuring bottle mass prior to and following the analysis period. Bottles can be weighed empty and again following refilling when multiple bottles were required. Urine and fecal output can be accounted for by measuring body mass prior to and following toilet breaks without the need to remove training attire. Respiratory and metabolic fluid losses are generally not accounted for in field settings, and it is assumed that 1 kg is equivalent to 1 L of sweat.

$$\text{Sweat Rate } (L \cdot h^{-1}) = (\text{Precompetition Body Mass} - \text{Postcompetition Body Mass} + \text{Fluid Consumption} - \text{Urine and Fecal Output}) / \text{Analysis Time (hours)}.$$

To complement sweat rate and dehydration, hydration status is routinely assessed through a variety of urinary indices (Kavouras 2002). The accuracy of urine hydration status diminishes during periods of rapid fluid turnover (Kovacs et al. 1999). Because athletes typically hydrate prior to competition and training, which generally requires the consumption of large volumes of fluid in hot climates, the validity of urine hydration indices could be jeopardized. Despite this limitation, urine specific gravity is a common measurement that can be compared with published standards (table 9.1) to permit interpretation of results and provide feedback to athletes and coaches.

Intense exercise in the heat can lead to high sweat rates (Brearley and Watkins 2007). Factors including gastrointestinal distress and limited access to fluids can reduce opportunities to match such sweat rates with fluid consumption. Hence, athletes generally develop a degree of dehydration during training or competition in the heat. Approximately 2% body mass is thought to be a criterion for dehydration-induced performance decrement (Below et al. 1995; Walsh et al. 1994), reflected in the consensus statement of the American College of Sports Medicine Roundtable on Hydration and Physical Activity (Casa et al. 2005). Through the estimation of sweat rates and dehydration, athletes can develop a schedule of fluid consumption to maintain their fluid balance within the 2% body mass window.

An alternative recommendation is for athletes to consume fluids ad libitum, thereby drinking to the dictates of thirst (Noakes 2007). Thirst sensation maintains serum osmolality and sodium concentration, increasing the drive to drink and reducing the intensity of exercise (Sawka and Noakes 2007). Hence, avoiding a reduction in power output can be achieved by avoiding thirst.

Despite the obvious differences between strategies, the common feature is that athletes require fluid during prolonged activity in the heat. Whether to consume fluid to nearly match sweat losses or to drink ad libitum is an individual consideration. Ad libitum consumption is an appropriate starting point for athletes. For short- to moderate-duration events, ad libitum fluid consumption may limit dehydration to less than 2% body mass because drinking to one's pleasure accounts for 50% to 70% of sweat losses during exercise (Cheuvront and Haymes 2001). As discussed for heat acclimatization and pacing, testing the fluid consumption strategy is required to optimize strategies for individual athletes. Fluid volume is only one drinking variable: beverage temperature, availability to fluids, scheduled breaks, and hydration beliefs are other factors to consider.

Acclimatization and fluid consumption are two strategies that potentially augment body heat loss in the heat. An alternative strategy is cooling, a method that demonstrates potential for increasing the ability to store heat.

Cooling

Precooling entails a range of methods that decrease skin and possibly core body temperature by the individual or combined application of ice, cold air, and cold or temperate water prior to competition. Specific precooling methods used to date include ice jackets, water immersion, crushed ice ingestion, ice towels, forearm immersion, fans, and cold air.

Ice Jackets

The cooling mode most widely used in sport science entails compressing ice against the skin in jackets or vests. Prior to the 1996 Atlanta Olympics, AIS scientists developed ice jackets that slightly blunted the core body temperature increase when worn during warm-up periods. Others have cooled athletes during warm-up periods to achieve cooler core and peripheral temperatures prior to performance. Arngrimsson and colleagues (2004) applied the AIS-designed ice jacket during a 38 min warm-up in tropical conditions to lower core body temperature by 0.2 °C. Using a cooling jacket during moderate cycling in hot conditions resulted in a lower rectal temperature (0.2 °C) following 45 min (Hasegawa et al. 2005). The limited cooling power of ice jackets probably explains the failure to lower core body temperature with a 5 min application before and during scheduled breaks of a simulated hockey protocol (Duffield et al. 2003). However, ice jackets have the potential to

TABLE 9.1 Urine Specific Gravity and Equivalent Level of Hydration

Condition	Urine specific gravity value
Well hydrated	<1.010
Minimal dehydration	1.010-1.020
Significant dehydration	1.021-1.030
Serious dehydration	>1.030

Adapted, by permission, from D.J. Casa et al., 2000, "National Athletic Trainers' Association position statement: Fluid replacement for athletes," *Journal of Athletic Training* 35(2): 212-224.

maintain a lower core body temperature achieved through other cooling methods (Quod et al. 2008).

The primary advantage of using garments containing ice for cooling is their practicality; however, this advantage is diminished for repeated bouts of use, because the garment requires recooling. Overall, commercially available ice jackets demonstrate a limited potential to curb heat storage and are only recommended for short warm-up periods. Combining the use of ice jackets with additional cooling modalities is required to achieve and maintain substantial decreases in core body temperature, because ice jackets are ineffective for precooling when used in isolation.

Water Immersion

There are two primary options for water immersion precooling: a relatively short immersion (<10 min) at cold water temperatures (<15 °C) or a longer immersion protocol of 20 to 30 min in temperate water (25-28 °C). The limited data available for short-duration immersions show that 5 min of exposure to 14 °C water is not effective for resting subjects (Peiffer et al. 2010a), which is not surprising given that short bouts of cooling are most effective for individuals following heat storage, as high internal temperatures limit the effect of local skin cooling on vasoconstriction, allowing greater contact time for blood with the cooled periphery (Casa et al. 2007). The same protocol can lower core body temperature by approximately 0.5 °C for warm or hot subjects (Peiffer et al. 2009a); however, there is minimal core body temperature difference when immersion in 14 °C is sustained for 5, 10, or 20 min (Peiffer et al 2009b) and minimal differences for 5 °C versus 14 °C cooling for 12 min (Clements et al. 2002). Although short-duration cold water protocols are effective in rapidly reducing core body temperature during hyperthermia, practicality issues arise when operating in tropical field conditions. The availability of the required volumes of ice to initially lower and subsequently maintain water temperature limits the utility of this method. For example, lowering the temperature of a 600 L water bath from 25 °C to 14 °C would require approximately 83 kg of ice (table 9.2).

In the absence of an appropriate mechanical cooling apparatus, immersion protocols that use temperate water are most likely to be of use in the field. Temperate water immersion is a popular means of cooling prior to competition, as athletes seek to lower their core body temperature without inducing shivering. To achieve this, athletes undergo approximately 30 min of 25 to 28 °C immersion, which can lower core body temperature by approximately 0.5 °C (Brearley and Finn 2006; Marino et al. 1998). Gradually lowering water temperature limits "cold

TABLE 9.2 Volume of Ice Required to Lower Water Temperature of a 600 L Bath From 25 °C to Desired Temperature

Desired water temperature	20 °C	14 °C	8 °C
Ice requirement, kg	37.5	82.5	127.5

shock" and allows transmission of body heat to the water over an extended period. Despite the logistical benefits of using temperate water as a cooling modality following heat storage, it has been the topic of less research than other cooling methods.

Water immersion shouldn't be limited to precooling. Immersion during scheduled breaks of competition can be achieved, even for sports that have short half-time periods (10 min), such as field hockey (figure 9.4).

Crushed Ice Ingestion

Consumption of cold fluid can modestly decrease core body temperature prior to performance and improve submaximal time to exhaustion in the heat (Lee et al. 2008), demonstrating some potential as an alternative to water immersion. Preliminary research suggests that the ingestion of crushed ice may be both effective and practical. Commonly referred to as a *slurpee* or *slushie*, crushed ice was identified as a possible cooling method for athletes prior to the 2004 Athens Olympics (Brearley and Finn 2003), because the mechanical cooling properties of ice are far greater than an equivalent volume of cold fluid. The Northern Territory Institute of Sport (NTIS) physiologists subsequently experimented with ice ingestion for athletes in the field, who provided positive feedback in terms of thermal comfort and performance. Preliminary research supports the feedback from field trials. For example, consuming 7.5 mL · kg^{-1} of crushed ice (−1 °C) lowered core body temperature by approximately 0.7 °C and improved submaximal running performance, whereas an equivalent volume of cold fluids (4 °C) reduced core body temperature by approximately 0.3 °C (Siegel et al. 2010). Ihsan and colleagues (2010) demonstrated that crushed ice ingestion lowers core body temperature prior to exercise and translates to improved time to complete a given volume of cycling compared with temperate fluid ingestion. The combination of a large bolus of crushed ice (14 mL · kg^{-1}) with the application of iced towels also improved physiological and performance responses to a simulated time trial based on characteristics of the 2008 Beijing Olympic time trial course (Ross et al. 2011).

Although crushed ice has been the subject of less research than have alternative modes of cooling, its

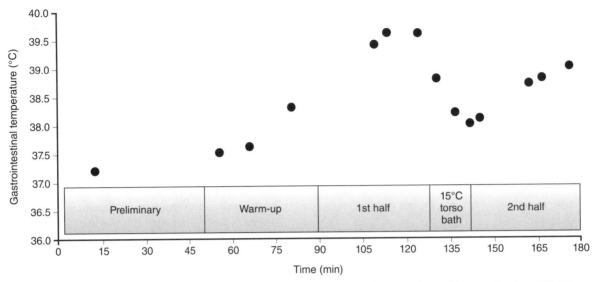

FIGURE 9.4 Gastrointestinal temperature response of an international field hockey athlete to 8 min of 15 °C torso immersion at half-time during a test match played in hot and humid conditions.

practicality and preliminary physiological and performance benefits warrant strong consideration for cooling in athletic settings. The potential for cooling at half-time and during breaks in competition ought to be explored in a research setting. Given the tropical field experiences of the NTIS, research is likely to support the use of crushed ice as a viable alternative to water immersion prior to and during endurance events in the heat (Siegel et al. 2012). This method requires refinement for each individual in terms of timing of ingestion and required volume. Making a slurpee or a slushie can involve elaborate, commercially available machines or simply cubed ice, a powered sports drink, and a blender.

Ice Towels

Ice towels have not been stringently researched but are nonetheless recognized as a treatment for hyperthermia (Roberts, 2007), best suited to field settings where core body temperature cooling rates for hyperthermic runners approximate half the rate observed for cold water immersion (Armstrong et al. 1996). In terms of precooling, ice towels have been tested as part of broader precooling strategies for athletes (Duffield et al. 2009; Myler et al. 1989; Ross et al. 2011). Based on the limited data available, alternative cooling techniques possess greater potential to achieve a substantial reduction of core body temperature, leading sport scientists to use ice towels in conjunction with other precooling methods (Duffield et al. 2009; Ross et al. 2011). Having limited cooling power, ice towels appear best suited to scheduled breaks in competition, and their use is frequently observed in elite team sports played in hot conditions.

Forearm and Hand Immersion

Another practical cooling method is immersion of the forearm and hand or of the hand alone in cold water, typically set at 10 °C. Forearm immersion relies on lowering core body temperature through convective heat transfer between the blood and forearm tissue (Ducharme and Tikuisis 1991) and can be effective for repeated work and cooling bouts. Giesbrecht and colleagues (2007) used three bouts of working for 20 min followed by 20 min of 10 °C forearm immersion, which lowered core body temperature by 0.6 °C more than control and 0.3 °C more than 20 °C forearm immersion. Hand immersion at 10 °C has also provided small physiological benefits when used for only 10 min following 50 min of work (Khomenok et al. 2008). Forearm immersion has been observed in some athletic settings despite the small physiological benefits observed to date. This method is much simpler to implement by using custom-made chairs with forearm cooling reservoirs and is useful in events where more powerful cooling modalities are not feasible. Using hand cooling may influence manual dexterity (Cheung et al. 2003), so testing this method in the training environment is recommended prior to implementation for competition.

Fans

Evaporative cooling is a simple method that can be effective and is often bypassed in favor of more technically demanding cooling modes. Evaporative cooling needs to be maximized as part of any heat management strategy. Movement of air should be the initial consideration for athlete management in the heat, with additional cooling strategies used to supplement the convection achieved by fans.

Cold Air

Cold air cooling is primarily reliant on convective heat loss for reductions in body temperature given the lack of mechanical cooling power exerted by air. Application of cold air to the skin causes vasoconstriction of the cutaneous circuit (Makinen et al. 2000), which limits the blood supply available to be cooled, negating convective heat transfer. This factor appears responsible for the inability of cold air (5-10 °C) cooling protocols to lower deep tissue temperature during precooling of moderately trained athletes (Lee and Haymes 1995; Oksa et al. 1993; Olschewski and Bruck 1988). However, 30 min of air cooling (5 °C) for lean, highly trained athletes can decrease core body temperature by approximately 0.5 °C (Kruk et al. 1990), suggesting that the lack of mechanical cooling power can be partially diminished by low subcutaneous fat levels. Overall, cold air precooling is unlikely to be an option for elite athletes because access to commercial portable "cold rooms" may be limited in hot conditions. Rooms that are cooled simply by standard air-conditioning are warmer (minimum temperature ~16 °C) than cold rooms and thus are unlikely to provide a substantial benefit during precooling; however, their use during scheduled breaks in competition is an option worth pursuing.

The different physiological responses between air precooling and the provision of air cooling during physical activity to alter body heat storage appear related to the temperature gradient between the air and skin and the evaporative potential of sweat. Maintenance of skin temperature above approximately 32 °C will minimize cutaneous vasoconstriction and maximize blood volume available to be cooled at the skin's surface (Veicsteinas et al. 1982). At higher metabolic rates that induce sweating, the provision of conditioned air augments sweat efficiency (Shapiro et al. 1982), thereby limiting body heat storage (Vallerand et al. 1991). Air conditioning also provides perceptual relief from the heat and is a preferred option for scheduled breaks in play, particularly for team sports in which many athletes potentially require cooling and other methods are deemed impractical.

Other Factors

Spending time between training sessions in cool conditions is considered important for acclimatizing athletes who are unused to the hotter conditions and assists in the adaptation process by allowing optimal recovery following training. Should the coach and athlete detect symptoms of unusual tiredness or problems associated with insufficient recovery between quality sessions or extended-duration ses-sions, the athlete should train during the coolest part of the day before resuming more stressful training in expected race conditions. This becomes increasingly important the closer it gets to the final taper period. Access to air-conditioned living quarters is essential between training to obtain respite from the heat and facilitate recovery. However, it is advisable to assist acclimatization by spending a certain amount of time in the natural environment. It is a good idea, perhaps for psychological as well as physical reasons, to train, at least on some occasions, at the same time of day as the competition. Of course, coaches and athletes must be very careful to avoid overexposure to any harsh conditions.

Athletes should avoid direct solar radiation where possible by using head protection as well as appropriate clothing and sunscreen. Clothing provides a barrier to heat dissipation during exercise and can interfere with evaporation of sweat and impair the cooling effect of sweating (Wendt et al. 2007). However, clothing provides protection from the radiant heat from the sun. Therefore, when exercising in the heat, athletes should wear minimal clothing if there is not a large degree of radiant heat from the sun (early morning, evening, or overcast) but should wear loose-fitting, light-colored clothing as well as a hat and sunglasses if exercising in the middle of the day to protect against the radiant heat from the sun. Sunscreen is recommended in summer to avoid sunburn.

Carbohydrate requirements for exercise are increased in the heat because of a shift in substrate utilization toward carbohydrate oxidation (Febbraio et al. 2001). The daily diet should focus on replacing glycogen stores after exercise. Competition strategies should include activities to enhance carbohydrate availability, such as building up glycogen stores in preparation for endurance events; pre-event carbohydrate intake; and intake of glucose or electrolytes in events lasting longer than 60 min. This can be done using water and carbohydrate gel preparations or using sports drinks. It has been reported that intake of carbohydrates prior to and during prolonged exercise in the heat enhances exercise performance (Burke et al. 2005).

Use of Heat and Humidity as a Training Intervention

Lower core body temperatures resulting from training in the heat as well as an increase in plasma volume resulting from both acute and chronic bouts of exercise in the heat may attenuate the magnitude of the thermoregulatory response (increased ventilation, circulation, and sweating) and reduce the increased energy requirements associated with heat stress (Svedenhag 2000). It follows that whole blood viscosity would be reduced from training in the heat, and a decreased viscosity has been shown to have positive effects on endurance performance (Telford et al. 1994). Adaptations from training in warm to hot conditions may also allow athletes to train at any given speed with a lower heart rate and core body temperature, both of which are associated with improved exercise economy (Thomas et al. 1995). These findings support the premise that training in moderate heat may improve economy and performance at normal temperatures. Training in the heat can be used as an additional stress, similar to altitude training, because the training load is higher at any given speed when training in the heat. A recent study demonstrated that 10 days of 90 min submaximal exercise training in a hot environment (40 °C and 30% humidity) improved cycling time trial performance in a cool environment (13 °C and 30% humidity) compared with a matched control group who did the same training in the cool environment (Lorenzo et al. 2010). The authors of this study suggested the increased performance was due to an ability to exercise at a higher power for reduced relative intensity (improved economy), an improved lactate threshold, and an improved cardiac performance, all a result of using heat as training intervention.

Core Body Temperature Measurement Site

No single internal temperature site represents core body temperature (Sawka and Wenger 1988); however, multisite internal temperature measurement is difficult in controlled settings and not possible during athletic training and competition. Given the proximity of the esophagus to the aorta, esophageal temperature is considered to represent the temperature of the blood perfusing the heart (Shiraki et al. 1986). Rectal temperature transition is more reliant on conduction of heat and therefore reacts more slowly to variations in temperature because of its deep position and lower blood flow (Mairiaux et al. 1983), whereas the gastrointestinal tract is not as well insulated as the rectum and receives higher blood flow (Mortensen et al. 1998). Direct comparisons between core body temperature measurement sites validated the use of gastrointestinal temperature as a core body temperature index, as it corresponds to increased heat storage in an intermediate manner to esophageal and rectal temperature (O'Brien et al. 1998). The valid measurement of core body temperature from the gastrointestinal tract has provided sport scientists access to data that were previously collected from the tympanic membrane. The cost-

effectiveness, ease of measurement, and relatively noninvasive nature of tympanic temperature measurement still promote the tympanum as a core body temperature measurement site. Although agreement between tympanic and other core body temperatures has been reported (Amoateng-Adjepong et al. 1999; Christensen and Boysen 2002), the data from athletes competing in the heat do not support such a claim. The NTIS has vast experience measuring athlete core body temperature. Several trials have analyzed tympanic temperature in the field referenced against gastrointestinal temperature. One such study examined the physiological responses of cricketers during competition in the tropics, where tympanic temperature averaged approximately 1.3 to 1.6 °C lower than gastrointestinal temperature, confirming that large discrepancies occur between tympanic and deep tissue temperature (Armstrong et al. 1994). It is reasonable to expect greater variation for tympanic temperature and gastrointestinal temperature in the field compared with the laboratory, because fluctuating environmental conditions would influence tympanic temperature (Fraden and Lackey 1991).

Improved results are likely through the use of tympanic probes (Sato et al. 1996) and insulation of the ear (Muir et al. 2001), but these are not practical in sporting settings given the requirement for exact positioning, irritability of placement proximal to tympanic membrane, and a decreased ability to detect aural cues. Additionally, the use of tympanic temperature is limited by the ability to straighten the auditory canal via the "ear tug" maneuver and permit access to the tympanic membrane. Therefore, it is strongly recommended that sport scientists refrain from interpreting tympanic temperature data of athletes in the field. If core body temperature is to be measured in field settings, gastrointestinal temperature is recommended due to its validity and practicality.

Heat-Associated Illness

The combination of physical activity and hot environments has produced observations of heat syncope, heat exhaustion, and general heat illness with collapse and a loss of consciousness a common feature. As Noakes (2008) describes, postexercise postural hypotension is the primary cause of heat illness, not heat storage per se. Although the hypotension is multifactoral, environmental conditions are not the primary cause. The term *heat illness* should be used to describe the excessive retention of body heat during exercise via abnormal heat production or failure to dissipate an adequate amount of body heat with a resultant severe elevation of core body temperature beyond 41 °C.

Normally, human behavior will modulate exercise intensity (and therefore metabolic heat production) in response to efferent feedback, including that from the central and peripheral thermoreceptors. The development of core body temperatures in excess of 41 °C is relatively rare, generally being observed in athletes treated by medical staff following mass participation events such as fun runs and marathons (Richards et al. 1979). Observations of athletes achieving core body temperatures of 38.5 to 40.5 °C in hot conditions are common, as these measurements can be considered operating temperature for athletes in such conditions. A more common symptom observed for those requiring medical attention is hypotension (Noakes 2008), most frequently observed immediately following exercise. Cessation of exercise removes the lower-limb muscle pump, instantly displacing central blood volume and rapidly reducing right atrial pressure (Noakes 2003). This in turn evokes the Barcroft-Edholm reflex, which results in paradoxical vasodilation and resultant hypotension. Sport scientists should be aware of this "finish line syncope" and understand that attaining a severely high core body temperature is extremely unlikely for healthy individuals who pace their effort according to physiological cues and have sufficient time in hot conditions to practice such pacing.

References

Adams, W.C., Fox, R.H., Fry, A.J., and MacDonald I.C. 1975. Thermoregulation during marathon running in cool, moderate, and hot environments. *Journal of Applied Physiology* 38(6):1030-1037.

Amoateng-Adjepong, Y., Del Mundo, J., and Manthous C.A. 1999. Accuracy of an infrared tympanic thermometer. *Chest* 115(4):1002-1005.

Aoyagi, Y., McLellan, T.M., and Shephard, R.J. 1994. Effects of training and acclimation on heat tolerance in exercising men wearing protective clothing. *European Journal of Applied Physiology and Occupational Physiology* 68(3):234-245.

Armstrong, L.E., and Maresh, C.M. 1991. The induction and decay of heat acclimatization in trained athletes. *Sports Medicine* 12(5):302-312.

Armstrong, C.G., and Kenney, W.L. 1993. Effects of age and acclimation on responses to passive heat exposure. *Journal of Applied Physiology* 75(5):2162-2167.

Armstrong, L.E., Epstein, Y., Greenleaf, J.E., Haymes, E.M., Hubbard, R.W., Roberts, W.O., and Thompson, P.D. 1996. American College of Sports Medicine position stand. Heat and cold illnesses during distance running. *Medicine and Science in Sports and Exercise* 28(12):i-x.

Armstrong, L.E., Maresh, C.M., Crago, A.E., Adams, R., and Roberts, W.O. 1994. Interpretation of aural temperatures during exercise, hyperthermia, and cooling therapy. *Medicine, Exercise, Nutrition and Health* 3:9-16.

Arngrimsson, S.A., Petitt, D.S., Borrani, F., Skinner, K.A., and Cureton, K.J. 2004. Hyperthermia and maximal oxygen uptake in men and women. *European Journal of Applied Physiology* 92(4-5):524-532.

Avellini, B.A., Kamon, E., and Krajewski, J.T. 1980. Physiological responses of physically fit men and women to acclimation to humid heat. *Journal of Applied Physiology* 49(2):254-261.

Bain, A.R., and Jay, O. 2011. Does summer in a humid continental climate elicit an acclimatization of human thermoregulatory responses? *European Journal of Applied Physiology* 111(6):1197-1205.

Ball, D., Burrows, C., and Sargeant, A.J. 1999. Human power output during repeated sprint cycle exercise: The influence of thermal stress. *European Journal of Applied Physiology and Occupational Physiology* 79(4):360-366.

Below, P.R., Mora-Rodríguez, R., González-Alonso, J., and Coyle, E.F. 1995. Fluid and carbohydrate ingestion independently improve performance during 1 h of intense exercise. *Medicine and Science in Sports and Exercise* 27(2):200-210.

Brearley, M.B., and Finn, J.P. 2003. Pre-cooling for Performance in the Tropics. Sportscience Perspectives: Training and Performance. www.sportsci.org/jour/03/mbb.htm.

Brearley, M., and Finn, J.P. 2006. Physiological, perceptual and performance responses to intermittent high intensity activity in a tropical environment following pre-cooling. In proceedings of the 2nd Australian Association of Exercise and Sport Science Conference and the 4th Sport Dieticians Australia Update, Sydney, September–October, p 48.

Brearley, M., and Watkins, M. 2007. Practical guidelines to minimize dehydration in the tropics. *Strength and Conditioning Coach* 15(4):7-11.

Burke, L.M., Wood, C., Pyne, D.B., Telford, R.D., and Saunders, P.U. 2005. Effect of carbohydrate intake on half-marathon performance of well-trained runners. *International Journal of Sport Nutrition and Metabolism* 15(6):573-589.

Casa, D.J., Armstrong, L.E., Hillman, S.K., Montain, S.J., Reiff, R.V., Rich, B.S., Roberts, W.O., and Stone, J.A. 2000. National Athletic Trainers Association position statement: Fluid replacement for athletes. *Journal of Athletic Training* 35(2):212-224.

Casa, D.J., Clarkson, P.M., and Roberts, W.O. 2005. American College of Sports Medicine Roundtable on Hydration and Physical Activity: Consensus statements. *Current Sports Medicine Reports* 4(3):115-127.

Casa, D.J., McDermott, B.P., Lee, E.C., Yeargin, S.W., Armstrong, L.E., and Maresh, C.M. 2007. Cold water immersion: The gold standard for exertional heatstroke treatment. *Exercise and Sport Sciences Reviews* 35(3):141-149.

Cheung, S.S., and McLellan, T.M. 1998. Heat acclimation, aerobic fitness, and hydration effects on tolerance during uncompensable heat stress. *Journal of Applied Physiology* 84(5):1731-1739.

Cheung, S.S., Montie, D.L., White, M.D., and Behm, D. 2003. Changes in manual dexterity following short-term hand and forearm immersion in 10 degrees C water. *Aviation, Space, and Environmental Medicine* 74(9):990-993.

Cheuvront, S.N., and Haymes, E.M. 2001. Thermoregulation and marathon running: Biological and environmental influences. *Sports Medicine* 31(10):743-762.

Cheuvront, S.N., Kolka, M.A., Cadarette, B.S., Montain, S.J., and Sawka, M.N. 2003. Efficacy of intermittent, regional microclimate cooling. *Journal of Applied Physiology* 94(5):1841-1848.

Christensen, H., and Boysen, G. 2002. Acceptable agreement between tympanic and rectal temperature in acute stroke patients. *International Journal of Clinical Practice* 56(2):82-84.

Clements, J.M., Casa, D.J., Knight, J., McClung, J.M., Blake, A.S., Meenen, P.M., Gilmer, A.M., and Caldwell, K.A. 2002. Ice-water immersion and cold-water immersion provide similar cooling rates in runners with exercise-induced hyperthermia. *Journal of Athletic Training* 37(2):146-150.

Dimri, G.P., Malhotra, M.S., Sen Gupta, J., Kumar, T.S., and Arora, B.S. 1980. Alterations in aerobic-anaerobic proportions of metabolism during work in heat. *European Journal of Applied Physiology* 45(1):43-50.

DuBois, E.F. 1951. The many different temperatures of the human body and its parts. *Western Journal of Surgery* 59(9):476-490.

Ducharme, M.B., and Tikuisis, P. 1991. In vivo thermal conductivity of the human forearm tissues. *Journal of Applied Physiology* 70(6):2682-2690.

Duffield, R., Dawson, B., Bishop, D., Fitzsimons, M., and Lawrence, S. 2003. Effect of wearing an ice cooling jacket on repeat sprint performance in warm/humid conditions. *British Journal of Sports Medicine* 37(2):164-169.

Duffield, R., Steinbacher, G., and Fairchild, T.J. 2009. The use of mixed-method, part-body pre-cooling procedures for team-sport athletes training in the heat. *Journal of Strength and Conditioning Research* 23(9):2524-2532.

Falk, B., Radom-Isaac, S., Hoffmann, J.R., Wang, Y., Yarom, Y., Magazanik, A., and Weinstein, Y. 1998. The effect of heat exposure on performance of and recovery from high-intensity, intermittent exercise. *International Journal of Sports Medicine* 19(1):1-6.

Febbraio, M.A. 2001. Alterations in energy metabolism during exercise and heat stress. *Sports Medicine* 31(1):47-59.

Febbraio, M.A., Snow, R.J., Stathis, C.G., Hargreaves, M., and Carey, M.F. 1994. Effect of heat stress on muscle energy metabolism during exercise. *Journal of Applied Physiology* 77(6):2827-2831.

Finn, J.P., Ebert, T.R., Withers, R.T., Carey, M.F., Mackay, M., Phillips, J.W., and Febbraio, M.A. 2001. Effect of creatine supplementation on metabolism and performance in humans during intermittent sprint cycling. *European Journal of Applied Physiology* 84(3): 238-243.

Fortney, S.M., Nadel, E.R., Wenger, C.B., and Bove, J.R. 1981. Effect of blood volume on sweating rate and body fluids in exercising humans. *Journal of Applied Physiology* 51(6):1594-1600.

Fraden, J., and Lackey, R.P. 1991. Estimation of body sites temperatures from tympanic measurements. *Clinical Pediatrics* 30(4):65-70.

Galloway, S.D., and Maughan, R.J. 1997. Effects of ambient temperature on the capacity to perform prolonged cycle exercise in man. *Medicine and Science in Sports and Exercise* 29(9):1240-1249.

Giesbrecht, G.G., Jamieson, C., and Cahill, F. 2007. Cooling hyperthermic firefighters by immersing forearms and hands in 10 degrees C and 20 degrees C water. *Aviation, Space, and Environmental Medicine* 78(6):561-567.

Gonzalez-Alonso, J., and Calbet, J.A. 2003. Reductions in systemic and skeletal muscle blood flow and oxygen delivery limit maximal aerobic capacity in humans. *Circulation* 107(6):824-830.

Gonzalez-Alonso, J., Calbet, J.A., and Nielsen, B. 1998. Muscle blood flow is reduced with dehydration during prolonged exercise in humans. *Journal of Physiology* 513(3):895-905.

Gonzalez-Alonso, J., Teller, C., Andersen, S.L., Jensen, F.B., Hyldig, T., and Nielsen, B. 1999. Influence of body temperature on the development of fatigue during prolonged exercise in the heat. *Journal of Applied Physiology* 86(3):1032-1039.

Hasegawa, H., Takatori, T., Komura, T., and Yamasaki, M. 2005. Wearing a cooling jacket during exercise reduces thermal strain and improves endurance exercise performance in a warm environment. *Journal of Strength and Conditioning Research* 19(1):122-128.

Ihsan, M., Landers, G., Brearley, M., and Peeling, P. 2010. Beneficial effects of ice ingestion as a precooling strategy on 40-km cycling time-trial performance. *International Journal of Sports Physiology and Performance* 5(2):140-151.

Jentjens, R.L., Wagenmakers, A.J., and Jeukendrup, A.E. 2002. Heat stress increases muscle glycogen use but reduces the oxidation of ingested carbohydrates during exercise. *Journal of Applied Physiology* 92(4):1562-1572.

Johnson, J.M., and Rowell, L.B. 1975. Forearm skin and muscle vascular responses to prolonged leg exercise in man. *Journal of Applied Physiology* 39(6):920-924.

Kamijo, Y., Okumoto, T., Takeno, Y., Okazaki, K., Inaki, M., Masuki, S., and Nose, H. 2005. Transient cutaneous vasodilatation and hypotension after drinking in dehydrated and exercising men. *Journal of Physiology* 56 (Pt 2):689-698.

Kampmann, B., Bröde, P., Schütte, M., and Griefahn, B. 2008. Lowering of resting core temperature during acclimation is influenced by exercise stimulus. *European Journal of Applied Physiology* 104(2):321-327.

Kavouras, S.A. 2002. Assessing hydration status. *Current Opinion in Clinical Nutrition and Metabolic Care* 5(5):519-524.

Khomenok, G.A., Hadid, A., Preiss-Bloom, O., Yanovich, R., Erlich, T., Ron-Tal, O., Peled, A., Epstein, Y., and Moran, D.S. 2008. Hand immersion in cold water alleviating physiological strain and increasing tolerance to uncompensable heat stress. *European Journal of Applied Physiology* 104(2):303-309.

Kovacs, E.M., Senden, J.M., and Brouns, F. 1999. Urine color, osmolality and specific electrical conductance are not accurate measures of hydration status during postexercise rehydration. *Journal of Sports Medicine and Physical Fitness* 39(1):47-53.

Kozlowski, S., Brzezinska, Z., Kruk, B., Kaciuba-Uscilko, H., Greenleaf, J.E., and Nazar, K. 1985. Exercise hyperthermia as a factor limiting physical performance: Temperature effect on muscle metabolism. *Journal of Applied Physiology* 59(3):766-773.

Kräuchi, K., and Wirz-Justice, A. 1994. Circadian rhythm of heat production, heart rate, and skin and core temperature under unmasking conditions in men. *American Journal of Physiology* 267(3 Pt 2):R819-R829.

Kraning, K.K., and Gonzalez, R.R. 1991. Physiological consequences of intermittent exercise during compensable and uncompensable heat stress. *Journal of Applied Physiology* 71(6):2138-2145.

Kruk, B., Pekkarinen, H., Harri, M., Manninen, K., and Hanninen, O. 1990. Thermoregulatory responses to exercise at low ambient temperature performed after precooling or preheating procedures. *European Journal of Applied Physiology and Occupational Physiology* 59(6):416-420.

Lee, C.F., Katsuura, T., Harada, H., and Kikuchi, Y. (1995. Different behaviour of forearm blood flow during intermittent isometric handgrip in a thermo-neutral and a hot environment. *Applied Human Sciences* 14(3):111-117.

Lee, D.T., and Haymes, E.M. 1995. Exercise duration and thermoregulatory responses after whole body precooling. *Journal of Applied Physiology* 79(6):1971-1976.

Lee, J.K., Shirreffs, S.M., and Maughan, R.J. 2008. Cold drink ingestion improves exercise endurance capacity in the heat. *Medicine and Science in Sports and Exercise* 40(9):1637-1644.

Lorenzo, S., Halliwill, J.R., Sawka, M.N., and Minson, C.T. 2010. Heat acclimation improves exercise performance. *Journal of Applied Physiology* 109(4):1140-1147.

MacDougall, J.D., Reddan, W.G., Layton, C.R., and Dempsey, J.A. 1974. Effects of metabolic hyperthermia on performance during heavy prolonged exercise. *Journal of Applied Physiology* 36(5):538-544.

Mairiaux, P., Sagot, J.C., and Candas, V. 1983. Oral temperature as an index of core temperature during heat transients. *European Journal of Applied Physiology and Occupational Physiology* 50(3):331-341.

Makinen, T., Gavhed, D., Holmer, I., and Rintamaki, H. 2000. Thermal responses to cold wind of thermoneutral and cooled subjects. *European Journal of Applied Physiology* 81(5):397-402.

Marino, F., Sockler, J.M., and Fry, J.M. 1998. Thermoregulatory, metabolic and sympathoadrenal responses to repeated brief exposure to cold. *Scandinavian Journal of Clinical Laboratory Investigation* 58(7):537-545.

Maw, G.J., Mackenzie, I.L., and Taylor, N.A. 1998. Human body-fluid distribution during exercise in hot, temperate

and cool environments. *Acta Physiologica Scandinavica* 163(3):297-304.

Maxwell, N.S., Aitchison, T.C., and Nimmo, M.A. 1996. The effect of climatic heat stress on intermittent supramaximal running performance in humans. *Experimental Physiology* 81(5):833-845.

Mitchell, J.B., McFarlin, B.K., and Dugas, J.P. 2003. The effect of pre-exercise cooling on high intensity running performance in the heat. *International Journal of Sports Medicine* 24(2):118-124.

Mittal, B.B., Sathiaseelan, V., Rademaker, A.W., Pierce, M.C., Johnson, P.M., and Brand, W.N. 1991. Evaluation of an ingestible telemetric temperature sensor for deep hyperthermia applications. *International Journal of Radiation, Oncology, Biology and Physiology* 21(5):1353-1361.

Morris, J.G., Nevill, M.E., Lakomy, H.K.A., Nicholas, C., and Williams, C. 1998. Effect of a hot environment on performance of prolonged, intermittent, high-intensity shuttle running. *Journal of Sports Sciences* 16:677-686.

Mortensen, F.V., Bayat, M., Friis-Andersen, H., Jespersen, S.M., Hoy, K., Christensen, K.O., Lindblad, B.E., and Hansen, E.S. 1998. Effect of moderate exercise on blood flow in the gastrointestinal tract in trained conscious miniature swine. *Digestive Surgery* 15(6):665-658.

Muir, I.H., Bishop, P.A., Lomax, R.G., and Green, J.M. 2001. Prediction of rectal temperature from ear canal temperature. *Ergonomics* 44(11):962-72.

Myler, G.R., Hahn, A.G., and Tumilty, D.M. 1989. The effect of preliminary skin cooling on performance of rowers in hot conditions. *Excel* 6:17-21.

Nadel, E.R., Cafarelli, E., Roberts, M.F., and Wenger, C.B. 1979. Circulatory regulation during exercise in different ambient temperatures. *Journal of Applied Physiology* 46(3):430-437.

Nadel, E.R., Fortney, S.M., and Wenger, C.B. 1980. Effect of hydration state of circulatory and thermal regulations. *Journal of Applied Physiology* 49(4):715-721.

Nielsen, B., Hales, J.R., Strange, S., Christensen, N.J., Warberg, J., and Saltin, B. 1993. Human circulatory and thermoregulatory adaptations with heat acclimation and exercise in a hot, dry environment. *Journal of Physiology* 460:467-485.

Noakes, T.D. 2003. The forgotten Barcroft/Edholm reflex: Potential role in exercise associated collapse. *British Journal of Sports Medicine* 37(3):277-278.

Noakes, T.D. 2007. Hydration in the marathon: Using thirst to gauge safe fluid replacement. *Sports Medicine* 37(4-5):463-466.

Noakes, T.D. 2008. A modern classification of the exercise-related heat illnesses. *Journal of Science and Medicine in Sport* 11(1):33-39.

Noakes, T.D. 2011. Time to move beyond a brainless exercise physiology: the evidence for complex regulation of human exercise performance. *Applied Physiology, Nutrition and Metabolism* 36(1):23-35.

Noakes, T.D., St Clair Gibson, A., and Lambert, E.V. 2005. From catastrophe to complexity: a novel model of integrative central neural regulation of effort and fatigue during exercise in humans: summary and conclusions. *British Journal of Sports Medicine* 39(2):120-124.

Nybo, L., and Nielsen, B. 2001a. Hyperthermia and central fatigue during prolonged exercise in humans. *Journal of Applied Physiology* 91(3):1055-1060.

Nybo, L., and Nielsen, B. 2001b. Perceived exertion is associated with an altered brain activity during exercise with progressive hyperthermia. *Journal of Applied Physiology* 91(5):2017-2023.

O'Brien, C., Hoyt, R.W., Buller, M.J., Castellani, J.W., and Young, A.J. 1998. Telemetry pill measurement of core temperature in humans during active heating and cooling. *Medicine and Science in Sports and Exercise* 30(3):468-472.

Oksa, J., Rintamaki, H., and Makinen, T. 1993. Physical characteristics and decrement in muscular performance after whole body cooling. *Annals of Physiological Anthropology* 12(6):335-359.

Olschewski, H., and Bruck, K. 1988. Thermoregulatory, cardiovascular, and muscular factors related to exercise after precooling. *Journal of Applied Physiology* 64(2):803-811.

Pandolf, K.B., Burse, R.L., and Goldman, R.F. 1977. Role of physical fitness in heat acclimatization, decay and reinduction. *Ergonomics* 20(4):399-408.

Parkin, J.M., Carey, M.F., Zhao, S., and Febbraio, M.A. 1999. Effect of ambient temperature on human skeletal muscle metabolism during fatiguing submaximal exercise. *Journal of Applied Physiology* 86(3):902-908.

Patterson, M.J., Warlters, D., and Taylor, N.A. 1994. Attenuation of the cutaneous blood flow response during combined exercise and heat stress. *European Journal of Applied Physiology and Occupational Physiology* 69(4):367-369.

Patterson, M.J., Stocks, J.M., and Taylor, N.A. 2004. Sustained and generalized extracellular fluid expansion following heat acclimation. *Journal of Physiology* 559:327-334.

Peiffer, J.J., Abbiss, C.R., Nosaka, K., Peake, J.M., and Laursen, P.B. 2009a. Effect of cold water immersion after exercise in the heat on muscle function, body temperatures, and vessel diameter. *Journal of Science and Medicine in Sport* 12(1):91-96.

Peiffer, J.J., Abbiss, C.R., Watson, G., Nosaka, K., and Laursen, P.B. 2009b. Effect of cold-water immersion duration on body temperature and muscle function. *Journal of Sports Sciences* 27(10):987-993.

Peiffer, J.J., Abbiss, C.R., Watson, G., Nosaka, K., and Laursen, P.B. 2010a. Effect of cold water immersion on repeated 1-km cycling performance in the heat. *Journal of Science and Medicine in Sport* 13(1):112-116.

Quod, M.J., Martin, D.T., Laursen, P.B., Gardner, A.S., Halson, S.L., Marino, F.E., Tate, M.P., Mainwaring, D.E., Gore, C.J., and Hahn, A.G. 2008. Practical precooling: Effect on cycling time trial performance in warm conditions. *Journal of Sports Sciences* 26(14):1477-1487.

Richards, D., Richards, R., Schofield, P.J., Ross, V., and Sutton, J.R. 1979. Management of heat exhaustion in Sydney's the Sun City-to-Surf run runners. *Medical Journal of Australia* 2(9):457-461.

Roberts, W.O. 2007. Exertional heat stroke in the marathon. *Sports Medicine* 37(4-5):440-443.

Ross, M.L., Garvican, L.A., Jeacocke, N.A., Laursen, P.B., Abbiss, C.R., Martin, D.T., and Burke, L.M. 2011. Novel precooling strategy enhances time trial cycling in the heat. *Medicine and Science in Sports and Exercise* 43(1):123-133.

Rowell, L.B. 1986. *Human Circulation: Regulation During Physical Stress*. New York: Oxford University Press.

Saboisky, J., Marino, F.E., Kay, D., and Cannon, J. 2003. Exercise heat stress does not reduce central activation to non-exercised human skeletal muscle. *Experimental Physiology* 88(6):783-790.

Sargeant, A.J. 1987. Effect of muscle temperature on leg extension force and short-term power output in humans. *European Journal of Applied Physiology and Occupational Physiology* 56(6):693-698.

Sato, K.T., Kane, N.L., Soos, G., Gisolfi, C.V., Kondo, N., and Sato, K. 1996. Re-examination of tympanic membrane temperature as a core temperature. *Journal of Applied Physiology* 80(4):1233-1239.

Sawka, M.N., Gonzalez, R.R., Young, A.J., Dennis, R.C., Valeri, C.R., and Pandolf, K.B. 1989. Control of thermoregulatory sweating during exercise in the heat. *American Journal of Applied Physiology* 257(2 Pt 2):R311-R316.

Sawka, M.N., Young, A.J., Latzka, W.A., Neufer, P.D., Quigley, M.D., and Pandolf, K.B. 1992. Human tolerance to heat strain during exercise: influence of hydration. *Journal of Applied Physiology* 73(1):368-375.

Sawka, M.N., and Noakes, T.D. 2007. Does dehydration impair exercise performance? *Medicine and Science in Sports and Exercise* 39(8):1209-1217.

Sawka, M.N., and Wenger, C.B. 1988. Physiological responses to acute exercise-heat stress. In: K.B. Pandolf, M.N. Sawaka, and R.R. Gonzalez (eds), *Human Performance Physiology and Environmental Medicine at Terrestrial Extremes*. Indianapolis, IN: Benchmark Press, Inc. pp 97-151.

Shapiro, Y., Pandolf, K.B., Sawka, M.N., Toner, M.M., Winsmann, F.R., and Goldman, R.F. 1982. Auxiliary cooling: Comparison of air-cooled vs. water-cooled vests in hot-dry and hot-wet environments. *Aviation, Space and Environmental Medicine* 53(8):785-789.

Shido, O., Sugimoto, N., Tanabe, M., and Sakurada, S. 1999. Core temperature and sweating onset in humans acclimated to heat given at a fixed daily time. *American Journal of Physiology* 276(4):R1095-R1101.

Shiraki, K., Konda, N., and Sagawa, S. 1986. Esophageal and tympanic temperature responses to core blood temperature changes during hyperthermia. *Journal of Applied Physiology* 61(1):98-102.

Siegel, R., Maté, J., Brearley, M.B., Watson, G., Nosaka, K., and Laursen, P.B. 2010. Ice slurry ingestion increases core temperature capacity and running time in the heat. *Medicine and Science in Sports and Exercise* 42(4):717-725.

Siegel, R., Maté, J., Watson, G., Nosaka, K., and Laursen, P.B. 2012. Pre-cooling with ice slurry ingestion leads to similar run times to exhaustion in the heat as cold water immersion. *Journal of Sports Sciences* 30(2):155-165.

Singh, M.V., Rawal, S.B., Pichan, G., and Tyagi, A.K. 1993. Changes in body fluid compartments during hypohydration and rehydration in heat-acclimated tropical subjects. *Aviation, Space and Environmental Medicine* 64(4):295-299.

Snow, R.J., Febbraio, M.A., Carey, M.F., and Hargreaves, M. 1993. Heat stress increases ammonia accumulation during exercise in humans. *Experimental Physiology* 78(6):847-850.

Svedenhag, J. 2000. Running economy. In: Bangsbo J, Larsen H, eds. *Running and Science*. Copenhagen: Munksgaard, pp 85-105.

Tatterson, A.J., Hahn, A.G., Martin, D.T., and Febbraio, M.A. 2000. Effects of heat stress on physiological responses and exercise performance in elite cyclists. *Journal of Science and Medicine in Sport* 3(2):186-193.

Telford, R.D., Kovacic, J.C., Skinner, S.L., Hobbs, J.B., Hahn, A.G., and Cunningham, R.B. 1994. Resting whole blood viscosity of elite rowers is related to performance. *European Journal of Applied Physiology and Occupational Physiology* 68(6):470-476.

Thomas, D.Q., Fernhall, B., Blanpied, P., and Stillwell, P. 1995. Changes in running economy and mechanics during a 5 km run. *Journal of Strength and Conditioning Research* 9(3):170-175.

Tucker, R., Rauch, L., Harley, Y.X., and Noakes, T.D. 2004. Impaired exercise performance in the heat is associated with an anticipatory reduction in skeletal muscle recruitment. *Pflugers Archives* 448(4):422-430.

Vallerand, A.L., Michas, R.D., Frim, J., and Ackles, K.N. 1991. Heat balance of subjects wearing protective clothing with a liquid- or air-cooled vest. *Aviation, Space and Environmental Medicine* 62(5):383-391.

Veicsteinas, A., Ferretti, G., and Rennie, D.W. 1982. Superficial shell insulation in resting and exercising men in cold water. *Journal of Applied Physiology* 52(6):1557-1564.

Walsh, R.M., Noakes, T.D., Hawley, J.A., and Dennis, S.C. 1994. Impaired high-intensity cycling performance time at low levels of dehydration. *International Journal of Sports Medicine* 15(7):392-398.

Wendt, D., van Loon, L.J., and Lichtenbelt, W.D. 2007. Thermoregulation during exercise in the heat: Strategies for maintaining health and performance. *Sports Medicine* 37(8):669-682.

Willis, W.T., and Jackman, M.R. 1994. Mitochondrial function during heavy exercise. *Medicine and Science in Sports and Exercise* 26(11):1347-1353.

Young, A.J., Sawka, M.N., Levine, L., Cadarette, B.S., and Pandolf, K.B. 1985. Skeletal muscle metabolism during exercise is influenced by heat acclimation. *Journal of Applied Physiology* 59(6):1929-1935.

Physiological Recovery

Joanna Vaile and Shona Halson

Optimal recovery provides numerous performance benefits during repetitive high-level training and competition. Adequate recovery has been shown to restore physiological and psychological processes, so that fatigue is minimized and the athlete can compete or train again at an appropriate level. Recovery from training and competition is complex and involves numerous factors, and it typically depends on the nature of the exercise performed and other external stressors to which the athlete may be exposed.

Research examining different recovery interventions and their effect on fatigue, muscle injury, recovery, and performance is increasing; however, there are still many unanswered questions. What remains clear is with the increasing professionalism in elite sport and increases in training loads, identifying optimal recovery strategies is an important aspect of elite performance.

This chapter examines the scientific basis of various popular recovery strategies. Some recovery interventions, although lacking scientific evidence, are still popular among elite athletes. The lack of scientific evidence is most likely the result of the relative infancy of this area of research, and thus these interventions should not be discounted as appropriate strategies for elite athletes. For practical guidelines and further information, see Vaile and colleagues (2010).

Compression Clothing

Historically, compression clothing has been used in a medical setting to treat various circulatory conditions; graduated compression generates a pressure gradient, with the level of compression decreasing from the distal to proximal portion of the limb. Graduated compression clothing has been reported to increase venous blood flow and venous return and reduce swelling in patients with venous insufficiencies (Agu et al. 2004) as well as in healthy subjects (Bringard et al. 2006). As a result of these findings and various manufacturers' claims that compression clothing can enhance recovery through improved peripheral circulation and venous return and via increased clearance of blood lactate and markers of muscle damage such as creatine kinase (CK), the use of commercially available compression garments for exercise and recovery has increased exponentially in recent years.

Duffield and Portus (2007) compared the effect of three different full-body compression garments and a control condition (no compression) on repeat sprint and throwing performance in club-level cricket players. The compression garments were worn during and for 24 h after exercise. Significantly lower blood CK concentrations and ratings of muscle soreness were observed 24 h after exercise in athletes who wore the compression garments; however, no differences were observed between the three different compression brands. The authors proposed that CK was reduced because the compression clothing limited the inflammatory response to acute muscle damage and reduced swelling (Duffield and Portus 2007).

Similarly, wearing a lower-limb compression garment for 12 h following an elite-level rugby match resulted in an 84.4% recovery in interstitial CK concentration 84 h after the match (compared with 39.0% recovery following passive recovery) (Gill et al. 2006). Furthermore, Ali and colleagues (2007) found that wearing a knee-length compression sock during a 10 km run significantly reduced perceived muscle soreness and frequency of reported soreness, especially in the lower extremities, 24 h after exercise. These results suggest that the potential ergogenic value of compression clothing may be the ability to assist recovery to improve or maintain subsequent exercise performance (Ali et al. 2007). However, none of these studies report any data on changes in exercise performance on subsequent days.

In an attempt to substantiate this claim, Duffield and colleagues (2008) investigated the effect of lower-limb compression clothing on recovery from a simulated team game exercise task, performed on 2 consecutive days. The compression clothing was worn during and for 15 h following the simulated team game protocol. No differences were observed in blood lactate or CK concentrations between the compression and control groups postexercise on day 1. Despite a reduction in perceived muscle soreness postexercise on day 1 in the compression group compared with the control group, there was no difference in exercise performance (speed and power) between the groups on day 2. The authors concluded that there appears to be a perceptual benefit from wearing compression clothing postexercise, reducing feelings of muscle soreness and potentially improving readiness to engage in training (Duffield et al. 2008). This in itself may be of benefit to athletes. Similarly, wearing full-length, lower-limb compression clothing for 12 h following resistance exercise was found to have no effect on subsequent exercise performance (speed, agility, power, and strength), flexibility, plasma CK and myoglobin concentrations, swelling, or perceived muscle soreness (French et al. 2008). Montgomery and colleagues (2008) investigated the effect of wearing full-length lower-limb compression clothing on recovery during and after a 3-day basketball tournament. Participants wore the compression clothing for an 18 h recovery period following each game. No effect was observed on exercise performance (agility, speed, and power), flexibility, swelling, or perceived muscle soreness (Montgomery et al. 2008).

A lower recovery blood lactate concentration has been observed when compression stockings were worn during and after exercise when compared with clothing worn during exercise only or not at all (Berry and McMurray 1987). However, as no plasma volume shifts were observed, it was suggested that the lower values were due to lactate being retained in the muscular bed rather than an enhanced rate of lactate removal (Berry and McMurray 1987). Kraemer and colleagues (2001) found the use of compression sleeves worn following muscle damage to reduce CK concentration 72 h postexercise, prevent the degree of loss of elbow extension, decrease subjects' perception of soreness, reduce swelling, and promote recovery of force production (Kraemer et al. 2001).

In summary, the limited amount of research on the use of compression specifically for recovery purposes has shown that wearing compression clothing during and after exercise may reduce postexercise blood lactate and CK concentrations as well as perceived muscle soreness. However, there is little evidence to suggest that wearing compression clothing is any more effective than other recovery interventions at maintaining or improving subsequent exercise performance. When interpreting these findings, we must consider the limitations of the research. None of the studies referred to here measured the compressive forces (in mmHg) applied to the underlying musculature by the compression clothing. Therefore, there are no data to suggest that the compression clothing worn by the participants exerted graduated compression on the musculature. For a compression garment to provide the ideal graduated compression, it needs to be custom fit to the contours of an individual's limbs. Furthermore, in the studies that observed a reduction in perceived muscle soreness postexercise, the placebo effect cannot be discounted, because a placebo condition was not included (i.e., wearing a garment that provides no compression or a lower level of compression).

This said, compression clothing is widely used for recovery by many elite athletes and teams around the world, and anecdotal evidence suggests compression clothing to be highly beneficial. Quality research needs to be completed before more specific guidelines can be provided.

Hydrotherapy

Despite the widespread integration of hydrotherapy into an athlete's postexercise recovery regime, information regarding these interventions is largely anecdotal. Although a number of physiological responses to water immersion are well researched, the underlying mechanisms related to postexercise recovery are poorly understood. The human body responds to water immersion with changes in cardiac response, peripheral resistance, and blood flow (Wilcock et al. 2006). In addition, both hydrostatic pressure and temperature of the immersion medium may influence the success of different hydrotherapy interventions (Wilcock et al. 2006).

Immersion of the body in water can result in an inward and upward displacement of fluid from the extremities to the central cavity due to hydrostatic pressure. As identified by Wilcock and colleagues (2006), the resulting displacement of fluid may increase the translocation of substrates from the muscle. Therefore, postexercise edema may be lessened and muscle function maintained. Another physiological response to water immersion is an increase in stroke volume, which has been shown to increase cardiac output.

Although the effects of hydrostatic pressure exerted on the body during water immersion may be beneficial, the temperature of water that the body is exposed to is also thought to influence the success of such recovery interventions. The main physiological effect of immersion in cold water is a reduction in blood flow due to peripheral vasoconstriction (Meeusen and Lievens 1986). In contrast, immersion in hot water increases blood flow via peripheral vasodilation (Bonde-Petersen et al. 1992; Knight and Londeree 1980).

Cold Water Immersion

Cryotherapy (meaning "cold treatment," often in the form of an ice pack) is the most commonly used treatment for acute soft tissue injuries, given its ability to reduce the inflammatory response and alleviate spasm and pain (Eston and Peters 1999; Meeusen and Lievens 1986; Merrick et al. 1999). Multiple physiological responses to various cooling methods have been observed, including a reduction in heart rate and cardiac output and an increase in arterial blood pressure and peripheral resistance (Sramek et al. 2000; Wilcock et al. 2006). Additional responses include a decrease in core and tissue temperature (Enwemeka et al. 2002; Lee et al. 1997; Merrick et al. 2003; Yanagisawa et al. 2007), acute inflammation (Yanagisawa et al. 2004), and pain (Bailey et al. 2007; Washington et al. 2000) and an improved maintenance of performance (Burke et al. 2000; Yeargin et al. 2006). Merrick and colleagues (1999) suggested that cryotherapy is an effective method for decreasing inflammation, blood flow, muscle spasm, and pain as well as skin, muscular, and intra-articular temperatures.

The use of cryotherapy for the treatment of muscle damage and exercise-induced fatigue has been investigated with varying findings. Eston and Peters (1999) investigated the effects of cold water immersion (of the exercised limb in 15 °C for 15 min) on the symptoms of exercise-induced muscle damage following strenuous eccentric exercise. The muscle-damaging exercise consisted of eight sets of five maximal isokinetic contractions (eccentric and concentric) of the elbow flexors of the dominant arm (0.58 rad · s^{-1} and 60 s rest between sets). The measures used to assess the presence of exercise-induced muscle damage included plasma CK concentration, isometric strength of the elbow flexors, relaxed arm angle, local muscle tenderness, and upper arm circumference. Eston and Peters (1999) found CK

activity to be lower and relaxed elbow angle to be greater for the cold water immersion group on days 2 and 3 following the eccentric exercise, concluding that the use of cold water immersion may reduce the degree to which the muscle and connective tissue unit becomes shortened after strenuous eccentric exercise (Eston and Peters 1999).

Bailey and colleagues (2007) investigated the influence of cold water immersion on indices of muscle damage. Cold water immersion (or passive recovery) was administered immediately following a 90 min intermittent shuttle run protocol; rating of perceived exertion (RPE), muscular performance (maximal voluntary contraction of the knee extensors and flexors), and blood variables were monitored prior to exercise, during recovery, and following recovery for 7 days. The authors concluded that cold water immersion was a highly beneficial recovery intervention, finding a reduction in muscle soreness, a reduced decrement of performance, and a reduction in serum myoglobin concentration 1 h following exercise (Bailey et al. 2007). However, further values across the 7-day collection period were not cited, and CK response was unchanged regardless of intervention. Lane and Wenger (2004) investigated the effects of active recovery, massage, and cold water immersion on repeated bouts of intermittent cycling separated by 24 h. Cold water immersion had a greater effect compared with passive recovery, active recovery, and massage on recovery between exercise bouts, resulting in enhanced subsequent performance (Lane and Wenger 2004). This is an important investigation, as most studies in the area of cold water immersion have been conducted using muscle damage models or recovery from injury. Despite these promising results, some studies have found negligible changes when investigating the recovery effects of cold water immersion (Paddon-Jones and Quigley 1997; Sellwood et al. 2007; Yamane et al. 2006).

In a randomized controlled trial, Sellwood and colleagues (2007) investigated the effect of ice-water immersion on delayed-onset muscle soreness (DOMS). Following a leg extension exercise task (5 ×10 sets at 120% concentric 1RM), participants performed either 3 × 1 min water exposure separated by 1 min in either 5 °C or 24 °C (control) water. Pain, swelling, muscle function (one-leg hop for distance), maximal isometric strength, and serum CK were recorded at baseline, 24, 48, and 72 h after damage. The only significant difference observed between the groups was lower pain in the sit-to-stand test at 24 h postexercise in the ice-water immersion group (Sellwood et al. 2007). In accordance with Yamane and colleagues (2006), only the exercised limb was immersed at a temperature of 5 °C. In this study, ice-water immersion was no more beneficial than tepid water immersion in the recovery from DOMS (Yamane et al. 2006). Paddon-Jones and Quigley (1997) induced damage in both arms (64 eccentric elbow flexion), and then one arm was immersed in 5 °C water for 5 × 20 min, with 60 min between immersions, while the other served as a control. No differences were observed between arms during the next 6 days for isometric and isokinetic torque, soreness, and limb volume (Paddon-Jones and Quigley 1997). In the aforementioned studies, cold water immersion appeared to be an ineffective treatment, specifically when immersing an isolated limb in 5 °C water.

Only one study has investigated the effect of cold water immersion on training adaptation. Yamane and colleagues (2006) investigated the influence of regular postexercise cold water immersion following cycling or handgrip exercise. Exercise tasks were completed 3 to 4 times per week for 4 to 6 weeks, with cooling protocols consisting of limb immersion in 5 °C (leg) or 10 °C (arm) water. The control group showed a significant training effect in comparison to the treatment group, and the authors concluded that cooling was ineffective in inducing molecular and humoral adjustments associated with specified training effects (e.g., muscle hypertrophy, increased blood supply, and myofibril regeneration).

Despite these findings, the majority of research supports the notion that cold water immersion is effective in reducing symptoms associated with DOMS (Eston and Peters 1999), repetitive high-intensity exercise (Bailey et al. 2007; Lane and Wenger 2004), and muscle injury (Brukner and Khan 1993). A more refined investigation into the individual components of a specific recovery protocol is needed to reveal the effect of varying the duration of exposure, the temperature, and the medium used, whether it is ice, air, or water. In addition, training studies are required to investigate the effectiveness of such interventions on training adaptations.

Hot Water Immersion

The use of heat as a recovery tool has been recommended to increase the working capacity of athletes (Viitasalo et al. 1995) and assist in the rehabilitation of soft tissue injuries and athletic recovery (Brukner and Khan 1993; Cornelius et al. 1992). The majority of hot water immersion protocols are performed in water warmer than 37 °C, resulting in an increase in muscle and core body temperature (Bonde-Petersen et al. 1992; Weston et al. 1987). The physiological effects of immersion in hot water remain to be elucidated. One of the main physiological responses associated with exposure to heat is increased periph-

eral vasodilation, resulting in increased blood flow (Bonde-Petersen et al. 1992; Wilcock et al. 2006).

The effect of hot water immersion on subsequent performance is also poorly understood. Only one study has investigated the effect of hot water immersion on postexercise recovery. Viitasalo and colleagues (1995) incorporated three 20 min warm (~37 °C) underwater water-jet massages into the training week of 14 junior track-and-field athletes. The results indicated an enhanced maintenance of performance (assessed via plyometric drop jumps and repeated bounding) following the water treatment, indicating a possible reduction in DOMS. However, significantly higher CK and myoglobin concentrations were observed following the water treatment, suggesting either greater damage to the muscle cells or an increased leakage of proteins from the muscle into the blood. Viitasalo and colleagues (1995) concluded that combining underwater water-jet massage with intense strength training increases the release of proteins from the muscle into the blood, while enhancing the maintenance of neuromuscular performance (Viitasalo et al. 1995).

Evidence to support these findings is lacking, and the use of hot water immersion for recovery has received minimal research attention. Despite the hypothesized benefits of this intervention, anecdotal evidence suggests that hot water immersion is not widely prescribed on its own or as a substitute for other recovery interventions. Speculation surrounds the possible effects, timing of recovery, and optimal intervention category (e.g., following which type or intensity of exercise) for the use of hot water immersion.

Contrast Water Therapy

During contrast water therapy, athletes alternate between heat exposure and cold exposure by immersion in warm and cold water, respectively. This therapy has frequently been used as a recovery intervention in sports medicine (Higgins and Kaminski 1998) and is commonly used within the sporting community. Although research investigating contrast water therapy as a recovery intervention for muscle soreness and exercise-induced fatigue is limited in comparison to that for cold water immersion, several researchers have proposed possible mechanisms that may support the use of contrast water therapy. Higgins and Kaminski (1998) suggested that contrast water therapy can reduce edema through a pumping action created by alternating peripheral vasoconstriction and vasodilation. Contrast water therapy may bring about other changes such as increased or decreased tissue temperature, increased or decreased blood flow, changes in blood flow distribution, reduced muscle spasm, hyperemia of superficial blood vessels, reduced inflammation, and improved range of motion (Myrer et al. 1994). Active recovery has traditionally been considered superior to passive recovery. Contrast water therapy may elicit many of the same benefits of active recovery and may prove to be more beneficial, given that contrast water therapy imposes fewer energy demands on the athlete (Wilcock et al. 2006).

Contrast water therapy has been found to effectively decrease postexercise lactate levels (Coffey et al. 2004; Hamlin 2007; Morton 2006; Sanders 1996). After conducting a series of Wingate tests, investigators found that blood lactate concentrations recovered at similar rates when using either contrast water therapy or active recovery protocols and that after passive rest blood lactate removal was significantly slower (Sanders 1996). Coffey and colleagues (2004) investigated the effects of three different recovery interventions (active, passive, and contrast water therapy) on 4 h repeated treadmill running performance. Contrast water therapy and active recovery reduced blood lactate concentration by similar amounts after high-intensity running. In addition, contrast water therapy was associated with a perception of increased recovery. However, performance during the high-intensity treadmill running task returned to baseline levels 4 h after the initial exercise task regardless of the recovery intervention performed.

In a more recent study investigating the effect of contrast water therapy on the symptoms of DOMS and the recovery of explosive athletic performance, recreational athletes completed a muscle-damaging protocol on two separate occasions in a randomized crossover design (Vaile et al. 2008a). The two exercise sessions differed only in recovery intervention (contrast water therapy or passive recovery/control). Following contrast water therapy, isometric force production was not significantly reduced below baseline levels throughout the 72 h data collection period; however, following passive recovery, peak strength was significantly reduced from baseline by 14.8% ± 11.4% (Vaile et al. 2008a). Strength was also restored more rapidly within the contrast water therapy group. Thigh volume measured immediately following contrast water therapy was significantly less than that following passive recovery, indicating lower levels of tissue edema. These results indicate that symptoms of DOMS and restoration of strength are improved following contrast water therapy compared with passive recovery (Vaile et al. 2007; Vaile et al. 2008a). However, Hamlin (2007) found contrast water therapy to have no beneficial effect on performance during repeated sprinting. Twenty rugby players performed two repeated sprint tests separated by 1 h; between trials subjects completed either contrast water therapy or active recovery. Although substantial decreases in blood lactate concentration and heart rate were observed following contrast water therapy, compared with the first exercise bout, performance in the second exercise bout was decreased regardless of intervention (Hamlin 2007). Therefore, although contrast water therapy appears to be beneficial in the treatment of DOMS, it may not hasten the recovery of performance following high-intensity repeated sprint exercise.

The physiological mechanisms underlying the reputed benefits remain unclear. Temperatures for contrast water therapy generally range from 10 to 15 °C for cold water and 35 to 38 °C for warm water. It is evident that contrast water therapy is being widely used; however, additional research needs to be conducted to clarify its optimal role and relative efficacy.

Pool Recovery

Pool or beach recovery sessions are commonly used by team sport athletes in an attempt to enhance recovery from competition. Almost all Australian Rules, Australian Rugby League, and Australian Rugby Union teams use pool recovery sessions to perform active recovery in a non–weight-bearing environment. These sessions are typically used to reduce muscle soreness and stiffness and therefore are thought to be effective in sports that involve eccentric muscle damage or contact. Sessions often include walking and stretching in the pool and occasionally some swimming.

Dawson and colleagues (2005) investigated the effect of pool walking as a recovery intervention immediately following a game of Australian Rules Football. Pool walking was compared with contrast water therapy, stretching, and passive recovery (control) to determine the effect on subjective ratings of muscle soreness, flexibility (sit and reach test), and power (6 s cycling sprint and vertical jumps) assessed 15 h after the game. For all four recovery interventions, muscle soreness was increased 15 h after the game; however, only pool walking resulted in a significant reduction in subjective soreness. There was a trend for lower flexibility and power scores; however, this was only significant in the control trial. Although there were no differences between the three recovery interventions with respect to flexibility and power, players subjectively rated pool walking as the most effective and preferable strategy. The authors speculated that the active, light-intensity exercise with minimal impact stress or load bearing, combined with the hydrostatic pressure, may have enhanced recovery (Dawson et al. 2005).

Cold Water Immersion Versus Hot Water Immersion Versus Contrast Water Therapy?

Only recently have the three aforementioned recovery interventions been compared with each other examining exercise-induced muscle damage and acute fatigue models.

A group of three independent studies examined the effects of three hydrotherapy interventions on the physiological and functional symptoms of DOMS. A total of 38 strength trained males completed two experimental trials separated by 8 months in a randomized crossover design. One trial involved passive recovery, the other a specific hydrotherapy protocol for a 72 h period postexercise: either (1) cold water immersion ($n = 12$), (2) hot water immersion ($n = 11$), or (3) contrast water therapy ($n = 15$) (Vaile et al. 2008a) (figure 10.1). For each trial, participants performed a DOMS-inducing leg press protocol followed by either passive recovery or one of the hydrotherapy interventions for a total of 14 min. Performance was assessed via weighted squat jump and isometric squat; perceived pain, thigh girths, and blood variables were measured prior to exercise, immediately afterward, and at 24, 48, and 72 h postexercise. Overall, cold water immersion and contrast water therapy were found to be effective in reducing the physiological and functional deficits associated with DOMS, including improved recovery of isometric force and dynamic power and a reduction in localized edema (Vaile et al. 2008a). Although hot water immersion was effective in the recovery of isometric force, it was ineffective for recovery of all other markers compared with passive recovery.

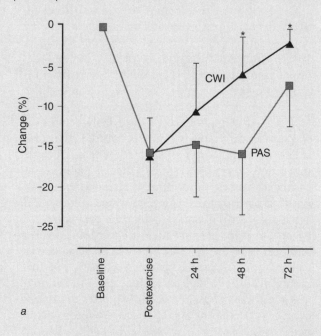

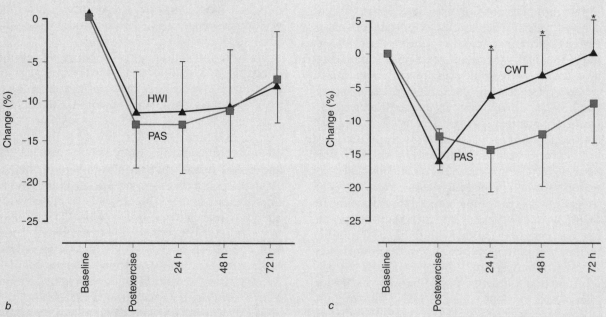

FIGURE 10.1 Percentage change in squat jump performance (peak power) following cold water immersion (CWI, *a*), hot water immersion (HWI, *b*), and contrast water therapy (CWT, *c*). Performance was assessed before and after muscle-damaging exercise as well as 24, 48, and 72 h following exercise (Vaile et al. 2008a). *Significant difference between hydrotherapy intervention and passive recovery (PAS).

Another study investigated the effects of three hydrotherapy interventions on next-day performance recovery following strenuous training. A total of 12 male cyclists completed four experimental trials differing only in recovery intervention: cold water immersion, hot water immersion, contrast water therapy, or passive recovery. Each trial comprised five consecutive exercise days (105 min duration, including 66 maximal effort sprints). Additionally, participants performed a total of 9 min of sustained effort (time trial). After completing each exercise session, participants performed one of the four recovery interventions (in a randomized crossover design). Performance (average power), core temperature, heart rate, and RPE were recorded throughout each session. Sprint (0.1%-2.2%) and time trial (0.0%-1.7%) performance were enhanced across the 5-day trial following both cold water immersion and contrast water therapy compared with hot water immersion and passive recovery. Differences in core body temperature were observed between interventions immediately and 15 min after recovery; however, no significant differences were observed in heart rate or RPE regardless of day of trial or intervention. Overall, cold water immersion and contrast water therapy improved recovery from high-intensity cycling compared with hot water immersion and passive recovery, enabling athletes to better maintain performance across a 5-day period (Vaile et al. 2008b).

Sleep

The relationship between sleep, recovery, and enhanced athletic performance is increasing in interest as the understanding of the function of sleep improves. A range of cognitive impairments and metabolic, immunological, and physiological processes are negatively affected by sleep deprivation (Samuels 2008).

Although there is almost no scientific evidence to support the role of sleep in enhancing recovery, elite athletes and coaches often identify sleep as a vital component of the recovery process. In a recent study, athletes and coaches ranked sleep as the most prominent problem when they were asked about the causes of fatigue (Fallon 2007). Sleep characteristics ranked first when athletes were asked about the aspects of the clinical history that they thought were important.

Sleep Deprivation

A limited number of studies have examined the effects of sleep deprivation on athletic performance. From the available data it appears that two phenomena exist. First, the sleep deprivation must be greater than 30 h to affect performance, and second, sustained or repeated bouts of exercise are affected to a greater degree than one-off maximal efforts (Blumert et al. 2007; Reilly and Edwards 2007).

Souissi and colleagues (2003) measured maximal power, peak power, and mean power before and after 24 and 36 h of sleep deprivation. Anaerobic power variables were not affected when athletes were awake for up to 24 h; however, these variables were impaired after 36 h without sleep (Souissi et al. 2003). Bulbulian and colleagues (1996) examined knee extension and flexion peak torque before and after 30 h of sleep deprivation in trained men. Isokinetic performance decreased significantly following sleep deprivation. In support of the contention that the effects of sleep deprivation are task specific, Takeuchi and colleagues (1985) reported that 64 h of sleep deprivation significantly reduced vertical jump performance and isokinetic knee extension strength; however, isometric strength and 40 m sprint performance were unaffected.

Although these studies provide some insight into the relationship between sleep deprivation and performance, most athletes are more likely to experience acute bouts of partial sleep deprivation where sleep is reduced for several hours on consecutive nights.

Partial Sleep Deprivation

A small number of studies have examined the effect of partial sleep deprivation on athletic performance. Reilly and Deykin (1983) reported decrements in a range of psychomotor functions after only one night of restricted sleep; however, gross motor functions such as muscle strength, lung power, and endurance running were unaffected (Reilly and Deykin 1983). Reilly and Hales (1988) reported similar effects in females following partial sleep deprivation; gross motor functions were less affected by sleep loss than were tasks requiring fast reaction times (Reilly and Hales 1988). It appears that submaximal prolonged tasks may be more affected than maximal efforts particularly, for the first two nights of partial sleep deprivation (Reilly and Percy 1994).

Evidence suggests that performance in maximal efforts may be unaffected by partial sleep deprivation but repeated submaximal efforts may be reduced. Reilly and Percy (1994) found a significant effect of sleep loss on maximal bench press, leg press, and dead lifts but not maximal bicep curl. Submaximal performance, however, was significantly affected on all four tasks and to a greater degree than maximal efforts. The greatest impairments were found later in the protocol, suggesting an accumulative effect of fatigue from sleep loss.

Napping

Athletes who suffer from some degree of sleep loss may benefit from a brief nap, particularly if a training session is to be completed in the afternoon or evening. Waterhouse and colleagues (2007) are among the few researchers to investigate the effects of a lunchtime nap on sprint performance following partial sleep deprivation. Following a 30 min nap, 20 m sprint performance was increased (compared with no nap), alertness was increased, and sleepiness was decreased. Napping may be beneficial for athletes who routinely have to wake early for training or competition and for athletes who are experiencing sleep deprivation (Waterhouse et al. 2007).

Extended Sleep

Another means of examining the effect of sleep on performance is to extend the amount of sleep an athlete receives and determine the effects on subsequent performance. Mah and colleagues (2007) instructed six basketball players to obtain as much extra sleep as possible following 2 weeks of normal sleep habits. Faster sprint times and increased free-throw accuracy were observed at the end of the sleep extension period. Mood was also significantly altered, with increased vigor and decreased fatigue (Mah et al. 2007). The same research group also increased the sleep time of swimmers to 10 h per night for 6 to 7 weeks. Following this period, 15 m sprint, reaction time, turn time, and mood all improved. The data from these small studies suggest that increasing the amount of sleep an athlete receives may significantly enhance performance.

Possible Mechanisms

A link between sleep deprivation and impaired neuroendocrine and immune function (Basta et al. 2007) has been proposed, as a number of critical metabolism and immune processes are known to occur during different stages of sleep. Daytime napping has recently been shown to result in beneficial changes in cortisol and interleukin-6, which are associated with inflammation. The neuroendocrine and immune systems are thought to be highly involved in the overtraining syndrome (Halson and Jeukendrup 2004), and thus a relationship between fatigue, sleep and underperformance most likely exists.

Massage

Massage is commonly used to enhance recovery from training and competition. Massage is defined as a mechanical manipulation of body tissues with rhythmical pressure and stroking for the purpose of promoting health and well being (Galloway and Watt 2004). Although massage is also used prior to training and competition and as a treatment for injury, for the purpose of this review, massage is discussed in the context of recovery only.

Massage is suggested to have numerous physiological benefits, including decreased muscle tension and stiffness, increased healing rate of injured muscles and ligaments, increased joint flexibility, increased range of motion, increased blood flow, decreased lactate concentrations, increased skin and muscle temperature, decreased anxiety, increased relaxation, enhanced immune and endocrine function, increased performance, and reduced muscle pain, swelling, and spasm (Arroyo-Morales et al. 2009; Best et al. 2008; Weerapong et al. 2005).

Despite the widespread use of massage and the anecdotally reported positive benefits, there is little high-quality scientific evidence to support or contest the claims.

Concentric Exercise

A small number of studies have investigated the effect of postexercise massage on performance measures. Forearm massage was shown to improve hand grip strength when applied immediately after 3 min of maximal hand exercise (Brooks et al. 2005). Other studies have examined massage in combination with other recovery strategies on performance outcomes (Lane and Wenger 2004; Monedero and Donne 2000). Lane and colleagues (2004) reported increased cycling power output following massage therapy when compared with passive recovery. The massage was 15 min in duration and was performed on the quadriceps, hamstring, and calf muscles. The massage techniques included deep effleurage (2.5 min per leg), compressions (1 min per leg), deep muscle stripping (2 min per leg), jostling (1 min per leg), and cross-fiber frictions (1 min per leg). However, active recovery and cold water immersion were superior to massage in terms of recovery from high-intensity cycling. Monedero and

Donne (2000) examined a combination of massage and active recovery in high-intensity cycling performance and reported no effect of massage alone on enhancing performance in cyclists. However, massage and active recovery significantly increased performance above passive recovery and isolated massage (Monedero and Donne 2000).

Robertson and colleagues (2004) investigated the effects of 20 min of leg massage on performance in a 30 s Wingate test, immediately following six high-intensity 30 s cycling efforts. When compared with supine passive recovery, there was no effect of massage on blood lactate concentration, heart rate, maximal power, and mean power. However, massage resulted in a lower fatigue index (percentage change in power between the first 5 s and the last 5 s of the 30 s exercise period) (Robertson et al. 2004). However, the participants in this study again were not elite athletes. Finally, Ogai and colleagues (2008) compared 10 min of leg massage to supine passive recovery between two bouts of high-intensity cycling. Massage did not alter lactate concentrations; however, power during the second exercise bout was higher in the massage treatment than in the control condition. Muscle stiffness and perceived fatigue were lower following massage, which may have contributed to the enhanced performance (Ogai et al. 2008).

Eccentric Exercise

Massage therapy following eccentric exercise that resulted in DOMS is a commonly used recovery treatment. Weber and colleagues (1994) investigated massage, aerobic exercise, microcurrent stimulation, and passive recovery on recovery of force from eccentric exercise. None of the treatment modalities had significant effects on soreness, maximal isometric contraction, or peak torque production (Weber et al. 1994). Hilbert and colleagues (2003) reported no effect of massage administered 2 h after a bout of eccentric exercise on peak torque produced by the hamstring muscle; however, muscle soreness ratings were decreased 48 h following exercise (Hilbert et al. 2003). Dawson and colleagues (2004) saw no effect of massage immediately following a marathon on quadriceps peak torque output or soreness. Farr and colleagues (2002) also reported no effect of 30 min of leg massage on muscle strength in healthy males, although soreness and tenderness ratings were lower 48 h following exercise. However, a significant improvement in vertical jump performance was reported after high-intensity exercise in college female athletes (Mancinelli et al. 2006).

Few investigations have examined the effect of massage on performance, and those investigations have examined a wide range of massage techniques and outcome measures. However, there may be some evidence to suggest that massage following eccentric exercise may reduce muscle soreness. It may also be possible that investigators have measured performance at inappropriate times and that benefits in performance may be seen 24 to 96 h following exercise, when inflammatory processes are at their peak. There is slightly more research regarding the effect of massage on recovery than on performance. Studies that focused on areas such as lactate concentrations and blood flow may have been ill-directed; muscle stiffness and muscle tone may be more important parameters for athlete recovery when considering the time-frame of when massage is performed following exercise.

Possible Mechanisms of Massage Therapy

It is highly likely that the effects of massage occur through more than one mechanism. Biomechanical, physiological, neurological, and psychophysiological mechanisms have been proposed (for comprehensive review, see Weerapong et al. 2005). Other research has examined immunological and endocrine parameters. Each of these areas is discussed briefly next.

Biomechanical Mechanisms

The biomechanical model primarily relates to increasing muscle–tendon compliance through mechanical pressure on the muscle tissue (Weerapong et al. 2005). This is thought to occur by mobilizing and elongating shortened or adhered connective tissue (Weerapong et al. 2005). One study examined the effect of massage on passive stiffness and reported no change when compared with passive rest (Stanley et al. 2001). However, two studies (Nordschow and Bierman 1962; Wiktorsson-Moller et al. 1983) reported increased range of motion in response to massage. Nordschow and Bierman (1962) reported an increase in lumbar range of motion following massage of the back and lower limbs, and Wiktorsson and colleagues (1983) reported increased ankle dorsiflexion following massage of the legs.

Physiological Mechanisms

It has been suggested that superficial skin friction may increase local heating and thus potentially blood flow. The mechanical pressure may stimulate the parasympathetic nervous system, altering hormone levels, heart rate, and blood pressure (Weerapong et al. 2005). A small amount of scientific literature has reported increased skin and muscle temperature (2.5 cm deep) after massage treatment (Drust et al. 2003; Longworth 1982). However, this increase does not appear to translate into increased blood flow. Shoemaker and colleagues (1997) reported no change in blood flow following massage of the forearm and quadriceps measured via pulsed Doppler ultrasound velocimetry. This same group also reported no change in arterial or venous blood velocity when comparing a massage and control leg in same individuals (Tiidus and Shoemaker 1995).

Wiltshire and colleagues (2010) investigated the effects of postexercise massage on muscle blood flow and lactic acid removal. Ten minutes of massage was performed immediately following 2 min of isometric handgrip exercise. The results suggest that massage impairs removal of lactic acid from the muscle because the massage impairs muscle blood flow (due to rhythmic compression of muscle tissue) (Wiltshire et al. 2010). However, handgrip exercise may not represent other forms of whole-body exercise, and measures of subsequent performance were not made.

Massage has been shown to reduce cortisol concentrations in dance students (Leivadi et al. 1999). Although there is almost no evidence in elite athletes, reductions in cortisol and serotonin are commonly reported following massage in a large number of patient populations (those with depression, anxiety, HIV, low back pain) (Weerapong et al. 2005). One study demonstrated that massage following high-intensity exercise had a positive effect on salivary immunoglobulin A rate, indicating that massage may negate some of the immunosuppressive effects of acute exercise (Arroyo-Morales et al. 2009). In this study, however, massage had no influence on cortisol concentrations.

Neurological Mechanisms

A reduction in neuromuscular excitability may occur with massage attributable to the stimulation of sensory receptors (Weerapong et al. 2005). One study reported decreased H-reflex amplitude during massage (Morelli et al. 1990). It has been suggested that this occurred as a result of decreased spinal reflex excitability through inhibition via muscle or other deep tissue mechanoreceptors (Weerapong et al. 2005).

The reduction in pain following massage may occur via the activation of the neural gating mechanism in the spinal cord (gate control theory of pain). Skin receptor information may block the information to the brain from the pain receptors. Additionally, massage is often used to reduce muscle spasm, which causes muscle pain. The spasm is thought to activate pain receptors or to compress blood flow, causing ischemia (Weerapong et al. 2005). A realignment of muscle fibers as a consequence of massage is thought to reduce the muscle spasms. However, although this theory is quite well accepted, there is no scientific evidence to support it.

Psychophysiological Mechanisms

Improved mood and reduced anxiety as a consequence of massage have been reported in several studies. Reduced anxiety and improvements in depressed mood were reported in dance students who received a 30 min massage twice a week for 4 weeks (Leivadi et al. 1999). Hemmings and colleagues (2000) compared massage, supine resting, and touching control during a period of training in boxers. Massage improved the Profile of Mood State questionnaire subscales of tension and fatigue (Hemmings et al. 2000).

Stretching

Stretching is one of the most commonly used post-training recovery interventions. However, there is limited evidence for the use of this modality in enhancing recovery immediately following exercise. The rationale for stretching during the recovery period is not clear, although a number of theories exist: to reduce muscle soreness and stiffness, to prevent injury, and to relax the muscle.

The majority of research has focused on the effects of static stretching on the signs and symptoms of DOMS. Jayaraman and colleagues (2004) examined the effects of static stretching and heat application on 32 untrained male subjects following eccentric knee extension exercise. Strength testing, pain ratings, and multiple-echo magnetic resonance imaging (MRI) were assessed to examine recovery of the thigh muscle. Both static stretching and heat application occurred 36 h after exercise to avoid any detrimental effects during the acute injury phase. The stretching treatment consisted of standing quadriceps, prone quadriceps, standing hamstring, seated hamstring, standing calf, and wall squat stretches. The findings demonstrated that neither heat application nor static stretching increased muscle recovery (indicated by changes in MRI). Additionally, there were no differences between treatments and control for swelling, pain, and recovery of strength. The authors concluded that static stretching did not increase recovery of damaged muscle fibers and that the prescription of static stretching did not enhance recovery following eccentric exercise (Jayaraman et al. 2004).

Similar results have been reported in studies examining the effects of postexercise stretching on signs and symptoms of DOMS (Buroker and Schwane 1989; McGlynn et al. 1979; Wessel and Wan 1994). Each of these investigations reported no effect of static stretching on enhancing recovery. However, two early studies reported a reduction in pain sensation following static stretching after eccentric exercise (Abraham 1977; DeVries 1966).

One investigation compared static stretching, active recovery, and passive recovery following fatiguing leg extension and leg flexion (Mika et al. 2007). Participants performed three sets of leg extension and leg flexion at 50% maximal voluntary contraction (MVC), where each set was performed to failure. Following a 5 min recovery period, isometric knee extensions were performed to fatigue. In the 4 min recovery period, participants either rested passively, performed light cycling at 10 W, or performed isometric quadriceps femoris contraction against resistance (applied by a physiotherapist). Participants were then asked to perform an isometric knee extension at 50% MVC until fatigue. The results suggest that the most effective strategy was active recovery. Subsequent MVCs were significantly greater in the trial in which active recovery was performed in comparison to stretching or passive recovery.

In summary, there is no evidence to suggest that stretching immediately following exercise enhances the recovery of performance; however, there is also no evidence to suggest a detrimental effect. A small amount of evidence suggests that stretching may reduce the sensation of pain after eccentric exercise.

Active Recovery

Active recovery (low-intensity exercise) is believed to be an integral component of physical recovery. Active recovery usually takes the form of walking, jogging, cycling, or swimming at a low submaximal intensity. Anecdotal evidence suggests that active recovery reduces postexercise muscle soreness and DOMS.

Although it is generally accepted that active recovery is superior to passive recovery; this assumption is based on the fact that enhanced blood lactate removal has been observed following active recovery compared with passive recovery (Ahmaidi et al. 1996; Dupont et al. 2004). Therefore, the literature suggests that active recovery is more efficient than passive recovery in improving the recovery process (Monedero and Donne 2000). However, as postexercise blood lactate concentrations will return to resting levels with passive rest in a time period much shorter than that typically available for recovery, the relevance of this research to elite athletes is questionable (Barnett 2006).

Literature appears to both support and oppose the use of active recovery following exercise. Sayers and colleagues (2000) used 26 male volunteers to investigate whether activity would affect the recovery of muscle function after high-force eccentric exercise of the elbow flexors. The exercise regime consisted of 50 maximal eccentric contractions of the elbow flexors of the nondominant arm (Sayers et al. 2000). Sayers and colleagues (2000) found that both light exercise and immobilization aided the recovery of maximal isometric force, also stating that recovery from soreness was aided by light exercise but delayed by immobilization. Gill and colleagues (2006) investigated the effect of four (active recovery, passive recovery, compression, and contrast water therapy) postmatch recovery interventions in 23 elite rugby players. For the active recovery component, the athletes completed 7 min of low-intensity cycling (80-100 revolutions per minute, ~150 W) followed by their normal postmatch routine (Gill et al. 2006). The authors concluded that the rate and magnitude of recovery were enhanced in the active recovery, contrast water therapy, and compression treatment groups, with an increase in CK clearance (Gill et al. 2006). Active recovery, contrast water therapy, and compression were all significantly superior to passive recovery, with no differences observed between the effective recovery strategies. However, a study investigating the effect of three different interventions (massage, electrical stimulation, and light exercise) found no support for the use of massage, microcurrent electrical stimulation, or light exercise to reduce or alleviate both soreness and force deficits associated with DOMS (Weber et al. 1994).

A combined recovery intervention including aspects of active recovery has demonstrated encouraging effects on lactate removal in a number of studies (Jemni et al. 2003; Monedero and Donne 2000; Watts et al. 2000). However, McAinch and colleagues (2004) examined the effect of active versus passive recovery on metabolism and performance during subsequent exercise. They found that when compared with passive recovery, active recovery between two bouts of intense aerobic exercise did not assist in the maintenance of performance nor did it alter either muscle glycogen content or blood lactate accumulation. The authors proposed there to be no rationale for the practice of active recovery following intense aerobic exercise (McAinch et al. 2004).

Carter and colleagues (2002) investigated the effects of mode of exercise recovery on thermoregulatory and cardiovascular responses; data suggested that mild active recovery may play an important role for postexertional heat dissipation. However the mechanisms behind these altered responses during active recovery are unknown, and further investigation is warranted (Carter et al. 2002).

Although the research on the benefits of active recovery beyond lactate removal is equivocal, the role of active recovery in reducing DOMS and enhancing range of motion after exercise warrants further investigation. This is anecdotally reported to be one of the most common forms of recovery and is used by the majority of athletes.

References

Abraham, W.M. 1977. Factors in delayed muscle soreness. *Medicine and Science in Sports* 9(1):11-20.

Agu, O., Baker, D., and Seifalian, A.M. 2004. Effect of graduated compression stockings on limb oxygenation and venous function during exercise in patients with venous insufficiency. *Vascular* 12(1):69-76.

Ahmaidi, S., Granier, P., Taoutaoum Z., Mercier, J., Dubouchaud, H., and Prefaut, C. 1996. Effects of active recovery on plasma lactate and anaerobic power following repeated intensive exercise. *Medicine and Science in Sports and Exercise* 28(4):450-456.

Ali, A., Caine, M.P., and Snow, B.G. 2007. Graduated compression stockings: Physiological and perceptual responses during and after exercise. *Journal of Sports Sciences* 25(4):413-419.

Arroyo-Morales, M., Olea, N., Ruiz, C., del Castilo Jde, D., Martinez, M., Lorenzo, C., and Diaz-Rodriguez, L. 2009. Massage after exercise—Responses of immunologic and endocrine markers: A randomized single-blind placebo-controlled study. *Journal of Strength and Conditioning Research* 23(2):638-644.

Bailey, D.M., Erith, S.J., Griffin, P.J., Dowson, A., Brewer, D.S., Gant, N., and Williams, C. 2007. Influence of cold-water immersion on indices of muscle damage following prolonged intermittent shuttle running. *Journal of Sports Sciences* 25(11):1163-1170.

Barnett, A. 2006. Using recovery modalities between training sessions in elite athletes: Does it help? *Sports Medicine* 36(9):781-796.

Basta, M., Chrousos, G.P., Vela-Bueno, A., and Vgontzas, A.N. 2007. Chronic insomnia and stress system. *Sleep Medicine Clinics* 2(2):279-291.

Berry, M.J., and McMurray, R.G. 1987. Effects of graduated compression stockings on blood lactate following an exhaustive bout of exercise. *American Journal of Physical Medicine* 66(3):121-132.

Best, T.M., Hunter, R., Wilcox, A., and Haq, F. 2008. Effectiveness of sports massage for recovery of skeletal muscle from strenuous exercise. *Clinical Journal of Sports Medicine* 18(5):446-460.

Blumert, P.A., Crum, A.J., Ernsting, M., Volek, J.S., Hollander, D.B.,. Haff, E.E., and Haff, G.G. 2007. The acute effects of twenty-four hours of sleep loss on the performance of national-caliber male collegiate weightlifters. *Journal of Strength and Conditioning Research* 21(4):1146-1154.

Bonde-Petersen, F., Schultz-Pedersen, L., and Dragsted, N. 1992. Peripheral and central blood flow in man during cold, thermoneutral, and hot water immersion. *Aviation, Space, and Environmental Medicine* 63(5):346-350.

Bringard, A., Denis, R., Belluye, N., and Perrey, S. 2006. Effects of compression tights on calf muscle oxygenation and venous pooling during quiet resting in supine and standing positions. *Journal of Sports Medicine and Physical Fitness* 46(4):548-554.

Brooks, C.P., Woodruff, L.D., Wright, L.L., and Donatelli, R. 2005. The immediate effects of manual massage on power-grip performance after maximal exercise in healthy adults. *Journal of Alternative and Complementary Medicine* 11(6):1093-1101.

Brukner, P., and Khan, K. 1993. *Clinical Sports Medicine.* Sydney: McGraw-Hill.

Bulbulian, R., Heaney, J.H., Leake, C.N., Sucec, A.A., and Sjoholm, N.T. 1996. The effect of sleep deprivation and exercise load on isokinetic leg strength and endurance. *European Journal of Applied Physiology and Occupational Physiology* 73(3-4):273-277.

Burke, D.G., MacNeil, S.A., Holt, L.E., MacKinnon, N.C., and Rasmussen, R.L. 2000. The effect of hot or cold water immersion on isometric strength training. *Journal of Strength and Conditioning Research* 14(1):21-25.

Buroker, K.C., and Schwane, J.A. 1989. Does post exercise static stretching alleviate delayed muscle soreness? *Physician and Sportsmedicine* 17(6):318-322.

Carter, R., 3rd, Wilson, T.E., Watenpaugh, D.E., Smith, M.L., and Crandall, C.G. 2002. Effects of mode of exercise recovery on thermoregulatory and cardiovascular responses. *Journal of Applied Physiology* 93(6):1918-1924.

Coffey, V., Leveritt, M., and Gill, N. 2004. Effect of recovery modality on 4-hour repeated treadmill running performance and changes in physiological variables. *Journal of Science and Medicine in Sport* 7(1):1-10.

Cornelius, W.L., Ebrahim, K., Watson, J., and Hill, D.W. 1992. The effects of cold application and modified PNF stretching techniques on hip joint flexibility in college males. *Research Quarterly for Exercise and Sport* 63(3):311-314.

Dawson, B., Cow, S., Modra, S., Bishop, D., and Stewart, G. 2005. Effects of immediate post-game recovery procedures on muscle soreness, power and flexibility levels over the next 48 hours. *Journal of Science and Medicine in Sport* 8(2):210-221.

Dawson, L., Dawson, K., and Tiidus, P.M. 2004. Evaluating the influence of massage on leg strength, swelling, and pain following a half marathon. *Journal of Sports Science & Medicine* 3:37-43.

DeVries, H.A. 1966. Quantitative electromyographic investigation of the spasm theory of muscle pain. *American Journal of Physical Medicine* 45:119-134.

Drust, B., Atkinson, G., Gregson, W., French, D., and Binningsley, D. 2003. The effects of massage on intra muscular temperature in the vastus lateralis in humans. *International Journal of Sports Medicine* 24(6):395-399.

Duffield, R., Edge, J., Merrells, R., Hawke, E., Barnes, M., Simcock, D., and Gill, N. 2008. The effects of compression garments on intermittent exercise performance and recovery on consecutive days. *International Journal of Sports Physiology and Performance* 3(4):454-468.

Duffield, R., and Portus, M. 2007. Comparison of three types of full-body compression garments on throwing and repeat-sprint performance in cricket players. *British Journal of Sports Medicine* 41(7):409-414.

Dupont, G., Moalla, W., Guinhouya, C., Ahmaidi, S., and Berthoin, S. 2004. Passive versus active recovery during high-intensity intermittent exercises. *Medicine and Science in Sports and Exercise* 36(2):302-308.

Enwemeka, C.S., Allen, C., Avila, P., Bina, J., Konrade, J., and Munns, S. 2002. Soft tissue thermodynamics before, during, and after cold pack therapy. *Medicine and Science in Sports and Exercise* 34(1):45-50.

Eston, R., and Peters, D. 1999. Effects of cold water immersion on the symptoms of exercise-induced muscle damage. *Journal of Sports Sciences* 17(3):231-238.

Fallon, K.E. 2007. Blood tests in tired elite athletes: Expectations of athletes, coaches and sport science/sports medicine staff. *British Journal of Sports Medicine* 41(1):41-44.

Farr, T., Nottle, C., Nosaka, K., and Sacco, P. 2002. The effects of therapeutic massage on delayed onset muscle soreness and muscle function following downhill walking. *Journal of Science and Medicine in Sport* 5(4):297-306.

French, D.N., Thompson, K.G., Garland, S.W., Barnes, C.A., Portas, M.D., Hood, P.E., and Wilkes, G. 2008. The effects of contrast bathing and compression therapy on muscular performance. *Medicine and Science in Sports and Exercise* 40(7):1297-1306.

Galloway, S.D., and Watt, J.M. 2004. Massage provision by physiotherapists at major athletics events between 1987 and 1998. *British Journal of Sports Medicine* 38(2):235-236; discussion 237.

Gill, N.D., Beaven, C.M., and Cook, C. 2006. Effectiveness of post-match recovery strategies in rugby players. *British Journal of Sports Medicine* 40(3):260-263.

Halson, S.L., and Jeukendrup, A.E. 2004. Does overtraining exist? An analysis of overreaching and overtraining research. *Sports Medicine* 34(14):967-981.

Hamlin, M.J. 2007. The effect of contrast temperature water therapy on repeated sprint performance. *Journal of Science and Medicine in Sport* 10(6):398-402.

Hemmings, B., Smith, M., Graydon, J., and Dyson, R. 2000. Effects of massage on physiological restoration, perceived recovery, and repeated sports performance. *British Journal of Sports Medicine* 34(2):109-114; discussion 115.

Higgins, D., and Kaminski, T.W. 1998. Contrast therapy does not cause fluctuations in human gastrocnemius. *Journal of Athletic Training* 33(4):336-340.

Hilbert, J.E., Sforzo, G.A., and Swensen, T. 2003. The effects of massage on delayed onset muscle soreness. *British Journal of Sports Medicine* 37(1):72-75.

Jayaraman, R.C., Reid, R.W., Foley, J.M., Prior, B.M., Dudley, G.A., Weingand, K.W., and Meyer, R.A. 2004. MRI evaluation of topical heat and static stretching as therapeutic modalities for the treatment of eccentric exercise-induced muscle damage. *European Journal of Applied Physiology* 93(1-2):30-38.

Jemni, M., Sands, W.A., Friemel, F., and Delamarche, P. 2003. Effect of active and passive recovery on blood lactate and performance during simulated competition in high level gymnasts. *Canadian Journal of Applied Physiology* 28(2):240-256.

Knight, K.L., and Londeree, B.R. 1980. Comparison of blood flow in the ankle of uninjured subjects during therapeutic applications of heat, cold, and exercise. *Medicine and Science in Sports and Exercise* 12(1):76-80.

Kraemer, W.J., Bush, J.A., Wickham, R.B., Denegar, C.R., Gomez, A.L., Gotshalk, A.L., Duncan, N.D., Volek, J.S., Newton, R.U., Putukian, M., and Sebastianelli, W.J. 2001. Continuous compression as an effective therapeutic intervention in treating eccentric-exercise-induced muscle soreness. *Journal of Sport Rehabilitation* 10:11-23.

Lane, K.N., and Wenger, H.A. 2004. Effect of selected recovery conditions on performance of repeated bouts of intermittent cycling separated by 24 hours. *Journal of Strength and Conditioning Research* 18(4):855-860.

Lee, D.T., Toner, M.M., McArdle, W.D., Vrabas, I.S., and Pandolf, K.B. 1997. Thermal and metabolic responses to cold-water immersion at knee, hip, and shoulder levels. *Journal of Applied Physiology* 82(5):1523-1530.

Leivadi, S., Hernandez-Reif, M., and Field, T. 1999. massage therapy and relaxation effects on university dance students. *Journal of Dance Medicine & Science* 3(3):108-112.

Longworth, J.C. 1982. Psychophysiological effects of slow stroke back massage in normotensive females. *ANS. Advances in Nursing Science* 4(4):44-61.

Mah, C., Mah, K., and Dement, W. 2007. *The effect of extra sleep on mood and athletic performance amongst collegiate athletes.* Paper presented at SLEEP 2008, 22nd Annual meeting of the Associated Professional Sleep Societies, Baltimore, Maryland, USA, June 9-12, 2007.

Mancinelli, C.A., Davis, D.S., Aboulhosn, L., Brady, M., Eisenhofer, J., and Foutty, S. 2006. The effects of massage on delayed onset muscle soreness and physical performance in female collegiate athletes. *Physical Therapy* 7:5-13.

McAinch, A.J., Febbraio, M.A., Parkin, J.M., Zhao, S., Tangalakis, K., Stojanovska, L., and Carey, M.F. 2004. Effect of active versus passive recovery on metabolism and performance during subsequent exercise. *International Journal of Sport Nutrition and Exercise Metabolism* 14(2):185-196.

McGlynn, G.H., Laughlin, N.T., and Rowe, V. 1979. Effect of electromyographic feedback and static stretching on artificially induced muscle soreness. *American Journal of Physical Medicine* 58(3):139-148.

Meeusen, R., and Lievens, P. 1986. The use of cryotherapy in sports injuries. *Sports Medicine* 3(6):398-414.

Merrick, M.A., Jutte, L.S., and Smith, M.E. 2003. Cold modalities with different thermodynamic properties produce different surface and intramuscular temperatures. *Journal of Athletic Training* 38(1):28-33.

Merrick, M.A., Ranin, J.M., Andres, F.A., and Hinman, C.L. 1999. A preliminary examination of cryotherapy and secondary injury in skeletal muscle. *Medicine and Science in Sports and Exercise* 31(11):1516-1521.

Mika, A., Mika, P., Fernhall, B., and Unnithan, V.B. 2007. Comparison of recovery strategies on muscle performance after fatiguing exercise. *American Journal of Physical Medicine & Rehabilitation* 86(6):474-481.

Monedero, J., and Donne, B. 2000. Effect of recovery interventions on lactate removal and subsequent performance. *International Journal of Sports Medicine* 21(8):593-597.

Montgomery, P.G., Pyne, D.B., Hopkins, W.G., Dorman, J.C., Cook, K., and Minahan, C.L. 2008. The effect of recovery strategies on physical performance and cumulative fatigue in competitive basketball. *Journal of Sports Sciences* 26(11):1135-1145.

Morelli, M., Seaborne, D.E., and Sullivan, S.J. 1990. Changes in h-reflex amplitude during massage of triceps surae in healthy subjects. *Journal of Orthopedic and Sports Physical Therapy* 12(2):55-59.

Morton, R.H. 2006. Contrast water immersion hastens plasma lactate decrease after intense anaerobic exercise. *Journal of Science and Medicine in Sport* 10(6):467-470.

Myrer, J.W., Draper, D.O., and Durrant, E. 1994. Contrast water therapy and intramuscular temperature in the human leg. *Journal of Athletic Training* 29:318-322.

Nordschow, M., and Bierman, W. 1962. The influence of manual massage on muscle relaxation: Effect on trunk flexion. *Journal of the American Physical Therapy Association* 42:653-657.

Ogai, R., Yamane, M., Matsumoto, T., and Kosaka, M. 2008. Effects of petrissage massage on fatigue and exercise performance following intensive cycle pedalling. *British Journal of Sports Medicine* 42(10):834-838.

Paddon-Jones, D.J., and Quigley, B.M. 1997. Effect of cryotherapy on muscle soreness and strength following eccentric exercise. *International Journal of Sports Medicine* 18(8):588-593.

Reilly, T., and Deykin, T. 1983. Effects of partial sleep loss on subjective states, psychomotor and physical performance tests. *Journal of Human Movement Studies* 9:157-170.

Reilly, T., and Hales, A. 1988. Effects of partial sleep deprivation on performance measures in females. In: E.D. McGraw (ed), *Contemporary Ergonomics*. London: Taylor & Francis.

Reilly, T., and Edwards, B. 2007. Altered sleep-wake cycles and physical performance in athletes. *Physiology & Behavior* 90(2-3):274-284.

Reilly, T., and Piercy, M. 1994. The effect of partial sleep deprivation on weight-lifting performance. *Ergonomics* 37(1):107-115.

Robertson, A., Watt, J.M., and Galloway, S.D. 2004. Effects of leg massage on recovery from high-intensity cycling exercise. *British Journal of Sports Medicine* 38(2):173-176.

Samuels, C. 2008. Sleep, recovery, and performance: The new frontier in high-performance athletics. *Neurology Clinics* 26(1):169-180; ix-x.

Sanders, J. 1996. *Effect of Contrast-Temperature Immersion on Recovery From Short-Duration Intense Exercise* (thesis). Canberra, Australia: University of Canberra.

Sayers, S.P., Clarkson, P.M., and Lee, J. 2000. Activity and immobilization after eccentric exercise: I. Recovery of muscle function. *Medicine and Science in Sports and Exercise* 32(9):1587-1592.

Sellwood, K.L., Brukner, P., Williams, D., Nicol, A., and Hinman, R. 2007. Ice-water immersion and delayed-onset muscle soreness: A randomized controlled trial. *British Journal of Sports Medicine* 41(6):392-397.

Shoemaker, J.K., Tiidus, P.M., and Mader, R. 1997. Failure of manual massage to alter limb blood flow: Measures by Doppler ultrasound. *Medicine and Science in Sports and Exercise* 29(5):610-614.

Souissi, N., Sesboue, B., Gauthier, A., Larue, J., and Davenne, D. 2003. Effects of one night's sleep deprivation on anaerobic performance the following day. *European Journal of Applied Physiology* 89(3-4):359-366.

Sramek, P., Simeckova, M., Jansky, L., Savlikova, J., and Vybiral, S. 2000. Human physiological responses to immersion into water of different temperatures. *European Journal of Applied Physiology* 81(5):436-442.

Stanley, S., Purdam, C., and Bond, T. 2001. *Passive tension and stiffness properties of the ankle plantar flexors: The effects of massage*. Paper presented at XVIIIth Congress of the International Society of Biomechanics, Zurich, Switzerland, July 9-13, 2001.

Takeuchi, L., Davis, G.M., Plyley, M., Goode, R., and Shephard, R.J. 1985. Sleep deprivation, chronic exercise and muscular performance. *Ergonomics* 28(3):591-601.

Tiidus, P.M., and Shoemaker, J.K. 1995. Effleurage massage, muscle blood flow and long-term post-exercise strength recovery. *International Journal of Sports Medicine* 16(7):478-483.

Vaile, J., Gill, N., and Blazevich, A.M. 2007. The effect of contrast water therapy on symptoms of delayed onset muscle soreness (DOMS) and explosive athletic performance. *Journal of Strength and Conditioning Research* 21(3):697-702.

Vaile, J., Halson, S., Gill, N., and Dawson, B. 2008a. Effect of hydrotherapy on the signs and symptoms of delayed onset muscle soreness. *European Journal of Applied Physiology* 102(4):447-455.

Vaile, J., Halson, S., Gill, N., and Dawson, B. 2008b. Effect of hydrotherapy on recovery from fatigue. *International Journal of Sports Medicine* 29(7):539-544.

Vaile, J., Halson, S., and Graham, S. 2010. Recovery review: Science vs. practice. *Journal of Australian Strength and Conditioning Supplement* 2(2):5-21.

Viitasalo, J.T., Niemela, K., Kaappola, R., Korjus, T., Levola, M., Mononen, H.V., Rusko, H.K., and Takala, T.E. 1995. Warm underwater water-jet massage improves recovery from intense physical exercise. *European Journal of Applied Physiology and Occupational Physiology* 71(5):431-438.

Washington, L.L., Gibson, S.J., and Helme, R.D. 2000. Age-related differences in the endogenous analgesic response to repeated cold water immersion in human volunteers. *Pain* 89(1):89-96.

Waterhouse, J., Atkinson, G., Edwards, B., and Reilly, T. 2007. The role of a short post-lunch nap in improving cognitive, motor, and sprint performance in participants with partial sleep deprivation. *Journal of Sports Sciences* 25(14):1557-1566.

Watts, P.B., Daggett, M., Gallagher, P., and Wilkins, B. 2000. Metabolic response during sport rock climbing and the effects of active versus passive recovery. *International Journal of Sports Medicine* 21(3):185-190.

Weber, M.D., Servedio, F.J., and Woodall, W.R. 1994. The effects of three modalities on delayed onset muscle soreness. *Journal of Orthopaedic and Sports Physical Therapy* 20(5):236-242.

Weerapong, P., Hume, P.A., and Kolt, G.S. 2005. The mechanisms of massage and effects on performance, muscle recovery and injury prevention. *Sports Medicine* 35(3):235-256.

Wessel, J., and Wan, A. 1994. Effect of stretching on the intensity of delayed-onset muscle soreness. *Clinical Journal of Sports Medicine* 4(2):83-87.

Weston, C.F., O'Hare, J.P., Evans, J.M., and Corrall. 1987. Haemodynamic changes in man during immersion in water at different temperatures. *Clinical Science (London, England: 1979)* 73(6):613-616.

Wiktorsson-Moller, M., Oberg, B., Ekstrand, J., and Gillquist, J. 1983. Effects of warming up, massage, and stretching on range of motion and muscle strength in the lower extremity. *American Journal of Sports Medicine* 11(4):249-252.

Wilcock, I.M., Cronin, J.B., and Hing, W.A. 2006. Physiological response to water immersion: A method for sport recovery? *Sports Medicine* 36(9):747-765.

Wiltshire, E.V., Poitras, V., Pak, M., Hong, T., Rayner, J., and Tschakovsky, M.E. 2010. Massage impairs postexercise muscle blood flow and "lactic acid" removal. *Medicine and Science in Sports and Exercise* 42(6):1062-1071.

Yamane, M., Teruya, H., Nakano, M., Ogai, R., Ohnishi, N., and Kosaka, M. 2006. Post-exercise leg and forearm flexor muscle cooling in humans attenuates endurance and resistance training effects on muscle performance and on circulatory adaptation. *European Journal of Applied Physiology* 96(5):572-580.

Yanagisawa, O., Homma, T., Okuwaki, T., Shimao, D., and Takahashi, H. 2007. Effects of cooling on human skin and skeletal muscle. *European Journal of Applied Physiology* 100(6):737-745.

Yanagisawa, O., Kudo, H., Takahashi, N., and Yoshioka, H. 2004. Magnetic resonance imaging evaluation of cooling on blood flow and oedema in skeletal muscles after exercise. *European Journal of Applied Physiology* 91(5-6):737-740.

Yeargin, S.W., Casa, D.J., McClung, J.M., Knight, J.C., Healey, J.C., Goss, P.J., Harvard, W.R., and Hipp, G.R. 2006. Body cooling between two bouts of exercise in the heat enhances subsequent performance. *Journal of Strength and Conditioning Research* 20(2):383-389.

PART III

Fundamental Assessment Principles and Protocols

Assessment of Physique

Gary Slater, Sarah M. Woolford, and Michael J. Marfell-Jones

A relationship between competitive success and physique traits has been identified in a range of sports, including football codes (Olds 2001), aesthetically judged sports (Claessens et al. 1999), swimming (Siders et al. 1993), track-and-field events (Claessens et al. 1994), and skiing (Larsson and Henriksson-Larsen 2008; Stoggl et al. 2010; White and Johnson 1991) as well as lightweight (Rodriguez 1986) and heavyweight rowing (Shephard 1998). The specific physique traits associated with competitive success vary with the sport. For athletes participating in aesthetically judged sports, maintenance of low body fat levels is associated with positive outcomes (Claessens et al. 1999; Faria and Faria 1989; Fry et al. 1991). A similar relationship exists in sports in which frontal surface area, power to weight ratio, and thermoregulation are important (Norton et al. 1996a). However, in sports demanding high force production, muscle mass may be more closely associated with performance outcomes (Brechue and Abe 2002; Kyriazis et al. 2010; Olds 2001; Siders et al. 1993; Stoggl et al. 2010), with specific distribution of muscle mass also important (Larsson and Henriksson-Larsen 2008; Stoggl et al. 2010). Likewise, in sports such as rowing, other physique traits like a shorter sitting height (relative to stature) and longer limb lengths are related to competitive success (DeRose et al. 1989), with such information being used successfully in talent identification (Hahn 1990). Because of these relationships, it has become common practice to monitor physique traits of athletes in response to growth, training, and dietary interventions.

Despite the association between physique traits and competitive success, the assessment of body composition among athletes, especially female athletes and dancers, has been questioned given the possibility that assessments promote anxiety and disordered eating (Carson et al. 2001). This comes despite recognition that evidence supporting a causal relationship between body composition assessments and disordered eating has yet to be established. However, current evidence indicates that when undertaken in conjunction with a suitably designed education program, assessments of physique can proceed without promoting adverse affective consequences (Whitehead et al. 2003). Thus, the concurrent education of athletes on the rationale for assessments makes good sense and should be actively promoted.

A wide array of techniques is available for the measurement of body composition, including anthropometric, radiographic (computed tomography [CT], dual energy x-ray absorptiometry [DXA]), and other medical imaging techniques (magnetic resonance imaging [MRI], ultrasound) as well as metabolic (creatinine, 3-methylhistidine), nuclear (total body potassium, total body nitrogen), and bioelectrical impedance analysis (BIA) techniques. To select the most appropriate technique, the tester must consider a range of factors, including technical issues such as the safety, validity, precision, and accuracy of measurement. Practical issues must also be considered, such as availability, financial implications, portability, invasiveness, time effectiveness, and technical expertise necessary to conduct the procedures. Consideration must be given to the ability of body composition assessment methods to accommodate the unique physique traits characteristic of some athletes, including particularly tall, broad, and muscular individuals or those with extremely low body fat levels. This chapter reviews the most common techniques used to assess the physique traits of athletes, with a focus on the technical or procedural aspects associated with their use. Brief consideration is given to emerging technologies that may have wider application to athletes into the future, such as three-dimensional photonic scanning.

Body Composition Models

A number of well-accepted two-compartment body composition assessment models are available to monitor body composition, including hydrodensitometry, air displacement plethysmography, and deuterium dilution. These methods are based on the premise that the body can be separated into two chemically distinct compartments: fat mass (FM) and fat-free mass (FFM) (Withers et al. 1999). However, each method carries some degree of error, most of which lies not in the technical accuracy of the measurements but in the biological variability of the assumptions associated with each technique used to generate body composition data from raw measures like body density and total body water. This is especially the case for FFM estimates. The combination of data from several of these two-compartment models into a multicompartment model reduces the number of assumptions made and is now recognized as the standard reference method in body composition assessment (Withers et al. 1999).

A three-compartment model will adjust the body density obtained from hydrodensitometry or air displacement plethysmography (Millard-Stafford et al. 2001) for FFM hydration or total body water using isotope dilution rather than assume a FFM total body water content of 73.72% (Brozek et al. 1963). Variation in FFM hydration from the assumed constant, as occurs in states of hypohydration and hyperhydration, will (respectively) increase and decrease FFM density with associated under- and overestimation of body fat percentage via hydrodensitometry by as much as 10% (Hewitt et al. 1993). Furthermore, measurement of FFM hydration is particularly relevant given that it has by far the lowest density of any FFM component yet occupies the largest percentage of the FFM.

The introduction of DXA has afforded the creation of a four-compartment model that controls for biological variability in both total body water and bone mineral content. Although this model, because it controls for biological variability, is theoretically more valid than the three-compartment model, work by Withers and associates (1998) indicates that the additional control for interindividual variation in bone mineral mass achieves little extra accuracy, at least in young untrained and trained males and females. This supports the original work of Siri (1993), indicating that the largest source of measurement error is related to FFM hydration.

Although this multicompartment approach to body composition assessment can be time-consuming and expensive, it is now widely recognized as the reference method in body composition assessment (Withers et al. 1999) and thus is recommended in research that uses changes in body composition as a key outcome measure. It should also be the criterion against which other body composition assessment techniques are validated. The resource-intensive nature of the multicompartment model limits its application in the routine monitoring of athletes. A detailed description of the method required is documented elsewhere (Withers et al. 1987b, 1998).

Air Displacement Plethysmography

Although air displacement plethysmography has been used to measure human body composition for some time, a viable system known by the trade name BOD POD (Life Measurement, Inc., Concord, California), has only been available commercially since the mid-1990s. This quick, comfortable, automated, noninvasive, and safe technique provides an estimate of fat percentage from body density without being submerged (as occurs with hydrodensitometry) and accommodates a range of subject types (Fields et al. 2002), including very tall and muscular athletes (Dempster and Aitkens 1995). Air displacement plethysmography uses basic gas laws to describe the inverse relationship between pressure and volume in two enclosed chambers, consequently allowing for the calculation of body density and body composition.

This two-compartment technique involves sitting quietly in an enclosed chamber while the volume of air displaced by the body is measured. Thereafter, lung functional residual capacity is measured using pulmonary plethysmography or, if this is not possible, is predicted based on age, gender, and height. Body density is then calculated by dividing the measured body mass by corrected body volume, with subsequent calculation of body fat percentage using either the Siri (Siri, 1993) or Brozek (Brozek et al. 1963) equations. As such, the BOD POD is constrained by the same issues as hydrodensitometry when converting a measure of body density into body composition.

The measurement of functional residual capacity using pulmonary plethysmography is both reliable and valid (Davis et al. 2007). However, functional residual capacity can change in response to significant adjustments in body composition and, as such, should be measured wherever possible and never used interchangeably with predicted thoracic gas volumes (Minderico et al. 2008). Furthermore, on an individual basis, body fat can deviate by as much as 3% depending on the use of measured versus predicted lung volume (Collins and McCarthy 2003).

The BOD POD compares favorably as a substitute for hydrodensitometry when a measure of body density is desired (Fields et al. 2000), although it has been observed to underestimate body fat slightly in males ($-1.2\% \pm 3.1\%$) and overestimate body fat in females ($1.0\% \pm 2.5\%$), independent of age, weight, or height (Biaggi et al. 1999). This effect is evident among athletic populations as well, with BOD POD–derived estimates of body fat consistently lower than those obtained via hydrodensitometry, DXA, and a three-compartment model for male collegiate football athletes (Collins et al. 1999). Among female athletes, the BOD POD overestimates body fat when compared against hydrodensitometry but either compares favorably (Ballard et al. 2004) or underestimates body fat compared with DXA (Bentzur et al. 2008). These differences may also be influenced by presenting body composition: The BOD POD, when compared against a multicompartment reference method, overestimates body fat as fat mass increases yet underestimates body fat in leaner athletes (Moon et al. 2009).

Although the ability to measure absolute characteristics is an important attribute of a physique assessment technique, equally important is the ability to identify small but potentially important changes in body composition in response to diet, training, or other interventions. Research using the artificial manipulation of body composition by the addition of 1 to 2 L of oil, water, or a combination of both substances to the BOD POD chamber among normal-weight individuals suggests that the technology has the ability to pick up changes in either component within the range of 2 kg (Le Carvennec et al. 2007; Secchiutti et al. 2007). When the BOD POD has been used in conjunction with DXA to monitor body composition changes in response to lifestyle interventions, BOD POD estimates of fat mass are typically lower, with concomitant higher estimates of FFM (Minderico et al. 2006; Weyers et al. 2002). However, agreement between techniques was high for identifying the changes in physique traits in response to lifestyle interventions (Frisard et al. 2005; Minderico et al. 2006; Weyers et al. 2002).

As with other assessment tools, subject presentation can influence results, and suitable protocols must be implemented to avoid the impact of these on the reliability of data. Specifically, uncompressed facial and scalp hair underestimate body fat attributable to trapped isothermal air in body hair (Higgins et al. 2001). Similarly, loose-fitting clothing worn during assessment influences body density mea-surements, underestimating body fat percentage by upward of 9% (Fields et al. 2000; Vescovi et al. 2002). Consequently, subjects are advised to wear standardized tight-fitting swimsuits (Fields et al. 2000; King et al. 2006; Vescovi et al. 2002) in conjunction with a silicone swimcap (Peeters and Claessens, 2011) and to remove excess facial hair (Higgins et al. 2001). Changes in body temperature and moisture content can also influence BOD POD data (Fields et al. 2004), suggesting that assessments should be undertaken independent of exercise. This is further supported by the fact that acute dehydration (within the range often experienced by athletes) influences BOD POD results, underestimating both body fat percentage and FFM, although the effect is within the range of 1% body fat (Utter et al. 2003).

Minimizing the influence of these variables enhances the ability of the BOD POD to track small but potentially important changes in body composition. Although the test–retest reliability of the BOD POD is excellent (Noreen and Lemon 2006), this does not provide insight into the biological variability evident between test measures. The between-day coefficient of variation for body fat using the BOD POD is within the range of 2.0% to 5.3% (Anderson 2007), although large discrepancies (up to 12%) between trials have been reported in a small percentage of individuals (Noreen and Lemon 2006), for reasons still yet to be determined (which could include variation in breathing patterns or transient change in pressure within the test room). In practice, technicians are encouraged to undertake repeat measurements, and if these two tests show a difference in body fat percentage greater than 0.5%, then a third test may be appropriate (Collins and McCarthy 2003). Although reliability between individual BOD POD systems is very good (Ball 2005), athletes should be assessed using the same machine each time when evaluating change.

Dual Energy X-Ray Absorptiometry

Dual energy x-ray absorptiometry (DXA) was originally developed for the diagnosis of osteoporosis and remains the reference method for this assessment (Lewiecki 2005). It may also have application as a screening tool in identifying athletes at risk of stress fractures (Kelsey et al. 2007; Prouteau et al. 2004). DXA technology is also able to measure soft tissue body composition and has rapidly gained popularity in recent years as one of the most widely used and accepted laboratory-based methods for body composition analysis. DXA not only provides a measure of FM and FFM but also provides information on regional body composition (e.g., arms, legs, trunk) and differences between left and right sides, making it unique among physique assessment tools and particularly appealing for athletes undertaking targeted training programs or undergoing periods of rehabilitation from injury. Furthermore, whole-body scans are rapid (~5 min) and noninvasive and are associated with very low radiation doses (~0.5 μSv or approximately 1/500th of annual natural background radiation), making the technology safe for longitudinal monitoring of body composition. However, individuals are cautioned against undertaking more than four scans per annum because of the cumulative radiation dose. Similarly, females of child-bearing age should be screened for pregnancy prior to scanning (Ackland et al. 2012). Because of its application in the assessment of bone mineral density, DXA technology is also becoming increasingly available for athletic measurement.

Although there are three manufacturers of DXA technology, including Hologic Inc. (Waltham, Massachusetts), Lunar Radiation Corp. (Madison, Wisconsin), and Norland Medical Systems (Fort Atkinson, Wisconsin), all models share in common an X-ray source, a scanning table, a detector, and a computer interface with complex algorithm software to convert raw data into estimates of body composition. The systems differ in analysis software and the geometry of scanning, using either fan-beam, narrow fan-beam, or pencil-beam technology, which ultimately determines scanning time, radiation dose, and accuracy (Genton et al. 2002). Because of these differences, longitudinal monitoring of athletes should be undertaken on the same machine (Hull et al. 2009; Soriano et al. 2004; Tothill et al. 2001) and using the same technician (Burkhart et al. 2009), especially if regional body composition changes are of interest.

DXA technology is based on the differential attenuation of transmitted photons at two energy levels by bone, fat, and lean tissue (Mazess et al. 1990). The differential attenuations are then expressed as a ratio, the outcome of which is specific to different molecular components, including fatty acids, protein, and bone. In theory, assessment of all three components would require measurement at three different photon energies. Thus, the DXA system can be used to estimate the fractional masses of only two components in any one pixel. That is, in bone-containing pixels, bone mineral and soft tissue can be measured, whereas in non–bone-containing pixels, fat and bone-mineral-free lean tissue can be measured (Pietrobelli et al. 1996). The proportion of fat and bone-mineral-free lean tissue in bone-containing pixels is assumed to be the same as the adjacent non–bone-containing pixels (Pietrobelli et

al. 1996), with the software subsequently converting individual pixel data into whole-body output. This assumed ratio of fat to bone mineral–free lean tissue in soft tissue pixels is applied to upward of one-third of pixels in a whole-body scan and is particularly evident in regions of low bone-free pixels, such as the thorax, arm, and head, resulting in less reliable identification of composition changes in these regions (Lands et al. 1996; Roubenoff et al. 1993).

DXA technology has been compared with the modern-day reference body composition assessment tool, the four-compartment model, which accounts for variation in the water and mineral fractions and the density of the FFM. Although some data suggest good agreement between DXA-derived measures of body composition and the four-compartment model in healthy young males and females (Prior et al. 1997), others have reported that DXA underestimates body fat (Deurenberg-Yap et al. 2001), especially among leaner individuals (Van Der Ploeg et al. 2003; Withers et al. 1998). This underestimation has been attributed to variation in FFM hydration (Deurenberg-Yap et al. 2001) or differences in anterior–posterior tissue thickness (Van Der Ploeg et al. 2003). Conversely, DXA has been shown to overestimate fat mass in females, and this bias increases as body fat levels increase (Moon et al. 2009; Silva et al. 2006). Despite this, when the primary focus is to monitor change in body composition, DXA appears to offer sufficient sensitivity to identify small changes in body composition, at least when a pencil-beam system is used (Houtkooper et al. 2000; Weyers et al. 2002). Fan-beam systems may not have the same resolution (Santos et al. 2010).

The precision of measurement for DXA in sedentary populations has been shown to be superior to hydrodensitometry and surface anthropometry (Pritchard et al. 1993), with a coefficient of variation of less than 1.0 kg for FM, FFM, and total mass (Mazess et al. 1990; Tothill et al. 1994). Any variability of results achieved by DXA can be divided into two categories: technical and biological error. In an effort to enhance precision of measurement, special consideration should be given to subject positioning (Lambrinoudaki et al. 1998; Lohman et al. 2009). Clothing should be kept to a minimum (Koo et al. 2004), with all metal objects removed. Although small amounts of food and fluid do not appear to influence results (Vilaca et al. 2009), larger volumes influence measurement of lean body mass (Horber et al. 1992), and thus measurements should be undertaken in a fasted state wherever possible, preferably soon after waking in a euhydrated state (Going et al. 1993), using a standardized scanning protocol (Nana et al. 2012). The reliability of regional measurements is inferior to total body results (Calbet et al. 1998; Lohman et al. 2009; Nana et al. 2012).

Of particular relevance to athletic populations is the defined scanning area available for assessment, typically within the range of 60 to 65 cm × 193 to 198 cm depending on the manufacturer (Genton et al. 2002). It is therefore difficult to perform whole-body DXA scans on athletes who are particularly tall or broad and very muscular, physique traits common to sports such as rowing, basketball, volleyball, and rugby union. There are several options when athletes are taller than the defined scanning area: exclude them from investigation altogether, scan the area of the body excluding the head or feet, have athletes bend their knees so as to fit within the scanned area (Silva et al. 2004), or sum data from two partial scans. The last option appears to be the method of choice; dividing the body at the neck for the two partial scans provides the most accurate estimates of bone and soft tissue composition (Evans et al. 2005). Until recently, very broad individuals were "mummy wrapped" in a sheet, bringing the arms forward, so as to fit within the scanning area. Although this allowed for a whole-body scan, the number of bone-containing pixels was significantly increased while limiting the ability to assess body composition at particular regions of interest such as the arms and torso. Newer DXA instruments like the iDXA from GE Lunar (Madison, Wisconsin) not only have larger scanning areas (66 cm wide) but also have software that allows an estimate of whole-body composition from a half-body scan (Rothney et al. 2009), a concept validated previously in obese individuals (Tataranni and Ravussin 1995). For particularly tall and broad athletes, two or three scans may be required to obtain whole-body composition data.

Bioelectrical Impedance Analysis

Bioelectrical impedance analysis (BIA) has become increasingly popular as a tool for assessing the physique traits of athletes given its relative ease of use, portability, and cost-effectiveness. It is a safe and noninvasive method to assess body composition that is based on the differing electrical conductivity of FM and FFM (Kyle et al. 2004a; National Institutes of Health 1996). FFM contains water and electrolytes and is a good electrical conductor, whereas anhydrous fat mass is not. The method involves measuring the resistance (R) to flow of a low-level current per second (Kyle et al. 2004a). Resistance is proportional to the length (L) of the conductor (in this case the human body) and inversely proportional to its cross-sectional area (A). A relationship then exists between the impedance quotient (i.e., L^2 / R) and the volume of water (total body water) containing electrolytes that conduct the electrical current. In practice, height in centimeters is substituted for length. Therefore, a relationship exists between FFM (approximately 73% water) and height $(cm)^2 / R$. FM is obtained from FFM by subtracting the value for FFM from total body mass (Kyle et al. 2004a).

Although the relationship between FFM and impedance is readily accepted, several assumptions are associated with the use of BIA to measure body composition. First, the human body is assumed to be a cylinder with a uniform length and cross-sectional area. In fact, the human body more closely resembles several cylinders. The body parts with the smallest FFM (the limbs) have the greatest influence on whole-body resistance. The trunk, which is a shorter, thicker segment, contains approximately 50% of body weight but contributes a minor amount to the overall resistance. Second, it is assumed that the conducting material within the cylinder is homogeneous, which it isn't. Third, the resistance to current flow per unit length of a specific conductor is assumed to be constant. However, this will vary depending on tissue structure, hydration status, and electrolyte concentration of the tissue (Kushner 1992).

Given the relevance of body water to conductivity of electrical current, there is substantial evidence that BIA is not valid for assessment of individuals with abnormal hydration such as visible edema, ascites, kidney, liver and cardiac disease, and pregnancy (Kyle et al. 2004b). Such medical conditions would be unlikely in an athletic population, but changes in hydration status as a consequence of exercise-induced hypohydration and acute food or fluid intake can still influence results. Exercise-induced hypohydration to a level of 3% body mass has been shown to decrease the estimate of FM via BIA by 1.7%. Conversely, acute rehydration increased estimates of FM by 3.2%, with a further increase in the estimate of FM as a state of hyperhydration was achieved (Saunders et al. 1998). Even the ingestion of relatively small volumes of fluid (591 ml) has been shown to increase estimates of FM (Dixon et al. 2009). Consequently, athletes should remain fasted for at least 8 h prior to assessment (Kyle et al. 2004b). Given this, it would be prudent to undertake assessments in the morning prior to breakfast wherever possible, with athletes encouraged to present in a well-hydrated state, which could be determined by collection of a first-morning urine sample.

Given the impact of acute changes in hydration status to resistance and thus estimates of total body weight (TBW) and body composition, it has been suggested that BIA could be used to assess acute changes in hydration status. However, preliminary results suggested that BIA is only able to predict half of the body water loss resulting from exercise (Koulmann et al. 2000). More recently, encouraging results have been reported on the ability of multi-frequency bioelectrical impedance spectroscopy to measure TBW (Moon et al. 2008) and exercise-induced changes in TBW (Moon et al. 2010) when contrasted against the deuterium dilution reference technique.

Other factors reported to influence BIA results include posture and positioning of the limbs, prior exercise, skin temperature (Kushner et al. 1996), and electrode placement (Dunbar et al. 1994). Adherence to the test protocol in this chapter will reduce measurement error but errors may still occur with this technique, prompting some to suggest that BIA may be most appropriate for estimating body composition of groups rather than individuals (Houtkooper et al. 1996). Kyle and associates (2004a) suggest that prediction of errors of less than 3.0 kg for males and 2.3 kg for females would be considered good.

Within an athletic population, BIA has been shown to overestimate FFM in smaller males and underestimate FFM in heavier athletes competing in a weight category sport (Utter and Lambeth 2010). Similarly, BIA has been shown to underestimate fat mass in team sport athletes (Svantesson et al. 2008). If BIA were used to calculate a minimum weight prediction for these athletes, errors of as much as 6 to 7 kg are possible, greater than those identified with other physique assessment techniques such as DXA, hydrodensitometry, and surface anthropometry (Clark et al. 2004). The selection of appropriate prediction equations to convert raw impedance results into FFM is an important consideration, with equations derived from untrained individuals likely to be inappropriate for an athletic population (Pichard et al. 1997). Kyle and associates (2004a) provide an excellent summary of prediction equations derived from healthy subjects and published since 1990 for adults for FFM, body fat, and TBW.

The measurement of difference in electrical properties of various body tissues is not a new concept. The original studies were conducted by Thomasset in the 1960s using two subcutaneously inserted needles (Thomasset, 1963). The technique was subsequently refined in the 1970s to include four surface electrodes and resulted in commercially available single-frequency analyzers (Kyle et al. 2004a). Since this time, there have been significant advances in BIA, with options of single-frequency, multifrequency, and segmental BIA.

Foot-to-foot BIA analyzers are the most readily available for public purchase (Jaffrin 2009). These inexpensive body fat "scales" are popular among the general public, being promoted as a simple, portable method of measuring body fatness. However, foot-to-foot devices have a number of limitations, because current is circulated only through the legs and lower part of the trunk and results are extrapolated to the whole body (Jaffrin, 2009). Furthermore, raw impedance data are rarely reported, making selection of population specific equation impossible. Although many systems come with an "athletic" mode option, this does not enhance the predictive accuracy of BIA compared with a reference method (Dixon et al. 2006).

Single-frequency BIA (50 kHz) is most commonly used as a field technique to measure body composition. It usually consists of four electrodes placed on the wrist, hand, ankle, and foot, although foot-to-foot and hand-to-hand analyzers are also available. The two source electrodes are placed on the dorsal surface of the right hand and foot proximal to the metacarpal–phalangeal and metatarsal–phalangeal joints, respectively. The two voltage electrodes are placed on the midpoint between the distal prominences of the radius and ulna of the right wrist and between the medial and lateral malleoli of the right ankle (Wagner and Heyward 1999).

Multifrequency BIA consists of measurement of impedance at various frequencies (0, 1, 5, 50, 100, 200, and 500 kHz). At a low single frequency, an electrical current will pass through extracellular water, not fully penetrating the cell membrane, whereas at high frequencies the current will penetrate the cell membrane (Cornish et al. 1996). By measuring various components across a number of frequencies, a mathematical model can be derived that can be used to predict TBW and subsequently FFM, eliminating the need for standard regression equations. This may be more appropriate for individuals who have variation in standard hydration status or body composition (Martinoli et al. 2003). The results of various studies comparing multifrequency BIA to single-frequency BIA and other methods such as DXA have reported mixed results (Pateyjohns et al. 2006; Sun et al. 2005).

Surface Anthropometry

For reasons of timeliness, practicality, and cost-effectiveness, the routine monitoring of body composition among athletic populations is most often undertaken using anthropometric characteristics such as body mass plus subcutaneous skinfold thicknesses and girths at specific anatomical landmarks. Unlike other techniques requiring expensive, laboratory-based equipment, surface anthropometry only requires relatively inexpensive equipment that is easily portable. However, highly skilled technicians are required if reliable data are to be collected. Technicians need to be meticulous in terms of both accurate site location and measurement technique. Measurements just 1 to 2 cm away from a defined site can produce significant differences in results (Hume and Marfell-Jones 2008; Ruiz et al. 1971). Furthermore, if repeat measurements are to be taken over time, it is important that the same technician collect the data (Hume and Marfell-Jones 2008).

The measurement of "skinfolds" or a double layer of skin and subcutaneous tissue as an index of whole body fat would appear to be reasonable. However, what is really being measured is the thickness of a double fold of skin and compressed subcutaneous adipose tissue (Martin et al. 1985). To infer from this the mass or percentage of total body fat requires a number of assumptions to be made:

- Constant compressibility of skinfolds across sites on the body
- Skin thickness at any one site that is negligible or a constant fraction of a skinfold
- Fixed adipose tissue patterning across the body
- A constant fat fraction in adipose tissue
- Fixed proportion of internal to external fat

When assessed via cadaver analysis, few of these assumptions hold true (Clarys et al. 2005). For example, skinfold compressibility is not constant between sites, and as a consequence, similar thicknesses of adipose tissue may yield different caliper values attributable to different degrees of tissue compressibility (Marfell-Jones et al. 2003). Furthermore, the patterning of adipose tissue varies markedly between individuals (Mueller and Stallones 1981), and so multiple skinfold sites should be used, including both upper- and lower-body landmarks (Eston et al. 2005). Similarly, although it is estimated that subcutaneous fat comprises one-third of total body fat, this can range from 20% to 70% depending on gender, degree of fatness, and age (Lohman 1981). Despite an obvious violation of these assumptions, a strong relationship does exist between subcutaneous adiposity and whole-body adiposity and between direct skinfold thickness measures and whole-body adiposity (Clarys et al. 2005).

Estimates of body density, FM, and FFM can be derived from raw skinfold data using one of many available regression equations. Altogether, more than 100 equations to predict body fat from skinfolds have been produced (Clarys et al. 1987, 1999; Lohman 1981; Martin et al. 1985). However, these equa-

tions are typically based on a single-measurement, between-subject, cross-sectional comparison of anthropometric parameters and laboratory-based techniques such as hydrodensitometry (Cisar et al. 1989), increasing the assumptions made. Because these equations are population specific, only equations derived from individuals with similarities in age, gender, body composition, and activity levels should be considered for use. Furthermore, compatibility in technical aspects of data collection, including anatomical landmarking and anthropometric equipment, is essential. Consequently, skinfold equations derived from athletic populations such as that of Withers and colleagues (1987a) are likely to offer a more accurate estimate of body composition (Reilly et al. 2009). However, the ability of these equations to track changes in physique traits in response to training or dietary interventions has not been widely assessed (Cisar et al. 1989; Wilmore et al. 1970). Preliminary data suggest that popular skinfold-based models, including those derived from athletes, lack the sensitivity to track small but potentially important changes in body composition (Silva et al. 2009; van Marken Lichtenbelt et al. 2004). As such, it seems unreasonable to introduce further error by transforming raw skinfold data into estimates of fat mass or body fat percentage. Thus, despite the advancement in physique assessment techniques and the notable desire of many athletes to know their body fat percentage, the conclusions of Johnston (1982) remain true to this date: Practitioners are better off using raw anthropometric data rather than attempting to estimate whole-body composition from available equations. Normative surface anthropometry data on a wide-ranging population of elite athletes across a range of sports are provided in table 11.1.

Although sums of skinfolds are highly correlated with body fat percentage, FFM correlates poorly with skinfolds (Roche 1996). It has been proposed that combining skinfolds with certain body circumferences leads to a better estimate of FFM (Forbes 1999). In theory, skinfold-corrected circumferences offer a more direct assessment of muscle mass, assuming that the skinfold thickness accurately partitions fat and lean components at a specific site (Reid et al. 1992). However, the skinfold-corrected girth estimates have been shown to be less accurate in monitoring changes in muscle mass than predictions using skinfolds alone (Cisar et al. 1989; Stewart and Hannan 2000). This imprecision may be explained, at least in part, by the fact that muscle hypertrophy does not occur uniformly throughout each body region (Abe et al. 2003), yet the anthropometric fractionation estimate of muscle mass places equal weighting on each of five girth measurements

TABLE 11.1 Anthropometric Test Data for Female and Male Sport Players (mean ± SD; range)

Sport	Gender	Squad/level	Age/position	n	Height, cm	Mass, kg	Σ7 skinfolds, mm
Athletics	Female	Elite	Sprints/hurdles (100-400 m)	63	172.5 ± 4.9 (162.1-181.1)	62.4 ± 4.6 (56.1-73.4)	52.9 ± 9.2 (36.8-81.9)
		Elite	Distance (800 m to marathon)	34	167.8 ± 6.2 (157.8-184.2)	55.6 ± 5.9 (46.2-67.5)	56.4 ± 14.3 (28.0-96.0)
		Elite	Walks (10-50 km)	42	167.9 ± 3.9 (159.2-174.0)	55.5 ± 3.1 (47.7-62.1)	77.3 ± 13.9 (43.0-111.7)
	Male	Elite	Sprints/hurdles (100-400 m)	175	182.4 ± 5.5 (170.7-192.0)	77.5 ± 6.6 (64.0-96.2)	39.4 ± 8.9 (25.9-90.6)
		Elite	Distance (800 m to marathon)	180	178.9 ± 6.0 (169.0-192.4)	66.2 ± 6.5 (54.0-82.7)	37.4 ± 5.4 (23.5-54.8)
		Elite	Walks (10-50 km)	66	180.0 ± 6.8 (166.2-191.7)	65.4 ± 6.9 (55.0-81.2)	38.2 ± 5.4 (30.3-55.5)
Australian Rules	Male	AIS	~18 years	284	186.0 ± 0.1 (167.0-202.0)	81.0 ± 8.0 (63.0-107.0)	56.0 ± 13.0 (31.0-113.0)
Basketball	Female	AIS	17.7 (14.2-27.7)	500	182.6 ± 7.6 (158.8-204.0)	75.0 ± 11.0 (47.0-118.0)	93.0 ± 21.0 (46.0-165.0)
		National	Senior	17	n/a	n/a	88.5 ± 21.2 (55.6-142.8)
	Male	AIS	17.8 (14.4-22.7)	500	197.7 ± 9.2 (174.1-221.8)	92.0 ± 13.0 (59.0-141.0)	58.0 ± 17.0 (34.0-220.0)
Cricket	Female	State squad - national	Open	246	167.6 ± 4.9 (153.6-176.0)	65.2 ± 8.0 (48.5-97.0)	109.1 ± 32.3 (49.5-213.5)
		CA contracted - international	Open	122	168.3 ± 5.7 (153.6-177.5)	65.4 ± 7.5 (55.2-86.0)	105.2 ± 29.5 (49.6-193.2)
	Male	National	Under 19	299	181.9 ± 6.6 (170.0-196.8)	84.0 ± 8.7 (61.5-103.5)	69.3 ± 17.1 (38.0-122.0)
		AIS	Open	632	183.2 ± 7.4 (170.0-203.0)	85.7 ± 8.3 (55.0-110.5)	71.6 ± 19.6 (30.2-149.2)
		State squad - national	Open	1302	183.8 ± 6.8 (169.6-203.0)	86.0 ± 8.7 (63.5-110.5)	70.8 ± 19.6 (30.2-152.8)
		CA contracted - international	Open	212	182.4 ± 7.3 (170.0-197.0)	85.8 ± 8.1 (72.7-102.0)	71.7 ± 16.8 (40.8-125.9)

Sport	Gender	Squad/level	Age/position	n	Height, cm	Mass, kg	Σ7 skinfolds, mm
Cycling	Female	Track sprint	Senior	3	n/a	70.4 ± 2.9 n/a	87.2 ± 15.9 n/a
		Road	Senior	9	n/a	58.2 ± 3.9 (50.9-64.9)	55.2 ± 15.5 (30.8-73.6)
		MTB	Senior	6	n/a	57.4 ± 4.3 (53.4-62.8)	68.0 ± 18.7 (41.6-92.2)
		MTB	Junior	5	n/a	58.2 ± 7.3 (49.9-65.2)	85.0 ± 9.7 (69.5-94.9)
		BMX	Senior	5	n/a	70.2 ± 2.9 (67.4-74.3)	109.2 ± 18.7 (77.8-130.0)
	Male	Track endurance	Senior	6	n/a	72.6 ± 4.4 (69.0-80.2)	38.1 ± 5.1 (30.5-53.4)
		Track sprint	Senior	5	n/a	89.8 ± 5.0 n/a	52.3 ± 10.0 n/a
		Road	Under 23	7	n/a	73.4 ± 6.8 (58.2-77.4)	45.1 ± 8.8 (32.5-56.2)
		Road	Junior	2	n/a	63.0 ± 0.6 (62.6-63.4)	37.9 ± 2.1 (36.4-39.3)
		MTB	Under 23	11	n/a	72.5 ± 7.2 (63.4-88.5)	42.9 ± 6.6 (32.5-53.9)
		MTB	Junior	9	n/a	67.0 ± 8.2 (56.7-80.5)	43.0 ± 7.3 (32.5-53.9)
		BMX	Senior	9	n/a	84.4 ± 6.7 (71.0-93.0)	47.6 ± 11.7 (31.8-72.2)
Football (soccer)	Female	National	Under 17	42	163.1 ± 7.6 (154.5-170.1)	52.9 ± 4.2 (44.6-68.4)	98.6 ± 11.1 (63.6-151.7)
		National	Senior	38	166.8 ± 6.1 (156.2-178.3)	61.3 ± 4.7 (52.5-72.7)	75.2 ± 7.2 (53.7-103.1)
		State	Under 18/senior	6	n/a	62.7 ± 6.0 (56.3-73.0)	100.7 ± 14.7 (76.6-114.7)
	Male	National	Under 17	74	176.7 ± 5.8 (163.3-190.8)	69.6 ± 4.8 (58.0-82.4)	56.4 ± 7.6 (43.2-79.6)
		Olympic squad	Under 23	54	181.8 ± 6.1 (165.0-192.2)	77.5 ± 5.1 (60.0-96.4)	57.8 ± 9.3 (37.2-78.1)
		National	Senior	44	182.3 ± 6.5 (166.1-197.8)	78.3 ± 6.6 (61.2-96.8)	47.5 ± 11.1 (32.3-73.6)

(continued)

TABLE 11.1 (*continued*)

Sport	Gender	Squad/level	Age/position	n	Height, cm	Mass, kg	Σ7 skinfolds, mm
Hockey	Female	National senior	Striker	4	n/a	65.2 ± 4.7 (58.9-70.1)	65.5 ± 13.6 (56.4-85.7)
		National senior	Midfield	9	n/a	60.4 ± 2.2 (57.1-62.3)	76.6 ± 13.5 (62.4-104.3)
		National senior	Defender	6	n/a	62.0 ± 3.2 (57.7-67.3)	68.4 ± 13.0 (49.5-82.5)
		National senior	Goalkeeper	2	n/a	65.0 ± 7.4 (59.8-70.2)	66.7 ± 17.3 (54.5-78.9)
		National	Under 21	23	n/a	61.2 ± 4.8 (54.9-71.7)	81.3 ± 13.4 (60.1-112.6)
		National	Under 17	17	n/a	61.9 ± 8.3 (50.5-79.8)	87.8 ± 26.2 (52.6-128.9)
		State	Senior/All positions	10	169.0 ± 4.9 (160.4-176.8)	69.2 ± 7.1 (60.4-80.2)	102.9 ± 21.6 (66.8-137.2)
	Male	National senior	Striker	8	n/a	80.3 ± 4.9 (74.9-89.3)	56.7 ± 9.6 (44.6-74.7)
		National senior	Midfield	12	n/a	76.0 ± 4.6 (67.0-85.4)	50.2 ± 11.6 (32.2-70.3)
		National senior	Defender	9	n/a	80.3 ± 9.3 (62.2-90.9)	60.0 ± 11.5 (42.9-79.3)
		National senior	Goalkeeper	4	n/a	83.5 ± 8.6 (74.5-95.1)	77.0 ± 26.2 (55.6-110.6)
		National	Under 21	20	n/a	74.9 ± 7.3 (58.9-86.5)	55.3 ± 14.4 (32.6-85.7)
		State	Senior/All positions	11	180.8 ± 7.3 (170.5-195.3)	79.6 ± 10.7 (64.2-102.2)	60.5 ± 19.8 (42.0-103.9)
Netball	Female	AIS	Under 21	46	181.5 ± 4.1 (170.5-186.6)	73.7 ± 5.4 (64.8-91.9)	99.8 ± 21.7 (62.6-169.9)
		National	Senior	30	182.4 ± 6.2 (170.0-196.0)	72.6 ± 8.0 (58.6-92.4)	82.9 ± 17.5 (46.4-136.6)
		State	Under 17	7	174.7 ± 7.4 (162.2-182.1)	63.8 ± 4.4 (57.7-69.1)	90.5 ± 21.6 (61.3-121.5)
		State	Under 19	15	175.2 ± 6.3 (161.2-183.7)	67.9 ± 8.7 (50.7-87.6)	96.6 ± 18.6 (62.7-133.5)
		State	Senior	23	178.3 ± 6.5 (168.5-193.9)	67.7 ± 6.2 (55.4-77.5)	92.7 ± 18.0 (54.8-123.8)
Rowing	Female	National	Lightweight	18	n/a	59.3 ± 1.8 (54.4-62.6)	60.2 ± 9.3 (49.2-84.0)
		National	Heavyweight	50	n/a	75.5 ± 5.3 (65.4-87.7)	82.4 ± 19.2 (49.0-139.4)
	Male	National	Lightweight	30	n/a	73.6 ± 1.2 (71.1-76.2)	37.6 ± 4.6 (30.3-47.7)
		National	Heavyweight		n/a	92.2 ± 6.7 (77.4-108.7)	54.0 ± 13.0 (36.5-90.4)

Sport	Gender	Squad/level	Age/position	n	Height, cm	Mass, kg	Σ7 skinfolds, mm
Rugby League	Male	Subelite	Junior	36	176.0 ± 6.0 (162.0-188.5)	74.0 ± 13.0 (48.0-104.0)	75.0 ± 32.0 (40.0-154.0)
		Elite	Junior	28	178.0 ± 5.9 (167.0-191.5)	78.0 ± 10.0 (64.0-104.0)	67.0 ± 15.0 (43.0-107.0)
		Semiprofessional	Senior	77	183.1 ± 6.5 (165.0-195.0)	96.0 ± 11.0 (74.0-119.0)	72.0 ± 23.0 (38.0-145.0)
		Professional	Senior	68	184.1 ± 5.5 (173.0-197.0)	96.0 ± 9.0 (82.0-121.0)	53.0 ± 12.0 (37.0-101.0)
Sailing	Female	National	470 Helm	6	165.1 ± 4.5 (159.9-173.1)	55.1 ± 4.9 (48.2-60.2)	86.3 ± 23.9 (55.8-116.4)
		National	470 Crew	5	175.5 ± 4.0 (170.5-179.5)	67.2 ± 4.0 (63.8-73.4)	76.9 ± 15.0 (60.5-94.0)
		National	RSX	3	170.1 ± 2.0 (168.1-172.1)	62.7 ± 7.7 (55.3-70.6)	87.3 ± 21.1 (69.5-110.7)
		National	Match racing	9	167.5 ± 4.3 (161.8-175.1)	66.4 ± 2.5 (61.8-70.5)	108.2 ± 15.4 (90.3-138.7)
	Male	National	470 Helm	9	173.0 ± 5.3 (167.1-182.9)	61.8 ± 4.2 (53.0-68.5)	50.7 ± 15.8 (37.9-89.1)
		National	470 Crew	7	186.5 ± 4.8 (179.1-194.4)	72.1 ± 4.8 (66.5-81.7)	47.7 ± 14.1 (31.4-73.8)
		National	49er Helm	6	178.8 ± 4.2 (173.0-185.7)	73.9 ± 3.8 (69.1-78.9)	54.5 ± 7.5 (39.8-60.4)
		National	49er Crew	6	184.5 ± 3.5 (179.8-187.6)	80.1 ± 2.3 (75.9-82.7)	53.9 ± 8.0 (43.0-66.9)
		National	Laser	8	182.4 ± 4.2 (176.9-189.3)	80.5 ± 3.5 (76.3-86.0)	60.5 ± 18.9 (37.7-87.0)
		National	Laser radial	4	172.0 ± 2.4 (169.7-174.6)	66.6 ± 3.8 (63.3-71.6)	89.1 ± 15.1 (76.7-110.5)
		National	Finn	3	186.3 ± 2.6 (184.6-189.3)	96.2 ± 4.3 (93.3-101.2)	80.6 ± 12.1 (71.9-94.4)
Sprint Kayak	Female	National	Senior	9	170.6 ± 6.2 (162-183)	67.4 ± 6.4 (56.2-76.5)	65.5 ± 13.4 (45.6-82.1)
		State	Junior/senior	6	n/a	66.0 ± 5.4 (57.9-72.4)	89.6 ± 16.6 (69.8-112.7)
	Male	National	Senior	15	185.1 ± 4.5 (178-193)	87.3 ± 5.7 (72.0-94.5)	42.6 ± 6.7 (31.8-59.9)
		State	Junior/senior	8	n/a	75.3 ± 4.1 (70.7-81.9)	45.9 ± 5.6 (40.5-57.0)

(continued)

TABLE 11.1 (*continued*)

Sport	Gender	Squad/level	Age/position	n	Height, cm	Mass, kg	Σ7 skinfolds, mm
Super rugby	Male	National	Prop	21	184.7 ± 2.8 (182.0-188.9)	112.5 ± 9.1 (106.4-128.0)	100.1 ± 22.0 (79.4-148.8)
		National	Hooker	8	181.5 ± 3.6 (179.2-186.2)	103.8 ± 4.2 (98.0-115.7)	83.7 ± 18.4 (66.9-97.2)
		National	Second row	14	194.2 ± 10.5 (193.3-202.0)	111.1 ± 3.7 (104.2-116.4)	78.9 ± 20.6 (52.7-116.1)
		National	Back row	25	187.9 ± 3.8 (175.5-196.0)	105.7 ± 7.3 (96.4-117.9)	75.2 ± 21.9 (52.6-104.0)
		National	Scrum half	12	180.0 ± 5.0 (174.6-183.7)	85.4 ± 9.5 (76.2-89.1)	60.5 ± 11.2 (37.0-64.8)
		National	Fly half	10	181.5 ± 5.1 (174.0 -188.7)	90.0 ± 5.8 (82.5-103.0)	57.1 ± 9.0 (42.7-99.7)
		National	Center	17	182.2 ± 5.1 (176.6-188.9)	94.4 ± 5.5 (83.2-107.8)	62.5 ± 13.4 (47.6-99.5)
		National	Wing/fullback	29	183.7 ± 5.5 (175.6-191.0)	93.6 ± 7.5 (77.1-105.9)	65.4 ± 15.0 (27.6-89.7)
Swimming	Female	National	Elite	31	171.5 ± 7.0 n/a	64.9 ± 9.0 (51.4-78.1)	67.6 ± 12.0 (49.9-100.5)
	Male	National	Elite	46	183.8 ± 7.1 n/a	82.1 ± 7.9 (66.1-99.6)	49.2 ± 9.0 (32.8-86.0)
Tennis	Female	AIS/national academy	16+ years	6	170.6 ± 6.4 (161.4-183.0)	62.7 ± 7.3 (47.1-77.0)	94.1 ± 20.0 (53.5-122.9)
		National academy	15-16 years	13	167.5 ± 5.7 (160.0-179.0)	57.4 ± 7.6 (46.7-73.0)	105.5 ± 28.9 (73.1-151.8)
		National academy	13-14 years	25	165.0 ± 7.0 (148.1-180.6)	54.8 ± 7.7 (43.7-73.0)	88.5 ± 23.6 (55.9-148.8)
		National academy	11-12 years	14	155.7 ± 7.7 (146.4-170.7)	43.1 ± 6.5 (34.8-52.3)	n/a
	Male	AIS/national academy	16+ years	15	183.6 ± 6.7 (174.6-199.2)	76.5 ± 8.8 (62.8-97.4)	46.5 ± 9.8 (35.9-67.5)
		National academy	15-16 years	22	180.4 ± 5.1 (174.0-189.8)	71.0 ± 5.4 (61.4-82.0)	59.5 ± 9.8 (52.1-74.0)
		National academy	13-14 years	34	171.7 ± 6.8 (152.4-183.2)	59.4 ± 9.0 (41.1-79.4)	55.1 ± 9.6 (38.9-73.9)
		National academy	11-12 years	15	159.8 ± 10.7 (149.0-191.3)	49.2 ± 10.4 (36.5-81.0)	n/a
Triathlon	Female	Elite	Junior	5	166.2 ± 4.9 (159.3-170.1)	52.0 ± 3.0 (49.0-58.0)	55.0 ± 15.0 (34.0-72.0)
		Elite	Under 23	10	165.7 ± 3.6 (159.2-170.5)	55.0 ± 4.0 (49.0-64.0)	52.0 ± 10.0 (40.0-70.0)
	Male	Elite	Junior	2	179.9 ± 11.1 (172.0-187.7)	67.0 ± 8.0 (62.0-73.0)	46.0 ± 11.0 (38.0-54.0)
		Elite	Under 23	18	178.0 ± 6.4 (169.3-193.8)	66.0 ± 7.0 (53.0-78.0)	38.0 ± 7.0 (27.0-50.0)

Sport	Gender	Squad/level	Age/position	n	Height, cm	Mass, kg	Σ7 skinfolds, mm
Volleyball: indoor	Female	National	Junior/senior	11	183.0 ± 6.4 (173.7-194.2)	73.0 ± 5.0 (65.0-84.0)	101.0 ± 21.0 (81.0-136.0)
	Male	National	Junior/senior	22	200.4 ± 5.0 (187.3-207.5)	95.0 ± 6.0 (82.0-109.0)	57.0 ± 9.0 (40.0-72.0)
Volleyball: beach	Female	National	Junior/senior	9	183.7 ± 4.3 (175.9-191.1)	71.0 ± 5.0 (63.0-82.0)	84.0 ± 22.0 (53.0-108.0)
	Male	National	Junior/senior	6	195.2 ± 3.76 (190.7-199.7)	90.0 ± 6.0 (84.0-99.0)	48.0 ± 13.0 (33.0-71.0)
Water polo	Female	National	Senior/center	7	177.0 ± 6.0 (170.0-183.0)	81.0 ± 7.0 (72.0-92.0)	102.0 ± 22.0 (66.0-127.0)
		National	Senior/perimeter	12	174.0 ± 6.0 (167.0-182.0)	70.0 ± 6.0 (62.0-84.0)	91.0 ± 20.0 (66.0-143.0)
		National	Senior/goalkeeper	4	174.0 ± 5.0 (169.0-183.0)	77.0 ± 5.0 (70.0-82.0)	99.0 ± 18.0 (76.0-114.0)
		State	Senior	4	173.1 ± 5.8 (167.8-178.4)	70.3 ± 5.1 (66.0-77.0)	107.0 ± 18.8 (82.9-122.8)
	Male	State	Senior	5	181.8 ± 6.6 (175.7-192.7)	79.1 ± 10.3 (66.4-91.9)	73.3 ± 18.8 (54.5-101.6)

n/a = not available; AIS = Australian Institute of Sport; CA = Cricket Australia; MTB = mountain bike; BMX = bicycle motocross; RSX = windsurfing discipline

(Drinkwater and Ross 1980). Consequently, anthropometric fractionation may not be appropriate when attempting to quantify skeletal muscle mass in athletes with significant muscle hypertrophy (Keogh et al. 2007).

Aside from the convenience of surface anthropometry for assessing physique traits of athletes, parameters such as skinfolds are very robust, not readily influenced by factors such as hydration status of the athlete (Norton et al. 1998) (see table 11.2).

However, measurement of body composition using surface anthropometry is typically undertaken in conjunction with the measurement of body mass, and body mass can be acutely influenced by an array of factors, independent of changes in FM or skeletal muscle mass. Consequently, body mass measurements should be made at the same time of day (preferably before breakfast or training but after voiding the bladder and bowel), and the athlete should wear minimal clothing (Maughan and Shirreffs, 2004) to

TABLE 11.2   Interpretation of Changes in Physique Traits Based on Skinfold and Body Mass Data

ANTHROPOMETRIC TRAIT			INTERPRETATION—PHYSIQUE TRAIT		
Body mass		Skinfolds	Muscle mass		Body fat
Increase	&	Stable	Gain	&	No change
Decrease	&	Stable	Loss	&	No change
Stable	&	Increase	Loss	&	Gain
Stable	&	Decrease	Gain	&	Loss
Increase	&	Increase	Potential gain	&	Gain
Increase	&	Decrease	Gain	&	Loss
Decrease	&	Increase	Loss	&	Gain
Decrease	&	Decrease	Potential loss	&	Loss

minimize the influence of extraneous factors that can affect body mass. Other issues to consider include consistency in the scales used (Stein et al. 2005), the phase of the menstrual cycle in females (Bunt et al. 1989), and hydration status.

Although surface anthropometry does not offer a direct measure of total fat mass or FFM, its robustness in the field, convenience, and low cost ensure that it remains a popular method of body composition monitoring among athletes. Newer technologies like DXA and the combination of technologies in a three- or four-compartment model offer an opportunity to better interpret surface anthropometry data. Preferably, this should be developed around interpretation of changes in physique traits over time rather than a one-off assessment. Such longitudinal investigations also create opportunities to examine the association between physique traits and competitive success.

Precise assessment of anthropometric traits, in particular skinfold thickness, can be difficult, and therefore extreme care in site location and measurement is required if meaningful results are to be obtained. Prior to assessment, the tester should develop the appropriate technique, which will reduce the level of error in repeated measurements and thus enhance the ability to detect small but potentially important changes. The standard assessment protocols of the International Society for the Advancement of Kinanthropometry (ISAK) (Stewart et al. 2011) are recommended. Professionals wishing to monitor the physique traits of athletes using surface anthropometry are strongly encouraged to undertake professional training through accredited courses.

Alternate Technologies

A range of alternate technologies are available for the assessment and monitoring of physique traits of athletes, including ultrasound and three-dimensional photonic scanning. Using ultrasound to measure subcutaneous fat is proposed to overcome some of the drawbacks of subcutaneous skinfold assessment using calipers, particularly error associated with the compressibility and elasticity of skinfolds (Ramirez 1992). The recent availability of higher-resolution, portable, and more affordable ultrasound equipment has created renewed interest in this technology for body composition assessment in field settings. A recent investigation in a mixed athlete population reported strong correlations for both males and females when body fat estimates measured by ultrasound were compared with DXA (Pineau et al. 2009). Similar estimates of FFM have been observed in wrestlers when assessed by ultrasound

and hydrodensitometry (Utter and Hager 2008). Consequently, ultrasound may also have application in tracking changes in skeletal muscle mass as a result of injury or training adaptations (Uremovic et al. 2004). More research is required to assess the reliability and validity of the method before it can be considered in routine athlete monitoring.

Three-dimensional photonic scanning (3DPS) was initially developed for use in the clothing industry to capture information about body surface topography, but it was soon recognized that the technology has potential application in medicine (Wells et al. 2000) and sports (Olds et al. 2008). It has several potential advantages over traditional anthropometric measurement techniques, capturing a much wider range of anthropometric dimensions in a more efficient and noninvasive manner (Schranz et al. 2010).

The method provides information on regional as well as whole-body volume and surface area plus various body dimensions and potentially body composition. Validation work on this approach for measurement of body composition is limited (Wang et al. 2006; Wells et al. 2000), but the technology has already been applied to contrast differences in physique traits between elite athletes and the general population (Schranz et al. 2010). Given the expense and poor portability of 3DPS systems, the technology is not practical for field use and is used only in research.

Factors Influencing Reliability of Methods

A wide range of tools are available to assess the physique traits of athletes. Although these tools are used to identify athletic talent and suitability for a given sport and player position, their primary use is for the longitudinal monitoring of physique traits. As such, techniques that are cost- and time-effective, portable, reliable, and safe and that provide insight into all physique traits, including both fat and muscle mass, are favored.

All physique assessment tools carry some degree of assumption and measurement error. Understanding this helps testers distinguish between measurement error and real changes in body composition, that is, documented change in physique traits that are greater than reported measurement error. Although measurement error is inherent in equipment and is often beyond the control of technicians, athlete presentation can also contribute to the error of repeat assessments. Factors such as time of day, hydration status, and gastrointestinal tract contents should be standardized whenever possible; fasted early morning assessments may be the most reliable. Minimizing the error or noise associated with a test enhances its reliability, making it easier to identify small but potentially important changes. Reliability of measurement also influences the frequency of assessment. In general, physique assessments should not be undertaken more often than every 4 to 8 weeks, depending on individual athletes and their body composition goals.

During data collection, the physical and emotional well-being of the athlete should remain a priority. Sensitivity should be shown to cultural beliefs and tradition. Testers should explain procedures to athletes who are unfamiliar with a given test, providing information on what testing is to be undertaken, the reason for profiling, what measurements are to be taken, and any specific requirements such as clothing to be worn. Where appropriate, the technician and athlete should be of the same gender,

and privacy in data collection and reporting should always be ensured. Administrators should consider using electronic databases that not only provide a secure means of data collection but also generate reports that provide invaluable historical data and compare results against previous assessments. Finally, where resources permit, data should be collected in duplicate, enhancing the reliability of measurement.

Understanding factors that influence the noise or error associated with body composition assessment tools will enable scientists to develop protocols that afford a much greater resolution of measurement, enabling the detection of small but potentially meaningful changes in body composition. This will allow coaches and scientists to assess interventions (e.g., diet, training) or unforeseen situations (e.g., injury, illness, detraining) that influence body composition, which will have application not only within sports but in the wider community as well.

TEST PROTOCOLS

TETRAPOLAR LEAD BIOELECTRICAL IMPEDANCE

The following protocol relates specifically to tetrapolar lead bioelectrical impedance equipment with two source and two sensor leads connected to four electrodes. For tetrapolar footpad style systems, issues such as electrode placement will be irrelevant. However, subject presentation should be in accordance with details specified next.

ATHLETE PREPARATION

The subject should present in a well-hydrated state (can be confirmed using an array of parameters, including urinary indices) and ideally after an overnight fast with no physical activity for the last 8 h. The bladder should be evacuated prior to assessment and all metal objects removed. The facility for assessment should offer privacy for the athlete and be maintained to ensure thermoneutral conditions.

TEST PROCEDURE

- Undertake calibration procedures in accordance with manufacturers' guidelines.
- Weigh the athlete on a calibrated scale in minimal clothing.
- Measure stretch stature using a wall-mounted stadiometer.
- Request relevant demographic data from athlete, including ethnicity, age, and gender,

ensuring that the most valid equations can be used in subsequent analysis.

- Place the athlete in a supine position on a flat, even surface free of any metal. The limbs should be abducted, arms at 30° from the trunk, and legs separated at 45°. This position should be maintained for at least 10 min prior to assessment to minimize the influence of body water shifts between compartments.
- Clean all electrode sites with alcohol swabs prior to electrode placement and allow them to dry.
- Place two electrodes on the right hand and wrist: one on the dorsal surface of the hand 1 cm proximal to the third metacarpophalangeal joint and the other centrally on the dorsal side of the wrist in line with the ulnar head.
- Place two electrodes on the right foot and ankle: one on the dorsal surface of the foot 1 cm proximal to the second metatarsophalangeal joint and other centrally on the dorsal surface of the ankle between the lateral and medial malleoli.
- Ensure all electrodes are placed at least 5 cm apart.
- Initiate data collection.

AIR DISPLACEMENT PLETHYSMOGRAPHY

ATHLETE PREPARATION

The subject should present in a well-hydrated state (can be confirmed using an array of parameters, including urinary indices), wearing a tight-fitting swimsuit (or similar clothing) in conjunction with a tight fitting silicone swimcap to compress hair. For males, facial hair should be removed prior to assessment or, alternatively, compressed similar to scalp hair. The bladder should be evacuated prior to assessment. The facility for assessment should offer privacy for the athlete and be maintained to ensure thermoneutral conditions.

TEST PROCEDURE

- Undertake a two-point calibration in accordance with manufacturer's guidelines.
- Weigh the athlete on a calibrated scale in minimal clothing.
- Request relevant demographic data from the athlete, including ethnicity, age, and gender.
- Have the athlete sit inside the chamber, closing the chamber door behind them.
- Initiate measurement, instructing the athlete to relax and breathe normally. This takes approximately 40 s.
- Following completion of the measurement, open the chamber door before repeating steps 3 and 4.
- If these two measurements are within 150 ml, the average will be used. If values differ by more than 150 ml, the software will prompt for a third measurement. If any two of the three measurements agree within 150 ml, the average of these will be used in subsequent calculations. If no two of the measurements meet this criteria, repeat the test procedure, including recalibration of equipment. Extra measurement error may be created if the athlete moves, talks, coughs, or sneezes.

- To account for the influence of thoracic gas volume on the measurement of body volume, this should be measured in accordance with manufacturer's instructions (Dempster and Aitkens 1995). Alternatively, thoracic gas volume can be predicted from gender, age, and height, although measured and predicted values should never be used interchangeably (Minderico et al. 2008).

- Body density is then calculated by dividing the measured body mass by corrected body volume, with subsequent calculation of body fat percentage using one of several equations including the Siri (Siri, 1993) or Brozek (Brozek et al. 1963) equations, specified below.

Siri equation	% Body fat = (495/Body Density) − 450.
Brozek equation	% Body fat = (457/Body Density) − 414.2.

DUAL ENERGY X-RAY ABSORPTIOMETRY WHOLE-BODY SCAN

ATHLETE PREPARATION

The subject should present wearing minimal clothing, in a well-hydrated state (can be confirmed using an array of parameters, including urinary indices), and ideally after an overnight fast or at least 4 h post prandial. The bladder should be evacuated prior to assessment and all metal objects removed. The facility for assessment should offer privacy for the athlete and be maintained to ensure thermoneutral conditions.

TEST PROCEDURE

- Prior to the first assessment of the day, undertake calibration procedures in accordance with manufacturers' guidelines.
- Given the radiation exposure associated with a DXA scan, ask all females whether there is a possibility they could be pregnant. If this is confirmed, a scan should not be undertaken.

- Weigh the athlete on a calibrated scale in minimal clothing.

- Measure stretch stature using a wall-mounted stadiometer to confirm that the athlete will fit within the scanning area of the machine being used. If the athlete is taller than the scanning area of the machine, the summed value of two separate scans may be required, one capturing data on the head and neck and the other capturing data on the neck and remainder of the body. Positioning of the head may be important in this situation; aligning the athlete into the Frankfort plane is encouraged.

- Position the athlete in a supine position, ensuring he or she is centered on the scanner with no part of the body excluded from the scanning area. The head should be placed in the Frankfort plane, with long hair let down and evenly distributed across the back. Arms should be slightly abducted from the torso (with hands in the midprone position) and legs separated (with feet at right angles to the leg, pointing upward) at a defined angle to ensure clarification of regions of interest such as the trunk, legs, and arms. The use of foam pads or similar material not recognized by the scanner can be particularly valuable in standardizing subject positioning while also minimizing movement during a scan. Particularly broad individuals who do not naturally fit on the scanner should be positioned on the scanner so that a minimum of half the body is within the scanning area. The half-body results can then be used to estimate whole-body composition, or the scan can be replicated with the other side of the body and results combined for both halves, depending on the scanner being used and associated software. If a single half-body scan is used, the same side of the body should be used for all subsequent scans on an individual. Particularly tall and broad individuals may need to undertake all of the above, ensuring two or three scans are undertaken to obtain a whole body measure.

- Initiate the scan, instructing the athlete to breathe normally and remain still throughout the scan. Using the report generated, visually confirm that no parts of the body have been occluded from the scan. If areas have been occluded, manual analysis alteration should be undertaken.

SURFACE ANTHROPOMETRY

The following protocol for the use of surface anthropometry has been endorsed by the International Society for the Advancement of Kinanthropometry (ISAK). It relates specifically to the measurement of subcutaneous fat thickness via the use of skinfolds. The reader is directed elsewhere for details on the procedure of other measures such as girths, lengths, and breadths (Stewart, et al. 2011).

ATHLETE PREPARATION

The subject should present in a well-hydrated state (can be confirmed using an array of parameters, including urinary indices), wearing comfortable clothing that allows access to most parts of the body, including the torso. The bladder should be evacuated prior to assessment. Measurements should not be undertaken after exercise, sauna, or showering, as heat promotes hyperemia in the skin with a concomitant increase in skinfold thickness.

The facility for assessment should offer privacy for the athlete and be maintained to ensure thermoneutral conditions. To eliminate the influence of diurnal variation on body mass, athletes should be assessed early in the morning or at least at the same time of day when serial measurements are undertaken over time.

TEST PROCEDURE

- Weigh the athlete on a calibrated scale in minimal clothing.

- Identify anatomical landmarks using the following description prior to taking any measurements. All sites should be marked and measured on the right side of the body.

- Prior to the measurement of skinfolds, ensure that the skinfold caliper is accurately measuring the distance between the center of its contact faces with the use of Vernier calipers.

- The skinfold is picked up at the marked site; the near edge of the thumb and finger are in line with the marked site. The back of the hand should be facing the measurer. It should be grasped and lifted so that a double fold of skin plus the underlying subcutaneous adipose tissue is held between the thumb and index finger of the left hand. The size of the fold to pick up should be the minimum necessary to ensure that the two skin surfaces of the fold are parallel.

- The nearest edge of the contact faces of the caliper is applied 1 cm away from the edge of the thumb and finger. The center of the caliper faces should be placed at a depth of approximately mid-fingernail.

- The caliper is held at 90° to the surface of the skinfold at all times. Ensure that the hand grasping the skin remains holding the fold the whole time the caliper is in contact with the skin.

- Measurement is taken 2 s after full pressure of the caliper is applied.
- Skinfold sites are measured in succession, reducing the effects of skinfold compressibility and measurer bias.
- Duplicate or triplicate measurements are taken where possible. A third measurement

should be taken when the second measurement is not within 5% of the first for skinfolds or within 1% for other measures.

SPECIFIED ANATOMICAL LANDMARKS

These landmarks are as defined by the International Society for the Advancement of Kinanthropometry (Stewart et al. 2011).

Acromiale landmark

Definition

The point on the superior aspect of the most lateral part of the acromion border

Subject position

The subject assumes a relaxed position with the arm hanging by the side. The shoulder girdle should be in a mid-position.

Location

Standing behind and on the right hand side of the subject, palpate along the spine of the scapula to the corner of the acromion. This represents the start of the lateral border, which usually runs anteriorly, slightly superiorly, and medially. Apply the straight edge of a pencil to the lateral and superior margin of the acromion to confirm the location of the most lateral part of the border. Mark this most lateral aspect. The acromion has an associated bone thickness. Palpate superiorly to the top margin of the acromion border in line with the most lateral aspect.

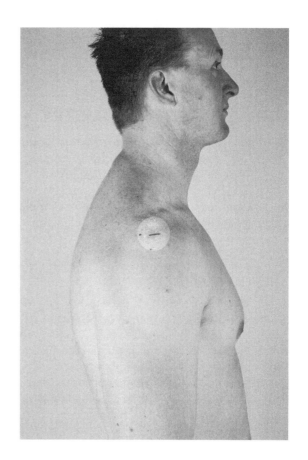

Radiale landmark

Definition

The point at the proximal and lateral border of the head of the radius

Subject position

The subject assumes a relaxed position with the arm hanging by the side and the hand in the mid-prone position.

Location

Palpate downward into the lateral dimple of the right elbow. It should be possible to feel the space between the capitulum of the humerus and the head of the radius. Move the thumb distally onto the most lateral part of the proximal radial head. Correct location can be checked by slight rotation of the forearm, which causes the head of the radius to rotate.

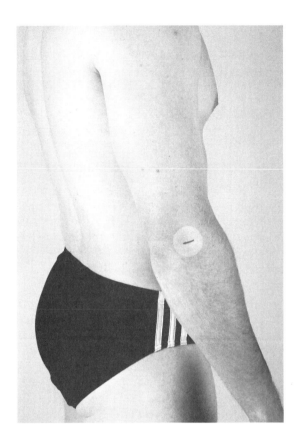

Mid-acromiale–radiale landmark

Definition

The midpoint of the straight line joining the acromiale and the radiale

Subject position

The subject assumes a relaxed position with the arms hanging by the sides.

Location

Measure the linear distance between the acromiale and radiale landmarks with the arm relaxed and extended by the side. The best way to measure this is with a segmometer or large sliding caliper. It is not acceptable to follow the curvature of the surface of the arm. If a tape must be used, hold it so that the perpendicular distance between the two landmarks is measured. Place a small mark at the level of the midpoint between these two landmarks. Project this mark around to the posterior and anterior surfaces of the arm as a horizontal line. This is required for locating the triceps and biceps skinfold sites.

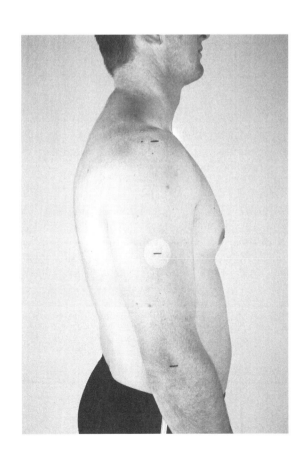

Triceps skinfold site

Definition

The point on the posterior surface of the arm, in the midline, at the level of the marked mid-acromiale–radiale landmark

Subject position

The subject assumes a relaxed standing position with the arm hanging by the side and the hand in the mid-prone position.

Location

This point is located by projecting the mid-acromiale–radiale site perpendicularly to the long axis of the arm around to the back of the arm and intersecting the projected line with a vertical line in the middle of the arm when viewed from behind.

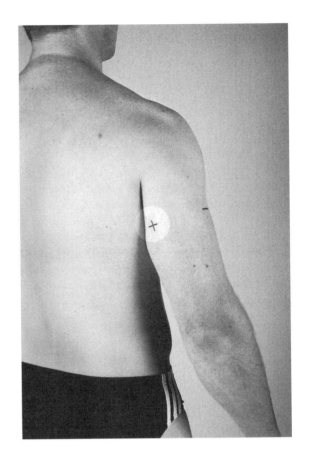

Biceps skinfold site

Definition

The point on the anterior surface of the arm in the midline at the level of the mid-acromiale–radiale landmark

Subject position

The subject assumes a relaxed standing position with the arm hanging by the side and the hand in the mid-prone position.

Location

This point can be located by projecting the mid-acromiale–radiale site perpendicularly to the long axis of the arm around to the front of the arm and intersecting the projected line with a vertical line in the middle of the arm when viewed from the front.

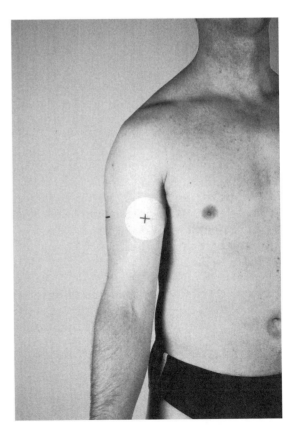

Subscapulare landmark

Definition

The undermost tip of the inferior angle of the scapula

Subject position

The subject assumes a relaxed standing position with the arms hanging by the sides.

Location

Palpate the inferior angle of the scapula with the left thumb. If there is difficulty locating the inferior angle of the scapula, have the subject slowly reach behind the back with the right arm. The inferior angle of the scapula should then be felt continuously as the subject's hand is again placed by the side of the body. A final check of this landmark should be made with the hand by the side in the relaxed position.

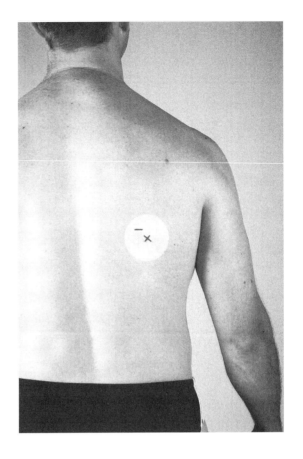

Subscapular skinfold site

Definition

The site 2 cm along a line running laterally and obliquely downward from the subscapulare landmark at a 45° angle

Subject position

The subject assumes a relaxed standing position with the arms hanging by the sides.

Location

Use a tape measure to locate the point 2 cm from the subscapulare in a line 45° laterally downward.

Iliocristale landmark

Definition

The point on the iliac crest where a line drawn from the midaxilla (middle of the armpit), on the longitudinal axis of the body, meets the ilium

Subject position

The subject assumes a relaxed position with the left arm hanging by the side and the right arm folded across her chest.

Location

Use your left hand to stabilize the subject's body by providing resistance on the left side of the pelvis. Find the general location of the top of the iliac crest with the palms of the fingers of the right hand. Once the general position has been located, find the specific edge of the crest by horizontal palpation with the tips of the fingers. Once identified, draw a horizontal line at the level of the iliac crest. Draw an imaginary line from the midaxilla down the midline of the body. The landmark is at the intersection of the two lines.

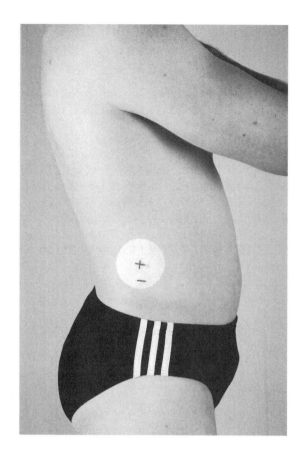

Iliac crest skinfold site

Definition

The site at the center of the skinfold raised immediately above the marked iliocristale

Subject position

The subject assumes a relaxed position with the right arm folded across the chest.

Location

This skinfold is raised superior to the iliocristale. To do this, place the left thumb tip on the marked iliocristale site, and raise the skinfold between the thumb and index finger of the left hand. Once the skinfold has been raised, mark its center with a cross (+). The fold runs slightly downwards anteriorly as determined by the natural fold of the skin.

The iliac crest skinfold is not included in the sum of seven skinfolds.

Iliospinale landmark

Definition

The most inferior or undermost part of the tip of the anterior superior iliac spine

Subject position

The subject assumes a relaxed position with the right arm folded across the chest.

Location

Palpate the superior aspect of the ilium and follow it anteriorly until the anterior superior iliac spine is reached. The landmark is marked at the lower margin or edge where the bone can just be felt. Difficulty in appraising the landmark can be eased by having the subject lift the heel of the right foot and rotate the femur outward. Because the sartorius muscle originates at the iliospinale, this movement of the femur enables palpation of the muscle and tracing to its origin.

Note: On females, the landmark is usually proportionally lower on the trunk, due to the flatter and broader shape of the female pelvis.

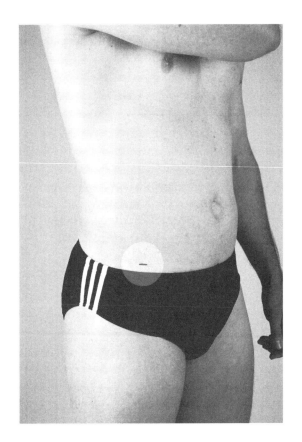

Supraspinale skinfold site

Definition

The point at the intersection of two lines:

- the line from the marked iliospinale to the anterior axillary border and
- the horizontal line at the level of the marked iliocristale

Subject position

The subject assumes a relaxed standing position with the arms hanging by the sides. The right arm may be abducted after the anterior axillary border has been identified.

Location

Run a tape from the anterior axillary border to the marked iliospinale and draw a short line along the side roughly at the level of the iliocristale. Then run the tape horizontally around from the marked iliocristale to intersect the first line.

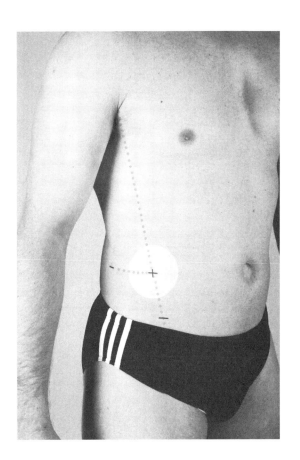

Abdominal skinfold site

Definition

The point 5 cm horizontally to the right hand side of the omphalion (midpoint of the navel)

Subject position

The subject assumes a relaxed standing position with the arms hanging by the sides.

Location

The site is identified by a horizontal measure of 5 cm, to the subject's right, from the omphalion. The skinfold taken at this site is a vertical fold.

Note: The distance of 5 cm assumes an adult height of approximately 170 cm. Where height differs markedly from this, the distance should be scaled for height. For example, if the stature is 120 cm, the distance will be 5 × 120/170 = 3.5 cm.

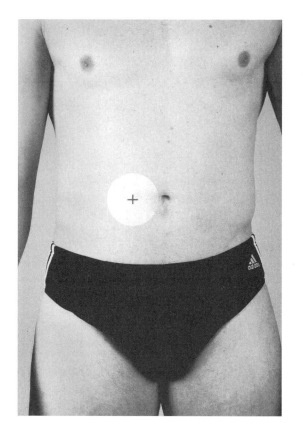

Medial calf skinfold site

Definition

The point on the most medial aspect of the calf at the level of the maximal girth

Subject position

The subject assumes a relaxed standing position with the arms hanging by the sides. The subject's feet should be separated with the weight evenly distributed.

Location

The level of the maximal girth is determined by trial and error. It is found by using the middle fingers to manipulate the position of the tape in a series of up or down measurements. Once the maximal level is located, the point is marked on the medial aspect of the calf with a small cross (+) or other suitable mark.

Note: For easier viewing, the photograph shows the medial aspect of the lower leg. However, the site is located with the subject standing.

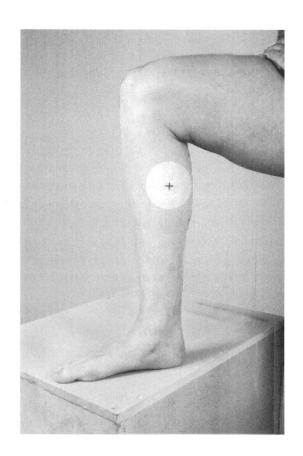

Front thigh skinfold site

Definition

The midpoint of the linear distance between the inguinal point and the patellare (the midpoint of the posterior, superior border of the patella)

Subject position

The subject assumes a seated position with the torso erect and the arms hanging by the sides. The knee of the right leg should be bent at a right angle.

Location

The measurer stands facing the right side of the seated subject on the lateral side of the thigh. If there is difficulty locating the inguinal fold, the subject should flex the hip to make a fold. Place a small horizontal mark at the level of the midpoint between the two landmarks. Now draw a perpendicular line to intersect the horizontal line. This perpendicular line is located in the midline of the thigh. If a tape is used, avoid following the curvature of the surface of the skin.

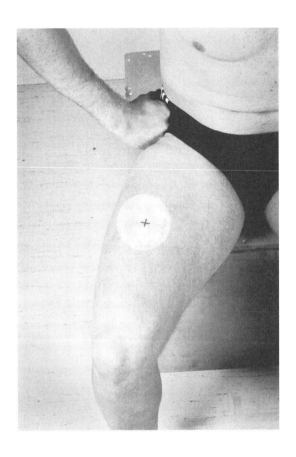

SPECIFIED SKINFOLD MEASUREMENTS

These measurements are as defined by the International Society for the Advancement of Kinanthropometry.

Triceps

Definition

The skinfold measurement taken parallel to the long axis of the arm at the triceps skinfold site

Subject position

The subject assumes a relaxed standing position with the right arm hanging by the side and the hand in the midprone position.

Biceps

Definition

The skinfold measurement taken parallel to the long axis of the arm at the biceps skinfold site

Subject position

The subject assumes a relaxed standing position with the right arm hanging by the side and the hand in the midprone position.

Subscapular

Definition

The skinfold measurement taken with the fold running obliquely downward at the subscapular skinfold site

Subject position

The subject assumes a relaxed standing position with the arms hanging by the sides.

Special note

The line of the skinfold is determined by the natural fold lines of the skin.

Iliac Crest

Definition

The skinfold measurement taken near horizontally at the iliac crest skinfold site

Subject position

The subject assumes a relaxed standing position. The right arm should be either abducted or placed across the trunk.

Special note

The line of the skinfold generally runs slightly downward posterior–anterior, as determined by the natural fold lines of the skin.

Supraspinale

Definition

The skinfold measurement taken with the fold running obliquely and medially downward at the supraspinale skinfold site

Subject position

The subject assumes a relaxed standing position with the arms hanging by the sides.

Special note

The fold runs medially downward and anteriorly at about a 45° angle as determined by the natural fold of the skin.

Abdominal

Definition

The skinfold measurement taken vertically at the abdominal skinfold site

Subject position

The subject assumes a relaxed standing position with the arms hanging by the sides.

Special note

It is particularly important at this site that the measurer's initial grasp is firm and broad since often the underlying musculature is poorly developed. This may result in an underestimation of the thickness of the subcutaneous layer of tissue. (Note: Do not place the fingers or caliper inside the navel.)

Medial Calf

Definition

The skinfold measurement taken vertically at the medial calf skinfold site

Subject position

The subject assumes a relaxed standing position with the right foot placed on the box. The right knee is bent at about 90°.

Special note

The subject's right foot is placed on a box with the calf relaxed. The fold is parallel to the long axis of the leg.

Front Thigh

Definition

The skinfold measurement taken parallel to the long axis of the thigh at the front thigh skinfold site

Subject position

The subject assumes a seated position at the front edge of the box with the torso erect, the arms supporting the hamstrings and the leg extended.

Because of difficulties with this skinfold, two methods are recommended. Be sure to document the method used as A or B. In both methods the leg is extended, and the subject supports the hamstrings.

Method A

The measurer stands facing the right side of the subject on the lateral side of the thigh. The skinfold is raised at the marked site and the measurement taken.

Method B

Subjects with particularly tight skinfolds are asked to assist by lifting the underside of the thigh (as in method A). The recorder (standing on the subject's left) assists by raising the fold, with both hands, at about 6 cm either side of the landmark. The measurer then raises the skinfold at the marked site and takes the measurement.

All photos reprinted, by permission, from M. Marfell-Jones et al., 2006, *International standards for anthropometric assessment* (Potchefstroom, SA: International Society for the Advancement of Kinanthropometry), 29-49.

References

Abe, T., Kojima, K., Kearns, C.F., Yohena, H., and Fukuda, J. 2003. Whole body muscle hypertrophy from resistance training: Distribution and total mass. *British Journal of Sports Medicine* 37(6):543-545.

Ackland, T. R., Lohman, T. G., Sundgot-Borgen, J., Maughan, R. J., Meyer, N. L., Stewart, A. D., and Muller, W. 2012. Current status of body composition assessment in sport: review and position statement on behalf of the ad hoc research working group on body composition health and performance, under the auspices of the I.O.C. Medical Commission. *Sports Medicine*, 42(3):227-249.

Anderson, D.E. 2007. Reliability of air displacement plethysmography. *Journal of Strength and Conditioning Research* 21(1):169-172.

Ball, S.D. 2005. Interdevice variability in percent fat estimates using the BOD POD. *European Journal of Clinical Nutrition* 59(9):996-1001.

Ballard, T.P., Fafara, L., and Vukovich, M.D. 2004. Comparison of Bod Pod and DXA in female collegiate athletes. *Medicine and Science in Sports and Exercise* 36(4):731-735.

Bentzur, K.M., Kravitz, L., and Lockner, D.W. 2008. Evaluation of the BOD POD for estimating percent body fat in collegiate track and field female athletes: A comparison of four methods. *Journal of Strength and Conditioning Research* 22(6):1985-1991.

Biaggi, R.R., Vollman, M.W., Nies, M.A., Brener, C.E., Flakoll, P.J., Levenhagen, D.K., et al. 1999. Comparison of air-displacement plethysmography with hydrostatic weighing and bioelectrical impedance analysis for the assessment of body composition in healthy adults. *American Journal of Clinical Nutrition* 69(5):898-903.

Brechue, W.F., and Abe, T. 2002. The role of FFM accumulation and skeletal muscle architecture in powerlifting performance. *European Journal of Applied Physiology* 86(4):327-336.

Brozek, J., Grande, F., Anderson, J.T., and Keys, A. 1963. Densitometric analysis of body composition: Revision of some quantitative assumptions. *Annals of the New York Academy of Sciences* 110:113-140.

Bunt, J.C., Lohman, T.G., and Boileau, R.A. 1989. Impact of total body water fluctuations on estimation of body fat from body density. *Medicine and Science in Sports and Exercise* 21(1):96-100.

Burkhart, T.A., Arthurs, K.L., and Andrews, D.M. 2009. Manual segmentation of DXA scan images results in reliable upper and lower extremity soft and rigid tissue mass estimates. *Journal of Biomechanics* 42(8):1138-1142.

Calbet, J.A., Moysi, J.S., Dorado, C., and Rodriguez, L.P. 1998. Bone mineral content and density in professional tennis players. *Calcified Tissue International*, 62(6):491-496.

Carson, J.D., Bridges, E., and Canadian Academy of Sport, M. 2001. Abandoning routine body composition assessment: A strategy to reduce disordered eating among female athletes and dancers. *Clinical Journal of Sport Medicine* 11(4):280.

Cisar, C.J., Housh, T.J., Johnson, G.O., Thorland, W.G., and Hughes, R.A. 1989. Validity of anthropometric equations for determination of changes in body composition in adult males during training. *Journal of Sports Medicine and Physical Fitness* 29(2):141-148.

Claessens, A.L., Hlatky, S., Lefevre, J., and Holdhaus, H. 1994. The role of anthropometric characteristics in modern pentathlon performance in female athletes. *Journal of Sports Sciences* 12(4):391-401.

Claessens, A.L., Lefevre, J., Beunen, G., and Malina, R.M. 1999. The contribution of anthropometric characteristics to performance scores in elite female gymnasts. *Journal of Sports Medicine and Physical Fitness* 39(4):355-360.

Clark, R.R., Bartok, C., Sullivan, J.C., and Schoeller, D.A. 2004. Minimum weight prediction methods crossvalidated by the four-component model. *Medicine and Science in Sports and Exercise* 36(4):639-647.

Clarys, J.P., Martin, A.D., Drinkwater, D.T., and Marfell-Jones, M.J. 1987. The skinfold: Myth and reality. *Journal of Sports Sciences* 5(1):3-33.

Clarys, J.P., Martin, A.D., Marfell-Jones, M.J., Janssens, V., Caboor, D., and Drinkwater, D.T. 1999. Human body composition: A review of adult dissection data. *American Journal of Human Biology* 11(2):167-174.

Clarys, J.P., Provyn, S., and Marfell-Jones, M.J. 2005. Cadaver studies and their impact on the understanding of human adiposity. *Ergonomics* 48(11-14):1445-1461.

Collins, A.L., and McCarthy, H.D. 2003. Evaluation of factors determining the precision of body composition measurements by air displacement plethysmography. *European Journal of Clinical Nutrition* 57(6):770-776.

Collins, M.A., Millard-Stafford, M.L., Sparling, P.B., Snow, T.K., Rosskopf, L.B., Webb, S.A., et al. 1999. Evaluation of the BOD POD for assessing body fat in collegiate football players. *Medicine and Science in Sports and Exercise* 31(9):1350-1356.

Cornish, B.H., Ward, L.C., Thomas, B.J., Jebb, S.A., and Elia, M. 1996. Evaluation of multiple frequency bioelectrical impedance and Cole-Cole analysis for the assessment of body water volumes in healthy humans. *European Journal of Clinical Nutrition* 50(3):159-164.

Davis, J.A., Dorado, S., Keays, K.A., Reigel, K.A., Valencia, K.S., and Pham, P.H. 2007. Reliability and validity of the lung volume measurement made by the BOD POD body composition system. *Clinical Physiology and Functional Imaging* 27(1):42-46.

Dempster, P., and Aitkens, S. 1995. A new air displacement method for the determination of human body composition. *Medicine and Science in Sports and Exercise* 27(12):1692-1697.

DeRose, E.H., Crawford, S.M., Kerr, D.A., Ward, R., and Ross, W.D. 1989. Physique characteristics of Pan American Games lightweight rowers. *International Journal of Sports Medicine* 10(4):292-297.

Deurenberg-Yap, M., Schmidt, G., van Staveren, W.A., Hautvast, J., and Deurenberg, P. 2001. Body fat measurement among Singaporean Chinese, Malays and Indians: A comparative study using a four-compartment model and different two-compartment models. *British Journal of Nutrition* 85(4):491-498.

Dixon, C.B., Deitrick, R.W., Cutrufello, P.T., Drapeau, L.L., and Lovallo, S.J. 2006. Effect of mode selection when using leg-to-leg BIA to estimate body fat in collegiate wrestlers. *Journal of Sports Medicine and Physical Fitness* 46(2):265-270.

Dixon, C.B., Ramos, L., Fitzgerald, E., Reppert, D., and Andreacci, J.L. 2009. The effect of acute fluid consumption on measures of impedance and percent body fat estimated using segmental bioelectrical impedance analysis. *European Journal of Clinical Nutrition* 63(9):1115-1122.

Drinkwater, D.T., and Ross, W.D. 1980. Anthropometric fractionation of body mass. In: M. Ostyn, G. Beunen, and J. Simons (eds), *Kinanthropometry II*, Vol IX. Baltimore: University Park Press, pp 178-189.

Dunbar, C.C., Melahrinides, E., Michielli, D.W., and Kalinski, M.I. 1994. Effects of small errors in electrode placement on body composition assessment by bioelectrical impedance. *Research Quarterly for Exercise and Sport* 65(3):291-294.

Eston, R.G., Rowlands, A.V., Charlesworth, S., Davies, A., and Hoppitt, T. 2005. Prediction of DXA-determined whole body fat from skinfolds: Importance of including skinfolds from the thigh and calf in young, healthy men and women. *European Journal of Clinical Nutrition* 59(5):695-702.

Evans, E.M., Prior, B.M., and Modlesky, C.M. 2005. A mathematical method to estimate body composition in tall individuals using DXA. *Medicine and Science in Sports and Exercise* 37(7):1211-1215.

Faria, I.E., and Faria, E.W. 1989. Relationship of the anthropometric and physical characteristics of male junior gymnasts to performance. *Journal of Sports Medicine and Physical Fitness* 29(4):369-378.

Fields, D.A., Goran, M.I., and McCrory, M.A. 2002. Body-composition assessment via air-displacement plethysmography in adults and children: A review. *American Journal of Clinical Nutrition* 75(3):453-467.

Fields, D.A., Higgins, P.B., and Hunter, G.R. 2004. Assessment of body composition by air-displacement plethysmography: Influence of body temperature and moisture. *Dynamic Medicine* 3(1):3.

Fields, D.A., Hunter, G.R., and Goran, M.I. 2000. Validation of the BOD POD with hydrostatic weighing: Influence of body clothing. *International Journal of Obesity and Related Metabolic Disorders* 24(2):200-205.

Forbes, G.B. 1999. Body composition: Overview. *Journal of Nutrition* 129(1S Suppl):270S-272S.

Frisard, M.I., Greenway, F.L., and Delany, J.P. 2005. Comparison of methods to assess body composition changes during a period of weight loss. *Obesity Research* 13(5):845-854.

Fry, A.C., Ryan, A.J., Schwab, R.J., Powell, D.R., and Kraemer, W.J. 1991. Anthropometric characteristics as discriminators of body-building success. *Journal of Sports Sciences* 9(1):23-32.

Genton, L., Hans, D., Kyle, U.G., and Pichard, C. 2002. Dual-energy X-ray absorptiometry and body composition: Differences between devices and comparison with reference methods. *Nutrition* 18(1):66-70.

Going, S.B., Massett, M.P., Hall, M.C., Bare, L.A., Root, P.A., Williams, D.P., et al. 1993. Detection of small changes in body composition by dual-energy x-ray absorptiometry. *American Journal of Clinical Nutrition* 57(6):845-850.

Hahn, A. 1990. Identification and selection of talent in Australian rowing. *EXCEL* 6(3):5-11.

Hewitt, M.J., Going, S.B., Williams, D.P., and Lohman, T.G. 1993. Hydration of the fat-free body mass in children and adults: Implications for body composition assessment. *American Journal of Physiology* 265(1 Pt 1):E88-E95.

Higgins, P.B., Fields, D.A., Hunter, G.R., and Gower, B.A. 2001. Effect of scalp and facial hair on air displacement plethysmography estimates of percentage of body fat. *Obesity Research* 9(5):326-330.

Horber, F.F., Thomi, F., Casez, J. P., Fonteille, J., and Jaeger, P. 1992. Impact of hydration status on body composition as measured by dual energy X-ray absorptiometry in normal volunteers and patients on haemodialysis. *British Journal of Radiology* 65(778):895-900.

Houtkooper, L.B., Going, S.B., Sproul, J., Blew, R.M., and Lohman, T.G. 2000. Comparison of methods for assessing body-composition changes over 1 y in postmenopausal women. *American Journal of Clinical Nutrition* 72(2):401-406.

Houtkooper, L.B., Lohman, T.G., Going, S.B., and Howell, W.H. 1996. Why bioelectrical impedance analysis should be used for estimating adiposity. *American Journal of Clinical Nutrition* 64(3 Suppl):436S-448S.

Hull, H., He, Q., Thornton, J., Javed, F., Allen, L., Wang, J., et al. 2009. iDXA, Prodigy, and DPXL dual-energy X-ray absorptiometry whole-body scans: A cross-calibration study. *Journal of Clinical Densitometry* 12(1):95-102.

Hume, P., and Marfell-Jones, M. 2008. The importance of accurate site location for skinfold measurement. *Journal of Sports Sciences* 26(12):1333-1340.

Jaffrin, M.Y. 2009. Body composition determination by bioimpedance: An update. *Current Opinion in Clinical Nutrition and Metabolic Care* 12(5):482-486.

Johnston, F.E. 1982. Relationships between body composition and anthropometry. *Human Biology* 54(2):221-245.

Kelsey, J.L., Bachrach, L.K., Procter-Gray, E., Nieves, J., Greendale, G.A., Sowers, M., et al. 2007. Risk factors for stress fracture among young female cross-country runners. *Medicine and Science in Sports and Exercise* 39(9):1457-1463.

Keogh, J.W., Hume, P.A., Pearson, S.N., and Mellow, P. 2007. Anthropometric dimensions of male powerlifters of varying body mass. *Journal of Sports Sciences* 25(12):1365-1376.

King, G.A., Fulkerson, B., Evans, M.J., Moreau, K.L., McLaughlin, J.E., and Thompson, D.L. 2006. Effect of clothing type on validity of air-displacement plethysmography. *Journal of Strength and Conditioning Research* 20(1):95-102.

Koo, W.W.K., Hockman, E.M., and Hammami, M. 2004. Dual energy X-ray absorptiometry measurements in small subjects: Conditions affecting clinical measurements. *Journal of the American College of Nutrition* 23(3):212-219.

Koulmann, N., Jimenez, C., Regal, D., Bolliet, P., Launay, J.C., Savourey, G., et al. 2000. Use of bioelectrical impedance analysis to estimate body fluid compartments after acute variations of the body hydration level. *Medicine and Science in Sports and Exercise* 32(4):857-864.

Kushner, R.F. 1992. Bioelectrical impedance analysis: A review of principles and applications. *Journal of the American College of Nutrition* 11(2):199-209.

Kushner, R.F., Gudivaka, R., and Schoeller, D.A. 1996. Clinical characteristics influencing bioelectrical impedance analysis measurements. *American Journal of Clinical Nutrition* 64(3 Suppl):423S-427S.

Kyle, U.G., Bosaeus, I., De Lorenzo, A.D., Deurenberg, P., Elia, M., Gomez, J.M., et al. 2004a. Bioelectrical impedance analysis—Part I: Review of principles and methods. *Clinical Nutrition* 23(5):1226-1243.

Kyle, U.G., Bosaeus, I., De Lorenzo, A.D., Deurenberg, P., Elia, M., Manuel Gomez, J., et al. 2004b. Bioelectrical impedance analysis—Part II: Utilization in clinical practice. *Clinical Nutrition* 23(6):1430-1453.

Kyriazis, T., Terzis, G., Karampatsos, G., Kavouras, S., and Georgiadis, G. 2010. Body composition and performance in shot put athletes at preseason and at competition.

International Journal of Sports Physiology and Performance 5(3):417-421.

Lambrinoudaki, I., Georgiou, E., Douskas, G., Tsekes, G., Kyriakidis, M., and Proukakis, C. 1998. Body composition assessment by dual-energy X-ray absorptiometry: Comparison of prone and supine measurements. *Metabolism: Clinical and Experimental* 47(11):1379-1382.

Lands, L.C., Hornby, L., Hohenkerk, J.M., and Glorieux, F.H. 1996. Accuracy of measurements of small changes in soft-tissue mass by dual-energy x-ray absorptiometry. *Clinical and Investigative Medicine* 19(4):279-285.

Larsson, P., and Henriksson-Larsen, K. 2008. Body composition and performance in cross-country skiing. *International Journal of Sports Medicine* 29(12):971-975.

Le Carvennec, M., Fagour, C., Adenis-Lamarre, E., Perlemoine, C., Gin, H., and Rigalleau, V. 2007. Body composition of obese subjects by air displacement plethysmography: The influence of hydration. *Obesity (Silver Spring)* 15(1):78-84.

Lewiecki, E.M. 2005. Clinical applications of bone density testing for osteoporosis. *Minerva Medica* 96(5):317-330.

Lohman, M., Tallroth, K., Kettunen, J.A., and Marttinen, M.T. 2009. Reproducibility of dual-energy x-ray absorptiometry total and regional body composition measurements using different scanning positions and definitions of regions. *Metabolism: Clinical and Experimental* 58(11):1663-1668.

Lohman, T.G. 1981. Skinfolds and body density and their relation to body fatness: A review. *Human Biology* 53(2):181-225.

Marfell-Jones, M., Clarys, J.P., Alewaeters, K., Martin, A.D., and Drinkwater, D.T. 2003. The hazards of whole body fat prediction in men and women. *Journal de Biometrie Humanaine et Anthropologie* 21:103-118.

Marfell-Jones, M.J., Olds, T., Stewart, A.D., and Carter, L. International Standards for Anthropometric Assessment (2006). International Society for the Advancement of Kinanthropometry (ISAK), Potchefstroom, South Africa, April 2006.

Martin, A.D., Ross, W.D., Drinkwater, D.T., and Clarys, J.P. 1985. Prediction of body fat by skinfold caliper: Assumptions and cadaver evidence. *International Journal of Obesity* 9(Suppl 1):31-39.

Martinoli, R., Mohamed, E.I., Maiolo, C., Cianci, R., Denoth, F., Salvadori, S., et al. 2003. Total body water estimation using bioelectrical impedance: A meta-analysis of the data available in the literature. *Acta Diabetologica* 40(Suppl 1):S203-S206.

Maughan, R., and Shirreffs, S. 2004. Exercise in the heat: Challenges and opportunities. *Journal of Sports Sciences* 22(10):917-927.

Mazess, R.B., Barden, H.S., Bisek, J.P., and Hanson, J. 1990. Dual-energy x-ray absorptiometry for total-body and regional bone-mineral and soft-tissue composition. *American Journal of Clinical Nutrition* 51(6):1106-1112.

Millard-Stafford, M.L., Collins, M.A., Evans, E.M., Snow, T.K., Cureton, K.J., and Rosskopf, L.B. 2001. Use of air displacement plethysmography for estimating body fat in a four-component model. *Medicine and Science in Sports and Exercise* 33(8):1311-1317.

Minderico, C.S., Silva, A.M., Fields, D.A., Branco, T.L., Martins, S.S., Teixeira, P.J., et al. 2008. Changes in thoracic gas volume with air-displacement plethysmography after a weight loss program in overweight and obese women. *European Journal of Clinical Nutrition* 62(3):444-450.

Minderico, C.S., Silva, A.M., Teixeira, P.J., Sardinha, L.B., Hull, H.R., and Fields, D.A. 2006. Validity of air-displacement plethysmography in the assessment of body composition changes in a 16-month weight loss program. *Nutrition & Metabolism* 3:32.

Moon, J.R., Tobkin, S.E., Roberts, M.D., Dalbo, V.J., Kerksick, C.M., Bemben, M.G., Cramer, J.T. and Stout, J.R. 2008. Total body water estimations in healthy men and women using bioimpedance spectroscopy: a deuterium oxide comparison. *Nutrition and Metabolism* 5:7.

Moon, J.R., Eckerson, J.M., Tobkin, S.E., Smith, A.E., Lockwood, C.M., Walter, A.A., et al. 2009. Estimating body fat in NCAA Division I female athletes: A five-compartment model validation of laboratory methods. *European Journal of Applied Physiology* 105:119-130.

Moon, J.R., Stout, J.R., Smith, A.E., Tobkin, S.E., Lockwood, C.M., Kendall, K.L., Graef, J.L., et al. 2010. Reproducibility and validity of bioimpedance spectroscopy for tracking changes in total body water: implications for repeated measurements. *British Journal of Nutrition* 104(9):1384-1394.

Mueller, W.H., and Stallones, L. 1981. Anatomical distribution of subcutaneous fat: Skinfold site choice and construction of indices. *Human Biology* 53(3):321-335.

Nana, A., Slater, G.J., Hopkins, W.G. and Burke, L.M. 2012. Effects of daily activities on dual-energy X-ray absorptiometry measurements of body composition in active people. *Medicine and Science in Sports and Exercise* 44(1):180-189.

National Institutes of Health. 1996. NIH Consensus statement. Bioelectrical impedance analysis in body composition measurement. National Institutes of Health Technology Assessment Conference Statement. December 12-14, 1994. *Nutrition* 12(11-12):749-762.

Noreen, E.E., and Lemon, P.W. 2006. Reliability of air displacement plethysmography in a large, heterogeneous sample. *Medicine and Science in Sports and Exercise* 38(8):1505-1509.

Norton, K., Hayward, S., Charles, S., and Rees, M. 1998. The effects of hypohydration and hyperhydration on skinfold measurements. Paper presented at the Sixth Scientific Conference of the International Society for the Advancement of Kinanthropometry, Adelaide, Australia, October 13-16, 1998.

Norton, K., Olds, T., Olive, S., and Craig, N. 1996. Anthropometry and sports performance. In: K. Norton and T. Olds (eds), *Anthropometrica*. Sydney: University of New South Wales Press, pp 287-364.

Norton, K., Whittingham, N.O., Carter, L., Kerr, D.A., Gore, C., and Marfell-Jones, M. 1996. Measurement techniques in anthropometry. In: K. I. Norton and T. Olds (eds),

Anthropometrica. Sydney: University of New South Wales Press, pp 25-76.

Olds, T. 2001. The evolution of physique in male rugby union players in the twentieth century. *Journal of Sports Sciences* 19(4):253-262.

Olds, T., Ross, J., Blanchonette, P., and Stratton, D. 2008. Virtual anthropometry. In: M. Marfell-Jones and T. Olds (eds), Kinathropometry X: Proceedings of the 10th International Society for the Advancement of Kinanthropometry Conference, held in conjunction with the 13th Commonwealth International Sport Conference. Oxon, UK: Routledge, pp 25-38.

Pateyjohns, I.R., Brinkworth, G.D., Buckley, J.D., Noakes, M., and Clifton, P.M. 2006. Comparison of three bioelectrical impedance methods with DXA in overweight and obese men. *Obesity (Silver Spring)* 14(11):2064-2070.

Peeters, M.W., and Claessens, A.L. 2011. Effect of different swim caps on the assessment of body volume and percentage body fat by air displacement plethysmography. *Journal of Sports Sciences* 29(2):191-196.

Pichard, C., Kyle, U.G., Gremion, G., Gerbase, M., and Slosman, D.O. 1997. Body composition by x-ray absorptiometry and bioelectrical impedance in female runners. *Medicine and Science in Sports and Exercise* 29(11):1527-1534.

Pietrobelli, A., Formica, C., Wang, Z., and Heymsfield, S. B. 1996. Dual-energy X-ray absorptiometry body composition model: Review of physical concepts. *American Journal of Physiology* 271(6 Pt 1):E941-E951.

Pineau, J.C., Filliard, J.R., and Bocquet, M. 2009. Ultrasound techniques applied to body fat measurement in male and female athletes. *Journal of Athletic Training* 44(2):142-147.

Prior, B.M., Cureton, K.J., Modlesky, C.M., Evans, E.M., Sloniger, M.A., Saunders, M., et al. 1997. In vivo validation of whole body composition estimates from dual-energy X-ray absorptiometry. *Journal of Applied Physiology* 83(2):623-630.

Pritchard, J.E., Nowson, C.A., Strauss, B.J., Carlson, J.S., Kaymakci, B., and Wark, J.D. 1993. Evaluation of dual energy X-ray absorptiometry as a method of measurement of body-fat. *European Journal of Clinical Nutrition* 47(3):216-228.

Prouteau, S., Ducher, G., Nanyan, P., Lemineur, G., Benhamou, L., and Courteix, D. 2004. Fractal analysis of bone texture: A screening tool for stress fracture risk? *European Journal of Clinical Investigation* 34(2):137-142.

Ramirez, M. 1992. Measurement of subcutaneous adipose tissue using ultrasound images. *American Journal of Clinical Nutrition* 89:347-357.

Reid, I.R., Evans, M.C., and Ames, R. 1992. Relationships between upper-arm anthropometry and soft-tissue composition in postmenopausal women. *American Journal of Clinical Nutrition* 56(3):463-466.

Reilly, T., George, K., Marfell-Jones, M., Scott, M., Sutton, L., and Wallace, J.A. 2009. How well do skinfold equations predict percent body fat in elite soccer players? *International Journal of Sports Medicine* 30:607-613.

Roche, A.F. 1996. Anthropometry and ultrasound. In: A.F. Roche, S.B. Heymsfield, and T.G. Lohman (eds), *Human Body Composition*. Champaign, IL: Human Kinetics, pp 167-190.

Rodriguez, F.A. 1986. Physical structure of international lightweight rowers. In: T. Reilly, J. Watkins and J. Borms (eds), *Kinanthropometry III*. London: E and F.N. Spon, pp 255-261.

Rothney, M.P., Brychta, R.J., Schaefer, E.V., Chen, K.Y., and Skarulis, M.C. 2009. Body composition measured by dual-energy X-ray absorptiometry half-body scans in obese adults. *Obesity (Silver Spring)* 17(6):1281-1286.

Roubenoff, R., Kehayias, J.J., Dawson-Hughes, B., and Heymsfield, S.B. 1993. Use of dual-energy x-ray absorptiometry in body-composition studies: Not yet a "gold standard." *American Journal of Clinical Nutrition* 58(5):589-591.

Ruiz, L., Colley, J.R., and Hamilton, P.J. 1971. Measurement of triceps skinfold thickness. An investigation of sources of variation. *British Journal of Preventive & Social Medicine* 25(3):165-167.

Santos, D.A., Silva, A.M., Matias, C.N., Fields, D.A., Heymsfield, S.B., and Sardinha, L.B. 2010. Accuracy of DXA in estimating body composition changes in elite athletes using a four compartment model as the reference method. *Nutrition & Metabolism* 7:22.

Saunders, M.J., Blevins, J.E., and Broeder, C.E. 1998. Effects of hydration changes on bioelectrical impedance in endurance trained individuals. *Medicine and Science in Sports and Exercise* 30(6):885-892.

Schranz, N., Tomkinson, G., Olds, T., and Daniell, N. 2010. Three-dimensional anthropometric analysis: Differences between elite Australian rowers and the general population. *Journal of Sports Sciences* 28(5):459-469.

Secchiutti, A., Fagour, C., Perlemoine, C., Gin, H., Durrieu, J., and Rigalleau, V. 2007. Air displacement plethysmography can detect moderate changes in body composition. *European Journal of Clinical Nutrition* 61(1):25-29.

Shephard, R.J. 1998. Science and medicine of rowing: A review. *Journal of Sports Sciences* 16:603-620.

Siders, W.A., Lukaski, H.C., and Bolonchuk, W.W. 1993. Relationships among swimming performance, body composition and somatotype in competitive collegiate swimmers. *Journal of Sports Medicine and Physical Fitness* 33(2):166-171.

Silva, A., Minderico, C., Rodrigues, A., Pietrobelli, A., and Sardinha, L. 2004. Calibration models to estimate body composition measurements in taller subjects using DXA. *Obesity Research* 12:A12-A12.

Silva, A.M., Fields, D.A., Quiterio, A.L., and Sardinha, L.B. 2009. Are skinfold-based models accurate and suitable for assessing changes in body composition in highly trained athletes? *Journal of Strength and Conditioning Research* 23(6):1688-1696.

Silva, A.M., Minderico, C.S., Teixeira, P.J., Pietrobelli, A., and Sardinha, L.B. 2006. Body fat measurement in adolescent athletes: Multicompartment molecular model comparison. *European Journal of Clinical Nutrition* 60(8):955-964.

Siri, W.E. 1993. Body composition from fluid spaces and density: Analysis of methods. 1961. *Nutrition* 9(5):480-491.

Soriano, J.M., Ioannidou, E., Wang, J., Thornton, J.C., Horlick, M.N., Gallagher, D., et al. 2004. Pencil-beam vs fan-beam dual-energy X-ray absorptiometry comparisons across four systems: Body composition and bone mineral. *Journal of Clinical Densitometry* 7(3):281-289.

Stein, R.J., Haddock, C.K., Poston, W.S., Catanese, D., and Spertus, J.A. 2005. Precision in weighing: A comparison of scales found in physician offices, fitness centers, and weight loss centers. *Public Health Reports* 120(3):266-270.

Stewart, A.D., and Hannan, W.J. 2000. Prediction of fat and fat-free mass in male athletes using dual X-ray absorptiometry as the reference method. *Journal of Sports Sciences* 18(4):263-274.

Stewart, A.D., Marfell-Jones, M.J., Olds, T., and De Ridder, J.H. International Standards for Anthropometric Assessment (2011). International Society for the Advancement of Kinanthropometry (ISAK), Lower Hutt, New Zealand, December 2011.

Stoggl, T., Enqvist, J., Muller, E., and Holmberg, H.C. 2010. Relationships between body composition, body dimensions, and peak speed in cross-country sprint skiing. *Journal of Sports Sciences* 28(2):161-169.

Sun, G., French, C.R., Martin, G.R., Younghusband, B., Green, R.C., Xie, Y.G., et al. 2005. Comparison of multifrequency bioelectrical impedance analysis with dual-energy X-ray absorptiometry for assessment of percentage body fat in a large, healthy population. *American Journal of Clinical Nutrition* 81(1):74-78.

Svantesson, U., Zander, M., Klingberg, S., and Slinde, F. 2008. Body composition in male elite athletes, comparison of bioelectrical impedance spectroscopy with dual energy X-ray absorptiometry. *Journal of Negative Results in Biomedicine* 7:1.

Tataranni, P.A., and Ravussin, E. 1995. Use of dual-energy X-ray absorptiometry in obese individuals. *American Journal of Clinical Nutrition* 62(4):730-734.

Thomasset, A. 1963. Bio-electric properties of tissues. Estimation by measurement of impedance of extracellular ionic strength and intracellular ionic strength in the clinic. *Lyon Médical* 209:1325-1350.

Tothill, P., Avenell, A., Love, J., and Reid, D.M. 1994. Comparisons between Hologic, Lunar and Norland dual-energy X-ray absorptiometers and other techniques used for whole-body soft-tissue measurements. *European Journal of Clinical Nutrition* 48(11):781-794.

Tothill, P., Hannan, W.J., and Wilkinson, S. 2001. Comparisons between a pencil beam and two fan beam dual energy X-ray absorptiometers used for measuring total body bone and soft tissue. *British Journal of Radiology* 74(878):166-176.

Uremovic, M., Bosnjak Pasic, M., Seric, V., Vargek Solter, V., Budic, R., Bosnjak, B., et al. 2004. Ultrasound measurement of the volume of musculus quadriceps after knee joint injury. *Collegium Antropologicum* 28(Suppl 2):227-283.

Utter, A., and Hager, M. 2008. Evaluation of ultrasound in assessing body composition of high school wrestlers. *Medicine and Science in Sports and Exercise* 40(5):943-949.

Utter, A.C., Goss, F.L., Swan, P.D., Harris, G.S., Robertson, R.J., and Trone, G.A. 2003. Evaluation of air displacement for assessing body composition of collegiate wrestlers. *Medicine and Science in Sports and Exercise* 35(3):500-505.

Utter, A.C., and Lambeth, P.G. 2010. Evaluation of multi-frequency bioelectrical impedance analysis in assessing body composition of wrestlers. *Medicine and Science in Sports and Exercise* 42(2):361-367.

Van Der Ploeg, G.E., Withers, R.T., and Laforgia, J. 2003. Percent body fat via DEXA: Comparison with a four-compartment model. *Journal of Applied Physiology* 94(2):499-506.

van Marken Lichtenbelt, W.D., Hartgens, F., Vollaard, N.B., Ebbing, S., and Kuipers, H. 2004. Body composition changes in bodybuilders: A method comparison. *Medicine and Science in Sports and Exercise* 36(3):490-497.

Vescovi, J.D., Zimmerman, S.L., Miller, W.C., and Fernhall, B. 2002. Effects of clothing on accuracy and reliability of air displacement plethysmography. *Medicine and Science in Sports and Exercise* 34(2):282-285.

Vilaca, K.H., Ferriolli, E., Lima, N.K., Paula, F.J., and Moriguti, J.C. 2009. Effect of fluid and food intake on the body composition evaluation of elderly persons. *Journal of Nutrition, Health & Aging* 13(3):183-186.

Wagner, D.R., and Heyward, V.H. 1999. Techniques of body composition assessment: A review of laboratory and field methods. *Research Quarterly for Exercise and Sport* 70(2):135-149.

Wang, J., Gallagher, D., Thornton, J., Yu, W., Horlick, M., and Pi-Sunyer, F. 2006. Validation of a 3-dimensional photonic scanner for the measurement of body volumes, dimensions, and percentage body fat. *American Journal of Clinical Nutrition* 83:809-816.

Wells, J., Douros, I., Fuller, N., Elia, M., and Dekker, L. 2000. Assessment of body volume using three dimensional photonic scanning. *Annals of the New York Academy of Sciences* 904: 247-254.

Weyers, A.M., Mazzetti, S.A., Love, D.M., Gomez, A.L., Kraemer, W.J., and Volek, J.S. 2002. Comparison of methods for assessing body composition changes during weight loss. *Medicine and Science in Sports and Exercise* 34(3):497-502.

White, A.T., and Johnson, S.C. 1991. Physiological comparison of international, national and regional alpine skiers. *International Journal of Sports Medicine* 12(4):374-378.

Whitehead, J.R., Eklund, R.C., and Williams, A.C. 2003. Using skinfold calipers while teaching body fatness-related concepts: Cognitive and affective outcomes. *Journal of Science and Medicine in Sport* 6(4):461-476.

Wilmore, J.H., Girandola, R.N., and Moody, D.L. 1970. Validity of skinfold and girth assessment for predicting alterations in body composition. *Journal of Applied Physiology* 29(3):313-317.

Withers, R.T., Craig, N. P., Bourdon, P. C., and Norton, K. I. 1987. Relative body fat and anthropometric prediction of body density of male athletes. *European Journal of Applied Physiology and Occupational Physiology* 56(2):191-200.

Withers, R. T., Laforgia, J., and Heymsfield, S.B. 1999. Critical appraisal of the estimation of body composition via two-, three-, and four-compartment models. *American Journal of Human Biology* 11(2):175-185.

Withers, R.T., LaForgia, J., Pillans, R.K., Shipp, N.J., Chatterton, B.E., Schultz, C.G., et al. 1998. Comparisons of two-, three-, and four-compartment models of body composition analysis in men and women. *Journal of Applied Physiology* 85(1):238-245.

Withers, R.T., Whittingham, N.O., Norton, K.I., La Forgia, J., Ellis, M.W., and Crockett, A. 1987. Relative body fat and anthropometric prediction of body density of female athletes. *European Journal of Applied Physiology and Occupational Physiology* 56(2):169-180.

Testing and Training Agility

Tim J. Gabbett and Jeremy M. Sheppard

Agility has traditionally been defined as the ability to change direction rapidly and accurately (Alricsson et al. 2001; Baechle 1994; Barrow and McGee 1971). The first edition of *Physiological Tests for Elite Athletes* emphasized this definition by only including protocols for the assessment of preplanned change of direction speed (e.g., zig-zag running). Consequently, agility has traditionally been viewed as a quality influenced predominantly by physical (e.g., leg muscular strength, reactive strength, and power) and biomechanical (e.g., running technique) qualities, and training programs designed to improve agility performance typically focus on enhancing these individual qualities.

Speed of movement both in a straight line and when changing direction is a clear determinant of performance in many team sports and therefore should be emphasized in the preparation of these athletes. However, speed qualities such as acceleration and acceleration with changes of direction are somewhat distinct from each other and likely require individual attention to maximize performance application to the sporting context (Young et al. 2001). Furthermore, an athlete's speed and ability to change direction are underpinned by a multifactorial model including strength, technique, and other related variables that should be addressed (Sheppard and Young 2006).

Factors Influencing Agility Performance

Although it is recognized that preplanned movements are important to team sport performance, effective agility movements are also commonly executed in response to a sport-specific stimulus (e.g.,

the offensive movements of an attacking player). It has been established that successful team sport performance depends on well-developed physical qualities (e.g., linear sprinting speed, change of direction speed) (Gabbett et al. 2009). However, higher-skilled athletes in team sports have also consistently shown perceptual skills that are superior to those of their less-skilled counterparts (refer to chapter 15, Perceptual–Cognitive and Perceptual–Motor Contributions to Elite Performance). Despite the wealth of evidence demonstrating differences between experts and novices in sport-specific anticipation and decision making tasks (see Williams et al. 2006 for review), the majority of agility testing protocols have used tests of preplanned change of direction speed. A limitation of tests that measure change of direction speed is that they fail to assess the perceptual components of agility. Figure 12.1 shows a multifactorial model of physical and perceptual qualities influencing agility performance. Although the physical and biomechanical qualities are important to enhance change of direction speed, perceptual qualities such as visual scanning, anticipation, situational knowledge, and pattern recognition are also critical factors discriminating the agility performances of higher- and lesser-skilled team sport athletes (Young et al. 2002).

Contemporary Definition of Agility

Because perceptual qualities are important for effective agility performance, and given that agility represents an integration of physical and perceptual factors, *agility* has recently been redefined as "a rapid, whole-body, change of direction or speed in response

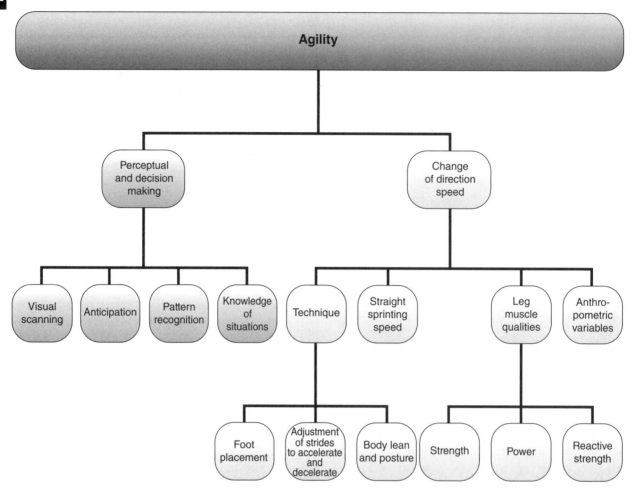

FIGURE 12.1 Theoretical model of agility components.

Adapted by permission of Edizioni Minerva Medica from: J Sports Med Phys Fitness, 2002 Sep;42(3);282-8.

to a sport-specific stimulus" (Sheppard and Young 2006, 919). This definition applies to agility tasks that are open skilled, in that they involve both temporal and spatial uncertainty.

We recently investigated the speed, change of direction speed, and reactive agility (using a test that relied on the ability of players to "read and react" to a game-specific stimulus) performances of higher- and lesser-skilled rugby league players (Gabbett et al. 2008b). Interestingly, significant differences between experts and novices were detected in the perceptual qualities of higher-skilled and less-skilled team sport athletes; however, no differences existed between groups in physical abilities (e.g., speed and change of direction speed) (table 12.1) (Gabbett et al. 2008b), with similar results found in netball (Farrow et al. 2005) and Australian Rules Football (Henry et al. 2011; Sheppard et al. 2006; Veale et al. 2010; Young et al. 2011). Equally important, the reactive agility test offered a reliable method of assessing the

perceptual components of agility performance (table 12.2) (Gabbett et al. 2008b; Sheppard et al. 2006).

The finding of superior anticipatory skill in the higher-skilled athletes suggests that these participants had a greater ability to extract relevant information earlier in the visual display by identifying relevant postural cues presented by the stimulus and disregarding irrelevant sources of information (Farrow et al. 2005). Collectively, these findings demonstrate the practical utility of the reactive agility test for assessing the perceptual components of agility and also raise questions about the discriminant validity of preplanned change of direction speed tests currently used by the majority of high-performance coaches.

Perhaps more important, among a squad of team sport athletes, the reactive agility test was able to distinguish four distinct classifications (table 12.3). Specifically, players were classified as requiring (1) decision-making and change of direction speed

TABLE 12.1 Speed, Change of Direction Speed, and Reactive Agility Results for First Grade and Second Grade Rugby League Players (mean ± SD)

	First grade	Second grade	Effect size	Difference
Speed				
5 m sprint, s	1.14 ± 0.06*	1.20 ± 0.11	0.68	Moderate
10 m sprint, s	1.90 ± 0.09*	2.00 ± 0.14	0.85	Large
20 m sprint, s	3.25 ± 0.16*	3.39 ± 0.21	0.75	Moderate
Change of direction speed				
5-0-5 agility test, s	2.34 ± 0.20	2.39 ± 0.15	0.28	Small
Modified 5-0-5, s	2.66 ± 0.14	2.71 ± 0.17	0.32	Small
L-run, s	6.36 ± 0.53	6.49 ± 0.40	0.28	Small
Reactive agility				
Movement time, s	2.48 ± 0.17*	2.60 ± 0.16	0.73	Moderate
Decision time, ms	55.3 ± 43.6	78.2 ± 40.4	0.54	Moderate
Response accuracy, %	89.3 ± 13.9	84.0 ± 17.3	0.34	Small

Effect size differences, <0.10 = trivial, 0.11-0.50 = small; 0.51-0.80 = moderate; >0.80 = large.

*Denotes significant difference ($p < .05$) between groups.

Adapted, by permission, from T.J. Gabbett, J.N. Kelly, and J.M. Sheppard, 2008, "Speed, change of direction speed, and reactive agility of rugby league players," *Journal of Strength and Conditioning Research* 22: 174-181.

TABLE 12.2 Test–Retest Reliability of Speed, Change of Direction Speed, and Reactive Agility Tests in a Sample of Rugby League Players (mean ± SD)

	Test 1	Test 2	ICC	%TE
Speed				
5 m sprint, s	1.20 ± 0.10	1.20 ± 0.10	0.84	3.2
10 m sprint, s	1.98 ± 0.13	1.99 ± 0.11	0.87	1.9
20 m sprint, s	3.39 ± 0.20	3.38 ± 0.20	0.96	1.3
Change of direction speed				
5-0-5 agility test, s	2.39 ± 0.17	2.37 ± 0.16	0.90	1.9
Modified 5-0-5, s	2.73 ± 0.17	2.72 ± 0.17	0.92	2.5
L-run, s	5.77 ± 0.69	5.63 ± 0.38	0.95	2.8
Reactive agility				
Movement time, s	2.50 ± 0.14	2.52 ± 0.18	0.92	2.1
Decision time, ms	71.5 ± 47.1	71.1 ± 46.0	0.95	7.8
Response accuracy, %	85.5 ± 16.3	85.4 ± 16.6	0.93	3.9

Data were collected on the same sample of rugby league players as in table 12.1. ICC = intraclass correlation coefficient; TE = typical error of measurement.

Adapted, by permission, from T.J. Gabbett, J.N. Kelly, and J.M. Sheppard, 2008, "Speed, change of direction speed, and reactive agility of rugby league players," *Journal of Strength and Conditioning Research* 22: 174-181.

training to further consolidate good physical and perceptual abilities, (2) decision-making training to improve below-average perceptual abilities, (3) speed and change of direction speed training to improve below-average physical attributes, or (4) a combination of decision-making and change of direction speed training to improve below-average physical and perceptual abilities.

Agility Testing Protocols

This section provides a sample of agility testing protocols. It is not designed to provide an exhaustive list of change of direction speed and reactive agility testing protocols but rather provides the reader with some of the options available to test preplanned change of direction speed and reactive agility.

TABLE 12.3 Interpretation and Training Prescription for Four Players With Different Results on the Reactive Agility Test

Player	Decision time, ms	Movement time, s	Interpretation	Prescription
Fast mover–fast thinker	58.75	2.31	Has speed and fast decision time, which contribute to above-average anticipation skills	Needs to continue to develop change of direction speed and decision-making skills.
Fast mover–slow thinker	148.75	2.33	Has speed but slow decision time, which contributes to below-average anticipation skills	Needs more decision-making training on (e.g., reactive agility training) and off (e.g., video-based perceptual training) the field.
Slow mover–fast thinker	28.75	2.85	Perceptually skilled but lacks change of direction speed	Needs more speed and change of direction speed training to improve physical qualities.
Slow mover–slow thinker	112.50	2.86	Has poor speed and slow decision time, which contribute to below-average anticipation skills	Needs more decision-making and speed and change of direction speed training to improve physical qualities and perceptual skill.

Fast movers–fast thinkers = good change of direction speed and good perceptual skill; fast movers–slow thinkers = good change of direction speed and below-average perceptual skill; slow movers–fast thinkers = below-average change of direction speed and good perceptual skill; slow movers–slow thinkers = below-average change of direction speed and below-average perceptual skill.

Adapted, by permission, from T.J. Gabbett, J.N. Kelly, and J.M. Sheppard, 2008, "Speed, change of direction speed, and reactive agility of rugby league players," *Journal of Strength and Conditioning Research* 22: 174-181.

Change of Direction Speed Testing Protocols

The following tests are examples of change of direction speed protocols.

L-Run

The L-run test was originally developed for rugby union (Webb and Lander 1983) but subsequently has been used with subelite (Gabbett et al. 2008b) and elite (Meir et al. 2001) rugby league players. In this preplanned running test, the athlete runs 5 m, turns 90° to the left, runs 5 m, makes a full 180° turn around a cone, returns to the original cone, makes another 90° turn, and returns to the start position (figure 12.2).

5-0-5 Agility Test

For the 5-0-5 agility test, 2 timing gates are placed 5 m from a designated turning point. Athletes assume a starting position 10 m from the timing gates (and therefore 15 m from the turning point). Athletes are instructed to accelerate as quickly as possible through the timing gates, pivot on the 15 m line, and return as quickly as possible through the timing gates (figure 12.3) (Draper and Lancaster 1985). Refer to chapter 14, Field Testing Principles and Protocols, for a more detailed description of the 5-0-5 agility test.

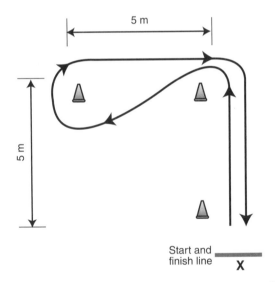

FIGURE 12.2 Schematic illustration of the L-run (Webb and Lander 1983).

Reprinted, by permission, from T.J. Gabbett, J.N. Kelly, and J.M. Sheppard, 2008, "Speed, change of direction speed, and reactive agility of rugby league players," *Journal of Strength and Conditioning Research* 22: 174-181.

T-Test

The T-test is a popular test of change of direction speed (Pauole et al. 2000; Semenick 1990). The athlete runs forward from point A to point B, shuffles

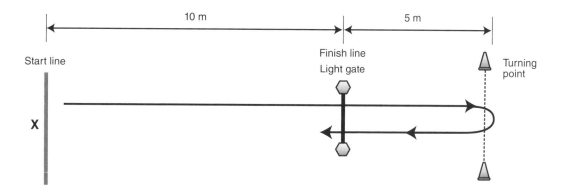

FIGURE 12.3 Schematic illustration of the 5-0-5 agility test (Draper and Lancaster 1985).

Reprinted, by permission, from T.J. Gabbett, J.N. Kelly, and J.M. Sheppard, 2008, "Speed, change of direction speed, and reactive agility of rugby league players," *Journal of Strength and Conditioning Research* 22: 174-181.

to the left (point C), shuffles to the right (point D), shuffles back to point B, and runs backward to the start position (point A) (figure 12.4). It should be emphasized that the T-test is clearly a preplanned change of direction speed test rather than a reactive agility test.

Reactive Agility Testing Protocols

Several sport-specific reactive agility testing protocols have been proposed for Australian football (Sheppard et al. 2006), rugby league (Gabbett et al. 2008b; Serpell et al. 2010), netball (Farrow et al. 2005), and softball (Gabbett et al. 2007). Faster decision times

and response accuracy have been demonstrated in response to lifelike video footage simulating sport-specific passages of play in higher-skilled athletes in comparison to their less-skilled counterparts in competitors from a wide range of team sports. Although the majority of these studies typically use expensive technology that involves projection of video images onto a large screen (Farrow et al. 2005; Gabbett et al. 2007; Serpell et al. 2010), the reactive agility test first described by Sheppard and colleagues (2006) offers an inexpensive and practical method of assessing reactive agility for a wide range of team sports. Although this reactive agility test was initially designed and validated for Australian football (Sheppard et al. 2006), it has recently gained popularity in rugby league (Gabbett et al. 2008b), rugby union (Sheppard et al. 2008), soccer (Sheppard, unpublished observations), and basketball (Gabbett et al. 2008a), demonstrating its practical utility for assessing the generic perceptual demands of a wide range of team sports.

The reactive agility test has been described in detail elsewhere (Sheppard et al. 2006). In brief, the athlete begins on a marked line, as illustrated in figure 12.5. Timing gates are placed 5 m to the left and right and 2 m forward of the start line. Therefore, the timing gates are placed 10 m from each other.

The tester (investigator) stands opposite and facing the athletes and is behind a set of timing lights. In each test trial the tester initiates movement, thereby beginning the timing. The athlete reacts to the movements of the tester by moving forward and then to the left or right in response to, and in the same direction as, the left or right movement of the tester. The timing stops when the athlete triggers the timing beam on either side.

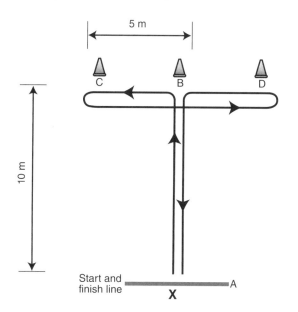

FIGURE 12.4 Schematic illustration of the T-test.

Adapted, by permission, from D. Semeick, 1990, "Tests and measurement: The T-test," *Strength and Conditioning Journal* 12(1): 36-37.

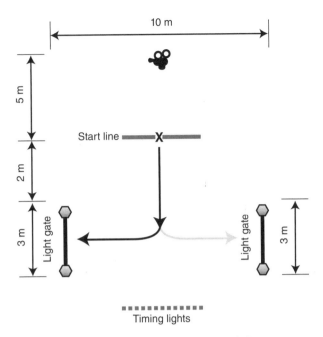

FIGURE 12.5 Schematic illustration of the reactive agility test.

Reprinted, by permission, from T.J. Gabbett, J.N. Kelly, and J.M. Sheppard, 2008, "Speed, change of direction speed, and reactive agility of rugby league players," *Journal of Strength and Conditioning Research* 22: 174-181.

The tester displays one of four possible scenarios for the athlete to react to. The four possible scenarios all involve steps of approximately 1/2 m and are presented in a random order that is different for each athlete:

- Step forward with right foot and change direction to the left.
- Step forward with the left foot and change direction to the right.
- Step forward with the right foot, then left, and change direction to the right.
- Step forward with the left foot, then right, and change direction to the left.

An equal number of each scenario is presented for each athlete. The test protocol involves randomized presentation of four different cues, for a total of eight trials. These cues create varying demands on the athletes (as in a game setting), resulting in intertrial variability. For this reason, the recorded score used represents the mean of all trials (8), which is an average of all trials to the left (4) and to the right (4).

The athletes sprint forward prior to any change of direction, in reaction to the forward movement of the tester. The athletes are instructed to recognize the cues as soon as possible (essentially while moving forward) and react by changing direction and sprint-

ing through the gates on the left or right in response. The athletes are instructed to emphasize accuracy (decision-making accuracy) and speed of movement. A high-speed video camera interfaced with a video recorder is usually positioned 5 m behind the subject to record the athlete's change of movement direction relative to the tester. Video is sampled at high frequency (e.g., 200 Hz) so that the number of frames between the tester's and athlete's movement initiation enables the athlete's decision-making time to be recorded to within ±5 ms for each trial.

Typical results of the reactive agility test are presented in table 12.3. Note that published results for the reactive agility test have been collected with different timing light systems, in different populations, and with different surfaces, all of which would likely contribute to differences in test results.

Individual Interpretation of Tests

Although it is clear that perceptual qualities contribute to effective agility performance, no study has used temporal or spatial occlusion techniques (which are traditionally used in anticipation studies) to identify the most relevant cues used by highly agile team sport players. Consequently, although reactive agility tests are acknowledged as important for informing agility training programs, the relevant cues to which athletes should attend are unclear. Identifying the information sources that are most relevant to specific sports offers an important research pursuit for applied sport scientists and is likely to provide great value to coaches attempting to train the perceptual components of agility.

Notwithstanding the current limitations associated with reactive agility tests, the inclusion of decision time and response accuracy information for the reactive agility test using high-speed video footage provides insight into the anticipatory strengths and weaknesses of team sport athletes. Table 12.3 shows the results of four rugby league players tested on the reactive agility test. The analysis of the high-speed video footage allows us to delineate the performance components of the reactive agility test by elucidating whether performance was limited by decision time or movement speed. Indeed, if movement time alone were used to classify reactive agility, some players would be incorrectly classified as having superior anticipatory skill (i.e., "fast thinkers"). Also note that a speed–accuracy trade-off exists in the majority of players, so that reductions in decision time (i.e., faster decision-making ability) result in reductions in response accuracy (Gabbett et al. 2008b).

Training Directions Based on Test Results

As stated previously in this chapter, the use of a traditional preplanned change of direction speed test, in combination with a reactive agility test, allows the coach to identify the individual factors limiting effective agility performance. For example, players with fast movement speed but slow decision time and low response accuracy clearly require additional perceptual training to improve their ability to recognize and respond to sport-specific cues. Conversely, players with fast decision times and high response accuracy but poor movement speed are likely to benefit from additional movement speed training that emphasizes speed in changing direction. To assist with the interpretation of change of direction speed and reactive agility tests, we have developed an agility training matrix (table 12.3). This matrix identifies players with fast change of direction speed, well-developed decision-making ability, or a combination or absence of both qualities so that coaches can provide players with specific agility training to meet their individual requirements.

References

Alricsson, M., Harms-Ringdahl, K., et al. 2001. Reliability of sports related functional tests with emphasis on speed and agility in young athletes. *Scandinavian Journal of Medicine and Science in Sports* 11(4):229-232.

Baechle, T.R. 1994. *Essentials of Strength and Conditioning.* Champaign, IL: Human Kinetics.

Barrow, H., and McGee, R. 1971. *A Practical Approach to Measurement in Physical Education.* Philadelphia: Lea & Febiger.

Draper, J.A., and Lancaster, M.G. 1985. The 505 test: A test for agility in the horizontal plane. *Australian Journal for Science and Medicine in Sport* 17(1):15-18.

Farrow, D., Young, W., and Bruce, L. 2005. The development of a test of reactive agility for netball: a new methodology. *Journal of Science and Medicine in Sport* 8(1):52-60.

Gabbett, T., Rubinoff, M., Thorburn, L., and Farrow, D. 2007. Testing and training anticipation skills in softball fielders. *International Journal of Sport Science and Coaching* 2(1):15-24.

Gabbett, T.J., Sheppard, J.M., Pritchard-Peschek, K.R., Leveritt, M.D., and Aldred, M.J. 2008a. Influence of closed skill and open skill warm-ups on the performance of speed, change of direction speed, vertical jump, and reactive agility in team sport athletes. *Journal of Strength and Conditioning Research* 22(5):1413-1415.

Gabbett, T.J., Kelly, J.N., and Sheppard, J.M. 2008b. Speed, change of direction speed, and reactive agility of rugby league players. *Journal of Strength and Conditioning Research* 22(1):174-181.

Gabbett, T., Kelly, J., Ralph, S., and Driscoll, D. 2009. Physiological and anthropometric characteristics of junior elite and sub-elite rugby league players, with special reference to starters and non-starters. *Journal of Science and Medicine in Sport* 12(1):215-222.

Henry, G., Dawson, B., Lay, B., and Young, W. 2011. Validity of a reactive agility test for Australian football. *International Journal of Sports Physiology and Performance.* 6(4):534-545.

Meir, R., Newton, R., et al. 2001. Physical fitness qualities of professional rugby league football players: determination of positional differences. *Journal of Strength and Conditioning Research* 15(4):450-458.

Pauole, K., Madole, K., Garhammer, J., Lacourse, M., and Rozenek, R. 2000. Reliability and validity of the t-test as a measure of agility, leg power, and leg speed in college-aged men and women. *Journal of Strength and Conditioning Research* 14(4):443-450.

Semenick, D. 1990. Tests and measurements: the T test. *National Strength and Conditioning Association Journal* 12(1):36-37.

Serpell, B.G., Ford, M., and Young, W.B. 2010. The development of a new test of agility for rugby league. *Journal of Strength and Conditioning Research* 24(12):3270-3277.

Sheppard, J. M., Barker, M., and Gabbett, T. 2008. Training agility in elite rugby players: a case study. *Journal of Australian Strength and Conditioning* 16(3):15-19.

Sheppard, J. M. and Young, W. B. 2006. Agility literature review: classifications, training and testing. *Journal of Sports Sciences* 24(9):919-932.

Sheppard, J. M., Young, W.B. Doyle, T.L., Sheppard, T., and Newton, R.U. 2006. An evaluation of a new test of reactive agility and its relationship to sprint speed and change of direction speed. *Journal of Science and Medicine in Sport* 9:342-349.

Veale, J.P., Pearce, A.J., and Carlson, J.S. 2010. Reliability and validity of a reactive agility test for Australian football. *International Journal of Sports Physiology and Performance,* 5(2):239-248.

Webb, P., and J. Lander 1983. An economical fitness testing battery for high school and college rugby teams. *Sports Coach* 7(3):44-46.

Williams, A.M., Hodges, N.J., North, J.S., and Barton, G. 2006. Perceiving patterns of play in dynamic sport tasks: investigating the essential information underlying skilled performance. *Perception* 35(3):317-332.

Young, W.B., McDowell, M.H., and Scarlett, B.J. 2001. Specificity of sprint and agility training methods. *Journal of Strength and Conditioning Research* 15(3):315-319.

Young, W.B., James, R., and Montgomery, I. 2002. Is muscle power related to running speed with changes of direction? *Journal of Sports Medicine and Physical Fitness* 42(3):282-288.

Young, W., Farrow, D., Pyne, D., McGregor, W., and Handke, T. 2011. Validity and reliability of agility tests in junior Australian football players. *Journal of Strength and Conditioning Research,* 25(12):3399-3403.

Strength and Power Assessment Protocols

Michael R. McGuigan, Jeremy M. Sheppard, Stuart J. Cormack, and Kristie-Lee Taylor

Both strength and power are critical components of athletic performance. Strength and power can represent specific or independent qualities of neuromuscular performance and therefore can be assessed and trained independently. Strength and power diagnosis is the process of determining an athlete's level of development of each of these distinct qualities. A variety of tests are used by practitioners to test their athletes and to describe the physical qualities of athletes. The information obtained from strength and power assessment can be used to monitor training, measure the response to training interventions, identify strengths and weaknesses, individualize training loads and programs, and compare the athlete with other athletes and with normative data.

Practitioners and researchers must consider the reliability and validity of the variables obtained from strength and power assessment. This chapter discusses the various strength and power qualities and the methods that are used to assess them.

Relationship Between Strength And Power

A robust relationship exists between strength and power. Strength is correlated to many variables of athletic performance (Pearson et al. 2009; Stone et al. 2003a, 2003b; Wisloff et al. 2004; Young 1995). Power output is an important attribute in determining athletic ability and predicting success in sport, such as vertical jump height (McBride et al. 2002) and sprint times (Cormie et al. 2010a). The magnitude of the relationship of strength to athletic performance

depends on the type of assessment that is performed. For example, isolated joint isokinetic measures show poor correlations to athletic performance variables (Pincivero et al. 1997; Pua et al. 2006). However, structural tests of strength such as free weight squats or isometric squat or isometric midthigh pull (IMTP) are consistently and strongly correlated to athletic performance (Kawamori et al. 2006; Nuzzo et al. 2008). Strength expressed in relative terms to body mass shows even stronger correlations to athletic performance (Nuzzo et al. 2008). Relatively low relationships have been reported between isokinetic knee extension torque and one-leg hop displacement ($r = .33$) (Pincivero et al. 1997). Another investigation also found low correlations between isokinetic knee extension torque and countermovement jump height ($r = .28-.47$) (Pua et al. 2006).

In contrast, isometric squat force has been strongly correlated with jump peak power and force ($r = .64-.71$), and one repetition maximum (1RM) squat strength has been strongly correlated with jump peak power and force ($r = .79-.84$) (Nuzzo et al. 2008). The IMTP has been strongly correlated with jump peak power and force ($r = .88$) by Haff and colleagues (2005). Power clean 1RM has been strongly correlated with jump peak power ($r = .86$) (Nuzzo et al. 2008) and ($r = .82$) (Haff et al. 2005). Squat 1RM has been significantly correlated with 10 m sprint time ($r = .940$) and 30 m sprint time ($r = .710$) (Wisloff et al. 2004). In addition, individuals who are stronger in the squat 1RM display lower sprint times at 10, 30, and 40 m (McBride et al. 2002). The relationship between strength and power and other measures of athletic performance is discussed in more detail later in the chapter.

13

Strength and Power Assessment Principles

A number of definitions have been used to describe strength. Commonly, it has been described as the "maximal force production capabilities of a muscle," but a more accurate definition is "the ability of a muscle to produce force." Rather than being constrained to maximal force production, the expression of strength exists on a continuum of force production capabilities, capped at one end by maximal strength production and at the other by the ability to repeatedly exert submaximal force (strength endurance).

The assessment of absolute maximal strength capabilities of a muscle or muscle group is conducted in the laboratory via superimposed electrical stimuli. The electrical stimulation is applied either to the innervating nerve or to the skin surface of a muscle group during a maximal voluntary isometric contraction, and the resultant force output is measured. Although knowledge of an individual's absolute maximal strength may be important, often it is more useful to know the maximal amount of the strength an athlete can produce voluntarily. Therefore, assessments of maximal voluntary strength are measured during a maximal isometric contraction without electrical augmentation. Maximal voluntary strength is affected by arousal and central nervous system activation, and therefore most people's absolute maximal strength is greater than the maximal voluntary strength achieved. Isometric testing for maximum voluntary strength has been used in a number of contexts with a variety of testing protocols. Historically, single-joint positions such as leg extension and elbow flexion have been used in research and clinical settings. A variety of hand-held dynamometers are available to assess isometric strength in hip abduction (Fenter et al. 2003), shoulder external rotation (Sullivan et al. 1988), back extension (Coldwells et al. 1994), and hand grip strength (McGuigan and Winchester 2008; McGuigan et al. 2006b; Schechtman et al. 2005). Multijoint body positions such as the squat and IMTP position have also been used extensively (McGuigan and Winchester 2008; McGuigan et al. 2006a, 2006b; Stone et al. 2003a, 2003b).

Isometric testing protocols are also used to measure the ability to develop force rapidly, referred to as the rate of force development (RFD). Both maximal isometric force and RFD have been shown to be important indicators of successful performance (Haff et al. 1997, 2005; Stone et al. 2003a), but many scientists have questioned the validity of isometric assessments given poor correlations between isometric test outcomes and dynamic performance (Murphy and Wilson 1997). As such, it has been suggested that tests of force production are better performed using dynamic movements that reflect the functional requirements of sporting performance.

Maximal Dynamic Strength

Dynamic strength can be assessed in a variety of ways using an assortment of testing equipment. Isokinetic dynamometry involves measuring force output, typically using open chain exercises, using a constant angular velocity throughout the test movement. Isokinetic strength testing is popular in clinical settings to assess bilateral force asymmetry and in research to compare changes in maximal dynamic strength during slow and fast velocity movements. Additionally, there have been efforts to design very specific isokinetic testing procedures that attempt to mimic athletic movements. However, similar to isometric testing protocols, isokinetic strength assessment has been criticized for a lack of performance specificity and is less commonly used in strength testing protocols for the elite athlete.

Isoinertial testing methods require a constant gravitational load throughout the movement. Traditionally, performance-related changes in maximal voluntary dynamic strength capabilities have been assessed using one repetition maximum (1RM) test protocols. This type of testing is conducted in a variety of core lifts and is operationally defined as the heaviest load that can be moved over a specified range of motion one time with correct performance. For this type of assessment, athletes need to be proficient in the movement patterns and able to handle large (maximal) loads. When this prerequisite is not met, 3 to 6 repetition maximum (RM) testing is often used, and the 1RM score is often predicted using equations (Reynolds et al. 2006). Repetition maximum tests can reveal the athlete's maximal strength capacities under dynamic conditions in movements similar to those performed in the sporting arena; however, the nature of the assessment requires high external loads to be moved, resulting in lower movement velocities. An additional aspect of dynamic strength assessment therefore involves measuring force production capabilities at higher movement speeds.

Speed-Strength

Speed-strength, also referred to as *power*, evaluates the athlete's ability to apply force rapidly, thus generating high levels of power (Abernethy et al. 1995; Cronin and Sleivert 2005). Considering that in the sporting context, power tasks can involve varying levels of force application and movement velocities, a strength and speed-strength continuum is useful for the practitioner to use in classifying movements and test protocols.

Speed-strength is generally tested using isoinertial protocols that use loads aimed at targeting specific aspects of the strength and speed-strength continuum or that involve a range of inertial loads to assess the spectrum of strength and speed-strength qualities (Sheppard et al. 2008b). Concentric-only squat jumps (SJ) (without countermovement) and countermovement jumps (CMJ) are commonly used for the lower body, and bench throws and bench pulls, with varying inertial loads, are commonly used for the upper body.

Strength Endurance

Strength endurance is at the end of the strength continuum opposite to maximal strength. Assessment is usually conducted using body weight or submaximal loads as the form of resistance, with the outcome variable being the maximum number of repetitions performed until failure or the maximum repetitions completed within a certain time interval. Strength endurance is also sometimes assessed as the time to fatigue in a submaximal isometric contraction. Strength endurance can be an important factor in a large variety of sports. A range of movement patterns are used in the assessment of strength endurance. The chosen exercises are largely dependent of the characteristics of the sport; however, the most common tests are those that include large muscle masses and compound movements. Examples include maximum number of push-ups, sit-ups, and chin-ups completed within 60 s. Isometric holds (planks, 90° squat against a wall) are also often performed, with the time to failure recorded.

Strength, Power, and Functional Performance

Elite athletes commonly undergo training programs designed to increase underlying strength and power qualities in an attempt to improve performance. Although the extent to which qualities of absolute strength, relative strength, strength endurance, or power are important to sports performance may vary depending on the activity, the associations between these qualities and performance have been well documented (Chelly et al. 2010; Gabbett et al. 2011; Gordon et al. 2009; Harris et al. 2008a; Hayes et al. 2011; Hennessy and Kilty 2001; Kyriazis et al. 2009; Lockie et al. 2011; Robbins et al. 2012; Markovic 2007; Nuzzo et al. 2008; Pearson et al. 2009; Peterson et al. 2006; Rønnestad et al. 2012; Wisloff et al. 2004; Young et al. 2011). In addition, the ability of strength and power tests to discriminate performance level and playing position in various sports has received

attention (Robbins et al. 2012; Alvarez-San Emeterio and Gonzalea-Badillo 2010; Baker 2009; Baker and Newton 2008; Tan et al. 2009). It appears that multiple factors influence the degree of association between underlying strength and power qualities and sporting movement as well as the degree to which training induced improvements in these qualities transfer to performance (Young et al. 2005; Campo et al. 2009; Cormie et al. 2010a; Harris et al. 2008b).

Relationships to Performance

Whilst the relative importance of strength and power to performance may be sport dependent, a large body of evidence supports a positive influence in a wide variety of activities. Specifically, numerous research projects have demonstrated a strong association between strength and power performance and explosive sporting movements (Comfort et al. 2012; Lockie et al. 2011; Nuzzo et al. 2008; Peterson et al. 2006; Robbins et al. 2012; West et al. 2011; Young et al. 2011). For example, West and colleagues (2011) found IMTP force at 100 ms was related to both 10 m sprint and CMJ performance, whilst Peterson and colleagues (2006) found 1RM squat strength to be highly correlated ($r = .76$-$.91$) with various jumping and sprinting tasks in college athletes. Similar associations have been found in various upper and lower body movements in sports such as team handball, soccer, Australian football, volleyball, basketball, and rugby league (Chaouachi et al. 2009a, 2009b; Chelly et al. 2010; Comfort et al. 2012; Sheppard et al. 2008; Wisloff et al. 2004; Young et al. 2011). In fact, weekly strength training may play an important role in maintaining sprint performance across competitive seasons in team sport athletes (Rønnestad et al. 2012). Interestingly, positive relationships between strength and power and sporting performance have also been demonstrated in skill-dominated activities. Recent examinations have revealed the importance of total body rotational power ($r = .54$) and chest strength ($r = .69$) to golf club head speed (Gordon et al. 2009) and a strong relationship between 1RM bench press and grinding performance ($r = .88$) in elite sailors (Pearson et al. 2009).

Although positive associations between measures of strength and sporting performance have been consistently reported, factors such as the movement similarity between the test and performance task (e.g., force, velocity, plane of movement, contraction type) are known to impact the strength of the relationship (Young et al. 2006). A potentially important consideration in these associations is the comparative importance of absolute strength, relative strength, and power. In some activities, absolute maximum strength appears to be more important whilst in others relative values are better predictors of performance (Nuzzo et al 2008; West et al. 2011; Young et al. 2011). Nuzzo and colleagues (2008) reported stronger correlations between relative and absolute squat strength when compared to numerous variables derived from a CMJ and Young et al. (2011) found relative CMJ peak power to be significantly correlated with maximum speed in Australian footballers (Young et al. 2011). Contraction type specificity also appears to influence the strength of the relationship, with evidence of CMJ performance being more closely related to maximum speed than acceleration (Young et al. 2011). Another factor to consider is the relative importance of strength and power. It appears that in some tasks, power is a more useful discriminator than strength. Kyriazis et al. (2009) found this to be the case with shot put performance. The current strength level of an athlete may also be important and it has been suggested that because of the similar short-term adaptations between maximal strength and explosive training, coupled with the potential long-term benefits of improved maximal strength, that strength training is a relatively more effective training modality for weaker individuals (Cormie et al. 2010a, b). However, power training may be more effective in improving explosive performance in stronger athletes (Cormie et al. 2010a, b). It should be noted that whilst increases in maximal strength have occurred in parallel to increases in sprint performance, this does not always occur (Comfort et al. 2012; Harris et al. 2008b). Task complexity may also affect the importance of strength and power to performance and perhaps not unsurprisingly, poor relationships have been reported between strength and agility performance (Markovic 2007). Critically, the role of strength and power to performance appears important in both male and female athletes (Hennessy and Kilty 2001).

Although much previous research has focused on relationships between strength and power and explosive functional performance such as sprinting and jumping, positive relationships between strength and power qualities and endurance performance are also evident (Hayes et al. 2011; Naclerio 2009). For example, muscular endurance has a positive impact on running economy and explosive strength training has been shown to improve 5 km run performance, also via changes in running economy (Hayes et al. 2011; Paavolainen et al. 1999). Maximal strength and power training has enhanced performance in other endurance activities such as cross-country skiing and cycling (Hoff et al. 2002; Mikkola et al. 2007; Rønnestad et al. 2010; Rønnestad et al. 2012) and these improvements may occur as a result of being able to generate more muscular force without a proportionate increase in metabolic cost (Saunders et al. 2004). It appears that there may be substantial performance advantages to including strength and power training in the program of athletes participating in endurance dominated events.

Performance Level Discriminations

There is evidence that strength and power qualities are useful discriminators of performance level in various sports. This has been shown in events as diverse as water polo, rugby league, alpine skiing, and rowing (Alvarez-San Emeterio and Gonzalea-Badillo 2010; Baker 2009; Baker and Newton 2008; Izquierdo-Gabarren et al. 2010; Tan et al. 2009).

Although various strength and power variables are able to discriminate between levels of performance, it appears that some measures may be more useful than others and that the usefulness of each measure may be sport dependent. For example, measures of leg power have been shown to be discriminators of starters compared to nonstarters in elite Australian Football players at the beginning of the season, whereas vertical jump height and upper and lower body strength were not significantly different between groups (Young et al. 2005).

Equipment Checklist

Isometric Midthigh Pull

- ☐ Force platform
- ☐ Personal computer for data collection and storage
- ☐ Immovable Olympic lifting bar (that can be adjusted for height)
- ☐ Goniometer
- ☐ Lifting straps (optional)

Speed Strength

- ☐ Force platform or position transducer
- ☐ Personal computer for data collection and storage (including data collection software)
- ☐ Known distance for position calibration
- ☐ Calibrated masses for force calibration
- ☐ Wooden or lightweight pole for body mass SJ and CMJ
- ☐ Barbell and weight plates for loaded movements
- ☐ Bench pull apparatus (for bench pull)
- ☐ Smith machine (for bench throw)
- ☐ Goniometer (for SJ joint angle assessment)

Maximum Strength or Strength Endurance

- ☐ Barbell and weight plates for loaded movements
- ☐ Bench pull apparatus (for bench pull)
- ☐ Smith machine (for bench throw)
- ☐ Chin up bar
- ☐ Leg press machine
- ☐ Stable boxes of various heights

Recommended Test Order

To maximize the ability to compare results from one test occasion to another, aspects such as training in the 72 h prior to testing and dietary and supplement intake (including alcohol and medications) should be standardized in the lead-up to testing. Testing should also be conducted at a similar time of day and in the same order as on previous occasions. Generally, it is suggested that explosive movements precede slower movements (e.g., CMJ prior to squat) and total body (complex) movements precede more isolated joint movements (e.g., squat before single-leg press).

ISOMETRIC MIDTHIGH PULL

RATIONALE

Isometric contractions are characterized by muscle actions that result in no change in joint angle. Maximal force production capabilities of a muscle are assessed by isometric strength tests, which require subjects to produce maximal force or torque against an immovable resistance; a strain gauge, cable tensiometer, force platform, or other devices are used to measure the applied force. Isometric tests are generally performed to quantify peak force application, as well as RFD, which is thought to be a critical component of successful athletic performance (Schmidtbleicher 1992b; Wilson et al. 1995).

Historically there has been much controversy in the scientific community over the validity of isometric testing in athletic populations. The previous edition of this text strongly suggested that isometric tests not be used for the purposes of athletic assessment, based on reports of low correlations between isometric tests and dynamic performance (Abe et al. 1992; Baker et al. 1994; Fry et al. 1991; Murphy and Wilson 1996; Murphy et al. 1994; Wilson et al. 1995). More recently, however, a growing body of evidence has shown positive correlations between peak force in an IMTP and dynamic performance (Haff et al. 1997; McGuigan and Winchester 2008; McGuigan et al. 2006; Stone et al. 2003b). In addition to having good predictive ability, these tests are popular because they can be easily administered and standardized, resulting in high reproducibility (Haff et al. 1997; Kawamori et al. 2006; Stone et al. 2003a); they are associated with minimal fatigue; they require relatively inexpensive equipment; and they carry a minimal risk of injury, even in populations with no previous resistance training history (Kilgallon 2008).

The IMTP requires the measurement of vertical ground reaction force via a force platform to obtain values for peak force and rate of force development (RFD). Total impulse and impulse at relevant time points throughout the pull are also commonly investigated.

TEST PROCEDURE

- Ensure that the force platform is correctly calibrated.
- Have the athlete stand on the force platform above an immovable bar, which should be set to a height where the hip angle and knee angles are approximately 155° to 165° and 125° to 135°, respectively.
- Lifting straps may be used to ensure a good grip is maintained throughout the lift.
- On the go command, instruct the athlete to pull "hard and fast" and maintain a maximal effort for 3 to 5 s.
- Ensure there is no change in hip or knee angle throughout the trial.
- Allow a minimum of 3 min rest between trials.

DATA ANALYSIS

Peak isometric force is highly related to maximal strength in a variety of dynamic lifts and therefore may be used as a substitute to traditional repetition maximum testing. This may be advantageous. Normative and reliability data are shown in table 13.1.

TABLE 13.1 Normative Data for Isometric Midthigh Pull Test

Author	Sport	Gender	Mass, kg	Peak force, N	Relative peak force, $N \cdot kg^{-1}$	Rate of force development, $N \cdot s^{-1}$
McGuigan et al. (2010)	Recreational athletes	Male (n = 26)	90 ± 10	1,217 ± 183	13.4 ± 2.4	5,729 ± 2,383
Stone et al. (2008)	Weightlifting	Male & female (n = 7)	98 ± 31	5,102 ± 1,536	52.6 ± 8.4	16,999 ± 5,878
McGuigan et al. (2008a)	Soccer	Male (n = 56)	72 ± 10	2,290 ± 528	31.8 ± 7.2	n/a
McGuigan et al. (2008b)	American football	Male (n = 22)	108 ± 23	2,159 ± 218	20.7 ± 3.5	13,489 ± 4,041
Nuzzo et al. (2008)	Various	Male (n = 12)	90 ± 15	3,144 ± 792	35.2 ± 6.8	3,555 ± 1,026*

n/a = not available. Reliability data: peak force intraclass correlation coefficient = .97, coefficient of variation = 2.4% (McGuigan et al. 2008).

*Average rate of force development from time of initiation to peak force.

MAXIMUM STRENGTH

RATIONALE

Maximum strength is generally defined as the maximum force that can be generated in a single effort against an external resistance (Cronin and Henderson 2004; Schmidtbleicher 1992a). The assessment of maximum force production capability does not include measurement of a "time component," and by definition (given the physics of the force–velocity relationship of muscle contraction), actions requiring maximum force production can only be performed at a relatively slow velocity (Schmidtbleicher 1992a).

To assess maximum strength in athletic populations, testers often use large muscle mass movements against an external resistance (e.g., squat, bench press) for a specified number of repetitions (RM testing). The general aim of these protocols is to determine the maximum load that can be lifted with the appropriate technique for a specified number of repetitions (often 1-3).

PREDICTION EQUATIONS

A number of prediction equations have been established to allow estimation of 1RM loads from other repetition ranges, and the reader is directed to the work of Reynolds and colleagues (2006), which contains the calculations for numerous prediction equations. This group developed regression equations to predict 1RM from 1 to 5, 10, and 20 repetitions in the bench press and leg press in a wide age range of men and women, although not elite performers. In an assessment of the ability of local muscular endurance to predict 1RM maximum bench press in four different athletic populations, Desgorces and colleagues (2010) developed and presented group-specific equations.

These equations are potentially useful when there are safety concerns over using low repetition range protocols in athletes who may lack technical skill in lifting movements or whose body type (e.g., exceptionally tall) makes it difficult to use maximum loads in some movement patterns (e.g., squat).

BODY WEIGHT CONSIDERATIONS

An important issue in the assessment of maximum strength is the role of body mass. It has been suggested that if body size is not taken into account, maximum strength may be overestimated and underestimated in heavier and lighter subjects, respectively, when using protocols involving various pushing, pulling, and lifting movements (Aasa et al. 2003; Atkins 2004). However, the opposite may be true in assessment tasks requiring athletes to overcome their own body weight (e.g., chin-ups) (Aasa et al. 2003). A common method used to account for differences in body size and to allow relative strength comparisons between athletes is to divide the lifted load by the mass of the athlete. This method is widely used and simple to calculate, but because the relationship between strength and body mass is not linear, such ratios may not be equivalent for athletes of different mass (Cleather 2006). In Olympic weightlifting, the Sinclair score based on the previous 4 years' Olympic and World records is used to scale the lifter's performance and provide a means of comparison across weight divisions (Cleather 2006). In powerlifting, the Wilks formula has been used to normalize lifting scores for different body weights, but recent work suggests that the Siff model is the best mathematical model to account for body weight differences in this population (Cleather 2006).

Although such equations may be useful for athletes for whom maximum force expression is the performance measure, the appropriateness of these formulas for assessing relative maximum strength in athletes from other sports is questionable given the different skill levels and variable importance of maximum strength to performance. Atkins (2004) determined that in elite rugby league players, the following equation provides an accurate assessment of strength, independent of body size: Corrected Strength = Maximum Strength/Body Mass ($kg^{0.67}$). The requirement for specific scaling values to allow comparison between athletes of different mass may depend on the requirements of the specific sport.

RELIABILITY

An important component of the assessment of maximum strength is the reliability of assessment tools. Unfortunately, such data are not always presented in published papers, which makes interpretation of the importance of observed changes somewhat difficult. Reliability data are shown in table 13.2.

PRETEST WARM-UP AND GUIDELINES

The National strength protocols in use across Australian institutes and academies of sport suggest a general aerobic warm-up, followed by a specific warm-up consisting of 40% to 60% of specified RM for a maximum of 10 repetitions, a minimum of 2 min recovery, 60% to 80% of specified RM for a maximum of 5 repetitions, another rest of at least 2 min, 90% of specified RM for a maximum of 3 repetitions and an extended rest period of at least 5 min prior to the test set.

TABLE 13.2 Reliability Data for Maximum Strength Tests

	1RM squat	1RM bench press	1RM power clean
Intraclass correlation coefficient	.97	.98	.94
Coefficient of variation	3.5%	2.8%	4.8%
Source	Sheppard et al. (2008c); national volleyball; n = 21	McGuigan et al. (2008b); college football; n = 22	Sheppard et al. (2008c); national volleyball; n = 21

In addition to specifications regarding pretest preparation and warm-up, the National protocols include the following guidelines:

- Lowering and lifting actions must be performed in a continuous manner with a single rest of no more than 3 s allowed between repetitions.
- Maximum of 5 min recovery is allowed between test sets.
- Minimum weight increments should be dictated by equipment availability; however, increments smaller than the typical error of the test may not allow determination of a biological change.
- The athlete should aim to achieve the highest RM score in a maximum of four test trials.
- A designated spotter should be used for safety.

TEST PROCEDURES

Bench Press

- Initially, athletes may choose the width of grip that they prefer, but this should remain consistent over consecutive attempts and tests. In the bottom position, the forearms should be perpendicular to the floor.
- Foot position should be recorded (both feet either on the floor or on the bench).
- Recommended assessor position is to the side of the athlete to allow observation of feet, shoulders, and buttocks and visibility of the bar touching the athlete's chest.

Technique

- A valid repetition is one in which the athlete lowers the bar to the highest point of the chest (above the bench) in a controlled movement prior to completing the lift to full elbow extension.

Technical Violations

If any of the following technical violations occur, the trial will be considered invalid and the athlete will perform a second trial at the same weight:

- Failing to make contact with or excessively bouncing the bar off the chest
- Lifting the shoulders or buttocks off the bench
- Raising either foot off the bench or ground so that it breaks contact
- Excessive deviation of bar from "normal" position (observed in warm-up)
- Uneven movement of the bar during the lift (shoulder elevation or uneven extension of arms during lift)
- Resting more than 3 s rest between repetitions

Bench Pull

- Bench height is set so that the athlete can comfortably take the desired grip while the weight is off the ground in the hang position.
- Athletes must start all lifts from hang position.
- Athletes may choose the width of grip that they prefer initially, but this should remain consistent over consecutive attempts and tests.
- The recommended assessor position is to the side of the athlete to allow observation of feet, knees, shoulders, and head and visibility of the bar touching the underside of the bench.

Technique

- A valid repetition is one in which the bar touches the underside of the bench (no bar pad) and the bar is lowered in a controlled manner to the hang position without touching the ground. Feet should remain off the ground throughout lift and in the same position throughout lifts.
- Use of abducted or adducted bench pull technique should be noted on testing results information (abducted bench pull: bar is lifted toward chest; adducted bench pull: bar is lifted toward navel).

Technical Violations

If any of the following technical violations occur, the trial will be considered invalid and the athlete will perform a second trial at the same weight:

- Moving the head or legs from chosen start position (i.e., athlete may start with head down or to the side, but it must remain in this position and in contact with the bench at all times)
- Moving the trunk away from bench or any hip flexion or extension
- Failing to make contact with bar on the bench
- Excessive deviation of bar from 'normal' position observed in warm-up (i.e., maintain abducted or adducted position)
- Uneven movement of the bar during the lift (shoulder depression, uneven flexion of elbows during lift)
- Resting more than 3 s rest between repetitions

Chin-Ups

- Chin-ups should be performed with a medium width pronated grip. The grip should be no wider than 1 hand width outside the shoulders while in the hang position. Athletes may choose the width of grip within limits, but this must remain consistent over consecutive attempts and tests.
- A straight bar should be used for testing chin-ups.
- Results should be recorded as body mass plus any external mass lifted.
- Recommended assessor position is at the athlete's side at eye level with bar.

Technique

Starting from a fully extended elbow position (hang position), the athlete is required to pull body up in one smooth action so that at the top of the lift, the top of the hands are level with the rear angle of the mandible (jaw). Legs can be held in semiflexed position or extended; however, they must not be moved in a way that increases momentum in the pulling phase of the lift. Athletes should be encouraged to complete the lift with minimal head movement.

Technical Violations

If any of the following technical violations occur, the trial will be considered invalid and the athlete will perform a second trial at the same weight:

- Not achieving correct height (top of hands level with rear angle of mandible)
- Movement at the hips or knees from start position during the lift
- Swinging body during lift
- Failing to extend elbow fully between repetitions
- Resting more than 3 s rest between repetitions

Incline Leg Press

- Set seat position at 90° to the slide. If adjustable, set footplate angle to 110°. If footplate angle is not adjustable, ensure that angle is consistent between tests (i.e., if using different machines for testing).
- With the heels as the reference, feet should be placed so that hips are flexed to 90°. Although toes and knees should remain in line during lowering, athletes may choose foot width initially, but this must remain consistent over consecutive attempts and tests.
- Results should be recorded as sum of weight plates plus mass of sled.
- Recommended assessor position is beside the athlete and at level of knees to assess depth.

Technique

A valid repetition is one in which the weight is lowered to a depth corresponding to 90° of knee flexion (anterior surface of the thigh is 90° with anterior surface of the shin) and then extended to full leg extension. (The footplate should be referenced at 90° knee flexion with tape or according to a pin position.)

Technical Violations

If any of the following technical violations occur, the trial will be considered invalid and the athlete will perform a second trial at the same weight:

- Not achieving required depth during lowering of weight
- Allowing hips or buttocks to lose contact with the seat
- Repositioning feet from start position
- Placing hands on thighs to assist lift
- Resting more than 3 s rest between repetitions

Single Leg Press

Technically, the single leg press test is completed as per incline leg press test (detailed previously) but with each leg on a separate occasion. Each set should be completed with both the left and right legs before moving to the next weight.

- Set seat position at 90° to the slide. If adjustable, set footplate angle to 110°. If footplate angle is not adjustable, ensure that angle is consistent between tests (i.e., if using different machines for testing).
- With the heel as the reference, foot should be placed on footplate so that hip is flexed to 90°. Although toes and knee should remain in line during lowering, athletes may choose foot placement.

- Results should be recorded as sum of weight plates plus mass of sled.

- Recommended assessor position is beside the athlete and at level of knees to assess depth.

Technique

- A valid repetition is one in which the weight is lowered to a depth corresponding to 90° of knee flexion (anterior surface of the thigh is 90° with anterior surface of the shin) and then extended to full leg extension (the footplate should be referenced at 90° knee flexion with tape or according to a pin position).

- The support foot can move during the lift as long as it is not moving in a way that is assisting the other leg.

- An appropriate warm-up should ensure that a safe and effective foot position is established before completing an RM in a testing situation.

Technical Violations

If any of the following technical violations occur, the trial will be considered invalid and the athlete will perform a second trial at the same weight:

- Not achieving required depth during lowering of weight

- Allowing hips or buttocks to lose contact with the seat

- Significantly repositioning the foot being tested from start position so as to change the nature of the lift

- Moving the support foot during the lift in a way that is assisting the other leg

- Placing hands on thighs to assist lift

- Resting more than 3 s rest between repetitions

Squat

This test requires a high level of technical proficiency and is recommended for athletes with a solid training base. A qualified and experienced strength coach or scientist must supervise this test.

Preparation and Test

- The safety bars should be set at the highest possible point without affecting the athlete's range of motion.

- Heel blocks should not be used unless anatomical structures limit the athlete's range of motion or prevent the exercise from being performed with correct technique. Use of heel blocks should be consistent between tests.

- The use of a weight belt is optional but should be consistent between tests.

- Athlete should assume a natural stance with feet approximately shoulder-width apart.

- Bar should be held in a "high" bar position on the trapezius during test. Hands should be held in a comfortable position as close to shoulders as possible.

- During the lowering action, knees should travel forward over toes. Heels must remain in contact with the floor at all times during test.

- Athletes are required to lower to a designated depth where crease of hips is level with the top of the knee.

- Recommended assessor position is beside the athlete to facilitate observation of hip and knee angle, back posture, and depth.

Technique

A valid repetition is one in which the weight is lowered to required depth and then extended to full leg extension with trunk as upright as possible.

Technical Violations

If any of the following technical violations occur, the trial will be considered invalid and the athlete will perform a second trial at the same weight:

- Moving forward or sideways excessively during test

- Losing controlled spinal position

- Lifting heels off the floor

- Not lowering to required depth

- Raising hips prior to shoulder elevation

- Resting more than 3 s rest between repetitions

Table 13.3 shows normative data for the various maximal strength tests.

TABLE 13.3 Elite Athlete Maximum Strength Test Norms

Sport	Gender	n	Body mass	Mean score	Range
1RM bench pull					
AIS water polo	Female	14	74.5 ± 7.5	61 ± 7	48-70
1RM power clean					
National volleyball (senior)	Male	11	n/a	104 ± 13	85-130
National volleyball (junior)	Male	9	86.9 ± 7.0	85 ± 13	70-105
National rowing (senior)	Male	8	94.6 ± 3.9	102 ± 10	85-115
3RM power clean					
Judo (elite)	Male	12	84.8 ± 13.3	83 ± 19	60-123
Judo (elite)	Female	11	66.0 ± 11.5	55 ± 10	43-85
3RM squat					
AIS basketball	Male	10	88.0 ± 4.2	115 ± 17	85-140
AIS basketball	Female	14	80.4 ± 10.6	75 ± 14	50-105
AIS rugby league	Male	14	89.7 ± 19.1	99 ± 24	40-130
National volleyball (senior) (90°)	Male	9	102.5 ± 2.9	139 ± 30	110-210
National volleyball (junior) (90°)	Male	7	82.6 ± 8.1	115 ± 25	80-155
AIS netball (parallel)	Female	12	73.7 ± 8.9	67 ± 15	38-90
Judo (elite)	Male	14	85.4 ± 12.6	121 ± 36	60-180
Judo (elite)	Female	11	66.0 ± 11.5	92 ± 14	70-110
Judo (NTID)	Male	16	65.2 ± 6.9	86 ± 33	60-140
Judo (NTID)	Female	14	59.7 ± 7.5	66 ± 15	40-100
National rowing (senior)	Male	10	96.4 ± 3.9	137 ± 14	110-160
3RM bench press					
AIS basketball	Male	11	89.5 ± 5.6	77 ± 13	55-93
AIS basketball	Female	14	80.4 ± 10.6	46 ± 9	30-68
AIS rugby league	Male	14	89.7 ± 19.1	89 ± 14	70-120
National volleyball (senior)	Male	13	95.7 ± 12.0	91 ± 13	68-110
National volleyball (junior)	Male	10	85.3 ± 8.3	71 ± 15	55-100
AIS netball	Female	13	73.7 ± 8.9	44 ± 10	30-60
Judo (elite)	Male	15	85.4 ± 12.6	92 ± 24	60-138
Judo (elite)	Female	10	66.0 ± 11.5	51 ± 16	40-90
Judo (NTID)	Male	16	65.2 ± 6.9	64 ± 14	40-80
Judo (NTID)	Female	14	59.7 ± 7.5	34 ± 10	20-50
National rowing (senior)	Male	13	94.6 ± 3.9	92 ± 7	85-103
Maximum repetitions chin-ups					
National volleyball (senior)	Male	9	102.5 ± 2.9	14 ± 3	8-17
National volleyball (junior)	Male	7	87.1 ± 8.8	8 ± 3	5-15
AIS water polo	Female	13	69.5 ± 7.1	6 ± 4	0-16
3RM chin-ups					
National volleyball (senior)	Male	13	95.7 ± 12.0	25 ± 8	14-43

(continued)

TABLE 13.3   (continued)

Sport	Gender	n	Body mass	Mean score	Range
1RM squat					
AIS water polo	Female	9	73.5 ± 6.0	72 ± 15	45-90
1RM bench press					
AIS water polo	Female	12	78.8 ± 8.0	60 ± 10	45-77
1RM bench pull					
AIS water polo	Female	14	74.5 ± 7.5	61 ± 7	48-70
1RM power clean					
National volleyball (senior)	Male	11	n/a	104 ± 13	85-130
National volleyball (junior)	Male	9	86.9 ± 7.0	85 ± 13	70-105
National rowing (senior)	Male	8	94.6 ± 3.9	102 ± 10	85-115
3RM power clean					
Judo (elite)	Male	12	84.8 ± 13.3	83 ± 19	60-123
Judo (elite)	Female	11	66.0 ± 11.5	55 ± 10	43-85

n/a = not available. NTID = national talent identification

SPEED-STRENGTH

RATIONALE

The assessment of speed-strength is an important consideration for a multitude of sports (Baker et al. 2001; Sheppard et al. 2008b; Stone et al. 2003a; Viitasalo 1983; Young 1995; Young et al. 1995). Speed-strength, also referred to as power, evaluates the athlete's ability to apply force rapidly in a dynamic movement, thus generating high levels of power (Abernethy et al. 1995; Cronin and Sleivert 2005). Speed-strength is generally tested using isoinertial protocols such as CMJ and squat jumps (SJ) for the lower body, whereas bench throws and bench pulls are commonly used for the upper body (Baker 2001a; Clark et al. 2008; Sheppard et al. 2008a).

Both upper-body (Baker 2001a; Baker and Newton 2006) and lower-body (Sheppard et al. 2008d) assessments of speed-strength have been shown to be reliable measures in athletic populations, particularly for peak and mean force platform kinetics, displacement and velocity kinematics, and derived power (table 13.4). Both upper-body (Baker 2002) and lower-body speed-strength (Sheppard et al. 2008d) measures have been shown to discriminate between higher and lower performers in a sporting context as well as detect training-induced changes (Baker and Newton 2006; Sheppard et al. 2008c) (table 13.5).

Speed-strength assessments often involve assessments against a spectrum of loads to assess performance against more than one inertial condition

TABLE 13.4   Reliability Data for the Incremental Load Power Profile Jump

Parameter	Technical error, absolute units	Change in mean, %	Percentage coefficient of variation	Intraclass correlation coefficient
Unloaded (body mass)				
Peak distance	0.03 m	1.7	7.2	.77
Peak velocity	0.24 m · s⁻¹	4.1	7.3	.25
Peak force	78.58 N	1.8	3.5	.96
Peak power	554.43 W	5.0	9.5	.80
Mean power	243.50 W	0.2	7.1	.89
Relative power	6.56 W · kg⁻¹	5.0	9.5	.74
Maximum rate of force development	1,067.74 N · s⁻¹	11.1	36.3	.43

Parameter	Technical error, absolute units	Change in mean, %	Percentage coefficient of variation	Intraclass correlation coefficient
BM + 25%				
Peak distance	0.03 m	3.8	8.3	.71
Peak velocity	0.10 m · s^{-1}	0.5	3.3	.83
Peak force	89.03 N	4.0	4.0	.95
Peak power	231.72 W	4.5	4.0	.95
Mean power	117.49 W	0.7	3.0	.98
Relative power	2.72 W · kg^{-1}	4.5	4.0	.94
Maximum rate of force development	2,033.10 N · s^{-1}	−9.0	47.4	−.04
BM + 50%				
Peak distance	0.01 m	−1.0	3.0	.95
Peak velocity	0.17 m · s^{-1}	−0.7	6.4	.71
Peak force	67.10 N	3.6	3.1	.97
Peak power	278.51 W	2.4	5.9	.90
Mean power	208.57 W	1.3	7.9	.86
Relative power	3.33 W · kg^{-1}	2.4	5.9	.87
Maximum rate of force development	1,097.68 N · s^{-1}	13.6	19.4	.79

n = 26 male subjects; 19.8 ± 2.6 years, 196.3 ± 9.6 cm, 88.6 ± 8.9 kg.

TABLE 13.5 Technical Error, Coefficients of Variation, and Intraclass Correlation Coefficients of Smith Machine and Free Weight–Loaded Jumps Squats in Elite Male Volleyball Players

Parameter	TECHNICAL ERROR, ABSOLUTE UNITS		INTRACLASS CORRELATION COEFFICIENT		PERCENTAGE COEFFICIENT OF VARIATION	
	Smith machine	Free weight	Smith machine	Free weight	Smith machine	Free weight
BM + 25%						
Peak displacement	0.01 m	0.01	.92	.97	4.23	2.65
Peak velocity	0.08 m · s^{-1}	0.07	.87	.92	3.07	2.37
Peak force	56.10 N	89.89	.92	.79	2.24	3.68
Mean force	7.29 N	17.76	.99	.99	0.6	1.16
Peak power	196.33 W	155.96	.90	.95	3.48	2.75
Mean power	89.34 W	180.34	.69	.65	11.06	16.45
BM + 50%						
Peak displacement	0.01 m	0.01	.96	.94	2.93	3.73
Peak velocity	0.07 m · s^{-1}	0.08	.88	.85	2.91	3.57
Peak force	79.62 N	108.98	.88	.76	3.02	4.22
Mean force	24.44 N	9.16	.99	.99	1.21	0.7
Peak power	278.01 W	267.69	.83	.87	4.47	4.94
Mean power	96.49 W	221.23	.70	.53	11.19	17.62

n = 18. BM = body mass.

(Clark et al. 2008; Hori and Andrews 2009; McBride et al. 2002; Sheppard et al. 2008b, 2009; Stone et al. 2003a). This testing concept appears to be very insightful, because it allows comparisons of speed-strength characteristics against varying isoinertial loads (Clark et al. 2008; Sheppard et al. 2008b). Based on the time in the training cycle, the athlete's developmental level, and the sport in question, this analysis could be useful in making decisions regarding the training needs of each athlete. The athlete's ability to accelerate a given load and achieve high power outputs appears to indicate the point on the force–velocity spectrum that the athlete's training program should emphasize to increase power.

Lower-body speed-strength is generally assessed via SJ and CMJ with the athlete's hands on the hips or using a bar across the shoulders, to restrict the assessment to lower body and trunk, without active arm swing. Upper-body laboratory assessments of speed-strength generally involve the bench throw for "pressing" speed-strength and the bench pull to assess "pulling" speed-strength qualities.

INSTRUMENTATION

Although speed-strength is a fundamental consideration in the testing of high-performance athletes, there are numerous methods by which data can be collected and varying instrumentation used (Cronin and Sleivert 2005). This can be problematic, in that direct comparisons between test results obtained with differing methods and instrumentation generally are not valid (Dugan et al. 2004). Central to the practitioner's consideration in assessing speed-strength is that the method produces reliable, valid, and insightful data. Practitioners should be aware of which variables are being obtained directly and which variables are being inferred, with particular reference to the validity of each measure.

If kinetic data are collected only (i.e., force platform only), kinematic variables such as displacement are calculated using the impulse-momentum approach, which is well accepted as a reliable and valid means of obtaining force kinetics and computing displacement kinematics (Kibele 1998). If positional data are collected only, velocity and acceleration data can be reliably determined through differentiation methods (Dugan et al. 2004; Hori and Andrews 2009). In turn, force can be inferred from the known mass (e.g., athlete's body mass) and the acceleration data, and then power can be calculated from the inferred force data and the velocity data. However, there may be concerns about the validity of these assessments, because any small inferential error that may be present in this process will be magnified during double differentiation processes and during multiplication processes (e.g., force × velocity).

It is becoming more common to collect data using a combined method (Dugan et al. 2004; Sheppard et al. 2008d) such that force data are collected directly from a force platform and position and velocity data are collected from a linear position transducer (figure 13.1). In terms of calculating power values, this method has the benefit of calculating power from force kinetics (directly collected from the force platform), and the velocity kinematic is obtained through single differentiation of displacement-time (i.e., velocity). The methods outlined in this section can be used with a force platform, a position transducer, or an integrated system, but testers should consider the differences between these methods.

In many cases, isoinertial assessment test protocols for both lower-body (e.g., CMJ) and upper-body (e.g., bench throw) speed-strength are performed in a Smith machine (McBride et al. 2002; Sleivert and Taingahue 2004; Wilson et al. 1993). With a Smith machine, the barbell is guided via an attachment to

FIGURE 13.1 Starting position of an unloaded jump squat using portable force plate and linear position transducer interfaced with software on a personal computer.

Photo courtesy of High Performance Sport New Zealand.

vertical supports, which restricts movement of the barbell to the vertical plane. For the bench throw, this is of practical importance, because spotters need to assist the athlete in catching the barbell on its descent, a task made considerably safer when conducted in the controlled environment of a Smith machine.

Whilst many practitioners may choose to perform SJ and CMJ testing using a Smith machine, perhaps believing that the vertical-only restriction allows for greater reliability and performance of the CMJ, low variability is found in free-weight jump squat movements (table 13.4), and in fact the variability is similar in both conditions (table 13.5) (Sheppard, Doyle, and Taylor 2008d). Another consideration is that the action of the body in the Smith machine may be dissimilar to sporting actions (Cotterman et al. 2005; Doyle et al. 2008). Furthermore, jump performance may in fact be superior with free-weight compared with the Smith machine (Sheppard et al. 2008d).

CALIBRATION AND PREPARATION OF EQUIPMENT

Prior to data collection, a linear position transducer should be calibrated against a known distance (e.g., 1 m), and force platforms should be calibrated using 2 disparate loads that span the likely scale of measurement (e.g., 20 kg and 300 kg). This is an important consideration, as all instrumentation will experience drift and require recalibration. Specific calibration procedures for each instrument brand are generally outlined by the manufacturer and should be followed by the practitioner.

TEST PROCEDURES

Squat Jump

Squat jumps (SJ) are a common test of concentric-only leg extensor speed-strength (Stone et al. 2003a; Sands et al. 2005; McGuigan et al. 2006; Sheppard and Doyle 2008). It is believed that by preceding the concentric jump action by a brief isometric hold lasting approximately 3 s, the athlete is unable to use the contribution from a stretch–shorten cycle (SSC), because the SSC is dependent on the rate of prestretch and a minimal duration between the prestretch and the concentric action, known as the amortization phase (Bobbert et al. 1996; Bosco and Komi 1981; Giovanni et al. 1968; Van Ingen Schenau et al. 1997; Walshe et al. 1998). Therefore, the SJ is seen as a valuable test to evaluate athletes who require powerful leg extension when a significant isometric hold precedes the concentric action (e.g., swim and athletics starts) and to compare concentric-only power with leg extensor power using a CMJ squat, which allows use of an SSC. This comparison is termed the *eccentric utilization ratio* (EUR) (McGuigan et al. 2006a).

When testing the SJ, researchers have observed that subjects require considerable familiarization to reduce or eliminate small eccentric actions (~1-2 cm change in center of gravity), which have been termed *small-amplitude countermovements* (SACM), prior to the initiation of the concentric jump (Hasson et al. 2004). This unloading occurs when an athlete lowers their center of gravity just prior to initiation of the propulsive phase; the force trace will show a decline from the steady state held prior to this movement (figure 13.2). Ideally, the presence of an SACM is determined by examining the force trace of the movement to identify an unloading of the force immediately prior to the concentric action. If a force plate is unavailable, the practitioner should use a position transducer to detect lowering of the center of gravity (Sheppard and Doyle, 2008).

The depth from which a SJ is initiated should be controlled for consistency between trials and among athletes. The majority of SJ methods use a 110° angle, measured at the knee, using a handheld goniometer (McGuigan et al. 2006a; Sheppard and Doyle 2008). A 3 s pause is held at this position before the athlete jumps vertical for maximum displacement. A typical trial procedure, when using an integrated force platform and position transducer system, can be performed as follows:

- Zero the force platform without the athlete on the platform (i.e., zero mass).

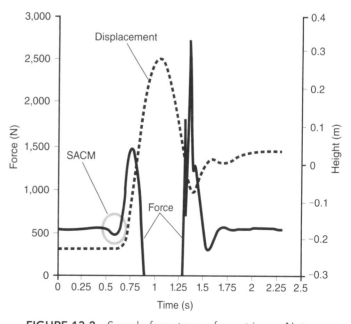

FIGURE 13.2 Sample force trace of squat jump. Note the small amplitude countermovement (SACM), circled, just prior to concentric phase of the jump. There is no observable SACM on the displacement trace in this example. This SACM represents a force unload of greater than 10% of body weight.

- Instruct the athlete to step onto the center of the platform, with a bar across their shoulders.

- Secure the position transducer cable to the side of the bar, such that the cable runs in parallel with the athlete's shoulders, hips, knee, and ankle (i.e., place it in line with the lateral malleolus).

- Zero the position transducer to set a reference for the takeoff and landing position.

- Instruct the athlete to descend to achieve the squat position (110° knee angle) with a practitioner providing feedback using a handheld goniometer.

- When the position is achieved, begin sampling data and instruct the athlete to hold this position for 3 s (i.e., counting down 3-2-1) before jumping vertically.

- Upon landing, instruct the athlete to stand in the landing position for 1 to 2 s or required time to complete the sampling period.

- Examine the force-trace for compliance (i.e., ensure no SACM was present).

- Save data, provide 1 min or more rest, and repeat for two to four trials.

Countermovement Jumps

Vertical jump squats or CMJs, as a speed-strength assessment, have been demonstrated to be a reliable method (Hori and Andrews 2009) that has a strong association with several sport performance measures and with improvements in performance with training (Baker and Nance 1999; Sands et al. 2005; Sheppard et al. 2007, 2009; Stone et al. 2003a). As such, the validity of jump squats as an assessment of lower-body speed-strength is well established.

The jump squat involves a countermovement prior to the propulsive phase and thus provides a means to assess speed-strength inclusive of the SSC. Athletes should be encouraged to perform a countermovement to a depth that will elicit the greatest jump height. With elite populations, this approach seems to produce more consistent and superior results compared with methods whereby the countermovement depth is dictated to the athlete.

A typical trial procedure, when using an integrated force platform and position transducer system, can be performed as follows:

- Zero the force platform without the athlete on the platform (i.e., zero-mass).

- Instruct the athlete to step onto the center of the platform with a bar across their shoulders.

- Secure the position transducer cable to the side of the bar, such that the cable runs in parallel with the athlete's shoulders, hips, knee, and ankle (i.e., place it in line with the lateral malleolus).

- Zero the position transducer to set a reference for the takeoff and landing position.

- Instruct the athlete to prepare for the vertical jump squat, count them in (i.e., counting down 3-2-1), and begin sampling data 1 to 2 s prior to completing the countdown.

- Upon landing, instruct the athlete to stand in the landing position for 1 to 2 s or required time to complete the sampling period.

- Examine the force-trace and position-trace data for normalcy.

- Save data, provide 1 min or more rest, and repeat for two to four trials.

Bench Throw

The purpose of the bench throw exercise is to explosively press the barbell from the chest as far away from the athlete as possible, hence the term *throw*. This differs from the bench press exercise, where the barbell is pressed to a lockout position, and thus involves a larger deceleration component (Clark et al. 2008).

The bench throw is a useful assessment of speed-strength for pressing movement in the upper body. Upper-body speed-strength is a critical component in several sports (Baker 2002), and assessment using bench throws is a practical and valid means to assess speed-strength and detect change (Baker 2001b; Baker and Newton 2006; Clark et al. 2008). The practitioner can use the bench throw with a variety of loads to target various aspects of the force–velocity continuum, and thus the bench throw is also used as a reliable and valid general assessment of upper-body explosive strength performance against different inertial loads (Alemany et al. 2005; Baker 2002; Baker and Newton 2006).

Bench throws are generally conducted in a Smith machine, enabling an easier catch of the bar after the athlete has displaced it. For heavier loads, one or two spotters should be present to assist in arresting the downward movement of the bar.

Data generally are collected with a linear position transducer attached to the barbell to obtain displacement-time data (see figure 13.1). Displacement values are of particular interest, as they define performance in the bench throw (i.e., the purpose is to displace the bar as far as possible). Practitioners will also find value in calculating power values (mean and peak instantaneous power output) through differentiation processes (Baker 2001a, 2002; Baker and Newton 2006). When linear position transducers are available, these procedures can also be used for other exercises such as bench pulls.

A typical trial procedure, when using a linear position transducer, can be performed as follows:

- Instruct the athlete to lie flat on the bench underneath the Smith machine bar with hands spaced on the bar such that they are either in the "strongest position" as selected by the athlete or are vertical to the athlete's shoulders.

- Secure the end of the position transducer to the barbell and ensure that the correct load is placed on the bar.

- Instruct the athlete to remove the barbell from the Smith machine supports.

- Zero the position transducer to set a reference for the release point of the bar during the throw.

- Instruct the athlete to prepare for the throw, count them in (i.e., counting down 3-2-1) and begin sampling data 1 to 2 s prior to completing the countdown.

- When the athletes (or spotters) catch the bar upon its descent, instruct the athlete to remain in position to complete the sampling period.

- Examine the position-trace data for normalcy.

- Save data, provide 1 min or more rest, and repeat for two to four trials.

DATA ANALYSIS

A key purpose of athlete assessment is to obtain insight into the training needs of an athlete. By comparing tests of differing isoinertial loads or different test types, as outlined previously, the practitioner can gain insight into the training needs of an athlete (McGuigan et al. 2006a; Sheppard et al. 2008b; Young 1995).

The load-spectrum speed-strength profile allows the coach to determine optimal load for average and peak power output, and this is believed to be an important consideration in designing power training programs (Baker et al. 2001; Driss et al. 2001; Dugan et al. 2004). Although the utility of training at the load in which peak power is achieved remains a contentious issue (Cronin and Sleivert 2005), it is likely a useful outcome of performing this test. The practitioner, in an aim to increase peak power capabilities, may design a training program using a range of loads that are above, at, and below the load that optimized peak power with that athlete. In other words, if during a testing bout, peak power in the loaded jump squat was observed to be achieved at 25 kg of additional mass, the strength and conditioning coach may design a program that involves an emphasis on loads ranging from 0 to 40 kg of additional mass. Follow-up

testing would determine not only whether increases in peak power occurred but also whether the load at which peak power occurred changed as a result of the training intervention. Figure 13.3 illustrates 2 athletes with differing speed-strength profiles in the bench throw. Athlete A achieved the highest power output in the 50 kg bench throw condition, whereas Athlete B did so in the 30 kg condition.

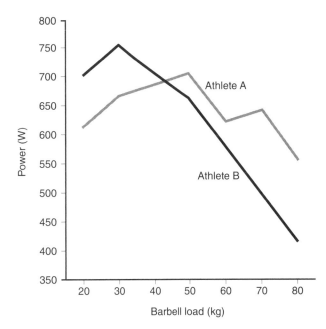

FIGURE 13.3 Comparison of mean concentric power (W) output in the bench throw across a spectrum of loads between two athletes.

Comparing the performance of an athlete under different conditions (e.g., SSC inclusive and exclusive) and with different loads can highlight what type of activity, and with which load upon the speed-strength continuum, the athlete can exploit. Figure 13.4 illustrates 2 athletes with similar power outputs in an unloaded jump squat. However, Athlete A produces similar power in initial (i.e., low) loading conditions, before any distinct decline (termed a *deflection point*) in power. By contrast, Athlete B exhibits a deflection point with the first external loading conditioning, suggesting poor low-load speed-strength development (i.e., low maximal strength). This analysis can highlight where deflection points occur, potentially indicating where an athlete should place his training emphasis (i.e., at the deflection point).

Another means by which to examine speed-strength profiles across a spectrum of loads is to compare the displacement values between conditions. For example, for a CMJ, comparisons of unloaded and loaded jump squats can be made, and ratios can

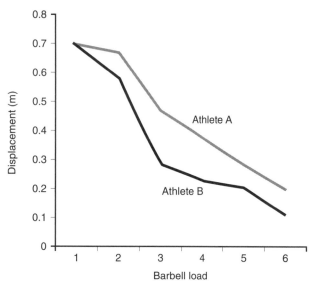

FIGURE 13.4 Comparison of peak displacement (m) in the jump squat (JS) across a spectrum of loads between two athletes. The unloaded/50% body-mass–loaded jump squat ratio for Athlete A is 67% and for Athlete B is 41%. The unloaded/100% body-mass–loaded jump squat ratio for Athlete A is 40% and for Athlete B is 29%.

be created such as the ratio of unloaded jump squat displacement and displacement from the 50% body-mass–loaded jump squat (UL/50%) as well as 100% body-mass loaded jump squat (UL/100%) (figure 13.4). Although ratios are likely context specific, a value of 65% ± 5% for the UL/50% jump squat comparison and 35% ± 5% for the UL/100% comparison may be a useful guide. In figure 13.4, Athletes A and B have nearly identical displacement scores for the unloaded jump squat. However, Athlete A has values within range for the UL/50% and UL/100%, whereas Athlete B has values well below range. Based on these results, Athlete B has poorly developed high-load speed-strength abilities and would benefit from further training in this area.

Isoinertial protocols can be used in conjunction with isometric assessments to compare speed-strength (i.e., dynamic strength) capabilities in relation to isometric (i.e., maximal strength) capabilities (Schmidtbleicher 1985; Young 1995; Zatsiorsky and Kraemer 2006) as well as in conjunction with isoinertial maximum volitional strength assessments (e.g., RM lifting). The combined use of isometric and isoinertial assessments as a means to evaluate the specific status of strength qualities is not entirely novel and may have been first used in Germany (Schmidtbleicher 1985) and adopted in varying forms elsewhere (Young 1995). Originally, this method used a maximum voluntary isometric contraction in a Smith machine in a squat position (immovable bar across the shoulders). More recent uses of maximum voluntary isometric contractions have involved an IMTP position rather than a squat position (Haff et al. 1997; Kawamori et al. 2006), perhaps with reasoning that the IMTP position may be a safer position in which to execute an isometric action, even though the IMTP likely elicits lower force application (Nuzzo et al. 2008).

To determine the dynamic strength index (DSI), peak force from a SJ is compared with the peak force from the IMTP. Although results are likely context specific, a DSI of 50% ± 5% for an isometric squat–SJ ratio and a DSI of 65% ± 5% for an IMTP–SJ is a typical range (Young 1995). Comparison of isometric test results and isoinertial test results may allow the strength and conditioning coach to determine the extent to which the athlete is able to apply force dynamically, in relation to their total maximal force capabilities. It is a valid and useful measure for guiding training prescription (Kawamori et al. 2006; Schmidtbleicher 1985; Young 1995). In other words, this assessment allows insight into the training status of the individual's strength qualities, providing coaches and sport scientists with a rationale from which to design individual strength training programs. If an athlete's DSI is below range but involves force application in the isometric movement, he likely would benefit from additional speed-strength training. Conversely, if the athlete's DSI is above range, additional maximum strength training may be of benefit.

STRENGTH ENDURANCE

RATIONALE

Athletes not only need to be strong but also need to have the ability to repeatedly develop a high level of force. Sports like wrestling, football, boxing, and strength competitions require high force and power production over an extended period of time. Therefore, testing specific strength-endurance in events such as this is important. Local muscular endurance is specific to the muscle or muscle group involved in the exercise, so more than one test may have to be performed if the goal is to determine muscular endurance in different areas of the body. Examples of local muscular endurance tests include body weight tests (sit-ups, push-ups, chin-ups) and weighted tests for maximum repetitions (bench press, leg press, bench pull).

One method, *absolute endurance*, is assessed as the number of repetitions performed with a specified constant load, for example, 40 kg. *Relative endurance* describes the number of repetitions performed with a load that is determined as a percentage of the athlete's maximum, for example, at 70% of 1RM. Values for absolute endurance will depend heavily on the maximum strength of an individual; however, values of relative endurance will not. By determining an individual's absolute and relative endurance, testers gain valuable information that can be used to design specific resistance training programs.

Absolute Strength Endurance

- The athlete is required to lift a predetermined weight (e.g., 40 kg) a set number of repetitions in the least amount of time.

- Cumulative repetitions are recorded along with the time taken to complete the test. Any repetition not performed with the correct technique is not included in the cumulative total.

Relative Strength Endurance

- The athlete is required to lift a load corresponding to a given percentage of RM weight as many times as possible until volitional fatigue or technical violation.

- If an athlete performs a valid repetition following one failed attempt, the test continues and the failed attempt is not recorded. Failing to complete two successive repetitions results in termination of test, and the last successful repetition is recorded as the performance.

All strength endurance tests must be completed with the same technical requirements and must be subject to the same technical violations as documented in the preceding sections. The reliability of the strength-endurance tests tends to be lower than for tests of strength or power. It is likely that strength-endurance tests are less robust than maximal strength assessments, given the need to standardize lifting tempos, test durations, and qualitative technical criteria (i.e., completed repetitions). Particular care needs to be taken to ensure that all repetitions are performed with correct technique and at a consistent rate. Practitioners could use a metronome to ensure that the lifting tempo is standardized. There are very few published data on the reliability of strength endurance tests.

HORIZONTAL JUMPS

A growing body of work has assessed the usefulness of various versions of the horizontal jump. Chaouachi and colleagues (2009b) used unilateral, bilateral, and five-jump horizontal tests in elite handball players and found that one-leg horizontal jump performance was a useful predictor of sprinting ability. This conclusion is supported by Holm and colleagues (2008), who found that one-leg drop jump performance could account for a high level of variance in male team sport athletes (Holm et al. 2008), and by Meylan and colleagues (2009), who suggested that the one-leg horizontal countermovement jump was the best choice for predicting sprint and change of direction performance (Meylan et al. 2009). From this work, it seems reasonable to conclude that horizontal jumps may provide a useful and, in some cases, potentially superior alternative to vertical jumps for the assessment of explosive qualities. As with other forms of assessment, coaches and sport scientists should calculate measures of validity and reliability for measures obtained in horizontal jumps in order to allow appropriate conclusions to be drawn from test results.

REACTIVE STRENGTH

Reactive strength is defined as the ability to develop maximal force in minimal time and is a requisite ability in most sports (Zatsiorsky and Kraemer 2006). Reactive strength is exhibited in movements consisting of a rapid eccentric contraction followed by a concentric muscle action. As with other stretch-shorten cycle activities, concentric force output is enhanced by the preceding eccentric contraction due to the storage and use of stored elastic energy in the muscle–tendon complex, the activation–loading dynamics, and the action of biarticular muscles (Komi and Bosco 1978; Voigt et al. 1995).

The most common test for assessing reactive strength is the drop or depth jump, which can be performed from varying drop heights. Numerous kinetic and kinematic variables can be assessed during the drop jump. A reactive strength index (RSI) can be calculated as well, which is a measure of the applied force and the time that is taken to develop it, describing an individual's ability to explosively transition from an eccentric to concentric muscular contraction. The RSI is calculated by dividing the ground contact time during the drop jump by the jump height achieved (Young 1995).

TEST PROTOCOLS

Two common techniques are used for the drop jump, the bounce drop jump (BDJ) and countermovement drop jump (CMDJ). In the BDJ, the downward movement after the drop is reversed as soon as possible into an upward push-off, whereas in the CMDJ this is done more gradually by increasing the amplitude of the downward movement after landing. Research has shown minimal differences in jump height between a

CMJ, slow SSC activities, and a CMDJ, indicating that these tests are influenced by similar neuromuscular qualities (Young 1995). In contrast, jump height was about 20% less when the athletes performed a BDJ, suggesting that it may be a better indicator of reactive strength under fast SSC conditions (Young 1995).

- Begin with the athlete standing on top of the drop box with the force plate or contact mat placed approximately 0.5 m in front of the box.

- Instruct the athlete to place their hands on their hips, step forward off the box without stepping down or jumping up, and, upon contact with the ground, jump as high as possible while minimizing time on the ground.

- Ensure that the takeoff and landing positions are the same for each trial. Common variations include landing with the feet dorsiflexed and knees or hips flexed.

- From the measurement device, obtain the jump height and contact times.

- Calculate RSI as jump height divided by ground contact time.

- Repeat this procedure from boxes of varying heights (e.g., 30, 45, 60, and 75 cm).

DATA ANALYSIS

The BDJ performed in the manner described here provides three sources of information. With repeated measures over time, improved reactive strength ability is indicated by an increase in RSI at a given drop height. A comparison across drop height conditions can provide insight into the capacity of the neuromuscular system to use the SSC because increasing the height of the drop increases the eccentric load. If the RSI increases or is at least maintained as the drop height is increased, it can be assumed that an individual's reactive strength capabilities are sound. Last, the optimal drop height (i.e., the drop height producing the best performance) can be useful in prescribing the height to be used in drop jump training because it is likely that the lower drop heights producing submaximal performances will produce an inadequate training stimulus, and drop heights above the optimal drop height may produce an excessive stretch load that is not well tolerated (Young 1995).

References

Aasa, U., Jaric, S., Barnekow-Bergkvist, and Johansson, H. 2003. Muscle strength assessment from functional performance tests: Role of body size. *Journal of Strength and Conditioning Research* 17(4):664-670.

Abe, T., Kawakami, Y., Ikegawa, S., Kanehisa, H., and Fukunaga, T. 1992. Isometric and isokinetic knee joint performance in Japanese alpine ski racers. *Journal of Sports Medicine and Physical Fitness* 32(4):353-357.

Abernethy, P., Wilson, G., and Logan, P. 1995. Strength and power assessment. Issues, controversies and challenges. *Sports Medicine* 19(6):401-417.

Alemany, J.A., Pandorf, C.E., Montain, S.J., Castellani, J.W., Tuckow, A.P., and Nindl, B.C. 2005. Reliability assessment of ballistic jump squats and bench throws. *Journal of Strength and Conditioning Research* 19(1):33-38.

Alvarez-San Emeterio, C., and Gonzalea-Badillo, J.J. 2010. The physical and anthropometric profiles of adolescent alpine skiers and their relationship with sporting rank. *Journal of Strength and Conditioning Research* 24(4):1007-1012.

Atkins, S.J. 2004. Normalizing expressions of strength in elite rugby league players. *Journal of Strength and Conditioning Research* 18(1):53-58.

Baker, D. 2001a. Comparison of upper-body strength and power between professional and college-aged rugby league players. *Journal of Strength and Conditioning Research* 15(1):30-35.

Baker, D. 2001b. A series of studies on the training of high-intensity muscle power in rugby league football players. *Journal of Strength and Conditioning Research* 15(2):198-209.

Baker, D. 2002. Differences in strength and power among junior-high, senior-high, college-aged, and elite professional rugby league players. *Journal of Strength and Conditioning Research* 16(4):581-585.

Baker, D., and Nance, S. 1999. The relationship between running speed and measures of strength and power in professional rugby league players. *Journal of Strength and Conditioning Research* 13(3):230-235.

Baker, D., Nance, S., and Moore, M. 2001. The load that maximizes the average mechanical power output during jump squats in power-trained athletes. *Journal of Strength and Conditioning Research* 15(1): 92-97.

Baker, D., and Newton, R.U. 2006. Adaptations in upper-body maximal strength and power output resulting from long-term resistance training in experienced strength-power athletes. *Journal of Strength and Conditioning Research* 20(3):541-546.

Baker, D., and Newton, R.U. 2008. Comparison of lower body strength, power, acceleration, speed, agility, and sprint momentum to describe and compare playing ranks among professional rugby league players. *Journal of Strength and Conditioning Research* 22(1):153-158.

Baker, D., Wilson, G., and Carlyon, B. 1994. Generality versus specificity: a comparison of dynamic and isometric measures of strength and speed-strength. *European Journal of Applied Physiology* 68(4):350-355.

Baker, D. G. 2009. Ability and validity of three different methods of assessing upper-body strength-endurance to endurance to distinguish playing rank in professional rugby league players. *Journal of Strength and Conditioning Research* 23(5):1578-1582.

Bobbert, M.F., Gerritsen, K.G.M., Litjens, M.C.A., and Van Soest, A.J. 1996. Why is countermovement jump height greater than squat jump height? *Medicine and Science in Sports and Exercise* 28(11):1402-1412.

Bosco, C., and Komi, P.V. 1981. Prestretch potentiation of human skeletal muscle during ballistic movement. *Acta Physiologica Scandinavica* 111(2):135-140.

Campo, S.S., Vaeyens, R., Philippaerts, R.M., Redondo, J.C., De Benito, A.M., and Cuadrado, G. 2009. Effects of lower-limb plyometric training on body composition, explosive strength, and kicking speed in female soccer players. *Journal of Strength and Conditioning Research* 23(6):1714-1722.

Chaouachi, A., Brughelli, M., Chamari, K., Levin, G.T., Abdelkrim, N.B., Laurenchelle, L., et al. 2009a. Lower limb maximal dynamic strength and agility determinants in elite basketball players. *Journal of Strength and Conditioning Research* 23(5):1570-1577.

Chaouachi, A., Brughelli, M., Levin, G., Boudhina, N.B., Cronin, J., and Chamari, K. 2009b. Anthropometric, physiological and performance characteristics of elite team-handball players. *Journal of Sports Sciences* 27(2):151-157.

Chelly, M. S., Hermassi, S., and Shephard, R. J. 2010. Relationships between power and strength of the upper and lower limb muscles and throwing velocity in male handball players. *Journal of Strength and Conditioning Research* 24(6):1480-1487.

Clark, R.A., Bryant, A.L., and Humphries, B. 2008. A comparison of force curve profiles between the bench press and ballistic bench throws. *Journal of Strength and Conditioning Research* 22(6):1755-1759.

Cleather, D. J. 2006. Adjusting powerlifting performances for differences in body Mass. *Journal of Strength and Conditioning Research* 20(2):412-421.

Coldwells, A., Atkinson, G., and Reilly, T. 1994. Sources of variation in back and leg dynamometry. *Ergonomics* 37(1):79-86.

Comfort, P., Haigh, A., and Matthews, M.J. 2012. Are changes in maximal squat strength during pre season training reflected in changes in sprint performance in rugby league players. *Journal of Strength and Conditioning Research* 26(3):772-776

Cormie, P., McGuigan, M.R., and Newton, R.U. 2010a. Adaptations in athletic performance after ballistic power versus strength training. *Medicine and Science in Sports and Exercise* 42(8):1582-1598.

Cormie, P., McGuigan, M.R., and Newton, R.U. 2010b. Influence of strength on magnitude and mechanisms of adaptation to power training. *Medicine and Science in Sports and Exercise* 42(8):1566-1581.

Cotterman, M.L., Darby, L.A., and Skelly, W.A. 2005. Comparison of muscle force production using the smith machine and free weights for bench press and squat exercises. *Journal of Strength and Conditioning Research* 19(1):169-176.

Cronin, J., and Sleivert, G. 2005. Challenges in understanding the influence of maximal power training on improving athletic performance. *Sports Medicine* 35(3):213-234.

Cronin, J.B., and Henderson, M. 2004. Maximal strength and power assessment in novice weight trainers. *Journal of Strength and Conditioning Research* 18(1):48-52.

Desgorces, F.D., Berthelot, G., Dietrich, G., and Testa, M.S. 2010. Local muscular endurance and prediction of 1 repetition maximum for bench in 4 athletic populations. *Journal of Strength and Conditioning Research* 24(2):394-400.

Doyle, T.L.A., Sheppard, J.M., and Sachlikidis, A. 2008. *Smith machine jump squats versus free weight jump squats: Which is safer?* Paper presented at the 2nd World Congress on Sports Injury Prevention, June 26-28, Tromso, Norway.

Driss, T., Vandewalle, H., Quievre, J., Miller, C., and Monod, H. 2001. Effects of external loading on power output in a squat jump on a force platform: a comparison between strength and power athletes and sedentary individuals. *Journal of Sports Sciences* 19(2):99-105.

Dugan, E.L., Doyle, T.L.A., Humphries, B., Hasson, C.J., and Newton, R. U. 2004. Determining the optimal load for jump squats: A review of methods and calculations. *Journal of Strength and Conditioning Research* 18(3):668-674.

Fenter, P.C., Beller, J., Pitts, T., and Kay, R. 2003. A comparison of 3 hand-held dynamometers used to measure hip abduction strength. *Journal of Strength and Conditioning Research* 17(2):531-535.

Fry, A., Kraemer, W., Weseman, C., Conroy, B., Gordon, S., Hoffman, J., and Maresh, C. (1991. The effects of an off-season strength and conditioning program on starters and non-starters in women's Intercollegiate volleyball. *Journal of Strength and Conditioning Research* 5(4):174-181.

Giovanni, A., Cavagna, A., Dusman, B., and Margaria, R. 1968. Positive work done by a previously stretched muscle. *Journal of Applied Physiology* 24:21-32.

Gordon, B.S., Moir, G.L., Davis, S.E., Witmer, C.A., and Cummings, D.M. 2009. An investigation into the relationship of flexibility, power, and strength to club head speed in male golfers. *Journal of Strength and Conditioning Research* 23(5):1606-1610.

Haff, G., Carlock, J., Hartman, M., Lon Kilgore, J., Kawamori, N., Jackson, J., Morris, R., Sands, W., and Stone, M. 2005. Force-time curve characteristics of dynamic and isometric muscle actions of elite women Olympic weightlifters. *Journal of Strength and Conditioning Research* 19(4):741-748.

Haff, G., Stone, M., O'Bryant, H., Harman, E., Dinan, C., Johnson, R., and Han, K-H. 1997. Force-time dependent characteristics of dynamic and isometric muscle actions. *Journal of Strength and Conditioning Research* 11(4):269-272.

Harris, N.K., Cronin, J.B., Hopkins, W.G., and Hansen, K.T. 2008a. Relationship between sprint times and the strength/power outputs of a machine squat jump. *Journal of Strength and Conditioning Research* 22(3):691-698.

Harris, N.K., Cronin, J.B., Hopkins, W.G., and Hansen, K.T. 2008b. Squat jump training at maximal power loads vs heavy loads: Effect on sprint ability. *Journal of Strength and Conditioning Research* 22(6):1742-1749.

Hasson, C.J., Dugan, E.L., Doyle, T.L., Humphries, B., and Newton, R.U. 2004. Neuromechanical strategies used to increase jump height during the initiation of the squat jump. *Journal of Electromyography and Kinesiology* 14(4):515-521.

Hayes, P.R., French, D.N, Thomas, K. 2011. The effect of muscular endurance on running economy. *Journal of Strength and Conditioning Research* 25(9):2464-2469.

Hennessy, L., and Kilty, J. 2001. Relationship of the stretch shortening cycle to sprint performance in trained female athletes. *Journal of Strength and Conditioning Research* 15(3):326-331.

Hoff, J., Gran, A., and Helgerud, J. 2002. Maximal strength training improves aerobic endurance performance. *Scandinavian Journal of Medicine & Science in Sports* 12(5):288-295.

Holm, D.J., Stalbom, M., Keogh, J.W., and Cronin, J. 2008. Relationship between the kinetics and kinematics of a unilateral horizontal drop jump to sprint performance. *Journal of Strength and Conditioning Research* 22(5):1589-1596.

Hori, G., and Andrews, W. 2009. Reliability of velocity, force, and power obtained from the Gymaware encoder with and without external loads. *Journal of Australian Strength and Conditioning* 17(1):12-17.

Hori, N., Newton, R.U., Andrews, W.A., Kawamori, N., McGuigan, M.R., and Nosaka, K. 2008. Does performance of hang power clean differentiate performance of jumping, sprinting, and changing of direction. *Journal of Strength and Conditioning Research* 22(2):412-418.

Izquierdo-Gabarren, M., Exposito, R., de Villarreal, E., and Izquierdo, M. 2010. Physiological factors to predict on traditional rowing performance. *European Journal of Applied Physiology* 108(1):83-92.

Kawamori, N., Rossi, S., Justice, B., Haff, E., Pistilli, E., O'Bryant, H., et al. 2006. Peak force and rate of force development during isometric and dynamic mid-thigh clean pulls performed at various intensities. *Journal of Strength and Conditioning Research* 20(3):483-491.

Kibele, A. 1998. Possibilities and limitations in the biomechanical analysis of countermovement jumps: a methodological study. *Journal of Applied Biomechanics* 14(1):105-117.

Kilgallon, M. 2008. Isometric training and the isometric pull: Applications for the modern strength and conditioning coach. *Journal of Australian Strength and Conditioning* 16(3):48-56.

Komi, P., and Bosco, C. 1978. Utilization of stored elastic energy in leg extensor muscles by men and women. *Medicine and Science in Sports and Exercise* 10(4):261-265.

Kyriazis, T. A., Terzis, G., Boudolos, K., and Georgiadis, G. 2009. Muscular power, neuromuscular activation, and performance in shot put athletes at pre season and competition period. *Journal of Strength and Conditioning Research* 23(6):1773-1779.

Lockie, R.G., Murphy, A.J., Knight T.J., Janse De Jong, X.A. 2011. Factors that differentiate acceleration ability in field sport athletes. *Journal of Strength and Conditioning Research* 25(10):2704-2714.

Markovic, G. 2007. Poor relationship between strength and power qualities and agility performance. *Journal of Sports Medicine and Physical Fitness* 47(3):276-283.

McBride, J.M., Triplett-McBride, T.N., Davie, A., and Newton, R.U. 2002. The effect of heavy-vs. light-load jump squats on the development of strength, power, and speed. *Journal of Strength and Conditioning Research* 16(1):75-82.

McGuigan, M.R., Newton, M.J., Winchester, J.B., and Nelson, A.G. 2010. Relationship between isometric and dynamic strength in recreationally trained men. *Journal of Strength and Conditioning Research* 24(9): 2570-2573.

McGuigan, M.R., Newton, M.J. and Winchester, J.B. 2008a. Use of isometric testing in soccer players. *Journal of Australian Strength and Conditioning* 16(4): 11-14.

McGuigan, M., and Winchester, J. 2008b. The relationship between isometric and dynamic strength in college football players. *Journal of Sports Science and Medicine* 7(1):101-105.

McGuigan, M.R., Doyle, T.L., Newton, M., Edwards, D.E., Nimphius, S., and Newton, R.U. 2006a. Eccentric utilization ratio: Effect of sport and phase of training. *Journal of Strength and Conditioning Research* 20(4):992-995.

McGuigan, M., Winchester, J., and Erikson, T. 2006b. The importance of isometric maximum strength in college wrestlers. *Journal of Sports Science and Medicine* 5(CSSI):108-113.

Meylan, C., McMaster, T., Cronin, J.B., Mohammad, N.I., Rogers, C., and DeKlerk, M. 2009. Single-leg lateral, horizontal, and vertical jump assessment: Reliability, interrelationships, and ability to predict sprint and change of direction performance. *Journal of Strength and Conditioning Research* 23(4):1140-1147.

Mikkola, J., Rusko, H., Nummela, A., Paavolainen, L., and Hakkinen, K. 2007. Concurrent endurance and explosive type strength training increases activation and fast force production of leg extensor muscle in endurance athletes. *Journal of Strength and Conditioning Research* 21(2):613-620.

Murphy, A., and Wilson, G. 1997. The ability of tests of muscular function to reflect training induced changes in performance. *Journal of Sports Sciences* 15(2):191-200.

Murphy, A., and Wilson, G. 1996. Poor correlations between isometric tests and dynamic performance: relationship to muscle activation. *European Journal of Applied Physiology* 73(3-4):353-357.

Murphy, A., Wilson, G., and Pryor, J. 1994. Use of the iso-inertial force mass relationship in the prediction of dynamic human performance. *European Journal of Applied Physiology* 69(3):250-257.

Naclerio, F.J., Colado, J.C., Rhea, M.R., Bunker, D., and Triplett, N.T. 2009. The influence of strength and power on muscle endurance test performance. *Journal of Strength and Conditioning Research* 23(5):1482-1488.

Nuzzo, J.L., McBride, J.M., Cormie, P., and McCaulley, G. O. 2008. Relationship between countermovement jump performance and multijoint isometric and dynamic tests of strength. *Journal of Strength and Conditioning Research* 22(3):699-707.

Paavolainen, L., Hakkinen, K., Hammalainen, I., Nummela, A., and Rusko, H. 1999. Explosive strength training improves 5km running time by improving running economy and muscle power. *Journal of Applied Physiology* 86(5):1527-1533.

Pearson, S.N., Hume, P.A., Cronin, J.B., and Slyfield, D. 2009. Strength and power determinants of grinding performance in America's Cup sailors. *Journal of Strength and Conditioning Research* 23(6):1883-1889.

Peterson, M.D., Alvar, B.A., and Rhea, M.R. 2006. The contribution of maximal force production to explosive movement among young collegiate athletes. *Journal of Strength and Conditioning Research* 20(4):867-873.

Pincivero, D., Lephart, S., and Karunakara, R. 1997. Relation between open and closed kinematic chain assessment of knee strength and functional performance. *Clinical Journal of Sport Medicine* 7(1):11-16.

Pua, Y., Koh, M., and Teo, Y. 2006. Effects of allometric scaling and isokinetic testing methods on the relationship between countermovement jump and quadriceps torque and power. *Journal of Sports Sciences* 24(4):423-432.

Reynolds, J.M., Gordon, T.J., and Roberts, R.A. 2006. Predictor of one repetition maximum strength from multiple repetition maximum testing and anthropometry. *Journal of Strength and Conditioning Research* 20(3):584-592.

Robbins, D.W., and Young, W.B. 2012. Positional relationships between various sprint and jump abilities in elite American football players. *Journal of Strength and Conditioning Research* 26(2):388-397.

Rønnestad, B., Hansen, E., and Raastad, T. 2010. Effect of heavy strength training on thigh muscle cross sectional area, performance determinants, and performance in well trained cyclists. *European Journal of Applied Physiology* 108(5):965-975.

Rønnestad, B., Hansen, E., and Raastad, T. 2012. Strength training affects tendon cross-sectional area and freely chosen cadence differently in non-cyclists and well trained cyclists. *Journal of Strength and Conditioning Research* 26(1):158-166.

Sands, W.A., Smith, S.L., Kivi, D.M.R., McNeal, J.R., Dorman, J.C., Stone, M.H., et al. 2005. Anthropometric and physical abilities profiles: US National skeleton team. *Sports Biomechanics* 4, 197-214.

Schechtman, O., Gestewitz, L., and Kimble, C. 2005. Reliability and validity of the DynEx dynamometer. *Journal of Hand Therapy* 18(3):339-347.

Saunders, P.U., Pyne, D.B., Telford, R.D., and Hawley, J.A. 2004. Factors affecting running economy in trained distance runners. *Sports Medicine* 34 (7): 465-485

Schmidtbleicher, D. 1985. Structural analysis of motor strength qualities and its application to training. *Science Periodical of Research and Technology in Sport*, 1-7.

Schmidtbleicher, D. 1992a. *Strength and Power in Sport*. Oxford, UK: Blackwell Scientific.

Schmidtbleicher, D. 1992b. Training for power events. In: P.V. Komi (ed), *Strength and Power in Sport*. Oxford, UK: Blackwell, pp 381-395.

Sheppard, J.M., Chapman, D., Gough, C., McGuigan, M., and Newton, R.U. 2008a. The association between changes in vertical jump and changes in strength and power qualities in elite volleyball players over 1 year. National Strength and Conditioning Association annual conference abstracts. *Journal of Strength and Conditioning Research* 22(6):1-115.

Sheppard, J.M., Chapman, D., Gough, C., McGuigan, M.R., and Newton, R.U. 2009. Twelve month training induced changes in elite international volleyball players. *Journal of Strength and Conditioning Research* 23(7):2096-2101.

Sheppard, J., Cormack, S., Taylor, K., McGuigan, M., and Newton, R. 2008b. Assessing the force–velocity characteristics of the leg extensors in well-trained athletes: The incremental load power profile. *Journal of Strength and Conditioning Research* 22(4):1320-1326.

Sheppard, J.M., Cronin, J., Extebarria, N., Gabbett, T., McGuigan, M., and Newton, R.U. 2007. The relative importance of strength and power for vertical jumping in elite male volleyball players. National Strength and Conditioning Association annual conference abstracts. *Journal of Strength and Conditioning Research* 21(4), E19.

Sheppard, J.M., Cronin, J.B., Gabbett, T.J., McGuigan, M.R., Etxebarria, N., and Newton, R.U. 2008c. Relative importance of strength, power, and anthropometric measures to jump performance of elite volleyball players. *Journal of Strength and Conditioning Research* 22(3):758-765.

Sheppard, J.M., and Doyle, T.L. 2008. Increasing compliance to instructions in the squat jump. *Journal of Strength and Conditioning Research* 22(2):648-651.

Sheppard, J.M., Doyle, T.L., and Taylor, K.L. 2008d. A methodological and performance comparison of Smith machine and free weight jump squats. *Journal of Australian Strength and Conditioning* 16(2):5-9.

Sleivert, G., and Taingahue, M. 2004. The relationship between maximal jump-squat power and sprint acceleration in athletes. *European Journal of Applied Physiology* 91(1):46-52.

Stone, M.H., O'Bryant, H.S., McCoy, L., Coglianese, R., Lehmkuhl, M., and Schilling, B. 2003a. Power and maximum strength relationships during performance of dynamic and static weighted jumps. *Journal of Strength and Conditioning Research* 17(1):140-147.

Stone, M., Sanborn, K., O'Bryant, K., Hartman, H., Stone, M., Proulx, C., et al. 2003b. Maximum strength-power-performance relationships in collegiate throwers. *Journal of Strength and Conditioning Research* 17(4):739-745.

Stone, M.H., Sands, W.A., Pierce, K.C., Ramsey, M.W., and Haff, G.G. 2008. Power and power potentiation among strength-power athletes: a preliminary study. *International Journal of Sports Physiology and Performance* 3(1):55-67.

Sullivan, S., Chelsey, A., Herbert, G., McFaull, S., and Scullion, D. 1988. The validity and reliability of hand-held dynamometry in assessing isometric external rotator performance. *Journal of Orthopaedic and Sports Physical Therapy* 10(6):213-217.

Tan, F.H.Y., Polglaze, T., Dawson, B., and Cox, G. 2009. Anthropometric and fitness characteristics of elite Australian female water polo Players. *Journal of Strength and Conditioning Research* 23(5):1530-1536.

Van Ingen Schenau, G. J., Bobbert, M.F., and De Haan, A. 1997. Does elastic energy enhance work and efficiency in the stretch-shortening cycle? *Journal of Applied Biomechanics* 13:389-415.

Viitasalo, J.T. 1983. Measurement of force–velocity characteristics for sportsmen in field conditions. *International Series of Biomechanics* IX(A):96-101.

Voigt, M., Simonsen, E.B., Dhyre-Poulsen, P., and Klausen, K. 1995. Mechanical and muscular factors influencing the performance in maximal vertical jumping after different prestretch loads. *Journal of Biomechanics* 28(3):294-307.

Walshe, A.D., Wilson, G.J., and Ettema, G.J. 1998. Stretch-shorten cycle compared with isometric preload: Contributions to enhanced muscular performance. *Journal of Applied Physiology* 84(1):97-106.

West, D.J., Owen, N J., Jones M. R., Bracken, R.M., Cook, C.J., Cunningham, D.J., Shearer, D.A., Finn, C.V., Newton, R.U., Crewther, B.T., and Kilduff, L.P. 2011. Relationship between force-time characteristics of the isometric midthigh pull and dynamic performance in professional rugby league players. *Journal of Strength and Conditioning Research* 25(11):3070-3075.

Wilson, G., Lyttle, A., Ostrowski, K., and Murphy, A. 1995. Assessing dynamic performance: a comparison of rate of force development tests. *Journal of Strength and Conditioning Research* 9(3):176-181.

Wilson, G.J., Newton, R.U., Murphy, A.J., and Humphries, B.J. 1993. The optimal training load for the development of dynamic athletic performance. *Medicine and Science in Sports and Exercise* 25(11):1279-1286.

Wisloff, U., Castagna, C., Helgerud, J., Jones, R., and Hoff, J. 2004. Strong correlation of maximal strength with sprint performance and vertical jump height in elite soccer players. *British Journal of Sports Medicine* 38(3):285-288.

Young, W. 1995. Laboratory strength assessment of athletes. *New Studies in Athletics* 10(1):89-96.

Young, W.B. 2006. Transfer of strength and power training to sports performance. *International Journal of Sports Physiology and Performance* 1:74-83

Young, W., Cormack, S., and Crichton, M. 2011. Which jump variables should be used to assess explosive leg muscle function? *International Journal of Sports Physiology and Performance* 6:51-57.

Young, W.B., McLean, B., and Ardagna, J. 1995. Relationship between strength qualities and sprinting performance. *Journal of Sports Medicine and Physical Fitness* 35(1):13-19.

Young, W.B., Newton, R., Doyle, T., Chapman, D., Cormack, S., and Stewart, G. 2005. Physiological and anthropometric characteristics of starters and non-starters and playing positions in elite Australian Rules Football: A case study. *Journal of Science and Medicine in Sport* 8(3):333-345.

Zatsiorsky, V., and Kraemer, W. 2006. *Science and Practice of Strength Training*, 2nd ed. Champaign, IL: Human Kinetics.

Field Testing Principles and Protocols

Sarah M. Woolford, Ted Polglaze, Greg Rowsell, and Matt Spencer

The physiological assessment of athletes in the field can provide a valid and reliable means of measuring the physiological characteristics that are important to sporting performance. Results provide an indication of an individual's strengths and weaknesses, their adaptation to training, and any changes in sport-related fitness over the course of a season. The testing regimes selected should use valid and reliable measures and reflect the movement patterns and physiological demands of the sport in which the athlete participates.

Field testing generally involves a series of tests that can be easily administered to multiple athletes using relatively inexpensive equipment. Some of the more common reasons for conducting field tests include these:

- To provide baseline measures
- To develop individual athlete and sport-specific team profiles
- To evaluate the effectiveness of the training stimulus
- To monitor rehabilitation status after injury
- To serve as a motivational tool
- To use for selection purposes

The following tests of jumping ability, speed, repeat sprint ability, agility, and aerobic power represent a general field testing battery.

Test Reliability

The typical error of measurement (TE) is our measure of test reliability. TE data are essential for the correct interpretation of results from one test to the next.

The test TE is used to determine whether a change in results between two testing sessions is due to an actual change in the physiological parameter being assessed or a technical error on the part of the tester and or equipment.

The field test TE data in table 14.1 indicate target TEs based on precision data provided by multiple laboratories in Australia accredited with the National Sport Science Quality Assurance (NSSQA) program. These can be used as a guide for those laboratories or groups developing their own test reliability data.

For more information on precision of measurement, please refer to chapter 3, Data Collection and Analysis.

Athlete Preparation

To ensure that any change in results can be attributed to changes in an athlete's physiological status, consistency across all areas of preparation is essential. Refer to chapter 2, Pretest Environment and Athlete Preparation, and sport specific chapters

TABLE 14.1 Target Typical Error (TE) Measurements for Physiological Field Tests

Test	Target TE
Vertical jump test	2 cm
Sprint tests (5, 10, 20 m)	0.03 s
Sprint tests (10, 30, 40 m)	0.04 s
5-0-5 agility test	0.05 s
Multistage shuttle run test	4 laps
Yo-Yo intermittent recovery test 1	80 m

for information relating to athlete and environment preparation. Further, the following principles should be applied.

To minimize the impact of test unfamiliarity on results, athletes should be familiar with test procedures and protocols before commencement. Where possible, athletes who are being tested for the first time should be provided familiarization sessions prior to the main testing block.

Changes in the fitness of the athlete should be the only variable that produces a change in the score on any test. Consistency in the condition in which an athlete presents for testing therefore should be encouraged. Athletes should observe the following:

- Avoid fatiguing training in the 24 h preceding a test to ensure athletes are tested in a relatively fresh state. Hence, no training or only light skills or low-volume training is recommended.

- Be free of injuries or niggles that will hinder performance.

- Be in good health.

- Follow normal dietary practices in the 24 to 48 h leading up to testing and arrive on test day well hydrated. Only a light meal 3 h preceding the test session is recommended. No food, cigarettes, or beverages containing alcohol or caffeine are to be consumed in the 3 h prior to testing.

- Be familiar with the testing battery.

- Perform the same warm-up before each test session.

- Wear appropriate clothing and footwear specific to the sport and testing surface.

Test Environment

Although the tests to be discussed are often referred to as field tests, there are advantages in performing as many of them as possible in a controlled environment, such as in a laboratory or indoor sports stadium. Unless stipulated in the sport-specific chapter, it is recommended that tests be conducted indoors so that climate and testing surface remain reasonably constant. The floor surface should be clean, free of dust, and in good condition.

For all field testing, testers should measure and document the environmental conditions (e.g., temperature, relative humidity, barometric pressure, wind speed and direction) and consider these when interpreting changes in results. Portable weather monitors (e.g., Kestral 4500 Pocket Weather Tracker) are typically used across the Australian Sport Institute network for this purpose.

Recommended Test Order

Testing sessions should be scheduled to coincide with the beginning and end of each training phase so that the effectiveness of the training phase can be assessed. Likewise, if a specific intervention has been programmed, pre- and posttesting is recommended to assess the impact of the intervention.

All test sessions should be scheduled at the same time of day to avoid fluctuations in physiological responses due to circadian rhythm (Winget et al. 1985) and to promote consistency with how the athlete presents herself for each test session.

It is recommended that the field tests be completed in a standardized order. This order should be determined in light of physiological considerations; that is, the completion of one test should not adversely affect performance in subsequent tests, thus promoting optimal performance and allowing for a valid comparison to previous test results. This test order will also require minimal recovery time between tests, thus allowing for an overall efficient testing session.

The recommended test order is as follows:

Day	Tests
1	Vertical jump
	Sprint
	Agility
	Multistage shuttle run
	Yo-Yo intermittent recovery
2	Repeat sprint ability

Equipment Checklist

The same tester should conduct each test occasion, and the same equipment should be used each time. Where possible, testing equipment should be calibrated prior to each testing session or at regular intervals throughout the testing year. Aspects such as the instruction and encouragement given to athletes and the rest period allowed between repetitions of a single test or between different tests should remain constant from one test occasion to the next.

Vertical Jump Test

- ☐ Yardstick jumping device (e.g., Swift Performance Yardstick)
- ☐ Testing Scales (accurate to ±0.05 kg) (optional; used for the prediction of peak power)
- ☐ Recording sheet (refer to the vertical jump data sheet template)
- ☐ Pen

Straight Line Sprint Tests

- ☐ Electronic light gate equipment (e.g., dual-beam light gates)
- ☐ Metal measuring tape
- ☐ Field marking tape
- ☐ Marker cones
- ☐ Recording sheet (refer to the sprint test data sheet template)
- ☐ Pen

Repeat Sprint Ability Tests

- ☐ Electronic light gate equipment (e.g., dual-beam light gates)
- ☐ Metal measuring tape
- ☐ Field marking tape
- ☐ Marker cones
- ☐ Stopwatch
- ☐ Recording sheet (refer to the repeat sprint ability test data sheet template)
- ☐ Pen

5-0-5 Agility Test

- ☐ Electronic light gate equipment (e.g., dual-beam light gates)
- ☐ Metal measuring tape
- ☐ Field marking tape
- ☐ Marker cones
- ☐ Recording sheet (refer to the agility test data sheet template)
- ☐ Pen

Multistage Shuttle Run Test

- ☐ Metal measuring tape
- ☐ Marking tape
- ☐ Marker cones
- ☐ Heart rate monitor
- ☐ Sound box or CD player
- ☐ Multistage Shuttle Run Test CD (or MP3 file)
- ☐ Recording sheet (refer to the multistage shuttle run test data sheet template)
- ☐ Pen

Yo-Yo Intermittent Recovery Test

- ☐ Metal measuring tape
- ☐ Marking tape
- ☐ Marker cones
- ☐ Heart rate monitor
- ☐ Sound box or CD player
- ☐ Yo-Yo IRT test CD, level 1 or 2 (or MP3 file)
- ☐ Recording sheet (refer to the Yo-Yo IRT data sheet template)
- ☐ Pen

VERTICAL JUMP TEST

RATIONALE

In many sports, jumping ability plays an integral part in the successful execution of sporting skills and overall team performance (e.g., soccer, Australian football, rugby union, rugby league, beach volleyball, volleyball, netball, basketball). Jumping activities in these sports typically incorporate an approach with countermovement and arm swing, which assists with overall jump height. Therefore, the vertical jump is recommended to assess jumping ability. However, to eliminate the variation and impact of the approach on the jump, only a standardized countermovement with arm swing is allowed.

TEST PROCEDURE

Standing Reach Height

- The athlete should stand with their feet together side-on to the Yardstick.

- Keeping the heels on the floor and looking straight ahead, the athlete reaches upward with their dominant hand as high as possible, fully elevating the shoulder to displace the vanes. This is recorded as the standing reach height in centimeters (e.g., vane 25 is displaced 25 cm).

- The absolute standing reach height from the floor may be calculated as the pole setting height (i.e., the height the zero vane is from the floor; either 160, 170, 180, 190, 200, or 210 cm) plus the standing reach height. Record this measure in centimeters.

Vertical Jump Height

- Move several of the lower vanes away before instructing the athlete to stand close to the Yardstick for their jump.

- The athlete uses an arm swing and countermovement to jump as high as possible in order to displace the vanes at the height of the jump.

- The takeoff must be from two feet with no preliminary steps or shuffling; however, feet can be comfortably apart.

- The athlete performs at least three trials and may continue as long as improvements are being made. The best trial, that is, the highest vane displaced, is recorded as the jump height.

- The difference between jump height and standing reach height is calculated to give the relative vertical jump result in centimeters.

- The absolute jump height from the floor may be calculated as the pole setting height (i.e., the height the zero vane is from the floor; 160, 170, 180, 190, 200, or 210 cm) plus the jump height (e.g., vane 80 = 80 cm). Record this measure in centimeters.

DATA ANALYSIS

Traditionally, vertical jump results have been presented as the difference between standing reach height and jump height; however, for some sports it may actually be more relevant to express these results as absolute jump height, that is, height from the ground to the peak of the jump. Indeed, athletes and coaches in many sports are interested in the absolute jump height because it is more relevant to performance in competition. For example, the higher a volleyball player can jump above the net, the greater their options for angle of attack in spiking as well as defensive abilities in blocking. For possession sports, an athlete's chances of "outreaching" his or her opponent and thus gaining possession are greater for athletes with a higher absolute jump height (e.g., basketball, netball, soccer, Australian football, rugby codes).

In addition to height jumped, the estimation of peak power may further assist in the evaluation of an athlete's explosive leg power. Sayers and colleagues (1999) established a valid and reliable formula for predicting peak power that can be used for both males and females. The TE for peak power using the following equation is 104 W or 2.0% (data supplied from South Australian Sports Institute, $n = 18$).

$$\text{Predicted Peak Power (W)} = (60.7 \times \text{Jump Height [cm]}) + (45.3 \times \text{Body Mass [kg]}) - 2{,}055.$$

Please refer to chapter 21 for normative data relevant to predicted peak power from vertical jump and body mass scores for hockey athletes.

LIMITATIONS

The test can be affected by player motivation and individual technique, particularly in relation to the athlete's coordination and ability to hit the vane at the peak of their jump.

It has been found that the arm swing in the vertical jump test contributes approximately 4 and 7 cm to vertical jump height for females and males, respectively (Harman et al. 1990; Walsh et al. 2007). Therefore improvements in vertical jump performance may reflect increases not only in lower-body power but also in jumping technique, timing, and coordination.

NORMATIVE DATA

Table 14.2 presents vertical jump test normative data for multiple sports players.

TABLE 14.2 Vertical Jump Test Data for Male and Female Sport Players

Sport	Gender	Squad/level	Age/position	n	JUMP HEIGHT, CM Mean ± SD	Range
Australian Football	Male	AIS	~18 years	284	60 ± 6	44-80
Basketball	Female	AIS	U20	500	47 ± 6	29-65
		State	U19	19	51 ± 5	39-62
		State	U17	14	48 ± 7	39-63
	Male	AIS	U20	500	65 ± 8	47-94
		State	U19	16	64 ± 7	54-78
		State	U17	16	62 ± 7	50-75
Cricket	Female	State	Open	163	42 ± 6	27-64
		National	Open	77	44 ± 6	31-63
	Male	National	U19	161	56 ± 7	36-75
		AIS	U23	266	58 ± 8	24-78
		State	Open	552	58 ± 8	23-82
		National	Open	151	60 ± 9	23-78
Football (soccer)	Female	National	U17	26	50 ± 4	42-60
		National	Open	27	54 ± 4	46-61
		State	U17	13	42 ± 4	36-50
	Male	National	U17	48	61 ± 5	54-69
		State	U16	n/a	n/a	n/a
		National	U23	47	62 ± 7	53-75
		National	Open	44	64 ± 6	55-76
Hockey	Female	State	Open	11	46 ± 4	37-53
	Male	National	Striker	6	65 ± 3	60-68
		National	Midfield	5	62 ± 5	56-69
		National	Defender	7	58 ± 5	52-66
		National	Goalkeeper	3	65 ± 5	59-69
		State	Open	10	61 ± 9	48-72
Netball	Female	National	U17	100	44 ± 5	31-58
		National	U19	67	44 ± 6	34-65
		National/AIS	U21	44	46 ± 6	34-57
		National	Open	55	46 ± 5	35-56
		State	U19	13	44 ± 4	38-52
Rugby League	Male	Junior (subelite)	U18	36	47 ± 7	33-58
		Junior (elite)	U18	28	52 ± 8	37-68
		Senior (semiprofessional)	Open	77	61 ± 7	49-75
		Senior (professional)	Open	68	63 ± 6	48-81
Rugby Union	Male	Junior talent squad	14-18 years	366	54 ± 6	n/a
		Provincial academy	15-26 years	289	58 ± 7	n/a
		Super Rugby	Open	163	62 ± 6	n/a

(continued)

TABLE 14.2 (*continued*)

Sport	Gender	Squad/level	Age/position	n	JUMP HEIGHT, CM Mean ± *SD*	Range
Tennis	Female	AIS/National academy	16+ years	22	40 ± 6	32-52
		National academy	15-16 years	28	39 ± 6	31-53
		National academy	13-14 years	52	37 ± 6	27-49
		National academy	11-12 years	23	36 ± 6	28-49
	Male	National academy	16+ years	56	54 ± 6	40-68
		National academy	15-16 years	46	51 ± 8	40-72
		National academy	13-14 years	62	43 ± 7	31-63
		National academy	11-12 years	26	37 ± 6	29-50

n/a = not available.

STRAIGHT LINE SPRINT TESTS

RATIONALE

The requirement to accelerate and sprint at maximal or near-maximal intensity is an important aspect of performance for many team sport athletes. Sprints over a short distance (i.e., 10 m) from a standing start are considered to reflect acceleration qualities, whereas longer sprints (30-40 m) also enable the measurement of maximal velocity when splits are recorded over the last 10 m (Young et al. 2001, 2008).

Athlete acceleration and speed qualities are ideally assessed using dual-beam light gates. For the purpose of control and standardization, it is recommended that sprint tests be conducted indoors and use a standing start.

Most court sports (netball, basketball, tennis) assess speed over 20 m and use 5 and 10 m split times to gauge acceleration. This distance is most relevant to these sports, because the court length and playing areas typically restrict players to shorter sprint distances.

Time–motion analyses during matches of various field sport athletes (rugby union, soccer, hockey, and Australian football) indicate that multiple sprints are performed with a mean sprint duration of 1.5 to 3 s (Bangsbo et al. 1991; Buchheit et al. 2010; Docherty et al. 1988; Dawson et al. 2004; Deutsch et al. 2007; Spencer et al. 2004), and the majority of these commence from a jogging or running motion and not from a standing start (Benton 2001). For this reason and the fact that longer duration sprints do occur, near-maximal or maximal running speeds can be achieved in competition (Duthie et al. 2006). For such sports it is appropriate to select longer sprint distances (e.g., 40 m) so that both acceleration and maximal running velocity can be assessed.

TEST PROCEDURE

- Measure specified distances with the measuring tape, checking that there are no twists in the tape when laid out. It is useful, where possible, to use a lane marker or sideline (straight line) to lay the measuring tape along.

- Mark each interval (i.e., 5, 10, 20 m or 10, 30, and 40 m) with masking tape including a start line (0 m) and a finishing line (20 m or 40 m).

- Place two cones approximately 4 m after the last set of light gates.

- Set the light gates at the appropriate intervals (0, 5, 10, and 20 m or 0, 10, 30, and 40 m).

- Set the light gates on the start line at a lower height to ensure capture of the start. Other light gates should be set at approximately torso height and approximately 2.0 m apart.

- The starting position is with front foot toe just touching the start line (0 m), heel up on back foot, body mass over the front foot, and shoulders and hips square in a crouched "ready" position.

- Once the athlete is in the ready position, all subsequent movement must be in a forward direction (i.e., no rocking is allowed).

- The athletes should be instructed to sprint as fast as possible, ensuring that they don't decelerate until they have passed the cones set 4 m after the final gate.

- The athlete may start in their own time once they have been advised that the timing system is ready.
- Record split times (at 5 and 10 m or 10 and 30 m) and final time (20 m or 40 m) for three trials to the nearest 0.01 s.
- Allow at least 2 min active recovery or rest between sprint trials.
- Use the best time for each split and final time as the final result, even if these times come from different trials.

DATA ANALYSIS

Results obtained from this test provide information on an athlete's ability to accelerate from a standing start and to cover a set distance quickly. The method for data analysis is presented next. Maximal speed will not be achieved during the 20 m sprint test from a standing start; however, it is possible to calculate a peak velocity if the intermediate split time for 10 m is known. A 40 m sprint with intermediate split distances allows the assessment of both acceleration and maximum velocity.

In sports such as rugby league and rugby union, where the ability to break a tackle is important, the momentum of the athlete in kilogram meters per second over the last 10 m of the 40 m sprint is a useful measure.

Improvements in 10 m sprint time will come from improvements in lower-body power and sprint starting technique from a stationary position. Improvements in total 40 m time and maximum velocity will come from speed drills focusing on acceleration and maximum velocity. Improvements in momentum will arise from increases in both body mass or running speed.

Data Analysis 20 m Sprint

- Acceleration score (s): fastest measured 0 to 5 m split time.
- Speed score (s): fastest measured 10 to 20 m split time.
- 20-m peak velocity score (m · s^{-1}): subtract the 10 m split time from the 20 m split time and divide 10 by this value.
- Combined acceleration/speed score (s): fastest measured 0 to 20 m time.

Data Analysis 40-m Sprint

- Acceleration score (s): fastest measured 0 to 10 m split time.
- Speed score (s): fastest measured 30 to 40 m split time.
- 40 m peak velocity score (m · s^{-1}): subtract the 30 m split time from the 40 m split time and divide 10 by this value.
- Momentum score (kg · m · s^{-1}): maximal velocity (m · s^{-1}) multiplied by body mass of player (kg).
- Combined acceleration/speed score (s): fastest measured 0 to 40 m time.

LIMITATIONS

To obtain valid results, athletes should be rested and relatively fresh for sprint testing. A well-structured warm-up is important to limit the risk of injury and promote optimal performance and may consist of light running, dynamic stretching, and some run-throughs building up to 100% efforts.

When sprint testing is conducted outdoors, the track should be positioned so that the wind direction is across the track. If this is not possible, wind velocity and direction should be recorded so that a time adjustment can be made. To do this, wind velocity is collected at the midpoint of the running track (i.e., 10 m mark for 20 m sprint test or 20 m mark for the 40 m sprint test) using a portable weather monitor. Wind velocity data collection commences when the athlete leaves the start and finishes when the athlete passes the final gate. The mean wind velocity over this time is recorded as a positive value for a tail wind and a negative value for a head wind. Split times can then be adjusted for zero wind conditions using the Linthorne (1994) formula:

$$\text{Males: } t_{ad} = t + [0.056 (w - w^2\{1/[2d/t]\})]$$

$$\text{Females: } t_{ad} = t + [0.067 (w - w^2\{1/[2d/t]\})]$$

where t_{ad} is the adjusted split time (s), t is the original split time (s), w is the mean wind velocity (m · s^{-1}), and d is the split distance (m).

NORMATIVE DATA

Table 14.3 presents 20 m sprint test normative data for female and male sport players. Table 14.4 presents 40 m sprint test normative data for female and male sport players.

TABLE 14.3   Data From 20 m Sprint Test for Female and Male Sport Players

Sport	Gender	Squad/level	Age/position	n	5 M TIME, S Mean ± SD	5 M TIME, S Range	10 M TIME, S Mean ± SD	10 M TIME, S Range	20 M TIME, S Mean ± SD	20 M TIME, S Range
Australian Football	Male	AIS	~18 years	284	1.09 ± 0.05	0.95-1.27	1.81 ± 0.07	1.65-2.03	3.04 ± 0.09	2.82-3.37
Basketball	Female	AIS	U20	500	1.17 ± 0.07	0.94-1.45	1.98 ± 0.09	1.73-2.35	3.41 ± 0.15	3.06-4.05
		State	U19	19	n/a	n/a	n/a	n/a	3.48 ± 0.13	3.29-3.78
		State	U17	14	n/a	n/a	n/a	n/a	3.56 ± 0.13	3.29-3.78
	Male	AIS	U20	500	1.10 ± 0.07	0.94-1.37	1.84 ± 0.09	1.43-2.25	3.13 ± 0.15	2.80-3.78
		State	U19	16	n/a	n/a	n/a	n/a	3.17 ± 0.09	3.06-3.34
		State	U17	16	n/a	n/a	n/a	n/a	3.24 ± 0.17	2.99-3.55
Cricket*	Female	State	Open	208	1.20 ± 0.08	0.95-1.47	2.02 ± 0.11	1.69-2.37	3.46 ± 0.18	2.59-3.91
		National	Open	90	1.19 ± 0.09	1.02-1.47	2.00 ± 0.11	1.81-2.37	3.43 ± 0.15	3.12-3.87
	Male	National	U19	197	1.05 ± 0.07	0.80-1.26	1.78 ± 0.10	1.01-2.01	3.04 ± 0.11	2.78-3.35
		AIS	U23	378	1.05 ± 0.06	0.87-1.26	1.78 ± 0.09	1.59-2.01	3.06 ± 0.11	2.76-3.37
		State	Open	808	1.06 ± 0.07	0.87-1.37	1.79 ± 0.09	1.59-2.20	3.06 ± 0.12	2.76-3.69
		National	Open	140	1.06 ± 0.07	0.90-1.28	1.79 ± 0.08	1.60-2.06	3.06 ± 0.11	2.77-3.70
Football (soccer)	Female	National	Open	27	1.18 ± 0.09	1.05-1.40	1.95 ± 0.10	1.77-2.21	3.32 ± 0.16	3.05-3.70
		State	U17	13	1.27 ± 0.05	1.21-1.39	2.08 ± 0.06	2.00-2.23	3.50 ± 0.10	3.39-3.72
	Male	National	U17	48	1.06 ± 0.06	0.95-1.18	1.76 ± 0.12	1.24-1.93	3.00 ± 0.10	2.79-3.19
		National	U23	47	1.05 ± 0.04	0.96-1.17	1.72 ± 0.05	1.72-1.91	2.91 ± 0.07	2.81-3.16
		National	Open	44	1.02 ± 0.04	1.03-1.14	1.69 ± 0.05	1.70-1.91	2.85 ± 0.07	2.72-3.12
		State	U16	18	1.15 ± 0.05	1.07-1.29	1.91 ± 0.07	1.82-2.11	3.20 ± 0.11	3.04-3.54
Netball	Female	National	U17	118	1.25 ± 0.09	1.09-1.50	2.07 ± 0.10	1.84-2.43	3.52 ± 0.16	3.15-4.15
		National	U19	76	1.24 ± 0.08	1.10-1.52	2.06 ± 0.09	1.87-2.40	3.51 ± 0.14	3.21-3.99
		National/AIS	U21	45	1.21 ± 0.07	1.03-1.36	2.03 ± 0.09	1.80-2.25	3.45 ± 0.14	3.14-3.74
		National	Open	55	1.22 ± 0.08	1.04-1.39	2.03 ± 0.10	1.78-2.25	3.46 ± 0.16	3.07-3.92
		State	U19	13	1.25 ± 0.06	1.16-1.35	2.08 ± 0.09	1.95-2.23	3.55 ± 0.15	3.30-3.80
Tennis	Female	AIS/National academy	16+ years	21	1.19 ± 0.05	1.10-1.28	2.01 ± 0.07	1.84-2.14	3.49 ± 0.14	3.27-3.88
		National academy	15-16 years	25	1.21 ± 0.07	1.10-1.34	2.03 ± 0.10	1.86-2.23	3.52 ± 0.16	3.24-3.83
		National academy	13-14 years	53	1.20 ± 0.08	1.06-1.36	2.05 ± 0.11	1.84-2.26	3.54 ± 0.18	3.16-3.92
		National academy	11-12 years	17	1.24 ± 0.08	1.12-1.38	2.11 ± 0.13	1.93-2.93	3.69 ± 0.25	3.30-4.24
	Male	AIS/National academy	16+ years	57	1.09 ± 0.06	0.97-1.25	1.81 ± 0.08	1.67-2.03	3.09 ± 0.10	2.80-3.36
		National academy	15-16 years	42	1.12 ± 0.07	0.91-1.24	1.88 ± 0.10	1.64-2.09	3.21 ± 0.15	2.85-3.69
		National academy	13-14 years	56	1.19 ± 0.08	1.01-1.42	2.00 ± 0.12	1.75-2.29	3.44 ± 0.20	3.00-3.93
		National academy	11-12 years	22	1.21 ± 0.09	1.05-1.45	2.06 ± 0.12	1.81-2.35	3.57 ± 0.19	3.14-3.98

*Cricket starts 50 cm behind the 0 m start gate; n/a = not available.

TABLE 14.4 Data From 40 m Sprint Test for Female and Male Sport Players

Sport	Gender	Squad/level	Age/position	n	0-10 M SPLIT, S Mean ± SD	Range	0-40 M SPLIT, S Mean ± SD	Range	30-40 M SPLIT, S Mean ± SD	Range
Hockey	Female	National	Striker	6	1.98 ± 0.04	1.90-2.03	5.92 ± 0.09	5.77-6.09	1.29 ± 0.03	1.23-1.34
		National	Midfield	8	1.96 ± 0.06	1.87-2.06	5.98 ± 0.22	5.54-6.26	1.32 ± 0.06	1.20-1.39
		National	Defender	4	2.01 ± 0.01	2.00-2.02	6.08 ± 0.13	5.94-6.19	1.34 ± 0.04	1.28-1.39
		National	Goalkeeper	2	2.08 ± 0.01	2.07-2.09	n/a	n/a	n/a	n/a
		National	U21	18	1.95 ± 0.11	1.74-2.13	5.95 ± 0.27	5.48-6.50	1.31 ± 0.07	1.20-1.45
		National	U17	15	1.94 ± 0.07	1.86-2.07	5.96 ± 0.22	5.68-6.47	1.31 ± 0.05	1.24-1.41
		State	Open	11	2.01 ± 0.09	1.89-2.22	6.10 ± 0.34	5.77-6.73	1.37 ± 0.10	1.26-1.55
	Male	National	Striker	8	1.82 ± 0.09	1.69-1.95	5.25 ± 0.18	5.02-5.54	1.09 ± 0.03	1.06-1.14
		National	Midfield	12	1.78 ± 0.06	1.68-1.88	5.29 ± 0.19	5.04-5.59	1.12 ± 0.05	1.07-1.19
		National	Defender	9	1.83 ± 0.06	1.78-1.94	5.43 ± 0.14	5.25-5.68	1.15 ± 0.04	1.07-1.20
		National	Goalkeeper	4	1.84 ± 0.08	1.86-2.07	n/a	n/a	n/a	n/a
		National	U21	20	1.80 ± 0.05	1.73-1.88	5.37 ± 0.15	5.68-6.47	1.14 ± 0.04	1.07-1.23
		State	Open	10	1.79 ± 0.09	1.60-1.86	5.36 ± 0.17	5.03-5.56	1.17 ± 0.05	1.08-1.24
Rugby League	Male	Junior (subelite)	U18	36	1.94 ± 0.11	1.77-2.23	5.83 ± 0.35	5.27-6.62	n/a	n/a
		Junior (elite)	U18	28	1.81 ± 0.08	1.63-1.93	5.56 ± 0.22	5.23-6.02	n/a	n/a
		Senior (semiprofessional)	Open	77	1.76 ± 0.08	1.63-2.00	5.29 ± 0.22	4.92-5.77	n/a	n/a
		Senior (professional)	Open	68	1.72 ± 0.08	1.59-1.97	5.23 ± 0.19	4.82-5.68	n/a	n/a
Rugby Union*	Male	Junior talent squad	14-18 years	368	1.92 ± 0.14	1.63-2.29	5.59 ± 0.25	4.89-6.42	n/a	n/a
		Provincial academy	15-26 years	292	1.93 ± 0.13	1.63-2.27	5.51 ± 0.25	4.81-6.47	n/a	n/a
		Super Rugby	Open	176	1.90 ± 0.12	1.63-2.28	5.47 ± 0.20	4.84-6.39	n/a	n/a

*Rugby Union: starts 30 cm behind the 0 m start gate; n/a = not available.

REPEAT SPRINT ABILITY TESTS

RATIONALE

Repeat sprint ability (RSA) is widely accepted as an important fitness component of team sports (Spencer et al. 2005). The assessment of RSA for team sports has gained considerable interest in recent years. Sport scientists working with sports such as soccer, hockey, Australian football, rugby league, and water polo have used various RSA protocols to investigate important aspects of this specific fitness quality with elite athletes. These assessments have included the correlations between RSA and tests of speed and aerobic fitness (Pyne et al. 2008); the use of RSA to discriminate between elite and subelite performance levels (Buchheit et al. 2010; Gabbett 2010; Impellizzeri et al. 2008); tracking changes in RSA throughout a competitive season (Impellizzeri et al. 2008); and determining the reliability and smallest worthwhile change in RSA (Gabbett 2010; Spencer et al. 2006b; Tan et al. 2010). However, despite the increased research and growing knowledge base, there is a real need for RSA tests that specifically assess the very intense or "worst-case" scenario of repeat sprint demands during competition for a particular sport. To achieve this, data from time–motion analysis are required to document the important variables of RSA (i.e., mean maximal sprint duration, maximal sprint number, mean recovery time between sprints, and likely recovery intensity). Variations in all of these aforementioned variables can have a considerable impact on test performance. For example, using an active versus a passive recovery between each sprint may result in a suboptimal effect on RSA (Spencer et al. 2006a). Game-specific tests of RSA have been developed for hockey (Spencer et al. 2006b), women's soccer (Gabbett 2010), and water polo (Tan et al. 2010).

TEST PROCEDURE

The specific RSA test procedure for a particular field-based sport should be designed on the principles discussed in the preceding rationale. In addition, some general procedures and important issues must be considered for all RSA testing. It is imperative that athletes give a maximal effort for all sprints and that no "pacing" is undertaken during the test. One method to reduce pacing involves monitoring the time of the first sprint. If the athlete does not obtain 95% or more of their best time from their individual one-off, straight line sprint conducted previously, the test is terminated. A subsequent attempt is performed after an appropriate rest period. Furthermore, it is suggested that the tester provide verbal encouragement to maximally motivate the athlete. The starting position of each sprint must be controlled, as described in the preceding section on straight line sprint tests.

Please refer to the individual sport chapters (e.g., hockey, Australian football, rugby league, tennis, soccer, and water polo) for sport-specific RSA testing protocols.

DATA ANALYSIS

The RSA results provide information on an athlete's ability to repeatedly perform short sprinting bursts. When an athlete scores poorly in this test, additional interval training specific to the work to rest ratios of repeat sprints for the sport is recommended.

When a test of RSA is expressed as the total sprint time, the reliability is excellent with reported TE values ranging from 0.7% to 1.5% (Gabbett 2010; Spencer et al. 2006b; Tan et al. 2010). However, the percentage sprint decrement is consistently reported to be less reliable (i.e., TE = 15%-30%; Gabbett 2010; Spencer et al. 2006b; Tan et al. 2010). Therefore, total sprint time should be reported for repeat sprint ability.

LIMITATIONS

The reliability and validity of RSA results rely on athletes' not pacing themselves over the course of the repeat efforts. Emphasizing the use of total sprint time and not percentage decrement will encourage athletes to perform each sprint as quickly as possible.

NORMATIVE DATA

Table 14.5 presents repeat sprint ability test normative data for female and male sport players.

TABLE 14.5 Repeat Sprint Ability (RSA) Test Data for Female and Male Sport Players

Sport	Gender	Squad/level	Age/position	n	TOTAL TIME, S Mean ± SD	Range
Australian Football*	Male	AIS	~18 years	229	25.63 ± 0.78	23.59-27.51
Football (soccer)†	Female	National	Open	n/a	20.9 ± 0.5	n/a
Hockey*	Female	National	All	15	29.93 ± 0.92	28.63-32.06
	Male	National	Striker	8	26.55 ± 0.58	26.02-27.71
		National	Midfield	12	26.31 ± 0.80	25.32-27.82
		National	Defender	6	27.02 ± 0.74	26.19-27.91
Rugby League‡	Male	Senior (semiprofessional)	Open	77	39.1 ± 3.3	33.8-46.7
		Senior (professional)	Open	68	38.4 ± 2.8	34.8-45.7
Tennis#	Female	AIS/National academy	16+ years	6	38.5 ± 2.4	35.0-42.2
		National academy	15-16 years	13	38.2 ± 2.4	35.1-42.3
		National academy	13-14 years	25	37.9 ± 2.0	33.7-41.3
		National academy	11-12 years	14	39.6 ± 2.6	35.3-44.7
	Male	AIS/National academy	16+ years	15	32.9 ± 1.2	31.0-35.1
		National academy	15-16 years	22	34.4 ± 1.7	32.2-38.4
		National academy	13-14 years	34	36.5 ± 2.6	32.1-45.5
		National academy	11-12 years	15	38.9 ± 1.6	35.3-41.4

*Australian Football and Hockey, 6 × 30 m RSA; †football, 6 × 20 m RSA; ‡Rugby League, 12 × 20 m RSA; #tennis, 10 × 20 m RSA; n/a = not available.

5-0-5 AGILITY TEST

RATIONALE

The basic movement patterns of many field and court sports require athletes to perform sudden changes in direction in response to the flight of the ball or movement of their opponents or teammates. Match agility permits an athlete to react to a stimulus, start quickly and efficiently, move in the correct direction, and be ready to change direction or stop quickly to make a play in a fast, smooth, efficient, and repeatable manner (Verstegen and Marcello 2001). Although new methods have been developed to measure the combination of visual processing, timing, reaction time, perception, anticipation, and decision making (Farrow et al. 2005; Gabbett et al. 2007; Sheppard et al. 2006), these reactive agility tests can be expensive and are not usually readily accessible. Therefore, the purpose of most agility tests is to simply measure the ability to rapidly change direction and position in the horizontal plane while performing a movement pattern already known to the athlete.

The 5-0-5 agility test is a relatively simple test that measures the time for a single rapid change of direction over a short "up and back" course with a running start. The test has been validated in the context of a team game (Draper and Pyke 1988) and is designed to minimize the influence of velocity while accentuating the effect of acceleration immediately before, during, and after the change of direction.

TEST PROCEDURE

- A metal tape measure and marking tape are used to mark out the points of the course as per figure 14.1.

- Light gates are set at 5 m approximately torso height and 2.0 m apart.

- The starting position is with front foot toe to the start line.

- After a signal that the light gates are ready, the athlete can start in their own time.

- The athlete should sprint from the start line through the light gates to the zero line, where they are required to turn and accelerate off the line, back through the light gates. One foot must be on or over the zero line at the turn for it to be a valid trial.

- The athlete may slow down only after passing through the light gates for the second time.

- The time taken to cover the 10 m distance is recorded to the nearest 0.01 s.
- The athlete completes three trials turning on their preferred foot. Alternatively, three trials on both sides may be given.
- The fastest time is recorded as the best score.

DATA ANALYSIS

Typically athletes who perform this test turning on both sides find they are faster on one side than the other. To improve imbalances between left and right sides and where the 5-0-5 test result is slow, it is recommended that athletes be given footwork drills, agility drills with and without the ball, and some acceleration and deceleration training.

LIMITATIONS

This test is purely an assessment of a planned movement pattern already known to the athlete. It does not require the athlete to react to any external stimuli such as ball movement or player movement, elements that are fundamental to team sport or open skill performance.

NORMATIVE DATA

Table 14.6 presents 5-0-5 agility test normative data for female and male sport players.

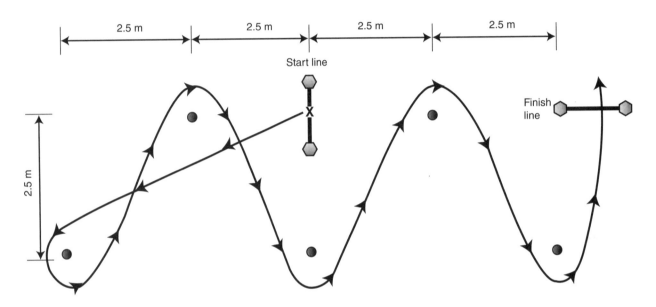

FIGURE 14.1 Equipment setup for the 5-0-5 agility test.

TABLE 14.6   Data From 5-0-5 Agility Test for Female and Male Sport Players

Sport	Gender	Squad/level	Age/position	n	TIME LEFT, S Mean ± *SD*	TIME LEFT, S Range	TIME RIGHT, S Mean ± *SD*	TIME RIGHT, S Range
Cricket	Female	State	Open	167	2.49 ± 0.11	2.20-2.77	2.48 ± 0.11	2.22-2.80
		National	Open	80	2.49 ± 0.36	2.26-2.77	2.45 ± 0.09	2.28-2.71
	Male	National	U19	158	2.27 ± 0.20	2.11-2.68	2.28 ± 0.11	2.06-2.61
		AIS	U23	280	2.28 ± 0.12	2.02-2.68	2.28 ± 0.11	2.02-2.71
		State	Open	589	2.29 ± 0.13	2.02-3.32	2.27 ± 0.12	2.02-2.81
		National	Open	98	2.27 ± 0.13	2.03-2.81	2.27 ± 0.13	2.02-2.68
Rugby League*	Male	Junior (subelite)	U18	36	2.38 ± 0.16	2.12-2.70	n/a	n/a
		Junior (elite)	U18	28	2.30 ± 0.13	2.13-2.65	n/a	n/a
		Senior (semiprofessional)	Open	77	2.32 ± 0.17	2.09-2.54	n/a	n/a
		Senior (professional)	Open	68	2.22 ± 0.21	1.86-2.55	n/a	n/a

Sport	Gender	Squad/level	Age/ position	n	TIME LEFT, S		TIME RIGHT, S	
					Mean ± SD	Range	Mean ± SD	Range
Tennis	Female	AIS/National academy	16+ years	2	2.48 ± 0.06	2.44-2.52	2.49 ± 0.03	2.47-2.51
		National academy	15-16 years	6	2.54 ± 0.12	2.38-2.64	2.53 ± 0.13	2.34-2.65
		National academy	13-14 years	24	2.54 ± 0.14	2.32-2.82	2.56 ± 0.15	2.29-2.82
		National academy	11-12 years	7	2.74 ± 0.11	2.59-2.87	2.68 ± 0.10	2.51-2.81
	Male	AIS/National academy	16+ years	21	2.33 ± 0.11	2.09-2.56	2.33 ± 0.07	2.19-2.46
		National academy	15-16 years	20	2.42 ± 0.13	2.20-2.59	2.42 ± 0.16	2.18-2.69
		National academy	13-14 years	21	2.49 ± 0.14	2.27-2.89	2.46 ± 0.19	2.13-2.91
		National academy	11-12 years	2	2.55 ± 0.26	2.36-2.73	2.49 ± 0.21	2.34-2.64

*Rugby League, left and right data combined; n/a = not available.

MULTISTAGE SHUTTLE RUN TEST

RATIONALE

The multistage shuttle run test (MSRT) has been shown to provide a valid estimate of maximal oxygen consumption (Brewer et al. 1988; Leger and Lambert 1982; Ramsbottom et al. 1988) and has been used extensively for the field assessment of maximal aerobic power for more than 20 years. The MSRT is a continuous time-efficient test capable of assessing a whole team or squad simultaneously. Given a maximal effort by an athlete, the MSRT also allows for the reliable measurement of maximal heart rate.

TEST PROCEDURE

- The 20 m distance is measured and clearly marked with marking tape and cones set at approximately 1.5 m intervals.
- Athletes line up along one of the lines, ready to start.
- Athletes listen carefully to the instructions on the CD. (Audio software is available from a variety of commercial and academic sources.) The commentary provides a brief explanation of the test leading to a 5 s countdown before the start of the test itself. Begin the test at level 1.
- The CD emits a single beep at various intervals. The athlete must try to be at the opposite end of the 20 m track by the time the following beep sounds. After approximately each minute, the time interval between beeps decreases and running speed is increased accordingly.

- The first running speed is referred to as level 1; the final speed is level 21.
- The athlete must place one foot on or over the 20 m mark at the sound of each beep.
- If the athlete arrives at the line before the beep sounds, they should turn (by pivoting) and wait for the beep before continuing to the next line.
- If an athlete fails to reach the line at the sound of the beep, the athlete must receive a warning that they will be eliminated if they are not at the opposite end of the 20 m track at the sound of the next beep.
- When near exhaustion, athletes falling short of the 20 m line twice in succession (one warning and a subsequent missed line) have their test terminated and their score recorded. Their score is the level and number of shuttles immediately prior to the beep on which they were eliminated.
- Additional indicators of maximal effort such as rating of perceived exertion, heart rate, and blood lactate concentration can be collected at the completion of the test.

- After completing the test, athletes should cool down by walking followed by stretching.

DATA ANALYSIS

There is a strong correlation ($r = 0.92$, $p < .01$) between MSRT score and maximal oxygen consumption, and it is possible to predict maximal oxygen consumption ($\dot{V}O_2$max) from MSRT final level scores using the following regression (Ramsbottom et al. 1988):

$$\text{Predicted } \dot{V}O_2\text{max (ml} \cdot \text{kg}^{-1} \cdot \text{min}^{-1}\text{)} = 14.4 + 3.48 \times$$
$$\text{MSRT (Shuttle Level); } SD \text{ 3.5 ml} \cdot \text{kg}^{-1} \cdot \text{min}^{-1}.$$

Athletes who score poorly on this test need additional endurance or aerobic training. Endurance fitness often improves substantially during the preseason period and is then maintained rather than improved during a long competitive season. Positional variation is common: Athletes who generally cover the most ground during competition score the best in this test.

Because the MSRT is a maximal test of approximately 1 min increments to exhaustion, it is possible to measure maximal heart rate values, generally during the final stages of this test. Unpublished TE data for this measure are 2 beats $\cdot$ min^{-1} (data provided by South Australian Sports Institute, $n = 40$). Maximal heart rate can then be used to calculate and prescribe aerobic training zones (refer to chapter 6, Blood Lactate Thresholds).

LIMITATIONS

The test can be affected by player motivation, running and turning technique, and environmental conditions. A limitation of this test is that athletes must know how to push themselves to their physiological limit.

NORMATIVE DATA

Table 14.7 presents MSRT normative data for female and male sport players.

TABLE 14.7   Multistage Shuttle Run Test Data for Female and Male Sport Players

Sport	Gender	Squad/level	Age/position	n	LEVEL/SHUTTLE Mean ± SD	LEVEL/SHUTTLE Range
Australian Football	Male	AIS	~18 years	214	13/5 ± 1/0	10/2 to 15/7
Hockey	Female	National	Striker	5	11/9 ± 0/6	11/1-12/4
		National	Midfield	8	12/0 ± 0/6	11/1-12/8
		National	Defender	6	12/6 ± 0/11	11/3-13/7
		National	U21	21	11/9 ± 1/1	9/2-13/9
		National	U17	13	10/8 ± 1/2	8/9-13/3
		State	Open	10	10/7 ± 0/9	9/4-12/1
	Male	National	Striker	6	14/9 ± 0/7	14/1-15/8
		National	Midfield	8	14/12 ± 0/7	13/11-15/11
		National	Defender	7	14/11 ± 1/1	13/2-16/4
		National	U21	18	14/4 ± 0/10	13/5-16/3
		State	Open	11	12/12 ± 2/0	10/5-15/2
Tennis	Female	AIS/National academy	16+ years	18	11/2 ± 1/2	8/9-12/8
		National academy	15-16 years	26	10/1 ± 1/5	7/2-12/8
		National academy	13-14 years	49	10/2 ± 1/4	6/7-13/2
		National academy	11-12 years	25	10/1 ± 1/4	6/2-12/2
	Male	AIS/National academy	16+ years	55	13/4 ± 1/0	11/1-15/4
		National academy	15-16 years	41	13/2 ± 1/1	8/9-15/5
		National academy	13-14 years	55	11/9 ± 1/1	9/1-14/5
		National academy	11-12 years	30	10/6 ± 1/8	6/9-13/9

YO-YO INTERMITTENT RECOVERY TEST

RATIONALE

The Yo-Yo intermittent recovery tests (IRTs) are specifically designed to mimic sports involving intensive exercise bursts followed by short recovery periods (i.e., soccer, Australian football, rugby, basketball, netball). Most team sport athletes must be able to recover quickly after hard exercise, because their ability to do so and "go again" can greatly affect the outcome of a competition.

The IRTs are a valid and practical means to evaluate the ability of an individual to repeatedly perform intense intermittent exercise while still requiring the maximal activation of the aerobic system (Bangsbo et al. 2008). Both tests can be used to determine maximal heart rate of an individual (Krustrup et al. 2003, 2006).

The Yo-Yo IRT has two levels. Both tests consist of 2 × 20 m shuttle runs at increasing speeds interspersed with a 10 s active recovery. The Yo-Yo IRT level 1 (IRT1) starts at a lower speed and has smaller increments in speed than the Yo-Yo IRT level 2 (IRT2). The Yo-Yo IRT1 is used to determine an individual's endurance capacity (Bangsbo et al. 2008) and may be more suited to younger athletes. The Yo-Yo IRT2 is used to determine the capacity of an athlete to recover from repeated exercise that requires a high contribution from the anaerobic systems.

The IRT tests involve a 5 to 15 s exercise period followed by a 10 s active recovery. Over the period of the test, the speed of running is progressively increased; the complete duration of the test varies from 5 to 20 min.

Currently, the Yo-Yo IRT1 test is the standard used across Australian NSSQA-accredited laboratories.

TEST PROCEDURE

- Using a measuring tape and marking tape, measure out a 20 m test course as per figure 14.2.

- Place markers 2 m apart at both ends of the 20 m test course (i.e., at start and turning lines).

- In addition to marking the 20 m line, measure out a 5 m distance behind the start line.

- Place a marker on the recovery line aligned to the middle of the two markers on the start line, as outlined in figure 14.2. Ensure there is one course setup per athlete being tested.

- Athletes assume a starting position on the start line.

- The Yo-Yo test CD is started. (Audio software is available from a variety of commercial and academic sources.)

- At the time of the first signal, athletes run forward to the turning line. At the sound of the second signal, athletes arrive and turn at the turning line and then run back to the start line arriving on the next beep. When the start marker is passed, the athletes continue forward at a reduced pace (jogging) toward the 5 m mark, where they then turn around the cone and return to the start line. At this point the athletes stop and wait for the next signal to sound. It is important that the athletes are stationary on the start line before the commencement of each sprint.

- Athletes are required to place one foot either on or over the start or turning lines at the sound of each beep.

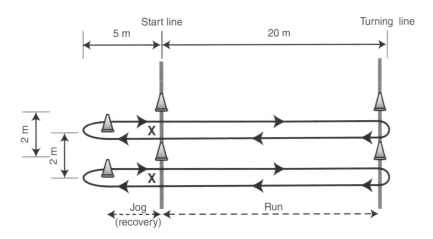

FIGURE 14.2 Setup for Yo-Yo intermittent recovery test.

- Athletes should continue running for as long as possible, until they are unable to maintain the speed as indicated by the CD (or MP3 file).

- The end of the test is indicated by the inability of an athlete to maintain the required pace for two trials. The first time the start line is not reached, a warning is given; the second time the athlete must withdraw.

- When the athlete withdraws, the last level and the number of 2 × 20 m intervals performed at this level are recorded on the appropriate recording sheet. (The last 2 × 20-m interval is included, even if the athlete did not complete at the right pace.)

- Additional indicators of maximal effort such as rating of perceived exertion, heart rate, and blood lactate concentration can be collected at the completion of the test.

- The Yo-Yo IRT is effort dependent, so for valid results athletes must attempt to reach the highest level possible before stopping.

- Verbal encouragement should be given to the athletes throughout the test.

- Upon completion of the test, all athletes should be encouraged to perform a warm down.

- The final Yo-Yo intermittent recovery speed and interval score obtained by each athlete are used to calculate the total distance covered by the athlete during the test.

Note: If an athlete is able to run faster than speed level 23 on IRT1, they should perform the IRT2 on the next occasion.

DATA ANALYSIS

The Yo-Yo IRTs provide a sensitive and reliable means of differentiating between players at different competitive levels and playing positions. These tests may also be used to monitor changes in performance after a training intervention or over the course of a competition season (Bangsbo et al. 2008). The total distance covered in the IRT has been shown to be related to match running performance, with athletes at higher levels of competition tending to run further in the tests and undertake more high-intensity running in matches. Although large variations in test results may be observed within a team, a sport may require a minimum level of fitness to perform at a high level (Bangsbo et al. 2008). Both Yo-Yo IRTs can be used to establish peak heart rate, which could then be used to prescribe aerobic training zones (refer to chapter 6 for aerobic training zone descriptions) and evaluate heart rate response to training or competition (Bangsbo et al. 2008). As with the MSRT, it is possible to theoretically predict maximal oxygen consumption ($\dot{V}O_2max$) from Yo-Yo IRT using the following regressions (Bangsbo et al. 2008). However, it should be noted that this practice is not considered accurate (Bangsbo et al. 2008) due to the scattered nature of the relationship between Yo-Yo IRT and $\dot{V}O_2max$.

Yo-Yo IRT1 test

$$\text{Predicted } \dot{V}O_2max \text{ (ml} \cdot \text{kg}^{-1} \cdot \text{min}^{-1}) = \text{IRT1 distance (m)} \times 0.0084 + 36.4$$

Yo-Yo IRT2 test

$$\text{Predicted } \dot{V}O_2max \text{ (ml} \cdot \text{kg}^{-1} \cdot \text{min}^{-1}) = \text{IRT2 distance (m)} \times 0.0136 + 45.3$$

LIMITATIONS

The test can be affected by player motivation, running and turning technique, and environmental conditions. The test is limited if athletes do not know how to push themselves to their physiological limit.

NORMATIVE DATA

Table 14.8 presents IRT1 normative data for female and male sport players.

TABLE 14.8 Yo-Yo Intermittent Recovery Test Data for Female and Male Sport Players

Sport	Gender	Squad/level	Age/position	n	LEVEL Mean ± SD	LEVEL Range	DISTANCE, M Mean ± SD	DISTANCE, M Range
Basketball	Female	AIS	U20	500	16.4 ± 1.3	12.9-20.2	1,240 ± 400	280-2,440
		State	U19	19	16.1 ± 0.8	14.1-17.6	1,083 ± 360	360-1,640
		State	U17	14	15.6 ± 1.1	14.1-17.5	985 ± 355	480-1,600
	Male	AIS	U20	500	18.8 ± 1.8	14.6-22.7	2,000 ± 562	680-3,280
		State	U19	16	18.1 ± 1.8	15.6-21.2	1,736 ± 598	1,000-2,760
		State	U17	16	16.2 ± 0.8	15.3-18.2	1,186 ± 264	880-1,800
Cricket	Female	State	Open	169	15.1 ± 1.3	12.3-20.1	1,155 ± 356	300-2,400
		National	Open	196	15.8 ± 1.2	12.1-20.1	970 ± 385	200-2,400
	Male	National	U19	190	17.7 ± 1.5	15.1-21.1	1,773 ± 454	800-2,720
		AIS	U23	575	17.9 ± 1.6	14.1-23.8	1,853 ± 467	480-3,640
		State	Open	845	18.1 ± 1.7	14.1-23.8	1,846 ± 454	480-3,640
		National	Open	148	18.1 ± 1.8	14.1-21.4	1,863 ± 444	480-2,840
Football (soccer)	Male	National	U17	18	20.6 ± 2.7	16.4-22.5	2,327 ± 358	1,240-3,200
		National	U23	24	20.8 ± 3.1	15.6-23.2	2,501 ± 434	1,000-3,400
		National	Open	18	21.4 ± 2.2	17.6-23.4	2,611 ± 421	1,640-3,480
		State	U16	18	18.4 ± 1.3	16.2-20.8	1,906 ± 450	1,160-2,680
Netball	Female	National	U17	25	15.7 ± 0.9	13.4-17.3	1,013 ± 286	440-1,520
		National/AIS	U21	22	16.6 ± 1.2	13.4-18.7	1,320 ± 372	440-2,000
		National	Open	27	17.2 ± 1.1	14.3-19.3	1,492 ± 358	560-2,160
		State	U19	12	15.2 ± 0.8	14.2-16.8	853 ± 323	520-1,400
Rugby League	Male	Junior Elite	U18	759	17.1 ± 1.0	13.4-21.5	1,440 ± 400	440-2,880
		Professional	Open	29	17.1 ± 0.9	15.4-20.4	1,789 ± 366	920-2,520
Rugby Union	Male	Junior talent squad	Backs	20	17.6 ± 1.1	15.8-19.5	1,654 ± 361	1,080-2,240
			Forwards	25	16.5 ± 1.2	14.3-19.2	1,291 ± 380	560-2,120
		Provincial academy	Backs	18	18.6 ± 1.2	15.4-20.4	1,951 ± 419	920-2,520
			Forwards	25	17.1 ± 1.3	14.4-21.1	1,424 ± 467	600-2,720

References

Bangsbo, J., Marcello, I., and Krustrup, P. 2008. The yo yo intermittent recovery test. A useful tool for the evaluation of physical performance in intermittent sports. *Sports Medicine* 38(1):37-51.

Bangsbo, J., Norregaard, L., and Thorso, F. 1991. Activity profile of competition soccer. *Canadian Journal of Sport Science* 16(2):110-115.

Benton, D. 2001. Sprint running needs of field sport athletes: a new perspective. *Sports Coach* 24(2):12-14.

Brewer J., Ramsbottom, R., and Williams, C. 1988. *Multistage Fitness Test*. Belconnen, ACT: Australian Coaching Council.

Buchheit M., Mendez-Villanueva, A., Simpson, B., and Bourdon, P. 2010. Repeated-sprint sequences during youth soccer matches. *International Journal of Sports Medicine* 31(10):709-716.

Dawson, B., Hopkinson, R., Appleby, B., Stewart, G., and Roberts, C. 2004. Player movement patterns and game activities in the Australian Football League. *Journal of Science and Medicine in Sport* 7(3):278-291.

Deutsch, M.U., Kearney, G.A., and Rehrer, N.J. 2007. Time-motion analysis of professional rugby union players during competition. *Journal of Sports Sciences* 25(4):461-472.

Docherty, D., Wenger, H.A., and Neary, P. 1988. Time motion analysis related to the physiological demands of rugby. *Journal of Human Movement Studies* 14:269-277.

Draper, J., and Pyke, F. 1988. Turning speed : A valuable asset in cricket run making. *Sports Coach* 11(3):30-31.

Duthie, G.M., Pyne, D.B., Marsh, D.J., and Hooper, S.L. 2006. Sprint patterns in rugby union players during competition. *Journal of Strength and Conditioning Research* 20(1):208-214.

Farrow, D., Young, W., and Bruce, L. 2005. The development of a test of reactive agility for netball: A new methodology. *Journal of Science and Medicine in Sport* 8(1):52-60.

Gabbett, T.J. 2010. The development of a test of repeated-sprint ability for elite women's soccer players. *Journal of Strength and Conditioning Research* 24(5):1191-1194.

Gabbett T., Rubinoff, M., Thorburn, L., and Farrow, D. 2007. Testing and training anticipation skills in softball fielders. *International Journal of Sport Science and Coaching* 2(1):15-24.

Harman, E.A., Rosenstein, M.T., Fryman, P.N., and Rosenstein, R.M. 1990. The effects of arm swing and counter movement on vertical jumping. *Medicine and Science in Sport and Exercise* 22(6):825-833.

Impellizzeri, F.M., Rampinini, E., Castagna, C., Bishop, D., Ferrari Bravo, D., Tibaudi, A., and Wisloff, U. 2008. Validity of a repeated-sprint test for football. *International Journal of Sports Medicine* 29(11):899-905.

Krustrup, P., Mohr, M., Amstrup, T., et al. 2003. The Yo-Yo intermittent recovery test: Physiological response, reliability and validity. *Medicine and Science in Sport and Exercise* 35(4):697-705.

Krustrup, P., Mohr, M., Nybo, L., et al. 2006. The Yo Yo IR2 test: Physiological response reliability, and application to elite soccer. *Medicine and Science in Sport and Exercise* 38(9):1666-1673.

Leger, L.A., and Lambert, J. 1982. A maximal multistage 20-m shuttle run test to predict $\dot{V}O_{2max}$. *European Journal of Applied Physiology and Occupational Physiology* 49(1):1-12.

Linthorne, N.P. 1994. Wind assistance in the 100-m sprint. *Modern Athlete and Coach* 32(1):6-9.

Pyne, D.B., Saunders, P.U., Montgomery, P.G., Hewitt, A., Sheehan, J., and Sheehan, K. 2008. Relationships between repeated sprint testing, speed and endurance. *Journal of Strength and Conditioning Research* 22(5):1633-1637.

Ramsbottom, R., Brewer, J., and Williams, C. 1988. A progressive shuttle run test to estimate maximal oxygen uptake. *British Journal of Sports Medicine* 22(4):141-145.

Sayers, S.P., Harackiewicz, D.V., Harman, E.A., Frykman, P.N., and Rosenstein, M.T. 1999. Cross-validation of three jump power equations. *Medicine and Science in Sports and Exercise* 31(4):572-577.

Sheppard, J.M., Young, W.B., Doyle, T.A., Sheppard, T.A., and Newton, R.U. 2006. An evaluation of a new test of reactive agility and its relationship to sprint speed and change of direction speed. *Journal of Science and Medicine in Sport* 9:342-349.

Spencer, M., Bishop, D., Dawson, B., and Goodman, C. 2005. Physiological and metabolic responses of repeated-sprint activities: Specific to field-based team sports. *Sports Medicine* 35(12):1025-1044.

Spencer, M., Bishop, D., Dawson, B., Goodman, C., and Duffield, R. 2006a. Metabolism and performance in repeated cycle sprints: Active versus passive recovery. *Medicine and Science in Sports and Exercise* 38(8):1492-1499.

Spencer, M., Fitzsimons, M., Dawson, B., Bishop, D., and Goodman, C. 2006b. Reliability of a repeated-sprint test for field-hockey. *Journal of Science and Medicine in Sport* 9(1-2):181-184.

Spencer, M., Lawrence, S., Rechichi, C., Bishop, D., Dawson, B., and Goodman, C. 2004. Time-motion analysis of elite field hockey, with special reference to repeated-sprint activity. *Journal of Sports Sciences* 22(9):843-850.

Tan, F., Polglaze, T., and Dawson, B. 2010. Reliability of an in-water repeated-sprint test for water polo. *International Journal of Sports Physiology Performance* 5(1):117-120.

Verstegen, M., and Marcello, B. 2001. In: B. Foran (ed), *High Performance Sports Conditioning*. Champaign, IL: Human Kinetics, pp 139-165.

Walsh, M.S., Boehm, H., Butterfield, M.M., and Santhosam, J. 2007. Gender bias in the effects of arms and counter-movement on jumping performance. *Journal of Strength and Conditioning Research* 21(2):362-366.

Winget, C.M., DeRoshia, C.W., and Holley, D.C. 1985. Circadian rhythms and athletic performance. *Medicine and Science in Sports and Exercise* 17(5):498-516.

Young, W., Benton, D., Duthie, G., and Pryor, J. 2001. Resistance training for short sprints and maximum-speed sprints. *Strength and Conditioning Journal* 23(2):7-13.

Young, W., Russell, A., Burge, P., Clarke, A., McCormack, S., and Stewart, G. 2008. The use of sprint tests for assessment of speed qualities of elite Australian rules footballers. *International Journal of Sports Physiology and Performance* 3:199-206.

Perceptual–Cognitive and Perceptual–Motor Contributions to Elite Performance

Damian Farrow

High-performance sport has always required athletes to develop a range of physical and mental capacities to succeed. However, traditionally, athlete testing has focused on the measurement of physiological capacities complemented with biomechanical assessment of technique. In recent times, coaches of open, fast-paced, interceptive sports in particular have sought to understand and in turn develop the perceptual and decision-making capacities of performance.

This chapter is divided into three main sections. First is a review of the defining perceptual–cognitive and perceptual–motor characteristics of sport performance and in particular the identification of those attributes that reliably distinguish the elite performer from those who are not. Data are drawn from a range of different Australian high-performance sports programs to illustrate each key finding. Second, a case study examining data collected from Netball Australia's skill acquisition testing battery is used to highlight some of the key issues of measurement and interpretation. Third, consideration of the concept of task representation is applied to the assessment and development of perceptual and decision-making tests.

Sport expertise research has evolved over the past 30 years to collectively detail a variety of characteristics that do and do not distinguish elite athletes from their lesser-skilled counterparts. The predominant approach has been empirical in nature, typically identifying and describing differences between expert and novice performers. The following sections outline the key perceptual–cognitive and perceptual–motor capacities that have been incorporated into high-performance sport testing regimes. More comprehensive reviews regarding these performance components and how elite and subelite performers differ are available elsewhere (see Abernethy 1994; Helsen and Starkes 1999; Starkes and Ericsson 2003).

Anticipating the Movement of an Opponent

In many sport tasks, especially fast ball and interceptive sports such as rugby, tennis, and cricket, the ability to anticipate the direction of an opponent's action is crucial. The temporal demands of the situation may make such anticipation a necessity if one is to intercept the ball in time (e.g., facing a fast bowler in cricket) or at the very least buy the receiving player time in which to prepare an appropriate response (e.g., returning a serve in tennis). The biomechanical properties of most sport skills ensure that performers must adhere to a relatively predetermined sequence of skill execution if they are to produce a biomechanically efficient action. For example, hitting a powerful tennis serve while maintaining an acceptable level of accuracy requires a player to produce hip to shoulder rotation that leads to a ballistic, sequential overarm action of the shoulder, arm, and wrist to serve the ball. This constraint on a server guarantees that at some point prior to racket–ball contact, a receiving player will be provided with invariant movement pattern information that is reliably predictive of the forthcoming service direction and spin.

Temporal Occlusion Measurement Approach

A variety of measurement approaches have been used to examine the anticipatory skill (i.e., advance information usage patterns) of elite and subelite performers involved in "open," interceptive sport skills. The most prominent method has been the temporal occlusion approach, which relies on the use of sport-specific dynamic visual images filmed from the perspective of a player while competing (e.g., a soccer goalkeeper facing a penalty kick is filmed from the perspective of the goalkeeper). These images are then selectively edited to provide differing amounts of advance and ball-flight information (see figure 15.1). Within this approach, participants are typically required to predict the opponent's action (e.g., kick

direction in soccer goalkeeping) from the information available to them under the different temporal occlusion conditions. A significant change in the prediction accuracy from one occlusion condition to the next is assumed to be indicative of information pickup from within the additional viewing period. For example, Farrow and colleagues (2005a) demonstrated that elite tennis players were able to pick up information to improve their prediction accuracy during the time period 300 ms before the ball was struck, whereas novices didn't extract any predictive information. By examining the key kinematic changes in the opponent's action during this time period (it coincided with the throwing action and ball toss of the opponent's action), elite players could determine the information usage patterns of the different skill groups. The use of the progressive temporal occlusion approach has repeatedly shown a visual–perceptual advantage in anticipation for elite players in comparison to their lesser-skilled counterparts that holds across a variety of sports, such as tennis (Goulet et al. 1989), badminton (Abernethy and Russell 1987), squash (Abernethy 1990), cricket (Müller et al. 2006), soccer (Savelsbergh et al. 2002), field hockey (Starkes 1987), and volleyball (Wright et al. 1990). Although obviously the information sources differ between

FIGURE 15.1 An example of the temporal occlusion conditions presented to a soccer goalkeeper attempting to intercept a penalty kick. The final four frames represent specific time points when the display would be occluded and the participant required to make a decision.

Reprinted, by permission, from D. Farrow and M. Raab, 2008, A recipe for expert decision making. In *Developing sport exercise: Research and coaches put theory into practice*, edited by D. Farrow, J. Baker, and C. Macmahon (United Kingdom: Taylor & Francis), 137-157.

sports, the pick-up of predictive movement pattern information from earlier time-windows remains a robust finding.

Visual Search Measurement

Another approach frequently used to assess information pickup in sports settings is to record players' eye movements as they view, and ideally respond to, a perceptual display identical, or at least very similar, to those they would encounter in the natural setting. It has been hypothesized that players' visual search patterns may provide a good indication of the location of the most informative features of an opponent's action that may be used for anticipatory purposes. In some instances, studies measuring visual search patterns have supported the occlusion-based research findings relating to the identification of what kinematic information sources experts use in their preparation of a response to an opponent's action (e.g., Savelsbergh et al. 2002; Singer et al. 1996; Williams et al. 1994). However, although many of these studies show some visual search differences between elite and lesser-skilled performers the differences are generally quite small, arguably trivial, leading at least some researchers (e.g., Abernethy 1990; Goulet et al. 1989) to conclude that visual search differences cannot alone explain information-processing differences and performance differences between elite and subelite performers.

Reactive Agility Measurement

In the previous edition of this text, the protocol for the assessment of agility performance in team sport athletes suggested that the ability to use agility movements "successfully in the actual game would depend on other factors such as visual processing, timing, reaction time, perception, and anticipation. Although all these factors combined are reflected in the player's on-field 'agility,' the purpose of most agility tests is simply to measure the ability to rapidly change body direction and position in the horizontal plane" (Ellis et al. 2002, pp. 132). Yet, ironically, at that time none of the typical agility tests actually attempted to measure any of these supposedly critical qualities. Rather, existing agility tests were based on planned movements initiated by the athlete (e.g., Draper and Lancaster 1985).

Farrow and colleagues (2005b) argued that agility tests must cater to this reactive aspect of performance through the addition of a visual–perceptual test component and accordingly examined the impact of such a stimulus to a planned netball agility task. Specifically, a near life-size video image of a skilled netball player was projected in front of the test participant (see figure 15.2). The projected player completed a

FIGURE 15.2 An example of a player performing the netball reactive agility test.

Adapted, by permission, from D. Farrow, 2010, "A multi-factorial examination of the development of skill expertise in high performance netball," *Talent Development and Excellence* 2(2): 123-135.

pass directed to either the left or right of the participant, who was instructed to attempt to intercept the simulated pass by adjusting the response and running through the corresponding timing gates. In addition to typical movement time measures, the decision-making time of the participant was recorded in the reactive test condition (refer to chapter 22, Netball Players, for the complete experimental protocol). A planned test condition was designed to replicate the movement requirements of the reactive condition with the key difference being that the participant knew the direction of travel before commencing the test and hence was not required to respond to a video stimulus (as per more typical agility tests).

The results demonstrated that although the planned agility test was completed faster than the reactive test, irrespective of skill level, it was the reactive test condition that was more likely to differentiate between the differing skill levels. In particular, the movement time component subsequent to the perceptual stimulus was the facet most likely to elicit time differences between the groups. In contrast, the planned components of agility performance contributed little to the between group differences found. Closer inspection of the reactive sprint time component revealed that decision-making time differed significantly between the elite and club level players examined. It is argued that the elite players' ability to anticipate the intended pass direction, as evidenced by a negative decision-making mean (–149 ms), allowed them to predict earlier their change of direction and hence complete the sprint component of the test with greater speed. In comparison, the lesser-skilled group's decision times (22 ms) indicated that they waited until the passer presented all available information before initiating their change of direction. This in turn slowed their sprint time component accordingly (see Farrow et al. 2005b for more details). Subsequently a number of researchers have used reactive agility tasks across a variety of sports, demonstrating findings consistent with those described above (e.g., Australian rules: Sheppard et al. 2006; Young et al. 2011; and rugby league: Gabbett et al. 2008; Gabbett and Benton 2009).

Recognizing Typical Patterns of Play in Team Sports

Sport-specific tests of pattern recognition and recall have been consistently used to demonstrate an elite advantage (e.g., Allard et al. 1980; Farrow et al. 2010; Starkes 1987; Williams and Davids 1995). Adapted from cognitive psychology examinations into chess expertise (Chase and Simon 1973), such tasks typically demonstrate that elite players can more accurately recall and recognize structured patterns of play from their sport, or within sports that are structurally similar, than can novices (Abernethy et al. 2005). The experimental task requires a player to view a typically occurring pattern of play from the sport before the vision is occluded at a key moment. The player is then provided a blank template of the scene and is required to recall the attacking and defensive structures of the two teams by plotting the location of each player as last presented before the trial was occluded (see figure 15.3). Scoring usually involves awarding a point for each player correctly recalled, which is then expressed as a percentage of total recall usually separated into attack and defensive recall (see table 15.1). Typically, elite athletes recall a higher

FIGURE 15.3 An AFL player completing a pattern recall task.

Photo courtesy of Damian Farrow.

TABLE 15.1 Percentage Recall (Mean ± *SD*) as a Function of Netball Skill Level

Group	Attack recall, %	Defense recall, %
Open	79.57 ± 6.55	68.18 ± 6.94
U21	66.64 ± 12.57	54.79 ± 12.13
U19	59.88 ± 8.85	51.28 ± 7.91
U17	63.69 ± 6.80	48.71 ± 5.54

Adapted, by permission, from D. Farrow, 2010, "A multi-factorial examination of the development of skill expertise in high performance netball," *Talent Development and Excellence* 2(2): 123-135.

proportion of players than do subelite performers (e.g., Farrow 2010; Williams and Davids 1995).

This advantage has been linked to the elite players' more sophisticated domain-specific knowledge structures (e.g., chunks) that are stored and retrieved efficiently from long-term memory (Ericsson and Kintsch 1995). The process of perceiving sport-specific patterns as chunks rather than individual items (i.e., as offensive or defensive patterns instead of individual players) allows an elite player to process the relevant patterns at a faster rate, an advantage often demonstrated in anticipation and decision-making tasks (Helsen and Starkes 1999). This is of particular interest to coaches and scientists alike; however, as suggested by Ward and colleagues (2006), "pattern recall requires individuals to invoke a process that they otherwise may not have used, at least explicitly, during a typical game. Accordingly, recall tasks may provide only limited insight into the mechanisms underlying actual performance compared to more representative tasks that simulate "real-world" constraints" (p. 245). Consequently, Farrow and colleagues (2010) examined the relationship between pattern recall and anticipation by requiring Australian team members and club-level and novice rugby players to view structured line-out patterns. Results reinforced the notion that the recall and anticipation tasks were able to differentiate the elite from the subelite. Pattern recall was found to be a strong predictor of anticipatory skill, explaining 40% of its variance. However, the nature of the link between pattern recall and anticipation is still generally considered equivocal and may even be sport-specific.

Possessing Superior Decision-Making Skills

Although the contribution of pattern recognition and recall to actual performance is still open to debate, the related process of decision making is not. Decision making is focused on how a performer elects to use the perceptual information picked up in the environment to select a response option. The quality of a decision is commonly evaluated relative to its accuracy and speed (faster usually being better). In field or team sports, players are continually confronted with complex tactical scenarios in which they must pick up relevant information from the ball, teammates, and opponents and then make an appropriate decision. These decisions are usually made under pressure as opponents try to limit the time and space available to perform; hence, the ability to process the perceptual information efficiently is an integral skill in such sports and highlights the important contribution this process makes to elite performance.

The most common method used to assess decision making is similar to the previously described film occlusion techniques used to examine anticipatory skill. Video-based game-specific scenarios are presented to the athlete, usually on a large screen, and the task of the performer is to respond to the scenario when the footage is occluded at specific tactical moment in time either by a physical or computer-based response (e.g., mouse click, touch screen) (see figure 15.4). Research typically reveals that elite performers make tactical decisions faster and more accurately than do their subelite counterparts (see table 15.2), attributable to their ability to selectively attend to and more effectively use relevant information in the environment than can subelite performers (Farrow 2010; Helsen and Starkes 1999; Vaeyens et al. 2007).

Consistent with the anticipation literature, the measurement of eye movement patterns has been used in an effort to better understand the informa-

TABLE 15.2 Decision-Making (DM) Accuracy and Speed (Mean ± *SD*) as a Function of Netball Skill Level

Group	DM accuracy, %	DM speed, ms
Open	64.71 ± 10.82	266.17 ± 209.76
U21	57.98 ± 10.03	196.80 ± 218.26
U19	46.00 ± 11.83	454.01 ± 302.30
U17	47.64 ± 12.39	452.72 ±334.40

Adapted, by permission, from D. Farrow, 2010, "A multi-factorial examination of the development of skill expertise in high performance netball," *Talent Development and Excellence* 2(2): 123-135.

tion sources contributing to skilled decision making. It is assumed that the location of each eye fixation represents an area of interest, and the number and duration of fixations provide an indication of the amount of information processed by the performer. Skilled performers are typically found to possess a more efficient visual search strategy, suggesting an enhanced selective attention process (Helsen and Pauwels 1993; Vaeyens et al. 2007; Williams and Davids 1998).

Helsen and Pauwels (1993) examined search patterns of elite (semiprofessional and professional) and novice (limited or no experience) soccer players as they interacted with video simulations of match-specific scenarios. At a specific point in the video, the ball appeared to be passed toward the screen as though it were directed to the participant. The participants then had to decide as quickly as possible whether they would shoot for goal, pass to a teammate, or dribble around an opponent. The elite soccer players, given their superior ability to recognize typical patterns of play, were found to make quicker and more accurate decisions than did the novice players. The eye movement data supported this conclusion, demonstrating that the elite soccer players made fewer fixations of longer duration on selected areas of the display. The elite performers focused on the key defender (or the "sweeper") and any areas of "free" space. Conversely, the subelite performers focused their attention on other areas of the display deemed to be of less importance, such as other attackers, the goal, and the ball.

Subsequent research (Mann et al. 2008; Vaeyens et al. 2007; Williams and Davids 1998) has expanded upon this early work and highlighted that visual search behavior is altered depending on the complexity of the scenario and nature of the display with which a player is confronted. For instance, Vaeyens and colleagues (2007) presented performers with a variety of scenarios (2 attackers vs. 1 defender, 3 vs. 2, 4 vs. 3, and 5 vs. 3) presented from an aerial or bird's eye perspective. Eye fixations increased and

FIGURE 15.4 An example of a decision making test for basketball involving a physical response.

Reprinted, by permission, from D. Farrow and M. Raab, 2008, A recipe for expert decision making. In *Developing sport exercise: Research and coaches put theory into practice*, edited by D. Farrow, J. Baker, and C. Macmahon (United Kingdom: Taylor & Francis), 137-157.

duration of each fixation decreased as the scenarios became more complex (i.e., with a greater number of players). Mann and colleagues (2008) explored the issue of how information pickup for decision making may be influenced by the viewing perspective afforded to the performer for the sport of soccer. The most widely used method has been that of an aerial perspective, where the camera is positioned above and behind the play, providing an aerial view of the scenario. This view has been found to provide the performer with greater access to information in the environment, namely detailed information regarding available space and the position of attackers and defenders in relation to each other. The alternative viewing perspective is that of a "first-person" or performer viewpoint, where the camera is located in the place of an actual player in the filmed scenario, and the play and ultimate decision-making response are structured around the location of this player view (see figure 15.5 for an example). When the decision-

making performances of Australian Institute of Sport (AIS) soccer players were evaluated under these two viewing conditions (identical tactical scenarios), Mann and colleagues (2008) found that the aerial view enhanced the quality of decision making relative to the player's view attributable to the player's greater access to key information, namely the time the player was able to observe open space. Such data highlight the importance of trying to create a representative task as much as possible when examining decision-making skill (see the final section of this chapter for more commentary on this issue).

Superior Anticipation of Event Probabilities

The previous sections have demonstrated that an extensive body of literature exists across a variety of sports demonstrating that elite performers are able to pick up useful information from both the kinematics of an opponent's movement pattern and the structure of common patterns of play to which subelite performers are not attuned. However, the contribution of situational probability information, such as sequential dependencies in an opponent's pattern of play, to anticipation and decision making is also critical.

In a fast-ball sport performance environment, the probability or expectancy of a particular response is affected by factors such as the respective court or field positions of players, relative stroke frequencies, and an opponent's strengths and weaknesses (Abernethy et al. 2001; Buckolz et al. 1988). Although it is difficult to simulate the experiential knowledge of an elite performer, some researchers have implemented simulation studies to manipulate the subjective probabilities available to a performer in an effort to better understand anticipatory performance. It is suggested that elite athletes have a greater understanding of probability information than do novices (i.e., elite athletes may understand better the possibilities an opponent has available and what the opponent's likely choice may be). Cohen and Dearnaley (1962) found that elite soccer players' subjective estimates matched more closely actual event probabilities than did estimates held by novice players, which were more likely to equate to the default situation of equiprobability. Abernethy and colleagues (2001) demonstrated, in simulated squash play, a greater capacity for elite players to move to the correct quadrant of the court to hit a return stroke in the complete absence of advance information from the hitting action of the opposing player. It is apparent that novices are not attuned to these same information sources (or,

FIGURE 15.5 Screen shot displaying (a) player perspective and (b) aerial perspective as seen by participants.

Photos courtesy of Damian Farrow.

at least, not attuned as well as elite performers) but tend to be more reliant on relatively late-occurring current information such as ball flight.

Farrow and Reid (2012) examined the contribution of probability information to the anticipatory responses of skilled tennis players who represented two different stages of development. Participants were required to predict the location of tennis serves presented to them on a plasma touch screen from the perspective of the receiver. Serves were sequenced into a series of games and sets, and a score was presented before each point, typical of a game of tennis. The game score (e.g., 15-0, 30-15) was manipulated to provide advance probability information. Specifically, the serve for the first point of each game was always directed to the same location. A total of 12 service games consisting of 96 points were presented to determine whether players would detect the rela-

tionship between the game score and resultant serve location as measured by response speed and accuracy. Results revealed that older and more skillful players picked up the occurrence of the first point service pattern after the ninth service game, whereas the younger players did not.

Improved Capacity to Dual Task

Demonstration of the elite performer's capacity to execute the primary skills of the sport more proficiently than subelite players is a valuable starting point for elite athlete testing (Pienaar et al. 1998). However, more sensitive approaches are available that have the potential to demonstrate differences between performers who on the surface seem to be of

Elite Athletes Do Not Possess Superior Generic Capacities

For some time research has been conducted to determine the contribution of selected visual parameters to the performance of sport skills (e.g., Blundell 1985). An underlying assumption of this body of research is that the physical capability of the visual system to receive information is a characteristic that may differentiate elite from novice athletes. Generic visual parameters such as static and dynamic visual acuity, peripheral vision, depth perception, visual phoria, and visual reaction time have all been examined to determine their relationship to sport performance (see Abernethy et al. 1998 for a review). The findings of these studies have been equivocal and have generally been unable to demonstrate a systematic, reproducible link between the visual parameter tested and sport performance, particularly as it pertains to the manifestation of expertise (e.g., Starkes 1987; Starkes and Deakin 1984; Ward and Williams 2003).

Hughes and colleagues (1993) found that although elite table tennis players had significantly better dynamic visual acuity, a wider visual field, and superior recognition of peripheral targets compared with lesser-skilled players, the differences accounted for less than 5% of the population variance. Applegate and Applegate (1992) investigated the relationship between degraded visual acuity and basketball set shot shooting performance and found no decline in shooting performance despite decreased visual acuity. Starkes (1987) investigated the relative importance of generic visual attributes compared with sport-specific cognitive attributes when explaining expertise differences within the sport of field hockey. Results demonstrated that none of the generic capacities (simple reaction time, dynamic visual acuity, and coincident anticipation) significantly contributed to the prediction of expertise. Likewise, the generalized visual performance of elite performers has proven not significantly different to others on tests of visual reaction time (Parker 1981), depth perception (Cockerill 1981), ocular dominance (Adams 1965), and color vision (Gavriysky 1970).

Helsen and Starkes (1999), using a multidimensional approach to predict performance differences among elite, intermediate, and novice soccer players, found that 84% of variance was accounted for by sport-specific capacities and that the only generic visual component to contribute even slightly (3%) was peripheral visual range in the horizontal dimension. In summary, as posited by Abernethy and Wood (2001), "Increasingly the consensus is that expert and novice athletes are not characterized by differences in basic visual function" (p. 204). One possible reason for this is that the variety of factors that contribute to sport performance limits the likelihood that vision alone correlates strongly with elite perceptual–motor performance. Any visual deficiency may be compensated for by superior performance in other aspects of performance such as psychological characteristics or physical strength. However, the most likely reason is that generalized (non-sport-specific) tests measure only the visual reception of information rather than the sport-specific perceptual interpretation of visual information; the latter appears to be the critical feature in distinguishing the visual–perceptual skill of expert and novice performers (Abernethy et al. 1998).

a similar skill level. This is important if one considers the demands of many sports, particularly team sports. Players are often required to simultaneously process numerous information sources while executing the skills of the game. For example, a netball player may receive the ball while running and within 3 s have to visually scan the court for passing options, decide between teammates running and calling for the ball, and prepare and execute the physical response of passing the ball accurately to her teammate. The way a player handles the complexity of such a game scenario is usually what separates the elite from subelite.

The most common method used to assess attentional processes as described here is via the dual-task testing paradigm. A participant is required to perform two tasks simultaneously: a primary task for which the attention demand is assessed (usually a core skill of the sport examined) and a secondary task (dual-task condition) for which performance changes are measured and implications drawn regarding the attentional demands of the primary task (Abernethy 1988). Experimental evidence typically reveals that expert players demonstrate dual-task performance superior to that of lesser-skilled players (Parker 1981; Smith and Chamberlain 1992). Such evidence suggests that through task-specific practice, the expert has automated the control of the primary skill and has spare attentional capacity to devote to secondary tasks.

A good example of how the dual task paradigm is applied in a high-performance test setting is described next for the sport of netball. The dual-task passing skill test consisted of three phases. In the primary task, the participant interacted with a near life-size video projection of four netball players (2 teammates and attackers and 2 opponents) involved in a 2v2 competition for the ball. The participant was required to shoulder pass as accurately as possible (instructed to "hit the hands") to the "free" teammate (not defended). A video record of the participant completing the task permitted post hoc analyses of decision-making accuracy and resultant passing accuracy. The

secondary task was a single-choice vocal reaction time test. Specifically, the participant was required say the word *tone* as quickly as possible into a lapel microphone upon hearing a target tone being emitted. The microphone was connected to a customized software program that recorded reaction time (RT) in milliseconds for each of 10 trials to provide a baseline measure to be used in comparison to the dual-task performance. In the dual-task test condition, the participant completed the same primary task, with the order of trials randomized, in addition to listening to a series of auditory tones. Participants were instructed to maintain their primary task performance but also monitor the tones and when a specified target tone was emitted (a tone of 660 Hz relative to 440 Hz) say the word *tone* as quickly as possible. Tones (500 ms in length) occurred once within every 2 s time interval, and the target tone occurred randomly (to the participant) on 10 occasions within the 100 s test; the participant's RT was recorded on each occasion. Secondary task performance was examined by comparing the baseline RT performance of the players to the RTs generated during the passing task (see table 15.3).

Constructing and Implementing a Testing Battery

The relative importance of the previously discussed components of performance will vary between sports. Similarly, pragmatic issues such as the time available to develop a test protocol and equally the time made available for its implementation will be key factors in determining which tests to implement. The following case study details the implementation of a netball-specific perceptual and decision-making test battery. Key features of this approach were that the sport remained committed to the test battery for more than 8 years and importantly sought testing to be completed across a variety of squads reflective of

TABLE 15.3 Netball Passing Skill Performance (Mean ± *SD*) as a Function of Skill Level

	PASS SELECTION, %		PASS ACCURACY, %		REACTION TIME, MS	
Group	Single	Dual	Single	Dual	Single	Dual
Open	86.49 ± 5.73	91.32 ± 5.16	75.55 ± 9.09	79.29 ± 7.41	259.95 ± 32.04	424.54 ± 68.75
U21	88.38 ± 6.54	85.86 ± 8.05	67.32 ± 7.98	66.82 ± 14.30	275.46 ± 43.59	464.80 ± 83.93
U19	85.09 ± 6.30	83.35 ± 5.42	63.33 ± 10.26	63.99 ± 10.08	264.28 ± 50.14	493.24 ± 109.11
U17	86.02 ± 8.33	88.51 ± 7.69	67.06 ± 10.62	73.26 ± 12.05	300.18 ± 42.83	511.43 ± 75.24

Adapted, by permission, from D. Farrow, 2010, "A multi-factorial examination of the development of skill expertise in high performance netball," *Talent Development and Excellence* 2(2): 123-135.

different stages of the developmental pathway (see Farrow 2010 for a more complete review). Similar initiatives have been adopted by other high-performance sport programs including the AIS-AFL Football Academy, AIS-Cricket Australia Centre of Excellence Program, and, more recently, the Tennis Australia High Performance Pathway.

Netball Australia was an early adopter of perceptual and decision-making testing. Since 2002, tests of pattern recall, decision making, netball passing (under a dual task load), and reactive agility have been implemented on a range of squads within the netball high-performance pathway. Eight years of testing permitted the examination of the development of perceptual–cognitive and perceptual–motor skill in highly-skilled netball players drawn from the under-17, under-19, and under-21 years National development squads and the Open National team. Two key objectives were addressed: (1) determine which perceptual–cognitive and perceptual–motor skills differentiated performers of differing skill levels, and (2) determine the relative contribution of these different test components in classifying the squad membership (skill level) of the players.

The tests all contained netball-specific stimuli and typically were administered to athletes over a 2-day period. All the tests were consistent with the tasks described in earlier sections (or see Farrow 2010 for full details). Consistent with previous research (e.g., Helsen and Starkes 1999; Ward and Williams 2003) performance on the pattern recall and decision-making tasks, in particular, demonstrated strong expertise effects. These variables and that of single task passing skill in turn accounted for 77.6% of between-group variance, despite all players' having amassed significant amounts of practice and being considered "skilled." The Open squad was distinguished from all of the remaining squads because of their superior capacity to perceive the netball-specific patterns of play as chunks rather than individual items, pointing to pattern recall as a critical element of elite netball performance. Decision-making results were also consistent with previous research reinforcing that elite performers could be distinguished from the subelite (e.g., Helsen and Starkes, 1999). The Open squad's accuracy was superior to the under-19 and under-17 squads' but not the under 21 squad's. Furthermore, the under-21 squad was discriminated from the under-19 and under-17 squads (who were only 1-2 years younger and less experienced) in relation to both decision-making accuracy and speed. These data suggest that decision-making skill becomes a critical discriminatory factor at a relatively specific stage of netball development and is not linked to biological age (McMorris 1999). Although no sig-

nificant between-group effects were found on the dual-task passing test, single-task passing accuracy was a significant contributor to the prediction of skill level differences, reinforcing the Open squad's general status as the elite group. In contrast to the previously reviewed study by Farrow and colleagues (2005b), the overall results of the reactive agility task did not reveal expertise differences. These findings suggest that reactive agility may not be a distinguishing factor at elite performance levels, although this is inconsistent with results from other research (e.g., Gabbett and Benton 2009; Sheppard et al. 2006). Furthermore, evidence suggested that the Open squad possessed faster decision-making times relative to the under-21 squad; however, this did not translate into faster movement time performance.

Task Representation

A key underpinning philosophy that needs to be considered when designing perceptual and decision-making tests for athletes is that of task representativeness (Brunswik 1956). Put simply, a test protocol needs to closely replicate the conditions occurring in the natural environment. Coupled to this point is the consistent finding in expertise research that elite performance is context specific and consequently it is unlikely that differences between elite and subelite performers will emerge on tasks that do not accurately reproduce the essential characteristics of the task (see Abernethy et al. 1993 and Williams and Ericsson 2005 for more complete reviews). Two key elements of the task that need to be considered from this perspective are the degree of perception–action coupling and the specificity of the stimuli presented. These two aspects of a task are discussed next to illustrate the importance of this principle.

Perception–Action Coupling

The separation of perception from action is a consistent feature within much of the sport expertise research. For example, a common test procedure used to examine the anticipatory skill of a performer in a time-stressed task such as cricket batting is to require the batter to sit and watch a video projection of the fast bowler and then respond verbally or via a paper and pencil response (termed a *decoupled response*). When this task is compared with the real-world setting, obvious differences include the use of a video display and the lack of any sport-specific movement-based response. Although research has demonstrated that elite performers may still maintain an advantage over their lesser-skilled counterparts, it is also appropriately argued that such test conditions have the potential to reduce or completely remove the elite performer advantage (Abernethy et al. 1993). Empirical support of this argument is provided by Mann and colleagues' (2007) meta-analysis of perceptual–cognitive expertise that reported a systematic relationship between display fidelity and perceptual skill. Increases in effect size were apparent as the mode of stimulus presentation became more representative of the "real-world" task. More recently, Mann and colleagues (2010), measuring anticipatory skill in cricket batting, demonstrated that elite performance increased as a function of the representativeness of the task presented. Specifically, the anticipatory skill of the elite batters, but not novices, increased as perception–action coupling increased from (1) verbal response only, (2) lower-body movement only, (3) full-body movement (but no bat), and (4) full-body movement with bat (i.e., the usual batting response) (see figure 15.6).

Generalized or Sport-Specific Stimuli

Consistent with the notion of task-representativeness is the need to consider the specificity of the test stimuli. As discussed in a previous section, some researchers have pursued an intuitive link between general processing capacities such as reaction time

and elite performance (see the section Elite Athletes Do Not Possess Superior Generic Capacities). However, empirical evidence supporting the link between these generic processes and elite performance is sparse. A similar issue exists when performance tests use non-sport-specific stimuli. For example, a common agility testing and training approach is for players to be presented with stimulus lights (typically arrows) that signal for the player to change direction to the left or right as rapidly as possible. It is argued this type of testing and training is analogous to the reactive agility tasks described in a previous section. However, such an approach obviously violates a key principle of task representation and removes a critical information source that an elite performer may use to make the agility responses. Recent evidence from Young and colleagues (2011) supports this argument demonstrating that although the sport-specific reactive agility responses of elite Australian rules footballers were superior to those of a lesser-skilled group, the elite players' responses to a nonspecific stimulus were not. In other words, the removal of sport-specific information, in this case an opposition player trying to dodge away from the participant, and replacement with an arrow signifying the change of direction required (nonspecific stimulus) removed the elite group's performance advantage.

References

Abernethy, B. 1988. Dual-task methodology and motor skills research: Some applications and methodological constraints. *Journal of Human Movement Studies* 14:101-132.

Abernethy, B. 1990. Expertise, visual search, and information pick up in squash. *Perception* 19(1):63-77.

Abernethy, B. 1994. The nature of expertise. In: Serpa, J. Alves, and V. Pataco (eds), *International Perspectives on Sport and Exercise Psychology*. Morgantown, VA: FIT Press, pp 57-68.

Abernethy, B., Baker, J., and Côté, J. 2005. Transfer of pattern recall skills as a contributor to the development of sport expertise. *Applied Cognitive Psychology* 19(6):705-718.

Abernethy, B., Gill, D., Parks, S.L., and Packer, S.T. 2001. Expertise and the perception of kinematic and situational probability information. *Perception* 30(2):233-252.

Abernethy, B., and Russell, D.G. 1987. Expert-novice differences in an applied selective attention task. *Journal of Sport Psychology* 9(4):326-345.

Abernethy, B., Thomas, K.T., and Thomas, J.T. 1993. Strategies for improving understanding of motor expertise (or mistakes we have made and things we have learned!!). In: J.L. Starkes and F. Allard (eds), *Cognitive Issues in Motor Expertise*. Amsterdam: Elsevier Science Publishers, B.V, pp 317-356.

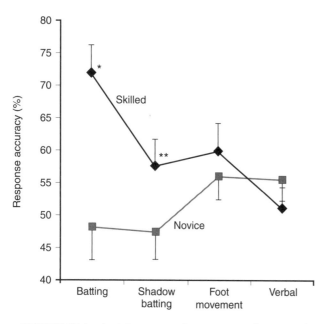

FIGURE 15.6 Anticipatory performance as a function of perception-action coupling in cricket batting.

Reprinted from *Acta Psychologica*, Vol 135(1), D.L. Mann, B. Abernethy, and D. Marrow, "Action specificity increases anticipatory performance and the expert advantage in natural interceptive tasks," pgs. 17-23, copyright 2010, with permission of Elsevier.

Abernethy, B., Wann, J., and Parks, S. 1998. Training perceptual–motor skills for sport. In: B. Elliott and J. Mester (eds), *Training in Sport: Applying Sport Science*. Chichester, UK: John Wiley, pp 1-68,.

Abernethy, B., and Wood, J. 2001. Do generalized visual training programs for sport really work? An experimental investigation. *Journal of Sport Sciences*, 19(3):203-222.

Adams, G.L. 1965. Effect of eye dominance on baseball batting. *Research Quarterly* 36:3-9.

Allard, F., Graham, S., and Paarsulu, M.E. 1980. Perception in sport: Basketball. *Journal of Sport Psychology* 2(1):14-21.

Applegate, R.A., and Applegate, R.A. 1992. Set shot shooting performance and visual acuity in basketball. *Optometry and Vision Science* 69(10):765-768.

Blundell, N. 1985. The contribution of vision to the learning and performance of sports skills: Part 1: The role of selected visual parameters. *Australian Journal of Science and Medicine in Sport* 17(3):3-11.

Brunswik, E. 1956. *Perception and the Representative Design of Psychological Experiments*, 2nd ed. Berkeley: University of California Press.

Buckolz, E., Prapavesis, H., and Fairs, J. 1988. Advance cues and their use in predicting tennis passing shots. *Canadian Journal of Sports Science* 13(1):20-30.

Chase, W.G., and Simon, H.A. 1973. Perception in chess. *Cognitive Psychology* 4:55-81.

Cockerill, I.M. 1981. Distance estimation and sports performance. In: M. Cockerill and W.W. MacGillivary (eds), *Vision and Sport*. London: Stanley Thornes, pp 116-125.

Cohen, J., and Dearnaley, E.J. 1962. Skill and judgement of footballers in attempting to score goals. *British Journal of Psychology* 53(1):71-88.

Draper, J.A., and Lancaster, M.G. 1985. The 505 test: A test for agility in the horizontal plane. *Australian Journal for Science and Medicine in Sport* 17(1):15-18.

Ellis, L., Gastin, P., Lawrence, S., et al. 2000. Protocols for the physiological assessment of team sport players. In: C.J. Gore (eds), *Physiological Tests for Elite Athletes*. Champaign, IL: Human Kinetics, pp 128-144.

Ericsson, K.A., and Kintsch, W. 1995. Long-term working memory. *Psychological Review* 102:211-245.

Farrow, D. 2010. A multi-factorial examination of the development of skill expertise in high performance netball. *Talent Development and Excellence* 2(2):123-135.

Farrow, D., Abernethy, B., and Jackson, R.C. 2005a. Probing expert anticipation with the temporal occlusion paradigm: Experimental investigations of some methodological issues. *Motor Control* 9(3):330-349.

Farrow, D., McCrae, J., Gross, J., and Abernethy, B. 2010. Revisiting the relationship between pattern recall and anticipatory skill. *International Journal of Sport Psychology* 41(1):91-106.

Farrow D, and Reid M. 2012. The contribution of situational probability information to anticipatory skill. *Journal of Science and Medicine in Sport*,, doi:10.1016/j.jsams.2011.12.007

Farrow, D., Young, W., and Bruce, L. 2005b. The development of a test of reactive agility for netball: A new methodology. *Journal of Science and Medicine in Sport* 8(1):52-60.

Gabbett, T., Kelly, J., and Sheppard, J. 2008. Speed, change of direction speed and reactive agility of rugby league players. *Journal of Strength and Conditioning Research* 22(1):174-181.

Gabbett, T., and Benton, D. 2009. Reactive agility of rugby league players. *Journal of Science and Medicine in Sport* 12(1):212-214.

Gavriysky, V. St. 1970. Vision and sporting results. *Journal of Sports Medicine and Physical Fitness* 10(4):260-264.

Goulet, C., Bard, C., and Fleury, M. 1989. Expertise differences in preparing to return a tennis serve: A visual information processing approach. *Journal of Sport and Exercise Psychology* 11(4):382-398.

Helsen, W.F., and Pauwels, J.M. 1993. The relationship between expertise and visual information processing in sport. In: J.L. Starkes and F. Allard (eds), *Cognitive Issues in Motor Expertise*. Amsterdam: Elsevier, pp 109-134.

Helsen, W.F., and Starkes, J.L. 1999. A multidimensional approach to skilled perception and performance in sport. *Applied Cognitive Psychology* 13(1):1-27.

Hughes, P., Blundell, N., and Walters, J. 1993. Visual and psychomotor performance of elite, intermediate and novice table tennis competitors. *Clinical and Experimental Optometry* 76(2):51-60.

Mann, D.L., Farrow, D., Shuttleworth, R., and Hopwood, M. 2008. The influence of viewing perspective on decision-making and visual search behaviour in an invasive sport. *International Journal of Sport Psychology* 40:546-564.

Mann, D.L., Abernethy, B., and Farrow, D. 2010. Action specificity increases anticipatory performance and the expert advantage in natural interceptive tasks. *Acta Psychologica* 135(1):17-23.

Mann, D.T.Y., Williams, A.M., Ward, P., and Janelle, C.M. 2007. Perceptual–cognitive expertise in sport: A meta-analysis. *Journal of Sport and Exercise Psychology* 29:457-478.

McMorris, T. 1999. Cognitive development and the acquisition of decision-making skills. *International Journal of Sport Psychology* 30(2):151-172.

Müller S., Abernethy, B., and Farrow, D. 2006. How do world-class cricket batsmen anticipate a bowler's intention? *Quarterly Journal of Experimental Psychology* 59(12):2162-2186.

Parker, H. 1981. Visual detection and perception in netball. In: I.M. Cockerill and W.W. MacGillivary (eds), *Vision and Sport*. London: Stanley Thornes, pp 42-53.

Pienaar, A.E., Spamer, M.J., and Steyn, H.S., Jr. 1998. Identifying and developing rugby talent among 10-year-old boys: A practical model. *Journal of Sports Sciences* 16(8):691-699.

Savelsbergh, G.J.P., Williams, A.M., Van Der Kamp, J., and Ward, P. 2002. Visual search, anticipation and expertise in soccer goalkeepers. *Journal of Sports Sciences* 20:279-287.

Sheppard, J.M., Young, W.B., Doyle, T.A., and Newton, R. 2006. An evaluation of a new test of reactive agility, and its relationship to sprint speed and change of direction speed. *Journal of Science and Medicine in Sport* 9:342-349.

Singer, R.N., Cauraugh, J.H., Chen, D., Steinberg, G.M., and Frehlich, S.G. 1996. Visual search, anticipation, and reactive comparisons between highly-skilled and beginning tennis players. *Journal of Applied Sport Psychology* 8(1):9-26.

Smith, M.D., and Chamberlain, C.J. 1992. Effect of adding cognitively demanding tasks on soccer skill performance. *Perceptual and Motor Skills* 75:955-961.

Starkes, J.L. 1987. Skill in field hockey: The nature of the cognitive advantage. *Journal of Sport Psychology* 9:146-160.

Starkes, J.L., and Deakin, J. 1984. Perception in sport: A cognitive approach to skilled performance. In: W.F. Straub and J.M. Williams (eds), *Cognitive Sport Psychology*. Lansing, NY: Sports Science Associates, pp 115-128.

Starkes, J.L., and Ericsson, K.A. (eds). 2003. *Expert Performance in Sports*. Champaign, IL: Human Kinetics.

Vaeyens, R., Lenoir, M., Williams, A.M., Mazyn, L., and Phillippaerts, R.M. 2007. The effects of task constraints on visual search behavior and decision-making skill in youth soccer players. *Journal of Sport and Exercise Psychology* 29(2):147-169.

Ward, P., and Williams, A.M. 2003. Perceptual and cognitive skill development in soccer: The multidimensional nature of expert performance. *Journal of Sport and Exercise Psychology* 25:93-111.

Ward, P., Williams, A.M., and Hancock, P.A. 2006. Simulation for performance and training. In: K.A. Ericsson, N. Charness, P.J. Feltovich, and R.R. Hoffman (eds), *The Cambridge Handbook of Expertise and Expert Performance*. Cambridge, UK: Cambridge University Press, pp 243-262.

Williams, A.M., and Davids, K. 1995. Declarative knowledge in sport: A by-product of experience or a characteristic of expertise? *Journal of Sport & Exercise Psychology* 17(3):259-275.

Williams, A.M., and Davids, K. 1998. Visual search in strategy, selective attention, and expertise in soccer. *Research Quarterly for Exercise and Sport*, 69(2):111-128.

Williams, A.M., Davids, K., Burwitz, L., and Williams, J.G. 1994. Visual search strategies in experienced and inexperienced soccer players. *Research Quarterly for Exercise and Sport* 65(2):127-135.

Williams, A.M., and Ericsson, K.A. 2005. Perceptual–cognitive expertise in sport: Some considerations when applying the expert performance approach. *Human Movement Science* 24(3):283-307.

Wright, D.L., Pleasants, F., and Gomez-Meza, M. 1990. Use of advanced visual cue sources in volleyball. *Journal of Sport and Exercise Psychology* 12(4):406-414.

Young, W., Farrow, D., Pyne, D., McGregor and Handke, T. (2011). Validity and reliability of agility tests in junior Australian football players. *Journal of Strength & Conditioning Research* 25(12): 3399-3403.

PART IV

Physiological Protocols for the Assessment of Athletes in Specific Sports

Australian Football Players

David B. Pyne and Stuart J. Cormack

This battery of testing protocols is used to assess the physiological status of Australian Football players. The tests have been selected to represent a range of physical and performance characteristics, including body composition, flexibility, lower-body power, speed, agility, anaerobic capacity, and aerobic endurance. These tests form part of the official AFL National Draft Combine Program and are used widely in the premier competition of the sport, the Australian Football League (AFL). Additional protocols have been suggested for tests that may be conducted where specialized equipment is available.

Australian Football is a physically demanding team sport requiring a combination of highly developed fitness capacities. Players at the elite level typically cover approximately 12 km per match, although this varies slightly depending on position (Coutts et al. 2010; Gray and Jenkins 2010; Pyne et al. 2006; Wisbey et al. 2010). The game is characterized by frequent accelerations, decelerations, changes in direction and heavy physical contact (Coutts et al. 2010; Wisbey et al. 2010). This combination of demands provides a challenge for both the assessment of physical attributes and design of training programs.

Match work rates and the level of development in various physical capacities in Australian Football players are well documented (Appleby and Dawson 2002; Brewer et al. 2010; Dawson et al. 2004a, 2004b; Wisbey et al. 2010). Much of this work relates to new smart sensor technologies that provide estimates of movement patterns and physical demands on game day (Wisbey et al. 2010) and also in training (Dawson et al. 2004a, 2004b). There are also substantial differences in match running performance between elite and subelite levels (Brewer et al. 2010). Collectively this information points to the highly intermittent

and intense nature of Australian Football games and training.

There are several potential benefits of physiological testing with junior and senior Australian Football players. Physiological testing can distinguish between starters and nonstarters at the beginning of an AFL season (Young et al. 2005) and the likely progression from junior to the elite level (Pyne et al. 2005). A number of recent studies have described capacities of elite junior players (Veale et al. 2008, 2009). Details on long-term power development in AFL players (McGuigan et al. 2009), the relationship between power and sprint performance (Young et al. 2011), and speed qualities in elite players (Young et al. 2008) provide useful information for coaches and sport scientists.

The highly demanding nature of Australian Football, the established links between various physical capacities and career progression, and data describing the level of development in various qualities at the elite level justify regular profiling of junior and senior players. Such testing allows objective decisions to be made regarding the design and implementation of programs aimed at improving Australian Football match performance.

Athlete Preparation

Standardized pretest preparation is recommended to enable the collection of reliable and valid physiological data. Refer to chapter 2, Pretest Environment and Athlete Preparation, for specific information relating to athlete and environment preparation.

- *Diet.* Athletes should be encouraged to follow their normal dietary practices in the 24 to

48 h preceding testing. A light breakfast should be consumed before testing conducted in the morning, and similarly a light lunch should be consumed 2 to 3 h before afternoon testing.

• *Training.* Athletes should be tested in a relatively fresh state so that they can give a full, unhindered effort in the maximal effort tests. It is preferred that athletes have only a light skill-oriented session on the evening before testing and that intensive training be conducted after testing.

• *Testing.* Athletes should be familiar with test procedures and protocols before commencement. Written and verbal instructions should be provided as necessary. Athletes being tested for the first time should undertake familiarization trials or sessions. Written informed consent may be required (from the player or parent or guardian) prior to testing in some circumstances.

Test Environment

Anthropometry testing can be conducted in the laboratory or a suitable indoor area (small room) where athletes can be measured in private. The 20 m sprint, vertical jump, agility run, multistage shuttle run, and 6 × 30 m repeat sprint ability tests should all be conducted indoors. The indoor venue should have a properly sprung wooden floor suitable for indoor sports. The surface should be clean, free of dust, and in good condition. The 3 km time trial should be conducted on a properly marked 400 m athletic track, preferably one with a synthetic surface. Environmental conditions should be recorded; if conditions are too adverse, testing should be postponed to another time.

Recommended Test Order

It is important that field and strength tests be completed in the same order to control the interference between tests. This order also allows valid comparison of results from different testing sessions. The order is as follows:

Day	Tests	Notes
1, morning	Anthropometry 20 m sprint AFL agility run Vertical jump	Anthropometry conducted in laboratory. For court testing, divide athletes into four groups and rotate on four stations (two stations for vertical jump testing).
1, afternoon	Repeat sprint ability	
2	Multistage shuttle run or Yo-Yo intermittent recovery test or 3 km time trial	Divide athletes into groups of ~8-10.
3	Strength, power, and competencies testing	Power tests should be undertaken before maximal strength tests.

Equipment Checklist

Anthropometry

- ☐ Stadiometer (wall mounted)
- ☐ Balance scales (accurate to ±0.05 kg)
- ☐ Anthropometry box
- ☐ Skinfold calipers
- ☐ Marker pen
- ☐ Anthropometric measuring tape
- ☐ Recording sheet (refer to the anthropometry data sheet template)
- ☐ Pen

20 m Sprint Test

- ☐ Electronic light gate equipment
- ☐ Measuring tape
- ☐ Field marking tape
- ☐ Witches hats
- ☐ Recording sheet (refer to the sprint data sheet template)
- ☐ Pen

Planned AFL Agility Run

- ☐ Electronic light gate equipment
- ☐ Measuring tape
- ☐ Field marking tape
- ☐ 5 × poles (see planned AFL agility run test procedure)
- ☐ Recording sheet (refer to the basic data sheet template)
- ☐ Pen

Vertical Jump Test

- ☐ Yardstick jumping device (e.g., Swift Performance Yardstick)
- ☐ Measuring tape
- ☐ Field marking tape
- ☐ Recording sheet (refer to the AFL vertical jump data sheet template)
- ☐ Pen

6 x 30 m Repeat Sprint Ability Test

- ☐ Electronic light gate equipment
- ☐ Measuring tape
- ☐ Witches hats
- ☐ Sound box or CD/MP3 player
- ☐ 6 × 30 m RSA test CD/MP3 file
- ☐ Stopwatch
- ☐ Recording sheet (refer to the repeat sprint ability test data sheet template)
- ☐ Pen

Multistage Shuttle Run Test or Yo-Yo Intermittent Recovery Test

- ☐ Measuring tape
- ☐ Field marking tape
- ☐ Witches hats
- ☐ Sound box or CD or MP3 player
- ☐ Multistage shuttle run or Yo-Yo intermittent recovery test CD (or MP3 file)
- ☐ Stopwatch
- ☐ Recording sheet (refer to the multistage shuttle run test or Yo-Yo intermittent recovery test data sheet template)
- ☐ Pen

3 km Time Trial

- ☐ 400 m synthetic running track
- ☐ Stopwatches
- ☐ Recording sheet (refer to the AFL 3 km time trial data sheet template)
- ☐ Pen

Strength and Power Testing

- ☐ Bench
- ☐ Chin-up bar
- ☐ Barbell (Olympic 20 kg)
- ☐ Weight plates (2.5-25 kg increments)
- ☐ Incline leg press or squat rack
- ☐ Recording sheet (refer to the AFL strength and power test data sheet template)
- ☐ Pen
- ☐ Force plate or linear position transducer connected to laptop computer (optional)

ANTHROPOMETRY

RATIONALE

Size and shape of players are important features of the game. Anthropometric measures are useful in talent identification and development and also in management of the training and dietary practices of senior players with mature physiques.

TEST PROCEDURE

Measurement of height, body mass, arm length, hand span, and skinfolds should be carried out prior to field testing protocols. Skinfolds are recorded over seven sites (triceps, biceps, subscapular, supraspinale, abdomen, thigh, and calf). The individual skinfold measures as well as the sum of the seven sites should be reported. Refer to anthropometry protocols outlined in chapter 11, Assessment of Physique, for a detailed description of all anthropometric test procedures.

Although the description of skinfold measurement procedures seems simple, a high degree of technical skill is essential for consistent results. It is therefore important that these measurements be taken by an experienced tester who has been trained in these techniques. It is also important that, where possible, the same tester conduct each retest to ensure adequate reliability.

DATA ANALYSIS

Anthropometric data can be used for serial monitoring (within-subject) of individual players during the preseason and competition season. For talent identification, anthropometric data can be used for cross-sectional (between-subject) comparison of the size and shape of players. Interpretation of data for younger players should account for variations in the time course of physical growth and maturation.

NORMATIVE DATA

Table 16.1 presents anthropometric normative data for Australian Football players.

TABLE 16.1 Normative Data for Australian Football Players

Normative data	Mean ± SD (range)	Typical error
Anthropometric data*		
Height, m	1.86 ± 0.06 (1.67-2.02)	0.01
Body mass, kg	81 ± 8 (63-107)	1.5
Σ7 skinfolds, mm	56 ± 13 (31-113)	1.5
Arm length, cm	82 ± 4 (69-89)	
Hand span, cm	22.9 ± 1.5 (18.9-27.5)	
Vertical jump data, cm*		1.4
Vertical jump: absolute	305 ± 11 (280-336)	
Vertical jump: relative	60 ± 6 (44-80)	
Running jump: right	70 ± 8 (50-95)	
Running jump: left	76 ± 7 (54-94)	
Sprint data, s*		0.03
5 m time	1.09 ± 0.05 (0.95-1.27)	
10 m time	1.81 ± 0.07 (1.65-2.03)	
20 m time	3.04 ± 0.09 (2.82-3.37)	
Other data		
Agility run data, s*	8.57 ± 0.28 (7.89-9.57)	0.13
Multistage shuttle run test (level/shuttle)†	13/5 ± 1/0 (10/2 to 15/7)	4 shuttles
3 km time trial data, min:s†	11:20 ± 0:45 (9:37-15:13)	24 s
Repeat sprint ability data, s†	25.63 ± 0.78 (23.59-27.51)	0.20

All normative data provided are for players of ~18 years of age and will vary for older players in senior football. Data source: *Pyne et al. 2005 (n = 284). †AFL National Draft Combine results for elite 18-year-old players, 2006-2009 (n = 214 players for MSRT; n = 196 players for 3 km time trial, n = 229 players for repeat sprint ability).

VERTICAL JUMP TEST

RATIONALE

The vertical jump is used to assess leg power and jumping ability, which are important elements in AFL when jumping for the ball in marking, scrimmages, or ruck contests.

TEST PROCEDURE

For Australian Football, two variations of the vertical jump test are implemented. The first is the standing vertical jump (with arm swing and countermovement). Refer to the vertical jump protocol in chapter 14, Field Testing Principles and Protocols, for setup details and test procedures for the standing vertical jump. The second variation is the running vertical jump test.

Running Vertical Jump

- Field marking tape is used to place a mark 5 m to the left and right side of the Yardstick (allowing for a straight-line approach).
- Standing reach height is measured as per standing vertical jump test.
- After several of the lower vanes are moved away, the athlete stands at the 5 m mark to commence the test.
- Using an approach run-up, the athlete jumps vertically off the outside leg and reaches as high as possible with the inside hand displacing the vanes at the height of the jump (the action is similar to a ruck contest).
- The takeoff must be from one foot. The left side is taken as the left leg takeoff–right hand jump, and the right side is taken as the right leg takeoff–left hand jump.

- The athlete performs three trials from each side.
- Jump height is recorded as the highest vane displaced for both the left and right sides.
- The difference between maximal jump height and standing reach height is calculated to give the relative jump result in centimeters for both sides.

DATA ANALYSIS

Absolute jump heights are likely to be more important for ruckmen and key position players in contested marking situations. The test results reflect the underlying jumping ability of players in a rested state. Another question is how jumping ability is affected by fatigue, and there is interest in implementing jumps tests to address this. In AFL players, the ratio of flight time to contraction time calculated during a countermovement jump is a sensitive measure of neuromuscular fatigue (Cormack et al. 2008). However, determination of this ratio requires use of a force plate and specialized software that might not be readily available in all programs. Another potentially useful variable for monitoring fatigue status in AFL players is average flight time calculated from five repeated countermovement jumps. Although this variable is a less sensitive fatigue marker, it can be calculated using a simple timing mat (Cormack et al. 2008).

NORMATIVE DATA

Table 16.1 presents vertical jump normative data for Australian Football players.

20 M SPRINT TEST

RATIONALE

This is a simple and easily administered test of the ability to sprint from a stationary start to 20 m. Over this short distance it is likely that the test is primarily a measure of acceleration qualities rather than maximal running velocity.

TEST PROCEDURE

For Australian Football, the sprint test is conducted over 20 m, with intermediate distances of 5 m and 10 m. Athletes start from a stationary, upright position with the front foot on the 0 m point, in line with the start gate. Refer to straight line sprint testing protocol in chapter 14, Field Testing Principles and Protocols, for a description of the procedures involved in sprint testing.

DATA ANALYSIS

Players should be rested and relatively fresh for sprint testing to obtain valid results. A well-structured warm-up is important to limit the risk of injury and promote optimal performance. The 20 m test has a long history in the Australian Football community, and many clubs and programs have established internal reference values. A common variation of this test is the 40 m sprint with timing gates at the 10, 20, 30, and 40 m marks. The longer distance should permit a better estimation of maximal running velocity, computed as the mean velocity between the 30 and 40 m gates, to complement the assessment of acceleration over 5, 10, and 20 m in the standard 20 m protocol.

PLANNED AFL AGILITY RUN

RATIONALE

The run is an AFL-specific planned agility test and used to indicate an athlete's overall agility and ability to change direction at speed around a short twisting course. This planned agility test and other variations are complementary to reactive agility testing.

TEST PROCEDURE

- Two sets of light gates are required, one at the start and one at the finish. The poles (obstacles) are typically made from PVC pipe with a circumference of ~35 cm and are ~1.5 m high. Field marking tape should be placed on the floor at two corners of the obstacle so it can be accurately repositioned if it is knocked over.

- Athletes start from a stationary, upright position with the front foot on the 0 m point, in line with the start gate. Athletes weave in and out of the obstacles or poles as per figure 16.1 and should avoid moving them in any way. If a pole is moved or knocked over completely, the trial is stopped and the athlete must start again (after a brief rest).

NORMATIVE DATA

Table 16.1 presents sprint test normative data for Australian Football players.

- Athletes should complete a short warm-up of light running, dynamic stretching, and some run-throughs. After instruction, athletes should have a practice trial at 50% effort to familiarize themselves with the course.

- Three 100% maximal effort trials are recorded, and the best time (seconds reported to two decimal places) taken as the score. Allow 2 to 3 min recovery between trials.

DATA ANALYSIS

This test is categorized as a planned agility test because it involves a fixed protocol in which athletes know the required movement pattern. The other variant is a reactive agility test whereby the player has to react to an external stimuli (e.g., light signal, another player, video screen). Reactive agility tests are thought to be more game-specific and possibly correlate better with actual performance. Refer to chapter 15, Perceptual–Cognitive and Perceptual–Motor Contributions to Elite Performance, for further detail. Athletes who are slow on the agility run should be given footwork drills, agility drills with and without the ball, and some acceleration and deceleration training. Consid-

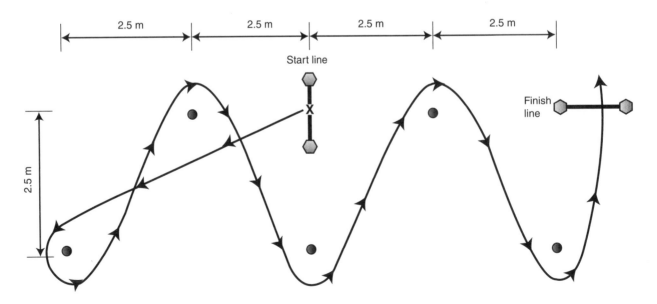

FIGURE 16.1 Schematic diagram of the AFL planned agility run.

eration should also be given to levels of underlying strength and power qualities that can be developed with appropriate resistance training programs.

MULTISTAGE SHUTTLE RUN TEST OR YO-YO INTERMITTENT RECOVERY TEST

RATIONALE

This test provides a simple means of quantifying endurance fitness.

TEST PROCEDURE

Refer to the multistage shuttle run test and Yo-Yo intermittent recovery test protocols in chapter 14, Field Testing Principles and Protocols, for detailed description of this test.

NORMATIVE DATA

Table 16.1 presents normative data for the AFL agility run for Australian Football players.

DATA ANALYSIS

Athletes who score poorly on these tests need additional endurance or aerobic training in the form of longer, slower intervals or shorter, faster intervals. Endurance fitness often improves substantially during the preseason period and is then maintained rather than improved during a long competitive season.

NORMATIVE DATA

Table 16.1 presents normative data for the multistage shuttle run test for Australian Football players.

3 KM TIME TRIAL

RATIONALE

The 3 km time trial is a field test that provides a measure of endurance fitness of AFL players. The test is self-paced, making it complementary to the externally paced multistage shuttle run test or Yo-Yo intermittent recovery tests often used to assess aerobic fitness.

TEST PROCEDURE

- A marked 400 m track is used (7.5 laps × 400 m). A synthetic surface is preferred, although a well-cut and maintained grass track could be a suitable alternative. The ambient temperature, relative humidity, wind speed, and direction should be recorded.
- The test protocol is a single maximal-effort 3 km time trial. Athletes should be given a short warm-up of light running and stretching and basic instructions on pacing.
- Athletes should be encouraged to adopt an even pacing with a fast finish strategy. An athlete expecting to run ~11 min for 3 km should run at ~88 s per lap, or ~92 and ~96 s per lap for 11:30 and 12 min, respectively. A common mistake is to go out too fast on the first lap.
- Athletes should be tested in small groups of 10 to 20 players. Staff members should be

assigned to the following roles: starter, lap counter, one or more spotters for finishing order of players, a spotter for stragglers who are lapped, one or more time keepers, and a recorder.
- Athletes should be given verbal instructions during the 3 km time trial including the number of laps completed or remaining, the elapsed time, and general encouragement on effort and performance.
- The total time in minutes and seconds for the 3 km for each player is reported.

DATA ANALYSIS

The 3 km time trial is a field test that provides a measure of endurance fitness. Some younger players have limited experience with pacing long efforts like this, so some familiarization and prior pacing efforts in training is suggested. The test is self-paced, making it complementary to the externally paced multistage shuttle run test. Athletes scoring poorly on this test need additional endurance (volume) training.

NORMATIVE DATA

Table 16.1 presents normative data for the 3 km time trial run for Australian Football players.

6 × 30 M REPEAT SPRINT ABILITY (RSA) TEST

RATIONALE

Repeated high-intensity efforts are a feature of AFL football. A number of different protocols have been developed in the last decade or so for assessment of the ability to repeat short sprints. The AFL has conducted the 6 × 30 m repeat sprint test at the National Combine camp since 2006. All AFL clubs use some form of repeat effort test to assess this attribute in their players.

TEST PROCEDURE

- The protocol is a 6 × 30 m maximal effort running sprint test on a 20 s cycle. The test is a measure of speed–endurance and repeated sprint ability. A short warm-up of light running and dynamic stretching should be undertaken prior to the test. If this test is conducted after other tests, adequate recovery should be provided.

- Testing is conducted indoors on a polished wooden (sprung) floor or indoor synthetic running track with sufficient space (approximately 70 m of track or floor).

- All athletes should be given appropriate warm-up and instructions for undertaking the test.

- Electronic light gates are used as the timing system. Generally, two or three instrumented lanes are set up in parallel when testing large numbers of athletes.

- Each player will complete one maximal effort trial involving 6 × 30 m sprints from a standing start. The start line is marked 1 m behind the line of the electronic timing gates at both ends of the track. The player is asked to complete each 30 m sprint as a maximal effort.

- The athletes are required to run through at least 10 m past the 30 m timing gate and then return by walking or with a slow recovery jog to the start gate ready for the next effort in the opposite direction. This procedure will apply at both ends of the testing lane.

- Athletes will be given a 10 s and 5 s warning (beep) at which time they must position themself at the starting line and wait for the countdown. The starting action is standardized with a standing start, front foot to the starting line. Athletes should be instructed and reminded not to prepare too early as this often puts them out of balance.

- A prerecorded MP3 track is used to conduct the test, with all timing instructions and signals included for the athlete and testing staff.

- The total time (in seconds) is used as the criterion score. Other measures include comparison of the fastest time against the primary 20 m sprint test time and the percentage decrement in velocity (time) over the test, but these are characterized by poorer reliability.

- The RSA test can be conducted in the same testing session as the other fitness tests (20 m sprint, jumps, and agility), although given its moderately fatiguing nature, it is routinely conducted as a separate test. It is recommended that if undertaken with other tests, the 6 × 30 m RSA test is conducted as the last test of the session.

DATA ANALYSIS

The results of the RSA test are primarily used for evaluation rather than prescription of training speeds. The total time for the six sprints (in seconds) is used as the criterion score rather than calculation of decrement indices such as percentage reduction from fastest to slowest sprint, first to last sprint, or a predicted target time. These indices are probably too noisy for routine use in testing of high level players.

NORMATIVE DATA

Table 16.1 presents normative data for the AFL repeat sprint ability test for Australian Football players.

STRENGTH AND POWER TESTING

RATIONALE

AFL players perform a variety of dynamic high-intensity actions and movements such as sprinting, jumping, and tackling that are influenced by underlying strength and power qualities. A detailed assessment of strength and power levels allows the design of programs to address specific limitations. The degree of complexity of strength and power testing is largely dependent on access to specialized equipment.

TEST PROCEDURE

Maximal Strength Testing: it is common for AFL programs to assess maximal strength with the use of compound (i.e., multijoint) lifts such as the bench press, leg press, chin-ups, and squats. Testing is generally conducted to determine the maximal load that can be lifted for a specified number of repetitions (usually 1-6). The specific exercises and precise repetition range should be determined by availability of equipment and experience of the group. Regardless of the repetitions and exercises selected, it is critical that athletes be completely familiar with the techniques required for each movement. Testing should be performed with athletes in a fatigue-free state and conditions standardized for each test occasion.

Specific guidelines for the conduct of upper- and lower-body maximal strength tests are contained in chapter 13, Strength and Power Assessment Protocols. These guidelines are applicable for the assessment of AFL players. In general, AFL clubs perform a lower-body test (e.g., leg press or squat) and an upper-body pushing (e.g., bench press) and pulling (e.g., chin-ups or bench pull) movement.

Power Testing: There is an important role for the direct measurement of power in AFL players (McGuigan et al. 2009; Sheppard et al. 2008; Young et al. 2011) using a force plate or position transducer. This type of equipment provides data on numerous force–time variables. Similar to maximal strength testing, power performance should be assessed in a fatigue-free state. A high degree of test familiarization and calibration of equipment is critical for this type of assessment.

Specific guidelines for the conduct of a range of power tests, protocols, and equipment considerations are provided in chapter 13. These guidelines are applicable for the assessment of AFL players. AFL clubs that use this form of testing are often interested in determining the ability of the athlete to generate power in various load conditions (e.g., body weight and body weight + extra load), in addition to assessing the ability to use the stretch–shorten cycle (SSC), for example, countermovement jump (CMJ) vs. squat jump (SJ). Although numerous qualities can be calculated (e.g., force, velocity, height), power (as measured in watts or watts per kilogram) may be one of the most useful.

At the elite level, power should be assessed with these tests:

- Countermovement jump at body weight
- Squat jump at body weight
- Countermovement jump with added load (suggested 40 kg)
- Squat jump with added load (suggested 40 kg)

During body weight conditions, the athlete should hold a fiberglass pole or broomstick across the shoulders to allow attachment of the position transducer. In the loaded condition, an Olympic bar with 10 kg weight plates on each side (total weight 40 kg) is used.

Athletes should commence with a dynamic warmup and perform at least three maximal body weight countermovement jumps prior to test trials. Three trials in each condition should be performed and the highest value from each condition used for analysis. The instruction to the athlete should be to jump as explosively as possible.

The squat jump is designed to assess explosive qualities without the contribution of the stretch shorten cycle. In both loaded and unloaded conditions, the athlete should squat to approximately 90° (alternatively, a self-selected knee angle may be used) and hold this position for 3 s before jumping.

The countermovement jump requires a self-selected rate of knee flexion followed by explosive knee extension.

Additional variables can be calculated from the various jump conditions:

- Eccentric utilization ratio: ratio of body weight CMJ to body weight SJ. Values less than 1.0 may suggest an inability to effectively use the SSC.
- Load Tolerance: ratio of loaded to unloaded power in both SJ and CMJ.

DATA ANALYSIS

Detailed strength and power profiling, including data obtained with specialized equipment, is used (particularly in the elite setting) to provide information on the underlying qualities that influence sprinting, jumping, and change of direction performance. It is common practice to conduct strength and power profiling in AFL players at the start and end of preparation phases and also during competitive seasons. The results against reference values (see table 16.2) can be used to identify particular performance limitations and help with design of a specific training program. Athletes must be thoroughly familiarized with test procedures to ensure acceptable levels of test–retest reliability. Values obtained may differ between phases of the year attributable to the different emphasis of training during preparation and competition periods. Results should be interpreted in light of the emphasis of the preceding training block and the possibility of fatigue influencing performance during the season.

NORMATIVE DATA

Table 16.2 presents normative data for maximal strength tests and power tests for Australian Football players.

TABLE 16.2   Maximal Strength and Power Test Data for Australian Football Players

| Normative data | MEAN ± *SD* (RANGE) | | |
	1RM (predicted and direct measure)	3RM (direct)	Typical error
Bench press, kg	~112-140 ± 15 (~1.0 kg per kg body mass)	~100 ± 15 (~1.1-1.4 kg per kg body weight)	2.5
Chin-up (load added to body mass), kg	~32 ± 15 (~0.3-0.4 kg per kg body mass)	~15-30 ± 8 (~0.3-0.4 kg per kg body weight)	1.25
Squat, kg	~100-160 ± 20 (~1.2-1.7 kg per kg body mass)	—	5.0
	Body mass	Body mass + 40 kg	
Squat jump, Watts/kg body mass	69-74 ± 8	52-58 ± 6	<3.5%
Countermovement jump, Watts/kg body mass	69-76 ± 7	59-60 ± 5	<3.5%

Data source: Senior elite AFL players; multiple AFL club data during preparation and competition phases 2006-2010 (*n* = 250 players).

References

Appleby, B., and Dawson, B. 2002. Video analysis of selected game activities in Australian Rules Football. *Journal of Science and Medicine in Sport* 5(2):129-142.

Brewer, C., Dawson, B., et al. 2010. Movement pattern comparisons in elite (AFL) and sub-elite (WAFL) Australian football games using GPS. *Journal of Science and Medicine in Sport* 13(6):618-623.

Cormack, S.J., Newton, R.U., et al. 2008. Neuromuscular and endocrine responses of elite players to an Australian Rules football match. *International Journal of Sports Physiology and Performance* 3(4):359-374.

Coutts, A.J., Quinn, J., et al. 2010. Match running performance in elite Australian Rules Football. *Journal of Science and Medicine in Sport* 13(5):543-548.

Dawson, B., Hopkinson, R., et al. 2004a. Comparison of training activities and game demands in the Australian Football League. *Journal of Science and Medicine in Sport* 7(3):292-301.

Dawson, B., Hopkinson, R., et al. 2004b. Player movement patterns and game activities in the Australian Football League. *Journal of Science and Medicine in Sport* 7(3):278-291.

Gray, A.J., and Jenkins, D.G. 2010. Match analysis and the physiological demands of Australian Football. *Sports Med* 40(4):347-360.

McGuigan, M.R., Cormack, S., et al. 2009. Long-term power performance of elite Australian Rules football players. *Journal of Strength and Conditioning Research* 23(1):26-32.

Pyne, D.B., Gardner, A.S., et al. 2005. Fitness testing and career progression in AFL football. *Journal of Science and Medicine in Sport* 8(3):321-332.

Pyne, D.B., Gardner, A.S., et al. 2006. Positional differences in fitness and anthropometric characteristics in Australian football. *Journal of Science and Medicine in Sport* 9(1-2):143-150.

Sheppard, J.M., Cormack, S., et al. 2008. Assessing the force-velocity characteristics of the leg extensors in well-trained athletes: the incremental load power profile. *Journal of Strength and Conditioning Research* 22(4):1320-1326.

Veale, J.P., Pearce, A.J., et al. 2009. The Yo-Yo intermittent recovery test (level 1) to discriminate elite junior Australian football players. *Journal of Science and Medicine in Sport* 13(3):329-331.

Veale, J.P., Pearce, A.J., et al. 2008. Performance and anthropometric characteristics of prospective elite junior Australian footballers: a case study in one junior team. *Journal of Science and Medicine in Sport* 11(2):227-230.

Wisbey, B., Montgomery, P.G., et al. 2010. Quantifying movement demands of AFL football using GPS tracking. *Journal of Science and Medicine in Sport* 13(5):531-536.

Young, W., Cormack, S., et al. 2011 Which jump variables should be used to assess explosive leg muscle function. *International Journal of Sports Physiology and Performance* 6(1):51-57.

Young, W., Russell, A., et al. 2008. The use of sprint tests for assessment of speed qualities of elite Australian rules footballers. *International Journal of Sports Physiology and Performance* 3(2):199-206.

Young, W.B., Newton, R., et al. 2005. Physiological and anthropometric characteristics of starters and non-starters and playing positions in elite Australian Rules football: a case study. *Journal of Science and Medicine in Sport* 8(3):333-345.

Basketball Players

David B. Pyne, Paul G. Montgomery, Markus J. Klusemann, and Eric J. Drinkwater

The testing protocols outlined in this chapter have been used extensively with elite junior players in programs of Basketball Australia. Various protocols have been established to assess the basketball-specific status of players preparing for national and international competitions. These performance, physiological, and physical tests include body composition, upper- and lower-body strength, lower-body power, physical competencies, speed, agility, repeat effort ability, and aerobic endurance. Similar tests and testing programs have evolved elsewhere in the international game, particularly in the United States, for example, the NBA Combine test data (Kamalsky 2010) and SPARQ (Wood 2010), and in Europe. Some of this material has been published in the scientific literature, but a substantial amount of work goes unpublished outside of the public domain. Other tests will evolve, and there are many variations of simple tests, giving coaches and scientists scope to develop more meaningful and useful tests. The comparative data on elite junior basketball players detailed here should provide a valuable resource for the basketball community.

The fundamental demands of basketball competition involve physiological contributions from both aerobic and anaerobic energy systems (Delextrat and Cohen 2008; Drinkwater et al. 2008; Ziv and Lidor 2009). A discrete set of physical competencies that translate into effective offensive and defensive performance during play are important as well. Basketball fitness also centers on strength (Drinkwater et al. 2008) to maintain the required musculoskeletal resilience to meet the demands of single and repeated games of tournament competition. The detrimental effect of fatigue experienced after basketball play has been assessed using many of the protocols described here (Montgomery et al. 2008a). A recent case study revealed that accumulated fatigue can negatively affect shooting performance (Erculi and Supej 2009). The relationships between fitness characteristics and game performance indicators are poorly understood, and the tests described in this chapter should be useful in this regard.

A comprehensive set of physiological and physical test protocols provides coaches and practitioners with information regarding the current fitness status of their players and short- to long-term improvement or impairment in key performance areas (Drinkwater et al. 2005, 2007; Montgomery et al. 2008b). Age and positional role differences in fitness performance in basketball are particularly evident in intermittent high-intensity endurance and agility performance (Abdelkrim et al. 2010b; Delextrat and Cohen 2009). It is clear that training should be individualized when dealing with positional roles in basketball. Interpretation of test data can influence decisions on training prescription to enhance fitness and performance but also career longevity and reduce the number of missed games through injury. Senior programs and teams need to establish their own normative data for interpreting test results.

The benefits of a testing program are underpinned by specificity of the tests that measure key attributes of basketball and the efficiency with which the tests are administered to the entire playing roster. The tests selected here reflect recent and contemporary practice in high-level junior basketball and published scientific research. For example, cycle ergometer tests used in the 1990s have been superseded by the line drill test, which has greater relevance to on-court demands and discriminates seasonal performance (Montgomery et al. 2008a). Similarly, for aerobic assessment, the intermittent nature of the Yo-Yo intermittent recovery test is representative of game play and correlates highly with laboratory-based aerobic testing (Bangsbo et al. 2008; Castagna et al. 2008; Krustrup et al. 2003). The agility drill has been revised to incorporate greater functional aspects of play and new international court markings. Although the lane agility test used in the NBA Combine also has extensive historical data to assess how quickly a player moves around the key, the use of backward running in that test is questionable with respect to specificity, particularly in defensive play around the basket. The revised basketball agility test incorporates selected movements implemented in a controlled environment emphasizing speed and footwork around the key.

Athlete Preparation

Standardized pretest preparation is needed to obtain reliable and valid physiological data. Refer to chapter 2, Pretest Environment and Athlete Preparation, for specific information relating to athlete and environment preparation.

- *Diet.* Athletes should be encouraged to follow their normal dietary practices in the 24 to 48 h preceding testing. A light breakfast should be consumed before testing conducted in the morning, and similarly a light lunch should be consumed 2 to 3 h before afternoon testing.

- *Training.* Athletes should be tested in a relatively fresh state so that they can give a full unhindered effort in the maximal effort tests. It is preferred that athletes have only a light skill-oriented session on the evening before testing and that intensive training be conducted after testing.

- *Testing.* Athletes should be familiar with test procedures and protocols before commencement. Written and verbal instructions should be provided as necessary. Athletes being tested for the first time should undertake familiarization trials or sessions, when possible. Written informed consent may be required (from the player or parent or guardian) prior to testing in some circumstances.

Test Environment

Anthropometry testing should be conducted in a suitable indoor area (small room) where athletes can be measured in private. The anaerobic capacity and power, flexibility, and aerobic performance tests should all be conducted indoors, preferably on a basketball court with a properly sprung wooden floor. The surface should be clean, free of dust, and in good condition. Environmental conditions should be recorded; if conditions are too adverse, testing should be postponed to another time.

Recommended Test Order

It is important that field and strength tests be completed in the same order to control the interference between tests. This order also allows valid comparison of different test occasions. The order is as follows:

Day	Tests
1 or morning session	Anthropometry
2 or afternoon session	Vertical jump
	20 m sprint
	Agility
	Sit and reach
	Basketball line drill
	Yo-Yo intermittent recovery level 1
3	Movement competencies
	Upper-body strength
	Lower-body strength

Equipment Checklist

Anthropometry

- [ ] Stadiometer (wall mounted)
- [ ] Balance scales (accurate to ±0.05 kg)
- [ ] Anthropometry box
- [ ] Skinfold calipers (Harpenden skinfold calipers)
- [ ] Marker pen
- [ ] Anthropometric measuring tape
- [ ] Recording sheet (refer to the anthropometry data sheet template)
- [ ] Pen

Vertical Jump Test

- [ ] Yardstick jumping device (e.g., Swift Performance Yardstick)
- [ ] Measuring tape
- [ ] Recording sheet (refer to the vertical jump data sheet template)
- [ ] Pen

20 m Sprint Test

- [ ] Electronic light gate equipment (e.g., dual-beam light gates)
- [ ] Measuring tape
- [ ] Field marking tape
- [ ] Witches hats
- [ ] Recording sheet (refer to the sprint test data sheet template)
- [ ] Pen

Basketball Agility Test

- [ ] Electronic light gate equipment (e.g., dual-beam light gates)
- [ ] Measuring tape
- [ ] Field marking tape
- [ ] Recording sheet (refer to the agility test data sheet template)
- [ ] Pen

Sit and Reach Test

- [ ] Sit and reach box
- [ ] Measuring slide
- [ ] Recording sheet (refer to the basic data sheet template)
- [ ] Pen

Basketball Line Drill

- [ ] Stopwatch
- [ ] Recording sheet (refer the basic data sheet template)
- [ ] Pen

Yo-Yo Intermittent Recovery Test

- [ ] Measuring tape
- [ ] Field marking tape
- [ ] Witches hats
- [ ] Sound box or CD player
- [ ] Yo-Yo intermittent recovery test CD
- [ ] Recording sheet (refer to the Yo-Yo IRT data sheet template)
- [ ] Pen

Functional Movement Screens

- [ ] Wooden dowel (approximately 2 m long)
- [ ] Adjustable hurdle or obstacle
- [ ] Field marking tape
- [ ] Measuring tape
- [ ] Recording sheet (refer to the basketball functional movement screening data sheet template)
- [ ] Pen

Strength Tests

- [ ] Bench press bench and rack
- [ ] Bench-pull bench
- [ ] Squat rack or power cage
- [ ] Barbell (Olympic 20 kg)
- [ ] Weight plates (2.5-25 kg increments)
- [ ] Recording sheet (refer to the strength and power test data sheet templates)
- [ ] Pen

ANTHROPOMETRY

RATIONALE

Anthropometry is a traditional method to assess an athlete's body composition. This testing enables coaches and support staff to track and evaluate changes in an athlete's body composition over time (a within-subject comparison) and to compare anthropometric data across a group of athletes (between-subject comparisons). A detailed overview of anthropometric data of basketball athletes is available (Drinkwater et al. 2007; Drinkwater et al. 2008).

TEST PROCEDURE

Measurement of height, body mass, arm length, hand span, and skinfolds should be carried out prior to court testing protocols. Skinfolds are recorded over seven sites (triceps, biceps, subscapular, supraspinale, abdominal, front thigh, and medial calf). The individual skinfold measures as well as the sum of the seven sites should be reported. Refer to anthropometry protocols outlined in chapter 11, Assessment of Physique, for a detailed description of all anthropometric test procedures. More advanced anthropometric assessment including muscle girths, bone breadths, and limb lengths can be conducted if required.

Although the description of skinfold measurement procedures seems simple, a high degree of technical skill is essential for consistent results. It is therefore important that these measurements be taken by an experienced tester who has been trained in these techniques. Where possible, the same tester should conduct each retest to ensure reliability.

DATA ANALYSIS

An anthropometric assessment gives a detailed overview of a player's body composition. Changes in body composition targeted through specific training can be tracked through repeated anthropometric measures. This approach evaluates the effect of various training interventions on the athlete's body composition. Depending on the training goal, the focus should be on changes in lean muscle mass or skinfolds rather than purely on weight gains or losses.

NORMATIVE DATA

Tables 17.1 and 17.2 present anthropometric normative data for female and male junior basketball players.

TABLE 17.1 Normative Data for Elite Junior (U20) Female Basketball Players

Normative data for females	Mean ± *SD* (range)	Typical error*
Anthropometric data[†]		
Height, m	182.8 ± 7.6 (158.8-204.0)	1.0 cm
Body mass, kg	75 ± 11 (47-118)	1.5
Σ7 skinfolds, mm	92 ± 20 (47-165)	~1.5
Vertical jump data[†]		
Distance, cm	47 ± 6 (29-65)	1.4
Sprint data, s[†]		0.04 s
5 m time	1.17 ± 0.07 (0.94-1.45)	
10 m time	1.98 ± 0.09 (1.73-2.35)	
20 m time	3.41 ± 0.15 (3.06-4.05)	

Normative data for females	Mean ± SD (range)	Typical error*
Basketball agility data, s[†]		
Right	5.83 ± 0.25 (5.58-6.08)	0.11
Left	5.80 ± 0.23 (5.57-6.03)	0.11
Line drill time	30.2 ± 1.3 (27.4-38.5)	0.25
Sit and reach		
Distance, cm	14 ± 7 (–3.5 to 30)	1.4
Yo-Yo test data		
Level	16.4 ± 1.3 (12.9-20.2)	~4 shuttles
Distance, m	1,240 ± 400 (280-2,440)	110
3RM squat		
kg	70 ± 13.9 (40-107.5) n = 147	4.5%
Ratio to body mass	0.91	
3RM bench press		
kg	45 ± 7.4 (30-72.5) n = 171	3.1%
Ratio to body mass	0.58	
3RM bench pull		
kg	49 ± 6.4 (35-72.5) n = 170	5.8%
Ratio to body mass	0.61	

*Typical error given in same units as normative data unless otherwise indicated.

[†]Data sources: 2010 AIS Basketball Physical Testing Database (500 female test results recorded 2003-2010): anthropometric data, 17.7 (14.2-27.7); vertical jump data, 17.3 (14.2-20.5); sprint data, 17.3 (14.2-20.5); agility test and line drill data—20 tests per group, 17.3 (14.2-20.5); sit and reach data, 17.3 (14.2-20.5); Yo-Yo test data, 17.3 (14.2-20.5); 3RM test data, 25 recreationally trained males.

TABLE 17.2 Normative Data for Elite Junior (U20) Male Basketball Players

Normative data for males	Mean ± SD (range)	Typical error*
Anthropometric data[†]		
Height, m	197.5 ± 9.1 (174.1-221.8)	1.0 cm
Body mass, kg	91 ± 12 (59-141)	1.5
Σ7 skinfolds, mm	57 ± 17 (34-220)	~1.5

(continued)

TABLE 17.2 *(continued)*

Normative data for males	Mean ± *SD* (range)	Typical error*
Vertical jump data†		
Distance, cm	65 ± 8 (47-94)	1.4
Sprint data, s†		**0.04**
5 m time	1.10 ± 0.07 (0.94-1.37)	
10 m time	1.84 ± 0.09 (1.43-2.25)	
20 m time	3.13 ± 0.15 (2.80-3.78)	
Basketball agility data, s†		
Right	5.30 ± 0.3 (5.00-5.60)	0.11
Left	5.24 ± 0.26 (4.98-5.50)	0.11
Line drill time	27.6 ± 1.3 (25.0-32.4)	0.25
Sit and reach		
Distance, cm	13 ± 7 (−15 to 29.5)	1.4
Yo-Yo test data		
Level	18.8 ± 1.8 (14.6-22.7)	~4 shuttles
Distance, m	2,000 ± 562 (680-3,280)	110
3RM squat		
kg	103 ± 23.0 (40-140) n = 61	4.5%
Ratio to body mass	1.1	
3RM bench press		
kg	72 ± 12.4 (40-100) n = 74	3.1%
Ratio to body mass	0.78	
3RM bench pull		
kg	49 ± 6.4 (50-100) n = 170	5.8%
Ratio to body mass	0.61	

*Typical error given in same units as normative data unless otherwise indicated.

†Data sources: 2010 AIS Basketball Physical Testing Database (500 male test results recorded 2003-2010): anthropometric data, 17.8 (14.4-22.7); vertical jump data, 17.6 (14.4-20.1); sprint data, 17.6 (14.4-20.1); agility test and line drill data = 20 tests per group, 17.6 (14.4-20.1); sit and reach data, 17.6 (14.4-20.1); Yo-Yo test data, 17.6 (14.4-20.1); 3RM test data, 25 recreationally trained males.

VERTICAL JUMP TEST

RATIONALE

Vertical jumping is a common feature of offensive and defensive movements in basketball competition (Abdelkrim et al. 2007, 2010a; McInnes et al. 1995; Narazaki et al. 2009). Vertical jump ability has been linked to an athlete's playing time (Hoffman et al. 1996) and is greater in national-level athletes than state-level athletes (Drinkwater et al. 2007). During basketball play overall jump height with arms extended is important, and a jump is often preceded by movement of one or both feet during situations of rebounding or jump-shot shooting. The vertical jump with countermovement and arm swing is a traditional test to assess an athlete's vertical jumping ability. Therefore this test is familiar to players and representative of game demands.

TEST PROCEDURE

Refer to vertical jump protocol in chapter 14, Field Testing Principles and Protocols, for setup details and test procedures.

DATA ANALYSIS

The vertical jump height assessed in this test is dependent on lower-limb power as well as the optimal coordination of all muscle groups involved in this movement. The results from this test can indicate a change in these two factors. The test does not involve a cognitive component often required in basketball competition (e.g., timing of jump, anticipation of ball flight). This limitation must be considered when interpreting results.

NORMATIVE DATA

Normative data for junior female and male basketball players are presented in tables 17.1 and 17.2.

20 M SPRINT TEST

RATIONALE

The 20 m sprint test determines an athlete's acceleration and speed over a short distance. Basketball athletes usually only conduct high intensity efforts for a very short distance (or duration) (McInnes et al. 1995; Bishop and Wright 2006; Abdelkrim et al. 2007; Narazaki et al. 2009). Similar to the vertical jump, an athlete's sprinting ability has been linked to playing time (Hoffman et al. 1996) and level of competition (Drinkwater et al. 2007).

TEST PROCEDURE

For basketball, the sprint test is conducted over 20 m, with intermediate distances of 5 m and 10 m. Athletes start from a stationary, upright position with the front foot on the 0 m point, in line with the start gate. Refer to straight line sprint testing protocol in chapter 14 for a description of the procedures involved in sprint testing.

DATA ANALYSIS

The 20 m sprint is dependent on an athlete's lower-body strength, power and sprinting technique. The 5 m and 10 m splits give an indication of the athlete's ability to accelerate quickly from a standing start position. The final time gives an overall measure of the athlete's sprinting ability specific to basketball (short distances). Maximal effort should be encouraged as motivation can influence the outcome of the sprint result. Providing immediate feedback (previous sprint times) can challenge and motivate the athlete to surpass the previous best result.

NORMATIVE DATA

Tables 17.1 and 17.2 present sprint test normative data for junior female and male basketball players.

BASKETBALL AGILITY TEST

RATIONALE

Although the commonly accepted definition of agility involves both mental and physical components (Sheppard and Young 2006), the agility test outlined in this section focuses solely on the physical component of agility. From a physical point of view, agility is the ability to conduct sudden changes in direction and velocity as quickly as possible (Young and Farrow 2006). This basketball agility test is derived from a "diamond agility drill" and has been modified to incorporate the new key format introduced by the international governing body of basketball (FIBA) in 2010.

The basketball agility test measures an athlete's ability to execute frequent changes in direction specific to the movement patterns seen in basketball competition. The athlete undertakes three maximal trials going toward both the left and the right sides. The test involves forward acceleration with four changes in direction. Electronic light gates are used to ensure accurate timing and are placed at the start–finish line of the agility test.

TEST PROCEDURE

- Using a measuring tape and field marking tape, mark the points for the start–finish line on the baseline as shown in figure 17.1. The marking points are placed a distance of 1.5 m from the intersection of the key with the baseline, leaving a 1.9 m start–finish line. Place the timing gates on the markers. Mark the center of the start–finish line.

- The markers for change of direction are 30 cm × 30 cm squares named *pivot boxes*. Using the field marking tape, place these boxes on either side of the key, 3.15 m from the baseline, and on points located at 1.8 m from the short elbow (where free-throw line and half-circle meet).

- Check the timing gates are operating correctly before the agility test proceeds.

- After providing instructions on the test procedures, allow each athlete to perform a 50% effort practice run to become familiar with the requirements of the test. Instruct the athlete to assume a stationary starting position, with the front foot up to and in the center of the starting line. The athlete should be instructed to "face the first pivot box" on their right or left and told that "the first movement is forward" to ensure a correct and standardized start.

- After a signal that the light gates are set, the athlete should start when ready. The light gate will start timing automatically as the athlete breaks the beam.

- The athlete sprints toward the first pivot box on the side of the key. A foot must be placed within each pivot box to ensure a valid trial. The athlete changes direction sprinting diagonally toward the upper pivot box, proceeds in a forward sprint to the opposite upper pivot box, sprints diagonally towards the final pivot box, and finishes with a sprint through the finish line (figure 17.1).

- Ensure that the athlete does not slow down before the finish gate and keeps sprinting right through.

- Record the total course time to the nearest 0.01 s on the appropriate recording sheet.

- Allow athletes three trials going towards the right and left side.

- Allow adequate rest, at least 30 s, between consecutive sprints. Rotate athletes one sprint effort at a time as part of a larger group.

- Record the criterion times as the fastest trial on the right side and left side reported to the nearest 0.01 s.

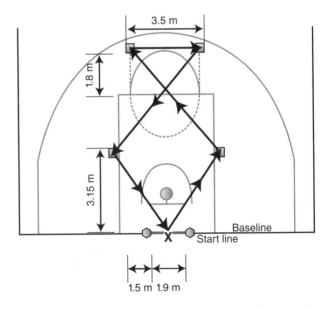

FIGURE 17.1 Schematic diagram of the basketball agility test.

DATA ANALYSIS

The basketball agility test is a practical method to assess an athlete's physical ability to accelerate, decelerate, and make rapid changes in direction. The test is a preplanned task and therefore does not involve a cognitive component, which should be considered when comparing results to on-court performances. Development of acceleration, deceleration technique, turning footwork, and lower-body power will improve agility test scores.

NORMATIVE DATA

Tables 17.1 and 17.2 present normative data for the basketball agility test for junior female and male basketball players.

SIT AND REACH TEST

RATIONALE

The sit and reach test measures the overall stretch length of the calf, hamstring, gluteal, and lower-back extensor muscles and passive tissues. This test gives an indication of an athlete's range of movement around the lower back and pelvis.

TEST PROCEDURE

- Prior to this test the athlete should stretch their hamstrings and lower back.
- The athlete sits on the floor in front of the sit and reach box with the legs fully extended, placing the bare feet against the vertical surface of the box.
- One hand should be placed over the top of the other with the palms facing down, fingertips overlapping, and fingers outstretched, and the elbows should be straight.
- The athlete leans forward as far as possible, sliding his or her hands along the ruler of the sit and reach box. It may be necessary for the tester to apply gentle pressure above the knees to ensure leg extension is maintained.

Full stretch must be held for at least 2 s to avoid bouncing.

- The distance the fingers pass beyond the toes is recorded as a positive score. If the fingers fall short of the toes or zero line, the score is recorded as a negative score. The best of three trials is recorded to the nearest 0.5 cm.

DATA ANALYSIS

The results from the sit and reach test should be used for within-athlete comparisons to evaluate changes in flexibility. Other factors not related to the athlete's flexibility will influence the test score. These factors include anthropometric characteristics of arm length, trunk length, and leg length. Consider any current or recent history of tightness, injury, or restrictions in the hamstring, gluteal, and lower-back muscle groups when interpreting results. Increasing range of motion and flexibility will improve test results.

NORMATIVE DATA

Tables 17.1 and 17.2 present normative data for the sit and reach test for junior female and male basketball players.

BASKETBALL LINE DRILL

RATIONALE

The basketball line drill is a sport-specific test to evaluate the anaerobic glycolytic capacity of an athlete. In performance terms, the basketball line drill tests the speed–endurance of a player in a basketball-specific context of transition play (getting up and down the court as quickly as possible).

TEST PROCEDURE

- Following a standard warm-up, the subject assumes a standing start position, at the chosen base line.

- On the start command, the stopwatch is started and the athlete is required to run as fast as possible to the closest free throw line, back to the base line, to the center line, back to the base line, to the distant free throw line, back to the base line, to the opposite base line, and finally to the base line where the test commenced.
- The stopwatch is stopped when the athlete crosses the base line on the last leg.
- The athlete must touch the lines with one foot and should run through the line on the last leg of the test without slowing.

- The time to completion is recorded to the nearest 0.01 s on the appropriate recording sheet.
- Each athlete should complete one trial.

DATA ANALYSIS

For the basketball line drill, the athlete is required to conduct multiple sprints with several rapid decelerations and accelerations. The results are therefore influenced by agility as well as straight line sprinting components. Having two athletes run at a time encourages the athletes to compete and give a maximal effort for this test. Compliance with the testing protocol (touching the line) needs to be monitored closely to ensure a valid result. If a line is missed, the test is terminated and the athlete can have a second trial after sufficient rest (minimum of 5 min). Strong verbal encouragement, especially toward the end of this test, can assist in the athlete pushing through associated fatigue. An enhanced anaerobic capacity and improved sprinting and turning technique will yield better results.

NORMATIVE DATA

Tables 17.1 and 17.2 present normative data for the basketball line drill for junior female and male basketball players.

YO-YO INTERMITTENT RECOVERY TEST

RATIONALE

The Yo-Yo intermittent recovery test (IRT) is now widely used in team sports to assess the endurance of team sport athletes. Performance in the Yo-Yo IRT level 1 has shown to significantly correlate with laboratory-based aerobic testing ($\dot{V}O_2$max) (Krustrup et al. 2003; Bangsbo et al. 2008) and basketball related endurance (Castagna et al. 2008).

TEST PROCEDURE

Refer to the Yo-Yo IRT protocol in chapter 14, Field Testing Principles and Protocols, for a detailed description of this test. This test should be completed at the end of the basketball testing session, because fatigue associated with this test may affect performance in the speed and power tests. Alternatives to the Yo-Yo IRT include the multistage shuttle run test and the cross-court sprint recovery (SPARQ).

DATA ANALYSIS

The aerobic fitness assessed in the Yo-Yo IRT involves elements of agility because it requires the athlete to accelerate, decelerate, and turn frequently. Even pacing should be encouraged to ensure maximal distances. A rapid start (quick 3 steps) to hit the turn line at the beep, a low turn, and then a second rapid acceleration to the finish line are useful cues for the athlete. Reminding athletes to turn on opposite legs is another useful cue to prevent unilateral leg fatigue. It is important that the athlete continues to perform the test until he or she is informed to stop by the instructor. Development of aerobic fitness and running and turning technique will result in improvements in this test.

NORMATIVE DATA

Tables 17.1 and 17.2 present normative data for the Yo-Yo intermittent recovery test for junior female and male basketball players.

FUNCTIONAL MOVEMENT SCREENS

RATIONALE

Poor motor control through a range of motion has been implicated in joint injuries as well as ongoing joint pain (Baratta et al. 1988; Wilson et al. 2009). Previous systems of screening athletes for potential risk of injuries have included static posture analysis or flexibility testing such as the sit and reach test. Unfortunately, neither of these systems assesses effective integration of the body segments during movement. That body segments do not work in isolation during functional movements is referred to as *regional interdependence* (Wainner et al. 2007). For example, the ankles, knees, and hips as well as the spinal stabilizers and mobilizers all work together to transfer maximal force into the ground for locomotion. A compromise anywhere in this kinetic chain will lead to dysfunction of the entire movement associated with a higher rate of injury, such as to the knee (Knapik et al. 1991) or lower back (Nadler et al. 1998). Although less obvious than ongoing pain or a sudden injury, even slightly impaired motor control has a negative influence on the accuracy, efficiency, and economy of most sporting movements. For example, individuals with a history of low back pain are known to have lasting impairment of their core muscles (Hodges and Richardson 1999). Such impairments have been linked to reduced running speed despite participants' being pain free (Nadler et al. 1998). Therefore, assessing an athlete's competency to perform movements fundamental to athletic performance is a higher-order examination of an athlete's readiness to train or compete than simple posture screening and flexibility testing.

TEST PROCEDURE

A series of tests have been developed by Cook et al. (2006a, b) to evaluate the quality of an athlete's movements through athletically relevant ranges of motion. These seven tests evaluate midline stability, mobility of the anterior and posterior chains, shoulder mobility, and trunk stability. The functional movement screens are performed in the following order: overhead squat, hurdle step, in-line lunge, shoulder mobility test, active straight-leg raise, trunk stability push-up, and rotary stability test.

Overhead Squat

- The athlete holds a wooden dowel overhead with the shoulders fully flexed and the elbows locked. The grip should be twice shoulder width.
- The athlete stands with feet approximately shoulder width apart and toes pointing forward or slightly out.
- The initial action is for the hips and knees to flex, gradually lowering the body. The heels remain in contact with the floor at all times.
- The lowering of the bar continues until the crease of hips is below the top of the knee.
- The athlete should be able hold this position with the torso remaining upright (parallel to the tibia) and the wooden dowel overhead comfortably.

Hurdle Step

- The athlete begins by standing in front of an obstacle set to knee height with a wooden dowel held across the trapezius in a high-bar back squat position.
- The athlete steps over the obstacle.
- During the step, the foot stepping over the obstacle must remain dorsiflexed.

- The hips, knees, and ankles remain sagittally aligned without pivoting the torso or breaking horizontal alignment of the wooden dowel.
- The action should be performed with both the left and right legs stepping first.

In-Line Lunge

- The athlete stands so that the body is parallel to a 1 m line of tape on the floor. The athlete takes a comfortable stride with both feet remaining on the line of tape.
- The athlete grasps a wooden dowel behind the back so that one arm is reaching overhead as though to touch the nape of the neck (palm prone), the other arm reaching behind the back as though to touch the inferior aspect of the scapula (palm supine).
- The athlete lowers the rear knee to the floor so that it touches the front heel in a lunging movement.
- The wooden dowel should remain in contact with the back of the head, thoracic spine, and sacrum throughout the movement.
- The hips, knees. and ankles remain sagittally aligned without pivoting the torso or breaking vertical alignment of the wooden dowel.
- The athlete should be able to hold this position with the torso remaining upright (parallel to the tibia).
- The action should be performed with both the left and right legs stepping first.

Shoulder Mobility Test

- The athlete reaches behind the back so that one arm is reaching overhead as though to touch the nape of the neck (palm prone), the other arm reaching behind the back as though to touch the inferior aspect of the scapula (palm supine).
- Both hands should form fists.
- The athlete cannot hold a stick or towel or receive any other aid in this position.
- The distance between the upper fist and the lower fist should not exceed one hand length as assessed by measuring tape.
- This distance should be assessed with both the left and right fists being the superior and inferior positions.
- The athlete should be able to comfortably hold this position while the measurements are taken.

Active Straight-Leg Raise

- The athlete lies supine on the ground in anatomical position.
- The athlete raises a leg off the ground, with both knees fully extended and both ankles dorsiflexed; the hips must remain flat (i.e., do not rotate).
- A wooden dowel is used as a vertical plumb line from the ankle (lateral malleolus) of the raised foot to assess the degree of hip flexion that can be achieved. The position of the wooden dowel in relation to the horizontal leg determines the achieved score. A score of 3 is given when the dowel resides between the hip and midthigh, a score of 2 between midthigh and midpatella, and a score of 1 below midpatella.

Trunk Stability Press-Up

- The athlete begins lying in the prone position as though about to perform a push-up. The athlete is supported on his or her elbows. A male athlete will begin with his thumbs in line with his forehead, whereas a female athlete will begin with her thumbs in line with her chin.
- The athlete will press his or her body up off the elbows to be supported on the hands.
- Male athletes unable to perform a repetition of this movement may lower their thumbs to be in line with the chin, whereas female athletes unable to perform a repetition may lower their thumbs to be in line with the clavicles.

Rotary Stability Test

- The athlete starts in a quadruped position with the body parallel to a 1 m piece of tape on the floor.
- The athlete should touch the unilateral knee to elbow without the spine deviating from parallel to the tape on the floor, the knee and elbow touching in line over the tape without hip or shoulder rotation.
- An athlete unable to perform the above movement should try touching the contralateral knee to elbow without the spine deviating from parallel to the tape on the floor, the knee and elbow touching in line over the tape without hip or shoulder rotation.

DATA ANALYSIS

Evidence suggests that validity ranges from substantial to excellent when comparing experienced testers with testers who have received basic instruction (Minick et al. 2010). Ideally, novice assessors should receive training from an experienced assessor, review the pertinent literature (Cook et al. 2006a, b), and practice extensively on a range of athletes. Test–retest agreement across the tests should exceed 85% before an assessor is considered competent. Considering that functional movement screens are scored on a 0 to 3 ordinal scale (described below), it is inappropriate to calculate an assessor's typical error of measurement (TE); error rates of ordinal scale data are best reported by the rate of agreement or weighted kappa scores (Minick et al. 2010).

There are seven functional movement screening tests, each test scored on a scale ranging from 0 to 3:

0   The athlete experiences pain when attempting the movement.

1   The athlete is unable to perform the movement or makes fundamental movement compensations.

2   The athlete performs the movement with minor movement compensations.

3   The athlete is able to perform the movement without the need for any movement compensations.

Because there are seven tests, each test having a maximal score of 3, it may seem that a "perfect" score on the functional movement screening test would be 21 whereas an athlete unable to perform any of the movements would score 0. However, because each functional movement screen assesses a different capacity, the sum of scores across all tests is just one method of interpreting the results. Test scores may also be assessed individually for each test. An athlete may score highly in all tests except one, but if that one test is a key movement pattern to the athlete's sport, the athlete's performance may be substantially impaired even though he appears to score overall quite well according to a summed score.

NORMATIVE DATA

The functional movement screens have not been in widespread use long enough to generate a representative sample of results. We therefore cannot present results that coaches should expect for basketball players. Although the maximal score on this test battery is 21, coaches should consider a summary score to have a lower limit of 14. Data from Kiesel and colleagues (2007) demonstrate that a score below 14 dramatically increased the odds of athlete injury in American football athletes.

STRENGTH AND POWER TESTING

RATIONALE

Based on the rules of the International Basketball Federation (FIBA), basketball is classified as a non-contact sport. There are, however, occasions when body contact between players occurs, such as when players form screens, players seal out an opponent, or centers compete for position under the basket. Therefore, strength plays an important role in a variety of basketball-specific skills (Chaouachi et al. 2009) and body mass plays an important role in basketball success (Drinkwater et al. 2007). Strength is also an integral part of the power equation; although high-force–low-velocity training can reduce contraction velocity (so-called velocity specificity) when that is the only training performed, athletes who maintain their speed training, such as in regular on-court training, improve running speed from strength training (Blazevich and Jenkins 2002).

Aside from tactical advantages, muscular strength can also impart stability around a joint, thereby protecting it from injury. Strength training can improve muscle coactivation and coordination, such as between the vastus lateralis and medialis or between the trunk flexors and extensors, thereby reducing the incidence of joint pain and noncontact injuries to the knee or low back (Baratta et al. 1988; Wilson et al. 2009). Muscular strength has also been strongly linked to bone density (Nilsson and Westlin 1971) and implicated in reducing skeletal injuries such as stress fractures (Bennell et al. 1999). Historically, there is concern regarding strength training in children and adolescents, but these risks are no greater than any other population using resistance training when adolescent athletes are properly supervised (Behm et al. 2008).

TEST PROCEDURE

A large range of different tests are available to assess strength. Given that muscle imbalances are linked with injury, it is important to assess muscular strength of both the upper and lower body as well as in the anterior and posterior planes bilaterally. Because fatigue accumulates quickly during maximal efforts, it is important that as few tests as possible are used. Therefore, the squat, bench press, and bench pull are the lifts to be tested.

Although the 1-repetition maximum (1RM) test is generally considered the best representative of strength, there are concerns about safety, reliability, and maintenance of technique when 1RM testing is conducted for individuals with low training age (Baechle et al. 2008). Therefore, 3RM testing is rec-ommended for athletes not specifically trained for strength.

The tests used in basketball are the 3RM squat, 3RM bench press, and 3RM bench pull. Tests should be completed in this order. Refer to chapter 13, Strength and Power Assessment Protocols, for a detailed description of strength and power tests. Warm-up and submaximal attempts of the bench press and squat should be performed in an appropriate rack with spotters present.

The following general guidelines must be adhered to for all tests:

- Strength testing should be performed on a separate day from the field tests. Strength and field tests should be separated by 48 h.

- The athlete should perform an appropriate warm-up. As a minimum, all athletes are required to perform a trial at approximately 90% of specified repetition maximum for each test. If tested for the first time, the athlete should perform an initial trial at approximately 90% of weight lifted in training.

- Lowering and lifting actions must be performed in a continuous manner. A single rest of no more than 2 s is allowed between repetitions.

- A maximum of 5 min recovery between trials is allowed.

- Minimum weight increments of 2.5 kg should be used between trials. However, increments should be guided by ease of each trial.

- Ideally, specified repetition maximum (RM) test should be completed within 4 trials (not including the warm-up).

- If the athlete is unable to complete tests as per protocol, this should be noted on testing results information, and values should not include any mathematical calculations (e.g., average, typical error).

- A spotter, other than the supervising coach, should be used when necessary.

DATA ANALYSIS

Many basketball athletes are strength tested without any prior experience in resistance training exercises. Furthermore, these players are assessed on movements that they lack the capability to perform and are impeded by lack of familiarity with the movement or lack of flexibility to attain the stipulated joint angles. As a result, athletes are often tested to the maximal range of motion they are capable of achieving rather than to the standards described. As a result, improvements with training may be observed as improvements in lifting technique and range of motion rather than as an increase in the load.

The data presented in tables 17.1 and 17.2 are not position-specific. Given the positive relationship between body mass and strength as well as the positional needs of strength, players in small positions (point guards, shooting guards, and small forwards) are likely to score slightly lower than the mean whereas those in big positions (power forwards and centers) should score above the mean (Drinkwater et al. 2008). To help interpret strength differences relative to size, the ratio of achieved RM to body mass (BM) is presented in tables 17.1 and 17.2. For the bench press and bench pull the ratio should be almost equal, thus indicating a balance between the musculature of pushing (bench press) and pulling (bench pull).

NORMATIVE DATA

Tables 17.1 and 17.2 present normative data for maximal strength tests and power tests for junior female and male basketball players.

References

Baechle, T.R., Earle, T.W., and Wathen, D. 2008. Resistance training. In: T.R. Baechle and R.W. Earle (eds). *Essentials of Strength Training and Conditioning*. Champaign, IL: Human Kinetics, pp 381-412.

Bangsbo, J., Iaia, F. and Krustrup, P. 2008. The Yo-Yo intermittent recovery test: A useful tool for evaluation of physical performance in intermittent sports. *Sports Medicine* 38(1):37-51.

Baratta, R., Solomonow, M., Zhou, B.H., et al. 1988. Muscular coactivation. The role of the antagonist musculature in maintaining knee stability. *American Journal of Sports Medicine* 16(2):113-122.

Behm, D.G., Faigenbaum, A.D., Falk, B., et al. 2008. Canadian Society for Exercise Physiology position paper: Resistance training in children and adolescents. *Applied Physiology Nutrition and Metabolism*, 33(3):547-561.

Ben Abdelkrim, N., Fazaa, S.E., and Ati, J.E. 2007. Time–motion analysis and physiological data of elite under-19-year-old basketball players during competition. *British Journal of Sports Medicine* 41(2):69-75.

Ben Abdelkrim, N., Castagna, C., Jabri, I., et al. 2010a. Activity profile and physiological requirements of junior elite basketball players in relation to aerobic-anaerobic fitness. *Journal of Strength and Conditioning Research* 24(9):2330-2342.

Ben Abdelkrim, N., Chaouachi, A., Chamari, K., et al. 2010b. Positional role and competitive-level differences in elite-level men's basketball players. *Journal of Strength and Conditioning Research* 24(5):1346-1355.

Bennell, K., Matheson, G., Meeuwisse, W., et al. 1999. Risk factors for stress fractures. *Sports Medicine* 28(2):91-122.

Bishop, D.C., and Wright, C. 2006. A time-motion analysis of professional basketball to determine the relationship between three activity profiles: High, medium and low intensity and the length of the time spent on court. *International Journal of Performance Analysis in Sport* 6(1):130-139.

Blazevich, A., and Jenkins, D. 2002. Effect of the movement speed of resistance training exercises on sprint and strength performance in concurrently training elite junior sprinters. *Journal of Sports Sciences* 20(12):981-990.

Castagna, C., Impellizzeri, F.M., Rampinini, E., et al. 2008. The Yo-Yo intermittent recovery test in basketball players. *Journal of Science and Medicine in Sport* 11(2):202-208.

Chaouachi, A., Brughelli, M., Chamari, K., et al. 2009. Lower limb maximal dynamic strength and agility determinants in elite basketball players. *Journal of Strength and Conditioning Research* 23(5):1570-1577.

Cook, E.G., Burton, L., and Hogenboom, B. 2006a. The use of fundamental movements as an assessment of function: Part 1. *North American Journal of Sport Physical Therapy* 1(2):62-72.

Cook, E.G., Burton, L., and Hogenboom, B. 2006b. The use of fundamental movements as an assessment of function: Part 2. *North American Journal of Sport Physical Therapy* 1(3):132-139.

Delextrat, A., and Cohen, D. 2008. Physiological testing of basketball players: Toward a standard evaluation of anaerobic fitness. *Journal of Strength and Conditioning Research* 22(4):1066-1072.

Delextrat, A., and Cohen, D. 2009. Strength, power, speed, and agility of women basketball players according to playing position. *Journal of Strength and Conditioning Research* 23(7):1974-1981.

Drinkwater, E.J., Hopkins, W.G., McKenna, M.J., et al. 2005. Characterizing changes in fitness of basketball players within and between seasons. *International Journal of Performance Analysis in Sport* 5(3):107-125.

Drinkwater, E.J., Hopkins, W.G., McKenna, M.J., et al. 2007. Modelling age and secular differences in fitness between basketball players. *Journal of Sport Sciences* 25(8):869-878.

Drinkwater, E.J., Pyne, D.B., and McKenna, M.J. 2008. Design and interpretation of anthropometric and fitness testing of basketball players. *Sports Medicine* 38(7):565-578.

Erculi, F., and Supej, M. 2009. Impact of fatigue on the position of the release arm and shoulder girdle over a longer shooting distance for an elite basketball player. *Journal of Strength and Conditioning Research* 23(3):1029-1036.

Hodges, P., and Richardson, C. 1999. Altered trunk muscle recruitment in people with low back pain with upper limb movement at different speeds. *Archives of Physical Medicine and Rehabilitation* 80(9):1005-1012.

Hoffman, J.R., Tenenbaum, G., Maresh, C.M., et al. 1996. Relationship between athletic performance tests and playing time in elite college basketball players. *Journal of Strength and Conditioning Research* 10(2):67-71.

Kamalsky, M. 2010. NBA Pre-Draft Measurements. www.draftexpress.com/nba-pre-draft-measurements.

Kiesel, K., Plisky, P., and Voight, M. 2007.Can serious injury in professional football be predicted by a preseason functional movement screen? *North American Journal of Sport Physical Therapy* 2(3):147-158.

Knapik, J.J., Bauman, C.L., Jones, B.J., et al. 1991. Preseason strength and flexibility imbalances associated with athletic injuries in female collegiate athletes. *American Journal of Sports Medicine* 19(1):76-81.

Krustrup, P., Mohr, M., Amstrup, T., et al. 2003. The Yo-Yo intermittent recovery test: Physiological response, reliability, and validity. *Medicine and Science in Sports and Exercise* 35(4):697-705.

McInnes, S., Carlson, J., Jones, C., et al. 1995. The physiological load imposed on basketball players during competition. *Journal of Sports Sciences* 13(5):387-397.

Minick, K.I., Kiesel, K.B., Burton, L., et al. 2010. Interrater reliability of the functional movement screen. *Journal of Strength and Conditioning Research* 24(2):479-486.

Montgomery, P.G., Pyne, D.B., Hopkins, W.G., et al. 2008a. The effect of recovery strategies on physical performance and cumulative fatigue in competitive basketball. *Journal of Sports Sciences* 26(11):1135-1145.

Montgomery, P.G., Pyne, D.B., Hopkins, W.G., et al. 2008b. Seasonal progression and variability of repeat-effort line drill performance in elite junior basketball players. *Journal of Sport Sciences*, 26(5):543-550.

Nadler, S.F., Wu, K.D., Galski, T., et al. 1998. Low back pain in college athletes: A prospective study correlating lower extremity overuse or acquired ligamentous laxity with low back pain. *Spine*, 23(7):828-833.

Narazaki, K., Berg, K., Stergiou, N., et al. 2009. Physiological demands of competitive basketball. *Scandinavian Journal of Medicine and Science in Sports*, 19(3):425-432.

Nilsson, B.E., and Westlin, N.E. 1971. Bone density in athletes. *Clinical Orthopaedics and Related Research* 77:179-182.

Sheppard, J., and Young, W. 2006. Agility literature review: Classifications, training and testing. *Journal of Sport Sciences* 24(9):919-932.

Wainner, R.S., Whitman, J.M., Cleland, J.A., et al. 2007. Regional interdependence: A musculoskeletal examination model whose time has come. *Journal of Orthopaedic and Sports Physical Therapy* 37(11):658-660.

Wilson, N.A., Press, J.M., and Zhang, L.Q. 2009. In vivo strain of the medial vs. lateral quadriceps tendon in patellofemoral pain syndrome. *Journal of Applied Physiology* 107(2):422-428.

Wood, R. 2010. SPARQ Testing for Basketball. www.topendsports.com/sport/basketball/basketball-sparq.htm.

Young, W., and Farrow, D. 2006. A review of agility: Practical applications for strength and conditioning. *Strength and Conditioning Journal* 28(5):24-29.

Ziv, G., and Lidor, R. 2009. Physical attributes, physiological characteristics, on-court performances and nutritional strategies of female and male basketball players. *Sports Medicine* 39(7):547-568.

Cricket Players

Marc Portus, Aaron Kellett, Stuart Karppinen, and Stephen Timms

In recent years the world of cricket has undergone significant change. Accordingly, and so as to maintain its international competitiveness, Australian cricket is rapidly moving on its journey from amateurism to professionalism. The advent of the new form of the game—Twenty/20 (T20)—and its associated competitions, such as the ICC Twenty/20 world cup, the Indian Premier League (IPL), and the Big Bash League (BBL), has placed many new physiological demands on players and presented new challenges for coaches and support staff. The longer forms of the game—One Day Internationals (ODIs), interstate matches (4 days), and test matches (5 days)—are still played in record numbers. For domestic and international cricketers this has resulted in an increasingly busy cricket calendar, with little time for rest and recovery. Consequently, the fitness of cricketers has never been more important to help them achieve long periods of high performance and reduce the risk of injury.

Research into the demands of all three forms of the game has demonstrated that elite cricketers undergo similar amounts of highly intermittent intense work bouts, but with increased rest intervals, when assessed against other field sports such as hockey, rugby, and soccer (Duffield and Drinkwater, 2008; Petersen et al. 2010). Fast bowlers have been reported to travel up to 25 km per day in multiday cricket. Multiday cricket presented a greater overall load than the T20 and ODI formats, but the latter formats involved higher work to rest ratios (Petersen et al. 2009, 2010). Average sprint distances for fast bowlers in all forms of the game were approximately 18 m and these sprints occurred from 42 times a day in the shorter formats up to 66 times a day for multiday cricket. These physiological demands, combined with extended periods of concentration and the requirement to display rapid decision making and explosive efforts (e.g., for a diving stop and throw in the field), present a strong case for the modern cricketer to be athletic.

To help address these demands, the Cricket Australia Sport Science Sport Medicine (SSSM) Unit has implemented a national standards fitness program in Australian cricket. The program consists of a national working group that includes strength and conditioning coaches from each state cricket association, the Cricket Australia Centre of Excellence, and the Australian teams working in a nationally coordinated approach. Similarly, national standards programs have been implemented in other SSSM disciplines (e.g., physiotherapy, medicine, sport psychology, and nutrition) to ensure that a holistic approach is taken to maximize the potential of professional Australian cricketers. A key aspect of the national standards fitness program is the development and implementation of a national protocol so all cricketers, regardless of their state, territory, or gender, have their physiological capacities assessed in a consistent and reliable manner (CANFWG, 2010).

National standards programs bring Australian cricket many benefits, including elimination of the duplication of efforts when a player transitions between various high-performance squads (e.g., State U19 side, Futures League, Centre of Excellence, Australia U19, Australia A), a continuity of service to the cricketers as they progress through the pathway, the ability to longitudinally track players, and a normative data set to establish developmental benchmarks for aspiring professional cricketers. This enables evidence-based, individualized training programs to be implemented to maximize each athlete's potential. This chapter outlines the physiological tests of

Cricket Australia's national fitness standards program and provides a summary of normative data from the 2007–2010 phase of the program.

Athlete Preparation

To ensure test reliability, the following specific pretest conditions should be observed in addition to those outlined in chapter 2, Pretest Environment and Athlete Preparation.

• *Diet.* Athletes should be encouraged to consume a series of high-carbohydrate meals in the 36 h prior to testing and arrive on test day well hydrated. No food, cigarettes, or beverages containing alcohol or caffeine are to be consumed in the 2 h prior to testing.

• *Training.* No training inducing severe physiological or neural fatigue should be undertaken in the 24 h prior to testing. This includes any high-intensity skills or physical training.

• *Testing.* All test sessions should be scheduled at the same time of day to avoid fluctuations in physiological response to the set protocols due to circadian rhythm (Winget et al. 1985). In addition, all tests must be completed in standardized order, with adequate and consistent recovery time between tests. This will promote optimal performance and allow for a valid comparison of results on different test occasions.

Test Environment

Environmental conditions will often influence field testing; therefore, documentation of ambient temperature, humidity, light, and wind is essential. All field tests should be conducted on a nonslip indoor surface.

Recommended Test Order

It is important that tests are completed in the same order to control the interference between tests. This order also allows valid comparison between different test occasions. The recommended order is as follows:

Day	Tests
1	Anthropometry
	20 m sprint
	5-0-5 agility
	Vertical jump
	Yo-Yo intermittent recovery

Equipment Checklist

Anthropometry

- ☐ Stadiometer (wall mounted)
- ☐ Balance scales (accurate to ±0.05 kg)
- ☐ Anthropometry box
- ☐ Skinfold calipers (Harpenden skinfold caliper)
- ☐ Marker pen
- ☐ Anthropometric measuring tape
- ☐ Recording sheet (refer to the anthropometry data sheet template)
- ☐ Pen

20 m Sprint Test

- ☐ Electronic light gate equipment (e.g., dual-beam light gates)
- ☐ Measuring tape
- ☐ Field marking tape
- ☐ Witches hats
- ☐ Recording sheet (refer to the sprint test data sheet template)
- ☐ Pen

5-0-5 Agility Test

- ☐ Electronic light gate equipment (e.g., dual-beam light gates)
- ☐ Measuring tape
- ☐ Field marking tape
- ☐ Witches hats
- ☐ Recording sheet (refer to the agility test data sheet template)
- ☐ Pen

Vertical Jump Test

- ☐ Yardstick jumping device (e.g., Swift Performance Yardstick)
- ☐ Measuring tape
- ☐ Recording sheet (refer to the vertical jump data sheet template)
- ☐ Pen

Yo-Yo Intermittent Recovery Test

- ☐ Measuring tape
- ☐ Witches hats
- ☐ Sound box or CD player
- ☐ Yo-Yo intermittent recovery test CD
- ☐ Recording sheet (refer to the Yo-Yo IRT data sheet template)
- ☐ Pen

ANTHROPOMETRY

RATIONALE

Anthropometric characteristics are easily overlooked aspects of a cricketer's physicality yet are crucial aspects to routinely monitor. This is particularly important for adolescent fast bowlers, for whom growth spurts represent a phase of heightened risk for serious injury such as spondylolysis (Duggleby and Kumar 1997). The nonfunctioning nature of excess body fat can also have detrimental effects on fast bowlers at the point of delivery, where large ground reaction forces have been well documented (e.g., Portus et al. 2004). Greater skinfolds mean greater mass, which transfers directly to greater forces that the body must overcome at the point of delivery. From a practical viewpoint, excessive skinfolds can negatively affect the rate at which athletes store and dissipate heat. Given that cricket is a summer sport played in some of the hottest countries in the world (e.g., India) this factor should not be overlooked.

TEST PROCEDURE

Measurement of height, body mass, and skinfolds should be carried out prior to field testing protocols. Skinfolds are recorded over seven sites (triceps, biceps, subscapular, supraspinale, abdominal, front thigh, and medial calf). The individual skinfold measures as well as the sum of the seven sites should be reported. Refer to anthropometry protocols outlined in chapter 11, Assessment of Physique, for a detailed description of all anthropometric test procedures.

More advanced anthropometric assessment including muscle girths, bone breadths, and limb lengths can be conducted if required.

Although the description of skinfold measurement procedures seems simple, a high degree of technical skill is essential for consistent results. It is therefore important that these measurements be taken by an experienced tester who has been trained in these techniques. It is also important that the same tester conduct each retest to ensure reliability.

DATA ANALYSIS

Skinfold results are commonly used as a practical basis for training and dietary interventions. Whether an event is primarily aerobic or anaerobic, increased fat mass will be detrimental to performance. Moreover, in sports requiring speed or explosive efforts, such as cricket, excess fat, which causes an increase in body mass, will decrease acceleration unless proportional increases in force are applied (Norton and Olds 1996). Excess fat will also increase the load that the musculoskeletal system must absorb during movement, thereby increasing the risk of injury. It is particularly important to monitor skinfolds during the season where physical conditioning is de-emphasized in favor of skills and competition.

NORMATIVE DATA

Table 18.1 presents anthropometric normative data for female and male high-performance Australian cricket players.

TABLE 18.1 Anthropometric Data for Female and Male High-Performance Australian Cricket Players (Mean ± *SD*; Range)

Players	Height, cm*	Body mass, kg[†]	Σ7 skinfolds, mm[‡]
Female			
State squad—national	167.6 ± 4.9 (153.6-176.0)	65.2 ± 8.0 (48.5-97.0)	109.1 ± 32.3 (49.5-213.5)
CA contracted—international	168.3 ± 5.7 (153.6-177.5)	65.4 ± 7.5 (55.2-86.0)	105.2 ± 29.5 (49.6-193.2)
Male			
Australia U19	181.9 ± 6.6 (170.0-196.8)	84.0 ± 8.7 (61.5-103.5)	69.3 ± 17.1 (38.0-122.0)
AIS program (emerging players)	183.2 ± 7.4 (170.0-203.0)	85.7 ± 8.3 (55.0-110.5)	71.6 ± 19.6 (30.2-149.2)
State squad (contracted)—national	183.8 ± 6.8 (169.6-203.0)	86.0 ± 8.7 (63.5-110.5)	70.8 ± 19.6 (30.2-152.8)
CA contracted—international	182.4 ± 7.3 (170.0-197.0)	85.8 ± 8.1 (72.7-102.0)	71.7 ± 16.8 (40.8-125.9)

Typical error: body mass = 1.5 kg; skinfolds = 1.5 mm. Data collected 2007-2010. Female State squad, n = 89*, 173[†], 246[‡]; female CA contracted, n = 46*, 94[†], 122[‡]; male Australia U19, n = 132*, 266[†], 299[‡]; male AIS program, n = 465*, 565[†], 632[‡]; male State squad, n = 445*, 1,213[†], 1,302[‡]; male CA contracted, n = 151*, 200[†], 212[‡].

20 M SPRINT TEST

RATIONALE

Testing for speed is relevant for cricketers given game demands that require players to be explosive and accelerate over short distances. These actions can typically be seen when players attempt to effect a run out in the field; take a quick 1, 2, or 3 runs while batting; stop a ball in the field; or perform the run-up phase of a fast bowler's delivery.

TEST PROCEDURE

For cricket, the sprint test is conducted over 20 m, with intermediate distances of 5 and 10 m. Athletes start the sprint from a standing position with the front foot at the horizontal start line marked 50 cm behind the start (0 m) timing gate. Athletes should be encouraged to start the sprint with their body mass over their front foot, shoulders and hips square in a crouch "ready" position, heel up on back foot, and no rocking. Refer to straight line sprint testing protocol in chapter 14, Field Testing Principles and Protocols, for a description of the procedures involved in sprint testing.

- Athletes must complete a maximum effort over 20 m, cutting the light gate beams at 0, 5, 10, and 20 m.

- Athletes should complete three maximum sprints, with the best split time (in seconds) for each gate and total time for 20 m recorded to the nearest 0.01 s.

- A rest of at least 2 min is required between trials.

DATA ANALYSIS

Cricket generally involves accelerations of distances less than 20 m (Petersen et al. 2010); therefore, understanding acceleration and maximal speed profiles of cricketers at multiple distances specific to match demands allows practitioners to specify their training interventions, based on normative data. Furthermore, these profiles may also allow coaches to best use a player's running speed in specific playing positions. In general terms players with faster speeds over 20 m are well suited to inner fielding positions, as they have an increased chance of fielding the ball in a shorter time to take a catch or run out the batter. The ability to read the play, agility, and throwing mechanics are also important factors to consider for inner field positions.

NORMATIVE DATA

Table 18.2 presents 20 m sprint normative data for female and male high-performance Australian cricket players.

TABLE 18.2 Sprint Data for Female and Male High-Performance Australian Cricket Players: Mean ± SD (Range)

Players	5 m time, s	10 m time, s	20 m time, s
Female			
State squad—national	1.20 ± 0.08 (0.95-1.47)	2.02 ± 0.11 (1.69-2.37)	3.46 ± 0.18 (2.59-3.91)
CA contracted—international	1.19 ± 0.09 (1.02-1.47)	2.00 ± 0.11 (1.81-2.37)	3.43 ± 0.15 (3.12-3.87)
Male			
Australia U19	1.05 ± 0.07 (0.8-1.26)	1.78 ± 0.10 (1.01-2.01)	3.04 ± 0.11 (2.78-3.35)
AIS program (emerging players)	1.05 ± 0.06 (0.87-1.26)	1.78 ± 0.09 (1.59-2.01)	3.06 ± 0.11 (2.76-3.37)
State squad (contracted)—national	1.06 ± 0.07 (0.87-1.37)	1.79 ± 0.09 (1.59-2.20)	3.06 ± 0.12 (2.76-3.69)
CA contracted—international	1.06 ± 0.07 (0.90-1.28)	1.79 ± 0.08 (1.60-2.06)	3.06 ± 0.11 (2.77-3.70)

Typical error: 20 m = 0.04 s. Data collected 2007-2010. Female State squad, $n = 208$; female CA contracted, $n = 90$; male Australia U19, $n = 197$; male AIS program, $n = 378$; male State squad, $n = 808$; male CA contracted, $n = 140$.

5-0-5 AGILITY TEST

RATIONALE

The ability to change direction rapidly is an important parameter in high-performance cricket. This is especially true for in-fielders and batters attempting to complete multiple runs between wickets. Elite cricketers usually specialize in a range of fielding positions (infield, outfield, and slips), so agility is an important factor in assessing suitability to these fielding positions.

TEST PROCEDURE

Refer to the 5-0-5 agility testing protocol in chapter 14 for standard setup details and test procedure.

DATA ANALYSIS

The 5-0-5 agility test is specific to the change of direction tasks that batters must perform when running between the wickets and fielding in the inner field. Profiling side-to-side turn differences can indicate the athlete's dominant turning side, which can drive specific training interventions to address this imbalance through a combination of physical and technical interventions. Examples of the utility of these interventions include alleviating a batter's tendency to turn blind when running under pressure between wickets and minimizing a player's movement time to their nondominant turning side to effect a run out in the field.

NORMATIVE DATA

Normative data for the standard 5-0-5 agility test for female and male high-performance Australian cricket players are presented in table 18.3.

TABLE 18.3 5-0-5 Agility Test Data for Female and Male High-Performance Australian Cricket Players: Mean ± *SD* (Range)

Players	5-0-5 time left, s	5-0-5 time right, s
Female		
State squad—national	2.49 ± 0.11 (2.20-2.77)	2.48 ± 0.11 (2.22-2.80)
CA contracted—international	2.49 ± 0.36 (2.26-2.77)	2.45 ± 0.09 (2.28-2.71)
Male		
Australia U19	2.27 ± 0.20 (2.11-2.68)	2.28 ± 0.11 (2.06-2.61)
AIS program (emerging players)	2.28 ± 0.12 (2.02-2.68)	2.28 ± 0.11 (2.02-2.71)
State squad (contracted)—national	2.29 ± 0.13 (2.02-3.32)	2.27 ± 0.12 (2.02-2.81)
CA contracted—international	2.27 ± 0.13 (2.03-2.81)	2.27 ± 0.13 (2.02-2.68)

Typical error: time = ~0.05 s. Data collected 2007-2010. Female State squad, $n = 167$; female CA contracted, $n = 80$; male Australia U19, $n = 158$; male AIS program, $n = 280$; male State squad, $n = 589$; male CA contracted, $n = 98$.

VERTICAL JUMP TEST

RATIONALE

It has been established that vertical jump test performance is a reliable measure of explosive muscular characteristics of the lower limbs (Markovic 2004). Fast muscular contraction is a key aspect of many modern cricket skills for wicketkeepers, slips catchers, infielders, batters, and pace bowlers. Two field methods for lower-body power assessment are used in elite Australian cricket: the static vertical jump and the standing (countermovement with arm swing) vertical jump. (Note that the term *power* is widely used in the practical strength and conditioning setting but its use is colloquial and technically incorrect. The true mechanical definition of the term "power" being the rate of doing work. See Knudson 2009 for a review of research and terminology problems in regard to the term "power" and the vertical jump). The static version assesses the concentric component only, whereas the countermovement version assesses the eccentric and concentric components together, thus providing an indication of the neuromuscular system's stretch–shorten cycle (SSC) capacity. The SSC is an important characteristic of many cricket movements, and understanding this quality can provide strength and conditioning practitioners with insight into the training needs and responses of their athletes (McGuigan et al. 2006).

TEST PROCEDURE

Refer to vertical jump protocol in chapter 14 for setup details and test protocol details for the standing (countermovement) vertical jump.

Specific protocol details are listed below:

Static Vertical Jump

- The tester measures standing reach height.
- The tester moves several of the lower vanes away before instructing the athlete to commence the test.
- Start position is with hands on hips. Starting from an upright position, the athlete slowly squats to 90° of knee flexion, pauses for 2 to 3 s, and then jumps for maximal height with no arm swing or countermovement.
- The athlete performs three maximal trials.
- Jump height is recorded as standing reach height subtracted from the highest vane displaced from the three trials.

Countermovement Jump

- The tester measures standing reach height.
- From the upright position, the athlete jumps for maximum vertical height with arm swing and a countermovement but no stride is permitted.
- The athlete needs to maintain the upper body in an upright position through the complete counter-movement and jump.
- The athlete performs three maximal trials.
- The difference between maximal jump height and standing reach height is calculated to give the relative jump result in centimeters.

DATA ANALYSIS

Assessing the differences between the static and countermovement jumps provides an insight into the athlete's ability to use the SSC. The SSC has been suggested as a useful indicator of dynamic performance in athletes (McGuigan et al. 2006) and is an important factor involved in many high-performance cricket skills. When scores indicate the athlete is not deriving an advantage from the SSC, that is, the static and countermovement scores are within 3 to 5 cm of each other, strength and conditioning professionals can implement training interventions to improve this important characteristic. This could include high-velocity resistance training and plyometric exercises challenging the neuromuscular system to enhance the use of the SSC, particularly in the lower body. This is on the proviso that the similar performance on the jumps is not due to a technique deficiency in the performance of the countermovement jump itself (e.g., poor arm swing contribution). These technique factors naturally should be corrected where possible.

NORMATIVE DATA

Vertical jump normative data for female and male high-performance Australian cricket players are presented in table 18.4.

TABLE 18.4 Vertical Jump Test Data for Female and Male High-Performance Australian Cricket Players: Mean ± SD (Range)

Players	Static jump height, cm	CMJ height, cm
Female		
State squad—national	41 ± 7 (25-59)	42 ± 6 (27-64)
CA contracted—international	41 ± 6 (25-55)	44 ± 6 (31-63)
Male		
Australia U19	51 ± 7 (31-63)	56 ± 6 (36-75)
AIS program (emerging players)	54 ± 8 (34-73)	58 ± 8 (24-78)
State squad (contracted)—national	55 ± 8 (31-74)	58 ± 8 (23-82)
CA contracted—international	57 ± 8 (40-73)	60 ± 9 (23-78)

Typical error: jump height = ~1.5 cm. Data collected 2007-2010. Female State squad, $n = 163$; female CA contracted, $n = 77$; male Australia U19, $n = 161$; male AIS program, $n = 266$; male State squad, $n = 552$; male CA contracted, $n = 151$.

YO-YO INTERMITTENT RECOVERY TEST

RATIONALE

There has long been debate regarding the need for aerobic fitness in cricketers. The exploits of some notable high-profile cricketers during the 1970s, 1980s, and to a lesser extent in the 1990s haven't helped portray an image of elite cricketers as being or needing to be fit to be successful. However, modern professional cricket entails sustained periods of international air travel and exposure to unfamiliar environments, hotel rooms, training grounds, and cricket grounds; many of these factors have been shown to be psychologically and physiologically taxing for athletes (e.g., Reilly et al. 2005). Compounded by three formats of the game, with each of the shorter formats influencing intensity levels in the longer forms, it becomes apparent the modern professional cricketer requires a good aerobic capacity. This will help athletes cope with physiological and psychological stress and expedite recovery between games and following acute intermittent effort, the latter being a feature of skill execution in modern cricket (Duffield and Drinkwater 2008; Petersen et al. 2010). Australian cricket uses the Yo-Yo intermittent recovery test (IRT) level 1 because it reflects the intermittent nature of cricket performance more closely than the multistage shuttle run test (the beep test), and provides a sound field-based measure of aerobic fitness (Bangsbo et al. 2008; Dupont et al. 2010; Krustrup et al. 2003; Rampinini et al. 2010).

TEST PROCEDURE

Refer to the Yo-Yo IRT protocol in chapter 14 for a detailed description of this test. It is recommended that this test be completed at the conclusion of the cricket testing battery, because fatigue associated with this test may affect performance in the speed and jump tests.

DATA ANALYSIS

The regular use of the Yo-Yo IRT allows strength and conditioning professionals to reliably monitor seasonal changes in intermittent running ability in response to preseason training interventions and in-season competition. The test is a useful field-based measure to assess cricketers against established norms in table 18.5. Athletes assessed with a score below the negative standard deviation for their player category must address their aerobic capacity to better cope with the chronic physiological and psychological effects of elite cricket. Due to travel, skills practice foci and the fact that team sport athletes notoriously lose aerobic conditioning during a playing season (e.g., Gabbett et al. 2008), concerted aerobic conditioning won't occur in many cases until the next off-season as part of an annually periodized program.

NORMATIVE DATA

Table 18.5 presents normative data for the Yo-Yo intermittent recovery test for female and male high-performance Australian cricket players.

TABLE 18.5   Yo-Yo Test Data for Female and Male High-Performance Australian Cricket Players: Mean ± *SD* (Range)

Players	Level	Distance, m
Female		
State squad—national	15.1 ± 1.3 (12.3-20.1)	1,155 ± 356 (300-2,400)
CA contracted—international	15.8 ± 1.2 (12.1-20.1)	970 ± 385 (200-2,400)
Male		
Australia U19	17.7 ± 1.5 (15.1-21.1)	1,773 ± 454 (800-2,720)
AIS program (emerging players)	17.9 ± 1.6 (14.1-23.8)	1,853 ± 467 (480-3,640)
State squad (contracted)—national	18.1 ± 1.7 (14.1-23.8)	1,846 ± 454 (480-3,640)
CA contracted—international	18.1 ± 1.8 (14.1-21.4)	1,863 ± 444 (480-2,840)

Typical error: distance = ~110 m. Data collected 2007-2010. Female State squad, n = 169; female CA contracted, n = 196; male Australia U19, n = 190; male AIS program, n = 575; male State squad, n = 845; male CA contracted, n = 148.

References

Bangsbo, J., Iaia, F., and Krustrup, P. 2008. The Yo-Yo intermittent recovery test: a useful tool for evaluation of physical performance in intermittent sports. *Sports Medicine* 38(1):37-51.

CANFWG. 2010. *Cricket Australia National Fitness Working Group Fitness Testing Protocol*. Brisbane, Australia: Sport Science Sport Medicine Unit, Cricket Australia Centre of Excellence.

Dupont, G. Defontaine, M., Bosquet, L., Moalla, W., and Berthoin, S. 2010. Yo-Yo intermittent recovery test versus the Universite de Montreal Track Test: relation with high intensity intermittent exercise. *Journal of Science and Medicine in Sport* 13(1):146-150.

Duffield, R., and Drinkwater, E. 2008. Time-motion analysis of Test and One-Day international cricket centuries. *Journal of Sports Sciences* 26(5):457-64.

Duggleby, T., and Kumar, S. 1997. Epidemiology of juvenile low back pain: a review. *Disability Rehabilitation* 19(12):505-512.

Gabbett, T., King, T., Jenkins, D. 2008. Applied physiology of rugby league. *Sports Medicine* 38(2):119-138.

Knudson, D. 2009. Correcting the use of the term "power" in the strength and conditioning literature. *Journal of Strength and Conditioning Research* 26(3):1902-1908.

Krustrup, P., Mohr, M., Amstrup, T., Rysgaard, T., Johansen, J., Pedersen, P., Bangsbo, J. 2003. The Yo-Yo intermittent recovery test: physiological response, reliability and validity. *Medicine and Science in Sports and Exercise* 35(4):697-705.

Markovic G., Dizdar D., Jukic I., Cardinale M. 2004. Reliability and factorial validity of squat and countermovement jump tests. *Journal of Strength and Conditioning Research* 18(3):551-555

McGuigan, M., Doyle, T., Newton, M., Edwards, D., Nymphius, S., Newton, R. 2006. Eccentric utilisation ratio: Effect of sport and phase of training. *Journal of Strength and Conditioning Research* 20(4):992-995.

Norton, K., and Olds, T. 1996. *Anthropometrica: A Textbook of Body Measurement for Sports and Health Courses*. Sydney, Australia: UNSW Press.

Rampinini, E., Sassi, A., Azzalin, A., Castagna, C., Menaspa, P., Carlomagno, D., and Impellizzeri, F. 2010. Physiological determinants of Yo-Yo intermittent recovery tests in male soccer players. *European Journal of Applied Physiology* 108(2):401-409.

Petersen, C., Pyne, D., Portus, M., Karppinen, S., and Dawson, B. 2009. Variability in movement patterns during One-Day internationals by a cricket fast bowler. *International Journal of Sports Physiology and Performance* 4(2):278-281.

Petersen, C., Pyne, D., Dawson, B., Portus, M., and Kellett, A. 2010. Movement patterns in cricket vary by both position and game format. *Journal of Sports Sciences* 28(1):45-52.

Portus, M., Mason, B., Elliott, B., Pfitzner, M., Done, R. 2004. Technique factors related to ball release speed and trunk injuries in high performance cricket fast bowlers. *Sports Biomechanics* 3(2):263-284.

Reilly, T., Waterhouse, J., Edwards, B. 2005. Jet lag and air travel: implications for performance. *Clinics in Sports Medicine* 4(2):367-80.

Winget, C.M., DeRoshia, C.W., Holley, D. 1985. Circadian rhythms and athletic performance. *Medicine and Science in Sports and Exercise* 17(5):498-516.

High-Performance Cyclists

Laura A. Garvican, Tammie R. Ebert, Marc J. Quod, Scott A. Gardner, John Gregory, Mark A. Osborne, and David T. Martin

Competitive track, road, BMX, and mountain bike cycling are physiologically demanding sports, spanning a wide time continuum from a 10 s flying 200 m track sprint to a 21-day grand tour where cyclists ride approximately 1400 km per week. At the 2008 Olympic Games in Beijing, 18 cycling events were contested (figure 19.1). Some events require a continuous paced effort (e.g., road time trial or individual pursuit), whereas other events are stochastic in nature, involving multiple intermittent high-intensity efforts (e.g., BMX, mountain bike cross country, points race). The physiological demands

associated with competitive cycling therefore vary greatly across the many different cycling events.

Although technological advances have led to lightweight, aerodynamic cycling equipment that can have a significant impact on cycling performance, it is clear that the greatest increases in cycling performance tend to be achieved by training-induced physiological adaptations that facilitate improvements in both cycling power and capacity. As a result, the physiological assessment of high-performance cyclists can have an important role at many stages of an athlete's development. Laboratory-based ergometer

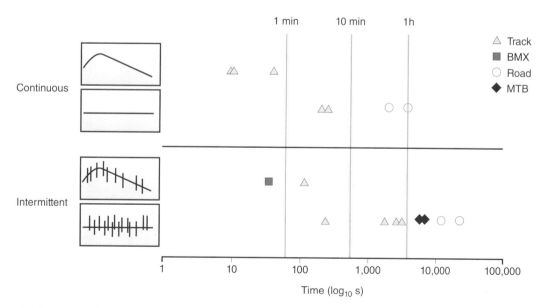

FIGURE 19.1 Competitive cycling events held at the 2008 Beijing Olympic Games: Cycling events can be either continuous or intermittent in nature, requiring either an all-out effort, with maximal power reached early and diminishing over time, or a paced effort, with an even effort throughout the event (y-axis). The duration of the event also varies from approximately 10 s (track sprint) to 5 to 6 hours (road race) (x-axis).

testing affords a controlled environment in which relevant fitness traits can be evaluated in isolation without the complicating influences of heat, altitude, technique, tactics, and other factors that can influence the cycling power output–speed relationship. Specifically, laboratory fitness testing may be used for the following reasons:

- Document physiological characteristics (e.g., $\dot{V}O_2$max, aerobic thresholds, cycling economy)
- Monitor physiological adaptations to a training intervention
- Quantify physiological effects of an ergogenic aid (e.g., enriched oxygen, caffeine)
- Define exercise intensity bands (training zones) relative to individual capabilities
- Assess exercise capacity over different durations and cadence ranges
- Evaluate the physiological demands of training sessions
- Characterize fatigue as revealed when physiological responses uncouple
- Troubleshoot when field-based performances are below expected values
- Construct a cyclist's individual profile of the key physiological indices contributing to track and road cycling success
- Augment information that is already available (e.g., track and road performances, field observations) when making decisions on competition readiness
- Aid in talent identification
- Predict cycling performance
- Help detect or confirm any acute or chronic overtraining syndrome

Track Cycling

Track cycling, which refers to all events performed in a velodrome (~250 m enclosed track), can be broadly categorized into sprint (<1,000 m) and endurance (>1,000 m) events, and the associated physiological demands have been shown to be vastly different (Craig and Norton 2001). Many of the shorter track events require the athlete to tax maximally both the anaerobic (i.e., oxygen independent) and aerobic (i.e., oxygen dependent) pathways (Jeukendrup et al. 2000), with the relative contribution of aerobic and anaerobic energy production primarily influenced by the duration of the effort (Craig and Norton 2001). For sprint cyclists, competitive events last a matter of

seconds, and until the advent of mobile power measuring devices such as the Schoberer Rad Meßtechnik (SRM), mathematical models were required to predict the bioenergetics associated with these events. The 2012 London Olympic cycling program includes three sprint events (match sprint, keirin, and team sprint) for both men and women.

Road Cycling

In contrast to track cycling, the majority of road cycling events are performed at submaximal intensities over prolonged durations. Therefore, elite road cyclists are categorized not only by a high maximal aerobic capacity but also by their ability to sustain a high percentage of their maximal power output for long periods of time (Mujika and Padilla 2001; Padilla et al. 2001). Road cycling races vary in length and terrain. A 1 h flat criterium race tends to favor powerful sprinters as opposed to 3-week stage races incorporating numerous mountain passes, which favor specialized hill climbers. Not surprisingly, specialists have evolved within the sport of road cycling in order to capitalize on individual physiological attributes. The single-day mass-start road race and the individual time trial have traditionally been included in the Olympic cycling program.

BMX

Bicycle motocross (BMX cycling), introduced to the Olympic program in 2008, consists of a series of time trials followed by up to seven races on route to the final. A single BMX race includes 6 to 8 riders and is of 30 to 40 s in duration, depending on the track layout. Recent competition analysis suggests that the race generally involves approximately 6 maximal sprints of between 1 and 3 s in length. In a 35 s race, riders will only complete 30 to 40 pedal revolutions (M.A. Osborne, personal communications, September 2010) and will need to generate high levels of peak power. In addition, the upper body contributes significantly to overcome the numerous jumps and obstacles on the track. The relatively short, high-intensity efforts associated with BMX place a high degree of importance on the alactic energy system because results are often determined by the riders who can repeatedly produce the highest power outputs without incurring any technical or tactical errors.

Mountain Bike

Mountain bike (MTB) cross country racing is an endurance event that requires bike handling skills, a

Laboratory Versus Field Testing

Although this chapter is primarily concerned with the assessment of cyclists within a laboratory environment, sport scientists and coaches should consider the efficacy of laboratory versus field performance testing (Paton and Hopkins 2001). The use of mobile power measuring devices is now common in many sporting institutes and trade teams. Accurate mobile cycling power meters can allow each training session to be a "testing" session, essentially providing feedback that has historically required laboratory testing. Indeed, the physiological assessment of high-performance track sprint cyclists in Australia has now largely moved to field-based activities that are incorporated into the yearly training program. Regardless of where and how fitness data are collected, the use of best practice procedures described throughout this book is paramount. The concept that field testing does not require careful attention to accuracy and reliability if it is not part of research or formal fitness assessment is not justifiable.

high degree of physical fitness, pacing, and tactical decision making. The cross country event is approximately 1.5 to 2.0 h in duration, and although it is comparable in effort to longer road time trials, power output profiles reveal a variable loading pattern. At the elite level, mountain bike cyclists possess similar physiological traits as road cyclists; however, the importance of these attributes relative to body mass appears to be greater in MTB (Lee et al. 2002). The Olympic program has included one-single-day mass start MTB cross country race for men and women since 1996.

Athlete Preparation

Standardized pretest preparation is recommended to enhance the reliability and validity of physiological data. Specific information relating to the preparation of the athlete and the testing environment is provided in chapter 2, Pretest Environment and Athlete Preparation. In addition, a number of considerations pertaining to performance testing of cyclists are outlined in this section.

Before scheduling a test, consider some questions. What is the reason for the test? What information are you hoping to collect? The time of year, current training load, and competitive level of the athlete should all be considered and documented whenever possible. As mentioned earlier, physiological performance testing should always be carried out with a clear goal and objective in mind.

- *Diet and Health Status.* Where possible, all athletes should consume a high-carbohydrate diet in the 24 h prior to testing. In the event that the cyclist's nutritional and health status cannot be standardized, it is useful to record the timing and content of prior meals, including caffeine intake.

- *Training.* Athletes should avoid very strenuous exercise or exercise that they are not accustomed to on the day prior to testing and on the testing day itself. Where possible, the time and type of training within 24 h of testing should be standardized. These training sessions should be of an intensity similar to light recovery (e.g., <50 km at <75% maximal heart rate [HRmax]). For highly trained cyclists it is likely that multiple high-intensity training and testing sessions can take place on successive days without noticeably influencing results. However, the lesser-fit cyclist may struggle to reproduce a maximal effort without a recovery day. An appreciation of the training phase and current training load of the athlete is also helpful for subsequent interpretation of the results.

- *Testing.* Where possible, time of day and order of testing should be standardized. Test familiarization is particularly important for all open-ended performance tests that include a pacing component. Ensure athletes are aware of the requirements and purpose of the test. In all cases athletes should be encouraged to produce their best effort on the day. After the cyclist finishes a maximal effort, it can be particularly informative to ask, "How much of yourself did you give?" using a scale from 0% to 100%. This question can reflect a different component of effort than the perceived exertion scale, which asks the cyclist to rate his or her perception of exercise intensity.

- *Pretest Warm-Up.* Cyclists should complete a thorough warm-up prior to testing; race-specific warm-ups are preferred before maximal time trials, whereas an abbreviated warm-up can be administered prior to graded exercise tests. Some athletes may wish to perform part of this warm-up on the road. Regardless of the modality of warm-up used, athletes should be given a few minutes to accustom themselves to the ergometer used for testing and make any adjustments to the riding position where necessary.

Test Environment

The purpose of the test will determine the protocol and ergometer used for testing. Many different types and models of cycle ergometer are available:

- Air-braked ergometers: e.g., Schwinn Velodyne (Schwinn, Dorel Industries Inc., Quebec, Canada) or custom-built ergometers (e.g., Hayes ergometer and Wombat, Australian Institute of Sport, Canberra, Australia)

- Electromagnetically braked ergometers (i.e., eddy current braked): e.g., Lode Excalibur Sport (Lode BV, Groningen, The Netherlands), SRM Ergometer (Schoberer Rad Meßtechnik, Julich, Germany), Velotron RacerMate (RacerMate Inc, Seattle, Washington)

- Friction braked ergometers: e.g., Monark ergometers (Vansbro, Sweden)

- Combination magnetic and air-braked ergometers: e.g., Wattbike (Wattbike Ltd., Nottingham, Great Britain)

- Portable trainer with a cycle power meter fitted to the athlete's bike: e.g., Cyclus2 (RBM Electronics, Leipzig, Germany); CycleOps Power (Saris Cycling Group, Madison, Milwaukee, USA)

Performance tests involving a predetermined fixed work load (e.g., graded exercise test) are most commonly performed on an electromagnetically braked ergometer, where the protocol can be programmed

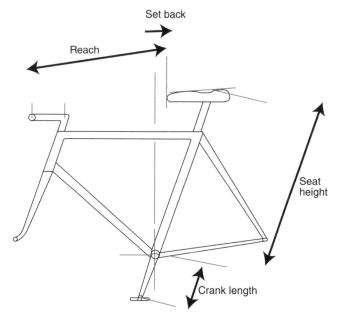

FIGURE 19.2 Bike setup diagram.

prior to the test. Self-paced maximal tests (e.g., the power profile or 30 min time trial), are best suited to air- or wind-braked ergometers, which have been specifically designed to replicate the kinetic energy and "feel" associated with riding on the road. However, in these types of tests it is important to ensure that the cyclist's capabilities are not limited by the choice of gear selection.

Test Protocols

The following procedures are common to all tests:

- Record cyclist's information on the appropriate recording sheet (i.e., time of test, diet, previous exercise, laboratory environmental conditions, body mass). If in doubt, record it as supplementary information; this is often useful for interpretation of results at a later date.

- Calibrate all equipment (e.g., metabolic cart and ergometer) prior to the testing session and record calibration results (refer to chapter 7, Determination of Maximal Oxygen Consumption). When using a metabolic cart, ensure that it is calibrated before each test. If using a portable power measuring device (e.g., an SRM), ensure that slope and sampling frequency are appropriate for the test protocol and record the zero offset prior to each test.

- Replicate the setup of the cyclist's own bike on the test ergometer (seat height, reach and set back, crank length, pedals) and record for future reference (figure 19.2).

- Place the electric fan in a standardized position (e.g., 45° to the front).

- Ensure that all equipment is correctly fitted to the cyclist (e.g., heart rate monitor, metabolic cart head piece) and is operating correctly.

- Fully explain the test procedure, have the cyclist sign an informed consent if required, and then have the cyclist complete the appropriate warm-up.

Equipment Checklist

Anthropometry

- ☐ Stadiometer (wall mounted)
- ☐ Balance scales (accurate to ±0.05 kg)
- ☐ Anthropometry box
- ☐ Skinfold calipers (Harpenden skinfold caliper)
- ☐ Bone calipers
- ☐ Anthropometer
- ☐ Marker pen
- ☐ Anthropometric measuring tape
- ☐ Recording sheet (refer to the anthropometry data sheet template)
- ☐ Pen

Capillary Blood Sampling

- ☐ Disposable lancets
- ☐ Alcohol swabs
- ☐ Heparinized capillary tubes (100 µL) or lactate strips
- ☐ Disposable rubber gloves
- ☐ Tissues
- ☐ Kidney or carry dish
- ☐ Sharps container
- ☐ Biohazard bag
- ☐ Blood lactate and blood gas analyzer

Bike Tool Kit for Bike Setup

- ☐ Full set of Allen keys
- ☐ Pedal spanner
- ☐ Measuring tape
- ☐ Plumb line

Oxygen Consumption ($\dot{V}O_2$max) and Maximal Heart Rate (HRmax): Blood Lactate Transition Threshold Test

- ☐ Calibrated, electromagnetically braked ergometer (e.g., Lode Excalibur Sport)
- ☐ Metabolic cart (and associated equipment)
- ☐ Heart rate monitoring equipment
- ☐ Electrode gel
- ☐ Electric fan
- ☐ Capillary blood sampling equipment (listed previously)
- ☐ Recording sheet (refer to the cycling short and long graded exercise test data sheet templates)
- ☐ Pen

Laboratory 30 Min Time Trial

- ☐ Calibrated, electromagnetically braked ergometer (e.g., Lode Excalibur Sport) *or* modified air-braked cycle ergometer with power measuring device (e.g., SRM crank, science version)
- ☐ Metabolic cart (and associated equipment)
- ☐ Heart rate monitoring equipment
- ☐ Electrode gel
- ☐ Electric fan
- ☐ Stopwatch
- ☐ Capillary blood sampling equipment (listed previously)
- ☐ Recording sheet (refer to the cycling 30 min time trial data sheet template)
- ☐ Pen

Power Profile Test

- ☐ Modified air-braked cycle ergometer with power measuring device (e.g., SRM crank, science version) *or* wind trainer with own road bike and SRM crank
- ☐ Metabolic cart (and associated equipment)
- ☐ Heart rate monitoring equipment
- ☐ Electrode gel
- ☐ Stopwatch
- ☐ Capillary blood sampling equipment (listed previously)
- ☐ Recording sheet (refer to the cycling power profile test data sheet template)
- ☐ Pen

Alactic Peak Power Test and BMX 6 × 3 s Repeat Sprint Test

- ☐ Modified air-braked cycle ergometer with power measuring device (e.g., SRM crank, science version)
- ☐ Stopwatch
- ☐ Recording sheet (refer to the cycling 6 s and BMX 6 × 3 s sprint test data sheet templates)
- ☐ Pen

When one is choosing the ergometer for the test, it is important to ensure that the riding position can be adjusted to replicate the cyclist's own bike as closely as possible. This should include the pedal system, crank length, and type of handle bar used. Measurements of the cyclist's own bicycle setup should be performed prior to testing and transferred to the ergometer to be used for testing (figure 19.2). The tester must be familiar with the operation of the chosen ergometer and the correct procedures for best practice. Regular calibration and maintenance of the ergometer are essential.

The laboratory setting should be controlled for temperature (18-23 °C), relative humidity (<70%), observers, noise, and other distractions. An electric fan is often useful to help cool cyclists who are completing protracted tests. For a testing session in the field, it is recommended to use a portable environmental sensor (e.g., Kestral3500 weather tracking device) to measure temperature, relative humidity, barometric pressure, and wind speed. The coach should be encouraged to be present at all testing sessions. Verbal encouragement is typically allowed as it replicates the normal training and competition environment.

Recommended Test Order

Day	Tests
1	$\dot{V}O_2$max or graded exercise test
3	Power profile test
5	30 min time trial

ANTHROPOMETRY

RATIONALE

Basic anthropometric characteristics of elite track, road, and mountain bike cyclists have been described in the literature (Craig and Norton 2001; Lee et al. 2002; Martin et al. 2001; Mujika and Padilla 2001) and document the extent that the body composition of cyclists can vary. Although a universal ideal anthropometric profile for successful performance does not exist, nonfunctional mass (either of the rider or bike) can be detrimental to performance. Added mass can inhibit the motion of a bike in the following ways (Olds et al. 1995b):

- Increase inertia and hence reduce the rate of acceleration
- Decrease the speed of the bike during hill climbing
- Increase rolling resistance
- Change frontal surface area

Mathematical models of cycling have been used to predict the effect of additional, nonfunctional mass on cycling performance (Martin et al. 2006; Olds et al. 1995a, 1995b). On the flat, adding an extra 1 kg may only result in an approximately 0.02% increase in performance time (Olds et al. 1995b), and even this small margin could determine a medal-winning performance in events such as the 4,000 m individual pursuit or individual time trial. During uphill cycling, the effect of added mass becomes much greater and a greater amount of energy is required to maintain a given speed (Olds et al. 1995b). For cyclists competing in events over hilly terrain (e.g., road and mountain bike races), a low, lean body mass may be desirable (Lee et al. 2002).

In light of the effects of body mass, longitudinal monitoring of a cyclist's body composition, in conjunction with their performance capabilities, can be useful. However, the significance and focus of a cyclist's anthropometric profile will depend on their level of development, the training phase, and the specific demands of the competitive event. Although a cyclist can control their diet and their training, body composition is heavily influenced by other environmental (e.g., life stress) and genetic factors. Therefore, focusing on target skinfolds can become a frustrating distraction compared with focusing on a healthy diet and consistent training.

TEST PROCEDURE

At a minimum, height, body mass, and skinfolds should be measured. Skinfolds are recorded over seven sites (triceps, biceps, subscapular, supraspinale, abdominal, front thigh, and medial calf). The individual skinfold measures as well as the sum of the seven sites should be reported. Refer to anthropometry protocols outlined in chapter 11, Assessment of Physique, for a detailed description of all anthropometric test procedures. More advanced anthropometric assessment including muscle girths, bone breadths, and limb lengths can be conducted if required.

Although the skinfold measurement procedures seem simple, a high degree of technical skill is essential for consistent results, and therefore it is important that an experienced, accredited, and trained tester carry out the measurements. Ideally the same tester should conduct each retest to ensure reliability.

If the opportunity arises, the restricted profile and four-compartment model may be useful for monitoring compartmental changes, especially during phases in which body composition has been targeted through dietary or training interventions.

DATA ANALYSIS

The application and limitations of anthropometric assessment are discussed in detail in chapter 11.

NORMATIVE DATA

Anthropometric data of Australian male and female cyclists from different cycling disciplines are presented in table 19.1. Substantial seasonal variation may occur in association with changes in training and competition load, and thus direct comparison across groups should be done with caution.

Table 19.1 Anthropometric Data for Female and Male High-Performance Cyclists: Mean ± *SD* (Range)

Athlete group	Body mass, kg	Σ7 skinfolds, mm
Female		
Senior track sprint (*n* = 3)	70.4 ± 2.9 n/a	87.2 ± 15.9 n/a
Senior road (*n* = 9)	58.2 ± 3.9 (50.9-64.9)	55.2 ± 15.5 (30.8-73.6)
Senior MTB (*n* = 6)	57.4 ± 4.3 (53.4-62.8)	68.0 ± 18.7 (41.6-92.2)
Junior MTB (*n* = 5)	58.2 ± 7.3 (49.9-65.2)	85.0 ± 9.7 (69.5-94.9)
Senior BMX (*n* = 5)	70.2 ± 2.9 (67.4-74.3)	109.2 ± 18.7 (77.8-130.0)
Male		
Senior track endurance (*n* = 6)	72.6 ± 4.4 (69.0-80.2)	38.1 ± 5.1 (30.5-53.4)
Senior track sprint (*n* = 5)	89.8 ± 5.0 n/a	52.3 ± 10.0 n/a
Under 23 road (*n* = 7)	73.4 ± 6.8 (58.2-77.4)	45.1 ± 8.8 (32.5-56.2)
Junior road (*n* = 2)	63.0 ± 0.6 (62.6-63.4)	37.9 ± 2.1 (36.4-39.3)
Under 23 MTB (*n* = 11)	72.5 ± 7.2 (63.4-88.5)	42.9 ± 6.6 (32.5-53.9)
Junior MTB (*n* = 9)	67.0 ± 8.2 (56.7-80.5)	43.0 ± 7.3 (32.5-53.9)
Senior BMX (*n* = 9)	84.4 ± 6.7 (71.0-93.0)	47.6 ± 11.7 (31.8-72.2)

n/a = not available. Typical error: Σ7 skinfolds = ~1 mm. Data source: sprint (August 2010); road (March-August 2010); MTB (September 2009); track endurance (March 2010).

OXYGEN CONSUMPTION ($\dot{V}O_2$MAX) AND MAXIMAL HEART RATE (HRMAX) (SHORT GRADED EXERCISE TEST)

RATIONALE

Elite track, road, and mountain bike cyclists are often characterized by a high maximal aerobic power ($\dot{V}O_2$max) (Craig and Norton 2001; Lee et al. 2002; Martin et al. 2001; Mujika and Padilla 2001). Although, undoubtedly, numerous factors contribute to performance outcomes during competitive events (e.g., tactics, drafting), the model of Olds and colleagues (1995a, b) suggests that $\dot{V}O_2$max is one of the main physiological parameters contributing to road cycling performance. In many instances, on both the road and the track, cyclists will be required to produce supramaximal efforts and thus place a demand on both the aerobic and anaerobic metabolic pathways. A high $\dot{V}O_2$max coupled with the ability to achieve it rapidly and maintain it would enable a large, rapid, and sustained aerobic energy release and reduce premature reliance on a large proportion of the finite oxygen deficit. Therefore, it is of no surprise that $\dot{V}O_2$max and associated indices have been strongly correlated to cycling performance (Craig et al. 1993).

A strong genetic component of $\dot{V}O_2$max (Martino et al. 2002) makes it a useful parameter for talent identification. However, $\dot{V}O_2$max can also be modulated to a degree by a variety of training interventions, such as altitude training (Levine and Stray-

Gundersen 1997) and high-intensity interval training (HIT) (Laursen et al. 2002); therefore, performance testing before and after the intervention period can offer insight into the success of the program and the responses evoked in the athlete.

The following test is often described as the 1 min ramp protocol or the short graded exercise test and enables $\dot{V}O_2$max to be achieved relatively quickly (10-15 min depending on the ability of the cyclist) in comparison to the long graded exercise test (see Blood Lactate Transition Thresholds: Long Graded Exercise Test) which may take up to 45 min in some circumstances. For this reason, the ramp protocol may be useful for talent identification or research purposes, if the maximal capabilities of the cyclist are the key variables of interest and steady-state metabolic analysis and blood lactate and heart rate responses are not required.

TEST PROCEDURE

- The tester records age, height, and body mass prior to the test.
- Common test procedures are completed as outlined previously.
- The cyclist is required to pedal at a constant workload for 1 min work durations. After each complete minute, the workload is increased. The test is continuous and is terminated when the cyclist can no longer maintain the required workload. This is deemed to be the point at which the desired cadence (90-100 revolutions per minute [rpm]) can no longer be maintained or the cyclist decides to terminate the test. The cyclist should attempt to complete a full minute or 30 s interval of the final workload.

- Table 19.2 provides information regarding the starting workloads, ramp increases, and desired cadence range. The respiratory valve mouth piece and nose clip must be worn for the entire duration of the test. It is often useful to place a piece of paper tape over the cyclist's nose before positioning the nose clip in order to stop the nose clip from slipping during the test.
- Maximal oxygen consumption ($\dot{V}O_2$max), maximal carbon dioxide production ($\dot{V}CO_2$max), respiratory exchange ratio (RER), ventilation ($\dot{V}_E$ in L · min^{-1} of expired gas), ventilatory equivalents (i.e., ratio of $\dot{V}_E$ to $\dot{V}O_2$, ratio of $\dot{V}_E$ to $\dot{V}CO_2$), and power output are recorded for each 30 s of the test.
- The $\dot{V}O_2$max is the sum of the highest two consecutive 30 s readings.
- Heart rate is recorded during the last 10 s of each work interval as well as at the end of the test.
- If peak blood lactate is required, blood samples should be taken at completion of the test and then at 1, 2, 5, and 7 min postexercise. In this instance the cyclist must stop pedaling and remain in a seated position, with no cooldown allowed for the first 7 min postexercise.

DATA ANALYSIS

When cycling performance is assessed, $\dot{V}O_2$max should not be interpreted without reference to the power output corresponding to $\dot{V}O_2$max. $\dot{V}O_2$max should be reported in both absolute (L · min^{-1}) and relative (ml · kg^{-1} · min^{-1}) units. Similarly, the final power output achieved (often described as maximal aerobic power MAP or Wmax) should be presented in watts and watts per kilogram (W and W · kg^{-1}).

TABLE 19.2 Protocol Selection for Assessing $\dot{V}O_2$max of Female and Male High-Performance Cyclists Using the Short Graded Exercise Test Protocol

Athlete group	Warm-up	Starting workload, W	Ramp increase, W	Cadence range selection, rpm	Step duration, min
Female senior	5 min at 100 W	125	25	90-105	1
Female junior	5 min at 50 W	75	25	90-105	1
Male senior	5 min at 150 W	175	25	90-105	1
Male junior	5 min at 125 W	150	25	90-105	1

The protocols suggested here are designed to assess the high-performance cyclist. Starting workloads, workload increments, and cadence ranges may not be applicable to the sub–state-level cyclist. Cadence ranges are based on use of a hyperbolic ergometer.

- Record the final workload and duration of the final workload completed for calculation of maximal aerobic power (in watts) as follows:
  - Maximal Aerobic Power (W) = Final Completed Workload (W) + [Workload Increment (W) · Fraction of Time Completed in Current Workload (min)].
  - Example: final completed workload = 300 W.
  - Workload increment = 25 W.
  - Time completed in current workload = 0.5 min of 1 min or 50%.
  - Hence, maximal aerobic power = 300 + (25 · 0.5) = 312.5 W.
- Having recorded maximal heart rate (HRmax) the following heart rate training zones can also be calculated:
  - Training zone 1 (T1): less than 65% to 75% HRmax
  - Training zone 2 (T2): 75% to 80% HRmax
  - Training zone 3 (T3): 80% to 85% HRmax
  - Training zone 4 (T4): 85% to 92% HRmax
  - Training zone 5 (T5): 92% to 100% HRmax

Although these estimated heart rate training zones lack the precision of those derived from a long graded exercise test with blood lactate measurements, they can provide a valuable training control for the cyclist only undertaking a short graded exercise test.

LIMITATIONS

The cyclist must be fully cooperative and motivated to cycle to exhaustion. When young cyclists are tested (e.g., talent identification programs), commitment and motivation may influence the likelihood of achieving maximal data. In addition, the type of equipment (e.g., mouth piece size) and protocol may need to be modified.

NORMATIVE DATA

Table 19.3 provides normative oxygen consumption and power output data collected from male road cyclists using the 1 min ramp protocol. Because of the additional information that can be obtained from a long graded exercise test (described subsequently), the use of the 1 min protocol within Australia to test national team cyclists is somewhat rare and thus the availability of normative data is limited.

TABLE 19.3 Oxygen Consumption and Power Output Data for Male High-Performance Cyclists, 1 Min Protocol: Mean ± SD (Range)

Athlete group	Body mass, kg	$\dot{V}O_2$max, $L \cdot min^{-1}$	$\dot{V}O_2$max, $ml \cdot kg^{-1} \cdot min^{-1}$	PO, W	PO, $W \cdot kg^{-1}$	HRmax, $beats \cdot min^{-1}$
Under 23 road (n = 9)	67.5 ± 5.4 (60.9-70.5)	4.94 ± 0.27 (4.5-5.4)	73.4 ± 5.2 (66.0-81.1)	468.7 ± 27.0 (425-512)	7.0 ± 0.5 (6.3-7.8)	189.4 ± 8.0 (175-200)
A grade (n = 10)	70.9 ± 5.5 (61.9-80.7)	5.2 ± 0.4 (4.4-5.6)	73.5 ± 4.3 (64.7-80.6)	452.9 ± 26.3 (425-481)	6.4 ± 0.4 (5.9-7.0)	183.3 ± 7.2 (170-193)

Typical error: absolute $\dot{V}O_2$max = 0.12 L · min⁻¹, % TE = 2.6; relative $\dot{V}O_2$max = 2.0 ml · kg⁻¹ · min⁻¹, % TE = 3.1; power output at $\dot{V}O_2$max = 14 W, % TE = 3.5; HRmax = 2 beats · min⁻¹, % TE = 1.2. Data source: Under 23 road (March 2010), A grade (November 2007).

BLOOD LACTATE TRANSITION THRESHOLDS (LONG GRADED EXERCISE TEST)

RATIONALE

Success in many cycling events depends not only on $\dot{V}O_2$max but also on the percentage of $\dot{V}O_2$max (%$\dot{V}O_2$max) that the cyclist can sustain for extended periods of time (e.g., during a time trial or while climbing) (Jeukendrup et al. 2000; Olds et al. 1995b). The method used to detect thresholds and the terms used to describe thresholds have been debated by the scientific community for more than 30 years. Very generally, a specific exercise intensity exists

below which endurance is mainly a function of fuel supply, body temperature, or both and above which there is a significant reduction in endurance time that is largely due to glycolytic stress (Joyner and Coyle 2008). The commonly referred to lactate threshold (power output that elicits ~4 mmol · L⁻¹ lactate) reflects an exercise intensity that signifies a marked transition from steady-state responses to an exercise intensity where heart rate, ventilation, and blood lactate accumulation progressively drift to higher and higher values even though the power output remains constant. Exercise at an intensity above threshold is therefore relatively short-lived compared with the exercise duration that can be endured at an exercise intensity below threshold. In road cycling, significant correlations between individual time trial performance and measured thresholds have been reported (Coyle et al. 1988, 1991), although in the context of track endurance cycling, power output at lactate threshold has been found to be a good predictor of 4,000 m individual pursuit time (Craig et al. 1993).

The long graded exercise test is often performed in preference to the short graded exercise test (i.e., the ramped $\dot{V}O_2$max test), because it enables blood lactate transition thresholds to be identified. The work intensity associated with these thresholds can subsequently be used to prescribe and monitor training programs as well as plan and monitor appropriate competition pacing strategies.

TEST PROCEDURE

- The tester records the subject's age, height, and body mass prior to the test.
- Common test procedures are completed as outlined previously.
- The test is very similar to the $\dot{V}O_2$max test described previously; however, the length of each work duration is increased to allow a steady-state to be reached.
- The cyclist is required to pedal at a constant workload for 3 (female) or 5 (male) min stages, with the work load increasing following the completion of each stage. The test is continuous; cyclists stop only when they can no longer maintain the required power output. Failure occurs when the desired cadence can no longer be maintained or the cyclist opts to terminate the test. The cyclist should be strongly encouraged to complete a full minute or 30 s interval of the final workload as this duration typically coincides with metabolic measurements.

- Table 19.4 for information regarding starting workloads, ramp increases, and the recommended cadence range. If the ergometer does not operate in hyperbolic mode (whereby power output is independent of cadence), it may be necessary to adjust the gearing in addition to changing the resistance to ensure that the cyclist's pedaling cadence stays within the required range.
- The test requires the mouthpiece and nose clip to be worn for the last half of each stage to ensure a stable reading for metabolic analysis (e.g., 1.5 min for 3 min stages, 2.5 min for 5 min stages). The cyclist is permitted to drink during the first half of each stage if required. However, toward the end of the test, the mouthpiece and nose clip should be in place continuously.
- The test should be completed within 18 to 30 min for females and 25 to 40 min for males.
- Oxygen consumption ($\dot{V}O_2$), carbon dioxide production ($\dot{V}CO_2$), respiratory exchange ratio (RER), ventilation ($\dot{V}_E$), ventilatory equivalents (i.e., ratio of $\dot{V}_E$ to $\dot{V}O_2$, ratio of $\dot{V}_E$ to $\dot{V}CO_2$) are recorded continuously, and the values recorded for the last 1 min of each 3 min stage and the last 2 min of each 5 min stage are averaged.
- The final workload and duration of the final workload completed are recorded for calculation of maximal aerobic power as follows:
  - Maximal Aerobic Power (W) = Final Completed Workload (W) + [Workload Increment (W) · Fraction of Time Completed in Current Workload (min)].
  - Example: final completed workload = 300 W.
- Workload increment = 25 W.
- Time completed in current workload = 1.5 min of 3 min or 50%.
  - Hence, maximal aerobic power (W) = 300 + (25 · 0.5) = 312.5 W.

TABLE 19.4 Protocol Selection for Assessing the Blood Lactate Transition Thresholds of High-Performance Cyclists Using Long Graded Exercise Test Protocol

Athlete group	Warm-up	Starting workload, W	Step increment, W	Cadence range selection (rpm)	Step duration, min
Female senior	5 min at 75-100 W	125	25	90-105	3
Female junior	5 min at 50-75 W	75	25	90-105	3
Male senior	5 min at 100 W	150	50	90-105	5
Male junior	5 min at 75-100 W	100	50	90-105	5

- The tester records heart rate during the last 15 s of each work load and heart rate at the end of the test.

- A pre-warm-up capillary blood sample is collected from the fingertip or ear lobe and analyzed for lactate concentration and (if required) pH, blood gases, and bicarbonate concentration.

- Throughout the test, capillary blood is collected from the fingertip or ear lobe during the last 30 s of each work load and analyzed in the same manner as the pre-exercise blood sample.

DATA ANALYSIS

- Report both $\dot{V}O_2$max (L · min^{-1} and ml · kg^{-1} · min^{-1}) and MAP (W and W · kg^{-1}). Note that the $\dot{V}O_2$max and MAP achieved during the graded exercise test may be slightly lower than those achieved during the 1 min ramp test.

- Lactate threshold 1 (LT1) and lactate threshold 2 (LT2: formerly known as the individual anaerobic threshold) are derived using customized software (e.g., ADAPT software, NSSQA program).

- The following indices should be reported for each threshold: $\dot{V}O_2$, % $\dot{V}O_2$max, HRmax,

%HRmax, power output (in watts), relative power output (W · kg^{-1}), and pH (if measured).

- The data collected during the graded exercise test can be used to prescribe training zones based on HR or power output. Table 19.5 provides some approximate guidelines (duration of effort, blood lactate levels, perceived exertion, power output) for the different training zones based on a five-zone model. A comparison of different training zone models commonly prescribed to cyclists is presented in table 19.6.

LIMITATIONS

The methods used to identify the blood lactate transition thresholds may affect the index values associated with each threshold. When comparing test data, use the same identification method. Because blood lactate concentration can be affected by diet and fatigue state, ensure that diet and training are standardized prior to testing.

NORMATIVE DATA

Normative data for the maximal values obtained by female and male cyclists of different disciplines during the graded exercise test, as well as indices associated with the LT1 and LT2 thresholds, are presented in tables 19.7a and b.

TABLE 19.5   Guidelines for Aerobic Training Zones Derived From Performance Testing of High-Performance Cyclists

Training zone	Heart rate, %HRmax	Duration of effort, min	Blood lactate, mmol · L^{-1}	Power output, % of 4 min max or maximal aerobic power	Perceived exertion
Recovery	<65%			40-50	Recovery
T1—training zone 1: aerobic	65-75%	>240	<1.5	50-65	Easy
T2—training zone 2: extensive endurance	75-80%	90-240	1.5-3.5	65-72.5	Comfortable
T3—training zone 3: intensive endurance	80-85%	15-90	3.3-3.6	72.5-80	Comfortable–uncomfortable
T4—training zone 4: threshold	85-92%	15-60	3.5-6.0	80-90	Uncomfortable
T5—training zone 5: $\dot{V}O_2$max	92-100%	3-7	>6.0	90-100	Stressful

TABLE 19.6 Comparison of Common Training Zones Prescribed to Cyclists Based on Maximal Heart Rate

%HRmax	Craig ENDURANCE ZONES	NSSQA TRAINING ZONES	Salzwedel ENDURANCE ZONES	Coggan LEVEL	McKenzie and Martin TRAINING ZONES
60		T1 (60%-75%)		L1 (<62%)	
61					
62				L2 (62%-75%)	
63					
64					
65	E1 (65%-75%)				T1 (65%-75%)
66					
67					
68					
69					
70			E1 (70%-80%)		
71					
72					
73					
74					
75	E2a (75%-80%)	T2 (75%-84%)			T2 (75%-80%)
76				L3 (76%-85%)	
77					
78					
79					
80	E2b (80%-85%)		E2 (80%-90%)		T3 (80%-85%)
81					
82		T3 (82%-89%)			
83					
84					
85	E3 (85%-92%)		E2b		T4 (85%-92%)
86				L4 (86%-95%)	
87					
88		T4 (88%-93%)			
89					
90					
91			E3 (90%-95%)		
92		T5 (92%-100%)			
93	E4 (92%-100%)				T5 (92%-100%)
94					
95					
96			E4 (95%-100%)	L5 (96%-100%)	
97					
98					
99					
100					

E = endurance zone; T = training zone; L= level, with 1 being the easiest zone level. Craig—Heart rate zones designed by Neil Craig, cycling sports science coordinator, and Charlie Walsh for use by Australian National team (1990-2000). NSSQA—Heart rate zones recommended by National Sport Science Quality Assurance program national guidelines. Salzwedel—Heart rate zones designed by Heiko Salzwedel, Australian National team coach (1990-2000). Coggan—Heart rate zones designed by Andy Coggan, author and sports scientist (Allen and Coggan 2006) and popularized by Training Peaks Software. McKenzie and Martin—Heart rate zones designed by David Martin, cycling sports science coordinator, and Ian McKenzie, AIS track endurance coach for use by Australian National team (2008 to present).

TABLE 19.7a   Long Duration Graded Exercise Test Data for Female and Male High-Performance Cyclists, Maximal Values: Mean ± *SD* (Range)

Athlete group	Body mass, kg	$\dot{V}O_2$max, $L \cdot min^{-1}$	$\dot{V}O_2$max, $ml \cdot kg^{-1} \cdot min^{-1}$	Power output, W	Power output, $W \cdot kg^{-1}$	HRmax, $beats \cdot min^{-1}$	Blood lactate, $mmol \cdot L^{-1}$
Female							
Senior road (*n* = 12)	58.8 ± 3.7 (50.9-64.9)	3.64 ± 0.24 (3.18-4.12)	62.1 ± 2.30 (59.2-67.7)	308.7 ± 17.9 (283-333)	5.3 ± 0.3 (4.8-6.0)	191.8 ± 8.5 (179-212)	13.8 ± 1.7 (10.6-16.0)
Senior MTB (*n* = 5)	59.5 ± 4.8 (54.2-65.2)	3.54 ± 0.4 (3.02-3.95)	59.4 ± 2.9 (55.0-62.9)	282.6 ± 39.8 (230-325)	4.7 ± 0.3 (4.2-5.2)	189.8 ± 7.9 (183-201)	12.7 ± 0.4 (12.2-13.1)
NTID (*n* = 9)	66.4 ± 6.8 (57.5-76.9)	3.72 ± 0.22 (3.40-4.07)	56.4 ± 4.8 (46.2-61.9)	302.0 ± 28.2 (262-354)	4.6 ± 0.3 (4.0-4.9)	188.3 ± 8.1 (179-206)	13.6 ± 2.5 (9.7-19.1)
Male							
Under 23 road (*n* = 13)	71.4 ± 6.9 (62.6-81.5)	4.77 ± 0.41 (4.02-5.28)	67.4 ± 7.8 (57.9-81.4)	385.0 ± 30.6 (350-435)	5.4 ± 0.6 (4.5-6.6)	193.1 ± 7.7 (182-208)	13.3 ± 3.0 (9.7-21.0)
Senior MTB (*n* = 4)	74.8 ± 9.3 (67.9-88.5)	5.25 ± 0.25 (4.92-5.53)	70.6 ± 5.4 (62.5-73.8)	417.5 ± 29.0 (390-455)	5.6 ± 0.4 (5.1-6.1)	189.0 ± 5.8 (182-196)	13.7 ± 1.1 (12.4-14.9)
Junior MTB (*n* = 8)	65.1 ± 7.4 (56.7-73.9)	4.24 ± 0.55 (3.42-4.85)	65.1 ± 2.3 (60.4-67.8)	342.0 ± 36.8 (265-405)	5.3 ± 0.4 (4.7-5.7)	189.6 ± 9.0 (176-201)	12.2 ± 1.1 (10.6-13.5)

HR = heart rate. Typical error: LT1: power output = 7 W, 3.4%; HR = 1 beats · min^{-1}, 1.0%; $\dot{V}O_2$ = 0.13 L · min^{-1}, 4.2%. LT2: power output = 5 W, 1.4%; HR = 2 beats · min^{-1}, 1.4%; $\dot{V}O_2$ = 0.08 L · min^{-1}, 2.0%. Data source: female road (January, February 2000), Under 23 road (May, 2006), NTID (December 2006), MTB (March 2009).

TABLE 19.7b   Lactate Transition Data for Female and Male High-Performance Cyclists Established During the Long Graded Exercise Test: Mean ± *SD* (Range)

Athlete group	Power output, W	Power output, $W \cdot kg^{-1}$	Blood lactate, $mmol \cdot L^{-1}$	HR, $beats \cdot min^{-1}$	%HRmax	$\dot{V}O_2$max, $L \cdot min^{-1}$	% $\dot{V}O_2$max
Female LT1							
Senior road (*n* = 12)	180 ± 27.9 (150-250)	3.0 ± 0.4 (2.5-3.9)	± 0.3 (0.6-1.7)	151 ± 1.0 (124-164)	79 ± 5.7 (68-86)	2.39 ± 0.30 (2.05-3.20)	65.8 ± 7.7 (55-84)
Senior MTB (*n* = 5)	198 ± 23.6 (175-230)	3.3 ± 0.2 (3.1-3.7)	1.8 ± 0.5 (1.2-2.4)	159 ± 6.8 (148-165)	82 ± 3.5 (78-87)	2.56 ± 0.28 (2.24-3.01)	71.7 ± 4.7 (66-78)
NTID (*n* = 9)	181 ± 20.8 (150-200)	2.7 ± 0.4 (2.1-3.4)	2.6 ± 0.8 (1.8-3.9)	157 ± 12.9 (138-175)	84 ± 5.1 (76-91)	2.48 ± 0.25 (2.07-2.75)	67.0 ± 7.8 (55-77)
Female LT2							
Senior road (*n* = 12)	245 ± 19.8 (217-289)	4.2 ± 0.2 (3.8-4.5)	4.2 ± 0.8 (3.0-5.5)	173 ± 9.0 (154-185)	90 ± 2.8 (86-94)	3.12 ± 0.26 (2.60-3.57)	85.6 ± 4.4 (79-94)
Senior MTB (*n* = 5)	233 ± 23.0 (206-265)	3.9 ± 0.2 (3.6-4.2)	3.8 ± 0.3 (3.5-4.2)	171 ± 5.2 (162-175)	87 ± 7.3 (75-93)	3.01 ± 0.33 (2.70-3.52)	87.8 ± 3.4 (84-93)
NTID (*n* = 9)	240 ± 29.6 (178-278)	3.6 ± 0.4 (2.8-4.1)	5.4 ± 1.7 (3.4-7.9)	175 ± 6.9 (165-184)	93 ± 2.3 (89-97)	3.13 ± 0.3 (2.47-3.51)	84.2 ± 6.7 (73-92)

Athlete group	Power output, W	Power output, $W \cdot kg^{-1}$	Blood lactate, $mmol \cdot L^{-1}$	HR, beats $\cdot$ min^{-1}	%HRmax	$\dot{V}O_2$max, $L \cdot min^{-1}$	% $\dot{V}O_2$max
Male LT1							
Under 23 road (n = 13)	201 ± 35.4 (150-250)	2.8 ± 0.4 (2.0-3.3)	1.5 ± 0.4 (0.7-2.6)	143 ± 11.6 (126-162)	74 ± 5.5 (63-82)	2.86 ± 0.36 (2.25-3.33)	60.0 ± 6.9 (47-70)
Senior MTB (n = 4)	263 ± 24.3 (250-300)	3.5 ± 0.2 (3.4-3.8)	2.1 ± 0.3 (1.8-2.6)	151 ± 6.2 (142-155)	80 ± 2.9 (77-83)	3.64 ± 0.19 (3.44-3.90)	70.5 ± 2.9 (67-74)
Junior MTB (n = 8)	229 ± 38.4 (170-287)	3.5 ± 0.4 (3.0-4.1)	2.2 ± 0.7 (1.6-3.5)	156 ± 6.6 (147-166)	90 ± 1.9 (80-88)	2.85 ± 0.40 (2.26-3.44)	66.6 ± 3.6 (63-72)
Male LT2							
Under 23 road (n = 13)	305 ± 31.0 (273-363)	4.3 ± 0.4 (3.6-5.1)	3.8 ± 0.7 (2.3-4.9)	173 ± 10.5 (150-187)	89 ± 4.1 (81-94)	4.04 ± 0.33 (3.47-4.48)	84.7 ± 3.9 (76-91)
Senior MTB (n = 4)	333 ± 18.8 (316-360)	4.5 ± 0.3 (4.1-4.7)	4.3 ± 0.2 (4.1-4.5)	176 ± 5.9 (162-174)	92 ± 3.1 (88-92)	4.33 ± 0.26 (4.10-4.60)	82.0 ± 6.1 (74-88)
Junior MTB (n = 8)	277 ± 47.3 (200-346)	4.3 ± 0.5 (3.5-4.8)	4.4 ± 0.3 (3.9-4.8)	171 ± 8.6 (160-186)	90 ± 1.9 (88-93)	3.43 ± 0.53 (2.58-4.14)	80.6 ± 3.6 (75-85)

HR = heart rate. Typical error: LT1: power output = 7 W, 3.4%; HR = 1 beats $\cdot$ min^{-1}, 1.0%; $\dot{V}O_2$ = 0.13 L $\cdot$ min^{-1}, 4.2%. LT2: power output = 5 W, 1.4%; HR = 2 beats $\cdot$ min^{-1}, 1.4%; $\dot{V}O_2$ = 0.08 L $\cdot$ min^{-1}, 2.0%. Data source: female road (January, February 2000), Under 23 road (May, 2006), NTID (December 2006), MTB (March 2009).

LABORATORY 30 MIN TIME TRIAL

RATIONALE

The time trial is often referred to as the "race of truth" because tactical decisions and drafting are not important; the goal is simply to cover the prescribed distance in the shortest time possible. The laboratory 30 min time trial serves as a standardized performance task that is not influenced by environmental conditions or aerodynamics. Assuming the athlete makes a maximal, well-paced effort, results from this test reliably and accurately reflect time trial cycling-specific fitness. In addition, when one is investigating the effects of an intervention on performance (e.g., training, diet, ergogenic aids), the laboratory time trial serves as a valid and reliable test that closely replicates the demands of competition (Jeukendrup et al. 1996).

TEST PROCEDURE

- The tester records the subject's age, height, and body mass prior to the test.
- Common test procedures are completed as outlined previously.
- Before commencing the test, an appropriate pacing strategy should be discussed with the cyclist. Many athletes tend to start too aggressively, and although cyclists are encouraged to treat the test as they would a real competitive effort, it is useful to remind them to pace themselves to achieve the highest average power output over 30 min.

- The test may be performed using either an air-braked or electromagnetically braked cycle ergometer. For all types of ergometers, the cyclist must be able to replicate a typical racing position (i.e., aerobars, bike position) and select a comfortable cadence during the test (typically 95-105 rpm). In the case of the Lode Excalibur Sport, a linear factor (LF) must be selected based on the following relationship:
  - Power Output (W) = LF $\cdot$ [Cadence (rpm)]^2

- For example, at a cadence of 100 rpm, an LF of 0.015 would result in a power output of 150 W. Refer to the ergometer's operating manual for further details on operation using the linear testing mode.

- If using an air-braked system fitted with an SRM, the SRM must be calibrated and the zero offset recorded before the start of the test. The athlete should select an appropriate gear to achieve the desired power output and cadence.

- Five minutes prior to starting the test, cyclists are allowed to become familiar with the desired power output and cadence by completing one or two 20 s efforts at the target power output. Within 2 min of starting the test, a pretest capillary blood sample is collected.

- The test begins from a stationary, standing start. If a Lode Excalibur Sport is used, the display of accumulated work needs to be easy to view or record continuously. It is important that cyclists are provided with feedback on both time and power output throughout the test.

- Expired air analysis may be performed from 0 to 5, 10 to 15, and 20 to 25 min if required.

- Heart rate, accumulated work (kJ), and cadence (rpm) are recorded every minute. If using an SRM, power output is recorded continuously at a 1 s sampling rate.

- Every 5 min and within 2 min of finishing the test, the athlete is asked to rate their perception of effort (RPE) using the Borg 6–20 scale. A capillary blood sample is collected for measurement of lactate and, if required, glucose, bicarbonate, and pH.

- Based on accumulated work, an average power output is calculated (e.g., 1 min accumulated work in kJ multiplied by 1,000 divided by 60 based on the relationship 1 watt = 1 J $\cdot$ s^{-1}). Power output is reflected as an average absolute and relative power output for 30 min (W and W $\cdot$ kg^{-1}). Alternatively, if an SRM is used, average power for the entire 30 min can be recorded as well as each 5 to 10 min segment.

- Oxygen uptake is presented as an average of all readings made over the 15 min of monitoring and is expressed as L $\cdot$ min^{-1}, ml $\cdot$ kg^{-1} $\cdot$ min^{-1}, and %$\dot{V}O_2$max if corresponding $\dot{V}O_2$max data are available.

- Other parameters collected during the 30 min test are reflected as an average score and a maximum (minimum in the case of pH) score.

DATA ANALYSIS

Of primary concern is the cyclist's mean power output expressed as both watts and watts per kilogram for the duration of the test. Although mean power output during a 1 h laboratory time trial has been correlated to 40 km time-trial performance (Coyle et al. 1991), changes in laboratory time trial performance may not

precisely parallel changes in time trial performance in the field. In general, an improved performance reflects improved cycling-specific fitness. However, other aspects of training and testing need to be considered in conjunction with the results:

- An improved performance may be attributable to a better pacing strategy, better motivation, increased aerobic fitness, increased anaerobic fitness, or less residual fatigue.

- A decrease in performance may be attributable to the pacing strategy used; many times an extremely aggressive start leads to a big decrease in power output during the middle third of the test. Other possibilities include a decrease in aerobic fitness, a decrease in anaerobic fitness, a decrease in motivation, and a large degree of residual fatigue. Fatigue can usually be distinguished from a loss of fitness because fatigued cyclists display a lower heart rate and sometimes a lower lactate for a given workload and perception of effort. A detrained athlete, in contrast, displays elevated heart rates and lactates for a given power output. A careful evaluation of training history prior to the performance test can help refine the explanation of a poor test.

- If performance is not dramatically different from previous tests, it is of interest to examine the RER, lactate, and pH to gain insights into anaerobic fitness. In some cases, performance is unchanged despite large increases in average heart rate, perceived exertion, and lactate. It is possible that this scenario reflects a detrained but very fresh and highly motivated athlete.

The heart rate and power output that can be sustained for 30 min can be used as a point of reference to guide future training sessions. If a cyclist can sustain 300 W for 30 min, then the following interval training session should be possible: 4 × 10 min at 305 W with 5 min recovery. This type of interval training represents an overload that is consistent with the cyclist's current level of fitness. Over time, this type of interval session should stimulate adaptations that will allow the cyclist to improve his 30 min maximal mean power output.

LIMITATIONS

As with any performance test it is very important that the athlete is sufficiently motivated to give a maximal effort and that pacing is appropriate. The RER and the RPE can be used as a guide to evaluate the relative exercise intensity. When RER is close to 1.00 and the RPE is greater than 17, it is likely that the athlete is producing a near-maximal, well-paced effort.

The best individual results tend to display a fairly flat power output profile over time with a gradual increase in power over the last 5 to 10 min. If there is a large power output decrease (more than 30-50 W) after 5 to 10 min, the cyclist may have been too aggressive at the start of the test. Similarly, if a large increase in power output is observed during the last 5 min of the test, the cyclist was likely too conservative.

NORMATIVE DATA

Normative data collected on male and female cyclists of different disciplines are presented in table 19.8.

TABLE 19.8 Data From 30 Min Time Trial for Female and Male High-Performance Cyclists: Mean ± SD (Range)

Athlete group	Body mass, kg	Power output, W	Power output, W · kg^{-1}	% MAP	$\dot{V}O_2$max, L · min^{-1}	% $\dot{V}O_2$max	HR, beats · min^{-1}	% HRmax	Blood lactate, mmol · L^{-1}	Cadence, rpm
Female										
Senior road (n = 12), lode ergo	58.8 ± 3.7 (50.9-64.9)	253 ± 14.8 (234-274)	4.3 ± 0.2 (4.0-4.7)	82.0 ± 4.0 (78-85)	3.28 ± 0.18 (2.98-3.49)	90.2 ± 4.0 (82-98)	179 ± 9.2 (163-200)	93 ± 1.4 (91-95)	7.9 ± 2.3 (4.7-12.2)	103.9 ± 3.6 (99-109)
Senior road (n =10), air-braked ergo	58.6 ± 5.1 (49.9-67.8)	253 ± 15.1 (237-283)	4.3 ± 0.4 (3.9-4.9)	81.0 ± 2.9 (78-86)	n/a	n/a	176 ± 6.8 (165-188)	n/a	8.8 ± 1.4 (7.1-11.7)	102.4 ± 2.5 (96-105)
Senior MTB (n = 5), air-braked ergo	62.2 ± 2.1 (60.3-65.2)	255 ± 20.7 (227-279)	4.1 ± 0.3 (3.7-4.5)	83.7 ± 3.8 (79-89)	3.31 ± 0.20 (3.12-3.56)	84.5 ± 2.6 (82-89)	177 ± 8.3 (169-188)	89 ± 5.4 (80-93)	6.6 ± 1.8 (4.1-9.1)	90.8 ± 5.1 (83-95)
Male										
Senior MTB (n = 4), air-braked ergo	74.8 ± 9.3 (67.9-88.5)	359 ± 24.7 (331-386)	4.9 ± 0.3 (4.7-5.3)	86.4 ± 1.1 (85-87)	4.73 ± 0.29 (4.41-5.11)	89.2 ± 1.8 (87-94)	176 ± 6.8 (167-181)	92 ± 3.6 (88-95)	9.6 ± 1.8 (7.3-11.1)	98.5 ± 3.0 (96-102)

n/a = not available; HR = heart rate; MAP = maximal aerobic power. Typical error: power output = 7 W (2.9%). Data source: Road (January/February 2000, March 2008), MTB (March 2009).

POWER PROFILE TEST

RATIONALE

Prior to the advent of mobile power-measuring devices, such as the SRM crank (SRM, Julich, Germany), quantification of the demands of cycling racing and the performance of cyclists during actual competition was limited to heart rate analysis and estimations of cycling power (Jeukendrup et al. 2000; Mujika and Padilla 2001). Unlike time trial performance, which has been correlated to laboratory fitness assessments (Coyle et al. 1991), finishing position in mass-start road races has not been signifi-cantly linked to traditional physiological indices; that is, due to the drafting and team tactics involved in road racing, the cyclist with the highest $\dot{V}O_2$max does not necessarily win. Rather, the decisive moment of a race may come down to how much power a cyclist can produce for a given period of time, for example, over 5 s in the final sprint to the line or for 5 min over a climb (Ebert et al. 2005; Quod et al. 2010).

Because of the widespread popularity of mobile power measuring devices, it has been possible to obtain data from cyclists during competition (Ebert et al. 2005, 2006; Quod et al. 2010). Analysis of these race data has enabled scientists not only to characterize the power output demands of particular competitions but also to identify the duration, number, and intensity of particular efforts that are required in order to be successful (Ebert et al. 2005).

The power profile test aims to determine the highest power output (maximal mean power: MMP) a cyclist can hold for a particular duration of effort when in a controlled environment and prepared for testing. The resulting time–power continuum determines the maximal capabilities of the cyclist at that phase of the season. The test has recently been shown to provide an ecologically valid assessment of a cyclist's power producing capabilities when compared with the efforts of the same duration obtained during actual competition (Quod et al. 2010). Therefore, the test may be useful for quantifying aspects of race performance as well as for talent identification and training prescription.

TEST PROCEDURE

- The tester records the subject's age, height, and body mass prior to the test.
- Complete common procedures as outlined previously.
- The power profile test may be performed on a calibrated air-braked cycle ergometer (e.g., Wombat, AIS, Canberra) or using a calibrated SRM mounted to the cyclist's bicycle, which is then placed in a wind trainer or rear wheel ergometer. When choosing the ergometer on which the test is to be performed, one should consider the kinetic energy associated with the system so as to replicate the feel associated with riding on the road as closely as possible. The ergometer's gearing should be adaptable to allow the cyclist to self-select a cadence, which will allow them to produce a maximal effort over the duration of each effort.

- The SRM should be set to a sampling rate of 1 s (or less) with the correct slope entered.
- A 5 min warm-up at 75 to 100 W is conducted to familiarize the cyclist with the bike setup and should include at least two 3 to 5 s sprints to ensure the correct gearing is selected for the test (this gearing should be recorded when possible). The gear selected should enable the cyclist to self-select a cadence that is appropriate to the duration of the effort and similar to when road riding.
- Following the warm-up, the SRM should be zeroed and the value recorded.
- The cyclist should produce as much power as possible for the duration of each effort. For the shorter efforts (6-30 s), the cyclist will typically engage in an all-out sprint. However, for the longer efforts (1-10 min), an element of pacing is required.
- The test protocol is outlined in table 19.9. Note that the first two sprints are conducted from a standing start, with the remaining

TABLE 19.9　Power Profile Test Protocol

Time, min:s	Effort	Power output	Heart rate	Metabolic analysis	Blood sample	Notes
0:00-0:06	6 s, small gear	*	*			Standing start
0:06-1:00	Active recovery (54 s)	*	*			50-100 W
1:00-1:06	6 s, big gear	*	*			Standing start
1:06-4:00	Active recovery (174 s)	*	*			50-100 W
4:00-4:15	15 s	*	*			Rolling start (70-80 rpm)
4:15-8:00	Active recovery (225 s)	*	*			50-100 W
8:00-8:30	30 s	*	*			Rolling start (70-80 rpm)
8:30-14:00	Active recovery (330 s)	*	*			50-100 W
14:00-15:00	1 min	*	*		†	Rolling start (70-80 rpm)
15:00-23:00	Active recovery (480 s)	*	*			50-100 W
23:00-27:00	4 min	*	*	*	†	Rolling start (70-80 rpm)
27:00-37:00	Active recovery (600 s)	*	*			50-100 W
37:00-47:00	10 min	*	*	*	†	Rolling start (70-80 rpm)

*Sampling during effort.

†Sampling before and after effort.

efforts completed from a rolling start between 70 and 80 rpm.

- Power output and cadence data are recorded throughout the test using a sampling rate of 1 s or less.

- Heart rate is recorded throughout the test using a sampling rate of 5 s. In addition, the heart rate at the end of each effort should be recorded.

- Oxygen consumption ($\dot{V}O_2$), carbon dioxide production ($\dot{V}CO_2$), respiratory exchange ratio (RER), ventilation ($\dot{V}_E$), ventilatory equivalents (i.e., ratio of $\dot{V}_E$ to $\dot{V}O_2$, ratio of $\dot{V}_E$ to $\dot{V}CO_2$) are recorded during the 4 and 10 min efforts.

- A capillary blood sample is collected from the fingertip or ear lobe 1 min prior to and 1 min after the 1 min, 4 min, and 10 min efforts. The sample is analyzed for lactate concentration and (if required) pH, blood gases, and bicarbonate concentration.

- On completion of the test, the zero offset of the SRM should be checked for drift.

DATA ANALYSIS

- Average power (MMP) and cadence for each effort are determined using commercially available software (e.g., SRM training software or Training peaks WKO+) (PeaksWare, USA). For the 2 × 6 s efforts, the highest average MMP for 5 s (MMP_{5s}) is reported. Peak power output for 1 s may also be of interest. MMP for each effort should be reported in W and $W \cdot kg^{-1}$.

- $\dot{V}O_2$peak (the highest recorded $\dot{V}O_2$ achieved during the test) produced during the 4 and 10 min effort is reported in $L \cdot min^{-1}$ and $ml \cdot kg^{-1} \cdot min^{-1}$. Providing the cyclist produces a well-paced, maximal effort during the 4 min piece, the $\dot{V}O_2$peak (the highest recorded $\dot{V}O_2$max achieved during the test) obtained during the effort provides a valid estimate of $\dot{V}O_2$max (Gore et al. 1998).

- Peak heart rate obtained during the 4 or 10 min effort or the highest average MMP for 4 min (MMP_{4min}) can be used to determine training zones based on a five-zone model (see table 19.5).

- The time–power relationship can be plotted and used to monitor changes over time or compare characteristics between athletes (figure 19.3). Typically, sprinters will display higher values over the shorter efforts, with

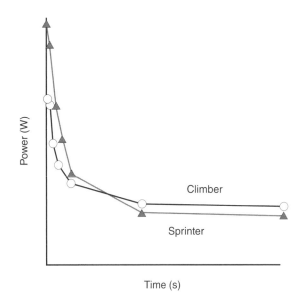

FIGURE 19.3 Examples of time–power relationship in two cyclists displaying different characteristics.

a steep decline in the curve as the duration of effort increases. Time trial specialists and hill climbers may exhibit flatter curves with minimal drop off in the tail of the curve (figure 19.3).

- If the cyclist has access to a mobile power measuring device, it is possible to relate testing performance to in-field race performance. An appreciation for measurement error associated with different power measuring devices is important here (Gardner et al. 2004; Paton and Hopkins 2001). In addition, it is unlikely that a maximal effort for each duration assessed during the power profile test will be produced during a single race, so analysis over series of races may be more appropriate (Quod et al. 2010).

LIMITATIONS

A familiarization trial on the test ergometer is recommended. Aspects of the test require an element of pacing, and cyclists may perform better on a second attempt at the test. Gear selection can have a limiting role on the power output produced. The average cadence of the effort may provide insight as to whether the gear selection was appropriate. Some cyclists may find that they produce lower power outputs during the shorter efforts compared to the field due to an inability to move the bike laterally in a laboratory setting (Quod et al. 2010).

NORMATIVE DATA

Normative data collected on male and female cyclists of different disciplines are presented in table 19.10.

TABLE 19.10   Power Profile Data for Female and Male High-Performance Cyclists: Mean ± *SD* (Range)

Parameter	Senior female road (n = 8)	Junior female (n = 4)	Under 23 male track endurance (n = 6)	Under 23 male road (n = 9)
Body mass, kg	63.6 ± 8.5 (50.2-76.3)	58.1 ± 1.9 (56.1-60.6)	76.4 ± 3.0 (72.4-80.2	72.6 ± 6.7 (64.1-81.9)
MMP_{5s}, W	877 ± 191.7 (659-1,222)	788 ± 63.4 (715-858)	1,126 ± 128.8 (984-1,330)	1,062 ± 103.1 (864-1,175)
MMP_{15s}, W	680 ± 128.6 (516-875)	664 ± 59.4 (576-705)	910 ± 53.8 (846-994)	836 ± 94.1 (677-945)
MMP_{30s}, W	526 ± 77.9 (438-658)	518 ± 59.7 (436-577)	739 ± 94.5 (644-905)	671 ± 73.0 (560-776)
MMP_{60s}, W	415 ± 54.6 (356-523)	368 ± 31.0 (334-398)	571 ± 47.8 (528-647)	536 ± 72.2 (427-651)
MMP_{4min}, W	302 ± 41.8 (243-361)	265 ± 5.3 (259-271)	423 ± 59.0 (376-523)	403 ± 64.5 (345-517)
MMP_{10min}, W	264 ± 29.6 (228-318)	239 ± 9.4 (228-249)	365 ± 45.6 (330-441)	354 ± 38.8 (314-433)
MMP_{5s}, $W \cdot kg^{-1}$	13.7 ± 1.8 (10.6-16.0)	13.6 ± 0.9 (12.5-14.7)	14.7 ± 1.3 (13.6-17.2)	14.6 ± 0.8 (13.2-15.9)
MMP_{15s}, $W \cdot kg^{-1}$	10.7 ± 1.0 (8.7-11.9)	11.4 ± 1.0 (10.1-12.4)	11.9 ± 0.5 (11.3-12.8)	11.5 ± 0.6 (10.3-12.6)
MMP_{30s}, $W \cdot kg^{-1}$	8.3 ± 0.6 (7.6-9.3)	8.9 ± 0.9 (7.6-9.5)	9.6 ± 0.9 (8.8-11.3)	9.2 ± 0.6 (8.3-9.9)
MMP_{60s}, $W \cdot kg^{-1}$	6.6 ± 0.5 (5.7-7.1)	6.3 ± 0.6 (5.8-7.1)	7.5 ± 0.6 (6.6-8.2)	7.4 ± 0.7 (6.1-8.2)
MMP_{4min}, $W \cdot kg^{-1}$	4.8 ± 0.5 (3.6-5.4)	4.6 ± 0.2 (4.3-4.7)	5.6 ± 0.7 (5.1-6.6)	5.5 ± 0.4 (4.9-6.3)
MMP_{10min}, $W \cdot kg^{-1}$	4.2 ± 0.4 (3.4-4.8)	4.1 ± 0.3 (3.8-4.4)	4.8 ± 0.5 (4.3-5.6)	4.9 ± 0.2 (4.5-5.3)
$\dot{V}O_2$peak 4 min, $L \cdot min^{-1}$	3.98 ± 0.44 (3.57-4.60)	3.46 ± 0.14 (3.35-3.66)	n/a	5.05 ± 0.39 (4.46-5.74)
$\dot{V}O_2$peak 4 min, $ml \cdot kg^{-1} \cdot min^{-1}$	63.0 ± 5.2 (52.9-71.1)	59.7 ± 3.2 (56.3-64.0)	n/a	70.31 ± 4.11 (65.9-76.7)
$\dot{V}O_2$peak 10 min, $L \cdot min^{-1}$	3.82 ± 0.40 (4.60-3.28)	n/a	n/a	4.90 ± 0.38 (4.48-5.63)
$\dot{V}O_2$peak 10 min, $ml \cdot kg^{-1} \cdot min^{-1}$	60.4 ± 3.8 (53.4-65.3)	n/a	n/a	68.33 ± 4.79 (62.0-76.9)

n/a = not available; MMP = maximal mean power. Typical error: power output: MMP_{5s} = 22 W, 2.0%; MMP_{15s} = 9 W, 1.6%; MMP_{30s} = 18 W, 2.4%; MMP_{1min} = 15 W, 2.4%; MMP_{4min} = 6 W, 2.7%; MMP_{10min} = 7 W, 2.8%; $\dot{V}O_2$peak = 0.04 $L \cdot min^{-1}$, 1.2%. Data source: senior female (January 2010), junior female (July 2008), Under 23 road (February 2007) track endurance (April 2010).

ALACTIC PEAK POWER TEST (6 S POWER TEST)

RATIONALE

Sprint track cycling requires a maximal rate of energy expenditure that must be matched by a rapid rate of energy resynthesis. In particular, the flying 200 m and matched sprint race are track cycling events lasting approximately 10 to 15 s for which the alactic anaerobic energy system is the dominant metabolic energy pathway (Craig and Norton 2001; Jeukendrup et al. 2000). This energy system is also relevant to other track and road cycling events, for example, the start of the 1,000 m time trial, individual and team pursuit, and the sprint to the finish line at the end of a road race. Similarly in BMX racing, athletes are required to produce approximately 6 to 8 maximal sprints of 1 to 3 s in duration over a 30 to 40 s period while negotiating various technical aspects of the track. As such, particular importance is placed on the alactic energy system, with race performance often determined by the riders who can repeatedly produce the highest power outputs without making any technical errors. The alactic peak power test described next can be used as a measure of alactic anaerobic performance in track sprint and BMX cyclists. When compared with similar efforts over 65 m in the field using world-class sprint cyclists, peak power output and power– and torque–pedaling relationships obtained during the 6 s power test are similar and may provide a valid means of modeling sprint performance (Gardner et al. 2007). In addition, the test may be a useful tool in the talent identification of potential BMX and track sprint cyclists. The tests described in this chapter will provide a useful estimation of an athlete's peak power producing capability. For a more diagnostic approach to determining maximal force or the power–cadence-producing capabilities, please refer to chapter 4, Ergometer-Based Maximal Neuromuscular Power.

TEST PROCEDURE

- The tester records the subject's age, height, and body mass prior to the test.
- Common test procedures are completed as outlined previously.
- The test should be performed on a calibrated air-braked ergometer (e.g., Wombat, AIS, Canberra).

- Depending on the ergometer used, the selected gearing should enable riders to approximate the cadences achieved during racing conditions: for example, 180 to 210 rpm for senior elite female and male BMX riders. When one is using an air-braked ergometer with a lightweight flywheel, a slightly heavier gear is recommended, which will slightly underestimate the peak cadences predominantly for safety reasons.
- Cyclists complete 2 or 3 warm-up sprints to familiarize themselves with the ergometer and ensure that the appropriate gear is selected.
- The test consists of a 6 s all-out maximal effort from a standing start. No pacing is allowed. The flywheel of the ergometer must be completely stationary prior to starting the test.
- If required, a second all-out sprint is performed in a larger gear, after a minimum 2 to 3 min recovery period.
- When BMX riders are tested, the test protocol is modified to accommodate the average length of each sprint in competition. BMX riders generally perform two trials using two different gear ratios. The duration of each trial is a maximum of 4 s or the point at which peak power begins to decrease, whichever comes first.
- Power output and cadence are recorded continuously a sampling rate of 5 Hz (0.2 s).

DATA ANALYSIS

- Peak power and maximal mean power for 5 s (MMP_{5s}) should be reported in watts and watts per kilogram.
- The results may be used to prescribe alactic interval training on the ergometer.

NORMATIVE DATA

Peak power normative data for track sprint cyclists collected in 2001–2002 are presented in table 19.11. More recent data are not available, because the majority of testing of Australian sprint cyclists is now conducted within training sessions using SRM cranks. Peak power data for BMX cyclists are presented in table 19.12.

TABLE 19.11 Peak Power Test Results for Female and Male High-Performance Track Sprint Cyclists: Mean ± *SD* (Range)

Parameter	Senior female (*n* = 5)	Senior male (*n* = 7)
Body mass, kg	65.7 ± 7.3 (56.7-76.6)	87.3 ± 6.2 (80.0-98.0)
Peak power, W	1,170 ± 100 (1,065-1,330)	1,840 ± 200 (1,615-2,282)
Peak power, W · kg^{-1}	17.9 ± 0.7 (17.0-19.0)	21.0 ± 1.0 (19.6-23.3)
Peak cadence, rpm	156 ± 17 (133-177)	156 ± 4 (151-160)

Typical error: peak power = 18.2 W (1.1%). Data collection: 2001-2002.

BMX 6 × 3 S REPEAT SPRINT TEST

RATIONALE

A single BMX race is typically 30 to 40 s in duration, depending on the track layout, and generally involves approximately 6 to 8 maximal sprints of between 1 and 3 s. In a 35 s race, the rider may only complete approximately 40 to 50 pedal revolutions under load; however, a significant upper-body contribution is needed to overcome the numerous jumps and obstacles on the track. In addition to the ability to generate high levels of peak power, the rider must be able to repeatedly accelerate into and out of the various corners and between particular jump obstacles.

TEST PROCEDURE

- The tester records the subject's age, height, and body mass prior to the start of the test.
- Common test procedures are completed as outlined previously.
- The test is performed on a calibrated air-braked cycle ergometer. The gear ratio should be consistent for all athletes and tests. Depending on the ergometer used, the selected gearing should approximate the cadences achieved during racing conditions. When an air-braked ergometer with a lightweight flywheel is used, a slightly heavier gear is recommended, which will slightly underestimate the peak cadences; hence, the work to rest ratio must be greater than that observed in racing conditions. The

19 tooth sprocket is recommended (e.g., 48/14/40/19).

- The test consists of 6 × 3 s maximal sprints on a 30 s cycle (i.e., 27 s of recovery between each 3 s maximal effort). The test begins from a standing start. The recovery periods are undertaken at a cadence of approximately 80 rpm and approximately 80 to 100 W depending on the ergometer and gearing, with the cyclist remaining in the saddle for the full length of the test. The test is completed in exactly 153 s.
- Power output and cadence are recorded continuously at a sampling rate of 5Hz (0.2 s).

DATA ANALYSIS

- Peak power for each 3 s sprint should be reported in watts and watts per kilogram.
- The decrement in peak power over the 6 efforts is reported in absolute terms (watts) and in percentages.
- The test protocol can also be used as an actual training session because it provides instantaneous feedback regarding progress.

NORMATIVE DATA

Normative data for the BMX 6 × 3 s repeat sprint test for male and female BMX cyclists are presented in table 19.12.

TABLE 19.12 Peak Power and 6 × 3 s Sprint Test Results for High-Performance Female and Male BMX Cyclists: Mean ± *SD* (Range)

Parameter	Senior female (*n* = 4)	Junior female (*n* = 4)	Senior male (*n* = 9)	Junior male (*n* = 4)
Body mass, kg	70.2 ± 2.9 (67.4-74.3)	68.9 ± 7.2 (60.5-78.2)	84.4 ± 6.7 (71.0-93.0)	85.1 ± 2.8 (81.4-87.6)
Peak power, W*	1,239 ± 67.5 (1,188-1,338)	1,161 ± 143.3 (1,005-1,346)	1,993 ± 142.6 (1,790-2,158)	1,721 ± 45.4 (1,679-1,783)
Peak power, W · kg^{-1}*	17.5 ± 1.2 (16.2-18.8)	16.8 ± 0.4 (16.8-17.2)	23.7 ± 1.9 (19.8-25.5)	20.2 ± 1.1 (19.2-21.2)
Peak power, W[†]	1,198 ± 71.4 (1,157-1,305)	1,090 ± 102.8 (971-1,215)	1,876 ± 120.9 (1,700-2,029)	1,579 ± 45.7 (1,537-1,638)
Peak power, W · kg^{-1}[†]	16.9 ± 1.1 (15.7-18.3)	15.8 ± 0.3 (15.5-16.2)	22.4 ± 2.0 (918.8-24.6)	18.6 ± 1.0 (17.6-19.5)
Sprint 1 power, W	1,219 ± 63.3 (1,170-1,311)	1,141 ± 127.1 (991-1,277)	2,000 ± 99.0 (1,837-2,092)	1,700 ± 45.3 (1,658-1,750)
Sprint 2 power, W	1,118 ± 53.6 (1,077-1,197)	1,063 ± 144.5 (868-1,217)	1,880 ± 132.0 (1,727-2,005)	1,574 ± 39.0 (1,520-1,613)
Sprint 3 power, W	1,018 ± 54.3 (946-1,042)	1,026 ± 126.8 (853-1,154)	1,807 ± 134.0 (1,646-1,907)	1,513 ± 35.7 (1,483-1,565)
Sprint 4 power, W	958 ± 66.9 (908-1,055)	1,020 ± 116.1 (867-1,136)	1,773 ± 114.0 (1,628-1,887)	1,484 ± 56.0 (1,426-1,556)
Sprint 5 power, W	963 ± 82.6 (904-1,085)	1,003 ± 141.0 (821-1,154)	1,749 ± 110.0 (1,633-1,863)	1,472 ± 55.8 (1,413-1,544)
Sprint 6 power, W	975 ± 57.0 (919-1,041)	1,001 ± 127.4 (826-1,121)	1,727 ± 89.0 (1,598-1,822)	1,448 ± 56.3 (1,410-1,531)
Decrement (sprint 1-6), %	19.9 ± 7.0 (14.0-28.6)	12.4 ± 3.3 (8.5-16.6)	13.7 ± 1.4 (12.1-15.8)	14.7 ± 4.9 (7.7-18.1)
Drop off (max-min), W	276 ± 98.1 (169-407)	146 ± 26.2 (109-170)	284 ± 27.0 (239-310)	254 ± 89.5 (127-324)

Chain rings: 48, 14, 40, 19. [†]Chain rings: 48, 14, 40, 21. Typical error: peak power = 18.2 W, 1.1%; 21.5 W, 1.3%[†]; 0.23 W · kg^{-1}, 1.1%*; 0.26 W · kg^{-1}, 1.3%[†]. % Decrement = 2.0%; drop off = 26.6 W, 23.4%. Data collection: July 2009, 2010.

References

Allen, H., and Coggan, A. 2006. *Training and Racing with a Power Meter*. Boulder, CO: VeloPress.

Coyle, E.F., Coggan, A.R., Hopper, M.K., and Walters, T.J. 1988. Determinants of endurance in well-trained cyclists. *Journal of Applied Physiology* 64(6):2622-2630.

Coyle, E.F., Feltner, M.E., Kautz, S.A., Hamilton, M.T., Montain, S.J., Baylor, A.M., Abraham, L.D., and Petrek, G.W. 1991. Physiological and biomechanical factors associated with elite endurance cycling performance. *Medicine and Science in Sports and Exercise* 23(1):93-107.

Craig, N.P., and Norton, K.I. 2001. Characteristics of track cycling. Sports Med 31(7):457-468.

Craig, N.P., Norton, K.I., Bourdon, P.C., Woolford, S.M., Stanef, T., Squires, B., Olds, T.S., Conyers, R.A., and Walsh, C.B. 1993. Aerobic and anaerobic indices contributing to track endurance cycling performance. *European Journal of Applied Physiology and Occupational Physiology* 67(2):150-158.

Ebert, T.R., Martin, D.T., McDonald, W., Victor, J., Plummer, J., and Withers, R.T. 2005. Power output during women's World Cup road cycle racing. *European Journal of Applied Physiology* 95(5-6):529-536.

Ebert, T.R., Martin, D.T., Stephens, B., and Withers, R.T. 2006. Power Output During a Professional Men's Road-Cycling Tour. *International Journal of Sports Physiology and Performance* 1(4):324-335.

Gardner, A.S., Martin, J.C., Martin, D.T., Barras, M., and Jenkins, D.G. 2007. Maximal torque- and power-pedaling rate relationships for elite sprint cyclists in laboratory and field tests. *European Journal of Applied Physiology* 101(3):287-292.

Gardner, A.S., Stephens, S., Martin, D.T., Lawton, E., Lee, H., and Jenkins, D. 2004. Accuracy of SRM and power tap power monitoring systems for bicycling. *Medicine and Science in Sports and Exercise* 36(7):1252-1258.

Gore, C.J., Hahn, A., Rice, A., Bourdon, P., Lawrence, S., Walsh, C., Stanef, T., Barnes, P., Parisotto, R., Martin, D., and Pyne, D. 1998. Altitude training at 2690m does not increase total haemoglobin mass or sea level VO_2max in world champion track cyclists. *Journal of Science and Medicine in Sport* 1(3):156-170.

Jeukendrup, A., Saris, W.H., Brouns, F., and Kester A,D. 1996. A new validated endurance performance test. *Medicine and Science in Sports and Exercise* 28(2):266-270.

Jeukendrup, A.E., Craig, N.P., and Hawley, J.A. 2000. The bioenergetics of World Class Cycling. *Journal of Science and Medicine in Sport* 3(4):414-433.

Joyner, M.J., and Coyle, E.F. 2008. Endurance exercise performance: The physiology of champions. *Journal of Physiology* 586(Pt 1):35-44.

Laursen, P.B., Shing, C.M., Peake, J.M., Coombes, J.S., and Jenkins, D.G. 2002. Interval training program optimization in highly trained endurance cyclists. *Medicine and Science in Sports and Exercise* 34(11):1801-1807.

Lee, H., Martin, D.T., Anson, J.M., Grundy, D., and Hahn, A.G. 2002. Physiological characteristics of successful mountain bikers and professional road cyclists. *Journal of Sports Sciences* 20(12):1001-1008.

Levine, B.D., and Stray-Gundersen, J. 1997. "Living high-training low": Effect of moderate-altitude acclimatization with low-altitude training on performance. *Journal of Applied Physiology* 83(1):102-112.

Martin, D.T., McLean, B., Trewin, C., Lee, H., Victor, J., and Hahn, A.G. 2001. Physiological characteristics of nationally competitive female road cyclists and demands of competition. *Sports Medicine* 31(7):469-477.

Martin, J.C., Gardner, A.S., Barras, M., and Martin, D.T. 2006. Modeling sprint cycling using field-derived parameters and forward integration. *Medicine and Science in Sports and Exercise* 38(3):592-597.

Martino, M., Gledhill, N., and Jamnik, V. 2002. High VO_2max with no history of training is primarily due to high blood volume. *Medicine and Science in Sports and Exercise* 34(6):966-971.

Mujika, I., and Padilla, S. 2001. Physiological and performance characteristics of male professional road cyclists. *Sports Medicine* 31(7):479-487.

Olds, T., Norton, K., Craig, N., Olive, S., and Lowe, E. 1995a. The limits of the possible: Models of power supply and demand in cycling. *Australian Journal of Science and Medicine in Sport* 27(2):29-33.

Olds, T.S., Norton, K.I., Lowe, E.L., Olive, S., Reay, F., and Ly, S. 1995b. Modeling road-cycling performance. *Journal of Applied Physiology* 78(4):1596-1611.

Padilla, S., Mujika, I., Orbananos, J., Santisteban, J., Angulo, F., and Jose Goiriena, J. 2001. Exercise intensity and load during mass-start stage races in professional road cycling. *Medicine and Science in Sports and Exercise* 33(5):796-802.

Paton, C.D., and Hopkins, W.G. 2001. Tests of cycling performance. *Sports Medicine* 31(7):489-496.

Quod, M.J., Martin, D.T., Martin, J.C., and Laursen, P.B. 2010. The power profile predicts road cycling MMP. *International Journal of Sports Medicine* 31(6):397-401.

Football (Soccer) Players

Darren J. Burgess and Tim J. Gabbett

Soccer, or football as it is known globally, is arguably the most widely played sport in the world. As evidence of its popularity, the World Cup, which is held every 4 years, attracts a larger television audience than the Olympic Games.

The demands of soccer match play have been studied for more than 40 years. Published literature using techniques ranging from hand notation to multicamera video analysis (Burgess et al. 2006; Carling et al. 2008; di Salvo et al. 2006; Ekblom 1986; Mayhew and Wenger 1985; Osgnach et al. 2010; Reilly and Thomas 1979) have determined the following requirements of the game at an elite level:

- Males typically travel between 10 and 13 km during a 90 min game; females cover between 9 and 11 km.
- Each player is in possession of the ball for, on average, less than 2 min per game.
- Sprinting, commonly defined as running faster than $20 \text{ km} \cdot \text{h}^{-1}$, contributes between 5% and 10% of total match distance.
- Each player performs a sprinting action, on average, every 70 to 90 s.
- Players travel approximately 5% further in the first half compared with the second half.

Given that most research suggests that the game is increasing in speed, these statistics indicate that the physical attributes of speed, agility, strength, power, and endurance are important qualities for soccer success.

Assessment of these physical qualities will enable the coach and fitness coach to determine player physical strengths and weaknesses and to assess player and team responses to training protocols.

Athlete Preparation

To ensure test reliability, the following pretest conditions should be observed in addition to those outlined in chapter 2, Pretest Environment and Athlete Preparation.

- *Diet*. It is not within the scope of this chapter to discuss specific dietary requirements for soccer. However, prior to testing athletes' diets should be, where possible, standardized. In general, players should not consume substantial quantities of food within 2 h of the test and should be well hydrated for all performance-based protocols.

- *Training*. Training prior to testing should be minimal and, where possible, standardized. Minimal training should occur on the day of field testing.

- *Testing*. Coaches and fitness coaches should agree on appropriate timing of physical fitness assessments. Typically this may occur at the start and conclusion of preseason, and some assessment protocols may occur at prearranged periods throughout the season. In-season assessment should be avoided during heavy training weeks and in the days surrounding games.

Test Environment

Ideally, tests should be conducted in a controlled, repeatable environment in order to ensure the reliability of test data. This controlled environment should include the following:

- Tests should be conducted at the same time of day so that the diurnal variation in physical attributes is minimized and variations of temperature and humidity are reduced.

- Particularly with running assessments, the surface should be standardized for each assessment. Ideally, sprinting and shuttle-based assessments should be conducted indoors on a nonslip surface.

- Environmental conditions will often influence field testing, so documentation of ambient temperature, humidity, light, and wind is essential.

Recommended Test Order

It is important that tests are completed in the same order to control the interference between tests. This order also allows valid comparison of different test occasions. The recommended order is as follows:

Day	Tests
1	Anthropometry
	Vertical jump
	Yo-Yo intermittent recovery
2	20 m sprint
	6 × 20 m repeat sprint ability

Equipment Checklist

Anthropometry

- ☐ Stadiometer (wall mounted)
- ☐ Balance scales (accurate to ±0.05 kg)
- ☐ Anthropometry box
- ☐ Skinfold calipers (Harpenden skinfold caliper)
- ☐ Marker pen
- ☐ Anthropometric measuring tape
- ☐ Recording sheet (refer to the anthropometry data sheet template)
- ☐ Pen

Vertical Jump Test

- ☐ Jump measuring device (e.g., Swift Performance Yardstick)
- ☐ Measuring tape
- ☐ Recording sheet (refer to the vertical jump data sheet template)
- ☐ Pen

20 m Sprint Test

- ☐ Electronic light gate equipment (e.g., dual-beam light gates)
- ☐ Measuring tape
- ☐ Field marking tape
- ☐ Witches hats
- ☐ Recording sheet (refer to the sprint test data sheet template)
- ☐ Pen

6 × 20 m Repeat Sprint Ability Test

- ☐ Electronic light gate equipment (e.g., dual-beam light gates)
- ☐ Measuring tape
- ☐ Field marking tape
- ☐ Marker cones
- ☐ Chalk
- ☐ Sound box or CD/MP3 player
- ☐ 6 × 30 m repeat sprint ability test CD/MP3
- ☐ Stopwatch
- ☐ Recording sheet (refer to the repeat sprint ability test data sheet template)
- ☐ Pen

Yo-Yo Intermittent Recovery Test and Heart Rate Recovery Test

- ☐ Measuring tape
- ☐ Witches hats
- ☐ Sound box or CD player
- ☐ Yo-Yo intermittent recovery test CD
- ☐ Recording sheet (refer to the Yo-Yo IRT data sheet template)
- ☐ Pen
- ☐ Heart rate monitoring equipment (heart rate recovery test only)

ANTHROPOMETRY

RATIONALE

Soccer is a high-intensity, intermittent sport requiring repeated explosive actions (sprinting, jumping, striking, tackling) over a 90 min game. To be explosive, players need to produce high amounts of force over a short period of time. Additional fat mass can be counterproductive to producing force and therefore soccer performance. An estimation of a player's fat mass is an important assessment tool. Subcutaneous skinfold assessment provides a reliable, valid, and relatively quick estimation of a player's total fat mass.

TEST PROCEDURE

Measurement of height, body mass, and skinfolds should be carried out prior to other field testing protocols. Skinfolds are recorded over seven sites (triceps, biceps, subscapular, supraspinale, abdominal, front thigh, and medial calf). The individual skinfold measures as well as the sum of the seven sites should be reported. Refer to anthropometry protocols outlined in chapter 11, Assessment of Physique, for a detailed description of all anthropometric test procedures. Measurement of height and body mass should also form a part of anthropometric assessment of soccer players. Traditionally, goalkeepers are the tallest players, followed by central defenders and strikers, with midfielders generally being the shortest players on a soccer team.

Although the description of skinfold measurement procedures seems simple, a high degree of technical skill is essential for consistent results. It is therefore important that these measurements be taken by an experienced tester who has been trained in these techniques. It is also important that the same tester conduct each retest to ensure reliability.

DATA ANALYSIS

Serial monitoring of anthropometry allows a coach or fitness coach to assess the effects of training programs on body type and indirectly assess players' diets. At the elite level, anthropometric assessment can assist in position-specific profiling, which may assist coaches in talent identification and development.

NORMATIVE DATA

Table 20.1 presents anthropometric normative data for female and male soccer players.

Table 20.1 Anthropometric Data for National Female and Male Soccer Players: Mean ± *SD* (Range)

Athlete group	Height, cm	Body mass, kg	Σ7 skin-folds, mm
Female			
National under 17s	163.1 ± 7.6 (154.5-170.1)	52.9 ± 4.2 (44.6-68.4)	98.6 ± 11.1 (63.6-151.7)
National squad	166.8 ± 6.1 (156.2-178.3)	61.3 ± 4.7 (52.5-72.7)	75.2 ± 7.2 (53.7-103.1)
Male			
National under 17s	176.7 ± 5.8 (163.3-190.8)	69.6 ± 4.8 (58.0-82.4)	56.4 ± 7.6 (43.2-79.6)
Olympic (under 23s) squad	181.8 ± 6.1 (165.0-192.2)	77.5 ± 5.1 (60.0-96.4)	57.8 ± 9.3 (37.2-78.1)
National squad	182.3 ± 6.5 (166.1-197.8)	78.3 ± 6.6 (61.2-96.8)	47.5 ± 11.1 (32.3-73.6)

Data source: female National under 17s, n = 42, 2005-2009; female National squad, n = 38, 2006-2009; male National under 17s, n = 74, 2003-2009; male Olympic squad, n = 54, 2003-2008; male National squad, n = 44, 2007-2010.

VERTICAL JUMP TEST

RATIONALE

Soccer players require well-developed muscular power for the many times they are required to jump and rapidly accelerate during the game. The vertical jump test is selected as the method for evaluating both explosive leg power and jumping ability. The test is easily performed in the field or laboratory and requires limited equipment, and substantial sport specific normative data are available.

Equipment required for the test can vary according to budget available and additional requirements of the equipment. The options are these:

- Force plate or platform: expensive yet comprehensive laboratory-based force assessment device that can assess numerous force and power parameters

- Contact mat: inexpensive, portable device designed to quickly and effectively assess ground contact time, flight time, and jump height

- Yardstick jumping device: device that directly measures jump height via the player touching the "vanes" attached to the device

TEST PROCEDURE

The standard vertical jump, with countermovement and arm swing, is used in soccer. Refer to vertical jump protocol in chapter 14, Field Testing Principles and Protocols, for setup details and test procedures.

DATA ANALYSIS

In general, vertical jump performance is expressed in two ways: as an absolute and a relative jump height. Absolute jump height is important when players are required to jump during a game (e.g., heading a ball), whereas relative jump height (i.e., the difference between standing reach height and jump height) provides an index of a player's lower-body muscular power and allows comparisons to be made between players. If desired, mechanical power can be calculated from vertical jump performances using the following equation: Peak Power (W) = 60.7 × [Jump Height (cm)] + 45.3 × [Body Mass (kg)] – 2,055 (Sayers et al. 1999).

NORMATIVE DATA

Vertical jump normative data for female and male soccer players are presented in table 20.2.

TABLE 20.2 Vertical Jump Data for National Female and Male Soccer Players: Mean ± *SD* (Range)

Athlete group	Jump height, cm
Female	
National under 17s	50 ± 4
	(42-60)
National squad	54 ± 4
	(46-61)
Male	
National under 17s	61 ± 5
	(54-69)
Olympic (under 23s) squad	62 ± 7
	(53-75)
National squad	64 ± 6
	(54-76)

Data source: female National under 17s, *n* = 26, 2007; female National squad, *n* = 27, 2006-2009; male National under 17s, *n* = 48, 2003-2009; male Olympic squad, *n* = 47, 2003-2008; male National squad, *n* = 44, 2007-2010.

20 M SPRINT TEST

RATIONALE

Acceleration and speed are crucial requirements in soccer; many studies conclude that these variables differentiate between players at different levels of the game. The average sprint distance in soccer in Australia is approximately 16 m (Burgess et al. 2006), so assessment of speed over this distance is appropriate. Acceleration is also important in soccer because it increases the player's chances of beating his or her opponent to the ball. Therefore, assessment over 5 m and 10 m is also recommended.

TEST PROCEDURE

For soccer, the sprint test is conducted over 20 m, with intermediate distances of 5 and 10 m. Athletes should be encouraged to start the sprint with their body mass over their front foot, shoulders and hips square in a crouch "ready" position, heel up on back foot, and no rocking allowed. Refer to straight line sprint testing protocol in chapter 14, Field Testing Principles and Protocols, for a description of the procedures involved in sprint testing.

DATA ANALYSIS

The fastest times achieved over the 5, 10, and 20 m distance are recorded, irrespective of trial number. To allow comparisons to be made between players and sports, sprint data are converted to acceleration and velocity scores. Although not a true maximal velocity, the 20 m to 10 m split is used to determine peak velocity achieved over the 20 m sprint distance. If coaches and sport scientists are interested in the maximal velocity achieved over the 20 m sprint distance, it is advisable to also include a split at 15 m. If coaches and sport scientists require information on the true maximal velocity of their players, they are encouraged to perform a traditional 40 m sprint test with splits recorded between 30 and 40 m.

NORMATIVE DATA

Table 20.3 presents 20 m sprint normative data for female and male soccer players.

TABLE 20.3 Sprint Data for National Female and Male Soccer Players: Mean ± *SD* (Range)

Athlete group	5 m time, s	10 m time, s	20 m time, s
Female			
National Squad	1.18 ± 0.09 (1.05-1.40)	1.95 ± 0.10 (1.77-2.21)	3.32 ± 0.16 (3.05-3.70)
Male			
National under 17s	1.06 ± 0.06 (0.95-1.18)	1.76 ± 0.12 (1.24-1.93)	3.00 ± 0.1 (2.79-3.19)
Olympic (under 23s) squad	1.05 ± 0.04 (0.96-1.17)	1.72 ± 0.05 (1.72-1.91)	2.91 ± 0.07 (2.81-3.16)
National squad	1.02 ± 0.04 (1.03-1.14)	1.69 ± 0.05 (1.70-1.91)	2.85 ± 0.07 (2.72-3.12)

Data source: female National squad, *n* = 27, 2006-2009; male National under 17s, *n* = 48, 2003-2009; male Olympic squad, *n* = 47, 2003-2008; male National squad, *n* = 44, 2007-2010.

6 × 20 M REPEAT SPRINT ABILITY TEST

RATIONALE

Repeated-sprint ability is widely accepted as a critical component of soccer match play. In soccer competition it has been shown that periods of fatigue follow the most intense bouts of high-intensity running (Mohr et al. 2005). In addition, repeated-sprint bouts are reported to occur immediately before a goal is scored or conceded, lending credence to the suggestion that the ability or inability to perform repeated sprints may prove critical to the outcome of the match (Spencer et al. 2004).

TEST PROCEDURE

Several repeat sprint protocols have been presented in the literature (Bishop and Edge 2006; Psotta et al. 2005; Spencer et al. 2006). Although several repeat sprint tests have been proposed for soccer, a major limitation of these tests is that few adequately assess the repeat sprint demands of competition in a game-specific manner. Of the repeat sprint protocols that have been developed for soccer, protocols either have been too strenuous or have been based on the overall exercise to rest ratios of competition (1:7-1:14) and have failed to take into account the most extreme repeat sprint demands of the sport.

Here, we recommend a repeat sprint test for soccer players (Gabbett 2010). The test involves maximal effort accelerations, decelerations, and active recovery periods that reflect the most extreme demands of international competition (Gabbett and Mulvey 2008). Players perform 6 × 20 m maximal effort sprints on a 15 s cycle. At the completion of each sprint, players perform a 10 m deceleration and 10 m active jog recovery (figure 20.1).

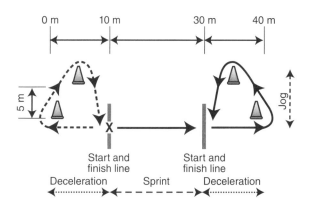

FIGURE 20.1 Schematic diagram of the 6 × 20 m repeat sprint test.

Adapted, by permission, from T.J. Gabbett, 2010, "The development of a text of repeated-sprint ability for elite women's soccer players," *Journal of Strength and conditioning Research* 24: 1191-1194.

DATA ANALYSIS

Repeat sprint test data are typically reported using the total repeat sprint time (i.e., the sum of all sprints) and the percentage decrement (or fatigue index). Percentage decrement has been calculated and reported in several different ways (Pyne et al. 2008); subtle differences in calculations can result in differences in the percentage decrement. If a fatigue index is of interest to the coach or sport scientist, we suggest the following calculation: (Total Time / Ideal Time) × 100, where total time represents the sum of the 6 individual sprints and the ideal time represents the theoretically ideal time if no decrement in performance occurred (i.e., the fastest sprint time multiplied by 6) (Spencer et al. 2006).

It has been shown that the total repeat sprint time is highly reproducible (intraclass correlation coefficient [ICC] = .91; typical error of measurement [TE] = 1.5%). However, the percentage decrement is less reliable (ICC = .14, TE = 19.5%) (Gabbett 2010). The relatively low TE for absolute repeat sprint time suggests good sensitivity of this test for tracking training induced changes in performance. However, the poor test–retest reliability for percentage decrement suggests that the calculation of a fatigue index is less reliable. Therefore, larger relative changes in percentage decrement would be required to be confident that a real change has occurred in the performance of the athlete. Given that the total time is a more reliable indicator than percentage decrement, the major measure of performance should be the total sprint time.

Studies have shown significant relationships among speed, maximal aerobic power, and repeat sprint ability (Bishop and Edge 2006; Gabbett 2010). Although it is clear that 20 m sprint tests, multistage shuttle run tests, and repeat sprint protocols assess distinct qualities, these results demonstrate that repeat sprint ability is influenced by the absolute speed and aerobic qualities of players. Some studies have demonstrated that repeat sprint ability is correlated with 20 m sprint times but not maximal aerobic power (Castagna et al. 2007; Pyne et al.

2008). However, others have shown that whereas the influence of maximal aerobic power on repeat sprint ability is minimal, subjects with moderate maximal aerobic power have a smaller decrement in repeat sprint performance than subjects with a low maximal aerobic power (Bishop and Edge 2006). Given that repeat sprint ability is likely to be influenced (at least in part) by both speed and maximal aerobic power, the repeat sprint test should be conducted and interpreted in combination with traditional speed (e.g., 20 m sprint with splits at 5 m intervals) and aerobic power tests.

NORMATIVE DATA

Table 20.4 presents repeat sprint ability test normative data for female soccer players.

TABLE 20.4 Repeat Sprint Ability Test Data for National Female Soccer Players: Mean ± SD (Range)

Athlete group	Initial 20 m time, s	Total time, s	% Decrement
National squad	3.3 ± 0.1 (3.0-3.7)	20.9 ± 0.5 (19.2-22.7)	6.2 ± 1.6 (2.4-7.9)

Typical error: total time = 0.3s; % Decrement = 19.5%.

YO-YO INTERMITTENT RECOVERY TEST

RATIONALE

The Yo-Yo intermittent recovery test (IRT) has been selected because it is a valid and reliable estimate of aerobic endurance, is simple and efficient to administer, and has been sufficiently researched and validated in professional and youth soccer players (Krustrup et al. 2003). It is intermittent in nature and therefore reflects the requirements of soccer more closely than other similar endurance-based assessments.

TEST PROCEDURE

Refer to the Yo-Yo IRT protocol in chapter 14 for a detailed description of this test. It is recommended that this test be completed at the conclusion of the testing session, as fatigue associated with this test may affect performance in the speed and power tests. Ideally this test should be conducted on a nonslip surface such as a basketball court or indoor training facility. The surface chosen should be identical upon subsequent testing with the same group of players.

DATA ANALYSIS

There are two forms of Yo-Yo IRT: level 1 (Yo-Yo IRT1) and level 2 (Yo-Yo IRT2). Studies have suggested a higher anaerobic contribution to the performance of Yo-Yo IRT2 (Bangsbo et al. 2008) than Yo-Yo IRT1. The Yo-Yo IRT1 has demonstrated reasonable correlations to high speed distance covered in matches (Krustrup et al. 2003), whereas performance in the Yo-Yo IRT2 has shown a positive relationship with peak distance of high-intensity running during a 5 min match period (Krustrup et al. 2006). The Yo-Yo IRT1 should be sufficient for most squads; however, some elite squads may elect to use the Yo-Yo IRT2.

Yo-Yo IRT results in preseason testing should assist coaches and fitness coaches in determining the specific level of conditioning players require. Any in-season results should be treated with caution, because results may be influenced by recent match play.

NORMATIVE DATA

Table 20.5 presents Yo-Yo intermittent recovery test normative data for male soccer players.

TABLE 20.5 Yo-Yo Intermittent Recovery Test Data for National Male Soccer Players: Mean ± *SD* (Range)

Athlete group	Level	Distance, m
National under 17s	20.6 ± 2.7 (16.4-22.5)	2,327 ± 358 (1,240-3,200)
Olympic (under 23s) squad	20.9 ± 3.1 (15.6-23.2)	2,501 ± 434 (1,000-3,400)
National squad	21.4 ± 2.2 (17.6-23.4)	2,611 ± 421 (1,640-3,480)

Data source: male National under 17s, *n* = 18, 2007-2009; male Olympic squad, *n* = 24, 2007; male National squad, *n* = 18, 2007-2009.

HEART RATE RECOVERY TEST: 4 MIN YO-YO

RATIONALE

Assessing players' fitness levels during a competitive season can be difficult. Because recovery and game preparation understandably take precedence over fitness assessment, the opportunities to assess players' readiness to play are very limited. One method of overcoming this is to have the players perform a submaximal assessment. This enables the fitness coach to monitor heart rate responses to the load as well as the rate of heart rate recovery following the assessment. This method is quick, is relatively inexpensive, and, because of its submaximal nature, will have very little impact on the training load of the team.

This type of assessment has been validated as a reliable fitness assessment tool (Lamberts et al. 2007) and therefore could be used to monitor players' response to training and match loads during the season.

TEST PROCEDURE

A number of submaximal assessment protocols are available. The most appropriate for soccer is to modify a currently validated intermittent protocol, such as the Yo-Yo IRT (Krustrup et al. 2003). The option chosen here is to limit the Yo-Yo IRT2 to 4 min and provide a standardized time after the 5 min has finished (in this case 3 min) to assess the magnitude of heart rate recovery from the test. Using this protocol, all players complete the same volume of work (5 min of the Yo-Yo IRT2) and have the same rest period (3 min) after the test.

The protocol for this assessment involves largely the same instructions as those outlined in the Yo-Yo IRT protocol in chapter 14. Players complete 4 min of the Yo-Yo IRT2 (finish at stage 20 level 5) while wearing a heart rate monitor. At the conclusion of stage 20 level 2, players are instructed to remain passive for a 3 min period during which their heart rate continues to be monitored.

This test can be performed monthly, fortnightly, or even weekly as is desired by the sport science and coaching staff.

DATA ANALYSIS

Performance during the test is best demonstrated by the average heart rate (HR) during the 5 min of the Yo-Yo IRT2 as well as the final recovery HR expressed as a percentage of the final test heart rate. For example, if the heart rate after the 5 min test is 180 and the heart rate after the 3 min recovery period is 120, the percentage recovery would be 66.7%.

NORMATIVE DATA

Table 20.6 presents heart rate recovery test normative data for male soccer players.

TABLE 20.6 Heart Rate Recovery Test Data for National Male Soccer Players: Mean ± *SD* (Range)

Athlete group	Average test HR, %HRmax	Heart rate recovered, %
National squad	85.9 ± 4.5 (71.1-97.1)	60.7 ± 3.9 (53.1-72.8)

Data source: male National squad, *n* = 53, 2007-2010.

References

Bangsbo, J., Iaia, F. M., & Krustrup, P. 2008. The Yo-Yo intermittent recovery test : a useful tool for evaluation of physical performance in intermittent sports. *Sports Medicine, 38*(1):37-51.

Bishop, D., and Edge, J. 2006. Determinants of repeated-sprint ability in females matched for single-sprint performance. *European Journal of Applied Physiology* 97(4):373-379.

Burgess, D., Naughton, G.A., and Norton, K.I. 2006. Profile of movement demands of National Football (Soccer) players in Australia. *Journal of Science and Medicine in Sport* 9(4):334-341.

Carling, C., Bloomfield, J., Nielsen, L., and Reilly, T. 2008. The role of motion analysis in elite soccer: Contemporary performance measurements techniques and work rate data. *Sports Medicine* 38(10):839-862.

Castagna, C., Manzi, V., D'Ottavio, S., Annino, G., Padua, E., and Bishop, D. 2007. Relation between maximal aerobic power and the ability to repeat sprints in young basketball players. *Journal of Strength and Conditioning Research* 21(4):1172-1176.

Di Salvo, V., Baron, R., Tschan, H., Calederon Montero, F.J., Bachl, N., and Pigozzi, F. 2006. Performance characteristics according to playing position in elite soccer. *International Journal of Sports Medicine* 28(3):222-227.

Ekblom, B. 1986. Applied physiology of football. *Sports Medicine* 3:50-60.

Gabbett, T.J., and Mulvey, M.J. 2008. Time-motion analysis of small-sided training games and competition in elite women football players. *Journal of Strength and Conditioning Research* 22(2):543-552.

Gabbett, T.J. 2010. The development of a test of repeated-sprint ability for elite women's football players. *Journal of Strength and Conditioning Research* 24(5):1191-1194.

Krustrup, P., Mohr, M., Amstrup, T., Rysgaard, T., Johansen, J., Steensberg, A., Pedersen, P.K., and Bangsbo, J. 2003. The Yo-Yo intermittent recovery test: Physiological response, reliability, and validity. *Medicine and Science in Sports and Exercise* 35(12):697-705.

Krustrup, P., Mohr, M., Nybo, L., Jensen, J.M., Nielsen, J.J., and Bangsbo, J. 2006. The Yo-Yo IR2 test: Physiological response, reliability, and application to elite soccer. *Medicine and Science in Sports and Exercise* 38(9):1666-1673.

Lamberts, R.P., Swart, J., Noakes, T.D., and Lambert, M.I. 2007. Changes in heart rate recovery after high-intensity training in well-trained cyclists. *European Journal of Applied Physiology* 105(5):705-713.

Mayhew, S.R., and Wenger, H.A. 1985. Time-motion analysis of professional soccer. *Journal of Human Movement Studies* 11:49-52.

Mohr, M., P. Krustrup, Bangsbo, J. 2005.Fatigue in soccer: A brief review. *Journal of Sports Sciences* 21(7): 593-599.

Osgnach, C., Poser, S., Bernardini, R., Rinaldo, R., and Di Prampero, P.E. 2010. Energy cost and metabolic power in elite soccer: A new match analysis approach. *Medicine and Science in Sports and Exercise* 42(1):170-178.

Psotta, R., Blahus, P., Cochrane, D.J., and Martin, A.J. 2005. The assessment of an intermittent high intensity running test. *Journal of Sports Medicine and Physical Fitness* 45(3):248-256.

Pyne, D.B., Saunders, P.U., Montgomery, P.G., Hewitt, A.J., and Sheehan, K. 2008. Relationships between repeated sprint testing, speed, and endurance. *Journal of Strength and Conditioning Research* 22(5):1633-1637.

Reilly, T., and Thomas, V. 1979. Estimated daily energy expenditures of professional association footballers. *Ergonomics* 22(5):541-548.

Sayers, S.P., Harackiewicz, D.V., Harman, E.A., Frykman, P.N., and Rosenstein, M.T. 1999. Cross-validation of three jump power equations. *Medicine and Science in Sports and Exercise* 31(4):572-577.

Spencer, M., Lawrence, S., Rechichi, C., Bishop, D., Dawson, B., and Goodman, C. 2004. Time-motion analysis of elite field hockey, with special reference to repeated-sprint activity. *Journal of Sports Sciences* 22(9):843-850.

Spencer, M., Fitzsimons, M., Dawson, B., Bishop, D., and Goodman, C. 2006. Reliability of a repeated-sprint test for field-hockey. *Journal of Science and Medicine in Sport* 9(1-2):181-184.

Tumilty, D. 1993. Physiological characteristics of elite football players. *Sports Medicine* 16(2):80-96.

Hockey Players

Claire Rechichi, Ted Polglaze, and Matt Spencer

Hockey is an Olympic sport played by both men and women. Although referred to as *field hockey* in North America, the official name for the sport, used worldwide and by the International Olympic Committee, is *hockey*. The game is played on a rectangular pitch measuring 91.4 by 55.0 m and consists of two 35 min halves with a 10 min break for half-time.

Each team consists of 11 players on the pitch, including the goalkeeper. In addition, there are 5 substitute players for each team, and the number of substitutions that can be made during a game is unlimited. For international matches, it is now common for teams to make approximately 20 to 30 substitutions per half of a game, which is a considerable increase from 5 to 10 years ago. Furthermore, the range in individual player game time is typically between 30 and 70 min, with the mean positional game time being approximately 45 to 55 min for elite competition. However, it is common for the attacking positions to be substituted more often and have less playing time than for the defending positions (table 21.1).

The physiological requirements of contemporary hockey involve a highly developed aerobic system, numerous high-intensity running bouts, and frequent sprint and agility efforts for both the men's and women's game (Gabbett 2010; Spencer et al. 2004). Although the mean playing time in hockey is considerably shorter than other field-based team sports such as soccer and Australian Rules football, the relative intensity of the game (as indicated by the meters covered per minute) is typically greater. For example, as presented in table 21.1, the meters covered per minute for the various playing positions

TABLE 21.1 Summary of Basic Time-Motion Analysis Data From International Men's and Women's Hockey Games

Position	Distance, m	Game time, min	$m \cdot min^{-1}$	$m >5 \ m \cdot s^{-1}$*
Female				
Striker	6,496	50	129	445
Midfield	6,451	51	126	393
Defender	6,691	56	120	394
Team mean	6,459	52	125	408
Male				
Striker	6,356	46	139	969
Midfield	7,366	52	142	1,189
Defender	6,956	53	132	716
Team mean	6,892	50	138	958

*Meters covered $>5 \ m \cdot s^{-1}$. Data source: male, 6 games; individual player files, $n = 68$, 2010. Female, 8 games; individual player files, $n = 93$, 2009-2010.

in elite men's hockey (i.e., 132-142 m · min⁻¹) is somewhat greater than for the reported distances in elite men's soccer (120-130 m · min⁻¹; Carling et al. 2010) and Australian Rules football (113-124 m · min⁻¹; Wisbey et al. 2010). This comparatively greater relative intensity appears to be also true for elite women's hockey, with the distance covered for the various playing positions being 120 to 129 m · min⁻¹ (table 21.1) compared with elite women's soccer (110-114 m · min⁻¹; Mohr et al. 2008).

There have been several changes to the testing protocols from the first edition of this chapter. The tests that have been removed from the protocol include the 10 s maximal cycle ergometer test, the 5-0-5 agility test, and the 5 × 6 s repeat effort test on the cycle ergometer. The repeat effort test has been replaced by the 6 × 30 m on-field repeat sprint ability test. This test was designed to simulate the intense repeated sprint efforts that occur during elite competition. Furthermore, the nature of the test is specific to hockey because it includes a deceleration aspect and an active recovery (Spencer et al. 2006).

Athlete Preparation

To ensure test reliability, the following specific pretest conditions should be observed in addition to those outlined in chapter 2, Pretest Environment and Athlete Preparation.

- *Diet*. Athletes should be encouraged to consume a series of high-carbohydrate meals in the 36 h prior to testing and arrive on test day well hydrated. No food, cigarettes, or beverages containing alcohol or caffeine are to be consumed in the 2 h prior to testing.

- *Training*. No training inducing severe physiological or neural fatigue should be undertaken in the 24 h prior to testing. This includes any high-intensity skills or physical training.

- *Testing*. All test sessions should be scheduled at the same time of day to avoid fluctuations in physiological response to the set protocols due to circadian rhythm (Winget et al. 1985). All tests must be completed in the recommended standardized order, with adequate and consistent recovery time between tests. This will promote optimal performance and allow for a valid comparison of results on different test occasions.

Test Environment

Environmental conditions will often influence field testing, so documentation of ambient temperature, humidity, light, and wind is essential. All field tests for hockey players should occur on a wet artificial grass surface, with the players wearing their normal playing footwear. For running tests, the preferred path is set down the baseline of a hockey field, where there is zero slope, compared with the center line or sidelines, which may have increasing gradients, potentially influencing performance.

Recommended Test Order

It is important that tests are completed in the same order to control the interference between tests. This order also allows valid comparison of different test occasions. The recommended order is as follows:

The 40 m sprint and the multistage shuttle run test are to be conducted during the same session but must be completed in that order. The repeat sprint ability test is to be conducted on a separate day, which can be a consecutive or subsequent day. The vertical jump test may be conducted on day 1 or in conjunction with other strength tests (if applicable) on a separate day.

Day	Tests
1	Anthropometry
	Vertical jump
	40 m sprint
	Multistage shuttle run
2	Repeat sprint ability

Equipment Checklist

Anthropometry

- ☐ Stadiometer (wall mounted)
- ☐ Balance scales (accurate to ±0.05 kg)
- ☐ Anthropometry box
- ☐ Skinfold calipers (Harpenden skinfold caliper)
- ☐ Marker pen
- ☐ Anthropometric measuring tape
- ☐ Recording sheet (refer to the anthropometry data sheet template)
- ☐ Pen

Vertical Jump Test

- ☐ Yardstick jumping device (e.g., Swift Performance Yardstick)
- ☐ Measuring tape
- ☐ Recording sheet (refer to the vertical jump data sheet template)
- ☐ Pen

40 m Sprint Test

- ☐ Electronic light gate equipment (e.g., dual-beam light gates)
- ☐ Measuring tape
- ☐ Field marking tape
- ☐ Witches hats
- ☐ Recording sheet (refer to the sprint test data sheet template)
- ☐ Pen

Multistage Shuttle Run Test

- ☐ Measuring tape
- ☐ Witches hats
- ☐ Sound box or CD player
- ☐ Multistage shuttle run test CD (or MP3 file)
- ☐ Stopwatch
- ☐ Recording sheet (refer to the multistage shuttle run test data sheet template)
- ☐ Pen

6 × 30 m Repeat Sprint Ability Test

- ☐ Electronic light gate equipment (e.g., dual-beam light gates)
- ☐ Measuring tape
- ☐ Field marking tape
- ☐ Marker cones
- ☐ Sound box or CD/MP3 player
- ☐ 6 × 30 m repeat sprint ability test CD/MP3
- ☐ Chalk
- ☐ Stopwatch
- ☐ Recording sheet (refer to the repeat sprint ability test data sheet template)
- ☐ Pen

ANTHROPOMETRY

RATIONALE

Anthropometric measurements of height, body mass, and sum of skinfolds provide a clear appraisal of the structural status of an individual at any given time (Ross et al. 1991). Detailed athletic profiles are also valuable in describing the characteristics of elite athletes across sports (Withers et al. 1987), by position within a sport (Morrow et al. 1980), and at various stages throughout a yearly training cycle.

TEST PROCEDURE

For senior athletes, body mass and sum of skinfolds are measured regularly throughout the year. For junior athletes, height should be assessed at least once per year. Skinfolds are recorded over seven sites (triceps, biceps, subscapular, supraspinale, abdominal, front thigh, and medial calf). The individual skinfold measures as well as the sum of the seven sites should be reported. Refer to anthropometry protocols outlined in chapter 11, Assessment of Physique, for a detailed description of all anthropometric test procedures. More advanced anthropometric assessment including muscle girths, bone breadths, and limb lengths can be conducted if required.

Although the description of skinfold measurement procedures seems simple, a high degree of technical skill is essential for consistent results. It is therefore important that these measurements be taken by an experienced tester who has been trained in these techniques. It is also important that the same tester conduct each retest to ensure reliability.

DATA ANALYSIS

Skinfold results are commonly used as a practical basis for training and dietary interventions. Whether an event is primarily aerobic or anaerobic, increased fat mass will be detrimental to performance. Moreover, in sports requiring speed or explosive power, such as hockey, excess fat, which causes an increase in body mass, will decrease acceleration unless proportional increases in force are applied (Norton et al. 1996). Excess fat will also increase the load that the musculoskeletal system must absorb during movement, thereby increasing the risk of injury.

NORMATIVE DATA

Table 21.2 presents anthropometric normative data for female and male hockey players by position.

TABLE 21.2 Anthropometric Data for National Female and Male Hockey Players: Mean ± *SD* (Range)

Position	Body mass, kg	Σ7 skinfolds, mm
Female		
Striker (n = 4)	65.2 ± 4.7 (58.9-70.1)	65.5 ± 13.6 (56.4-85.7)
Midfield (n = 9)	60.4 ± 2.2 (57.0-62.3)	76.6 ± 13.5 (62.4-104.3)
Defender (n = 6)	62.0 ± 3.2 (57.7-67.3)	68.4 ± 13.0 (49.5-82.5)
Goalkeeper (n = 2)	65.0 ± 7.4 (59.8-70.2)	66.7 ± 17.3 (54.5-78.9)
Under 21 (n = 23)	61.2 ± 4.8 (54.9-71.7)	81.3 ± 13.4 (60.1-112.6)
Under 17 (n = 17)	61.9 ± 8.3 (50.5-79.8)	87.8 ± 26.2 (52.6-128.9)
Male		
Striker (n = 8)	80.3 ± 4.9 (74.9-89.3)	56.7 ± 9.6 (44.6-74.7)
Midfield (n = 12)	76.0 ± 4.6 (67.0-85.4)	50.2 ± 11.6 (32.2-70.3)
Defender (n = 9)	80.3 ± 9.3 (62.2-90.9)	60.0 ± 11.5 (42.9-79.3)
Goalkeeper (n = 4)	83.5 ± 8.6 (74.5-95.1)	77.0 ± 26.2 (55.6-110.6)
Under 21 (n = 20)	74.9 ± 7.3 (58.9-86.5)	55.3 ± 14.4 (32.6-85.7)

Typical error: Σ7 skinfolds = ~1 mm. Data source: National senior women's squad, n = 21, July 2010. National under 21 women's squad, n = 23, May 2009. National under 17 women's squad, n = 17, September 2009. National senior men's squad, n = 33, January 2010. National under 21 men's squad, n = 20, May 2009.

VERTICAL JUMP TEST

RATIONALE

The vertical jump test is an established measure of explosive and anaerobic power of the lower limbs and hips that is easy to perform, requires limited equipment, and is common to many power-related sports, allowing easy comparison.

TEST PROCEDURE

Refer to vertical jump protocol in chapter 14, Field Testing Principles and Protocols, for setup details and test procedures.

DATA ANALYSIS

Although vertical jump height is generally accepted as a measure of leg power, a more valid power score is obtained when both jump height and body mass are considered. Sayers and colleagues (1999) developed an equation to calculate power output from these two variables:

Peak Power (W) = [60.7 × Jump Height (cm)] + [45.3 × Body Mass (kg)] − 2,055.

Both jump height and peak power, calculated from this equation, should be reported when assessing vertical jump.

LIMITATIONS

The test is affected by player motivation and individual technique (particularly in relation to the athlete hitting the vane at the peak of the jump).

NORMATIVE DATA

Vertical jump normative data for male hockey players and goalkeepers are presented in table 21.3.

TABLE 21.3 Vertical Jump Data for National Male Hockey Players: Mean ± *SD* (Range)

Position	Jump height, cm	Peak power, W
Striker (*n* = 6)	65 ± 3 (60-68)	5,628 ± 292 (5,349-6,030)
Midfield (*n* = 5)	62 ± 5 (56-69)	5,073 ± 431 (4,440-5,476)
Defender (*n* = 7)	58 ± 5 (52-66)	4,964 ± 343 (4,561-5,371)
Goalkeeper (*n* = 3)	65 ± 5 (59-69)	5,640 ± 474 (5,168-6,115)

Typical error: jump height = 1.0 cm. Data source: National senior men's squad, *n* = 21, October 2007.

40 M SPRINT TEST

RATIONALE

The requirement to accelerate and sprint at maximal or near-maximal intensity is an important aspect of hockey. Time–motion analysis of international men's hockey indicates that field players perform on average 30 sprints per game, with a mean sprint duration of approximately 2 s (Spencer et al. 2004). However, occasionally field players are required to perform sprints of 30 to 40 m. Therefore, speed (i.e., 30-40 m time split) should be measured in addition to acceleration (i.e., 0-10 m time split). Warm-up should include at least two maximal sprints over 30 to 40 m, because pilot testing has shown the best sprint to be produced on numbers 3 to 5.

Note: For goalkeepers, only acceleration (i.e., 0-10 m split) is required.

TEST PROCEDURE

For hockey, the sprint test is conducted over 40 m, with intermediate distances of 10 and 30 m. Athletes should be encouraged to start the sprint with their body mass over their front foot, shoulders and hips square in a crouch "ready" position, heel up on back foot, and no rocking allowed. Refer to straight line sprint testing protocol in chapter 14 for a description of the procedures involved in sprint testing.

- The preferred sprint path is set down the baseline of a hockey field, because this sprint path has a zero slope. Sprint paths along the center line or sidelines may have increasing gradients which will influence performance.

- Split times (at 10 m and 30 m) and final time (40 m) for minimum two (preferably three) trials are recorded to the nearest 0.01 s.

- At least 2 min active recovery and rest is allowed between sprint trials.

DATA ANALYSIS

Data reduction for the 40 m sprint test is as follows:

- Fastest measured 0 to 10 m time is recorded as the acceleration score.

- Fastest measured 30 to 40 m time is recorded as the speed score.
- Fastest measured 0 to 40 m time is recorded as a combined acceleration and speed score.

NORMATIVE DATA

Table 21.4 presents 40 m sprint normative data for female and male hockey players by position.

TABLE 21.4 Sprint Data for National Female and Male Hockey Players: Mean ± *SD* (Range)

Position	0-10 m split, s	0-40 m split, s	30-40 m split, s
Female			
Striker (*n* = 6)	1.98 ± 0.04 (1.90-2.03)	5.92 ± 0.09 (5.77-6.09)	1.29 ± 0.03 (1.23-1.34)
Midfield (*n* = 8)	1.96 ± 0.06 (1.87-2.06)	5.98 ± 0.22 (5.54-6.26)	1.32 ± 0.06 (1.20-1.39)
Defender (*n* = 4)	2.01 ± 0.01 (2.00-2.02)	6.08 ± 0.13 (5.94-6.19)	1.34 ± 0.04 (1.28-1.39)
Goalkeeper (*n* = 2)	2.08 ± 0.01 (2.07-2.09)	n/a	n/a
Under 21 (*n* = 18)	1.95 ± 0.11 (1.74-2.13)	5.95 ± 0.27 (5.48-6.50)	1.31 ± 0.07 (1.20-1.45)
Under 17 (*n* = 15)	1.94 ± 0.07 (1.86-2.07)	5.96 ± 0.22 (5.68-6.47)	1.31 ± 0.05 (1.24-1.41)
Male			
Striker (*n* = 8)	1.82 ± 0.09 (1.69-1.95)	5.25 ± 0.18 (5.02-5.54)	1.09 ± 0.03 (1.06-1.14)
Midfield (*n* = 12)	1.78 ± 0.06 (1.68-1.88)	5.29 ± 0.19 (5.04-5.59)	1.12 ± 0.05 (1.07-1.19)
Defender (*n* = 9)	1.83 ± 0.06 (1.78-1.94)	5.43 ± 0.14 (5.25-5.68)	1.15 ± 0.04 (1.07-1.20)
Goalkeeper (*n* = 4)	1.84 ± 0.08 (1.78-1.95)	n/a	n/a
Under 21 (*n* = 20)	1.80 ± 0.05 (1.73-1.88)	5.37 ± 0.15 (5.12-5.63)	1.14 ± 0.04 (1.07-1.23)

n/a = not available. Typical error: 0-10 m = 0.03 s; 0-40 m = 0.05 s; 30-40 m = 0.02 s. Data source: National senior women's squad, *n* = 18, July 2010. National under 21 women's squad, *n* = 18, May 2009. National under 17 women's squad, *n* = 15, September 2009. National senior men's squad, *n* = 33, January 2010. National under 21 men's squad, *n* = 20, May 2009.

6 × 30 M REPEAT SPRINT ABILITY TEST

RATIONALE

Studies that have investigated the time motion analysis of sprinting during team-sports competition have generally reported that an intense sprint (i.e., 2-3 s) is performed every 90 to 120 s during a game (Dawson et al. 2004; Di Salvo et al. 2007; Duthie et al. 2005; Spencer et al. 2004). However, this information does not provide an insight into the typical movement patterns of repeat sprint activity, which is suggested to be an important aspect of team sports. During an international men's hockey competition, it was reported that a typical bout of repeated sprinting consisted of approximately 4 sprints with a mean recovery time between sprints being approximately 15 s (Spencer et al. 2004). Furthermore, it was reported that approximately 95% of the recovery between sprints was of an active nature. Although intense periods of repeated sprinting do not occur frequently during a game, they may be important in determining the result. Therefore, it is suggested that an assessment of hockey-specific, repeat sprint ability (RSA) is important to include in the routine field testing of elite hockey players. It is not necessary for goalkeepers to complete this test.

TEST PROCEDURE

- The overground RSA test requires the athletes to perform 6 maximal 30 m sprints departing every 25 s for males and 30 s for females. The preferred sprint path is set down the baseline of a hockey field, because this sprint path has a zero slope.

- The tester measures 50 m along the baseline of the field and marks out each 10 m section. The timing gates are set up at the 10, 20, 30, and 40 m marks. A cone is placed at 0 and at 50 m to mark out the deceleration zone. From both the 5 m and 45 m mark, a cone is placed at 19.5. This marks out the 40 m active recovery path.

- The electronic light gate equipment must be correctly set up prior to the commencement of the test. Lap mode is selected and 24 splits are entered (this is the total amount of split times required for the test). It is probably easier to record the trial number from the timing unit and recall the data after the testing session than to write down data during the test.

- Pretest preparation should include an active running warm-up of jogging to striding intensity for 5 to 10 min in addition to approximately 3 sprints of 20-30 m. It is also suggested that athletes undertake two submaximal repetitions of the RSA test to familiarize themselves with the test procedures (especially the jog intensity for the active recovery component). A recovery period of approximately 5 min should be given before commencing the test.

- The start position for each sprint is the same as for the 40 m sprint test. Each sprint must be performed maximally. As a check of pacing, athletes must obtain 95% of their first 10 m time from the acceleration and speed test. For example, if the 100% 10 m time is 1.80 s and the first 10 m split from the repeat sprint test is 1.86 s (i.e., 96%), then the test is continued. If the target time is not achieved, a minimum rest period of 3 min is required before starting again.

- The test starts with a 30 m sprint. The athlete is then required to decelerate to a walk by the 10 m cone, turn sharply, and commence the jog recovery. A distance of 40 m jogging is then completed within approximately 12 to 13 s (i.e., 3.1-3.3 m · s^{-1}) to allow 4 to 5 s of passive recovery before the commencement of the next sprint repetition. Constant verbal feedback is required during the jogging component to ensure adequate pacing. A 5 s warning is given and the athlete assumes the ready position and awaits the go command to commence the next sprint.

- Once the final sprint has been completed, the tester presses the recall button to retrieve the data. The "recall event" is the trial number. The tester presses the up or down arrow to select the appropriate trial number and presses enter. The first split time will be displayed S01; if the left or right arrow is pressed, this will change to L01. Now each of the individual spilt times will be displayed, rather than the cumulative time for the whole test. Split times (0-10, 10-20, and 20-30 m) are entered into the data spreadsheet, and the total 30 m time is calculated for each sprint.

- The test produces two scores, the absolute (total time in seconds) and relative (percentage decrement) scores. The calculations for percentage decrement are explained below. The total sprint time has been shown to be very reliable, with a typical error (TE) of

0.7%. However, the percentage decrement was less reliable, with a TE of 14.9% (Spencer et al. 2004). Therefore, the major measure of performance should be the total sprint time.

Figure 21.1 describes the equipment setup for the 6 × 30 m repeat sprint ability test.

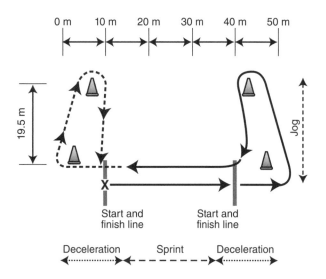

FIGURE 21.1 Schematic diagram of the 6 × 30 m overground repeat sprint test.

Adapted from *Journal of Science and Medicine in Sport*, Vol. 9, M. Spence et al., "Reliability of a repeated-sprint test for field-hockey," pgs. 181-184, copyright 2006, with permission from Elsevier.

DATA ANALYSIS

The following is an example of the methods used to calculate the absolute (total time in seconds) and relative (percentage decrement) scores for the 6 × 30 m RSA test (modified from Fitzsimons et al. 1993).

Repetition	Time (s)
1	4.28
2	4.48
3	4.52
4	4.58
5	4.65
6	4.75

Total time = 27.26 s.

$$\text{Ideal Time (s)} = \text{Best Time} \times 6$$
$$= 4.28 \times 6$$
$$= 25.68.$$

$$\text{Decrement (\%)} = (\text{Total/Ideal} \times 100)$$
$$= (27.26/25.68 \times 100)$$
$$= 106.2$$
$$= 106.2 - 100$$
$$= 6.2\%.$$

NORMATIVE DATA

Table 21.5 presents repeat sprint ability test normative data for female and male hockey players by position.

TABLE 21.5 Repeat Sprint Ability Test Data for National Female and Male Hockey Players: Mean ± *SD* (Range)

Position	Total time, s	% Decrement
Female		
All (*n* = 15)	29.93 ± 0.92 (28.63-32.06)	4.8 ± 2.5 (1.8-11.5)
Male		
Striker (*n* = 8)	26.55 ± 0.58 (26.02-27.71)	4.3 ± 1.4 (2.3-7.1)
Midfield (*n* = 12)	26.31 ± 0.80 (25.32-27.82)	4.3 ± 0.8 (3.1-5.4)
Defender (*n* = 6)	27.02 ± 0.74 (26.19-27.91)	4.6 ± 2.3 (1.4-7.0)

Typical error: total time = 0.7 s; % Decrement = 14.9%. Data source: National senior women's squad, *n* = 15, 2006. National senior men's squad, *n* = 26, January 2010.

MULTISTAGE SHUTTLE RUN TEST

RATIONALE

The multistage shuttle run test (MSRT) has been selected as the test of aerobic power for hockey for a number of reasons. First and most important, the test has been found to be a sufficiently accurate estimate of aerobic power (Leger and Lambert 1982; Ramsbottom et al. 1988). Second, the activity is similar to that of many team sports with respect to the stop, start, and change-of-direction movement patterns. Third, it is a very time-efficient test with which a whole team or squad can be assessed simultaneously.

TEST PROCEDURE

Refer to the multistage shuttle run test protocol in chapter 14 for detailed description of this test. The test should be performed in a corner of the hockey field to ensure the test area is as flat as possible.

LIMITATIONS

The test is affected by player motivation, technique, and environmental conditions. Also, although the test has a strong correlation with $\dot{V}O_2max$ ($r = .92$), a degree of error exists (i.e., 20% of the variance between tests is not explained by $\dot{V}O_2max$), which needs to be acknowledged when multistage shuttle run test results are used to predict $\dot{V}O_2max$ (Ramsbottom et al. 1988).

NORMATIVE DATA

Table 21.6 presents multistage shuttle run test normative data for female and male hockey players by position.

TABLE 21.6 Multistage Shuttle Run Test Data for National Female and Male Hockey Players: Mean ± *SD* (Range)

Position	Level/shuttle
Female	
Striker (*n* = 5)	11/9 ± 0/6 (11/1-12/4)
Midfield (*n* = 8)	12/1 ± 0/6 (11/1-12/8)
Defender (*n* = 6)	12/6 ± 0/11 (11/3-13/7)
Under 21 (*n* = 21)	11/9 ± 1/1 (9/2-13/9)
Under 17 (*n* = 13)	10/8 ± 1/2 (8/9-13/3)
Male	
Striker (*n* = 6)	14/9 ± 0/7 (14/1-15/8)
Midfield (*n* = 8)	14/12 ± 0/7 (13/11-15/11)
Defender (*n* = 7)	14/11 ± 1/1 (13/2-16/4)
Under 21 (*n* = 18)	14/4 ± 0/10 (13/5-16/3)

Typical error: shuttle decimal = 0.3. Data source: National senior women's squad, *n* = 19, July 2010. National under 21 women's squad, *n* = 21, May 2009. National under 17 women's squad, *n* = 13, September 2009. National senior men's squad, *n* = 21, October 2009. National under 21 men's squad, *n* = 18, May 2009.

References

Carling, C., Espié, V., Le Gall, F., Bloomfield, J., and Jullien, H. 2010. Work-rate of substitutes in elite soccer: A preliminary study. *Journal of Science and Medicine in Sport* 13(2):253-255.

Dawson, B., Hopkinson, R., Appleby, B., Stewart, G., and Roberts, C. 2004. Comparison of training activities and game demands in the Australian Football League. *Journal of Science and Medicine in Sport* 7(3):292-301.

Di Salvo, V., Baron, R., Tschan, H., Calderon Montero, F.J., Bachl, N., and Pigozzi, F. 2007. Performance characteristics according to playing position in elite soccer. *International Journal of Sports Medicine* 28(3):222-227.

Duthie, G., Pyne, D., and Hooper, S. 2005. Time motion analysis of 2001 and 2002 Super 12 Rugby. *Journal of Sports Sciences* 23(5):523-530.

Fitzsimons, M., Dawson, B., Ward, D., and Wilkinson, A. 1993. Cycling and running tests of repeated sprint ability. *Australian Journal of Science and Medicine in Sport* 25(4):82-87.

Gabbett, T.J. 2010. GPS analysis of elite women's field hockey training and competition. *Journal of Strength and Conditioning Research* 24(5):1321-1324.

Leger, L.A., and Lambert, J. 1982. A maximal multistage 20 m shuttle run test to predict $\dot{V}O_{2max}$. *European Journal of Applied Physiology* 49(1):1-12.

Mohr, M., Krustrup, P., Andersson, H., Kirkendal, D., and Bangsbo, J. 2008. Match activities of elite women soccer players at different performance levels. *Journal of Strength and Conditioning Research* 22(2):341-349.

Morrow, J.R., Hosler, W.W., and Nelson, J.K. 1980. A comparison of women intercollegiate basketball players, volleyball players and non-athletes. *Journal of Sports Medicine and Physical Fitness* 20(4):435-440.

Norton, K., Olds, T., Olive, S., and Craig, N. 1996. Anthropometry and sport performance. In: K. Norton and T. Olds (eds), *Anthropometrica*. Sydney, Australia: University of NSW Press, pp 287-364.

Ramsbottom, R., Brewer, J., and Williams, C. 1988. A progressive shuttle run test to estimate maximal oxygen uptake. *British Journal of Sports Medicine* 22(4):141-144.

Ross, W.D., and Marfell-Jones, M.J. 1991. Kinanthropometry. In: J.D. MacDougall, H.A. Wenger, and H.J. Green (eds), *Physiological Testing of the High Performance Athlete*, 2nd ed. Champaign, IL: Human Kinetics, pp 223-308.

Sayers, S.P., Harackiewicz, D.V., Harman, E.A., Frykman, P.N., and Rosenstein, M.T. 1999. Cross-validation of three jump power equations. *Medicine and Science in Sports and Exercise* 31(4):572-577.

Spencer, M., Fitzsimons, M., Dawson, B., Bishop, D., and Goodman, C. 2006. Reliability of a repeated-sprint test for field-hockey. *Journal of Science and Medicine in Sport* 9(1-2):181-184.

Spencer, M., Lawrence, S., Rechichi, C., Bishop, D., Dawson, B., and Goodman, C. 2004. Time-motion analysis of elite field-hockey: special reference to repeated-sprint activity. *Journal of Sports Sciences* 22(9):843-850.

Winget, C.M., DeRoshia, C.W., and Holley, D.C. 1985. Circadian rhythms and athletic performance. *Medicine and Science in Sports and Exercise* 17(5):498-516.

Wisbey, B., Montgomery, P.G., Pyne, D.B., and Rattray, B. 2010. Quantifying movement demands of AFL football using GPS tracking. *Journal of Science and Medicine in Sport* 13(5):531-536.

Withers, R.T., Craig, N.P., Bourdon, P.C., and Norton, K.I. 1987. Relative body fat and anthropometric prediction of body density of male athletes. *European Journal of Applied Physiology* 56(2):191-200.

Netball Players

Kristie-Lee Taylor, Darrell L. Bonetti, and Rebecca K. Tanner

The game of netball is an international sport based on throwing and catching. It is played by two teams of seven players and takes place over four 15 min quarters. It is played on a court (double sprung wooden timber) measuring 30.5 m long by 15.25 m wide. The court is divided into three thirds measuring 10.17 m each. The area of the court accessible to each player is determined by the player position. Given the confined court area, players are constantly involved in offensive and defensive maneuvers.

The objective of the game is to advance the ball down the court and have the two shooting players (goal shooter and goal attack) score goals from within a defined area (goal circle) by throwing the ball through a ring that is attached to a 3.05 m (10 feet) high post. Netball is the most popular women's sport in Australia with an estimated one million players nationwide. At an international level, it is played in more than 40 countries. Australia has claimed the World Championship title in 10 of the 13 World Netball Championships held since 1963.

Netball, like many team sports, requires players to perform game-related skills with the ball (e.g., throwing, catching, shooting, offensive and defensive rebounds) as well as game-related activity without the ball (e.g., jumping, lunging, defending). These skills are performed in combination with movement patterns of various intensities and duration. Success in netball at an elite level requires players to possess an exceptional level of skill in combination with high levels of aerobic and anaerobic fitness. Netball has been described as a game reliant on rapid acceleration and sudden and rapid changes in direction in combination with jumping or leaping movements (Otago 1983). It is an interval-type game involving a combination of short work intervals (e.g., sprints,

jumps, shuffling movements) interspersed with short recovery periods (e.g., jogging, shooting, and passive defense). Aerobic conditioning has been identified as a major contributor to performance (Allison 1978; Woolford and Angove 1992).

Studies of movement patterns and work intensities of netball players provide valuable information relating to the key physical qualities related to the game. All of these qualities should be considered in the physical preparation and conditioning of netball players. Key activity patterns and motion variables for netball include the following:

- The average work to rest ratio is 1:3 (Davidson and Trewartha 2008; Loughran and O'Donoghue 1999; Otago 1983).

- There is a greater work to rest ratio for center players as opposed to goalkeepers and goal shooters (Davidson and Trewartha 2008).

- Center players spend less time per standing and walking event but greater amount of time per jog and sprint events relative to all other positions (Davidson and Trewartha 2008; Loughran and O'Donoghue 1999; Steele and Chad 1992).

- On average, center players cover approximately 8 km during a game compared with 4.2 km covered by goalkeepers and goal shooters (Davidson and Trewartha 2008).

- The majority of work periods are less than 10 s (Otago 1983), and the average time spent per activity is typically less than 5 s (Steele and Chad 1992).

- On average, players change direction every 4 s (Davidson and Trewartha 2008).

- Mean sprint duration is approximately 1.4 s (Allison 1978; Davidson and Trewartha 2008; Loughran and O'Donoghue 1999); sprint events equate to a quarter of the total distance covered by center players (Davidson and Trewartha 2008).

This chapter contains guidelines for assessing the physiological abilities of netball players. The tests are those currently used by Australian national squads (e.g., National team, under-21 squad, under-19 and under-17 talent squads) and include field tests and strength tests.

The field tests have been selected to measure some of the major physical abilities required for playing netball, namely, acceleration, agility or footwork, jumping ability, and endurance. Resistance training is an important component of the netball player's physical preparation, and the selected strength tests enable basic diagnosis of strength and changes related to training. The field tests and strength tests are two separate procedures and should be conducted on separate occasions.

Athlete Preparation

Standardized pretest preparation is recommended to enable reliable and valid physiological data to be obtained. Refer to chapter 2, Pretest Environment and Athlete Preparation, for specific information relating to athlete and environment preparation.

In general, field tests and strength tests should be carried out at the end of each appropriate macrocycle to help determine the effectiveness of the program completed and structure of the next program to be implemented.

Test Environment

- The tests should be carried out at the same time of day for each testing session.

- Testing should not be conducted in extreme environmental conditions, particularly hot and humid conditions and extreme cold conditions. Although it is sometimes impossible to control the environmental conditions for field testing, it is possible to test at times of the day that are the least extreme.

- The time of day, temperature (°C) and relative humidity (%) should be recorded.

Recommended Test Order

It is important that the field and strength tests are completed in the same order to control the interference between tests. This order also allows valid comparison of different test occasions. Generally strength tests are completed on a day following the field-based tests, although this is not always practical. The order of testing within the strength testing session should be decided in consultation with the strength and conditioning specialist conducting the assessments.

The reactive agility test is not included in the regular testing battery for netball. This test requires specialized equipment and a detailed setup; testing is usually only conducted annually and occurs on a separate day from the rest of the regular testing battery.

Day	Tests
1	Anthropometry
	20 m sprint
	Planned agility
	Vertical jump
	Yo-Yo intermittent recovery
2	Muscular strength

Equipment Checklist

Anthropometry

- [ ] Stadiometer (wall mounted)
- [ ] Balance scales (accurate to ±0.05 kg)
- [ ] Anthropometry box
- [ ] Skinfold calipers (Harpenden skinfold caliper)
- [ ] Marker pen
- [ ] Anthropometric measuring tape
- [ ] Recording sheet (refer to the anthropometry data sheet template)
- [ ] Pen

20 m Sprint Test

- [ ] Electronic light gate equipment (e.g., dual-beam light gates)
- [ ] Measuring tape
- [ ] Field marking tape
- [ ] Marker cones
- [ ] Recording sheet (refer to the sprint test data sheet template)
- [ ] Pen

Planned Agility Test

- [ ] Electronic light gate equipment (e.g., dual-beam light gates)
- [ ] Measuring tape
- [ ] Field marking tape
- [ ] 5 obstacles or poles
- [ ] Recording sheet

Reactive Agility Test

- [ ] Electronic light gate equipment (e.g., dual-beam light gates)
- [ ] Measuring tape
- [ ] Field marking tape
- [ ] Marker cones
- [ ] Digital video camera
- [ ] Tripod
- [ ] Data projector (including power lead and computer connector lead)
- [ ] Computer
- [ ] Projection screen
- [ ] Recording sheet
- [ ] Pen

Vertical Jump Test

- [ ] Yardstick jumping device (e.g., Swift Performance Yardstick)
- [ ] Measuring tape
- [ ] Field marking tape
- [ ] Recording sheet (refer to the vertical jump data sheet template)
- [ ] Pen

Yo-Yo Intermittent Recovery Test

- [ ] Measuring tape
- [ ] Witches hats
- [ ] Sound box or CD player
- [ ] Yo-Yo intermittent recovery test CD
- [ ] Recording sheet (refer to the Yo-Yo IRT data sheet template)
- [ ] Pen

Muscular Strength Tests

- [ ] Squat rack with safety bars
- [ ] Olympic bench press
- [ ] Barbell (Olympic 20 kg)
- [ ] Weight plates (2.5-25 kg increments)
- [ ] Recording sheet (refer to the strength and power test data sheet templates)
- [ ] Pen

ANTHROPOMETRY

RATIONALE

Height, body mass, and sum of seven skinfolds provide objective measures of the athlete's body structure. These measurements are important in quantifying differential growth and training influences (Ross and Marfell-Jones 1987). In particular, skinfold measurement can be used to monitor and understand the effect of diet and training intervention on body composition. For example, skinfold measurements may be used to determine whether changes in body mass are due to changes in body fat levels or lean body mass (Norton et al. 1994). Furthermore, increased body fat levels may have an adverse effect on netball performance because excess fat has no functional role in activities on the court and can be regarded as dead weight. Excess fat will decrease acceleration and is detrimental to the player's ability to rebound or leap for the ball (Norton et al. 1994). The development of anthropometric profiles of elite netballers is also of importance. Analysis of such information highlights the physical requirements of particular netball positions (Bale and Hunt 1986).

TEST PROCEDURE

Measurement of height, body mass, and skinfolds should be carried out prior to field testing protocols. Skinfolds are recorded over seven sites (triceps, biceps, subscapular, supraspinale, abdominal, front thigh, and medial calf). The individual skinfold measures as well as the sum of the seven sites should be reported. Refer to anthropometry protocols outlined in chapter 11, Assessment of Physique, for a detailed description of all anthropometric test procedures.

Although the description of skinfold measurement procedures seems simple, a high degree of technical skill is essential for consistent results. It is therefore important that these measurements be taken by an experienced tester who has been trained in these techniques. It is also important that the same tester conduct each retest to ensure reliability.

DATA ANALYSIS

Anthropometric data can be used for serial monitoring (within-subject) of changes in body composition of individual players across the course of a competitive season. In the younger age groups, body mass and height are the only anthropometric measures used for between-subject comparisons within the national talent identification pathway.

NORMATIVE DATA

The anthropometry of netball players can be influenced by player position, age group, and competitive level. Normative data for national squads are presented in table 22.1.

TABLE 22.1 Anthropometric Data for U21 and National Squad Netball Players: Mean ± *SD* (Range)

Squad/ level	Height, cm	Body mass, kg	Σ7 skinfolds, mm
U21	181.5 ± 4.1	73.7 ± 5.4	99.8 ± 21.7
	(170.5-186.6)	(64.8-91.9)	(62.6-169.9)
National team	182.4 ± 6.2	72.6 ± 8.0	82.9 ± 17.5
	(170.0-196.0)	(58.6-92.4)	(46.4-136.6)

Data source: AIS Netball and National team database 2010; U21 (*n* = 46, age 17-21 years); National team (*n* = 30, age 20-33 years).

20 M SPRINT TEST

RATIONALE

Although netballers rarely sprint at maximal speed during the game, a large number of efforts involve acceleration from a jog, shuffle, or stationary position (Steele and Chad 1992). These running efforts average less than 2 s, but the acceleration involved in making position, evading an opponent, or intercepting a pass is an important requirement of the game. Therefore, the ability of the netballer to accelerate is best indicated by the times to 5 and 10 m, with 20 m being the maximal distance any player might travel.

TEST PROCEDURE

For netball, the sprint test is conducted over 20 m, with intermediate distances of 5 and 10 m. Athletes should be encouraged to start the sprint with their body mass over their front foot, shoulders and hips square in a crouch "ready" position, heel up on back foot, and no rocking allowed. Refer to straight line sprint testing protocol in chapter 14, Field Testing Principles and Protocols, for a description of the procedures involved in sprint testing.

Ideally, the 20 m sprint test should be conducted on a wooden floor that is used for indoor court play. It is important to ensure that the surface used is consistent for each test—results can be affected by dust or dirt on the floor boards or the use of a different playing surface.

Athletes must warm up adequately in preparation for the sprint test. A poor warm-up may result in unreliable or slower results. The warm-up should include a period of easy running and several maximal sprints and start efforts (e.g., 10 min of easy running and dynamic exercises for increasing range of motion, approximately six "stride-throughs" over 30 m building in intensity from 85% to 100% maximum with jog recovery and four sprints from a stationary start over 10 m building in intensity from 90% to 100%). A minimum of 2 min recovery should be given between each sprint trial.

DATA ANALYSIS

The assessment of maximal 20 m sprint time can assist the practitioner in monitoring within-athlete changes in netball relevant speed and the ability of the athlete to accelerate from a stationary start. The assessment of changes after specific training interventions can assist in determining the appropriateness of the training program.

NORMATIVE DATA

Normative data for the 20 m sprint for under 17, under 19, under 21, and National squad netball players are presented in table 22.2.

TABLE 22.2 Sprint Data for U17, U19, U21, and National Squad Netball Players: Mean ± SD (Range)

Squad/ level	5 m time, s	10 m time, s	20 m time, s
U17 TID	1.25 ± 0.09 (1.09-1.50)	2.07 ± 0.10 (1.84-2.43)	3.52 ± 0.16 (3.15-4.15)
U19 TID	1.24 ± 0.08 (1.10-1.52)	2.06 ± 0.09 (1.87-2.40)	3.51 ± 0.14 (3.21-3.99)
U21	1.21 ± 0.07 (1.03-1.36)	2.03 ± 0.09 (1.80-2.25)	3.45 ± 0.14 (3.14-3.74)
National team	1.22 ± 0.08 (1.04-1.39)	2.03 ± 0.10 (1.78-2.25)	3.46 ± 0.16 (3.07-3.92)

Data source: AIS Netball database 2010; U17 TID (n = 118); U19 TID (n = 76); U21 (n = 45); National team (n = 55).

PLANNED AGILITY TEST

RATIONALE

The basic movement patterns of netball players involve numerous sideways movements, sudden changes in direction, and quick stops and starts. A player will perform shuffling or sideways movement of the body at speed and full effort between 100 and 300 times a game depending on the playing position (Steele and Chad 1992). The 5-0-5 agility test was used for the assessment of agility in all national and state level programs up until 2010, after which the planned agility test protocol was introduced. The planned agility test involves maximal effort trials around a predetermined course of cones consisting of three left and two right 90° turns. This has been used extensively in the assessment of agility in Australian Football players (Pyne et al. 2005).

TEST PROCEDURE

- Two sets of light gates are required, one at the start and one at the finish. The poles (obstacles) are typically made from PVC pipe with a circumference of approximately 35 cm and are approximately 1.5 m high. Field marking tape should be placed on the floor at two corners of the obstacle so it can be accurately repositioned if it is knocked over.

- Athletes start from a stationary, upright position with the front foot on the 0 m point, in line with the start gate. Athletes weave in and out of the obstacles or poles as per figure 22.1 and should avoid moving them in any way. If this occurs, or a pole is knocked over completely, the trial is stopped and the athlete must start again (after a brief rest).

- Athletes should complete a short warm-up of light running, dynamic stretching, and some run-throughs. After instruction, athletes should have a practice trial at 50% effort to familiarize themselves with the course.

- Three 100% maximal effort trials are recorded, and the best time (seconds reported to two decimal places) taken as the score. Allow 2 to 3 min of recovery between trials.

DATA ANALYSIS

Similar to the assessment of maximal 20 m sprint time, monitoring within-athlete change of direction ability can assist the practitioner in determining the

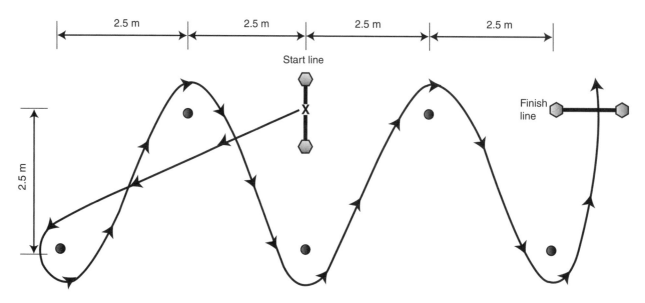

FIGURE 22.1 Schematic diagram of the planned agility test.

appropriateness of conditioning stimuli. In addition to monitoring athletes' ability to negotiate a predetermined agility course in minimal time, the practitioner should consider assessing reactive agility and decision-making time (see following section on assessment of reactive agility). Athletes who are slow on the agility run should be given footwork drills, agility drills with and without the ball, and acceleration and deceleration training.

NORMATIVE DATA

Planned agility test normative data for under 17 and National squad netball players are presented in table 22.3.

TABLE 22.3   Planned Agility Test Data for U17 and National Squad Netball Players: Mean ± *SD* (Range)

Squad/level	Planned agility test, s
U17 TID	9.05 ± 0.26
	(8.6-9.5)
National team	8.92 ± 0.34
	(8.3-9.7)

Data source: AIS Netball database 2010; U17 TID (*n* = 95); U19 TID (*n* = 77); U21 (*n* = 45); National team (*n* = 50).

REACTIVE AGILITY TEST

RATIONALE

The reactive agility test combines both planned and reactive movements to a sport-specific stimulus. This reflects the actual game of netball, because the majority of footwork patterns initiated by the players (particularly defensive players) are triggered by an opponent's movement. It is the aim of this test to measure the movement speed of players performing a defensive agility movement when the movement is (a) planned and (b) in reaction to a sport-specific stimulus. The additional measures of movement direction accuracy and movement initiation time are recorded within the sport-specific display condition.

This test requires specialized equipment, including a custom-designed software package developed by the Australian Institute of Sport (AIS). An overview is presented here; however for more detail readers

are referred to chapter 15, Perceptual–Cognitive and Perceptual–Motor Contributions.

SETUP

- Figure 22.2 shows the dimensions and setup of the reactive agility test. To begin, use a tape measure and field marking tape to place a mark 5 m in front of the projection screen to indicate the position of the projector. Place another mark 3 m directly behind the projector to indicate where timing gate 2 will be placed. Place a third mark 1 m directly behind the 2nd mark indicating the line of the defensive shuffle. Extend this mark 2 m to the left and 2 m to the right to indicate the start and end points of the defensive shuffle.

- Assemble 4 sets of timing gates on tripods with the bottom segment extended only. Set up the gates as shown in figure 22.2. Gate 1 is positioned at the start of the defensive shuffle (to the right hand side of the projector looking at the screen). Gate 2 is positioned at the mark 1 m in front of the defensive shuffle line, gate 3 is positioned 4.1 m diagonally left of the cone marking the point of change of direction (looking at the screen), and gate 4 is positioned 4.1 m diagonally right of the cone marking the point of change of direction.

- Connect gate 4 to gate 2, gate 2 to gate 1, and gate 1 to gate 3. Connect the timing gate control box to gate 4.

- Connect the timing gate control box to the computer.

- Once the data projector is positioned as per the setup diagram, mount the projection screen on the wall. Adjust the position and focus of the projector so that the image is displayed clearly on the projection screen and ensure that the resultant video image size is as realistic as possible (i.e., height and width). To do this, zoom the projector out as far as possible and then focus.

- Set up a digital camera behind the defensive shuffle line, directly in line with the projector. The camera should be placed on a tripod set at the lowest possible height. Ensure that the camera has the closest, unobstructed view of the test screen and the athlete's lower body and feet as possible. Place a labeled, blank tape in the camera and set the camera to stand-by mode.

TEST PROCEDURE

- Prior to the commencement of testing, inform the athlete of the test protocol. There will be 2 test conditions—reactive agility test condition and planned agility test condition. The reactive agility test condition will be performed first, followed by the planned agility test condition.

- At the commencement of testing, switch the camera into the record mode. Get the athlete to face the camera and say her name and the date of testing. There is no need to record practice trials or the test trials of the planned test condition. It is only necessary to record the test trials of the reactive test condition.

Reactive Test Condition

- The test requires an athlete to shuffle (sidestep), from gate 1, for 4 m to the left and then 2 m to the right, while facing forward. The athlete then sprints forward for 1 m before changing direction and running for 4.1 m in a straight sprint on a 45° angle. When the athlete begins the sidestepping movement, this triggers the video to play a near-life-size video image of an opponent passing the ball, which indicates whether the participant has to run to the right or the left direction.

- The athlete performs 2 practice trials and 12 test trials.

- When the athlete is ready to begin the test trials, turn the camera into record mode.

- Before the commencement of each trial, the timing gates must be armed. To do this the athlete must pass through timing gate 1 from the left of the gate (looking at the screen) to the starting position on the right of the gate. It is essential that this is done quickly and cleanly so that the gate beeps once only. The most effective way of arming the gate is to instruct the athlete to hold their arms straight

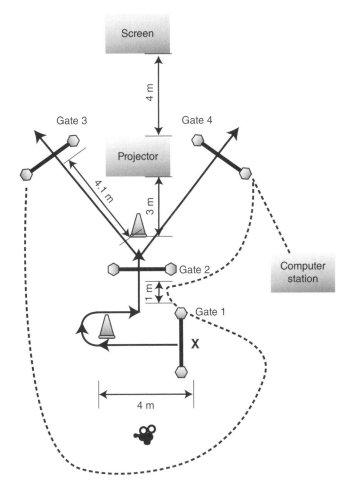

FIGURE 22.2 Netball reactive agility test setup.

up above their head and to jump through quickly. If the gate beeps more than once, you must begin the trial again.

- Arming the gates will begin the timer on the timing gate control box and the timing for display of the video image on the projection screen.

- After arming the gates, the athlete should commence the test as soon as possible. Facing the screen, the athlete should begin the sidestep shuffle to the left through timing gate 1 to the end of the 4 m defensive shuffle line.

- When the athlete reaches the end of the defensive shuffle line, she remains facing the screen but changes direction and side-shuffles back along the line to the right. At the mark on the halfway point of the defensive shuffle line, the athlete sprints forward through timing gate 2. Upon passing through timing gate 2, the athlete must respond as though she were about to intercept the pass viewed on the screen and sprints diagonally left or right through timing gate 3 or 4. Remind the athlete that it is important to continue sprinting until she has passed through the final gate and not to slow down until after the final gate has been cleared.

- Record the trial number, the direction of movement, and whether the trial was correct or incorrect on the athlete's manual data sheet.

- Perform the second trial in the same way, ensuring to arm the timing gates.

- Repeat the procedure as above for 12 test trials. Remember to reset the timing gates between trials and to record all of the trials on the manual data sheet.

Planned Test Condition

- The planned condition is designed to replicate agility tests where the direction of travel is known to the subject before commencing the test. In this case the player is required to move around a cone that is situated 50 cm from the end of the 1 m sprint and 25 cm to either the right or the left of the center line (dependent on which way the subject was required to run).

- The test is performed in a similar way to the reactive test condition, and setup is the same. However, the projector is not necessary for the planned test condition.

- During the directional sprints, the athlete runs to the left and through timing gate 3 for the first 6 trials, and then to the right and through timing gate 4 for the final 6 trials

(i.e., the direction of movement is planned). To ensure that the athlete does not cut corners as she passes through timing gate 2, she must pass the cone situated 50 cm in front of timing gate 2 before turning in the predetermined direction.

- Perform 1 practice trial for each direction followed by 6 trials to the left and 6 trials to the right.

- Proceed as for the reactive agility test, remembering to reset the timing gates and record the trials on the manual data sheet following each individual trial.

- There is no need to film the planned test condition trials.

DATA ANALYSIS

Outcome measures from the reactive agility test include the following:

- Shuffle time (seconds)—time taken from the start of the test until the completion of the shuffle movement (from gate 1 until gate 2).

- Sprint time (seconds)—time from the end of the shuffle until the completion of the test. This measures the athlete's straight sprinting ability (from gate 2 until either gate 3 or 4, depending on left or right movement direction).

- Total time (seconds)—the addition of the shuffle and sprint time components.

- The shuffle and sprint time components are also separated into left and right changes of direction.

- Decision-making time (meters per second)—defined as the first initiation of a movement toward the required direction of travel relative to the moment of ball release from the athlete passing in the video image. The video image is occluded at this moment (ball leaving the hand of the passer) providing a definitive frame from which to start this calculation. Typically the athlete's heel is the first signal of a change in movement direction. A negative score indicates anticipation of the pass direction. Frame-by-frame analysis of video footage should be used for timing calculations.

For a more detailed discussion surrounding the interpretation and practical recommendations resulting from the reactive agility test for netball, please refer to chapter 15.

NORMATIVE DATA

See chapter 15, Perceptual–Cognitive and Perceptual–Motor Contributions to Elite Performance.

VERTICAL JUMP TEST

RATIONALE

The vertical jump test is selected as the method for evaluating both explosive leg power and jumping ability. The test is easily performed in the field and requires limited equipment, and substantial sport-specific normative data are available.

TEST PROCEDURE

Refer to vertical jump protocol in chapter 14 for setup details and test procedures.

In addition, the following points apply to the measurement of relative and absolute jump height:

- Reach height must be recorded to help ensure test reliability for relative height.
- The absolute jump height is recorded to the nearest 1 cm.
- The relative jump height is the difference between the two points (reach height and jump height) recorded to the nearest 1 cm.
- The jump measurement device used to conduct the test is noted.

DATA ANALYSIS

The test provides two jump heights: the relative vertical jump height and absolute jump height. The absolute jump height is the highest point reached or touched when jumping, whereas the relative jump height is the difference between the standing reach and absolute jump height. Absolute jump height provides an objective measure of the specific skill and physical ability to rebound; the height reached indicates an important performance component for goal shooters and defenders. Relative jump height can be used to monitor the effects of strength and power training programs and make comparisons between players.

The vertical jump test result can be affected by the coordination of trunk and arms, and the interpretation of any changes due to strength training may be obscured by changes in the jump action used (Young 1994). It is assumed that the jump action of elite level netballers will not change substantially from test to test; however, junior netballers may improve their vertical jump result due to skill and growth related changes.

NORMATIVE DATA

Normative data for the vertical jump tests for under 17, under 19, under 21, and National squad netball players are presented in table 22.4.

TABLE 22.4 Vertical Jump Data U17, U19, U21, and National Squad Netball Players: Mean ± *SD* (Range)

Squad/level	Standing reach height, cm	Absolute jump height, cm	Relative jump height, cm
U17 TID	234.9 ± 8.7 (213-254)	278.8 ± 9.2 (258-300)	43.9 ± 5.0 (31-58)
U19 TID	235.1 ± 7.3 (216-251)	279.2 ± 7.9 (262-295)	44.1 ± 5.5 (34-65)
U21	238.5 ± 9.5 (218-263)	284.3 ± 9.8 (262-299)	45.8 ± 5.5 (34-57)
National team	237.0 ± 9.5 (217-263)	283.4 ± 9.5 (262-301)	46.4 ± 5.4 (35-56)

Data source: AIS Netball database 2010; U17 TID (*n* = 100); U19 TID (*n* = 67); U21 (*n* = 44); National team (*n* = 55).

YO-YO INTERMITTENT RECOVERY TEST

RATIONALE

The contribution of the aerobic energy system to netball play will depend on factors such as playing position, how closely the competing teams are matched, and the distribution or pattern of court play (Chad and Steele 1990; Woolford and Angove 1992). Because athletes may be required to sustain high levels of intensity with little opportunity for substantial recovery, a high level of aerobic fitness is required (Chad and Steele 1990). Furthermore, a high level of aerobic fitness will enable the netballer to play and practice longer at higher intensities.

Historically, the multistage shuttle run has been used to assess aerobic power in netball. In 2009–2010 the multistage shuttle run was replaced with the Yo-Yo intermittent recovery test (IRT). This intermittent test involves a 5 to 15 s exercise period followed by a 10 s active recovery. Over the period of the test, the speed of running is progressively increased. The complete duration of the test may vary between 2 and 15 min. This test is specifically designed to mimic sports involving intensive exercise bursts followed by short recovery periods.

TEST PROCEDURE

Refer to the Yo-Yo IRT protocol in chapter 14 for detailed description of this test. It is recommended that this test be completed at the conclusion of the netball testing battery, as fatigue associated with this test may affect performance in the speed and power-type tests.

Note: There are two levels for this Yo-Yo IRT: level 1 and level 2. Currently, the standard testing criteria for netball is level 1. If an athlete is able to run faster than level 1 speed, she should perform the level 2 test on the next occasion.

DATA ANALYSIS

The Yo-Yo IRT is effort dependent; thus, for valid results athletes must attempt to reach the highest level possible before stopping. Providing verbal encouragement throughout the test can assist in ensuring this outcome. The results of the Yo-Yo IRT are primarily used for evaluation of changes in intermittent running endurance capacity rather than prescription of training speeds and estimation of $\dot{V}O_2$max.

NORMATIVE DATA

Yo-Yo IRT normative data for under 17, under 21, and National squad netball players are presented in table 22.5.

TABLE 22.5 Yo-Yo Test Data for U17, U21, and National Squad Netball Players Mean ± *SD* (Range)

Squad/level	Level	Distance, m
U17 TID	15.7 ± 0.9	1,013 ± 286
	(13.4-17.3)	(440-1,520)
U21	16.6 ± 1.2	1,320 ± 372
	(13.4-18.7)	(440-2,000)
National team	17.2 ± 1.1	1,492 ± 358
	(14.3-19.3)	(560-2,160)

Data source: AIS Netball database 2010; U17 TID (*n* = 25); U21 (*n* = 22); National team (*n* = 27).

MUSCULAR STRENGTH TESTS

RATIONALE

Netball play involves explosive actions such as leaping for intercepts, rebounding, and hard changes of direction. These explosive actions require high levels of force to be exerted over a relatively short period of time. Maximal force and rate of force development have been demonstrated to be important to the performance of such activities (Hakkinen et al. 1986; Schmidtbliecher 1992). Furthermore, resistance and plyometric training effectively develop the strength qualities of maximal force and rate of force development, respectively (Hakkinen et al. 1985; Schmidtbliecher 1992). Resistance training, therefore, is an important component of the netball player's physical preparation. The bench press, bench pull, and squat lifts are used to assess maximal force production capabilities for netball because they test multijoint function, use the major muscle groups of the upper and lower body, and are commonly used training exercises that are familiar to the high-level player.

TEST PROCEDURE

Strength tests should be administered by a coach or instructor who is qualified and experienced in teaching strength training methods. The athlete's technique should be closely scrutinized during the warm-up, particularly for the squat. Instruction and coaching tips should be given if these are required. If the tester believes that the athlete's technique is very poor and cannot perform the tests safely, the test should not be conducted. Before leaving the weight training facility, the athlete should perform a number of light sets of each exercise under instruction from the tester. Follow-up technical coaching or testing should be arranged at this time.

For the back squat and bench press tests, the bar should be loaded such that no more than five sets are performed in achieving the 3 repetition maximum (3RM). A minimum of 3 min of recovery should be given to the athlete between sets. Please refer to chapter 13, Strength and Power Assessment Protocols, for detailed protocols for the assessment of the squat, bench press, and bench pull exercises.

DATA ANALYSIS

The selected 3RM strength tests enable basic diagnosis of strength levels and changes related to training for lower- and upper-body pushing and pulling movements. As well as providing information on force production capabilities under high loads, the assessment of absolute changes after specific training interventions can assist in determining the appropriateness of the training program. Because maximal strength is largely influenced by body size and muscle mass, between-athlete comparisons of maximal strength scores are often divided by body mass, as presented in table 22.6.

NORMATIVE DATA

Table 22.6 presents 3RM strength test normative data for AIS Scholarship holders (under 19) and National squad netball players.

TABLE 22.6 Data From 3RM Test for AIS Scholarship Holders (U19) and National Squad Netball Players: Mean ± SD (Range)

	3RM SQUAT		3RM BENCH PRESS		3RM BENCH PULL	
Squad/level	kg	Ratio to body mass	kg	Ratio to body mass	kg	Ratio to body mass
AIS squad	71.4 ± 12.6 (37.5-90.0)	1.0 ± 0.2 (0.6-1.3)	43.5 ± 7.2 (30.0-60.0)	0.6 ± 0.1 (0.4-1.0)	47.4 ± 6.7 (37.5-60.0)	0.6 ± 0.1 (0.5-0.9)
National team	74.2 ± 11.0 (60.0-100.0)	1.0 ± 0.1 (0.8-1.2)	46.7 ± 5.0 (35.0-55.0)	0.6 ± 0.1 (0.5-0.8)	n/a	n/a

n/a = not available. Data source: AIS Netball database 2010; AIS squad (n = 26); National team (n = 18).

References

Allison, B. 1978. A practical application of specificity in netball training. *Sports Coach* 2(2):9-13.

Bale, P., and Hunt, S. 1986. The physique, body composition and training variables of elite and good netball players in relationship to playing position. *Australian Journal of Science and Medicine in Sport* 18(4):16-19.

Chad, K., and Steele, J. 1990. Relationships between physiological requirements and physiological responses to match play and training in skilled netball players: Basis of tailor-made training programs. A report presented to the Australian Sports Commission's Applied Sports Research Program. Canberra, Australia: Australian Sports Commission.

Davidson, A., and Trewartha, G. 2008. Understanding the physiological demands of netball: a time-motion investigation. *International Journal of Performance Analysis in Sport* 8(3):1-17.

Hakkinen, K., Komi, P.V., and Alen, M. 1985. Effect of explosive type strength training on isometric force- and relaxation-time, electromyographic and muscle fibre characteristics of leg extensor muscles. *Acta Physiologica Scandinavica* 125(4):587-600.

Hakkinen, K., Komi, P.V., and Kauhanan, H. 1986. Electromyographic and force production characteristics of leg extensor muscles of elite weightlifters during isometric, concentric and various stretch shortening cycle exercises. *International Journal of Sports Medicine* 7(3):144-151.

Loughran, B.J., and O'Donoghue, P.G. 1999. Time-motion analysis of work-rate in club netball. *Journal of Human Movement Studies* 36(1):37-50.

Norton, K.I., Craig, N.P., Withers, R.T., and Whittingham, N.O. 1994. Assessing the body fat of athletes. *Australian Journal of Science and Medicine in Sport* 26(1 & 2):6-13.

Otago, L. 1983. A game analysis of the activity of netball players. *Sports Coach* 7(1):24-28.

Pyne, D., Gardner, A.S., Sheehan, K., and Hopkins, W.G. 2005. Fitness testing and career progression in AFL football. *Journal of Science and Medicine in Sport* 8(3):331-332.

Ross, W.D., and Marfell-Jones, M.J. 1987. Kinathropometry. In: J.D. MacDougall, H.A. Wenger, and H.J. Green (eds), *Physiological Testing of the Elite Athlete*. Champaign, IL: Human Kinetics, pp 223-308.

Schmidtbleicher, D. 1992. Training for power events. In: P. Komi (ed), *Strength and Power in Sport*. Oxford, UK: Blackwell, pp 381-395.

Steele, J., and Chad, K. 1992. An analysis of the movement patterns of netball players during matchplay: implications for designing training programs. *Sports Coach* 15(1):21-28.

Woolford, S., and Angrove, M. 1992. Games intensities in elite level netball: position specific trends. *Sports Coach* 15(2):28-32.

Young, W. 1994. A simple method for evaluating the strength qualities of the leg extensor muscles and jumping abilities. *Strength and Conditioning Coach* 2(4):5-8.

Rowers

Anthony J. Rice and Mark A. Osborne

Traditional rowing is a maximal intensity event over 2,000 m with a duration of between 5:30 and 8:30 min:s depending on the category (i.e., male, female, heavyweight, and lightweight), boat type (scull or sweep oar), boat size (1, 2, 4, or 8 rowers), environmental conditions, and depth, temperature, and salinity of the water. Elite rowers typically have long limbs relative to their height (Hahn 1990; Rodriguez 1986; Ross et al. 1982; Shephard 1998), high maximal aerobic powers (both in absolute and relative terms; Secher et al. 1982), and the ability to sustain high blood lactate concentrations (Hagerman 1984). A typical rowing event can be completed with 70% to 90% of the energy requirements met by aerobic sources with the remainder coming from anaerobic sources; the percentage mix is determined primarily by the time required to complete the 2,000 m distance (Steinacker 1993).

Adaptive rowing is a recent development in the sport of rowing. Athletes race over 1,000 m with race durations typically between 3 and 5 min. Rowers compete in four boat classes, which are adapted for different types of disabilities: (1) LTAMix4+ legs, trunk, and arms, in which the crew of four must consist of both genders in a normal rowing shell with a sliding seat; (2) LTAIDMix4+ mixed coxed four, which is like the LTAMix4+ except that the crew is intellectually disabled; (3) TAMix2x, in which athletes can use their trunk and arms in a mixed double scull on a fixed seat; and (4) the men's and women's ASW1x, in which a single scull is used and athletes are limited to using their arms and shoulders on a fixed seat. The hull of the adaptive rowing boat is identical to able-bodied boats. However, adaptive rowing boats are equipped with seats that vary according to the disability of the rower. The LTA4+

has a standard sliding seat, whereas the other three boat classes have fixed seats. The TA2x has a seat that provides "complementary support." The AW1x and AM1x are equipped with a seat that offers postural support to those individuals with compromised sitting balance (i.e., spinal cord injury, cerebral palsy). This ensures that the upper body is strapped in and supported while kept in a fixed position. Smaller boats are equipped with pontoons to provide additional buoyancy and are attached to the boats' riggers to provide additional lateral balance.

For the coach and scientist, the aims of physiological testing include these:

- Establish and quantify the current fitness status of the rower across a range of physiological capacities
- Profile each rower relative to established benchmarks as well as the current elite group means
- Measure the effectiveness of the previous training phase and plan for future training phases
- Establish training intensities and zones (based on heart rate, power output, stroke rate, and rating of perceived exertion) that can be used for training prescription

The test battery provides an opportunity to demonstrate a rower's current maximal capabilities over a number of test distances or time increments, which can help athletes regularly "recalibrate" their perceptions of maximal performances across the distances and time periods that emphasize the different physiological capacities. The submaximal component of the incremental protocol carried out in the laboratory is independent of the athlete's motivation and helps provide direct evidence on current physiological status and changes in rowing specific efficiency induced by training or detraining.

Other reasons to undertake physiological testing of rowers include these:

- Quantify physiological effects of an ergogenic aid (i.e., beta alanine, caffeine, sodium bicarbonate)
- Trial the effect of different warm-up strategies on subsequent race performance
- Characterize fatigue as revealed when physiological responses uncouple
- Troubleshoot when race performances are below expected values
- Assist with other talent identification measures

In the current Olympic cycle, elite Australian rowers undertake a number of physiological and performance tests as part of their preparation for domestic and international seasons. These include strength and power measures, anthropometry, an assessment of the rower's maximal power-generating capabilities over four test distances (power profile: 100 m, 500 m, 2,000 m, and 6,000 m), and a seven-stage 4 min discontinuous incremental ergometer test, with the final 4 min stage being an all-out effort. These tests are outlined in the sections following.

Athlete Preparation

Standardized pretest preparation is recommended to enhance the reliability and validity of physiological data. Specific information relating to the preparation of the athlete and the testing environment is outlined in chapter 2, Pretest Environment and Athlete Preparation. In addition, a number of considerations pertaining to performance testing of rowers are outlined next.

- *Training.* Prior to the commencement of ergometer power profile testing, athletes should not train on the day preceding the first test. Standardized training and recovery are prescribed prior to and between ergometer tests to ensure comparability of results. The Recommended Test Order highlight box later in the chapter outlines the time (or distance), modality, and intensity of the training in the sessions between each ergometer test.

Prior to laboratory testing, athletes must not train in the 12 h preceding the test. On the day before the test, the afternoon training session should consist of no more than 12 km on the water and should be of low intensity (T3-T2 range). There should be no heavy weight training or exercise to which the athlete is not accustomed. It is suggested that the athlete replicate as closely as possible similar training loads in the 24 h leading into each testing session.

- *Diet.* A normal meal (incorporating a high-carbohydrate component) should be consumed on the evening preceding each test and, if scheduling allows, also on the day of the test. No alcohol should be consumed in the 24 h preceding the test. The athlete should give special attention to ensuring good hydration in the lead-up to each test.

- *Ergogenic aids.* To make accurate comparisons, no ergogenic aids should be permitted prior to undertaking any of the ergometer tests. Testing ergogenic aids (e.g., caffeine, sodium bicarbonate) or perfecting previous ergogenic strategies should occur only during training sessions, alternative ergometer sessions, or regattas.

Test Environment

In Australia, Concept 2 rowing ergometers have been the stable testing ergometer for the past 20 years. The Gjessing ergometer was used previously to test rowers, but this model was found to not simulate the real demands of on-water rowing. In the last 10 years, 4 models of the Concept rowing ergometer (Concept IIB, IIC, IID and IIE) have been developed and used. The Concept IID and IIE ergometers are currently used for the testing of Australian rowers. Research comparing the IIC and IID ergometers (Vogler et al. 2007) indicates that the physiological responses to the IIC and IID ergometers are similar. As such, either ergometer can be used for testing. The Concept IIE uses the same flywheel as the Concept IID and has similar airflow characteristics around the flywheel; as a result, the physiological response to rowing on a model IIE is likely to be similar to the IIC and IID.

To better replicate the movement patterns of on-water rowing and therefore the feel of ergometer rowing, Concept 2 developed slides as an accessory for rowing its ergometers. Two slides are required to support the ergometer from each end. Each slide consists of a base with two tracks and a carriage with wheels that run along the tracks. A bungee cord is attached at either end of the base frame to maintain tension to keep the carriage centered on the base.

When using slides, the ergometer moves back and forth under the rower, so that the rower does not move his own body mass along the ergometer seat rail. This action is similar to the movement of a boat when rowing on the water. Because rowers are required to move the smaller mass of the ergometer underneath them rather than their own usually larger body mass on a stationary ergometer, most athletes, particularly heavier individuals, find that the slides enable them to maintain higher stroke rates because only the mass of the ergometer is in motion (not the mass of the athlete and the ergometer seat, handle, and chair). This also enables rowers to maintain stroke rates closer to those experienced when rowing on the water and typically higher than those previously achievable during maximal efforts on a fixed ergometer (unpublished data, Marsden 2006).

Sport scientists had to consider whether a move to testing protocols using the ergometer on slides would have a measurable effect on the physiological variables associated with current testing protocols. Unpublished data from Marsden and colleagues (2006) as well as published data from Holsgaard-Larsen and Jensen (2010) and Mahoney and colleagues (1999) suggest that although the stroke rate and time to complete 2,000 m may be different on a sliding ergometer versus a fixed ergometer, the physiological variables associated with the two ergometers as well as the calculated blood lactate transition thresholds are similar. As a result of these investigations, rowing ergometer testing and the majority of ergometer training in Australia have shifted to Concept II ergometer models (IID and IIE) on slides.

Recommended Test Order

It is important that the ergometer power profile tests are completed in the same order to control the interference. This order also allows valid comparison of results from different testing sessions. The order is as follows (with day 1 normally being a Saturday).

Day	Tests
1, afternoon	90 min aerobic training
2, all day	No training—only active recovery and stretching or walking
3, morning	100 m all-out followed in 20 min by 500 m all-out (Able-bodied rowers) or 250 m all-out (adaptive rowers)
	10 km low-intensity aerobic row or cross-training equivalent
3, afternoon	Able-bodied rowers—6,000 m all-out
	Adaptive rowers—3,000 m all-out
	Normal gym training (no new exercise or changes in weights program)
4, morning	15 km low-intensity aerobic rowing
4, afternoon	10-12 km low-intensity aerobic row (technique) or ergometer preparation
5, morning	Able-bodied rowers—2,000 m all-out
	Adaptive rowers—1,000 m all-out
	Resume normal training

Equipment Checklist

Anthropometry

- ☐ Stadiometer (wall mounted)
- ☐ Balance scales (accurate to ±0.05 kg)
- ☐ Anthropometry box
- ☐ Skinfold calipers (Harpenden skinfold caliper)
- ☐ Bone calipers
- ☐ Anthropometer
- ☐ Anthropometric measuring tape (Lufkin, Model W606PM)
- ☐ Anthropometric measuring tape
- ☐ Recording sheet (refer to the anthropometry data sheet template)
- ☐ Marker pen
- ☐ Pen

Strength and Power Testing

- ☐ Bench press bench
- ☐ Squat rack or power cage
- ☐ Barbell (Olympic 20 kg)
- ☐ Weight plates (2.5-25 kg increments)
- ☐ Recording sheet (refer to the strength and power test data sheet templates)
- ☐ Pen

Ergometer Power Profile

- ☐ Concept IID or IIE rowing ergometer on Concept II sliders
- ☐ Heart rate monitor (if desired)
- ☐ Stopwatch
- ☐ Lactate analyzer and associated consumables
- ☐ Recording sheet (refer to the rowing power profile data sheet template)
- ☐ Concept 2 PM3/4 interface software if available
- ☐ Pen

7 × 4 min Incremental Test

- ☐ Concept IIC, IID, or IIE rowing ergometer
- ☐ Concept II slides
- ☐ Heart rate monitoring system
- ☐ Metabolic cart (and associated equipment)
- ☐ Stopwatch
- ☐ Capillary blood sampling equipment (as described subsequently)
- ☐ Rating of perceived exertion (RPE) chart
- ☐ Recording Sheet (refer to the rowing incremental test data sheet template)
- ☐ Pen

Adaptive Rower Tests

- ☐ 2 towels (to chock up either end of the slider arrangement for fixed-seat rowers)
- ☐ Chest and leg straps for arms only and trunk and arms only

Capillary Blood Sampling

- ☐ Disposable lancets
- ☐ Alcohol swabs
- ☐ Heparinized capillary tubes (100 µL) or lactate strips
- ☐ Disposable rubber gloves
- ☐ Tissues
- ☐ Kidney or carry dish
- ☐ Sharps Container
- ☐ Biohazard bags
- ☐ Blood lactate/blood gas analyzer

ANTHROPOMETRY

RATIONALE

Regular anthropometric assessment of elite rowers is conducted throughout both domestic and international seasons. Depending on the category of rower (gender, age group, heavyweight, or lightweight), anthropometric assessment is completed every 3 months but may occur more frequently as competition approaches.

TEST PROCEDURE

At a minimum, height, body mass, and skinfolds should be measured. Skinfolds are recorded over seven sites (triceps, biceps, subscapular, supraspinale, abdominal, front thigh, and medial calf). The individual skinfold measures as well as the sum of the seven sites should be reported. Refer to anthropometry protocols outlined in chapter 11, Assessment of Physique, for a detailed description of all anthropometric test procedures. More advanced anthropometric assessments including muscle girths, bone breadths, limb lengths, arm span, and sitting height can be conducted if required.

Although the description of skinfold measurement procedures seems simple, a high degree of technical skill is essential for consistent results, and therefore it is important that an experienced, accredited, and trained tester carries out the measurements. Ideally the same tester should conduct each retest to ensure reliability.

If the opportunity arises, the restricted profile and four-compartment model may be useful for monitoring compartmental changes, especially during phases where body composition has been targeted through dietary or training interventions.

DATA ANALYSIS

The application and limitations of anthropometric assessment are discussed in detail in chapter 11.

NORMATIVE DATA

Anthropometric data of Australian male and female, lightweight and heavyweight rowers are presented in table 23.1. Substantial seasonal variation may occur in association with changes in training and competition load, and thus direct comparison across groups should be made with caution.

TABLE 23.1 Anthropometric Data for Female and Male, Lightweight and Heavyweight Rowers: Mean ± *SD* (Range)

Athlete group	Body mass, kg	Σ7 skinfolds, mm
Female		
Lightweight (*n* = 18)	59.3 ± 1.8 (54.4-62.6)	60.2 ± 9.3 (49.2-84.0)
Heavyweight (*n* = 50)	75.5 ± 5.3 (65.4-87.7)	82.4 ± 19.2 (49.0-139.4)
Male		
Lightweight (*n* = 30)	73.6 ± 1.2 (71.1-76.2)	37.6 ± 4.6 (30.3-47.7)
Heavyweight (*n* = 76)	92.2 ± 6.7 (77.4-108.7)	54.0 ± 13.0 (36.5-90.4)

Typical error: Σ7 skinfolds = ~1 mm. Data source: Senior National–caliber athletes, December 2009.

STRENGTH AND POWER TESTING

RATIONALE

Strength is an important factor in rowing. General strength, power development, and strength endurance should be the ultimate goal of a rowing training program. Excessive hypertrophy can be detrimental to rowers because they have to change the direction of the moving body mass without disturbing boat velocity, and excessive hypertrophy can also affect boat drag, compounding the difficulties faced by lightweight competitors who often have to manage their weight.

Strength training for rowing has evolved from traditional circuit-based strength-endurance programs to methods originally developed for weightlifting and track-and-field athletes. Strength training during the preseason is generally designed to develop overall strength, whereas training during the competition season is designed to develop power and ensure that maximal force can be applied to the oar handle.

Elite rowers generate maximal force during the first stroke of a race and can reach more than 1,350 N for men and 1,000 N for women (Hartmann et al. 1993). Isometric testing has also been used to assess maximal force generation throughout the rowing stroke (Secher 1975). Results showed that international standard rowers were able to generate significantly more force than national-level rowers, who in turn were better than club-level rowers. However, isometric testing generally requires the construction of specially built equipment and requires additional resources to measure handle force. As a result, a number of traditional gym-based lifts are widely used by rowers to assess basic strength.

TEST PROCEDURE

For rowers who have achieved appropriate technical competency, lifts include 3 repetition maximum (3RM) bench press, bench pull, and parallel squat tests along with a 1RM clean. These exercises represent a generic pushing activity (bench press), pulling activity (bench pull), and assessment of leg strength (parallel squat). A clean is also regarded as a useful indicator of rowing strength because it approximates the rowing action although in a vertical plane instead of horizontal plane.

Refer to chapter 13, Strength and Power Assessment Protocols, for a detailed description of strength and power tests. Warm-up and submaximal attempts of the bench press and squat should be performed in an appropriate rack with spotters present.

The following general guidelines must be adhered to for all tests (as per chapter 13):

- Strength testing should be performed on a separate day from the ergometer tests. Strength and ergometer tests should be separated by 48 h.

- The athlete must perform an appropriate warm-up. As a minimum, all athletes are required to perform a trial at approximately 90% of specified repetition maximum for each test. If it is the first test, the athlete should perform an initial trial at approximately 90% of the weight lifted in training.

- Lowering and lifting actions must be performed in a continuous manner. A single rest of no more than 2 s is allowed between repetitions.

- A maximum of 5 min recovery between trials is allowed.

- Minimum weight increments of 2.5 kg should be used between trials. However, increments should be guided by ease of each trial.

- Ideally, specified repetition maximum (RM) test should be completed within 4 trials (not including the warm-up).

- If the athlete is unable to complete tests as per protocol, this should be noted on testing results information, and values should not include any mathematical calculations (e.g., average, typical error).

- It is recommended that a spotter, other than the supervising coach, be used where necessary.

DATA ANALYSIS

Results should be reported in absolute and relative terms. Relative strength is generally more important than absolute strength for the rower because the total mass of the rower and boat affects the drag through water. Continuing to develop absolute strength is of minimal benefit if the mass gained offsets the strength gains by increasing drag. With improvements in relative strength as opposed to increasing mass, it is generally easier to accelerate the boat because drag is not compromised. Because relative strength is more important to rowers than absolute strength, the strength gains should be expressed as a percentage of body mass.

NORMATIVE DATA

Strength data of Australian male and female, heavyweight and lightweight rowers are presented in table 23.2. Substantial seasonal variation may occur in association with changes in training and competition load and body mass. An athlete's skill in executing these lifts will also affect results obtained.

TABLE 23.2 Strength Test Data for Female and Male, Lightweight and Heavyweight Rowers: Mean ± *SD* (Range)

Athlete group	3RM bench press	3RM bench pull	3RM squat	1RM clean
Female				
Lightweight (n = 10)	43.6 ± 9.6 (25.0-55.0)	53.0 ± 6.7 (37.5-60.0)	75.3 ± 16.4 (50.0-102.5)	50.0 ± 9.8 (30.0-60.0)
Heavyweight (n = 10)	52.5 ± 5.0 (45.0-60.0)	62.3 ± 6.5 (47.5-72.5)	86.7 ± 14.6 (67.5-110.0)	61.3 ± 6.8 (47.5-70.0)
Male				
Lightweight (n = 13)	68.5 ± 6.2 (57.5-80.0)	77.5 ± 7.2 (67.5-90.0)	109.4 ± 9.5 (95.0-130.0)	84.8 ± 10.8 (70.0-97.5)
Heavyweight (n = 20)	93.8 ± 10.8 (70.0-110.0)	95.5 ± 7.8 (85.0-110.0)	133.9 ± 19.3 (110.0-170.0)	101.3 ± 11.2 (80.0-120.0)

Typical error: 3RM tests = ~3%. Data source: Senior National caliber–athletes, December 2009/January 2010.

ERGOMETER POWER PROFILE TEST

RATIONALE

The ergometer power profile test protocol aims to provide detailed physiological information about the rower's submaximal capacity and efficiency as well as measuring maximal performance parameters as accurately as possible.

Unfortunately, given the relatively, time-consuming nature of the laboratory-based 7 × 4 min incremental protocol (i.e., only a single athlete is able to complete the test every 45 min) it is undertaken only a few times each year. The information from the 7 × 4 min protocol provides an accurate assessment of the training and performance status of the rower, but the protocol is not conducive to measurements of a large group of individuals on a regular basis nor can it be conducted in the field environment or when a rower does not have access to a laboratory environment. As a result, a battery of ergometer performance tests has been developed.

The ergometer power profile test battery includes maximal 100 m, 500 m, 2,000 m, and 6,000 m efforts completed over a 3-day period. Data from the maximal efforts provide useful information on the power–time continuum of each individual rower, which can be used to infer changes in anaerobic capacity and maximal aerobic power as well as show direct changes in ergometer performance. Figure 23.1 illustrates the contribution of the various energy systems to maximal ergometer performance over the prescribed distances and approximate time span. The group data can be used by coaches to objectively assess the effectiveness of their training programs since the previous series of tests.

For consistency and comparability, it is vital that the power–distance ergometer tests are completed in an identical manner each time. This includes the training undertaken on the days between ergometer tests as well as the order in which the tests are completed (see recommended test order). It is imperative that the training for the 2 days prior to the testing week and the majority of the testing week itself is prescribed. The testing should occur within a rest and recovery week, and frequency and duration of activities are programmed accordingly. To ensure comparability, both within and across programs, all rowers and training groups should be asked to present to the testing week in similar physical states each time (see recommended test order).

TEST PROCEDURE

- Record relevant information on the test recording sheet (i.e., time of test, laboratory environmental conditions).

- Measure and record athlete's body mass (especially lightweights).

- Attach a heart rate monitor (if being used) and ensure that it is working correctly.

- Ask the athlete to adjust positioning of foot stretchers.

- Instruct the athlete to take a few strokes to set the appropriate drag factor (see drag factor settings, table 23.3).

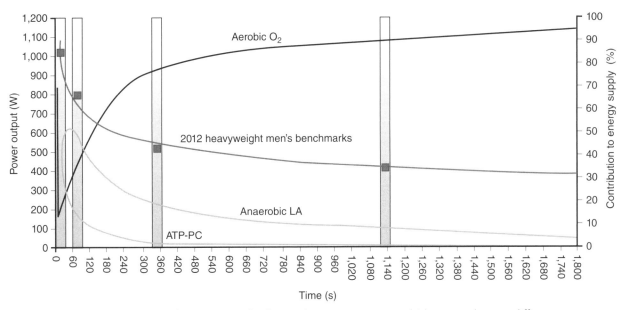

FIGURE 23.1 The energy supply continuum displaying the energy systems which are used across different ergometer distances.

TABLE 23.3 Sliding Ergometer Drag Factor Settings

Category	Drag factor
Able-bodied athletes	
Junior female	95
Lightweight female	95
Heavyweight female	105
Junior male	105
Lightweight male	105
Heavyweight male	115
Adaptive athletes	
Arms-only male (fixed seat)	120
Arms-only female (fixed seat)	120
Trunk and arms male (fixed seat)	120
Trunk and arms female (fixed seat)	120
Legs, trunk, and arms male (sliding seat)	115
Legs, trunk, and arms female (sliding seat)	105

- Allow the athlete to complete individualized warm-up. Athletes are allowed to individualize their warm-up prior to each ergometer test session, but this must be replicated as closely as possible prior to subsequent test sessions.

- Set the monitor on the rowing ergometer for appropriate test distance (i.e., 100 m, 500 m, 2,000 m, or 6,000 m).

- Fully explain the test procedure, ensuring that the rower understands that they are required to complete the set distance in the shortest possible time.

- Ensure that the flywheel is completely stopped before commencing the test.

- Instruct the athlete to commence the test when they are ready.

- At the end of the test, record the following data from the work monitor:
  - Average 500 m pace (min:s.s)
  - Average power output (watts)
  - Average stroke rate (strokes per minute)
  - Split data can also be recalled for various intervals if desired (usually average pace and stroke rate over each 500 m interval).

- Alternatively, save each test effort using appropriate PM3/4 interface software.

DATA ANALYSIS

Figure 23.2 shows a power output (W)–distance (m) relationship (vertical axis and horizontal axis, respectively) for the four test distances (100 m, 500 m 2,000 m, and 6,000 m), and the subsequent curve fit demonstrates how the power–time continuum is established.

The analysis of the data obtained from a power–distance continuum can be very powerful because it shows how different fitness traits can change with specific training but how these traits in some cases have only a small, if any, transfer to other distances (i.e., 100 m improvement will have little effect on 6,000 m performance).

Figure 23.3 shows an upward shift in the entire curve, suggesting that all fitness traits (alactic, lactic,

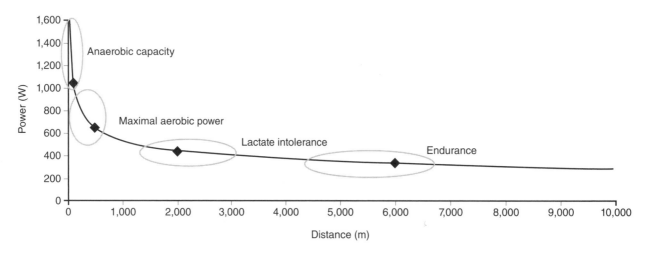

FIGURE 23.2 The relationship between ergometer distance and average power output.

and aerobic) have shown a significant improvement as a result of training. In the next example (figure 23.4), there has been a greater upward shift in the longer distances than in the shorter distances, suggesting an important improvement in aerobic and lactic energy systems but not so in the alactic system. This may be representative of what most athletes see through the early preparation phases of training prior to race specific training and speed work.

In the final example (figure 23.5), there has been no shift in the medium- and long-distance power outputs but an important upward shift in the power outputs that can be exhibited over the shorter distances. This would suggest an improvement in alactic and lactic systems but no improvement in aerobic pathways. This example would represent improvements in rowing-specific fitness that we may see later in the international season.

These examples show how regular measurement of ergometer performance can guide and individualize training prescription as well as assist with evaluation of the success of a specific training block. The test battery is essentially a simple strength and weakness assessment across the specific fitness traits required by rowers that does not require rowers to undertake laboratory tests or time out from the routine training environment. This battery has the ability to display changes that are expected to occur in a rower's fitness throughout both domestic and international preparations. The rower and coach are able to immediately gauge their progress, and the data will allow coaching staff to determine each athlete's specific improvement as well as benchmark athletes against traits that are believed to be gold medal standard at World Championships and Olympic Games.

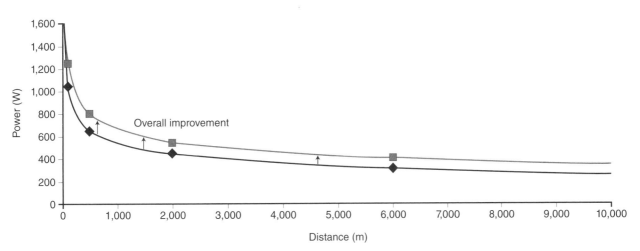

FIGURE 23.3 The expected shift in the relationship between ergometer distance and average power output when all components of fitness have been addressed.

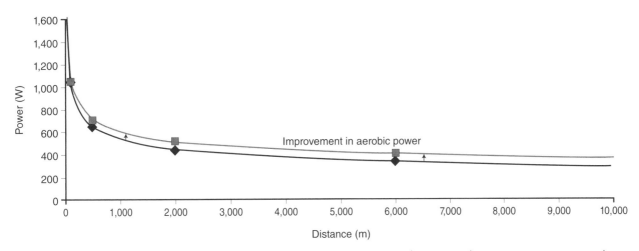

FIGURE 23.4 The expected shift in the relationship between ergometer distance and average power output when the training emphasis has been on aerobic development.

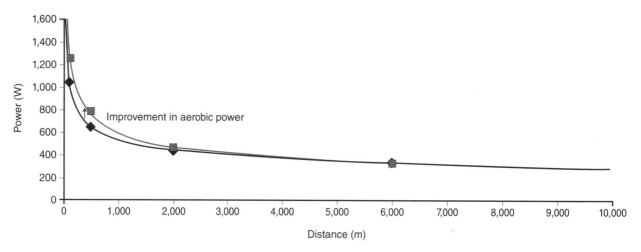

FIGURE 23.5 The expected shift in the relationship between ergometer distance and average power output when the training emphasis has been on anaerobic energy system development.

NORMATIVE DATA

Normative data of Australian male and female, heavyweight and lightweight rowers are presented in table 23.4. These data are extracted from tests conducted on senior National caliber athletes during the December 2009 testing block. Substantial seasonal variation occurs in association with changes in training and competition load, and thus direct comparison across groups should be made with caution.

TABLE 23.4 Power Profile Test Data for Female and Male, Lightweight and Heavyweight Rowers: Mean ± *SD* (Range)

Athlete group	Time, mm:s.s	Power, W	Stroke rate, strokes · min⁻¹
Female lightweight (*n* = 19)			
100 m	00:19.2 ± 00:00.7 (00:18.4-00:20.4)	401 ± 41 (330-449)	54 ± 6 (37-65)
500 m	01:40.1 ± 00:03.8 (01:34.3-01:45.8)	351 ± 40 (296-417)	45 ± 4 (36-51)
2,000 m	07:19.3 ± 00:13.5 (06:57.4-07:50.0)	266 ± 24 (216-308)	36 ± 2 (32-41)
6,000 m	23:17.3 ± 00:48.1 (22:14.2-25:28.7)	223 ± 22 (169-255)	30 ± 3 (25-35)
Female heavyweight (*n* = 48)			
100 m	00:17.3 ± 00:00.8 (00:15.6-00:19.0)	544 ± 75 (408-738)	57 ± 5 (44-68)
500 m	01:32.7 ± 00:03.2 (01:27.2-01:40.9)	443 ± 46 (341-528)	46 ± 3 (39-51)
2,000 m	06:57.8 ± 00:15.6 (06:33.8-07:38.5)	310 ± 33 (232-367)	35 ± 3 (27-41)
6,000 m	22:23.5 ± 00:53.7 (21:11.1-25:33.0)	252 ± 28 (168-294)	30 ± 3 (22-35)

Athlete group	Time, mm:s.s	Power, W	Stroke rate, strokes · min⁻¹
Male lightweight (n = 42)			
100 m	00:17.4 ± 00:09.3	692 ± 63	57 ± 4
	(00:15.0-00:18.0)	(560-830)	(55-61)
500 m	01:25.2 ± 00:02.1	568 ± 42	43 ± 2
	(01:20.6-01:30.7)	(469-668)	(39-48)
2,000 m	06:21.7 ± 00:11.9	405 ± 35	36 ± 2
	(06:05.5-06:58.0)	(307-459)	(33-38)
6,000 m	20:30.0 ± 00:38.4	327 ± 28	29 ± 2
	(19:36.0-22:27.8)	(247-372)	(26-30)
Male heavyweight (n = 70)			
100 m	00:14.7 ± 00:01.0	892 ± 132	61 ± 5
	(00:13.6-00:20.6)	(320-1,113)	(48-77)
500 m	01:18.9 ± 00:04.9	725 ± 96	49 ± 3
	(01:12.9-01:44.3)	(308-903)	(40-57)
2,000 m	05:59.6 ± 00:21.8	489 ± 62	37 ± 2
	(05:40.7-07:53.3)	(211-566)	(30-43)
6,000 m	19:37.2 ± 01:15.1	377 ± 49	30 ± 2
	(18:38.5-26:05.9)	(158-432)	(24-33)

Data source: Senior National–caliber athletes, December 2009.

7 × 4 MIN INCREMENTAL TEST

RATIONALE

This ergometer test protocol aims to provide detailed physiological information of the rower's submaximal capacity and efficiency and to measure maximal performance parameters in a time efficient manner. In the 7 × 4 min incremental test, athletes complete six submaximal efforts at predetermined workloads separated by 1 min rest periods. The six submaximal workloads are followed by a maximal 4 min effort, where athletes are required to complete as much work (cover as much distance) as possible in the 4 min period. To more closely simulate the rowing motion, it is recommended the test be undertaken using Concept II slides. The main aims of using these are to more accurately reflect the stroke rate and drive–recovery ratio of on-water rowing and minimize the risk of injury that may result from spending considerable time on the ergometer with significant low back load.

TEST CONSIDERATIONS

The starting workload and increment between submaximal workloads is individualized and based on the athlete's best 2,000 m time within the last 12 months. The range of times for the 2,000 m ergometer tests have been categorized based on 10 s increments, with the fastest 2,000 m times having the highest starting workload and largest increment (see table 23.5).

To ensure that individual athletes complete an identical amount of work prior to beginning the seventh step (4 min at maximal pace), every 7 × 4 min incremental test undertaken by the athlete for a seasonal year should use the same starting workload and increment. There should be no change to the starting workload during a given season. Only a significant change in 2,000 m time will warrant a change in workload and increment.

The workloads and increments have been designed such that the sixth step (i.e., the step immediately preceding the maximal step) produces a blood lactate value in the range of 5 to 8 mmol · L⁻¹ and corresponds to a pace very close to the athlete's 6,000 m pace from the power profile. This will obviously change depending on the time of year and the athlete's current training status, but in a single season the athlete will complete an identical amount of work leading into the maximal performance component of the test, and thus the workloads remain constant.

A decision often has to be made about whether moving to the next 10 s increment is valuable. An example is when an athlete has 4 years of data all

TABLE 23.5 Ergometer Workloads (W and mm:s.s) Categorized With Respect to the Athlete's Best 2,000 m Time in the Previous Year

Previous years selection ergometer time	<5:50.0	5:50.0-6:00.0	6:00.0-6:10.0	6:10.0-6:20.0	6:20.0-6:30.0	6:30.0-6:40.0
Work load increments, W	45	45	40	35	35	30
Workload 1	200	170	170	160	140	140
Workload 2	245	215	210	195	175	170
Workload 3	290	260	250	230	210	200
Workload 4	335	305	290	365	245	230
Workload 5	380	350	330	300	280	260
Workload 6	425	395	370	335	315	290
Workload 7	max	max	max	max	max	max
Workload 1	02:00.7	02:07.4	02:07.4	02:10.0	02:15.9	02:15.9
Workload 2	01:52.7	01:57.8	01:58.7	02:01.7	02:06.2	02:07.4
Workload 3	01:46.6	01:50.5	01:52.0	01:55.2	01:58.7	02:00.7
Workload 4	01:31.6	01:44.8	01:46.6	01:49.8	01:52.7	01:55.2
Workload 5	01:37.4	01:40.1	01:42.1	01:45.4	01:47.8	01:50.5
Workload 6	01:33.8	01:36.1	01:38.2	01:41.6	01:43.7	01:46.6
Workload 7	max	max	max	max	max	max

starting at the same workload and increment but finally betters his 2,000 m time from 5:50.6 to 5:49.8. Taking the protocol to its exact description would mean the athlete would change his starting workload and increment. However, the gain in doing this is far outweighed by the inability to compare the athlete across the 4 years of data using the previous starting workload and increment. In cases such as this it is left to the discretion of the coach or scientist and athlete to decide what should be the appropriate starting workload and increment for that athlete.

The 2,000 m category times used in table 23.5 are based on the athlete's best 2,000 m time from the previous year and not his all-time personal best 2,000 m test. Thus it is possible that workloads could change slightly from one year to the next. If this is the case, comparison between years for the same individual, or in the same year between individuals, must only be done using variables such as heart rate, blood lactate, $\dot{V}O_2$, and perceived exertion at LT1, LT2, and the maximal step. Distance covered in the final maximal step can only be used as a comparison between athletes when those athletes have completed the same amount of work prior to the maximal step (i.e., identical starting workload and increment).

TEST PROCEDURE

- Ensure that the rowing ergometer is clean and fully operational (e.g., chain is oiled, slide is clean, seat moves freely, ergometer display unit is working). The ergometer should be positioned at least 1 m from any other fixed structure.

- A fan can be used to cool the rower during the test, but the position and speed of the fan should remain a constant factor across testing blocks. The fan must be placed to the side of the rower (to prevent the air flow from pushing the rower away from the fan while she is sitting on the slides) on the opposite side of the flywheel (to avoid corrupting the drag factor and resultant power output measurements).

- Ensure that the metabolic cart has been adequately warmed up and has been calibrated for ventilation and gas composition. It is recommended that three primary gravimetric-standard gases (±0.02%) spanning the physiological range be used for oxygen and carbon dioxide analyzer calibration.

- Ensure that relevant informed consent and pretest forms are completed.

- Record relevant information on the test recording sheet (i.e., time of test, laboratory environmental conditions).

- Measure and record athlete's body mass.

- Attach a heart rate monitor and ensure it is working correctly.

6:40.0-6:50.0	6:50.0-7:00.0	7:00.0-7:10.0	7:10.0-7:20.0	7:20.0-7:30.0	7:30.0-7:40.0	7:40.0-7:50.0
25	25	25	20	20	15	15
150	130	115	125	110	120	110
175	155	140	145	130	135	125
200	180	165	165	150	150	140
225	205	190	185	170	165	155
250	230	215	205	190	180	170
275	255	240	225	210	195	185
max	max	max	max	max	max	max
02:12.8	02:19.3	02:25.2	02:21.2	02:27.3	02:23.1	02:27.3
02:06.2	02:11.4	02:15.9	02:14.3	02:19.3	02:17.6	02:21.2
02:00.7	02:05.0	02:08.7	02:08.7	02:12.8	02:12.8	02:15.9
01:56.0	01:59.7	02:02.7	02:03.8	02:07.4	02:08.7	02:11.4
01:52.0	01:55.2	01:57.8	01:59.7	02:02.7	02:05.0	02:07.4
01:48.5	01:51.3	01:53.5	01:56.0	01:58.7	02:01.7	02:03.8
max	max	max	max	max	max	max

- Place a piece of tape across the bridge of the athlete's nose. This is used to help prevent the nose clip from slipping during the test.
- Ask the athlete to adjust positioning of foot stretchers.
- Position the headset and mouthpiece on the athlete to ensure maximal comfort.
- Instruct the athlete to take a few strokes to set the appropriate drag factor (see drag factor settings table 23.3) and to allow for adjustment of respiratory hosing or other test apparatus so that it is not interfering at any stage of the stroke.
- Collect a pretest capillary blood sample from the fingertip or ear lobe for determination of blood lactate concentration.
- Set the monitor on the Concept II Rowing ergometer to display watts for each stroke and set the workload time (4 min) and rest interval (1 min).
- Prior to commencement of the test, ensure that the nose clip is positioned to block nasal air flow and that the flywheel is completely stopped.
- Start the metabolic cart and provide the athlete with a countdown to commence the first workload of the test.
- During each workload record heart rate and stroke rate during the last 30 s of each min. Record distance covered, average power output, and 500 m pace at the end of each workload.
- On the completion of each workload, a blood sample should be drawn from the fingertip or ear lobe and analyzed for blood lactate concentration.
- Ask the athlete to identify a rating of perceived exertion (RPE) from Borg chart, and record this value on the data sheet.
- During the 1 min rest period between workloads, athletes are permitted to remove the mouthpiece attached to the metabolic cart to have a drink. However, the breathing apparatus must be back in position well before the start of the next work bout (approximately 10-15 s).
- Prior to commencement of the final maximal workload, remind the athlete that they should aim to complete as much work as possible in the 4 min period. Athletes should be encouraged to split the 4 min evenly rather than starting conservatively and then coming home strong. Athletes should be able to hold or better their average 2,000 m split for the entire 4 min period.
- Record maximal heart rate in the last minute of the maximal workload. On completion of the maximal effort record average stroke

rate, power output, 500 m pace, and distance covered. Coincident with the maximal performance assessment will be the attainment of maximal heart rate, oxygen consumption, and blood lactate concentration.

- Immediately on completion of the test, help the athlete to remove the breathing apparatus as soon as possible.

- Collect a capillary blood sample from the fingertip or ear lobe on completion of the test and 4 min post effort. Analyze the sample for determination of peak blood lactate concentration. The highest value from the two readings should be used as the peak blood lactate.

- Ask the athlete to identify RPE from Borg chart and record this value on the data sheet.

ADAPTIVE ROWERS

The 7 × 4 min incremental test can also be used to assess the physiological capacity of adaptive rowers. The test is administered as per that for able-bodied rowers; however, the following points of difference should be noted.

Test Administration

The athlete's best 1,000 m time within the last 12 months is used to determine the starting workload and increment between submaximal workloads. As with the protocol for able-bodied rowers, only a significant change in 1,000 m time will warrant a change in workload and increment. The starting workload and increments outlined in the table 23.6 have been based on the able-bodied protocol used by Australian rowers prior to the 2000 Sydney Olympics. Some trial and error may be required to individualize the workloads for rowers given their physical limitations; however, once a working protocol has been established it must remain that rower's individual protocol.

Similar to the able-bodied protocol, the workloads and increments have been designed such that the sixth step (i.e., the step immediately preceding the maximal step) produces a blood lactate value in the range of 5 to 8 mmol · L^{-1} and corresponds to a pace very close to the rower's 3,000 m pace from the performance test.

Slide Ergometer Arrangement for Fixed-Seat Adaptive Athletes

The slide arrangement for fixed-seat rowers (arms only and trunk and arms) allows the ergometer to slide freely but remain centered on the apparatus. The slide setup shown in figure 23.6 ensures an "on-water" feel of the ergometer and a reduced risk of injury but also prevents the ergometer from hitting

TABLE 23.6 Ergometer Workload Setting for Adaptive Rowers

Workload stage	Pace per 500 m, mm:s.0
Step 1	+ 34 s per 500 m on Step 7
Step 2	+ 28 s per 500 m on Step 7
Step 3	+ 22 s per 500 m on Step 7
Step 4	+ 16 s per 500 m on Step 7
Step 5	+ 10 s per 500 m on Step 7
Step 6 (~3,000 m pace)	+ 6 s per 500 m on Step 7
Step 7 (maximal step)	Average 1,000 m pace from power profile

either end of the slide. Each end of the slide arrangement should be elevated by using a folded towel. The normal slide setup (i.e., no towel) is used for legs, trunk, and arms rowers.

Chest and Leg Strap Placement

For arms-only and trunk and arms rowers, correct strap placement is essential. The chest strap for arms-only rowers must sit slightly below the nipple line. The leg strap for both arms-only and trunk and arms rowers must be slightly above the knees.

DATA ANALYSIS

Submaximal oxygen values are calculated by averaging the readings recorded during the final min of each submaximal workload. Maximal oxygen consumption ($\dot{V}O_2$max) is recorded as the highest value actually attained during a full minute. Thus, if the metabolic cart is based on 30 s sampling periods, $\dot{V}O_2$max is the sum, or if all results are expressed in L · min^{-1} the average of the highest two consecutive readings. If 15 s sampling periods are used, $\dot{V}O_2$max is the highest value obtained on the basis of any four consecutive readings.

Submaximal heart rates are the values for the final 30 s of each submaximal workload. The maximal heart rate is the highest value recorded over a 5 s sampling period during the entire test.

Data from the final maximal step are included as the final workload values (including peak lactate) and used in combination with values from the submaximal workloads for calculation of the blood lactate thresholds and related measures.

Data collected during the incremental test can be used for the determination of various blood lactate transition thresholds (refer to chapter 6, Blood Lactate Thresholds) and prescription of training zones based on heart rate, blood lactate concentration, and power output. Table 23.7 provides some approximate

FIGURE 23.6 *(a)* Slide ergometer arrangement for fixed seat adaptive athletes, with towel elevation and *(b)* with straps for preventing the ergometer from hitting slide end points.

Photos courtesy of Anthony Rice.

TABLE 23.7 Guidelines for Blood Lactate Transition Thresholds and Training Zones

Training zone	Prescriptive description	Blood lactate threshold relationship	Blood lactate concentration, mmol · L⁻¹	%HRmax	%V̇O₂max	Rating of perceived exertion	Duration capable, h:min:s
T1	Light aerobic	<LT1	<2.0	60-75	<60	Very light	>3 h
T2	Moderate aerobic	Lower half between LT1 and LT2	1.0-3.0	75-84	60-72	Light	1-3 h
T3	Heavy aerobic	Upper half between LT1 and LT2	2.0-4.0	82-89	70-82	Somewhat hard	20 min⁻¹ h
T4	Threshold	LT2	3.0-6.0	88-93	80-85	Hard	12-30 min
T5	Maximal aerobic	>LT2	>5.0	92-100	85-100	Very hard	5-8 min

guidelines (duration of effort, blood lactate levels, power output) for the different training zones based on a five-zone model.

NORMATIVE DATA

Normative data for Australian male and female, heavyweight and lightweight rowers for the maximal step of the 7 × 4 min incremental test are presented in table 23.8. These data are extracted from tests conducted on senior National caliber athletes during the December 2009 testing block. Substantial seasonal variation occurs in association with changes in training and competition load, and thus direct comparison across groups should be made with caution. Normative data for adaptive rowers for the maximal step of the 7 × 4 min incremental test are presented in table 23.9.

TABLE 23.8 Data From 7 × 4 min Incremental Test for Female and Male, Lightweight and Heavyweight Rowers: Mean ± *SD* (Range)

Athlete group	Lightweight	Heavyweight
Female*		
Distance covered, m	1,105 ± 40 (1,035-1,169)	1,145 ± 28 (1,074-1,208)
Peak power, W	273.9 ± 29.4 (224.6-323.6)	304.7 ± 22.5 (250.9-357.0)
$\dot{V}O_2$peak, L · min^{-1}	3.61 ± 0.2 (3.20-4.11)	4.19 ± 0.3 (3.55-4.86)
$\dot{V}O_2$peak, ml · kg^{-1} · min^{-1}	60.8 ± 4.0 (55.3-69.6)	55.2 ± 4.4 (43.7-65.1)
HRmax, beats · min^{-1}	190 ± 9 (177-208)	191 ± 8 (169-205)
Peak blood lactate, mmol · L^{-1}	12.4 ± 2.8 (6.8-16.8)	11.6 ± 2.5 (5.9-16.9)
LT1 power, W	167.7 ± 24.3 (125.0-215.0)	172.4 ± 21.8 (140.0-216.0)
LT2 power, W	217.9 ± 23.2 (182.7-259.0)	238.2 ± 17.9 (189.0-275.2)
Male†		
Distance covered, m	1,267 ± 26 (1,215-1,319)	1,314 ± 40 (1,206-1,397)
Peak power, W	412.5 ± 25.2 (363.3-464.8)	460.8 ± 41.5 (355.3-552.2)
$\dot{V}O_2$peak, L · min^{-1}	5.12 ± 0.3 (4.58-5.65)	6.05 ± 0.5 (4.67-7.17)
$\dot{V}O_2$peak, ml · kg^{-1} · min^{-1}	69.7 ± 3.6 (63.4-77.2)	65.6 ± 4.5 (55.8-76.9)
HRmax, beats · min^{-1}	188 ± 8 (175-199)	190 ± 8 (168-205)
Peak blood lactate, mmol · L^{-1}	13.7 ± 2.7 (7.6-17.6)	12.5 ± 2.4 (6.4-16.6)
LT1 power, W	231.3 ± 24.9 (174.0-290.0)	237.3 ± 33.1 (170.0-468.0)
LT2 power, W	312.0 ± 20.1 (277.0-354.0)	348.6 ± 39.1 (272.0-468.0)

Typical error: power ~1.5%. Data source: Senior National–caliber athletes, December 2009.

*Lightweight (*n* = 18), heavyweight (*n* = 50).

†Lightweight (*n* = 30), heavyweight (*n* = 76).

TABLE 23.9 Maximal Step Test Data for Six Adaptive Rowers: Mean ± *SD* (Range)

Test	Mean ± *SD* (range)
Distance covered, m	989 ± 76 (909-1,087)
Peak power, W	198.4 ± 46.4 (152.1-260.1)
$\dot{V}O_2$peak, L · min^{-1}	3.36 ± 0.5 (3.00-4.20)
$\dot{V}O_2$peak, ml · kg^{-1} · min^{-1}	46.5 ± 10.0 (35.7-60.4)
HRmax, beats · min^{-1}	185 ± 10 (169-193)
Peak blood lactate, mmol · L^{-1}	12.1 ± 1.9 (9.4-14.0)
LT1 power, W	122.6 ± 36.9 (75.0-178.0)
LT2 power, W	157.3 ± 35.8 (120.0-211.0)

Typical error: power ~2.75%. Data source: Senior National–caliber athletes, December 2009/January 2010.

References

Hagerman, F.C. 1984. Applied physiology of rowing. *Sports Medicine* 1(4):303-326.

Hahn, A. 1990. Identification and selection of talent in Australian rowing. *Excel* 6(3):5-11.

Hartmann, U., Mader, A., Wasser, K., and Klauer, I. 1993. Peak force, velocity, and power during five and ten maximal rowing ergometer strokes by world class female and male rowers. *International Journal of Sports Medicine* 14(Suppl 1):S42-S45.

Holsgaard-Larsen, A., and Jensen, K. 2010. Ergometer rowing with and without slides. *International Journal of Sports Medicine* 31(12):870-874.

Mahoney, N., Donne, B., and O'Brien, M. 1999. A comparison of physiological responses to rowing on friction-loaded and air-braked ergometers. *Journal of Sports Sciences* 17(2):143-149.

Marsden J. 2006. Presentation at the Applied Physiology Conference, November 2006, Sydney, New South Wales.

Rodriguez, F.A. 1986. Physical structure of international lightweight rowers. In: T. Reilly, T. Watkins, and J. Borms (eds). *Kinanthropometry III*. London: E and FN Spon, pp 255-261.

Ross, W., Ward, R., Leahy, R., and Day, J. 1982. Proportionality of Montreal athletes. In: Carter J. (ed), *Physical Structure of Olympic Athletes. Part 1: The Montreal Olympic Games Anthropological Project*. Basel, Switzerland: Karger, pp 81-106.

Shephard, R.J. 1998. Science and medicine of rowing: A review. *Journal of Sports Sciences* 16(7):603-620.

Secher, N.H. 1975. Isometric rowing strength of experienced and inexperienced oarsmen. *Medicine and Science in Sports and Exercise* 7(4):280-283.

Secher, N.H., Vaage, O., and Jackson, R. 1982. Rowing performance and maximal aerobic power of oarsmen. *Scandinavian Journal of Sport Sciences* 4(1):9-11.

Steinacker, J.M. 1993. Physiological aspects of training in rowing. *International Journal of Sports Medicine* 14: S3-S10.

Vogler, A.J., Rice, A.J., and Withers, R.T. 2007. Physiological responses to exercise on different models of the Concept II rowing ergometer. *International Journal of Sports Physiology and Performance* 2(4):360-370.

Rugby League Players

Tim J. Gabbett and Grant M. Duthie

Rugby league is a collision sport played at junior and senior levels by amateur (Gabbett 2000), semiprofessional (Gabbett 2002a), and professional (Brewer et al. 1994; Larder, 1992; Meir, 1993b; Meir et al. 1993; O'Connor, 1995) competitors. The game is physically demanding, requiring players to participate in frequent bouts of high-intensity activity (e.g., sprinting, physical collisions, and tackles), separated by short bouts of low-intensity activity (e.g., walking and jogging) (Gabbett, 2011, 2012; Gabbett et al., 2008, 2012; Waldron et al. 2011). A rugby league team consists of 13 players and 4 replacement players. Professional rugby league teams are permitted 10 interchange movements during the course of an 80 min (2 × 40 min halves) match. The objective is to advance the ball down the field into the opposition's territory and score a try (analogous to an American football touchdown). The ball must be passed backward but can be carried or kicked into the opposition's territory. At the completion of a set of six tackles, the ball is immediately given to the opposition team to commence its set of six tackles. Therefore, the same players are involved in both attack and defense.

Although variations exist among studies (Austin et al. 2011; King et al. 2009; Sirotic et al. 2009; Sykes et al. 2009), the mean total distance covered by players is reported to range from 3.5 to 6.8 km (Gabbett et al. 2012), with average work to rest ratios reported to be approximately 1:5 (King et al. 2009). Defending has been associated with lower work rates than attacking (Sykes et al. 2009). Depending on the standard of competition, the mean intensity of rugby league matches is reported to be in the range of 78% to 87% of maximal heart rate (Coutts et al. 2003; Estell et al. 1996; Gabbett 2003), with reported mean blood lactate concentrations of 5.2 to 7.2 mmol · L^{-1} (Coutts et al. 2003; Gabbett 2003).

As a result of the high physiological demands, rugby league players require well-developed muscular strength and power, speed, agility, and aerobic power (Gabbett 2002, 2002b, 2004a, 2004b; Gabbett and Herzig 2004; Gabbett et al. 2011a; Meir 1993a, 2001; O'Connor 1996). Indeed, the physical qualities of rugby league players have been reported to progressively increase with increases in playing standard (Gabbett 2002b). Furthermore, previous studies have reported significant differences between elite and subelite players (Gabbett et al. 2009, 2011b), and starters and non-starters (Gabbett 2002a, Gabbett et al. 2009, 2011b) for speed, change of direction speed, lower-body muscular power, and maximal aerobic power; better players typically have better developed physical qualities.

With this in mind, the physiological testing of rugby league players has several aims:

1. To assess the physiological characteristics believed to be the primary determinants of playing performance
2. To produce guidelines for the establishment of appropriate training intensities
3. To ensure that players are coping with and adapting to the prescribed training programs
4. To provide normative data and performance standards for junior and senior players

Athlete Preparation

Standardized pretest preparation is recommended to enable reliable and valid physiological data to be obtained. Refer to chapter 2, Pretest Environment and Athlete Preparation, for specific information relating to athlete and environment preparation.

In general, field tests and strength tests should be carried out at the end of each appropriate macrocycle to help determine the effectiveness of the program completed and the structure of the next program to be implemented.

Test Environment

- The tests should be carried out at the same time of day for each testing session.
- Testing should not be conducted in extreme environmental conditions, particularly hot and humid conditions and extreme cold conditions. Although it is sometimes impossible to control the environmental conditions for field testing, it is possible to test at times of the day that are the least extreme.

- The time of day, temperature (°C), and humidity (%) should be recorded.

Coaches are encouraged to supervise, help organize the athletes during the test session, and ensure that athletes perform the tests correctly. Coaches should avoid being involved as a tester or scorer if possible. Supervision of the athletes includes ensuring that the correct warm-up is performed effectively and that correct and effective starting techniques are used for sprint tests.

Recommended Test Order

Because of the number of tests required, and given that some tests induce considerable fatigue (e.g., Yo-Yo intermittent recovery test, multistage shuttle run), it may be necessary to conduct the physical testing protocol over a number of days. It is important that field and strength tests are completed in the same order to control the interference between tests. This order also allows valid comparison of different test occasions. The order is as follows:

Day 1	Tests
Morning	Vertical jump
	40 m sprint
	5-0-5 agility
	Yo-Yo intermittent recovery or multi-stage shuttle run
Afternoon or evening	1RM strength
Day 2	
Morning	Anthropometry
	Repeat sprint

Equipment Checklist

Anthropometry

- [ ] Stadiometer (wall mounted)
- [ ] Balance scales (accurate to ±0.05 kg)
- [ ] Anthropometry box
- [ ] Skinfold calipers (Harpenden skinfold caliper)
- [ ] Marker pen
- [ ] Anthropometric measuring tape
- [ ] Recording sheet (refer to the anthropometry data sheet template)
- [ ] Pen

Vertical Jump Test

- [ ] Yardstick jumping device (e.g., Swift Performance Yardstick)
- [ ] Measuring tape
- [ ] Field marking tape
- [ ] Recording sheet (refer to the vertical jump data sheet template)
- [ ] Pen

40 m Sprint Test

- [ ] Electronic light gate equipment (e.g., dual-beam light gates)
- [ ] Measuring tape
- [ ] Field marking tape
- [ ] Witches hats
- [ ] Recording sheet (refer to the sprint test data sheet template)
- [ ] Pen

5-0-5 Agility Test

- [ ] Electronic light gate equipment (e.g., dual-beam light gates)
- [ ] Measuring tape
- [ ] Field marking tape
- [ ] Witches hats
- [ ] Recording sheet (refer to the agility test data sheet template)
- [ ] Pen

12 × 20 m Repeat Sprint Ability Test

- [ ] Electronic light gate equipment (e.g., dual-beam light gates)
- [ ] Measuring tape
- [ ] Field marking tape
- [ ] Marker cones
- [ ] Recording sheet (refer to the repeat sprint ability test data sheet template)
- [ ] Pen

Yo-Yo Intermittent Recovery Test or Multistage Shuttle Run Test

- [ ] Measuring tape
- [ ] Witches hats
- [ ] Sound box or CD player
- [ ] Yo-Yo intermittent recovery test or multistage shuttle run CD
- [ ] Recording sheet (refer to the Yo-Yo IRT or multistage shuttle run test data sheet template)
- [ ] Pen

1RM Strength Tests

- [ ] Olympic bench press
- [ ] High bench
- [ ] Squat rack with safety bars
- [ ] Olympic bar and plates
- [ ] Recording sheet (refer to the strength and power test data sheet templates)
- [ ] Pen

ANTHROPOMETRY

RATIONALE

Measurement of height, body mass, and sum of seven skinfolds should be carried out prior to field testing protocols.

Height, body mass, and sum of skinfolds provide objective measures of the athlete's body structure. These measurements are important in quantifying differential growth and training influences (Ross and Marfell-Jones 1987). In particular, skinfold measurement can be used to monitor and understand the effect of diet and training intervention on body composition. For example, skinfold measurements may be used to determine whether changes in body mass are due to changes in body fat levels or lean body mass (Norton et al. 2000). If one is recommending a particular skinfold level, the playing position or role of the player and the extent to which excess body fat may affect performance should be considered.

Skinfolds are recorded over seven sites (triceps, biceps, subscapular, supraspinale, abdominal, front thigh, and medial calf). The individual skinfold measures as well as the sum of the seven sites should be reported.

TEST PROCEDURE

Refer to anthropometry protocols outlined in chapter 11, Assessment of Physique, for a detailed description of anthropometric test procedures.

NORMATIVE DATA

Normative data for junior (subelite and elite) and senior (semiprofessional and professional) players are presented in table 24.1.

TABLE 24.1 Anthropometric Data for Junior and Senior Rugby League Players: Mean ± *SD* (Range)

Athlete group	Height, cm	Body mass, kg	Σ7 skinfolds, mm
Junior subelite	176.0 ± 6.0 (162.0-188.5)	74 ± 13 (48-104)	75 ± 32 (40-154)
Junior elite	178.0 ± 5.9 (167.0-191.5)	78 ± 10 (64-104)	67 ± 15 (43-107)
Senior semiprofessional	183.1 ± 6.5 (165.0-195.0)	96 ± 11 (74-119)	72 ± 23 (38-145)
Senior professional	184.1 ± 5.5 (173.0-197.0)	96 ± 9 (82-121)	53 ± 12 (37-101)

Data source: junior (subelite): Gold Coast junior development players (*n* = 36, age 15-17 years); junior (elite): National Rugby League junior development players (*n* = 28, age 15-16 years); senior (semiprofessional): Queensland Cup players (*n* = 77, age 17-30 years); senior (professional): National Rugby League first grade players (*n* = 68, age 18-32 years).

VERTICAL JUMP TEST

RATIONALE

The vertical jump test is used to assess leg muscle power. The test is easily performed in the field, it requires limited equipment, and substantial sport-specific normative data are available.

Vertical jump height can be used to monitor the effects of strength training programs that target the legs and to make comparisons between players.

TEST PROCEDURE

Refer to vertical jump protocol in chapter 14, Field Testing Principles and Protocols, for setup details and test procedures.

NORMATIVE DATA

Normative data for junior (subelite and elite) and senior (semiprofessional and professional) players are presented in table 24.2.

40 M SPRINT TEST

RATIONALE

Rugby league players require the ability to move quickly in order to position themselves in attack and defense (Meir et al. 2001). However, time–motion studies have shown that rugby league players are rarely required to sprint distances greater than 40 m in a single bout of intense activity (Gabbett, 2012).

TEST PROCEDURE

For Rugby League, the sprint test is typically conducted over 40 m, with intermediate distances of 10 m, 20 m, and 30 m. However, there are two variations to this test, chosen at the discretion of the coach and scientist:

- 20 m test, with split times at 5 m and 10 m
- 40 m test, with split times at 10 m, 20 m and 30 m

Athletes start from a stationary, upright position with the front foot on the 0 m point, in line with the start gate. Refer to straight line sprint testing protocols in chapter 14 for a general description of the procedures involved in sprint testing.

NORMATIVE DATA

Normative data for junior (subelite and elite) and senior (semiprofessional and professional) players are presented in table 24.2.

5-0-5 AGILITY TEST

RATIONALE

Rugby league players require the ability to rapidly accelerate, decelerate, and change direction. Indeed, players are required to display agility both in attack and defense (Meir et al. 2001).

TEST PROCEDURE

For rugby league, change of direction speed is assessed using the 5-0-5 test. Athletes start from a stationary, upright position with the front foot on the 0 m point. Refer to the 5-0-5 agility testing protocol in chapter 14 for setup details and test procedure.

NORMATIVE DATA

Normative data for junior (subelite and elite) and senior (semiprofessional and professional) players are presented in table 24.2.

TABLE 24.2 Vertical Jump, Speed, and Change of Direction Data for Junior and Senior Rugby League Players: Mean ± SD (Range)

Athlete group	Vertical jump, cm	10 m sprint, s	40 m sprint, s	5-0-5 test, s
Junior subelite	47 ± 7 (33-58)	1.94 ± 0.11 (1.77-2.23)	5.83 ± 0.35 (5.27-6.62)	2.38 ± 0.16 (2.12-2.70)
Junior elite	52 ± 8 (37-68)	1.81 ± 0.08 (1.63-1.93)	5.56 ± 0.22 (5.23-6.02)	2.30 ± 0.13 (2.13-2.65)
Senior semiprofessional	61 ± 7 (49-75)	1.76 ± 0.08 (1.63-2.00)	5.29 ± 0.22 (4.92-5.77)	2.32 ± 0.17 (2.09-2.54)
Senior professional	63 ± 6 (48-81)	1.72 ± 0.08 (1.59-1.97)	5.23 ± 0.19 (4.82-5.68)	2.22 ± 0.21 (1.86-2.55)

Source: junior (subelite): Gold Coast junior development players ($n = 36$, age 15-17 years); junior (elite): National Rugby League junior development players ($n = 28$, age 15-16 years); senior (semiprofessional): Queensland Cup players ($n = 77$, age 17-30 years); senior (professional): National Rugby League first grade players ($n = 68$, age 18-32 years).

12 × 20 M REPEAT SPRINT ABILITY TEST

RATIONALE

Repeat sprinting and tackling are common requirements of rugby league match play. Time–motion studies have shown that players are frequently required to perform three or more repeated efforts with minimal recovery between efforts (Austin et al. 2010). Furthermore, the majority of these repeated-effort bouts have been shown to occur immediately before a try is scored or conceded (Austin et al. 2010). As a result, the ability or inability to perform repeated-effort bouts may prove critical to the outcome of the game. It has also been shown that the vast majority of sprint efforts in rugby league are of short duration (King et al. 2009; Sirotic et al. 2009), with very few sprints exceeding 30 m (Austin et al. 2010). Despite the importance of this physical quality, very few studies have documented the repeat sprint qualities of rugby league players (Clark 2003; O'Connor, 1996). Moreover, of the studies that have been performed, repeat sprint protocols have typically used sprint distances of 40 m, which may not be representative of the repeat sprint demands of competition. Consequently, we recommend the use of repeat sprinting protocols that replicate the high-intensity repeat sprinting demands (both in terms of duration and recovery) of competition. However, we acknowledge that any protocol that only uses sprinting will likely underestimate the true repeat sprinting and tackling (i.e., repeated effort) demands of competition.

Although investigators acknowledge the importance of repeated-effort ability in rugby league (Gabbett et al. 2008), a game-specific repeated-effort testing protocol currently does not exist. Standardizing the intensity of contact and wrestling in a repeated-effort protocol has commonly been viewed as the major obstacle preventing the development of a repeated-effort test. It has also been suggested that although a repeated-effort test would allow sport scientists to replicate the specific sprinting and tackling demands placed on players during competition, the reliability and validity of such a protocol are questionable.

Despite this perceived limitation, we have recently found (Johnston and Gabbett 2011) that a repeated-effort protocol (using 12 × 20 m sprint efforts, with intermittent tackling and wrestling on a 20 s cycle) offered greater reliability than a repeat sprint protocol (using 12 × 20 m sprint efforts on a 20 s cycle). Consequently, the development of a game-specific repeated-effort testing protocol should be considered a valuable research pursuit for rugby league investigators.

TEST PROCEDURE

The repeat sprint ability of rugby league players is assessed using a repeated 20 m sprint test. Players perform 12 maximal effort sprints over a 20 m distance, with each sprint performed on a 20 s cycle. The players' total sprint time and percentage decrement are calculated and used as the repeat sprint score.

NORMATIVE DATA

Normative data for junior (subelite and elite) and senior (semiprofessional and professional) players are presented in table 24.3.

TABLE 24.3   Repeat Sprint and Aerobic Qualities for Junior and Senior Rugby League Players: Mean ± SD (Range)

Athlete group	Repeat sprint ability, s	$\dot{V}O_2$max, ml · min⁻¹ · kg⁻¹*	Yo-Yo test distance, m†
Junior subelite	n/a	43.3 ± 5.4 (32.9-55.4)	n/a
Junior elite	n/a	48.2 ± 4.6 (39.6-57.9)	1,440 ± 400 (440-2,880)
Senior semiprofessional	39.1 ± 3.3 (33.8-46.7)	53.2 ± 3.9 (43.3-59.3)	n/a
Senior professional	38.4 ± 2.8 (34.8-45.7)	55.7 ± 2.9 (49.3-63.0)	1,789 ± 366 (920-2,520)

n/a = not available. Data source: junior (subelite): Gold Coast junior development players (n = 36, age 15-17 years); junior (elite): National Rugby League junior development players (n = 28, age 15-16 years); senior (semiprofessional): Queensland Cup players (n = 77, age 17-30 years); senior (professional): National Rugby League first grade players (n = 68, age 18-32 years).

*Estimated from multistage shuttle run. †Yo-Yo intermittent recovery test level 1; junior (elite), n = 759, age 15-17 years; senior (professional), n = 29, age 17-33 year.

YO-YO INTERMITTENT RECOVERY TEST OR MULTISTAGE SHUTTLE RUN

RATIONALE

Given the importance of aerobic qualities to rugby league playing performance, it is recommended that either the Yo-Yo intermittent recovery test (IRT) or multistage shuttle run test (MSRT) be included in any physical testing battery for this sport. Many coaches have extensively used the MSRT. However, the Yo-Yo IRT offers a reliable and valid assessment of aerobic fitness and has been suggested to more closely replicate the intermittent demands of high-intensity, intermittent team sport competition. Therefore, the decision to include the Yo-Yo IRT or MSRT will depend largely on the philosophies of the sport scientist responsible for the physical assessment of players.

TEST PROCEDURE

Refer to the Yo-Yo intermittent recovery test or multistage shuttle run protocols in chapter 14 for detailed description of this test. It is recommended that this test be completed at the conclusion of the testing battery, because fatigue associated with this test may affect performance in the speed and power type tests.

NORMATIVE DATA

Normative data for junior (subelite and elite) and senior (semiprofessional and professional) players are presented in table 24.3.

STRENGTH TESTING

All of the strength tests must be supervised by an accredited strength and conditioning coach or scientist. All exercises are to be performed in a controlled manner; any noted technical violations will result in the trial being invalid, and a second attempt at the same weight will be performed.

The tests used in Rugby League include the bench press, bench pull, and back squat. The equipment and protocols required for these strength tests are described in chapter 13, Strength and Power Assessment Protocols.

The following general guidelines must be adhered to for all tests:

- Strength testing should be performed on a separate day from the field tests. Strength and field tests should be separated by 48 h.

- The athlete must perform an appropriate warm-up. As a minimum, all athletes are required to perform a trial at approximately 90% of the specified repetition maximum for each test. If this is the first test, the athlete should perform an initial trial at approximately 90% of the weight lifted in training.

- Lowering and lifting actions must be performed in a continuous manner. A single rest of no more than 2 s is allowed between repetitions.

- A maximum of 5 min recovery between trials is allowed.

- Minimum weight increments of 2.5 kg should be used between trials. However, increments should be guided by ease of each trial.

- Ideally, the specified repetition maximum (RM) test should be completed within 4 trials (not including the warm-up).

- If the athlete is unable to complete tests as per protocol, this should be noted on testing results information, and values should not include any mathematical calculations (e.g., average, typical error of measurement).

- A spotter, other than the supervising coach, should be used where necessary.

DATA ANALYSIS

All data collected from the testing protocols are interpreted via comparison with previous test results as well as with group averages. Sport scientists should develop their own population-specific test–retest reliability data, with test results interpreted using the typical error of measurement of each test.

NORMATIVE DATA

Normative data for senior (professional) players are presented in table 24.4.

TABLE 24.4 Strength Test Data for Senior Rugby League Players: Mean ± SD (Range)

Group	Bench press, kg	Bench pull, kg	Back squat, kg
Senior professional	130 ± 17 (105-170)	107 ± 18 (90-115)	133 ± 18 (100-180)

Data source: senior (professional): National Rugby League first grade players (n = 32, age 17-32 years).

References

Austin, D., Gabbett, T., and Jenkins, D. 2011. Repeated high-intensity exercise in a professional rugby league. *Journal of Strength and Conditioning Research* 25:1898-1904.

Brewer, J., Davis, J., and Kear, J. 1994. A comparison of the physiological characteristics of rugby league forwards and backs. *Journal of Sports Sciences* 12:158.

Clark, L. 2003. A comparison of the speed characteristics of elite rugby league players by grade and position. *Strength and Conditioning Coach* 10:2-12.

Coutts, A., Reaburn, P., and Abt, G. 2003. Heart rate, blood lactate concentration and estimated energy expenditure in a semi-professional rugby league team during a match: a case study. *Journal of Sports Sciences* 21:97-103.

Estell, J., Lord, P., Barnsley, L., Shenstone, P., and Kannongara, S. 1996. Physiological demands of rugby league. In proceedings of the Australian Conference of Science and Medicine in Sport. Sports Medicine Australia, Canberra, ACT, 338-339.

Gabbett, T.J. 2000. Physiological and anthropometric characteristics of amateur rugby league players. *British Journal of Sports Medicine* 34:303-307.

Gabbett, T.J. 2002a. Influence of physiological characteristics on selection in a semi-professional rugby league team: a case study. *Journal of Sports Sciences* 20:399-405.

Gabbett, T.J. 2002b. Physiological characteristics of junior and senior rugby league players. *British Journal of Sports Medicine* 36:334-339.

Gabbett, T. 2003. Do skill-based conditioning games simulate the physiological demands of competition? *Rugby League Coaching Manuals* 32:27-31.

Gabbett, T.J. 2004a. Changes in physiological and anthropometric characteristics of rugby league players during a competitive season. *Journal of Strength and Conditioning Research* 19:400-408.

Gabbett, T.J. 2004b. Science of rugby league football: A review. *Journal of Sports Sciences* 23:961-976.

Gabbett, T.J. 2011. Activity cycles of National Rugby League and National Youth Competition matches. *Journal of Strength and Conditioning Research.* In press.

Gabbett, T.J. 2012. Sprinting patterns of National Rugby League competition. *Journal of Strength and Conditioning Research* 26:121-130.

Gabbett, T.J., and Herzig, P.J. 2004. Physiological characteristics of junior elite and sub-elite rugby league players. *Strength and Conditioning Coach* 12:19-24.

Gabbett, T.J., Jenkins, D.G., and Abernethy, B. 2012. Physical demands of professional rugby league training and competition using microtechnology. *Journal of Science and Medicine in Sport* 15:80-86.

Gabbett, T.J., Jenkins, D.G., and Abernethy, B. 2011a. Relationships between physiological, anthropometric, and skill qualities and playing performance in professional rugby league players. *Journal of Sports Sciences* 29:1655-1664.

Gabbett, T.J., Jenkins, D.G., and Abernethy, B. 2011b. Relative importance of physiological, anthropometric, and skill qualities to team selection in professional rugby league. *Journal of Sports Sciences,* 29:1453-1461.

Gabbett, T., Kelly, J., Ralph, S., and Driscoll, D. 2009. Physiological and anthropometric characteristics of junior elite and sub-elite rugby league players, with special reference to starters and non-starters. *Journal of Science and Medicine in Sport,* 12:215-222.

Gabbett, T., King, T., and Jenkins, D. 2008. Applied physiology of rugby league. *Sports Medicine* 38:119-138.

Johnston, R., and Gabbett, T. 2011. Repeated-sprint and effort ability in rugby league players. *Journal of Strength and Conditioning Research* 25:2789-2795.

King, T., Jenkins, D., and Gabbett, T. 2009. A time-motion analysis of professional rugby league match-play. *Journal of Sports Science* 27:213-219.

Larder, P. 1992. *The Rugby League Coaching Manual.* 2nd ed. London: Kingswood Press.

Meir, R. 1993a. Evaluating players fitness in professional rugby league: Reducing subjectivity. *Strength and Conditioning Coach* 1:11-17.

Meir, R. 1993b. Seasonal changes in estimates of body composition in professional rugby league players. *Sport Health* 11:27-31.

Meir, R., Arthur, D., and Forrest, M. 1993. Time motion analysis of professional rugby league: A case study. *Strength and Conditioning Coach* 1:24-29.

Meir, R., Newton, E., Curtis, M., Fardell, M., and Butler, B. 2001. Physical fitness qualities of professional rugby league football players: determination of positional differences. *Journal of Strength and Conditioning Research* 15:450-458.

Norton, K., Marfell-Jones, M., Whittingham, N., Kerr, D., Carter, L., Saddington, K., and Gore, C. 2000. Anthropometric assessment protocols. In: *Physiological tests for elite athletes.* C.J. Gore, ed. Champaign, Illinois: Human Kinetics, pp. 66-85.

O'Connor, D. 1995. Fitness profile of professional rugby league players (Abstract). *Journal of Sports Sciences* 13:505.

O'Connor, D. 1996. Physiological characteristics of professional rugby league players. *Strength and Conditioning Coach* 4:21-26.

Ross, W.D. and Marfell-Jones, M.J. 1987. Kinanthropometry. In: *Physiological testing of the elite athlete.* MacDougall, J.D., Wenger, H.A., and Green, H.J. (eds). Champaign, Illinois: Human Kinetics, pp. 223-308.

Sirotic, A., Coutts, A., Knowles, H., and Catterick, C. 2009. A comparison of match demands between elite and semi-elite rugby league competition. *Journal of Sports Science* 27:203-211.

Sykes, D., Twist, C., Hall, S., Nicholas, C., and Lamb, K. 2009. Semi-automated time-motion analysis of senior elite rugby league. *International Journal of Performance Analysis of Sport* 9:47-59.

Waldron, M., Twist, C., Highton, J., Worsfold, P., and Daniels, M. 2011. Movement and physiological match demands of elite rugby league using portable global positioning systems. *Journal of Sports Sciences,* 29:1223-1230.

Rugby Union Players

Dean G. Higham, David B. Pyne, and John A. Mitchell

Rugby union is a full-contact, field-based team sport played by men and women in many countries worldwide. Matches are played over two 40 min halves with a short break (not more than 15 min) between halves. The physiological demands of rugby union are broadly characterized by the high frequency of physical contacts and repeated bouts of short-duration, high-intensity efforts (Duthie et al. 2005). Although there is individual variation in the competition demands relative to each playing position, all rugby union players require a high degree of strength, power, speed, and anaerobic and aerobic endurance (Duthie et al. 2003).

Testing in rugby union is valuable for a variety of reasons, including monitoring long- and short-term training adaptations, prescribing individualized programs, selecting players, and predicting performance outcomes. The testing protocols outlined in this chapter are approved by the Australian Rugby Union and used at junior and senior levels of the game. The field-based testing protocols can largely be administered in basic indoor testing venues and generally do not require specialist sport science laboratory facilities.

Athlete Preparation

Standardized pretest preparation is essential to enable reliable and valid physiological data to be obtained. The following areas require particular attention when testing rugby union players. These pretest conditions should be observed in addition to those outlined in chapter 2, Pretest Environment and Athlete Preparation. Many of the considerations relevant to laboratory testing are equally important for field testing and data analysis and interpretation when athletes are tested in several different locations.

- *Diet.* Athletes should present for testing in a hydrated state. Hydration status may be assessed by checking the urine specific gravity of the midstream, first bladder void upon waking.

- *Training.* To maximize the validity of test results in representing the current physiological status of athletes, testing should be scheduled a minimum of 1 day after intense activity (training or match). This practice will allow players sufficient recovery to ensure test results are not confounded by residual fatigue or soreness.

- *Testing.* To help prevent injury and give athletes the opportunity to produce a maximal effort during tests, ensure athletes complete a thorough and standardized warm-up consisting of running and stretching before all tests (excluding anthropometry). Gym-based assessments of strength and power require an additional movement-specific warm-up. Tests should be scheduled at a similar time of day on each testing occasion or across testing locations to maintain consistency in athletes' physiological state. Diurnal variations in an athlete's circadian rhythm can influence test performance. The order of tests should be consistent between testing sessions (see Recommended Test Order). Athletes should be allowed sufficient recovery time between all tests and between individual trials within each test.

Test Environment

Whenever possible, athletes should be tested under similar environmental conditions (temperature and humidity). Consistent environmental conditions are important when comparing results between testing occasions: testing can be conducted in a wide variety of conditions from midwinter to midsummer.

Running surface and athletes' footwear can have a significant effect on test performance and should be consistent between testing sessions. Testing should preferably be conducted indoors on a polished wooden floor in a sports hall or basketball stadium rather than outdoors on a grass surface with variable surface conditions, winds, and temperatures. Environmental conditions and running surface should be recorded for each testing session.

Recommended Test Order

It is important that tests are completed in the same order to control the interference between tests. This order also allows valid comparison of testing results from different testing sessions. The recommended order is as follows:

Day	Tests
1	Anthropometry
	Vertical jump
	40 m sprint
	Yo-Yo intermittent recovery (level 1)
2	Physical competency screening
	1RM strength
3	6 × 30 m repeat sprint ability

Equipment Checklist

Anthropometry

- [ ] Stadiometer (wall mounted)
- [ ] Balance scales (accurate to ±0.05 kg)
- [ ] Anthropometry box
- [ ] Skinfold calipers (Harpenden skinfold caliper)
- [ ] Marker pen
- [ ] Anthropometric measuring tape
- [ ] Recording sheet (refer to the anthropometry data sheet template)
- [ ] Pen

Vertical Jump Test

- [ ] Jump measuring device (e.g., Swift Performance Equipment Yardstick)
- [ ] Tape
- [ ] Recording sheet (refer to the vertical jump data sheet template)
- [ ] Pen

40 m Sprint Test

- [ ] Electronic timing gate equipment (e.g., dual-beam light gates)
- [ ] Measuring tape
- [ ] Field marking tape
- [ ] Marker cones
- [ ] Hand-held environmental monitor (if testing outdoors)
- [ ] Recording sheet (refer to the sprint test data sheet template)
- [ ] Pen

Yo-Yo Intermittent Recovery Test

- [ ] Measuring tape
- [ ] Marker cones
- [ ] Sound box or CD/MP3 player
- [ ] Yo-Yo intermittent recovery level 1 test CD/MP3
- [ ] Recording sheet (refer to the Yo-Yo IRT data sheet template)
- [ ] Pen

6 × 30 m Repeat Sprint Ability Test

- [ ] Electronic timing gate equipment (e.g., dual-beam light gates)
- [ ] Measuring tape
- [ ] Field marking tape
- [ ] Marker cones
- [ ] Sound box or CD/MP3 player
- [ ] 6 × 30 m repeated-sprint ability test CD/MP3
- [ ] Stopwatch
- [ ] Recording sheet (refer to the repeat sprint ability test data sheet template)
- [ ] Pen

Physical Competency Screening

- [ ] Bench press bench
- [ ] Squat rack or power cage
- [ ] Box or step (15-50 cm in height)
- [ ] Barbell (Olympic 20 kg)
- [ ] Broomstick or lightweight barbell
- [ ] Recording sheet (refer to the rugby union physical competency screening data sheet template)
- [ ] Pen

Strength Tests

- [ ] Bench press bench
- [ ] Squat rack or power cage
- [ ] Chin-up bar
- [ ] Barbell (Olympic 20 kg)
- [ ] Weight plates (2.5-25 kg increments)
- [ ] Recording sheet (refer to the strength and power test data sheet templates)
- [ ] Pen

ANTHROPOMETRY

RATIONALE

Rugby union players are often characterized by the heterogeneity of their physical attributes. Unlike athletes in many other team sports, rugby players in different positions frequently have distinct physiques. For example, a player's height is an important characteristic for several positions in which a greater height may offer a competitive advantage (Duthie et al. 2003). However, the development of lean mass is desirable for rugby union players of all positions to enhance speed, strength, and power (Duthie et al. 2006a). Similarly, high levels of body fat are detrimental to performance by increasing energy expenditure and decreasing a player's acceleration and power to weight ratio. The lean mass index (LMI) was developed as a practical method to track proportional body mass changes adjusted for skinfold thickness (Slater et al. 2006).

TEST PROCEDURE

Skinfolds are recorded over seven sites (triceps, biceps, subscapular, supraspinale, abdominal, front thigh, and medial calf). The individual skinfold measures as well as the sum of the seven sites should be reported. Refer to anthropometry protocols outlined in chapter 11, Assessment of Physique, for a detailed description of all anthropometric test procedures. More advanced anthropometric assessment including girths, bone breadths, and limb lengths can be conducted if required.

Although the description of skinfold measurement procedures appears simple, a high degree of technical skill is essential for consistent results. The measurements should be taken by an experienced tester who has been trained in these techniques. It is also important, where possible, that the same tester conduct each retest to ensure reliability.

DATA ANALYSIS

Based on body mass and skinfolds, a different LMI coefficient is indicated for forwards and backs. The LMI is calculated as M/S^x, where M is body mass (kg), S is sum of seven skinfolds (mm), and x is an exponent (0.132 for forwards and 0.139 for backs).

Anthropometric results should be interpreted primarily on an individual player basis using change scores and an athlete's previous test results. For junior athletes, changes in height, body mass, body fat, and lean mass typically reflect the underlying growth and maturation in the adolescent years and the initial impact of dietary and strength and conditioning programs. There might be some interest in a between-player comparison of results, but this exercise needs to be position-specific (see tables 25.1, 25.2, and 25.3) and account for differences in rates of physical maturation. For older, more physically mature senior athletes, changes in anthropometric measures typically reflect the balance of diet, training demands, and strength and conditioning programs.

NORMATIVE DATA

Table 25.1, 25.2, and 25.3 present normative anthropometry data for male Australian junior talent squad players aged 14 to 18 years, provincial academy players aged 15 to 26 years, and Super Rugby players aged 17 to 35 years.

TABLE 25.1 Anthropometric Data for Male Australian Junior Talent Squad Players: Mean ± *SD* (Range)

Positional group	Height, cm	Body mass, kg	Σ7 skinfolds, mm	Lean mass index, mm · kg$^{-0.14}$
Prop (*n* = 45)	181.3 ± 5.1 (167.7-194.1)	102.0 ± 13.2 (60.2-142.2)	118.6 ± 28.8 (57.3-193.1)	55.3 ± 5.0 (49.0-71.0)
Hooker (*n* = 16)	176.3 ± 3.4 (169.0-182.0)	91.1 ± 9.9 (76.8-111.3)	108.2 ± 16.5 (78.1-130.6)	51.2 ± 4.2 (45.8-57.3)
Second row (*n* = 37)	188.2 ± 6.0 (170.2-199.5)	91.0 ± 9.6 (71.0-111.5)	82.7 ± 25.1 (44.5-136.2)	53.1 ± 3.5 (47.9-59.2)
Back row (*n* = 92)	182.2 ± 5.0 (169.7-194.1)	87.8 ± 10.8 (66.0-117.8)	78.0 ± 22.7 (43.8-150.2)	50.3 ± 4.9 (39.1-61.1)
Scrum half (*n* = 29)	173.8 ± 6.6 (159.7-186.2)	73.5 ± 9.3 (55.0-92.3)	62.4 ± 18.0 (40.7-111.3)	42.6 ± 4.2 (32.9-50.7)
Fly half (*n* = 29)	176.1 ± 5.5 (164.6-185.4)	76.7 ± 6.8 (62.2-93.9)	59.2 ± 11.8 (42.6-92.3)	44.0 ± 2.6 (40.2-48.4)
Center (*n* = 56)	180.0 ± 6.0 (162.3-193.5)	82.5 ± 7.8 (67.9-100.3)	64.8 ± 15.3 (38.8-108.6)	47.0 ± 3.3 (40.7-52.3)
Wing/fullback (*n* = 52)	179.0 ± 5.1 (165.8-187.5)	80.3 ± 8.8 (63.7-102.1)	59.6 ± 15.0 (34.0-92.6)	46.7 ± 4.4 (36.5-59.1)

Typical error: height = 1 cm; body mass = 1 kg; Σ7 skinfolds = ~1.5 mm; lean mass index = 0.3 mm · kg$^{-0.14}$. Data source: Australian Institute of Sport rugby union testing database 2001-2010; age 14-18 years.

TABLE 25.2 Anthropometric Data for Male Australian Provincial Academy Players: Mean ± *SD* (Range)

Positional group	Height, cm	Body mass, kg	Σ7 skinfolds, mm	Lean mass index, mm · kg$^{-0.14}$
Prop (*n* = 50)	182.3 ± 5.1 (173.6-193.1)	112.8 ± 11.0 (90.7-139.3)	116.7 ± 26.9 (55.1-179.1)	59.9 ± 4.3 (50.3-70.6)
Hooker (*n* = 24)	177.6 ± 3.5 (171.2-183.5)	100.4 ± 8.4 (79.3-119.2)	99.3 ± 27.1 (58.7-163.0)	54.7 ± 4.2 (43.7-61.7)
Second row (*n* = 33)	193.6 ± 5.7 (176.0-204.4)	104.7 ± 7.5 (85.6-118.2)	85.9 ± 21.7 (46.9-129.4)	58.4 ± 3.5 (49.4-63.8)
Back row (*n* = 66)	185.9 ± 4.9 (173.0-196.0)	96.2 ± 9.2 (75.7-122.6)	79.9 ± 26.2 (42.3-179.6)	54.6 ± 4.4 (43.0-64.8)
Scrum half (*n* = 22)	174.5 ± 6.2 (161.9-187.4)	77.8 ± 4.7 (68.8-87.1)	54.3 ± 15.1 (33.8-93.6)	44.7 ± 2.5 (38.5-50.6)
Fly half (*n* = 22)	179.1 ± 5.5 (171.2-188.5)	85.3 ± 7.3 (69.7-104.6)	62.4 ± 13.6 (41.2-88.0)	47.7 ± 3.8 (39.8-52.9)
Center (*n* = 50)	180.2 ± 5.2 (170.0-194.6)	87.9 ± 8.5 (71.2-108.1)	72.6 ± 18.5 (46.5-116.0)	48.9 ± 4.2 (41.2-59.1)
Wing/fullback (*n* = 52)	181.8 ± 5.5 (172.4-194.6)	86.6 ± 7.0 (75.0-104.4)	63.3 ± 15.6 (38.3-116.7)	48.9 ± 3.4 (42.3-57.0)

Typical error: height = 1 cm; body mass = 1 kg; Σ7 skinfolds = ~1.5 mm; lean mass index = 0.3 mm · kg$^{-0.14}$. Data source: Australian Institute of Sport rugby union testing database 2001-2010; age 15-26 years.

TABLE 25.3   Anthropometric Data for Male Australian Super Rugby Players: Mean ± *SD* (Range)

Positional group	Height, cm	Body mass, kg	Σ7 skinfolds, mm	Lean mass index, mm · kg^{-0.14}
Prop (*n* = 21)	184.7 ± 2.8 (182.0-188.9)	112.5 ± 9.1 (106.4-128.0)	100.1 ± 22.0 (79.4-148.8)	61.4 ± 4.7 (55.0-69.9)
Hooker (*n* = 8)	181.5 ± 3.6 (179.2-186.2)	103.8 ± 4.2 (98.0-115.7)	83.7 ± 18.4 (66.9-97.2)	58.0 ± 2.4 (53.8-63.2)
Second row (*n* = 14)	194.2 ± 10.5 (193.3-202.0)	111.1 ± 3.7 (104.2-116.4)	78.9 ± 20.6 (52.7-116.1)	62.7 ± 2.3 (57.9-65.8)
Back row (*n* = 25)	187.9 ± 3.8 (175.5-196.0)	105.7 ± 7.3 (96.4-117.9)	75.2 ± 21.9 (52.6-104.0)	60.0 ± 2.7 (54.4-66.4)
Scrum half (*n* = 12)	180.0 ± 5.0 (174.6-183.7)	85.4 ± 9.5 (76.2-89.1)	60.5 ± 11.2 (37.0-64.8)	48.3 ± 4.6 (46.1-51.1)
Fly half (*n* = 10)	181.5 ± 5.1 (174.0-188.7)	90.0 ± 5.8 (82.5-103.0)	57.1 ± 9.0 (42.7-99.7)	51.4 ± 3.8 (47.7-58.6)
Center (*n* = 17)	182.2 ± 5.1 (176.6-188.9)	94.4 ± 5.5 (83.2-107.8)	62.5 ± 13.4 (47.6-99.5)	53.7 ± 4.7 (47.3-61.1)
Wing/fullback (*n* = 29)	183.7 ± 5.5 (175.6-191.0)	93.6 ± 7.5 (77.1-105.9)	65.4 ± 15.0 (27.6-89.7)	52.5 ± 3.9 (43.1-60.4)

Typical error: height = 1 cm; body mass = 1 kg; Σ7 skinfolds = ~1.5 mm; lean mass index = 0.3 mm · kg^{-0.14}. Data source: Australian Institute of Sport rugby union testing database 2001-2010; age 17-35 years.

VERTICAL JUMP TEST

RATIONALE

Muscular power is fundamental for success in rugby union, particularly during tackles, scrums, line-outs, rucks, and mauls. The vertical jump test provides a useful field-based measure of leg power. Force produced during the maximal vertical jump has been shown to be related to scrummaging force (Robinson and Mills 2000).

TEST PROCEDURE

Refer to vertical jump protocol in chapter 14, Field Testing Principles and Protocols, for setup details and test procedures.

DATA ANALYSIS

Vertical jump results can be interpreted against position standards (see table 25.4) or against a player's previous test results. Improvements in jump height should reflect increases in lower body power but also in jumping technique, timing, and coordination. Familiarization and prior practice can influence test results, particularly for junior athletes.

NORMATIVE DATA

Table 25.4 presents normative vertical jump data for male Australian junior talent squad players aged 14 to 18 years, provincial academy players aged 15 to 26 years, and Super Rugby players aged 18 to 35 years.

TABLE 25.4 Vertical Jump Data for Male Australian Rugby Players: Mean ± SD (Range)

Positional group	Junior talent squad players	Provincial academy players	Super Rugby players
Prop (n = 46; 45; 21)	49 ± 7 (32-66)	53 ± 9 (36-75)	54 ± 6 (38-64)
Hooker (n = 15; 22; 11)	49 ± 8 (35-64)	56 ± 6 (47-68)	61 ± 6 (51-71)
Second row (n = 37; 33; 20)	52 ± 7 (34-69)	56 ± 8 (39-70)	60 ± 7 (46-75)
Back row (n = 94; 60; 25)	55 ± 7 (36-73)	58 ± 6 (40-72)	62 ± 6 (53-77)
Scrum half (n = 29; 19; 15)	56 ± 5 (47-68)	61 ± 5 (49-70)	65 ± 6 (57-74)
Fly half (n = 29; 20; 11)	56 ± 6 (41-70)	60 ± 6 (51-72)	66 ± 6 (57-78)
Center (n = 61; 44; 20)	57 ± 7 (45-74)	58 ± 7 (41-71)	64 ± 6 (54-76)
Wing/fullback (n = 55; 46; 40)	59 ± 6 (46-74)	61 ± 7 (46-78)	64 ± 5 (51-73)

Jump height given in centimeters. Typical error: jump height = 1 cm. Data source: Australian Institute of Sport rugby union testing database 2001-2010; junior talent squad players, age 14-18 years; provincial academy players, age 15-26 years; Super Rugby players, age 18-35 years.

40 M SPRINT TEST

RATIONALE

Speed is an essential quality for high-level rugby union players. Speed during sprinting is often classified into two related phases: initial acceleration and maximal velocity. Although the mean duration of sprints observed during competition is approximately 3 s (Deutsch et al. 2007), players also regularly achieve speeds in excess of 90% maximal velocity (Duthie et al. 2006b). These data suggest that both characteristics of speed are important in rugby union. The 40 m sprint test with intermediate split distances allows the assessment of both acceleration and maximal velocity in the same test.

TEST PROCEDURE

For rugby union, the sprint test is conducted over 40 m, with intermediate split distances of 10, 20, and 30 m. The athlete starts from a stationary standing position with the front foot at the horizontal start line marked 30 cm behind the start (0 m) timing gate. Typically two or three 40 m trials are undertaken at the discretion of the coach or athlete.

If testing is conducted outdoors on a synthetic running track or a grass surface, wind velocity and direction must be recorded and the times adjusted accordingly. Wind velocity is collected in the direction of the sprint at the 20 m mark using a handheld environmental monitor. Wind velocity data collection commences when the athlete leaves the start and finishes when the athlete passes the 40 m gate. The mean wind velocity over this time is recorded.

Split times for the sprints are adjusted using the Linthorne (1994) formula:

$$t_{ad} = t + [a \, (w - w^2 \{1/[2(d/t)]\})]$$

where t_{ad} is the adjusted split time (s), t is the original split time (s), $a = 0.056$ for males and 0.067 for females, w is the wind velocity (m · s^{-1}), and d is the split distance (m).

An estimation of maximal velocity (m · s^{-1}) is obtained by dividing 10 by the value obtained by subtracting the 30 m split time from the 40 m split time. The momentum of the athlete (kg · m · s^{-1}) over the last 10 m of the 40 m sprint is then calculated as the maximal velocity (m · s^{-1}) multiplied by the mass of the player (kg).

DATA ANALYSIS

In rugby union, the fastest 40 m time is typically reported as the test score, although some coaches might prefer to use the mean 40 m time over the two or three trials. Improvements in 10 m sprint time will come from enhancements in lower-body power and the technique of sprint starts from a stationary position. Advice from a specialist sprint coach is useful in this process. Improvements in total 40 m time and maximal velocity should come from speed drills focusing on acceleration and maximal veloc-ity. Speed-resisted and speed-assisted drills may also be useful. Momentum is often associated with the ability to break tackles and get over the advantage line. Improvements in momentum will arise from increases in both body mass and running speed.

NORMATIVE DATA

Tables 25.5, 25.6, and 25.7 present normative 40 m sprint data for male Australian junior talent squad players aged 14 to 18 years, provincial academy players aged 15 to 26 years, and Super Rugby players aged 18 to 35 years.

TABLE 25.5   Sprint Data for Male Australian Junior Talent Squad Players: Mean ± *SD* (Range)

Positional group	10 m time, s	40 m time, s	Maximal velocity, $m \cdot s^{-1}$	Momentum, $kg \cdot m \cdot s^{-1}$
Prop (*n* = 44)	1.98 ± 0.14 (1.71-2.27)	5.82 ± 0.28 (5.20-6.41)	8.0 ± 0.4 (7.2-9.1)	808 ± 100 (533-1,085)
Hooker (*n* = 15)	1.99 ± 0.13 (1.80-2.20)	5.87 ± 0.27 (5.38-6.42)	7.9 ± 0.4 (6.9-8.7)	727 ± 85 (634-896)
Second row (*n* = 38)	1.94 ± 0.15 (1.71-2.24)	5.70 ± 0.28 (5.25-6.28)	8.2 ± 0.5 (7.4-9.3)	753 ± 86 (559-947)
Back row (*n* = 95)	1.92 ± 0.14 (1.69-2.29)	5.58 ± 0.24 (5.10-6.16)	8.5 ± 0.4 (7.5-9.3)	742 ± 104 (502-990)
Scrum half (*n* = 29)	1.89 ± 0.14 (1.65-2.26)	5.49 ± 0.26 (5.02-6.21)	8.6 ± 0.4 (7.6-9.3)	635 ± 92 (448-781)
Fly half (*n* = 28)	1.88 ± 0.12 (1.70-2.10)	5.49 ± 0.22 (5.15-6.09)	8.6 ± 0.4 (7.1-9.1)	653 ± 69 (530-809)
Center (*n* = 61)	1.88 ± 0.14 (1.65-2.22)	5.40 ± 0.21 (5.03-5.79)	8.8 ± 0.3 (8.1-9.5)	731 ± 72 (579-851)
Wing/fullback (*n* = 58)	1.89 ± 0.14 (1.63-2.14)	5.38 ± 0.22 (4.89-5.91)	8.9 ± 0.4 (7.8-9.7)	718 ± 81 (541-879)

Typical error: 10 m time = 0.4 s; 40 m time = 0.4 s; maximal velocity = 0.1 $m \cdot s^{-1}$, momentum = 10 $kg \cdot m \cdot s^{-1}$. Data source: Australian Institute of Sport rugby union testing database 2001-2010; age 14-18 years.

TABLE 25.6   Sprint Data for Male Australian Provincial Academy Players: Mean ± *SD* (Range)

Positional group	10 m time, s	40 m time, s	Maximal velocity, m · s⁻¹	Momentum, kg· m · s⁻¹
Prop (n = 45)	2.02 ± 0.13 (1.80-2.27)	5.83 ± 0.28 (5.37-6.47)	8.1 ± 0.5 (7.0-8.8)	940 ± 88 (755-1,123)
Hooker (n = 24)	1.96 ± 0.14 (1.73-2.19)	5.64 ± 0.27 (5.26-6.22)	8.4 ± 0.4 (7.4-8.9)	853 ± 71 (695-985)
Second row (n = 31)	1.99 ± 0.13 (1.74-2.21)	5.74 ± 0.26 (5.26-6.32)	8.3 ± 0.4 (7.3-8.9)	867 ± 64 (648-975)
Back row (n = 60)	1.92 ± 0.12 (1.68-2.23)	5.47 ± 0.20 (5.10-5.89)	8.7 ± 0.3 (8.0-9.4)	846 ± 78 (644-1,022)
Scrum half (n = 21)	1.91 ± 0.15 (1.64-2.08)	5.39 ± 0.28 (4.93-5.86)	8.9 ± 0.4 (8.1-9.4)	694 ± 52 (573-770)
Fly half (n = 21)	1.85 ± 0.12 (1.64-2.06)	5.33 ± 0.21 (4.91-5.89)	9.0 ± 0.5 (8.1-10.3)	771 ± 63 (676-931)
Center (n = 48)	1.93 ± 0.14 (1.67-2.16)	5.42 ± 0.25 (4.97-6.00)	8.9 ± 0.4 (8.0-9.6)	786 ± 76 (647-918)
Wing/fullback (n = 42)	1.86 ± 0.13 (1.63-2.06)	5.28 ± 0.22 (4.81-5.71)	9.2 ± 0.4 (8.4-10.1)	784 ± 72 (644-932)

Typical error: 10 m time = 0.4 s; 40 m time = 0.4 s; maximal velocity = 0.1 m · s⁻¹, momentum = 10 kg · m · s⁻¹. Data source: Australian Institute of Sport rugby union testing database 2001-2010; age 15-26 years.

TABLE 25.7   Sprint Data for Male Australian Super Rugby Players: Mean ± *SD* (Range)

Positional group	10 m time, s	40 m time, s	Maximal velocity, m · s⁻¹	Momentum, kg · m · s⁻¹
Prop (n = 24)	2.00 ± 0.13 (1.80-2.26)	5.86 ± 0.23 (5.53-6.39)	8.1 ± 0.4 (7.4-8.8)	925 ± 57 (823-1,032)
Hooker (n = 12)	1.92 ± 0.12 (1.77-2.08)	5.60 ± 0.18 (5.32-5.81)	8.5 ± 0.2 (8.2-8.8)	870 ± 59 (823-956)
Second row (n = 22)	1.90 ± 0.10 (1.75-2.10)	5.57 ± 0.21 (5.16-5.93)	8.5 ± 0.4 (7.8-9.3)	939 ± 61 (874-1,029)
Back row (n = 29)	1.90 ± 0.14 (1.73-2.28)	5.46 ± 0.23 (5.09-6.01)	8.7 ± 0.4 (7.8-9.6)	923 ± 53 (848-1,057)
Scrum half (n = 18)	1.86 ± 0.14 (1.66-2.13)	5.32 ± 0.21 (4.96-5.68)	9.1 ± 0.4 (8.5-9.6)	751 ± 45 (676-818)
Fly half (n = 11)	1.84 ± 0.10 (1.68-1.98)	5.33 ± 0.15 (5.03-5.55)	9.0 ± 0.2 (8.7-9.4)	779 ± 43 (743-851)
Center (n = 18)	1.88 ± 0.12 (1.70-2.05)	5.35 ± 0.19 (4.96-5.72)	9.0 ± 0.4 (8.0-9.7)	863 ± 59 (768-937)
Wing/fullback (n = 41)	1.85 ± 0.11 (1.63-2.06)	5.22 ± 0.15 (4.84-5.54)	9.3 ± 0.3 (8.5-10.0)	848 ± 88 (659-1,028)

Typical error: 10 m time = 0.4 s; 40 m time = 0.4 s; maximal velocity = 0.1 m · s⁻¹, momentum = 10 kg · m · s⁻¹. Data source: Australian Institute of Sport rugby union testing database 2001-2010; age 18-35 years.

YO-YO INTERMITTENT RECOVERY TEST

RATIONALE

Players may cover in excess of 7 km in an 80 min match of rugby union (Cunniffe et al. 2009). Aerobic endurance is related to fatigue resistance and promotes recovery from high-intensity activity, including sprints, tackles, rucks, and mauls (Tomlin and Wenger 2002). The Yo-Yo intermittent recovery test (IRT) level 1 involves acceleration, deceleration, and change of direction, making it more specific to the work demands of rugby union than continuous running endurance tests.

TEST PROCEDURE

Refer to the Yo-Yo IRT protocol in chapter 14, Field Testing Principles and Protocols, for a detailed description of this test. It is recommended that the Yo-Yo IRT level 1 test is completed at the conclusion of the rugby union testing battery, as fatigue associated with this test may impact on performance in the speed and power-based tests.

DATA ANALYSIS

Test results presumably reflect the endurance fitness of players. Players seeking improved aerobic fitness should undertake additional endurance-type training including lower-intensity interval running, cross-training, and circuit training. Skill-based conditioning games have also been shown to improve markers of endurance (Gamble 2004). Strength and conditioning coaches need to be mindful of not compromising strength, speed, and power attributes in the quest for improved aerobic fitness. Endurance will be improved by a reduction in fat mass in certain players exhibiting elevated sum of skinfolds. In some players, Yo-Yo IRT results can be improved by attention to the efficiency of starting and turning technique.

NORMATIVE DATA

Table 25.8 presents normative Yo-Yo IRT level 1 data for male Australian junior talent squad players aged 14 to 18 years and provincial academy players aged 15 to 26 years.

TABLE 25.8 Level 1 Yo-Yo Intermittent Recovery Test Data (Distance Covered) for Male Australian Rugby Players: Mean ± *SD* (Range)

Positional group	JUNIOR TALENT SQUAD PLAYERS		PROVINCIAL ACADEMY PLAYERS	
	Meters	Level · shuttle	Meters	Level · shuttle
Backs (*n* = 20; 18)	1,654 ± 361 (1,080-2,240)	17.6 ± 1.1 (15.8-19.5)	1,951 ± 419 (920-2,520)	18.6 ± 1.2 (15.4-20.4)
Forwards (*n* = 25; 25)	1,291 ± 380 (560-2,120)	16.5 ± 1.2 (14.3-19.2)	1,424 ± 467 (600-2,720)	17.1 ± 1.3 (14.4-21.1)

Typical error: distance = 90 m. Data source: Australian Institute of Sport rugby union testing database 2001-2010; junior talent squad players, *n* = 45, age 14-18 years; provincial academy players, *n* = 43, age 15-26 years.

6 × 30 M REPEAT SPRINT ABILITY TEST

RATIONALE

Rugby union players frequently perform repeated high-intensity efforts with only short recovery between bouts (Duthie et al. 2005). These periods do not allow players to achieve complete recovery. The 6 × 30 m repeat sprint ability (RSA) test is designed to evaluate a player's speed-endurance qualities and ability to resist fatigue under time and distance demands similar to those experienced during a match (Pyne et al. 2008).

TEST PROCEDURE

- The protocol is a 6 × 30 m maximal effort running sprint test on a 20 s cycle. All players are given appropriate instruction on the test procedures and a short warm-up of light running and stretching prior to the test. If this test is conducted after other tests, adequate recovery should be provided.

- Testing is conducted indoors on a synthetic running track surface or a polished wooden (sprung) floor with sufficient space (~60-70 m in length) to allow for decelerations and turnarounds. Record the type of running surface. Each repeat sprint test takes 2 min; allow another 30 to 60 s for changeover to the next player or group of players.

- Electronic timing gates are set up according to the manufacturer's specifications at 0 m and 30 m. An amplifier with the test protocol loaded onto a MP3 or CD player is positioned halfway along the testing lane. A marker cone is positioned 10 m beyond the timing gates at each end of the testing lane to signify the specified 10 m recovery distance (turnaround point). The start line at both ends of the sprint lane is marked exactly 1 m back from the 0 m (start) timing gate.

- Each player will complete one maximal effort trial involving 6 × 30 m running sprints from a stationary standing start. The player is asked to complete each 30 m sprint as a maximal effort.

- The player is required to run (decelerate) in a straight line at least 10 m past the 30 m timing gate (deceleration zone) and then return by walking or slowly jogging to the start line ready for the next effort. This procedure applies at both ends of the testing area. That is, the 30 m timing gate becomes the start timing gate for the next sprint.

- The player is given a 5 s warning and then a verbal command of "ready" at approximately 0.5 s before a starting signal. The player commences the maximal sprint effort at the starting signal. The starting action is standardized in the same manner as for the 40 m sprint test (i.e., crouch start, front foot at the starting line, self-start). The players should be instructed not to assume the crouch start position too early; they only need to adopt this position immediately prior to the starting signal. A common mistake is for the player to adopt the starting position too early and lose balance.

- One submaximal repetition of the test should be undertaken to familiarize players with the test procedures (especially the jog intensity for the active recovery component). A recovery period of approximately 5 min should be given before commencing the test.

- The time for each of the six sprints is recorded accurate to 0.01 s and the total time (in seconds) used as the criterion score.

- The RSA test can be conducted in the same testing session as other speed and power fitness tests (40 m sprint, vertical jump), although given its fatiguing nature, should be the last test of the session.

- If sufficient timing equipment and staff resources are available, two or more athlete tests should be run in parallel to reduce the time needed to test all players.

DATA ANALYSIS

It is important players perform each sprint with a maximal effort. A common limitation is that a player pulls up short of the full 30 m sprint, particularly toward the end of the test when fatigue is apparent. Results should be interpreted with regard to previous results and by playing position. In general, backs are expected to have a total time approximately 1 to 2 s quicker than forwards. Training of both speed and endurance qualities is necessary to improve players' RSA. Attention should be given to drills that emphasize both attributes in isolation and combination.

NORMATIVE DATA

At the time of publication, insufficient data were available to compile reference values for the 6 × 30 m RSA test.

PHYSICAL COMPETENCY SCREENING

RATIONALE

Physical competency can change throughout the career of a rugby union player because of several factors, including balance and coordination, strength and muscle hypertrophy, injury, tightness, and muscular imbalances. Dysfunctional or limited movement patterns and asymmetries have been associated with an increased risk of injury (Kiesel et al. 2007).

Changes in a player's movement competency will influence the requirements of the training program and must be identified by the strength and conditioning coach. A qualitative understanding of the movement patterns of a player provides the foundation for individualized exercise programming and prescription (Kiesel et al. 2011).

The physical competency screen consists of eight basic movement tests. Physical competency is specific to each movement pattern and region of the body. The selected range of tests is designed to assess the functional movement dynamics over the player's entire body. The screening allows coaches to track the long-term progression of players as well as monitor for short-term, injury-related weaknesses in specific areas.

TEST PROCEDURE

The physical competency movement screen is a qualitative assessment and should be conducted by an experienced practitioner. The screen takes approximately 8 to 10 min per player. The athlete is required to complete the set number of repetitions or duration for each test and based on the movement cues is deemed to be competent or not (pass/fail). The aim of the physical competency screen is to assess movement proficiency, not muscular strength. Junior athletes or those without previous strength training experience should substitute the 20 kg barbell for a broomstick or lightweight barbell when attempting the overhead squat and Romanian deadlift. The movement cues for each test are:

Barbell Overhead Squat (5 repetitions)

- Head is centered.
- Movement is initiated through hips.
- Elbows are locked; bar is aligned with midfoot.
- Lumbar spine maintains a neutral position.
- Thighs attain parallel position to the ground.
- Heels remain on the ground.

Photos courtesy of the Australian Rugby Union.

A-Lunge With Twist (3 repetitions each side)

- Arms are held across shoulders with elbows up.
- Lunge begins at static "A" position.
- Stride length equates 90° knee bend at front and back leg.
- Front knee is behind line of toes.
- Back foot is positioned on forefoot.

- Head is centered and trunk is stable.
- Athlete rotates toward front leg through thoracic spine.
- Lumbar position is neutral, resisting rotation.
- Balance is maintained throughout movement.

Photos courtesy of the Australian Rugby Union.

Barbell Romanian Deadlift (5 repetitions)

- Head is centered.
- Trunk is straight, scapulae set, chest up.
- Bend occurs at hips, not the lumbar spine.
- Knees maintain a slightly bent position.

Photos courtesy of the Australian Rugby Union.

Single Leg Squat to Bench (5 repetitions each side)

- Movement is initiated through hips.
- Lumbar spine maintains a neutral position.
- Knees are aligned over toes.
- Pelvis remains parallel to the ground.
- Athlete lowers in a controlled matter until buttocks touch bench.
- Heel remains on the ground.
- Trunk integrity is maintained throughout movement.

Photos courtesy of the Australian Rugby Union.

Push-Up (5 repetitions)

- Head is centered and held stable.
- Shoulders are held down and away from the ears.
- Elbows are at 90° at bottom position.
- Lumbar spine is in neutral position.
- Athlete has obvious scapulae control.
- Gluteals are activated.
- Legs are straight and stable.

Photos courtesy of the Australian Rugby Union.

45° Pull-Up (5 repetitions)

- Head is centered and held stable.
- Shoulders are held down and away from the ears.
- Head, shoulders, hips, and feet maintain alignment.
- Lumbar spine is in neutral position.
- Elbows are approximately 90° at top, bar to nipple line.
- Athlete has obvious scapulae control.
- Athlete pulls up and lowers in a controlled manner.

Photos courtesy of the Australian Rugby Union.

Single Leg Calf Raise (20 repetitions each side)

- Athlete stands upright on edge of box or step.
- Heel is down, leg straight.
- Athlete performs 1 s concentric contraction (up), 1 s eccentric contraction (down).
- Athlete uses full, controlled range.
- Pelvis remains parallel to the ground.

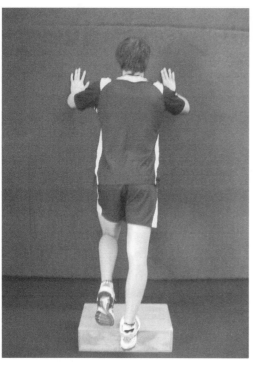 

Photos courtesy of the Australian Rugby Union.

Trunk Bridge (prone and lateral, 60 s each)

Prone

- Athlete begins on forearms and toes.
- Head is centered, looking down.
- Shoulders are held down and away from the ears.
- Lumbar spine is in neutral position and gluteals are activated.
- Body maintains alignment.

Lateral

- Athlete begins on forearm and foot.
- Free arm is in line with body.
- Shoulders are held down and away from ears.
- Support arm is held at 90° to the body.
- Lumbar spine is in neutral position and gluteals are activated.
- Body maintains alignment.

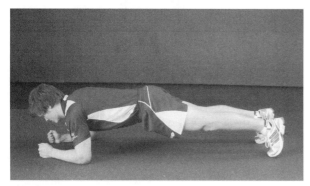 

Photos courtesy of the Australian Rugby Union.

DATA ANALYSIS

The physical competency screening requires minimal equipment and is therefore relatively simple to implement with limited resources. It is important that all tests within the competency screening are attempted. The screening should be conducted on two or three occasions over various stages of the training year. Previous results should be kept as a reference to monitor the progress of each player. If a player fails to complete a test or their physical competency regresses, a review of the player's current physical state is indicated. The strength and conditioning coach should determine whether there is any obvious reason for the result (e.g., injury) and modify the training or rehabilitation program as required.

STRENGTH TESTING

RATIONALE

Strength is a crucial quality for success in rugby union during both attack and defense. Absolute strength and power (regardless of body mass) are required to apply high forces quickly in contact situations and during the scrum (Quarrie and Wilson 2000). Players' running velocity and ability to change direction are also related to their strength and power relative to body mass (Nimphius et al. 2010; Young et al. 1995). The back squat, bench press, and prone grip chin-up provide a useful index of whole body strength.

TEST PROCEDURE

The strength tests used in rugby union include the 1RM back squat, bench press, and prone grip chin-up. Refer to chapter 13, Strength and Power Assessment Protocols, for a detailed description of strength and power tests. Warm-up and submaximal attempts of the bench press and squat should be performed in an appropriate rack with spotters present.

The following general guidelines must be adhered to for all tests:

- Strength testing should be performed on a separate day from the field tests. Strength and field tests should ideally be separated by 48 h.
- The athlete must perform an appropriate warm-up. As a minimum, the athlete is required to perform a trial at approximately 90% of specified repetition maximum (RM) for each test. If the athlete is being tested for the first time, he or she should perform an initial trial at approximately 90% of weight lifted in training.
- Lowering and lifting actions must be performed in a continuous manner. A single rest of no more than 2 s is allowed between repetitions.
- A maximum of 5 min recovery between trials is allowed.
- Minimum weight increments of 2.5 kg should be used between trials. However increments should be guided by ease of each trial.
- Ideally, the specified RM test should be completed within four trials (not including the warm-up).
- If the athlete is unable to complete tests as per protocol, this fact should be noted on the recording sheet. The test scores should not be included in any mathematical calculations (e.g., mean, typical error of measurement).
- It is recommended that a spotter, other than the supervising coach, be used where necessary.

DATA ANALYSIS

The use of the three standard strength tests will allow the coach to gain insight into the relative strength deficiencies of each player as they apply to his or her particular stage of physical development and prior training experience. The coach can identify imbalances between the upper and lower segments of the body as well as between the upper pushing (bench press) and upper pulling (prone grip chin-up) muscle groups. This analysis will assist individual program design and potentially reduce the likelihood of injuries associated with muscular imbalances.

NORMATIVE DATA

At all levels of rugby union, there are recommended standards of physical strength (see table 25.9). The strength program for each individual athlete should address the relative strength standards appropriate to the level of competition in which they are participating. If a player is unable to complete a single chin-up, loads of less than body mass can be achieved by testing the athlete using a standard lat pull-down machine.

TABLE 25.9 Australian Rugby Union Strength Standards (1RM or Estimated)

Lift	Junior	Advanced Junior	Intermediate	Senior	International
Back squat	1-1.2 × BM	1.2-1.4 × BM	1.4-1.6 × BM	1.6-1.8 × BM	1.8-2.0 × BM
Bench press	0.5-0.7 × BM	0.7-0.9 × BM	0.9-1.1 × BM	1.1-1.3 × BM	1.3-1.5 × BM
Prone grip chin-up	0.5-0.7 × BM	0.7-0.9 × BM	0.9-1.1 × BM	1.1-1.3 × BM	1.3-1.5 × BM

BM = body mass.

Used with permission from the Australian Rugby Union.

References

Cunniffe, B., Proctor, W., Baker, J.S., and Davies, B. 2009. An evaluation of the physiological demands of elite rugby union using global positioning system tracking software. *Journal of Strength and Conditioning Research* 23(4):1195-1203.

Deutsch, M.U., Kearney, G.A., and Rehrer, N.J. 2007. Time-motion analysis of professional rugby union players during match-play. *Journal of Sports Sciences* 25(4):461-472.

Duthie, G., Pyne, D., and Hooper, S. 2003. Applied physiology and game analysis of rugby union. *Sports Medicine* 33(13):973-991.

Duthie, G., Pyne, D., and Hooper, S. 2005. Time motion analysis of 2001 and 2002 super 12 rugby. *Journal of Sports Sciences* 23(5):523-530.

Duthie, G.M., Pyne, D.B., Hopkins, W.G., Livingstone, S., and Hooper, S.L. 2006a. Anthropometry profiles of elite rugby players: quantifying changes in lean mass. *British Journal of Sports Medicine* 40(3):202-207.

Duthie, G.M., Pyne, D.B., Marsh, D.J., Livingstone, S., and Hooper, S.L. 2006b. Sprint patterns in rugby union players during competition. *Journal of Strength and Conditioning Research* 20(1):208-214.

Gamble, P. 2004. A skill-based conditioning games approach to metabolic conditioning for elite rugby football players. *Journal of Strength and Conditioning Research* 18(3):491-497.

Kiesel, K., Plisky, P., and Voight, M. 2007. Can serious injury in professional football be predicted by a preseason functional movement screen? *North American Journal of Sports Physical Therapy* 2(3):147-158.

Kiesel, K., Plisky, P., and Butler, R. 2011. Functional movement test scores improve following a standardized off-season intervention program in professional football players. *Scandinavian Journal of Medicine and Science in Sports* 21(2):287-292.

Linthorne, N.P. 1994. Wind assistance in the 100-m sprint. *Modern Athlete and Coach* 32(1):6-9.

Nimphius, S., McGuigan, M.R., and Newton, R.U. 2010. Relationship between strength, power, speed and change of direction performance of female softball players. *Journal of Strength and Conditioning Research* 24(4):885-895.

Pyne, D.B., Saunders, P.U., Montgomery, P.G., Hewitt, A.J., and Sheehan, K. 2008. Relationships between repeated sprint testing, speed, and endurance. *Journal of Strength and Conditioning Research* 22(5):1633-1637.

Quarrie, K.L., and Wilson, B.D. 2000. Force production in the rugby union scrum. *Journal of Sports Sciences* 18(4):237-246.

Robinson, P.D., and Mills, S.H. 2000. Relationship between scrummaging strength and standard field tests for power in rugby. In: Y. Hong and D.P. Johns (eds), *Proceedings of the XVIII International Symposium on Biomechanics in Sports.* Hong Kong: The Chinese University of Hong Kong, pp 980.

Slater, G.J., Duthie, G.M., Pyne, D.B., and Hopkins, W.G. 2006. Validation of a skinfold based index for tracking proportional changes in lean mass. *British Journal of Sports Medicine* 40(3):208-213.

Tomlin, D.L., and Wenger, H.A. 2002. The relationships between aerobic fitness, power maintenance and oxygen consumption during intense intermittent exercise. *Journal of Science and Medicine in Sport* 5(3):194-203.

Young, W., McLean, B., and Ardagna, J. 1995. Relationship between strength qualities and sprinting performance. *Journal of Sports Medicine and Physical Fitness* 35(1):13-19.

Runners and Walkers

Philo U. Saunders and Daniel J. Green

Competitive running and walking are physiologically demanding, with events ranging from the 100 m sprint lasting approximately 10 s to the marathon and 50 km race walk, which take several hours to complete. The physiological characteristics of elite performers will vary drastically between events, with both aerobic and anaerobic capabilities being important.

Testing of runners and walkers will help coaches achieve several objectives:

- Construct event- and athlete-specific physiological profiles
- Monitor and assess effectiveness of training programs
- Prescribe and implement training intensities for specific training sessions
- Identify talent and determine event specialization
- Detect acute or chronic overreaching or overtraining syndromes
- Assess mechanisms responsible for improved performance after various interventions

The laboratory testing protocols in this chapter include anthropometry, an incremental treadmill test for middle- and long-distance runners, an incremental treadmill test for race walkers, and a maximal accumulated oxygen deficit (MAOD) treadmill test for 400 to 800 m runners. The main aim of the laboratory-based testing is to provide information on physiological parameters associated with optimal running and walking performance.

For the endurance-based events (middle- and long-distance running as well as race walking), test-ing needs to quantify the major components that contribute to best performance. Some 20 years ago it was demonstrated that endurance running performance is a function of an athlete's maximal volume of oxygen consumption ($\dot{V}O_2max$), the maximal fraction of $\dot{V}O_2max$ that can be maintained for the duration of a run, and the energy cost of running per unit distance (di Prampero 1986). The distance running and walking protocols quantify all three measures so that changes in all areas can be monitored and improvements can be made through training and other interventions. The other key outcome of physiological testing in endurance athletes is to determine training zones based on speed or heart rates so athletes can perform sessions in specific zones. When one is assessing 400 m runners, the anaerobic energy contribution becomes more important; and similar to endurance running events, equations accounting for utilization of anaerobic energy stores have been used to predict performances for different running distances (Capelli 1999; di Prampero et al. 1993; Lacour et al. 1990). The laboratory tests developed for assessing 400 m athletes have been designed to quantify the characteristics identified as critical for anaerobic energy production and to quantify aerobic energy contribution.

Athlete Preparation

Standardized pretest preparation is recommended to enable reliable and valid physiological data to be obtained. To standardize test conditions, the following specific guidelines should be met. Refer to chapter 2, Pretest Environment and Athlete Preparation, for specific information relating to athlete and environment preparation.

- *Diet and Health Status.* There should be no radical changes to diet in the days prior to testing. Most athletes already follow a high-carbohydrate diet, but strategies should be used to ensure that high-carbohydrate meals are eaten in the final 2 meals prior to testing to ensure optimal energy stores. Athletes should refrain from consuming food 3 h prior to testing because this can affect the rate of substrate utilization, particularly during the initial stages of the testing protocol. It is important for athletes to present for testing in a euhydrated state, having consumed adequate fluid in the 12 h prior to testing. Furthermore, athletes should be tested under a consistent health status; if any major illness or injury is present, testing should be postponed or cancelled. It is good practice to record the diet for the preceding 24 h and replicate this for all testing sessions.

- *Training.* Where possible, time and the type of training performed within the previous 48 h should be standardized. These training sessions should be familiar and not too demanding to ensure that athletes are in an optimal state to give a maximal performance in the laboratory testing. It is recognized that training programs differ between athletes so not all athletes will be doing the same training, but each athlete should try to present in a similar manner for all laboratory testing sessions especially when trying to quantify the effects of training or a specific intervention.

- *Footwear.* The same or a similar pair of shoes should be worn for each test session. Shoes of vary-ing mass may have an influence on the economy and time to exhaustion during the progressive treadmill test. The type of shoes used during testing should be recorded. Lightweight racing flats are recommended for best results.

- *Familiarity.* The athlete should be familiar with all test procedures prior to data collection. This is crucial for performance tests and those for which submaximal heart rates (HR) are recorded. Where possible, a practice session should be conducted on a separate day; at the very minimum, a run on the treadmill to familiarize athletes is essential.

Test Environment

The environmental and psychological conditions surrounding the athlete should be standardized. The laboratory should be controlled for temperature (18-23 °C), relative humidity (<70%), noise, verbal encouragement, and onlookers. An electric fan should be used to cool subjects to prevent excessive heat stress or dehydration during the tests. Finally, the athlete's coach should be encouraged to attend.

Athletes are encouraged to fill out the standard testing questionnaire, which will help them meet the pretest conditions and help the test coordinator interpret the results. Questionnaires should include laboratory conditions (temperature, humidity, and barometric pressure), description of last 48 h of training, diet for last 24 h, health status, and type and mass of running shoes.

Recommended Test Order

It is important that the tests are completed in the same order to control the interference. This order also allows valid comparison of results from different testing sessions. The order is as follows:

Day	Tests	Notes
1	Anthropometry	Height, body mass, skinfolds
1	Treadmill testing	Incremental test or maximal accumulated oxygen deficit (MAOD) test

Equipment Checklist

Anthropometry

- [ ] Stadiometer (wall mounted)
- [ ] Balance scales (accurate to ±0.05 kg)
- [ ] Anthropometry box
- [ ] Skinfold calipers (Harpenden skinfold caliper)
- [ ] Marker pen
- [ ] Anthropometric measuring tape
- [ ] Recording sheet (refer to the anthropometry data sheet template)
- [ ] Pen

Capillary Blood Sampling

- [ ] Disposable lancets
- [ ] Alcohol swabs
- [ ] Heparinized capillary tubes (100 μL) or lactate strips
- [ ] Disposable rubber gloves
- [ ] Tissues
- [ ] Kidney or carry dish
- [ ] Sharps container
- [ ] Biohazard bags
- [ ] Blood lactate or blood gas analyzer

Incremental Treadmill and MAOD Tests

- [ ] Treadmill (calibrated for velocity and elevation)
- [ ] Metabolic cart (and associated equipment)
- [ ] Heart rate monitoring equipment
- [ ] Electrode gel
- [ ] Electric fan
- [ ] Stopwatch
- [ ] Capillary blood sampling equipment (as described previously)
- [ ] Rating of perceived exertion (RPE) chart
- [ ] Recording sheet (refer to the treadmill test and MAOD data sheet templates)

ANTHROPOMETRY

RATIONALE

Body composition is an important part of high-performance running and walking, and low body fat levels are important. It has been demonstrated that lower skinfold measurements are associated with better distance running performance (Bale et al. 1985, 1986; Legaz and Eston 2005; Legaz et al. 2005). Therefore, it is important to regularly measure skinfolds for individual athletes. Another important parameter to monitor is body mass, because this provides an indication of how muscle and fat mass have changed (Bale et al. 1985, 1986).

TEST PROCEDURE

Measurement of height, body mass, and skinfolds should be carried out prior to further testing protocols. Skinfolds are recorded over seven sites (triceps, biceps, subscapular, supraspinale, abdominal, front thigh, and medial calf) for both males and females using techniques endorsed by the International Society for the Advancement of Kinanthropometry. The individual skinfold measures as well as the sum of the seven sites should be reported. Refer to anthropometry protocols outlined in chapter 11, Assessment of Physique, for a detailed description of anthropometric test procedures.

DATA ANALYSIS

Although the description of skinfold measurement procedures seems simple, a high degree of technical skill is essential for consistent results. It is therefore important that these measurements be taken by an experienced tester who has been trained in these techniques. It is also important that the same tester conduct each retest to ensure reliability. Depending on the requirements of the athlete, coach, and program, anthropometric measurements can be obtained quite frequently. Results given to athletes and coaches should take into account inherent measurement errors by reporting the range corresponding to a 68% confidence interval (value $\pm \sqrt{2} \times$ TE) or 95% confidence interval (value $\pm 2 \times \sqrt{2} \times$ TE).

NORMATIVE DATA

Height and body mass results are evaluated in terms of the chronological and biological age of the runner. However, the most appropriate evaluation of an individual's anthropometric results is a comparison with his or her previous results. Comparisons with group means and normal ranges may also be of interest. Tables 26.1 and 26.2 present anthropometric normative data for elite female and male runners visiting the Australian Institute of Sport (AIS) from 2001 to 2010.

The sum of skinfold values recorded by runners tend to show a degree of variation. Consideration should therefore be given to the phase of the training cycle, because skinfold readings can vary markedly throughout the training season. The minimal level of skinfolds that can be maintained with good health varies among individuals.

TABLE 26.1   Anthropometric Data for Elite Female Runners: Mean ± *SD* (Range)

Athlete group	Height, cm	Body mass, kg	Σ7 skinfolds, mm
Sprints/hurdles (100-400 m)	172.5 ± 4.9 (162.1-181.1) n = 61	62.4 ± 4.6 (56.1-73.4) n = 63	52.9 ± 9.2 (36.8-81.9) n = 63
Distance (800 m to marathon)	167.8 ± 6.2 (157.8-184.2) n = 25	55.6 ± 5.9 (46.2-67.5) n = 33	56.4 ± 14.3 (28.0-96.0) n = 34
Walks (10-50 km)	167.9 ± 3.9 (159.2-174.0) n = 31	55.5 ± 3.1 (47.7-62.1) n = 42	77.3 ± 13.9 (43.0-111.7) n = 42

Data source: Nationally ranked Australian runners and walkers, 2001-2010.

TABLE 26.2 Anthropometric Data for Elite Male Runners: Mean ± *SD* (Range)

Athlete group	Height, cm	Body mass, kg	Σ7 skinfolds, mm
Sprints/hurdles (100-400 m)	182.4 ± 5.5 (170.7-192.0) n = 154	77.5 ± 6.6 (64.0-96.2) n = 174	39.4 ± 8.9 (25.9-90.6) n = 175
Distance (800 m to marathon)	178.9 ± 6.0 (169.0-192.4) n = 80	66.2 ± 6.5 (54.0-82.7) n = 172	37.4 ± 5.4 (23.5-54.8) n = 180
Walks (10-50 km)	180.0 ± 6.8 (166.2-191.7) n = 53	65.4 ± 6.9 (55.0-81.2) n = 66	38.2 ± 5.4 (30.3-55.5) n = 66

Data source: Nationally ranked Australian runners and walkers, 2001-2010.

INCREMENTAL TREADMILL TEST FOR MIDDLE- AND LONG-DISTANCE RUNNERS

RATIONALE

The incremental running test is designed to measure the physiological parameters that determine distance running performance: maximal oxygen consumption ($\dot{V}O_2$max), running economy (i.e., the $\dot{V}O_2$ consumed at a given running speed or the slope of the speed vs. $\dot{V}O_2$ regression), and the fractional utilization of $\dot{V}O_2$max, which can be determined as the $\dot{V}O_2$ consumed at the lactate threshold as a percentage of $\dot{V}O_2$max. This test can also be used to determine the velocity at $\dot{V}O_2$max ($v\dot{V}O_2$), which is highly correlated with performance and takes into account running economy and $\dot{V}O_2$max.

The submaximal portion of the test is used to measure running economy (i.e., the $\dot{V}O_2$ at a given running speed), measure the HR and lactate responses at certain speeds, and thereby construct HR–velocity and lactate–velocity curves. This can be used to provide an assessment of aerobic fitness without relying on the motivation for maximal effort. The submaximal portion of the test involves subjects completing four or five stages of 4 min duration and constant speed between 15 and 21 km · h⁻¹ for male runners and 13-19 km · h⁻¹ for female runners. All submaximal stages are completed on a treadmill at 0% gradient with a 1 min rest period between stages for collection of a capillary blood sample, and each subsequent stage is performed at a speed 1 km · h⁻¹ faster than the previous stage. A general guide for calculating test speeds is to complete the fifth stage at the runner's current 10 km race pace. That is, if a subject's current estimated 10 km time is 30 min, the fourth stage should be set at 19 km · h⁻¹.

This subject's submaximal test would then consist of four stages completed at 16, 17, 18, and 19 km · h⁻¹, with the possibility of a fifth and final stage at 20 km · h⁻¹. Ideally the aim is to establish a range of blood lactate values from baseline (1-2 mmol · L⁻¹) to just above 4 mmol · L⁻¹. During the test, a heart rate (HR) monitor is worn and the average HR for the last minute of each 4 min stage recorded. At the completion of each stage, a blood lactate measure is taken and recorded. The fifth stage is conducted if blood lactate concentration ([La⁻]) remains under 4 mmol · L⁻¹ at the completion of the fourth stage; however, if [La⁻] has already exceeded 4 mmol · L⁻¹ at the completion of the fourth stage, the fifth stage is not required. Expired gas is collected throughout the test for determination of ventilation, $\dot{V}O_2$, $\dot{V}CO_2$, and respiratory exchange ratio (RER). Similar to HR measurements, all metabolic measures are taken from the final min of each 4 min stage to ensure a plateau has been reached.

$\dot{V}O_2$max is measured during an incremental test to volitional exhaustion performed 5 min after the completion of the final stage of the submaximal test. The initial stages of the test are performed with the treadmill set at a gradient of 0% with increases in speed up until a critical speed, where further increases in intensity are then achieved through increases in treadmill gradient. This is to ensure that the test is suitable for a wide range of athletes but that athletes are not "outrun" because the test speed increases too quickly before a $\dot{V}O_2$max is reached. The critical speed is the last speed completed in the submaximal test. The incremental protocol commences at a treadmill speed 4 km · h^{-1} lower than the critical speed and increases by 0.5 km · h^{-1} every 30 s up until the critical speed. After 30 s at the critical speed and 0% gradient, the treadmill gradient increases by 0.5% every 30 s until volitional exhaustion is reached, with subjects instructed to attempt to finish the test on a 30 s increment if possible. With 10 s to go in each 30 s increment, the subject is asked whether he can complete another 30 s workload and is required to signal with a thumbs-up that he is physically able to begin the next increment. The maximal HR is recorded and a blood lactate sample collected 1-3 min after completion of the test. Expired gas is collected throughout the test for determination of ventilation, $\dot{V}O_2$, $\dot{V}CO_2$, and respiratory exchange ratio. $\dot{V}O_2$max is reported in ml · kg^{-1} · min^{-1} and is the highest 30 s value multiplied by 2, because the protocol increases in 30 s increments. If a breath-by-breath system is used, 30 s increments are an ideal duration to provide stable measurements in $\dot{V}O_2$, increase test sensitivity, and allow for practical implementation.

TEST PROCEDURE

- Ensure that the treadmill is clean and has been recently calibrated for velocity and gradient. Run the treadmill through the range of velocities required in the test before the athlete arrives to verify proper operation.

- Ensure that the metabolic cart has been adequately warmed up and has been calibrated for ventilation and gas composition. It is recommended that three primary gravimetric-standard reference gases (±0.02%) spanning the physiological range are used for oxygen and carbon dioxide analyzer calibration.

- When the athlete arrives, make sure the guidelines for athlete preparation have been met and complete the pretest questionnaire.

- Measure the athlete's body mass (±0.05 kg) and the mass of one of the athlete's running shoes.

- Attach HR monitoring device to the athlete and ensure that the device is operating correctly. Electrode gel will help with electrical conduction.

- Place a piece of tape across the bridge of the athlete's nose. This is used to help prevent the nose clip slipping during the test.

- Instruct the athlete to take a position at the front of the treadmill and, when ready, allow the athlete a brief warm-up and familiarization period on the treadmill. During this period, outline the test procedures to the athlete (i.e., the athlete will spend 4 min at each workload with 1 min rest; each workload will increase in velocity by 1 km · h^{-1}; capillary blood samples will be taken after each workload).

- Position the headset and mouthpiece on the athlete, ensuring maximal comfort. Instruct the athlete to position the nose-clip comfortably to block nasal air flow.

- Prior to commencement of the test, have the athlete hold the handrails. Start the treadmill while simultaneously starting the metabolic cart and time devices. During each workload, indicate the time remaining and check with the athlete about his or her ability to continue. Stop the treadmill at the end of the 4 min, giving the athlete prior warning with a countdown over the last 5 s.

- Record the athlete's HR at the end of each minute of every submaximal stage; this serves as a backup in case there are difficulties in downloading data from the HR device following the test.

- Immediately on stopping exercise, a capillary blood sample should be drawn from the finger or earlobe and analyzed for blood lactate concentration.

- Ask the athlete to identify on the rating of perceived exertion chart (Borg 1962) their subjective assessment of exertion and record this value on the data sheet.

- During the rest interval, allow the athlete to remove the nose clip and mouthpiece; however, the headset should remain in place. With approximately 30 s before the start of the next workload, instruct the athlete to replace the mouthpiece and nose clip, if removed, and assume the starting position.

- Proceed as outlined above, with the athlete completing four or five different workloads.

- To assess the appropriateness of a further stage, observe the athlete's [La$^-$] and keep going until he records a value over 4 mmol · L^{-1}. If you are not sure whether the value is going to be above 4 mmol · L^{-1}, wait an extra 30 s (90 s total) to see what the value is and whether another stage is necessary.

- During the maximal portion of the test, have the athlete continue exercising until exhaustion. With 10 s left in each 30 s period during the expected final workload, ask the athlete if he is capable of running for at least a further 30 s (i.e., thumbs-up is yes; thumbs-down is no).

- Stop the test when the athlete voluntarily presses the emergency stop button or when the athlete indicates that he cannot complete the next 30 s period.
- At the end of the test, remove the nose clip and headset and obtain a 1-3 min posttest blood sample.
- If there are no more tests scheduled, ensure that the metabolic cart, the blood analyzers, and all other equipment used for testing are shut down and packed away in accordance with normal procedures. Ensure that the respiratory apparatus of the $\dot{V}O_2$max system is thoroughly sterilized, rinsed, and hung up to dry.

DATA ANALYSIS

- Once the HR and blood lactate data are collected from the submaximal test they can be graphed, and the speeds and heart rates relating to the 2 and 4 mmol · L^{-1} lactate thresholds can be calculated. Speeds are determined by solving the regression equation for lactate versus speed for lactate values of 2 and 4 mmol · L^{-1}. Threshold heart rates are then determined by solving the regression equation for heart rate versus speed for the speeds calculated previously. Fractional utilization can be calculated by dividing the $\dot{V}O_2$ at 4 mmol · L^{-1} lactate threshold by $\dot{V}O_2$max.
- An index of running economy and performance can be determined by calculating the $\dot{V}O_2$ (ml · kg^{-1} · min^{-1}) consumed at a fixed running speed (14 km · h^{-1} for females and 16 km · h^{-1} for males) expressed as a percentage of $\dot{V}O_2$max.

For prescription of training intensities, HR and lactate zones should be provided using the following classifications based on the prediction of the 2 and 4 mmol · L^{-1} lactate thresholds.

- *Easy running* is done to establish base mileage and improve running economy. Runs should be for longer than 30 min at an HR or speed under the 2 mmol · L^{-1} lactate.
- *Steady running* consists of varying durations but usually between 30 and 90 min at a pace to maintain lactate levels between 2 and 4 mmol · L^{-1}.
- *Threshold running* is a key intensity for long-distance training because it is closely related to exercise tolerance and replicates race intensity. Threshold is the speed or HR required to produce blood lactate levels at, or just above, 4 mmol · L^{-1}. Training at or around this intensity improves performance over 5 km and longer distance events. Threshold training can be either continuous or in the form of interval sessions, which are usually 20 to 50 min in duration.
- *High-intensity training* is for fast interval work at speeds or heart rates above 4 mmol · L^{-1} lactate. The interval duration, speed, and rest interval can be manipulated depending on the desired outcome of the session.

Table 26.3 presents a summary of the HR ranges for a five training zone model based on the athlete's 4 mmol · L^{-1} lactate threshold or maximal HR.

NORMATIVE DATA

Normative data for lactate threshold (4 mmol · L^{-1} speed), velocity at $\dot{V}O_2$max ($v\dot{V}O_2$), $\dot{V}O_2$max, and submaximal $\dot{V}O_2$ (running economy) for high-performance runners tested at the AIS between 2001-2010 are presented in table 26.4.

TABLE 26.3 Heart Rate (HR) Training Zones (5 Training Zone Model)

Zone	HR using % of 4 mmol · L^{-1} lactate threshold HR	HR using % of maximum HR
T1—recovery running	<75%	60%-70%
T2—aerobic endurance	75%-85%	70%-80%
T3—intense aerobic	85%-95%	80%-90%
T4—threshold	95%-102%	90%-95%
T5—lactate tolerance	>102%	95%-100%

TABLE 26.4 Incremental Treadmill Test Data for High-Performance Female and Male Runners: Mean ± *SD* (Range)

Parameter	Female	Male
Lactate threshold (4 mmol · L^{-1} speed), mmol · L^{-1}	16.2 ± 1.4 (13.7-18.4) $n = 25$	18.5 ± 1.1 (15.9-20.7) $n = 50$
Velocity at $\dot{V}O_2$max, km · h^{-1}	18.0 ± 1.4 (15.8-20.8) $n = 25$	20.5 ± 1.0 (18.2-22.5) $n = 49$
$\dot{V}O_2$max, ml · kg^{-1} · min^{-1}	61.72 ± 5.78 (46.17-72.44) $n = 82$	70.84 ± 4.57 (54.29-84.08) $n = 276$
$\dot{V}O_2$ 12 km · h^{-1}, ml · kg^{-1} · min^{-1}	41.31 ± 4.39 (33.27-50.21) $n = 52$	n/a
$\dot{V}O_2$ 13 km · h^{-1}, ml · kg^{-1} · min^{-1}	44.90 ± 1.97 (42.16-47.24) $n = 10$	n/a
$\dot{V}O_2$ 14 km · h^{-1}, ml · kg^{-1} · min^{-1}	48.28 ± 3.66 (38.96-56.70) $n = 83$	45.96 ± 3.90 (32.37-56.52) $n = 370$
$\dot{V}O_2$ 15 km · h^{-1}, ml · kg^{-1} · min^{-1}	50.94 ± 2.51 (42.40-54.81) $n = 25$	50.53 ± 3.22 (43.18-57.57) $n = 19$
$\dot{V}O_2$ 16 km · h^{-1}, ml · kg^{-1} · min^{-1}	54.48 ± 3.05 (46.17-61.91) $n = 74$	53.63 ± 3.89 (40.53-66.79) $n = 428$
$\dot{V}O_2$ 17 km · h^{-1}, ml · kg^{-1} · min^{-1}	58.23 ± 2.68 (52.65-61.37) $n = 17$	58.19 ± 3.03 (52.25-64.89) $n = 49$
$\dot{V}O_2$ 18 km · h^{-1}, ml · kg^{-1} · min^{-1}	61.85 ± 3.05 (56.36-65.49) $n = 9$	61.05 ± 3.94 (48.04-71.97) $n = 373$
$\dot{V}O_2$ 19 km · h^{-1}, ml · kg^{-1} · min^{-1}	n/a	65.23 ± 3.24 (55.84-72.19) $n = 36$
$\dot{V}O_2$ 20 km · h^{-1}, ml · kg^{-1} · min^{-1}	n/a	68.46 ± 1.93 (65.68-71.64) $n = 12$

n/a = not applicable. Typical error: $\dot{V}O_2$max = 1.8%-2.4%; lactate threshold = 3%-6%; velocity at $\dot{V}O_2$ = 2.4%. Data source: Nationally ranked Australian runners and walkers, 2001-2010.

INCREMENTAL TREADMILL TEST FOR RACE WALKERS

RATIONALE

Similar to the running test, the submaximal portion of the test is used to measure walking economy (i.e., the $\dot{V}O_2$ at a given walking speed) and to measure the HR and blood lactate responses at certain speeds and thereby construct HR–velocity and lactate–velocity curves. The submaximal economy test is performed on a treadmill and involves exercising for four or five progressively increasing workloads for 4 min, with each workload separated by a 1 min rest interval (capillary blood sample taken for measurement of $[La^-]$). The initial workload is typically 9 to 13 km · h^{-1} for female/junior athletes and 11 to 15 km · h^{-1} for male athletes, with successive workloads increased by 1 km · h^{-1}. During the test an HR monitor is worn and the average HR for the last minute of each 4 min stage recorded. At the completion of each stage, a blood sample is collected and $[La^-]$ recorded. Expired gas is collected throughout the test for determination of ventilation, $\dot{V}O_2$, $\dot{V}CO_2$, and respiratory exchange ratio. All metabolic measures are reported as the last min of each 4 min stage to ensure a plateau has been reached.

$\dot{V}O_2$max is measured during an incremental test to volitional exhaustion performed 5 min after the completion of the final stage of the submaximal test. The initial stages of the test are performed with the treadmill set at a gradient of 0% with increases in speed up until a critical speed, where further increases in intensity are then achieved through increases in treadmill gradient. This is to ensure that the test is suitable for a wide range of athletes but that athletes are not "outwalked" because the test speed increases too quickly before a $\dot{V}O_2$max is reached. The critical speed is the last speed completed in the submaximal test. The incremental protocol commences at a treadmill speed 4 km · h^{-1} lower than the critical speed

and increases by 0.5 km · h^{-1} every 30 s up until the critical speed. After 30 s at the critical speed and 0% gradient, the treadmill gradient increases by 0.5% every 30 s until volitional exhaustion is reached, with subjects instructed to attempt to finish the test on a 30 s increment if possible. With 10 s to go in each 30 s increment, the athlete is asked whether he can complete another 30 s workload and is required to signal with a thumbs-up that he is physically able to begin the next increment. The maximal HR is recorded and a blood lactate sample collected 1-3 min after completion of the test.

$\dot{V}O_2$max (ml · kg^{-1} · min^{-1}) is defined as the highest 30 s value multiplied by 2, because the protocol increases in 30 s increments. The main criterion for terminating the test is the walker's indication that he or she is unable to continue; the tester should carefully observe the walker's control, because a fall might result in injury or unwillingness to participate in further tests.

TEST PROCEDURE

The protocol is as described in incremental treadmill test for middle- and long-distance runners.

DATA ANALYSIS

Analysis is as described in incremental treadmill test for middle and long distance runners, however use of different speeds for walking economy.

NORMATIVE DATA

Normative data for lactate threshold (4 mmol · L^{-1} speed), velocity at $\dot{V}O_2$max ($v\dot{V}O_2$), $\dot{V}O_2$max, and submaximal $\dot{V}O_2$ (walking economy) for high-performance race walkers tested at the AIS between 2001-2010 are presented in table 26.5.

TABLE 26.5 Incremental Treadmill Test Data for High-Performance Female and Male Walkers: Mean ± *SD* (Range)

Parameter	Female	Male
Lactate threshold (4 mmol · L⁻¹ speed), mmol · L⁻¹	12.3 ± 1.1 (8.7-14.3) $n = 41$	13.9 ± 0.7 (12.2-15.1) $n = 40$
Velocity at $\dot{V}O_2$max, km · h⁻¹	13.9 ± 0.9 (11.9-16.6) $n = 39$	15.8 ± 0.8 (14.3-17.3) $n = 40$
$\dot{V}O_2$max, ml · kg⁻¹ · min⁻¹	56.14 ± 4.95 (46.42-66.60) $n = 49$	66.76 ± 4.16 (59.04-77.76) $n = 59$
$\dot{V}O_2$ 8 km · h⁻¹, ml · kg⁻¹ · min⁻¹	29.66 ± 1.81 (27.81-32.91) $n = 6$	n/a
$\dot{V}O_2$ 9 km · h⁻¹, ml · kg⁻¹ · min⁻¹	32.00 ± 3.21 (22.45-36.59) $n = 35$	n/a
$\dot{V}O_2$ 10 km · h⁻¹, ml · kg⁻¹ · min⁻¹	36.67 ± 3.46 (27.55-42.24) $n = 51$	37.94 ± 2.39 (34.87-42.15) $n = 7$
$\dot{V}O_2$ 11 km · h⁻¹, ml · kg⁻¹ · min⁻¹	41.51 ± 3.52 (33.13-49.30) $n = 50$	71.72 ± 3.51 (31.20-46.80) $n = 41$
$\dot{V}O_2$ 12 km · h⁻¹, ml · kg⁻¹ · min⁻¹	46.95 ± 3.64 (40.35-56.39) $n = 43$	47.23 ± 3.40 (39.28-57.00) $n = 60$
$\dot{V}O_2$ 13 km · h⁻¹, ml · kg⁻¹ · min⁻¹	51.08 ± 4.25 (45.10-61.72) $n = 21$	52.58 ± 3.58 (43.72-60.04) $n = 60$
$\dot{V}O_2$ 14 km · h⁻¹, ml · kg⁻¹ · min⁻¹	n/a	57.15 ± 3.39 (50.17-65.63) $n = 35$
$\dot{V}O_2$ 15 km · h⁻¹, ml · kg⁻¹ · min⁻¹	n/a	61.25 ± 3.12 (56.30-66.16) $n = 20$

n/a = not applicable. Typical error: $\dot{V}O_2$max = 1.8%-2.4%; lactate threshold = 3%-6%; velocity at $\dot{V}O_2$ = 2.4%. Data source: Nationally ranked Australian runners and walkers, 2001-2010.

MAXIMAL ACCUMULATED OXYGEN DEFICIT (MAOD) TEST

RATIONALE

The MAOD test is designed to determine the aerobic and anaerobic contributions of a maximal exercise bout of approximately 90 to 180 s in duration. The test also measures running economy, $\dot{V}O_2$max, $v\dot{V}O_2$, MAOD, and maximal [La$^-$] in the maximal exercise bout. As with other laboratory tests, the information collected during this test gives an indication of the different capacities of the athlete and how these change with training and other interventions. Training can then be modified to work on specific areas, for example, aerobic capacity, blood lactate production, or blood lactate tolerance. The test is divided into 2 sections like the incremental treadmill tests for distance runners and walkers.

The first part of the test consists of 4 or 5×4 min submaximal stages at 0% gradient to determine a speed versus $\dot{V}O_2$ curve. The speeds are individualized depending on the athlete and training history (i.e., an 800 m runner with an extensive aerobic training history will be able to run at much faster speeds than a 400 m runner who has completed limited aerobic training). The aim is to obtain a wide range of blood lactate measurements (1-8 mmol $\cdot$ L^{-1}) without fatiguing the runners too much before the maximal run to exhaustion. The ranges for 400 m runners are 10 to 14 km $\cdot$ h^{-1} for females and 12 to 16 km $\cdot$ h^{-1} for males; 800 m runners with more distance running history will do similar speeds as described in the distance running protocol.

The second part of the test involves a timed run to exhaustion at 21 km $\cdot$ h^{-1} (females) or 24 km $\cdot$ h^{-1} (males) at a 0% gradient. Metabolic analysis is measured during both sections of the test. Several steps are required to calculate the contribution of aerobic and anaerobic energy systems to the maximal test. The regression line of speed versus $\dot{V}O_2$ obtained from the submaximal test is used to estimate the $\dot{V}O_2$ at 21 km $\cdot$ h^{-1} (females) or 24 km $\cdot$ h^{-1} (males). It is important to ensure that data points used to establish the regression line have an RER less than 1; an assumption of this calculation is that the energy cost of running is linearly related to speed up to these maximal test values. Aerobic contribution to the maximal test is calculated as the $\dot{V}O_2$max obtained during the maximal run, and the MAOD (ml $\cdot$ kg^{-1} $\cdot$ min^{-1}) is calculated as the difference between the estimated $\dot{V}O_2$ required to run at 21 or 24 km $\cdot$ h^{-1} and the actual $\dot{V}O_2$max measured during the maximal portion of the test. The percentage of aerobic and anaerobic energy contribution can then be calculated.

TEST PROCEDURE

- Equipment preparation is as described in the incremental treadmill test for middle and long distance running test.

- When the athlete arrives, make sure the guidelines for athlete preparation have been met and have the athlete complete the pretest questionnaire.

- Measure the athlete's body mass (±0.05 kg) and the mass of one of the athlete's running shoes.

- Attach HR monitoring device to the athlete and ensure that the device is operating correctly. Electrode gel will help with electrical conduction.

- Place a piece of tape across the bridge of the athlete's nose. This is used to help prevent the nose clip slipping during the test.

- Instruct the athlete to take a position at the front of the treadmill and, when ready, allow the athlete a brief warm-up and familiarization period on the treadmill. During this period, outline the test procedures for the athlete (i.e., the athlete will perform 4 min at each workload with 1 min rest; each workload will increase in velocity by 1 km $\cdot$ h^{-1} capillary blood samples will be taken after each workload).

- Position the headset and mouthpiece on the athlete ensuring maximal comfort. Instruct the athlete to position the nose clip comfortably to block nasal air flow.

- Prior to commencement of the test, have the athlete hold the handrails. Start the treadmill while simultaneously starting the metabolic cart and time device. During each workload indicate the time remaining and check with the athlete about her ability to continue. Stop the treadmill at the end of the 4 min, giving the athlete prior warning with a countdown over the last 5 s.

- Record the athlete's HR at the end of each minute of every submaximal stage; this serves as a backup in case there are difficulties in downloading data from the HR device following the test.

- Immediately on stopping exercise, a blood sample should be drawn from the finger or earlobe and analyzed for [La$^-$].

- Ask the athlete to identify on the rating of perceived exertion chart (Borg 1962) her subjective assessment of exertion and record this value on the data sheet.

- During the rest interval allow the athlete to remove the nose-clip and mouth piece; however, the headset should remain in place. With approximately 30 s before the start of the next workload, instruct the athlete to replace the mouthpiece and nose clip, if removed, and assume the starting position.

- Proceed as outlined above, with the athlete completing four or five different workloads.

- To assess the appropriateness of a further stage, observe the athlete's lactate value and keep going until she records a value over 4 mmol · L^{-1}. If you are not sure whether the value is going to be above 4 mmol · L^{-1}, wait an extra 30 s (90 s total) to see what the value is and whether another stage is necessary. For many 400 m runners, the lactate values can easily be as high as 6 to 8 mmol · L^{-1}, so the aim is to ensure that 4 speeds are performed and a good lactate curve is obtained.

- Give athletes a 30 min recovery before the maximal portion of the test to ensure they are fully recovered and ready for a maximal run to exhaustion.

- Prior to commencement of the maximal run to exhaustion, give the runner the opportunity to perform a couple of 5 to 10 s runs at the test speed (21 or 24 km · h^{-1}) to familiarize her with how the treadmill gains speed.

- Because all 30 s values are used in the calculation of MAOD, the metabolic cart needs to be started at least 1 min prior to the start of the test to rid the system of all of the dead space and obtain an accurate $\dot{V}O_2$ from the first second of the test onward.

- Start the treadmill and time devices simultaneously with a 30 s increment of the already-running metabolic cart.

- Instruct the runner to run as long as she can and to grab hold of the side rails when finished. For safety, a "catcher" is required on each side of the treadmill in case the athlete suddenly falls at the end of the test, with another person operating the treadmill and a fourth person monitoring the time of the test. As soon as the runner is no longer able to keep a good position at the front of the treadmill or if he or she grabs the rails, the treadmill is instantly stopped and the time of the test recorded.

- At the end of the test, remove the nose clip and headset and obtain a 3 min posttest blood sample.

- If there are no more tests scheduled, ensure that the metabolic cart, the blood analyzers, and all other equipment used for testing are shut down and packed away in accordance with normal procedures. Ensure that the respiratory apparatus of the $\dot{V}O_2$max system is thoroughly sterilized, rinsed, and hung up to dry.

DATA ANALYSIS

- A speed versus $\dot{V}O_2$ relationship is graphed from the submaximal portion of the test and a regression equation calculated.

- The speed of the maximal portion of the test (21 or 24 km · h^{-1}) is put into the regression equation to determine the $\dot{V}O_2$ required to run at that speed (predicted $\dot{V}O_2$).

- The aerobic contribution of the maximal test is the maximal $\dot{V}O_2$ measured during the test (i.e., sum of all 30 s values in the case of 30 s collection period).

- MAOD = Predicted $\dot{V}O_2$ – Aerobic Contribution.

- Aerobic Contribution as a Percentage = (Measured $\dot{V}O_2$/Predicted $\dot{V}O_2$) × 100.

- Anaerobic Contribution = (MAOD/Predicted $\dot{V}O_2$) × 100.

NORMATIVE DATA

Normative data for elite female and male 400 m runners tested at the AIS are listed in table 26.6.

TABLE 26.6 Maximal Accumulated Oxygen Deficit (MAOD) Test Data for High-Performance Female and Male 400 m Runners: Mean ± SD (Range)

Parameter	Female	Male
Test time, min:s	2:02 ± 0:17 (1:32-2:30) n = 13	2:17 ± 0:25 (1:32-3:14) n = 13
$\dot{V}O_2$max, ml · kg^{-1} · min^{-1}	51.31 ± 2.66 (47.02-55.40) n = 13	56.96 ± 4.52 (47.42-64.46) n = 13
MAOD, ml · kg^{-1} · min^{-1}	30.53 ± 5.06 (22.95-40.45) n = 12	37.88 ± 6.43 (28.47-48.77) n = 13
Aerobic %	57 ± 5 (46-64) n = 12	55 ± 6 (47-63) n = 13
Anaerobic %	43 ± 5 (36-54) n = 12	45 ± 6 (37-53) n = 13
Velocity at $\dot{V}O_2$, km · h^{-1}	14.6 ± 0.9 (13.6-16.4) n = 12	16.0 ± 1.2 (13.6-18.0) n = 13
Maximal lactate concentration, mmol · L^{-1}	15.2 ± 1.5 (11.2-17.6) n = 13	16.3 ± 1.0 (14.8-17.9) n = 12

Typical error: test time = 8.6%, $\dot{V}O_2$max = 4.0%, MAOD = 6.5%, aerobic % = 3.2%, anaerobic % = 3.5%, velocity at $\dot{V}O_2$ = 4.0%, maximal lactate concentration = 6.3%. Data source: Nationally ranked Australian runners and walkers, 2001-2010.

References

Bale, P., Bradbury, D., and Colley, E. 1986. Anthropometric and training variables related to 10km running performance. *British Journal of Sports Medicine* 20(4):170-173.

Bale, P., Rowell, S., and Colley, E. 1985. Anthropometric and training characteristics of female marathon runners as determinants of distance running performance. *Journal of Sports Sciences* 3(2):115-126.

Borg, G. 1962. Physical performance and perceived exertion. Studia Psychogia et Paedagogia. Geerup vol 11, pp 1-35.

Capelli, C. 1999. Physiological determinants of best performances in human locomotion. *European Journal of Applied Physiology and Occupational Physiology* 80(4):298-307.

di Prampero, P.E. 1986. The energy cost of human locomotion on land and in water. *International Journal of Sports Medicine* 7(2):55-72.

di Prampero, P.E., Capelli, C., Pagliaro, P., Antonutto, G., Girardis, M., Zamparo, P., and Soule, R.G. 1993. Energetics of best performances in middle-distance running. *Journal of Applied Physiology* 74(5):2318-2324.

Lacour, J.R., Padilla-Magunacelaya, S., Barthelemy, J.C., and Dormois, D. 1990. The energetics of middle-distance running. *European Journal of Applied Physiology and Occupational Physiology* 60(1):38-43.

Legaz, A., and Eston, R. 2005. Changes in performance, skinfold thicknesses, and fat patterning after three years of intense athletic conditioning in high level runners. *British Journal of Sports Medicine* 39(11):851-856.

Legaz Arrese, A., Gonzalez Badillo, J.J., and Serrano Ostariz, E. 2005. Differences in skinfold thicknesses and fat distribution among top-class runners. *Journal of Sports Medicine and Physical Fitness* 45(4):512-517.

Sailors

Andrew Verdon, Hamilton Lee, and Michael Blackburn

Sailing is primarily a tactical, strategic, and technical sport. However, as wind strength increases so does the physiological demand because of the greater loads on the sails, and thus greater effort is required to control the boat proficiently.

Fitness requirements vary with the class of boat sailed, but the primary actions for the Olympic dinghies and keelboat classes include steering, hiking (leaning back over the side of the boat), and sheeting (pulling on the ropes controlling the sails) and for the Olympic sailboard class, pumping (repeated fanning of the sail).

There have been several changes in the sport of sailing over the three Olympic Games since the first edition of this book. Since 2000, the Soling, Europe, Tornado, Mistral, and Yngling classes were removed and the Laser Radial (female single-handed dinghy), Elliot 6 m (Women's match racing), and RSX (male and female sailboard) included. More recently, the RSX Sailboard has been replaced with a Kiteboard event. In May 2011 further changes were made for the 2016 Rio Olympic Games. A double-handed Female skiff (49er FX) will be added and a multihull (Nacra 17) will return as a mixed (male and female) event. In 2016, for the first time since 1900, there will be no keel boat class in the Olympic Games, as the Star class and the Elliot 6 m (Women's match racing) will be dropped.

There appears to be a push from the International Sailing Federation (ISAF) and the International Olympic Committee (IOC) toward more athletic classes, shorter races, and a greater number of races over the entire regatta period, using boats that are worldwide and easily transported to reduce costs.

In a typical 5- to 7-day championship sailing regatta, crews generally sail two races per day lasting about 60 min with a 15 min or longer break between races (depending on the weather conditions). For every race, approximately 2 h are spent on the water including sailing to the course area, warming up, racing, and sailing back to the marina (i.e., 3-6 h a day on the water). A major change to the format of Olympic Regattas occurred in 2008 with the introduction of a medal race. This is a final race worth double points, in which only the top 10 boats, determined from the initial races, compete. The results and points score including this medal race are used to determine the final overall results. The medal race is usually a short race of about 20 min.

The Women's match racing (WMR) event follows a different regatta format consisting of 12 to 20 min races. Sailors may complete as many as five races per day over a 6-day regatta. They progress through a round-robin series of races within a group before moving to a knockout series of quarter finals, semifinals, and a final to decide the medals.

Variability in wind and water conditions makes it difficult to conduct regular on-water field testing of sailors to monitor changes in fitness. However, several studies of the physiological demands of dinghy sailing have provided useful reference data (Castagna and Brisswalter 2007; Spurway 2007; Spurway et al. 2007).

This chapter contains the basic physical and physiological fitness tests adopted in Australia across the classes and quantifies general fitness qualities relating to sailing (lifting, pulling, aerobic fitness, and body composition management) across most of the current Olympic classes. These tests provide a good general assessment of the capacities that are important to Olympic class sailors.

Many national- and international-level Australian sailors have become proficient through many years of mainly on-water training and practice to develop their sailing skills. Many do not have a physical training background and are unfamiliar with structured resistance or cardiovascular training; This lack of training is combined with a lack of general fitness traits. The following tests are used with elite and developmental Australian sailing athletes. Ideally these tests are implemented after athletes have had some basic strength training (a minimum of a few weeks) prior to testing to establish baseline values.

Hiking is a major performance characteristic of some classes; however, it is beyond the scope of this chapter to describe the specialist hiking tests and the equipment that has been developed. Measurement of hip and trunk endurance and subsequent interventions in this area are better dealt with by a qualified physiotherapist or strength and conditioning specialist.

Athlete Preparation

Standardized pretest preparation is recommended to enable reliable and valid physiological data to be obtained. In order to standardize test conditions, the following specific guidelines should be met. Refer to chapter 2, Pretest Environment and Athlete Preparation, for specific information relating to athlete and environment preparation.

- *Meal Content and Timing.* When using testing as a monitoring tool to evaluate athlete progress as opposed to an investigative research design, individual comfort levels should dictate actual meal size, volume, and timing.

- *Stimulants.* The use of caffeinated sports drinks or commercial beverages 2 h prior to testing should be discouraged.

- *Training.* Athletes should not incur training-induced severe physiological or neural fatigue in the 72 h prior to testing (Abernethy 1995). This will exclude athletes from participating in any maximal physiological testing or maximal physical training prior to strength testing.

No unaccustomed exercise should be performed 72 h prior to strength testing that may result in sarcomere damage or decreased activation of motor units. Unaccustomed exercise includes a change in resistance exercise selection, increases in training volume (number of sets, exercises, or resistance sessions), or the performance of high-volume eccentric contractions.

Athletes should following these guidelines prior to testing:

- Rise a minimum of 1 h prior to testing.
- Eat lightly in relation to normal food habits, and eat nothing in the 1 h prior to testing.
- Consume fluid as needed; observe caffeine guidelines as given previously.
- Consume fluid during testing as required.
- Complete a 5 min stationery bicycle ride at a moderate intensity.
- Complete a 2 min row on the Concept II ergometer at approximately 2:30/500 m pace.
- Stretch as desired prior to testing.
- Be ready to test at the allocated time.

Test Environment

The physical and physiological fitness tests used for sailing were designed so that athletes could test themselves in an informal training environment. However, for formal testing it is recommended that tests be conducted in standardized conditions (i.e., in laboratory conditions on a well-maintained ergometer). It is recommended that athletes test using the same ergometer.

Equipment Checklist

Anthropometry

- ☐ Stadiometer (wall mounted)
- ☐ Balance scales (accurate to ±0.05 kg)
- ☐ Anthropometry box
- ☐ Skinfold calipers (Harpenden skinfold caliper)
- ☐ Marker pen
- ☐ Anthropometric measuring tape

- ☐ Recording sheet (refer to anthropometry data sheet template)
- ☐ Pen

And for fractionation:
- ☐ Small sliding caliper (bone caliper)
- ☐ Large sliding caliper
- ☐ Wide-spreading caliper

Strength Tests

- ☐ Bench (with barbell rack)
- ☐ Chin-up bar
- ☐ Barbell (Olympic 20 kg)
- ☐ Weight plates (2.5-25 kg increments)

- ☐ Squat rack
- ☐ Recording sheet (refer to sailing tests data sheet template)
- ☐ Pen

Anaerobic—Rowing 30 s Sprint Test

- ☐ Concept II rowing ergometer
- ☐ Recording sheet (refer to sailing tests data sheet template)

- ☐ Pen

Aerobic Endurance—O'Neill 4 min Rowing Test

- ☐ Concept II rowing ergometer
- ☐ Heart rate monitoring equipment
- ☐ Borg RPE scale (6-20)

- ☐ Recording sheet (refer to sailing tests data sheet template)
- ☐ Pen

Recommended Test Order

It is important that strength and field tests are completed in the same order to control the interference between tests. This order also allows valid comparison of different test occasions. The order is as follows:

DAY	TESTS
1	All physical and strength tests:
	Chin-ups
	Squat
	Bench press
	Bench pull
2	Anthropometry
	Ergometer

Physiologically, the result of a "general warm-up" or aerobic preparation is an increase in circulation, an increase in oxygen delivery to muscle tissue, and an elevation in body core temperature. Therefore, muscle temperature rises and viscosity is reduced, resulting in greater efficiency of muscle contraction while also reducing the potential for muscle pulls and skeletal muscle injury (Burke 1977).

Physiologically, a "specific warm-up" results in a spectrum of advantages. An increase in muscle and blood temperature improves both contractile force and contractile speed because muscles contract and relax faster and contract with greater efficiency as a result of lower viscous resistance (Burke 1977). A specific warm-up also warms and circulates lubricating fluid within the joints (synovial fluid), which facilitates movement of the associated joints, increases range of motion, and decreases the potential for injury (Harris and Elbourn 2002). The final advantage of the specific warm-up is that it prepares the athlete mentally for a maximal effort while also enhancing the transmission of nerve impulses for the specific movement or exertion (Burke 1977; Harris and Elbourn 2002).

In summary, an effective warm-up includes these advantages:

- Increases the heart rate and blood circulation gradually
- Increases body core temperature and muscle temperature
- Permits freer movement in joints
- Prepares the joints and associated muscles to function through their full range of motion
- Improves the efficiency of muscular actions
- Reduces risk of injury
- Improves the transmission of nerve impulses
- Aids psychological preparation

The warm-up process should be broken into two sections. First is the general warm-up, consisting of a light aerobic format. This is followed by a test-specific warm-up that incorporates the action of the particular test.

The intensity and duration of the warm-up should be adjusted to the individual's level of strength and fitness or training age. This is the tester's or coach's duty. It is at the athlete's discretion to use stretching of any form as part of a general warm-up and in recovery periods of the specific warm-up.

Table 27.1 details the various tests that are currently used with elite and developmental Australian sailing athletes.

TABLE 27.1 Test Selection for Various Sailing Categories

Class	Anthro-pometry	STRENGTH TESTS				ANAEROBIC	AEROBIC
		Chin-ups	Squat	Bench press	Bench pull	30 s	O'Neill
470 Helm	✓	✓			✓		✓
470 Crew	✓	✓			✓	✓	✓
49er Helm	✓	✓			✓		✓
49er Crew	✓	✓			✓	✓	✓
Laser	✓	✓			✓	✓	✓
Laser Radial	✓	✓			✓	✓	✓
Finn	✓	✓	✓	✓	✓	✓	✓
RSX	✓	✓	✓	✓	✓	✓	✓
Women's Match Racing	✓	✓	✓	✓	✓	✓	✓

ANTHROPOMETRY

RATIONALE

Body mass is important in Olympic sailing because the body is used as a righting moment to provide leverage against the force of the wind. Because sailing is a weight-supported and weight-dependent activity, specific height and body mass ranges for each Olympic class have been associated with success. All Olympic classes have narrowed in on a small body mass band of generally 3 to 4 kg for the sailor or crew (outside of any rule requirements).

These ranges have emerged naturally as a function of the design of the equipment, particularly the power that can be generated by the sails. Depending on the class of boat sailed, crews deliberately try to gain or lose weight, which often involves adjustment of body fat levels. Therefore, assessment of sailors' body size and composition is important.

TEST PROCEDURE

Measurement of height, body mass, and skinfolds should be carried out prior to ergometer testing. Skinfolds are recorded over seven sites (triceps, biceps, subscapular, supraspinale, abdomen, thigh, and calf). The individual skinfold measures as well as the sum of the seven sites should be reported. Refer to anthropometry protocols outlined in chapter 11, Assessment of Physique, for a detailed description of all anthropometric testing procedures. If significant physical alterations are anticipated, either through diet and exercise interventions or because of growth, it is recommended that additional anthropometric assessment (extended profile for fractionation) including muscle girths and bone breadths be conducted. These measures should be made in accordance with protocols of the International Society for the Advancement of Kinanthropometry (ISAK).

A high degree of technical skill is essential for consistent anthropometry results. It is therefore important that these measurements be taken by an experienced tester who has been trained in these techniques. It is also important that the same tester conduct each retest to ensure reliability.

NORMATIVE DATA

Table 27.2 presents anthropometric normative data for female and male national- and international-level Australian sailing athletes.

TABLE 27.2 Anthropometric Data for Female and Male Australian Sailing Athletes: Mean ± *SD* (Range)

Class	Height, cm	Body mass, kg	$\sum$7 skinfolds, mm
470 Male—Helm (n = 9)	173.0 ± 5.3	61.8 ± 4.2	50.7 ± 15.8
	(167.1-182.9)	(53.0-68.5)	(37.9-89.1)
470 Male—Crew (n = 7)	186.5 ± 4.8	72.1 ± 4.8	47.7 ± 14.1
	(179.1-194.4)	(66.5-81.7)	(31.4-73.8)
470 Female—Helm (n = 6)	165.1 ± 4.5	55.1 ± 4.9	86.3 ± 23.9
	(159.9-173.1)	(48.2-60.2)	(55.8-116.4)
470 Female—Crew (n = 5)	175.5 ± 4.0	67.2 ± 4.0	76.9 ± 15.0
	(170.5-179.5)	(63.8-73.4)	(60.5-94.0)
49er—Helm (n = 6)	178.8 ± 4.2	73.9 ± 3.8	54.5 ± 7.5
	(173.0-185.7)	(69.1-78.9)	(39.8-60.4)
49er—Crew (n = 6)	184.5 ± 3.5	80.1 ± 2.3	53.9 ± 8.0
	(179.8-187.6)	(75.9-82.7)	(43.0-66.9)
Laser (n = 8)	182.4 ± 4.2	80.5 ± 3.5	60.5 ± 18.9
	(176.9-189.3)	(76.3-86.0)	(37.7-87.0)
Laser Radial (n = 4)	172.0 ± 2.4	66.6 ± 3.8	89.1 ± 15.1
	(169.7-174.6)	(63.3-71.6)	(76.7-110.5
Finn (n = 3)	186.3 ± 2.6	96.2 ± 4.3	80.6 ± 12.1
	(184.6-189.3)	(93.3-101.2)	(71.9-94.4)
RSX Female (n = 3)	170.1 ± 2.0	62.7 ± 7.7	87.3 ± 21.1
	(168.1-172.1)	(55.3-70.6)	(69.5-110.7)
Match Racing Female (n = 9)	167.5 ± 4.3	66.4 ± 2.5	108.2 ± 15.4
	(161.8-175.1)	(61.8-70.5)	(90.3-138.7)

Typical error: height = 1 cm; body mass = 1 kg; $\sum$7 skinfolds = ~ 1.5 mm. Data source: Australian Institute of Sport, 2006-2010.

STRENGTH TESTS

RATIONALE

Olympic sailors need power, strength (including core strength), aerobic fitness, agility, and balance. The amounts of these required will depend on not only the class sailed (or position within the boat) but also the wind strength and the sea state for that race. Generally for dinghies, the higher the wind strength and the larger the waves, the greater are the physical demands placed on the sailor. This may differ slightly for keel boats (Star and Elliot 6 m) and boards (RSX).

Strength is an important quality for Olympic sailors for several reasons. As the strength of the wind increases, so do the forces acting on the boat and the load in the sheets (control ropes form the sails). As these forces increase the sailor needs to use more force to steer, control, and trim the boat and the sails. The sailor also acts as a lever arm via hiking or trapezing to increase the righting moment. This requires strength through the lower body and the trunk and is crucial to performance and boat speed. Hiking elicits a high heart rate and blood pressure response but imposes only a moderate aerobic demand as the strength of the static contractions of the quadriceps and other active muscles impinge on local blood vessels and restrict flow. This mechanism has been referred to in the literature as quasi-isometric (Gallozzi et al. 1993, Vogiatzis et al. 1995).

To trim the sails for optimal power, the sailor pulls on the sheets and engages in powerful and vigorous pumping actions to increase the apparent wind over the sails and promote the hull to plane, reducing drag in the water and increasing boat speed. In certain classes sailors are required to rapidly hoist and lower sails as they round marks on the course to increase or reduce sail area by using a spinnaker. The sailor needs to repeatedly do this around the course over the length of the race and each race of the day over successive days.

Strength also plays a role in the agility and balance needed to dynamically sail the hull efficiently in choppy water and waves, because the hull is always unstable and constantly moving due to the forces of the wind and the water acting upon it.

TEST PROCEDURE

The strength tests used in Australia include maximal chin-ups and 3RM squat, bench press, and bench pull. Refer to chapter 13, Strength and Power Assessment Protocols, for a detailed description of these tests. Warm-up and submaximal attempts of the bench press and squat should be performed in an appropriate rack with spotters present.

The following general guidelines must be adhered to for all tests:

- Strength testing should be performed on a separate day from other tests. Strength and ergometer tests should ideally be separated by 48 h.

- The athlete must perform an appropriate warm-up. As a minimum, all athletes are required to perform a trial at approximately 90% of specified repetition maximum (RM) for each test. If it is the first test, the athlete should perform an initial trial at approximately 90% of weight lifted in training.

- Lowering and lifting actions must be performed in a continuous manner. A single rest of no more than 2 s is allowed between repetitions.

- A maximum of 5 min recovery between trials is allowed.

- Minimum weight increments of 2.5 kg should be used between trials. However, increments should be guided by ease of each trial.

- Ideally, the specified RM test should be completed within four trials (not including the warm-up).

- If the athlete is unable to complete tests as per protocol, this should be noted in the testing results, and values should not be included in any mathematical calculations (e.g., average, typical error of measurement).

- It is recommended that a spotter, other than the supervising coach, be used where necessary.

NORMATIVE DATA

Table 27.3 presents normative data for female and male national- and international-level Australian sailing athletes for maximal chin-ups. At the time of publication, limited data were available for other strength tests.

TABLE 27.3 Strength Test Data for Female and Male Australian Sailing Athletes: Mean ± *SD* (Range)

Class	Chin-ups, count
470 Male—Helm (*n* = 6)	15 ± 2 (12-18)
470 Male—Crew (*n* = 6)	14 ± 2 (12-17)
470 Female—Helm (*n* = 4)	6 ± 5 (0-10)
470 Female—Crew (*n* = 5)	6 ± 4 (2-10)
49er—Helm (*n* = 5)	12 ± 2 (10-15)
49er—Crew (*n* = 7)	14 ± 4 (10-21)
Laser (*n* = 7)	13 ± 4 (7-17)
Laser Radial (*n* = 3)	6 ± 2 (5-8)
Finn (*n* = 3)	13 ± 5 (8-18)
RSX Female (*n* = 3)	10 ± 3 (7-12)
Match Racing Female (*n* = 9)	7 ± 3 (3-11)

Data source: testing period 2006-2010.

ANAEROBIC—ROWING 30 S SPRINT TEST

RATIONALE

The anaerobic system is called upon in sailing for high-intensity efforts such as occur during starts, when pumping the sails, during top and bottom mark roundings, and in close race finishes to generate maximal speed through the water.

TEST PROCEDURE

A Concept IID or IIE rowing ergometer (using a PM2 or later monitor) should be used for the 30 s sprint test.

- Instruct the athlete to sit on the ergometer and take a few strokes to set the appropriate drag factor. Ensure that the drag factor is set according to the following athlete body mass guidelines:

  - Women ≤59 kg = 110
  - Women >59 kg = 120
  - Men ≤72 kg = 120
  - Men >72 kg = 130

- Position the foot stretchers on the ergometer for comfort of the athlete; legs should be flat at the finish or fully extended position.

- Allow the athlete to complete a 5 min general warm-up on a cycle ergometer followed by 2 min of light to moderate rowing then rest or stretch for approximately 2 min.

- Set the monitor on the rowing ergometer for 30 s.

- Ensure that the flywheel is completely stopped before commencing the test.

- Instruct the athlete to row as hard as possible for 30 s to achieve the highest average power output for the full 30 s duration of the test.
- At the end of the test, record distance covered (m), average power output (W), and average 500 m pace (min:s.0).
- Allow athletes to complete an active recovery following the test and after relevant data have been collected from the work monitor.

- If the athlete is to complete the O'Neill 4 min test in the same session, allow 10 min active recovery and rest between tests.

NORMATIVE DATA

No normative data for the 30 s sprint test were available at the time of publication.

AEROBIC ENDURANCE—O'NEILL 4 MIN ROWING TEST

RATIONALE

Aerobic capacity is an important quality for Olympic class sailors. It serves various purposes:

- It improves recovery after high-intensity efforts in racing.
- It increases hiking ability in strong wind racing. Hiking is a dynamic constant activity that requires contribution from the aerobic system.
- It enhances the sailor's ability to perform sheeting and pumping in windy conditions. These actions place large demands on the aerobic system because of the constant nature of effort and the loads on the hull and sails.

TEST PROCEDURE

A Concept IID or IIE rowing ergometer (using a PM2 or later monitor) should be used for the O'Neill 4 min rowing test.

- Attach heart rate (HR) monitoring device to the athlete and ensure that the device is operating correctly. Electrode gel will help with electrical conduction.
- Instruct the athlete to sit on the ergometer and take a few strokes to set the appropriate drag factor. Ensure the drag factor is set according to the following athlete body mass guidelines:
  - Women ≤59 kg = 110
  - Women >59 kg = 120
  - Men ≤72 kg = 120

  - Men >72 kg = 130
- Position the foot stretchers on the ergometer for comfort of the athlete; legs should be flat at the finish or fully extended position.
- Set the monitor on the rowing ergometer for 4 min.
- Ensure that the flywheel is completely stopped before commencing the test.
- Instruct the athlete to row as hard as possible for 4 min with aim of achieving the greatest possible distance.
- Record HR immediately upon completion of test (i.e., last 10-15 s) and recovery HRs at 1 and 2 min posttest.
- At the end of the test, record distance covered (m) and average 500 m pace (min:s.0).
- Ask the athlete to use the rating of perceived exertion chart (Borg 1962) to indicate his or her subjective assessment of exertion; record this value on the data sheet.
- Have athletes complete an active recovery following the test and after relevant data have been collected from the work monitor.

NORMATIVE DATA

Table 27.4 presents O'Neill test normative data for female and male national- and international-level Australian sailing athletes.

TABLE 27.4 O'Neill 4 Min Rowing Test Data for Female and Male Australian Sailing Athletes: Mean ± *SD* (Range)

Class	Distance, m	Average pace, min:s
470 Male—Helm (*n* = 6)	1,103 ± 27 (1,069-1,134)	1:48 ± 0:02 (1:52-1:46)
470 Male—Crew (*n* = 6)	1,127 ± 30 (1,093-1,162)	1:46 ± 0:02 (1:49-1:43)
470 Female—Helm (*n* = 4)	925 ± 40 (880-952)	2:09 ± 0:05 (2:16-2:06)
470 Female—Crew (*n* = 5)	1,035 ± 29 (1,009-1,066)	1:53 ± 0:04 (1:57-1:49)
49er—Helm (*n* = 5)	1,149 ± 25 (1,110-1,172)	1:44 ± 0:02 (1:48-1:42)
49er—Crew (*n* = 7)	1,180 ± 26 (1,150-1,218)	1:41 ± 0:02 (1:44-1:39)
Laser (*n* = 7)	1,173 ± 47 (1,101-1,227)	1:42 ± 0:04 (1:49-1:38)
Laser Radial (*n* = 3)	1,047 ± 32 (1,028-1,084)	1:51 ± 0:04 (1:56-1:47)
Finn (*n* = 3)	1,175 ± 42 (1,144-1,223)	1:41 ± 0:03 (1:44-1:38)
RSX Female (*n* = 3)	1,026 ± 58 (990-1,093)	1:57 ± 0:06 (2:01-1:50)
Match Racing Female (*n* = 9)	1,039 ± 18 (1,008-1,066)	1:55 ± 0:01 (1:58-1:53)

Data source: testing period 2006-2010.

References

Abernethy, P., Wilson, G., and Logan, P. 1995. Strength and power assessment: Issues, controversies and challenges. *Sports Medicine* 19(6):401-417.

Borg, G. 1962. Physical performance and perceived exertion. *Studia Psychologia et Paedagogia* Geerup vol 11, pp 1-35.

Burke, E.J. 1977. *Toward an Understanding of Human Performance*. Ithaca, NY: Movement Publications.

Castagna, O., and Brisswalter, J. 2007. Assessment of energy demand in Laser sailing: Influences of exercise duration and performance level. *European Journal of Applied Physiology* 99(2):95-101.

Gallozzi, C., Fanton, F., De Angelis, M., and Dal Monte, A. 1993. The energetic cost of sailing. *Medical Science Research* 21:851-853.

Harris, J., and Elbourn, J. 2002. *Warming Up and Cooling Down*, 2nd ed. Champaign, IL: Human Kinetics.

Spurway, N., Legg, S., and Hale, T. 2007. Sailing physiology. *Journal of Sports Sciences* 25(10):1073-1075.

Spurway, N.C. 2007. Hiking physiology and the "quasi-isometric" concept. *Journal of Sports Sciences* 25(10):1081-1093.

Vogiatzis, I., Spurway, N.C., Wilson, J., and Boreham, C. 1995. Assessment of aerobic and anaerobic demands of dinghy sailing at different wind velocities. *Journal of Sports Medicine and Physical Fitness* 35(2):103-107.

Sprint Kayak Athletes

Nicola Bullock, Sarah M. Woolford, Peter Peeling, and Darrel L. Bonetti

Flatwater sprint kayaking has been an Olympic sport since 1936, with athletes competing over race distances of 500 m (both men and women) and 1,000 m (men only). In 2009, changes were made to the Olympic program, resulting in the introduction of the 200 m (K1 for women, K1 and K2 for men). This addition resulted in the loss of the men's K1 and K2 500 m events. The shortest competitive flatwater race for men is the K2 200 m (~31 s to complete) and the longest the K1 1,000 m (~3 min 30 s). The shortest competitive women's race is the K1 200 m (~40 s to complete) and the longest is the K1 500 m (~1 min 50 s).

The differences in the physiological demands of the 500 m and 1,000 m events are considered to be subtle (Fry and Moreton 1991; Tesch et al. 1976), clearly demonstrated by athletes successfully competing over both distances. At the 2008 Beijing Olympic Games the men's K1 1,000 m gold medalist won the bronze medal in the K1 500 m and the K1 500 m gold medalist won bronze in the K1 1,000 m. Furthermore, variations to the type of training undertaken by athletes for each of these competition distances are minimal. It is well established that the 500 m and 1,000 m events are predominately aerobic (62% and 82%, respectively) in nature (Byrnes and Kearney 1997); however, a more recent review concluded that successful kayakers over 500 m and 1,000 m require not only a high level of aerobic power but also a high anaerobic energy yield and great upper-body muscle strength (Michael et al. 2008).

In contrast to the 500 m and 1,000 m events, the shorter duration of the 200 m is expected to impose different physiological demands on the paddler, resulting in training programs with a bigger emphasis on speed, speed-endurance, and acceleration. As such, there is likely to be a shift in the focus toward enhancing the power to weight ratio of 200 m kayak athletes. Research suggests that a superior capacity of the anaerobic energy pathway may be essential for successful 200 m racing (Van Someren and Palmer 2003). However, the aerobic contribution for the 200 m is approximately 37% (Byrnes and Kearney 1997), which suggests that the aerobic energy contribution to the 200 m cannot be ignored. As a result, both the aerobic and anaerobic energy systems should be targeted in the 200 m paddler's training plan.

Regardless of the distance of the event in which an elite kayak athlete competes, it is essential that the progression of the athlete's training program be monitored via regular physiological testing throughout the annual training plan. Testing is not meant to be the only indicator of on-water performance, because many other factors are involved with performance and cannot be measured from this test. The best measure of performance is the race itself; however, this does not provide a clear insight into the physiological capacities that contribute to performance.

To this end, the physiological testing of kayakers has the following aims:

- To assess the characteristics believed to be directly related to elite kayaking performance
- To compare an athlete's current physiological capabilities with prior tests at the same time of year and identify progression and deficiencies
- To provide meaningful data that can be used to design an effective training program
- To provide data that may indicate an athlete's adaptation to a prescribed training program, which may include training-induced changes in submaximal capabilities and maximal performance parameters.

The normative data given in this chapter are from male 1,000 m paddlers and female 500 m paddlers. To date, not enough information has been collected to present normative data for 200 m paddlers.

Athlete Preparation

Prior to testing, a standardized pretest preparation is recommended to allow for reliable and valid physiological data to be obtained. Refer to chapter 2, Pretest Environment and Athlete Preparation, for specific information relating to athlete and laboratory preparation. Athletes should be injury free, well hydrated, and have consumed adequate prior nutrition.

On the day preceding testing, any training sessions that are completed should consist of no more than 12 km on the water and should be of low intensity. The athlete should undertake no heavy weight training or exercise to which they are not accustomed. Where possible the athlete should replicate (as closely as possible) similar training loads in the 24 h prior to each testing session attended.

Test Environment

The Dansprint Pro kayak ergometer (Dansprint ApS, Hvidovre, Denmark) is currently used for kayak ergometer laboratory testing in Australia; however, other ergometers can be used. Dynamic calibration of the Dansprint ergometer should occur at least every 12 months or ideally immediately before each block of athlete testing (refer to chapter 7, Determination of Maximal Oxygen Consumption).

Calibration data against a first-principles torque meter indicate that it is necessary to establish a calibration regression or correction that takes into account the total true load exerted by the paddler. The power output displayed by the Dansprint ergometer is approximately 22% higher than the output on the calibration rig, and it is likely there will be minor variations between ergometers with regard to flywheel characteristics and resistance to flywheel rotation (Tom Stanef, South Australian Sports Institute, unpublished findings 2007-2011). Therefore, all tests performed on a particular athlete during a competitive season should be conducted on the same well-maintained ergometer.

Prior to testing, the Dansprint ergometer resistance should be standardized by selecting the "user calibration" function on the ergometer digital display. The flywheel resistance should then be adjusted as per the settings outlined in table 28.1. These recommendations are based on applied anecdotal evidence from

TABLE 28.1 Ergometer Flywheel Settings for Laboratory Testing

Athlete category	Drag resistance setting
Junior female	30
Senior female	30
Junior male	35
Senior male	40

the South Australian Sports Institute. It is suggested that these settings are associated with a resistance generated on the ergometer that simulates those experienced on water.

To standardize the paddle tension, the bungee cord used to retract the rope drive system via the ergometers paddle shaft must be calibrated. It is important that the load factor of this bungee cord is standardized using a digital scale (e.g., Kern hanging scales) to a tension value of 1.5 ± 20 g (see figure 28.1). Tension adjustments are made when the drive ropes are extended at a length of 210 cm (figure 28.2), which is representative of the extension achieved in typical use. Adjustment of the tension is done by resetting the length of the bungee cord using the lock-

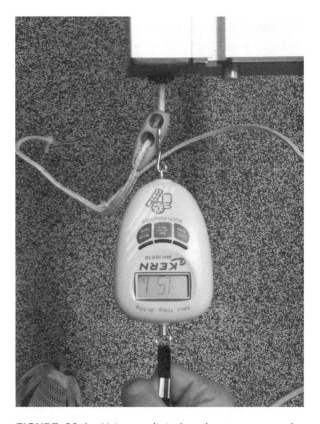

FIGURE 28.1 Using a digital scale to measure the bungee cord tension.

Photo courtesy of Nicola Bullock.

ing device and simple knots. It is recommended that the drive ropes on Dansprint ergometers be replaced after approximately 20 maximal tests to avoid wearing that may influence test results. Following the replacement of the drive ropes the ergometer should be recalibrated.

In addition to adjusting the flywheel resistance and bungee tension, the tester should ensure that the paddler's body mass on the day of testing is entered into the Dansprint ergometer digital display via the "user setup" function. The manufacturer's software analysis of the distance travelled during each testing stage is influenced by the body mass of the athlete.

The following factors are also important for the standardization of testing:

- The back of the seat base is positioned 19 cm from the end of the kayak ergometer frame to keep the calibration consistent.

- The seat height and type (swivel vs. flat), the footrest position, and the paddle shaft length are consistent between tests of the same athlete.

- The gas analyzers are turned on at least 45 min before the scheduled start of the test.

- The gas analyzers are calibrated against reference standards shortly before the commencement of testing and again at the completion of the test to account for any analyzer drift.

- The ergometer is set up no closer than 1 m to the closest wall or barrier for safety reasons.

The laboratory environment should adhere to recommendations detailed in chapter 2. It is recommended that during testing, the thermal environment of the laboratory is kept between 18 and 23 °C with a relative humidity less than 70%.

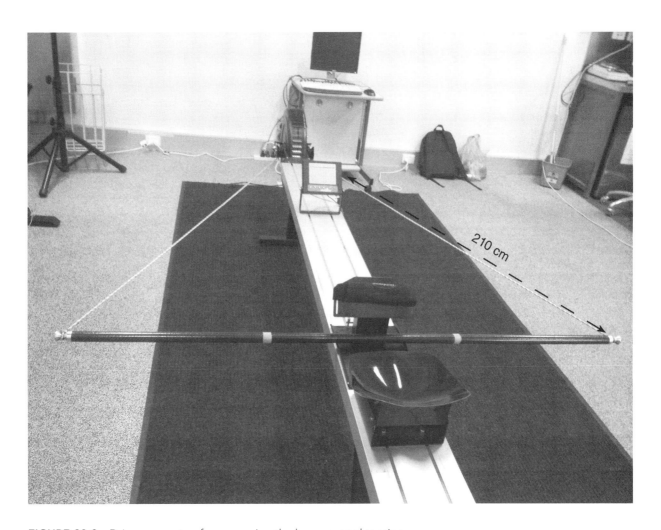

FIGURE 28.2 Drive rope setup for measuring the bungee cord tension.

Photo courtesy of Nicola Bullock.

Equipment Checklist

Anthropometry

- [ ] Stadiometer (wall mounted)
- [ ] Reference scales (accurate to ±0.05 kg)
- [ ] Anthropometry box
- [ ] Skinfold calipers (Harpenden skinfold caliper)
- [ ] Marker pen
- [ ] Anthropometric measuring tape
- [ ] Recording sheet (refer to the anthropometry data sheet template)
- [ ] Pen

7 × 4 min Incremental Test and Accumulated Oxygen Deficit Test

- [ ] Kayak ergometer
- [ ] Ergometer calibration sheets (refer to example calibration sheet in chapter 7, Determination of Maximal Oxygen Consumption)
- [ ] Heart rate monitoring system
- [ ] Metabolic cart
- [ ] Rating of perceived exertion scale (e.g., 6-20 Borg scale)
- [ ] Reference scales (accurate to ±0.05 kg)
- [ ] Lactate analyzer (and consumables as required)
- [ ] Sterile alcohol swabs
- [ ] Lancets and sharps container
- [ ] Biohazard bag
- [ ] Tissues
- [ ] Disposable gloves
- [ ] Recording sheet (refer to the sprint kayak 7 × 4 min data sheet template)
- [ ] Pen

On-Water Incremental Test

- [ ] Lactate analyzer (and consumables as required)
- [ ] Sterile alcohol swabs
- [ ] Lancets and sharps container
- [ ] Biohazard bag
- [ ] Tissues
- [ ] Disposable gloves
- [ ] Heart rate monitoring system
- [ ] Rating of perceived exertion scale (e.g., 6-20 Borg scale)
- [ ] Reference scales (accurate to ±0.05 kg)
- [ ] GPS system that measures speed or pace
- [ ] Stroke rate watch
- [ ] Recording sheet (refer to the kayak on-water incremental test data sheet)
- [ ] Pen

Recommended Test Order

Where possible, anthropometric and laboratory testing should be completed on the same day. If possible, on-water testing should be conducted within a week of the laboratory tests. The recommended test order is as follows:

Day	Tests	Notes
1	Anthropometry	Undertaken prior to kayak laboratory test
2	Laboratory testing	Following anthropometry
3	On-water testing	Optional, but doing the laboratory test and on-water test in the same week enables a cross-check of prescribed training zones.

ANTHROPOMETRY

RATIONALE

Tracking an athlete's body composition over time allows the sport scientist to understand the interaction of diet, training status, and other lifestyle factors that can affect body composition. To date, few investigations have shown a positive relationship between anthropometry and kayak performance. Although Sitkowski (2002) and van Someren and Howatson (2008) demonstrated that anthropometric parameters were not associated with 1,000 m performance, it was shown that chest circumference and humeral breadth correlated strongly with successful 500 m and 200 m performances. Additionally, Ackland and colleagues (2003) showed that Olympic-level sprint kayak athletes have a lean body composition, with a proportionally large upper-body girth and narrow hips (for males). Such characteristics are unique and are not commonly observed in the general population. From a state-level perspective, team-selected athletes have been shown to be taller and heavier than nonselected paddlers (Fry and Moreton 1991). Furthermore, bicep and forearm girths of the state team members were significantly greater than those nonselected athletes, suggesting that a high degree of muscularity may be a prerequisite for success at a higher level (Fry and Moreton 1991). Despite these confounding outcomes, it is likely that the long-term tracking of body composition may indicate the current training status of an athlete and where they are positioned relative to previous peak performances.

TEST PROCEDURE

Measurement of height, body mass, and skinfolds should be carried out prior to ergometer testing. Skin-folds are recorded over seven sites (triceps, biceps, subscapular, supraspinale, abdominal, front thigh, and medial calf). The individual skinfold measures as well as the sum of the seven sites should be reported. Refer to anthropometry protocols outlined in chapter 11, Assessment of Physique, for a detailed description of all anthropometric test procedures. More advanced anthropometric assessment including muscle girths, bone breadths, and limb lengths can be conducted if required.

Although the description of skinfold measurement procedures seems simple, a high degree of technical skill is essential for consistent results. It is therefore important that these measurements be taken by an experienced tester who has been trained in these techniques. It is also important that the same tester conduct each retest to ensure reliability.

DATA ANALYSIS

An anthropometric assessment gives a detailed overview of an athlete's body composition. Changes in body composition targeted through specific training can be tracked through repeated anthropometric measures. This allows an evaluation of the training effect on the athlete's body composition. Depending on the training goal, the focus should be on changes in lean muscle mass or skinfolds rather than purely on total body mass gains or losses.

NORMATIVE DATA

Table 28.2 presents anthropometric normative data for female and male national senior team kayak athletes.

TABLE 28.2 Anthropometric Data for Female and Male National Senior Team Kayak Athletes: Mean ± SD (Range)

Gender	Height, cm	Body mass, kg	Σ7 skinfolds, mm
Female	170.6 ± 6.2 (162.0-183.0)	67.4 ± 6.4 (56.2-76.5)	65.5 ± 13.4 (45.6-82.1)
Male	185.1 ± 4.5 (178.0-193.0)	87.3 ± 5.7 (72.0-94.5)	42.6 ± 6.7 (31.8-59.9)

Typical error: height 1.0 cm; body mass 1 kg; Σ7 skinfolds ~1.5 mm. Data source: National senior team during international competition; female n = 9; male n = 15, 2008-2009.

7 × 4 MIN INCREMENTAL TEST

RATIONALE

The 7 × 4 min incremental test is designed to provide reliable information regarding the athlete's submaximal and peak physiological capabilities.

The 7 × 4 min incremental test provides a good indication of the physiological efficacy of the on-water training program in athletes over the different mesocycles within a periodized training program. By using standardized workloads and amounts of work done prior to the final step, one can compare both efficiency and maximal performance improvements among junior, developing, and senior elite athletes during a season and over an Olympic cycle. These results then allow the coach or sport scientist to set benchmarks to ascertain the minimal standards each gender and age group should achieve.

TEST PROCEDURE

The first six stages of the 7 × 4 min incremental test are submaximal and should be performed at fixed intensities depending on the gender, age, and performance ability of the athlete. These workloads are outlined in table 28.3. Each workload is followed by a 60 s rest to allow for the collection of a blood lactate sample and the recording of average power, average stroke rate, and ratings of perceived of exertion (RPE).

Submaximal Workloads

When the athlete arrives at the laboratory, the tester should ensure that the guidelines for athlete preparation have been met and that all pretest consent forms have been signed. The protocol is designed such that the initial two workloads act as a warm-up. Therefore, no pretest warm-up needs to be conducted prior to testing, as this could influence the subsequent heart rate and blood lactate readings recorded during the early stages of the test.

Because the starting workload and subsequent work increment will vary depending on the athlete's gender, age, and performance level, it is suggested that for all testing sessions completed throughout a given competition year, the athlete's starting workload and the subsequent work increments during the first six steps be kept consistent. This will help to ensure that a similar workload is completed between testing sessions during the submaximal stages. Unless there has been a big performance change, only in the subsequent competitive season should a coach or sport scientist move the athlete up to a new category, which will influence the starting workload.

The designated workloads and step increments of the 7 × 4 min incremental test have been designed so that the sixth step produces a blood lactate value in the range of 5 to 8 mmol · L^{-1}. This range of blood lactate concentration ([La$^-$]) is often seen at training intensities of anaerobic threshold and slightly above. These values will vary depending on the time of year and the athlete's current training status. If an athlete is struggling to maintain power output (PO) in the fifth step or if [La$^-$], HR, or RPE are higher than they should be, you can stop the test after the fifth step and go straight to the maximal effort.

Once the testing equipment has been set up and calibrated, the test protocol proceeds as follows:

- Collect all necessary details for the athlete to be tested and enter them into the software associated with the testing equipment. These

TABLE 28.3 Submaximal Workloads for the 7 × 4 Min Incremental Test and Accumulated Oxygen Deficit Test

	MALE				FEMALE			
Step	World class K1 senior	Elite senior	Senior	Junior	World class K1 senior	Elite senior	Senior	Junior
1	150	125	100	100	100	80	60	45
2	175	150	125	120	120	100	80	60
3	200	175	150	140	140	120	100	75
4	225	200	175	160	160	140	120	90
5	250	225	200	180	180	160	140	105
6	275	250	225	200	200	180	160	120
7	Maximum	Maximum	Maximum	Maximum	Maximum	Maximum	Maximum	Maximum

All values are power output in watts. World class K1 senior – K1 A final at World Championship level; Elite senior – team boat paddlers making A finals or K1 paddlers making B finals at international level; Senior – athletes in national team but not making finals at the international level or athletes who have just transitioned from junior-senior level; Junior – athletes 18 years and under.

details might include such variables as height, body mass, and sum of skinfolds.

- Attach heart rate (HR) monitoring chest transmitter to the athlete and ensure that the device is synchronized to its associated data receiver.

- Ask the athlete to be seated on the kayak ergometer. If necessary, adjust the position of the ergometer footrest according to the athlete's preference and record for future reference.

- Position the metabolic cart apparatus (e.g., respiratory valve) appropriately and ensure that the athlete is comfortable. After the athlete takes several light paddle strokes, make any necessary adjustment to the respiratory hoses or other apparatus, ensuring that the paddle shaft is not impeded at any stage during the stroke.

- Program the ergometer for 4 min workloads. It is suggested that the recovery period is not programmed so that delays that may be encountered in the recovery period do not impact on the workload time.

- If the athlete is using a mouth piece, place a small piece of tape on the athlete's nose and secure the nose clip. Alternatively, a face mask may be used.

- Ensure that the athlete's feet are secured in position on the foot rest, and ask the athlete to prepare the paddle shaft in anticipation of the test commencement.

- Activate the metabolic cart for test commencement and instruct the athlete when to begin paddling the first workload of the test.

- As the athlete works through the test protocol, provide continual information from the calibration sheets produced for the specific ergometer concerning target power outputs. It is recommended the tester count the athlete into each step with a "ready, set, go" to ensure synchronization of the work effort to the acquisition of oxygen consumption data.

- Stroke ratings and heart rates should be recorded during the last 30 s of each workload. At the end of each workload record average PO, then correct to the true PO using the calibration regression equation generated from the first principles calibration of the ergometer.

- On completion of each workload, request an RPE from the athlete. An earlobe blood sample should be collected and analyzed for [La⁻] as soon as possible.

- During the rest period between each workload, permit the athlete to remove the gas collection apparatus in order to have a drink. Ensure that the breathing apparatus is back in position well before the start of the next work bout (approximately 10-15 s).

Maximal Workloads

The final (usually seventh) 4 min step of the test protocol is a maximal effort and is completed after 60 s recovery from the previous step. This stage is where the athlete is required to perform at a perceived time-trial intensity. The results of this time trial can be compared with previous maximal efforts from the same testing protocol. This final stage will also provide the sport scientist with information about an athlete's ability to perform at their maximal intensity and will differentiate performance between athletes who have completed the same submaximal steps. Several factors are important during this seventh stage:

- The athletes should be informed that they must complete as much work as possible over the 4 min period.

- Most athletes tend to overpace rather than underpace the maximal stage, and variations in pacing can have a substantial effect on the test result. Therefore, the athletes should adopt an even pacing strategy during this 4 min time-trial period. This can be aided by limiting the start phase to 10 s and then providing a guideline on what power output to hold for the first half of the test based on an athlete's previous test results. The power output attained by the athlete during the final 2 min is then optional.

- On completion of the final step, a blood lactate sample should be collected 4 min posttest.

LIMITATIONS

The limitations of the 7 × 4 min incremental test may include, but are not limited to, the following. A substantial learning effect can be present between test sessions, particularly with athletes who are not familiar with the ergometer or test protocols. Therefore, new athletes should be tested in duplicate fashion to minimize these learning effects. Most athletes will be able to technically use the ergometer at submaximal intensities; however, some athletes may experience technical difficulties during the final maximal stage.

DATA ANALYSIS

The following points should be considered when interpreting oxygen consumption and blood lactate results from the 7 × 4 min incremental test:

- Submaximal oxygen consumption should be reported as the average value for the final 1 min of each 4 min submaximal stage (L · min⁻¹ and ml · kg⁻¹ · min⁻¹).
- Maximal oxygen consumption ($\dot{V}O_2$max) is reported as the highest value attained over a period of one full minute or sum of the two highest consecutive 30 s values (L · min⁻¹ and ml · kg⁻¹ · min⁻¹).
- Submaximal heart rates are the values for the final 30 s of each submaximal workload.
- Maximal heart rate is the highest value recorded over a 5 s sampling period during the entire test. This will generally be achieved during the final stage of the maximal 4 min effort.
- Data from the maximal 4 min effort are included as the final workload values of the test and are used in combination with values from the submaximal workloads for the calculation of blood lactate thresholds and related measures (refer to chapter 6, Blood Lactate Thresholds). Guidelines for sprint kayak training zones are presented in table 28.4.
- The tester should report power, $\dot{V}O_2$, and HR data for the first and second lactate thresholds (i.e., LT1 and LT2) with the option of using the fixed points of 2, 4, and 6 mmol · L⁻¹ of lactate.
- Power to weight ratio of an athlete can be measured using the average power output from the maximal stage.

NORMATIVE DATA

Normative data for aerobic power parameters for female and male national senior team kayak athletes are presented in table 28.5. Lactate transition threshold data are presented in table 28.6.

TABLE 28.4   Guidelines for Sprint Kayak Training Zones

Training zone	Prescriptive description	Blood lactate threshold relation- ship	Blood lactate concen- tration, mmol · L⁻¹	%HRmax	%$\dot{V}O_2$max	Kayak stroke rate, strokes · min⁻¹*	RPE	Exercise time to exhaustion
T1	Light aerobic	< LT1	<2.0	60-75	<60	<60	Very light	>3 h
T2	Moderate aerobic	Lower half between LT1 and LT2	1.0-3.0	75-84	60-75	56-72	Light	1-3 h
T3	Heavy aerobic	Upper half between LT1 and LT2	2.0-4.0	82-89	75-85	70-82	Some-what hard	20 min to 1 h
T4	Threshold	LT2	3.0-6.0	88-93	85-90	78-92	Hard	12-30 min
T5	Maximal aerobic (1,000 m race pace)	> LT2	>5.0	92-100	90-100	90-110	Very hard	5-8 min
T6	500 m race pace	> LT2	>8.0	100	n/a	106-120	Very, very hard	1.5-2 min
T7	200 m race pace	> LT2	>6.0	x	n/a	115-140	Almost maximal	30-50 s
T8	Sprints	x	x	x	n/a	>130	Maximal	10-15 s

These are guidelines only for training zones. n/a = not available; x = not applicable

*Stroke rates will vary throughout the year with athlete's aerobic fitness.

TABLE 28.5 Aerobic Power Data From 7 × 4 min Incremental Test for Female and Male National Senior Team Kayak Athletes: Mean ± *SD* (Range)

Gender	Body mass, kg	$\dot{V}O_2$max, L · min⁻¹	$\dot{V}O_2$max, ml · kg⁻¹ · min⁻¹	PO at $\dot{V}O_2$max, W	PO, W · kg⁻¹	HRmax, beats · min⁻¹	Blood lactate, mmol · L⁻¹
Female	66.2 ± 3.8 (61.5-71.7)	3.69 ± 0.24 (3.50-3.96)	56.1 ± 0.9 (55.3-57.1)	205.4 ± 8.4 (193-215)	3.1 ± 0.2 (2.9-3.4)	174.2 ± 10.8 (157-186)	9.3 ± 1.6 (7.1-11.6)
Male	89.6 ± 6.1 (80.7-100.4)	5.17 ± 0.12 (5.00-5.30)	56.5 ± 3.3 (52.0-62.3)	313.6 ± 23.9 (291-357)	3.6 ± 0.3 (3.1-4.1)	180.1 ± 4.0 (175-187)	10.2 ± 2.3 (6.2-14.0)

HR = heart rate; PO = power output. Typical error: $\dot{V}O_2$max ±0.06 L · min⁻¹, ±1.0 ml · kg⁻¹ · min⁻¹; PO at $\dot{V}O_2$max ±3.0 W; HRmax ±1.7 beats · min⁻¹; blood lactate ± 0.6 mmol · L⁻¹ (*n* = 9). Data source: National senior team during domestic competition; female *n* = 5; male n = 9, 2008-2009.

TABLE 28.6 Lactate Transition Data From 7 × 4 min Incremental Test for Female and Male National Senior Team Kayak Athletes: Mean ± *SD* (Range)

Parameter	LT1	LT2
Female		
$\dot{V}O_2$, L · min⁻¹	2.46 ± 0.07 (2.38-2.51)	3.24 ± 0.18 (3.12-3.45)
%$\dot{V}O_2$max	69.3 ± 2.9 (66.0-71.0)	91.6 ± 5.5 (88.0-98.0)
PO, W	113 ± 12 (100-122)	158 ± 13 (148-174)
HR, beats · min⁻¹	143 ± 12 (132-160)	166 ± 12 (148-180)
%HRmax	82 ± 4 (76-86)	95 ± 2 (93-97)
Blood lactate, mmol · L⁻¹	1.2 ± 0.1 (1.1-1.4)	3.7 ± 0.7 (2.8-4.5)
Male		
$\dot{V}O_2$, L · min⁻¹	3.25 ± 0.23 (2.90-3.60)	4.41 ± 0.30 (4.06-4.90)
%$\dot{V}O_2$max	63.8 ± 6.6 (56.0-75.0)	86.3 ± 6.5 (75.0-94.0)
PO, W	160 ± 20 (135-200)	226 ± 30 (184-283)
HR, beats · min⁻¹	139 ± 6 (129-146)	166 ± 7 (159-178)
%HRmax	77 ± 3 (72-80)	92 ± 3 (87-96)
Blood lactate, mmol · L⁻¹	1.2 ± 0.4 (0.8-1.9)	3.7 ± 0.5 (3.3-4.6)

HR = heart rate; PO = power output. Typical error: LT2 $\dot{V}O_2$ 0.15 L · min⁻¹; % $\dot{V}O_2$max 2.2%; PO 4 W; HR 5 beats · min⁻¹; %HRmax 2.1%; Blood lactate 0.4 mmol · L⁻¹ (*n* = 9). Data source: National senior team during domestic competition; female *n* = 5; male *n* = 9, 2008-2009.

KAYAK ACCUMULATED OXYGEN DEFICIT TEST

RATIONALE

The kayak accumulated oxygen deficit test (AOD) allows the sport scientist to determine an athlete's submaximal workload (LT1 and LT2) and maximal oxygen consumption and to estimate anaerobic capacity (i.e., accumulated oxygen deficit). This testing protocol requires the athlete to paddle at progressively ramped submaximal 4 min work increments, each separated by a 1 min recovery period. The submaximal component is completed when the athlete produces a lactate 4 mmol · L^{-1} or greater. Following the completion of the submaximal work stages, the athlete is given a 20 min break before performing a maximal 4 min effort.

TEST PROCEDURE

The submaximal stages of the AOD test are the same as the 7 × 4 min incremental test. However, athletes are only required to complete as many 4 min submaximal workloads as required to elicit a blood lactate concentration of 4 mmol · L^{-1} or greater. Once this has been achieved, the submaximal component of the test is stopped to avoid fatiguing the athlete prior to the ensuing maximal 4 min maximal effort.

The 20 min Rest Interval

On completion of submaximal workloads, a 20 min active recovery period is allowed to prepare the athlete for the maximal 4 min maximal effort test. During this recovery period, the following should occur:

- The athlete should be provided the opportunity to drink and actively recover as needed. The active recovery component should incorporate at least 10 min of work at an intensity corresponding to LT1 to help clear any lactate accumulated during the submaximal component of the test. This active period can be performed using an exercise modality of the athlete's choice (e.g., cycling, rowing, jogging).

- Calibration of the gas analyzers should be checked and verified.

- A countdown to the commencement of the maximal 4 min maximal effort test should be provided regularly to the athlete.

- If AOD is to be determined, ensure the collection period for your metabolic cart is set to 30 s. Ergometer data will also need to be collected by an external computer so that average power for each 30 s of the 4 min maximal effort can be determined.

- At an elapsed recovery time of 17 min, the athlete should return to the kayak ergometer to prepare for commencement of the maximal 4 min effort.

The Maximal 4 min Effort

This final component of the AOD testing protocol is treated the same as the 7 × 4 min maximal workload. At the 18 min mark of the active recovery period, ask the athlete to be seated on the kayak ergometer. Position the gas collection apparatus appropriately and ensure that the athlete's feet are secured in position on the footrest. At the 20 min mark of the recovery period, provide a 10 s count-in for the athlete to commence the 4 min maximal effort.

DATA ANALYSIS

The blood lactate data collected from both the submaximal and 4 min maximal efforts during the AOD testing protocol can be analyzed in the same manner as that indicated in the 7 × 4 min incremental test.

A measure of anaerobic capacity as calculated by the AOD method first described by Medbo et al., (1988) can be determined from the 4 min maximal effort of the AOD test. The AOD is calculated by extrapolation from the linear relationship between submaximal work intensity (power output or meters per minute) and oxygen consumption. This relationship can be determined from data obtained during the submaximal component of the test. Once the relationship has been established, work intensities (power output) can be quantified in terms of their oxygen consumption equivalents for each 30 s period of the maximal 4 min performance test. Measured oxygen consumption is subtracted from the estimated oxygen requirement to give the oxygen deficit for each 30 s. The oxygen deficits over the entire period of the test (up to the end of the 30 s collection period preceding test completion) are then summed to obtain the AOD. Please refer to chapter 5, Anaerobic Capacity, for a comprehensive description of this method.

An alternative to the AOD full test is to collect data using the 7 × 4 min incremental test, before following up 1 or 2 days later with a maximal 4 min effort (following a standardized warm-up), as mentioned previously. If one method is chosen over another, then the tester cannot compare data between the different tests because different amounts of work prior to the AOD test will have been performed.

NORMATIVE DATA

Table 28.7 presents normative data for the AOD test.

TABLE 28.7   Accumulated Oxygen Deficit Test Results From 4 min Maximal Effort for Female and Male State Institute Athletes: Mean ± *SD* (Range)

Parameter	Female	Male
Body mass, kg	69.7 ± 3.6 (66.3-73.4)	81.6 ± 11.1 (73.5-94.3)
4 min average power, W	196 ± 4 (193-200)	296 ± 54 (263-359)
HRmax, beats · min^{-1}	175 ± 15 (157-186)	184 ± 8 (176-192)
$\dot{V}O_2$max, L · min^{-1}	3.8 ± 0.2 (3.5-3.9)	5.3 ± 0.2 (5.1-5.6)
$\dot{V}O_2$max, ml · kg^{-1} · min^{-1}	54.0 ± 6.2 (47.3-59.6)	62.3 ± 7.7 (56.8-71.1)
Anaerobic capacity, L	2.3 ± 1.0 (1.2-2.9)	2.4 ± 1.0 (1.3-3.3)
Total $\dot{V}O_2$, L	13.0 ± 0.8 (12.1-13.7)	18.9 ± 0.9 (18.2-19.9)
Aerobic potential, %	86.5 ± 0.8 (85.6-87.0)	89.2 ± 0.2 (89.0-89.4)
Aerobic contribution, %	85.4 ± 5.1 (82.2-91.3)	89.1 ± 3.7 (86.0-93.2)
Anaerobic contribution, %	14.6 ± 5.1 (8.7-17.8)	10.9 ± 3.7 (6.8-14.0)
Peak blood lactate, mmol · L^{-1}	10.1 ± 2.9 (7.1-12.9)	10.7 ± 1.3 (9.4-11.9)

Data source: State Institute athletes during domestic competition; female *n* = 3; male *n* = 3, 2008-2009.

ON-WATER INCREMENTAL TEST

RATIONALE

Following is one example of on-water testing that can be undertaken with sprint kayak athletes. By conducting this test with an emphasis on consistency, the coach and sport scientist can determine the number of steps, the velocity of each step, and the amount of recovery time or the "cycle time."

TEST PROCEDURE: 6 × 1,000 M

The on-water test entails the completion of an incremental test performed on a marked course with no flowing water. If weather permits (i.e., wind velocity is <1.0 m · s^{-1} and therefore is not affecting boat speed), a straight 1,000 m course should be used, with the athletes paddling up and back on subsequent steps.

Athletes are required to complete 6 × 1,000 m efforts of increasing speeds on a 7 min cycle (i.e., athletes start each subsequent effort on an elapsed work plus rest time of 7 min). The velocity for each 1,000 m effort is calculated as a percentage of the athlete's season's best 1,000 m time (i.e., 60%, 66%, 73%, 81%, 90%, and 100%), with the first increment being the slowest and the sixth being a maximal effort. Even pacing of these 1,000 m efforts is crucial to the success of the testing session.

Administration of the test protocol should proceed as follows:

- Record all necessary details for the athlete being tested (e.g., body mass) and environmental conditions (e.g., test location, ambient temperature, relative humidity, water temperature, water depth, water currents, wind speed and direction).

- Attach heart rate (HR) monitoring chest transmitter to the athlete and ensure that the device is synchronized to its associated data receiver.

- Ensure that a global positioning system (GPS) watch is attached to the boat so the athlete can view the elapsed time and speed.

- Line the athlete up on the start of the marked course, ensuring that he knows the pace of the ensuing 1,000 m effort.

- Ask the athlete to start the GPS watch.

- The athlete should commence the first 1,000 m effort when the GPS signals 1 min of elapsed time, and the subsequent efforts should then start on 8, 15, 22, 29, and 36 min of elapsed time.

- As the athlete works through the test protocol, provide continual information concerning speed and pace outputs.

- During each workload, record time to complete the 1,000 m, stroke ratings, and heart rate (averaged over the last 150 m).

- At the completion of each of the first five 1,000 m efforts, the athlete should paddle immediately toward the blood collector and provide an RPE; an earlobe blood sample should be collected to determine [La$^-$].

- During the remaining rest period prior to commencement of the next 1,000 m effort, the athlete should remain active at a light intensity.

Maximal Component

The sixth 1,000 m effort of the on-water test is completed on 36 min of elapsed time from the GPS watch. The final stage can provide the sport scientist with information about an athlete's ability to perform at maximal intensity and can differentiate performance between athletes who have completed the same submaximal steps. Several factors are important during the sixth effort:

- Athletes should be informed that they must complete the 1,000 m as fast as possible and without the aid of a GPS giving time or velocity feedback during the effort.

- Athletes should be encouraged to adopt an even pacing strategy during the final trial.

- On completion of the sixth step, a blood lactate sample should be collected 4 min posttest.

LIMITATIONS

One consideration of the on-water incremental kayak test is the ambient temperature in which the test is conducted. High ambient temperatures ($\sim$30 °C) have been shown to increase blood lactate concentrations (Chen et al. 2006), heart rate and RPE (Marino et. al. 2001), and lactate threshold (Flore et al. 1992) compared with similar workloads in normothermic temperatures (10-15 °C). Such findings are potentially of major importance when taking blood lactate measures in the field and should be considered when interpreting the test results. The water temperature, water depth, and direction of current will also affect boat speed and should be recorded during the testing session for future reference.

Laboratory vs. On-Water Testing

Advantage of laboratory-based tests:

- Tests are performed in a controlled, stable environment.
- Physiological data are easier to gather.
- Tests provide greater accuracy in the control of workloads.

Disadvantages of laboratory-based tests:

- Specific paddling technique may be questionable due to ergometer design.
- Tests may be affected by the ability of the ergometer to cope with high stroke rates.
- Because of equipment limitations, generally only one athlete can be tested at a time.

Advantages of field-based tests:

- Tests are extremely specific to actual athletic performance.
- Tests are perceived as more practically significant to the athlete and coach.
- Tests enable the scientific staff to assess more than one athlete at a time.

Disadvantages of field-based tests:

- Tests are effected by environmental conditions (e.g., ambient and water temperature, wind speed and direction, water surface conditions).
- Tests are limited by the equipment the athlete can use.
- Tests have a limited number of physiological parameters that can be measured and monitored

References

Ackland, T.R., Ong, K.B., and Ridge, B. 2003. Morphological characteristics of Olympic sprint canoe and kayak paddlers. *Journal of Science and Medicine in Sport* 6(3):285-294.

Byrnes, W.C., and Kearney, J.T. 1997. Aerobic and anaerobic contributions during simulated canoe/kayak events. *Medicine and Science in Sports and Exercise* 29:S220.

Chen, S., Wang, J., Lee, W., Hou, C., Chen, C., Laio, Y., Lin, C. and Kuo, C. 2006. Validity of the 3 min step test in moderate altitude: Environmental temperature as a confounder. *Applied Physiology, Nutrition, and Metabolism* 31(6):726-730.

Flore, P., Therminarias, A., Oddou-Chirpaz, M.F., and Quirion, A. 1992. Influence of moderate cold exposure on blood lactate during incremental exercise. *European Journal of Applied Physiology and Occupational Physiology* 64(3):213-217.

Fry, R.W., and Moreton, A.R. 1991. Physiological and kinanthropometric attributes of elite flatwater kayakists. *Medicine and Science in Sports and Exercise* 23(11):1297-1301.

Marino, F.E., Mbambo, Z., Kortekass, E., Wilson, G., Lambert, M.I., Noakes, T.D., and Dennis, S.C. 2001. Influence of ambient temperature on plasma ammonia and lactate accumulation during prolonged submaximal and self paced running. *European Journal of Applied Physiology* 86(1):71-78.

Medbo, J.I., Mohn, A.C., Tabata, I., Barh, R., Vaage, O., and Sejersted, O.M. 1988. Anaerobic capacity determined by maximal accumulated O2 deficit. *Journal of Applied Physiology* 64(1): 50-60

Michael, J.S., Rooney, K.B., and Smith, R. 2008. The metabolic demands of kayaking: A review. *Journal of Sports Science and Medicine* 7:1-7.

Sitkowsi, D. 2002. Some indices distinguishing Olympic or World Championship medalist in sprint kayaking. *Biology of Sport* 19(2):133-147.

Tesch, P., Piehl, K., Wilson, G., and Karlsson, J. 1976. Physiological investigations of Swedish elite canoe competitors. *Medicine and Science in Sports and Exercise* 8(4):214-218.

Van Someron, K.A., and Palmer, G.S. 2003. Prediction of 200-m sprint kayaking performance. *Canadian Journal of Applied Physiology* 28(4):505-517.

Van Someren K.A., and Howatson, G. 2008. Prediction of flatwater kayaking performance. *International Journal of Sports Physiology and Performance* 3(2):207-18.

Swimmers

Bernard Savage and David B. Pyne

Physiological testing is a feature of many high-level swimming programs. The evaluation of physiological and sport-specific performance measures provides fundamental information for the coach, athlete, and sport scientist about the athlete's response to the training program (Smith et al. 2002). Given that performance in swimming is closely related to the physiological adaptations induced by the athlete's training program (Edelmann-Nusser et al. 2002), the assessment of various components of performance can provide important information on training progress and likely competition potential. The most practical tests for elite athletes are generally those that can be easily administered in the training environment. Ease of testing is especially important in swimming, given that its physiological and movement demands cannot be easily replicated in a laboratory setting.

Progressive incremental tests are commonly used to assess the physiological adaptations of swimmers; blood lactate concentration ($[La^-]$) and heart rate (HR) are measured over a range of intensities culminating in a maximal effort (Anderson et al. 2006, 2008; Pyne et al. 2001). Incremental swimming tests can also provide feedback on performance measures such as pacing and stroke and movement characteristics at increasing speeds. Relationships between $[La^-]$ and velocity have been widely used as a means to monitor the training state of swimmers by identifying an intensity (velocity) at a fixed $[La^-]$ or lactate threshold (Hein et al. 1989). The main premise for this type of testing is that the lactate threshold is a useful measure of submaximal endurance fitness and assumed to reflect training-induced adaptations occurring in skeletal muscle (Pyne et al. 2001). Changes in the lactate threshold through a training

season have been previously documented in swimmers, but there is limited information available on international-level swimmers or on measures taken consistently within and between seasons.

Beyond the traditional pool-based incremental tests, many other applications of physiological testing hold potential benefits for elite swimmers (Pyne et al. 2001). The most common but perhaps the least studied in the published literature is routine HR and blood lactate profiling of training sets and sessions. Physiological testing has also been used during competitions for postrace lactate testing and monitoring the effectiveness of swim-downs and recoveries (Toubekis et al. 2008). In recent years there has been substantial interest into altitude training by swimmers and changes in physiological and performance measures (Robertson et al. 2010). Swimming coaches have also shown interest in enhancing training and competitive performance via use of dietary manipulations, ergogenic aids, and other training interventions such as respiratory muscle training. In these and other areas, physiological testing provides objective evidence for the suitability and effectiveness (or otherwise) of training and competition interventions.

Practical Applications of Physiological Testing

A common application of physiological testing in higher-level swimming programs is prescription and monitoring of training. Prescription of training sets should involve elements of physical load and technical characteristics of the stroking parameters. The distance–time relationship from short interval swimming can be used to develop training sets that elicit an intense physiological, mechanical, and perceptual stimulus (Ribeiro et al. 2010). Although measurements of maximal oxygen consumption are logistically challenging with most elite swimmers, they can provide useful research information on relationships between cardiorespiratory responses and various training intensity domains (Dekerle et al. 2009). Several methods have evolved to quantify training load in swimming, including HR and [La⁻] responses; subjective ratings of perceived exertion (RPE); self-reported measures of fatigue, soreness, health, and well-being; and observations from the coach or scientist (Borresen and Lambert 2009; Smith et al. 2002). However, no single physiological marker can quantify the relationship between training load and performance. The physiological demands of various swimming sets can be assessed with HR, [La⁻], and RPE methods (Borresen and Lambert 2009; Pyne and Telford 1988; Smith et al. 2002; Wallace et al. 2009). A key consideration is the relative longitudinal stability of the relationship between various power (velocity) threshold values and the associated values for HR and RPE used as practical markers of training intensity (Foster et al. 1999). Given this stability, serial monitoring of HR and [La⁻] training markers should yield useful training information. A number of training classification systems for swimming were developed and promoted in the 1980s and 1990s (Pyne and Telford 1988). Although these remain popular and are used widely in contemporary swimming, they have some inherent limitations that should be considered during routine monitoring of swim training sets and sessions (Counsilman and Counsilman 1991).

Apart from pool and laboratory testing of anthropometric, fitness, and performance characteristics of swimmers, physiological testing for elite athletes has several other applications. Physiological testing of athletes has broadened substantially in the last decade or two, and several areas have been the focus of interest for coaches, scientists, and researchers. Physiological testing is also useful in planning or modeling training (Mujika et al. 1996), quantifying short- and long-term training loads, and assessing effects of the taper phase immediately prior to major competition (Bosquet et al. 2007; Mujika et al. 2004). The effects of simulated and real altitude exposure in elite swimmers have been investigated by several groups in parallel with significant investment of time, resources, and funding by leading swimming nations (Robertson et al. 2010; Trujens et al. 2008). Physiology testing is also useful to evaluate the effects of different recovery modalities specific to swimming training and competitive performance (Parouty et al. 2010). There is a longstanding practice of profiling the blood lactate response to competitive swimming and monitoring the recovery of swimmers via the rate of blood lactate disappearance (Greenwood et al. 2008; Toubekis et al. 2008). Respiratory monitoring may be useful in tracking the effects of inspiratory muscle training on competitive or time-trial swim performance (Kilding et al. 2010). Finally, physiological testing has a role in evaluating the physiological and performance effects of various ergogenic aids such as caffeine and bicarbonate (Pruscino et al. 2008) and beta-alanine (Baguet et al. 2010).

Athlete Preparation

Prior to testing, a standardized pretest preparation is recommended to enable reliable and valid physiological data to be obtained. Refer to chapter 2, Pretest Environment and Athlete Preparation, for specific information relating to athlete and laboratory and field environment preparation.

• *Diet.* A normal high-carbohydrate–low-fat diet should be followed on the days before and the day of testing. Swimmers should be instructed to abstain from food and beverages containing caffeine or alcohol in the 2 h prior to testing. Adequate hydration with either water or a sports drink should be encouraged. This practice is particularly important when testing is conducted outdoors or indoors in conditions of high temperature and humidity.

• *Training.* No highly stressful swimming training should be undertaken within the previous 18 h. It is also recommended that no heavy weight training or exercise to which the swimmer is not accustomed be undertaken within the preceding 24 h.

• *Testing.* Swimmers should be reasonably well rested and free of illness and injury prior to testing. If there is any doubt whether the swimmer is ready to undertake testing, it may be appropriate to postpone testing to another day. Results of testing with swimmers who are not adequately prepared or motivated can be difficult to interpret.

A standardized warm-up of 1,000 to 1,500 m, consisting primarily of low- to moderate-intensity aerobic swimming and some pace work, should be completed prior to pool testing. The warm-up is normally undertaken in freestyle with some elements of pull, kick, and the main stroke used in the subsequent testing (if this is not freestyle). Some swimming at the speed of the first swim in an incremental test set is recommended. Coaches may elect to have their swimmers perform their standard prerace warm-up. A swimmer should complete the same warm-up prior to the same test from session to session.

Test Environment

Most pools have closely monitored water temperature (usually 27.0 ± 1.0 °C; mean ± *SD*) and water quality levels. These conditions should be checked with pool staff if there is any doubt over the suitability of the pool. Poor results may be obtained if the water temperature is outside comfortable limits. Pool testing can be completed either indoors or outdoors. If testing is conducted outdoors, ambient air temperature and wind may affect swimmers who are waiting to be tested. Close cooperation with coaches and swimmers is needed to ensure appropriate timing of warm-up procedures and testing. The tests described here are typically completed in a long course (50 m) pool; a short course (25 m) pool can be used, but the results between long course and short course are not directly interchangeable.

Recommended Test Order

Anthropometry should be conducted before any training or testing session. Where possible the timing of anthropometric testing should be standardized so body mass measurements are always taken with the swimmers in a fasted state (before breakfast) or after a meal (e.g., after breakfast or after lunch).

The two variations of the step test (5 ×200 m and 8 × 100 m step test) should be conducted in different weeks or as determined by the coach and scientist. The sequence of testing through a training season should be planned in advance. Some coaches prefer to alternate the 200 m and 100 m variants monthly throughout a training season or in conjunction with the training emphasis at the time of testing.

Day	Tests
Week 1 day 1, morning	Anthropometry
Week 1 day 1, afternoon	5 × 200 m
Week 1 day 2, afternoon	6 × 50 m stroke efficiency
Week 2 day 1, afternoon	8 × 100 m
By arrangement	Other physiological applications

Equipment Checklist

Anthropometry

- [ ] Stadiometer (wall mounted)
- [ ] Balance scales (accurate to ±0.05 kg)
- [ ] Anthropometry box
- [ ] Skinfold calipers (Harpenden skinfold caliper)
- [ ] Marker pen
- [ ] Anthropometric measuring tape
- [ ] Recording sheet (refer to the anthropometry data sheet template)
- [ ] Pen

Blood Sampling

- [ ] Lactate analyzer (and consumables as required)
- [ ] Sterile alcohol swabs
- [ ] Lancets and sharps container
- [ ] Biohazard bag
- [ ] Tissues
- [ ] Disposable rubber gloves
- [ ] Surgical tape

Aerobic Step Test (5 × 200 m)

- [ ] 50 m pool (long course)
- [ ] Stopwatches with stroke rate facility
- [ ] Heart rate monitoring equipment
- [ ] Clipboard
- [ ] Rating of perceived exertion (RPE) chart
- [ ] Pre–step test questionnaire
- [ ] Recording sheet (refer to the swimming 5 × 200 m step test data sheet template)
- [ ] Blood sampling equipment (as above)
- [ ] Pen

8 × 100 m Step Test

- [ ] 50 m pool (long course)
- [ ] Stopwatches with stroke rate facility
- [ ] Heart rate monitoring equipment
- [ ] Clipboard
- [ ] Rating of perceived exertion (RPE) chart
- [ ] Pre–step test questionnaire
- [ ] Recording Sheet (refer to the swimming 8 × 100 m step test data sheet template)
- [ ] Blood sampling equipment (as described previously)
- [ ] Pen

Stroke Mechanics Test (6 × 50 m)

- [ ] 50 m pool (long course)
- [ ] Stopwatch with stroke rate facility
- [ ] Tape measure for checking distance markers
- [ ] Clipboard
- [ ] Recording sheet (refer to the swimming 6 × 50 m stroke mechanics data sheet template)
- [ ] Pen

ANTHROPOMETRY

RATIONALE

The size and shape of swimmers are important features of the competitive performance in high-level swimming. Anthropometric measures are important in talent identification, development of junior or age group swimmers, and management of the training and dietary practices of senior swimmers.

TEST PROCEDURE

Measurement of height, body mass, and skinfolds should be carried out prior to swim testing. Skinfolds are recorded over seven sites (triceps, biceps, subscapular, supraspinale, abdominal, front thigh, and medial calf). The individual skinfold measures as well as the sum of the seven sites should be reported. Refer to anthropometry protocols outlined in chapter 11, Assessment of Physique, for a detailed description of all anthropometric test procedures.

DATA ANALYSIS

Anthropometric data can be used for serial monitoring of individual swimmers (within-subject) during the preseason and competitive season. Interpretation

of data for younger swimmers should account for variations in the time course of physical growth and maturation.

A comprehensive anthropometric assessment was undertaken on the 1991 World Swimming Championships (Carter and Ackland 1994). Although the data are somewhat dated, the descriptive analysis of international swimmers provides many benchmarks that are still relevant today. A more recent analysis detailed within-season and season-to-season changes in anthropometric characteristics in highly trained Australian swimmers (Pyne et al. 2006).

Interpretation of results should account for previous test results, recent swimming training history, dietary practices, and any strength and conditioning (dry land) training. Where possible, the same accredited anthropometrist should perform serial monitoring of individual swimmers.

NORMATIVE DATA

Table 29.1 presents anthropometric normative data for elite female and male swimmers.

TABLE 29.1 Anthropometric Data for Elite Female and Male Swimmers: Mean ± *SD* (Range)

Gender	Height, cm*	Body mass, kg[†]	Σ7 skinfolds, mm[†]
Female	171.5 ± 7.0	64.9 ± 9.0	67.6 ± 12.0
	(n/a)	(51.4-78.1)	(49.9-100.5)
	n = 170	n = 31	n = 31
Male	183.8 ± 7.1	82.1 ± 7.9	49.2 ± 9.0
	(n/a)	(66.1-99.6)	(32.8-86.0)
	n = 231	n = 46	n = 46

Typical error: height = 1.0 cm, mass = 1.0 kg, skinfolds = 1.5 mm. n/a = not available

*Data from Carter and Ackland 1994. [†]Data from Pyne et al. 2006.

AEROBIC STEP TEST (5 × 200 M)

RATIONALE

The aim of this test is to provide objective information on aerobic or endurance fitness capacity of the swimmer (Anderson et al. 2008; Smith et al. 2002). The protocol involves a progressive incremental test in which cardiovascular (HR), metabolic ([La⁻]), and mechanical (stroke rate and stroke count) responses to increasing speeds of swimming are measured. The premise of this test is that HR and [La⁻] responses to submaximal exercise are sensitive indicators of endur-

ance fitness and that [La⁻] responses are related to training-induced adaptations occurring within skeletal muscle (Foster et al. 1993). Step tests using 100 or 200 m intervals are more specific to the training and competitive requirements of sprint and middle-distance swimmers. The 200 m interval represents a compromise between achieving a steady state in metabolism and reaching swimming speeds that are more specific to competition levels. Longer intervals are not well tolerated by sprint swimmers.

TEST PROCEDURE

- Five evenly paced swims graded from easy to maximal on a 5 min cycle (a 6 min cycle for breaststroke swimmers is optional) are undertaken.

- Swimmers are prepared using the guidelines discussed earlier in the sections "Athlete Preparation" and "Test Environment." Swimmers should complete the pre–step test questionnaire (form 29.1) in consultation with the coach before testing.

- Target times for each swimmer are calculated before the session in consultation with the swimmer, coach, and testing staff. Times should be calculated using the guidelines shown in table 29.2.

- A 5 s differential is added to account for push start and training setting to estimate the time for the final swim (step 5) (e.g., 1:50 + 0:05 = 1:55).

- In reverse order from the fifth and final swim, 5 s are added to each subsequent interval to establish the full test protocol.

- The test should be conducted using small groups of three or four swimmers at a time (each group takes approximately 30 min). This number of swimmers requires three or four testing staff plus the coach to collect all the necessary data.

- All swims use a push start. Emphasis on even pace or even splits and the target time should be given before each swim. A common mistake is to go too fast on the first swim.

- The 100 m split time and the total 200 m time are recorded. With manual timing, the first movement is used as the starting time and the hand or foot touch for turns and the finish. The stroke rate is measured on the third lap and the stroke count on the fourth lap.

- Immediately upon completion of each swim, HR is measured by the swimmer with an HR monitor and the perceived effort rated using the Borg scale of perceived exertion chart (either the 6-20 or 1-10 scale variants).

- The swimmer exits the pool after each swim and has a capillary blood sample taken as soon as possible from the earlobe or fingertip.

- Samples need to be drawn rapidly (with the exception of the final effort) to ensure that the swimmer can adhere to the 5 min (or 6 min) cycle. Swimmers should enter the pool at least 15 s prior to the push start.

- Samples are processed with a calibrated blood lactate analyzer. Where possible, the same analyzer should be used for a given swimmer throughout the step test.

- On completion of the final swim, the blood sample is drawn 3 min postexercise. Some coaches and scientists take additional samples for up to 10 min to catch the peak postexercise lactate concentration.

- All results are recorded on the aerobic test (5 × 200 m) data sheet.

DATA ANALYSIS

A typical data set for the aerobic step test (5 × 200 m) is presented in the table 29.3. The swimming velocity is represented directly as velocity (m · s^{-1}) or indirectly as time per 100 m of swimming (s). The measure of time per 100 m is preferred by coaches.

VARIANTS

There are two common variants for this step test. The first variant, the 7 × 200 m, is merely an extension of the 5 × 200 m aerobic step test with smaller increments between swims (from low-intensity aerobic [first swim] to maximal intensity [seventh swim]). The second variant is 5 × 200 m, 2 × 100 m, and 4 × 50 m as a series in the same session. This test set uses two broken swims as the sixth and seventh efforts: the sixth effort is swum as 2 × 100 m on a 1:45 cycle (freestyle, backstroke) or a 2:15 cycle for breaststroke. The seventh and final effort is completed as 4 × 50 m on a 40 s cycle for freestyle and backstroke and a 50 s cycle for breaststroke. The idea here is to use short rest intervals to assist the swimmer in reaching race speeds in the test.

INTERPRETATION

The recent training background and health status of each individual swimmer should be considered when interpreting data: recent or current illness, recent or current injury, recent training absences, the number of weeks completed in the current training cycle, and any other information relevant to the interpretation of test performances and results. Plotting the HR–velocity and lactate–velocity relationships is a common means of analyzing and interpreting step test data. The HR–velocity and lactate–velocity curves shift *upward, leftward, or both* if fitness has deteriorated. In contrast, a shift *downward, rightward, or both* is evidence of improved aerobic fitness. A rightward shift (to a higher velocity) in lactate threshold indicates an improvement in aerobic endurance—of particular importance to middle- and long-distance swimmers. The choice of a simple line chart, an exponential curve, or a second- or third-order polynomial

FORM 29.1   Swimming Pre–Step Test Questionnaire

Name: _____

Date of Birth: _____

Date of Test: _____

Place of Test: _____

Coach: _____

Scientist: _____

☐ Short Course   or   ☐ Long Course

Are you suffering from an injury about which you have consulted a doctor or physiotherapist in the last week?

☐ Yes   ☐ No

Please provide details _____

Are you suffering from an illness or have you consulted a doctor in the last week?

☐ Yes   ☐ No

Please provide details _____

What phase are you in?

☐ Aerobic   ☐ Quality   ☐ Taper   ☐ Competition   ☐ Recovery

How many kilometers did you swim last week?

☐ <30   ☐ 31-40   ☐ 41-50   ☐ 51-60   ☐ >60

Please circle your answer the following questions, with 1 = poor, 2 = fair, 3 = average, 4 = good, and 5 = excellent.

How was your quality of sleep in the last week?	1	2	3	4	5
How was your general health in the last week?	1	2	3	4	5
How were your general energy levels in the last week?	1	2	3	4	5
How is your confidence in your swimming at the moment?	1	2	3	4	5
How good has your training been in the last week?	1	2	3	4	5
How has your overall mood been in the last week?	1	2	3	4	5

From Australian Institute of Sport, 2013, *Physiological tests for elite athletes*, 2nd ed. (Champaign, IL: Human Kinetics).

TABLE 29.2   Criteria for Establishing Target Speeds for the Aerobic Step Test (5 × 200 m)

Criterion	STEP				
	1	2	3	4	5
Approximate % of target time	80%	85%	90%	95%	100%
Approximate heart rate below maximum, beats · min⁻¹	–40	–30	–20	–10	10-0
Seconds slower than final target time	–20	–15	–10	–5	0
Example (personal best 1:50)	2:15	2:10	2:05	2:00	1:55

TABLE 29.3   Typical Results From Aerobic Step Test (5 × 200 m) for a Male Freestyle Swimmer

Result	STEP				
	1	2	3	4	5
200 m, min:s	2:21.40	2:18.10	2:11.15	2:05.13	1:59.47
1st 100 m, s	71.0	69.3	65.1	62.5	59.5
2nd 100 m, s	70.4	68.9	66.1	62.6	60.0
Mean 100 m, s	70.7	69.1	65.6	62.6	59.7
Heart rate, beats · min⁻¹	120	133	142	156	172
Blood lactate, mmol · L⁻¹	1.8	1.9	3.3	4.7	8.8
Stroke rate, strokes · min⁻¹	35.5	35.4	38.9	39.5	42.5
Stroke count, strokes/50 m	38	37	38	38	38
Rating of perceived exertion	8	11	12	15	17

A typical set of results from the 5 × 200 m step test showing a male freestyle swimmer completing the test with a 1:59.47 effort. The final swim was relatively evenly paced and associated with a heart rate of 172 beats · min⁻¹ and blood lactate of 8.8 mmol · L⁻¹. The derived 4 mmol · L⁻¹ lactate threshold speed was estimated by linear regression as 64.0 s/100 m.

plot is made at the discretion of the scientist. Additional information such as stroke rate, stroke count, stroke efficiency index, and the rating of perceived exertion (RPE) can be graphed in a similar manner.

The following questions should assist the interpretation of testing results:

- Was the final time (i.e., the seventh 200 m) substantially faster or slower than before and how does it compare to the athlete's personal best time for the 200 m?

- What were the magnitudes of increments in speed between each of the five swims?

- Were the swims completed with even or near-even splits?

- Have the HR–velocity and blood lactate–velocity curves moved and, if so, in which direction?

- What are the direction and magnitude of change in the derived velocity at the lactate threshold?

- Were changes in performance and physiological measures greater than the typical error of measurement for each of these measures (table 29.4)?

- Were any stroke measurements made (e.g., stroke rate or stroke count) and, if so, are they consistent with the physiological data?

- Was any biomechanical or subjective evaluation of the swimmer's technique undertaken?

- What was the swimmer's perception of her test performance?

The other main analytical approach is interpretation of changes in estimates of lactate threshold during serial monitoring of a swimmer within or

TABLE 29.4 Test–Retest Typical Error of Measurement Expressed as a Coefficient of Variation

Parameter	Typical error, %
Maximal (final 200 m step)	
200 m time, s	0.8
Lactate, mmol · L^{-1}	16
Heart rate, beats · min^{-1}	2.3
Submaximal (4 mmol · L^{-1} lactate)	
Time per 100 m, s	0.8
Stroke rate, strokes · min^{-1}	3.3
Heart rate, beats · min^{-1}	3.2

Adapted, by permission, from M.E. Anderson et al., 2006, "Monitoring test performance in elite swimmers," *European Journal of Sports Science* 6(3): 145-154.

between training seasons (Anderson et al. 2006). Table 29.5 shows a typical set of data for a leading male 200 m freestyle swimmer. Although the performance times for the fifth and maximal effort 200 m swim were fairly stable at approximately 1:55, there were substantial improvements in the 4 mmol · L^{-1} threshold velocity in February (prior to the summer national championships) and July (prior to the main international competition for the season). The HR at the 4 mmol · L^{-1} threshold was relatively stable at approximately 176 beats · min^{-1}.

Other factors can influence step test results, including glycogen availability and swimming efficiency. A swimmer who is glycogen depleted might show an attenuated lactate response due to a lack of available substrate (Tegtbur et al. 1993). This response will generally be accompanied by slower times during the final stages of the test and a decrease in the ratio of [La$^-$] to the swimmer's perceived level of exertion. Elevations in HR may reflect poor technique rather than poor fitness. Monitoring stroke rates, counts, and efficiency can be used to distinguish between technical and fitness (physiological) changes. However, changes in technique could be elicited by fatigue or fitness, and further investigations may be necessary.

NORMATIVE DATA

Table 29.3 presents a typical data set for the 5 × 200 m aerobic step test. In practice, all results are interpreted in a within-subject (individual athlete) framework given the wide range of strokes and distances swum. Consequently, few normative data sets are available for testing of high level swimmers. Reliability data are shown in Table 29.4.

TABLE 29.5 Typical Summary of Changes in the Lactate Threshold (at 4 mmol · L^{-1}) and Related Parameters

	FIFTH (MAXIMAL) EFFORT				ESTIMATED 4 MMOL · L^{-1}			
Date	Time, min:s	HR, beats · min^{-1}	[La$^-$], mmol · L^{-1}	SR, strokes · min^{-1}	Time, s/100 m	HR, beats · min^{-1}	SR, strokes · min^{-1}	SC, s/50 m
January	1:54.0	184	11.1	36.0	62.2	172	29.5	28
February	1:55.7	186	8.9	39.0	60.9	177	32.9	29
May	1:54.5	185	9.1	35.3	61.8	173	30.9	28
July	1:55.2	186	8.0	35.7	60.9	173	32.3	27
Best	1:54.0	184	11.1	36.0	62.2	172	29.5	28
Mean	1:55.8	62	9.0	35.9	61.7	176	30.4	28
SD	0:01.3	4	1.3	1.5	0.5	4	2.7	2

HR = heart rate; [La$^-$] = blood lactate concentration; SC = stroke count; SR = stroke rate.

8 × 100 M STEP TEST

RATIONALE

The aim of this test is to provide objective information on the fitness of 50 m and 100 m sprinters. The protocol involves a graded incremental test in which cardiovascular (HR), metabolic ([La$^-$]), and mechanical (stroke rate and stroke count) responses to increasing speeds of swimming are measured.

TEST PROCEDURE

- Swimmers are prepared using the guidelines discussed in the sections "Athlete Preparation" and "Test Environment." Swimmers should complete the pre–step test questionnaire (form 29.1) in consultation with the coach before testing.
- The protocol for the 8 × 100 m test is as follows:
  - 3 × 100 m at A1 on 1:45/2:00, 200 m swim-down
  - 2 × 100 m at A2 on 2:00/2:15, 200 m swim-down
  - 1 × 100 m at LT, 200 m swim-down
  - 1 × 100 m at MVO$_2$, 400 m swim-down
  - 1 × 100 m dive max

  where

  A1 = low-intensity aerobic work approximately 80% of personal best (PB) time.

  A2 = moderate-intensity aerobic work approximately 85% of PB time.

  LT = lactate threshold 4 mmol · L^{-1} velocity approximately 90% of PB time.

  MVO$_2$ = maximal aerobic work approximately 95% of PB time.

- All testing is done in a long course (50 m pool). Target times for each swimmer are calculated before the session and discussed with the swimmer, coaches, and testing staff on the pool deck.
- A 2 s differential is added to account for training situation to estimate the time for the final swim (step 8) (e.g., 1:02 + 0:02 = 1:04 s). In reverse order from the eighth and final swim, 4 or 5 s are added for each subsequent interval to establish the full test protocol (see previous example).
- The test is stroke-specific for freestyle, butterfly, backstroke, and breaststroke swimmers.

Some coaches have their butterfly swimmers perform this test as 50 m freestyle–50 m butterfly; in this case the split time needs to be timed as "feet leave" for the second 50 m split. Individual medley swimmers should use their weakest stroke.

- A results sheet is prepared to record all pertinent information for each individual swimmer, including recent or current illness, recent or current injury, recent training absences, the number of weeks completed in the current training cycle, and any other information relevant to the interpretation of test performances and results.
- The test should be conducted using small groups of three or four swimmers at a time (each group takes approximately 40 min). This arrangement requires three or four testing staff plus the coach to collect all the necessary data.
- All swims, with the exception of the final swim, use a push start. Emphasis on even pace and the target time should be given before each swim. A common mistake is to go too fast on the first swim.
- The 50 m split and total time for the 100 m are recorded using the 8 × 100 m aerobic test data sheet. With manual timing, the first movement is used as the starting time and the hand touch for turns and the finish. The stroke rate and stroke count are measured on the second lap.
- Immediately upon completion of each swim, HR is measured by the swimmer with a HR monitor.
- The swimmer exits the pool after each group of swims (i.e., after the first three, then after the second two, and after each subsequent swim) and has a blood sample taken as soon as possible (with the exception of the final swim) from the earlobe or fingertip.
- On completion of the final swim, the blood sample is drawn 3 min postexercise. Some coaches and scientists take additional samples for up to 10 min to quantify the peak postexercise lactate concentration.
- Samples are processed with a calibrated blood lactate analyzer.
- The swimmer then swims down in accordance with the protocol before commencing the next series of swims.

DATA ANALYSIS

Interpretation of the results from the 8 × 100 m step test broadly follow the same process as that used for the 5 × 200 m step test outlined previously. Given that the speeds here are closer to 100 m race pace, the coach and scientist should compare results of the step test with competition performances (or the race model) for the 100 m freestyle and form events. Particular focus should be given to comparing the 50 m split times and stroke mechanics of the last two 100 m swims in this step test with the details contained in the swimmer's race model. For example, in table 29.6 the splits of 28.3 and 29.4 s (57.7 s) and 26.6 and 27.3 s (53.9 s) should be compared with the swimmer's personal best 100 m race performance and his race model for the next major competition.

NORMATIVE DATA

No specific reliability data have been obtained. The reliability data for the aerobic step test (5 × 200 m) can be used as a guide.

Example times for a female 100 m backstroke swimmer with a personal best time of 1:02 are shown next:

- 3 × 100 m at 1:25 on 2:00 cycle, [La⁻], 200 m swim
- 2 × 100 m at 1:20 on 2:15 cycle, [La⁻], 200 m swim
- 1 × 100 m at 1:15, [La⁻], 200 m swim
- 1 × 100 m at 1:10 sec, [La⁻], 400 m swim
- 1 × 100 m dive max at 1:05, 3-5 min [La⁻]

TABLE 29.6 Typical Results From 8 × 100 m Step Test for a Male Freestyle Swimmer

| | STEP | | | | |
| | 3 × 100 | 2 × 100 | 1 × 100 | 1 × 100 | 1 × 100 |
Result	A1	A2	LT	MVO₂	Maximal Dive
50 m, s	35.2	33.3	31.5	28.3	26.6
100 m, s	70.3	66.9	63.5	57.7	53.9
Heart rate, beats · min⁻¹	131	140	153	158	163
Blood lactate, mmol · L⁻¹	1.1	1.0	3.3	5.8	11.4
Stroke rate, strokes · min⁻¹	27.5	28.1	28.8	40.4	45.6
Stroke count, strokes/50 m	27.0	28.0	27.0	28.0	35.0

A1 = low-intensity aerobic work; A2 = moderate-intensity aerobic work; LT = lactate threshold; MVO₂ = maximal aerobic work.

STROKE MECHANICS TEST (6 × 50 M)

RATIONALE

Swimming is a technically demanding sport, and a substantial proportion of training time is devoted to the refinement of a swimmer's technique. A series of 6 × 50 m swims of progressively increasing speed are used to establish the relationship between swimming velocity (V), stroke rate (SR), and distance per stroke (DPS). These relationships can be summarized as follows:

$$\text{Velocity (V)} = \text{Stroke Rate (SR)} \times \text{Distance Per Stroke (DPS)}.$$

$$\text{Velocity (m} \cdot \text{s}^{-1}) = \text{Stroke Rate (strokes} \cdot \text{s}^{-1}) \times \text{Distance Per Stroke (m} \cdot \text{stroke}^{-1}).$$

$$\text{Distance Per Stroke (m)} = [\text{V (m} \cdot \text{s}^{-1}) \times 60]/\text{SR (strokes} \cdot \text{min}^{-1}).$$

In practice, distance per stroke is difficult to measure in the pool without sophisticated biomechanical analysis. Simply counting strokes per lap may be inaccurate as this does not account for how much distance was travelled underwater and whether the lap finished on a complete stroke. Distance per stroke can be calculated from velocity and stroke rate by recording stroke characteristics over a known distance in the pool (Maw and Volkers 1996).

TEST PROCEDURE

- Swimmers are prepared using the guidelines discussed earlier in the sections "Athlete Preparation" and "Test Environment." Swimmers should complete the pre–step test questionnaire (form 29.1) in consultation with the coach before testing.

- The protocol for this test is 6 × 50 m swims on a 2 min cycle. All swimmers use their main stroke. Individual medley swimmers should use butterfly (their lead-off stroke). A 50 m pool is mandatory for this test.

- The target times for the test are determined as follows. The slowest swim (i.e., step 1) is undertaken approximately 10 s slower than the predicted best time on the day. The following swims are then undertaken approximately 2 s faster than the preceding swim, until the sixth and final (and maximal effort) swim is completed. In colloquial terms, the protocol can be summarized as "add 10 s to your predicted 50 m time and then descend by 2." A common mistake is for the swimmer to start too fast on the first swim.

- All swims use a push start.

- When timing is manual, the first observed movement is used as the starting time and hand touch at the 50 m as the finishing time.

- All times are recorded to a tenth of a second using the 6 × 50 m stroke mechanics data sheet template.

- Stroke rate and distance per stroke are recorded for each repeat using the following procedures.

- Stroke data (DPS) should be recorded from the 15 m to the 45 m of each 50 m swim. The time to swim this segment is taken with a stopwatch. The stroke rate is taken midpool, and the time for three complete stroke cycles is recorded. The time that the head breaks the 15 and 45 m marks is used.

- Stroke rate (strokes per minute) is measured using the base 3 stroke rate facility on a manual stopwatch. The stopwatch is started as the swimmer's hand enters the water to commence a stroke. At the completion of three complete stroke cycles, the stopwatch is stopped as the same hand enters the water for the fourth time. Alternatively, the stroke cycles can be timed and stroke rate calculated using the following equation (Maw and Volkers 1996):

$$SR = (60 \times 3) / \text{Time for 3 Strokes (in seconds)}.$$

For example, if three consecutive strokes take 4.25 s, then SR = $(60 \times 3)/4.08$ = 44.1 strokes per minute, whereas for three strokes in 3.90 s, the SR = $(60 \times 3)/3.90$ = 46.2 strokes per minute.

- For breaststroke, it is often easier to use the point where the head comes up rather than the hand entry.

- Distance per stroke is calculated using the equation DPS = $(V \times 60)/SR$ (strokes · min^{-1}) giving a results in units of meters per stroke.

- Stroke rate, distance per stroke, and stroke count (y-axis) against swimming speed (x-axis) are plotted on separate graphs.

DATA ANALYSIS

The premise of this test is to provide a qualitative analysis of stroke mechanics during a series of progressively faster swims. This information should be used in conjunction with the coach's subjective assessment of the technical quality of the stroke. Each swimmer will have a different combination of distance per stroke and stroke rate for her particular stroke.

Good technique should be maintained from the slowest to fastest swim. Better-performing swimmers are able to "hold the stroke together" at the fastest speeds. In contrast, less-skilled performers can lose control evidenced by nonlinear changes in stroke rate or distance per stroke. Inspection of the graph should indicate the speed at which control of stroke mechanics starts to deteriorate or when stroke efficiency is optimized.

NORMATIVE DATA

No normative or reliability data are available. A typical set of results from the stroke mechanics test (6 × 50 m) for a female butterfly swimmer is provided in table 29.7.

TABLE 29.7 Results From Stroke Mechanics Test (6 × 50 m) for a Female Butterfly Swimmer

Result	STEP					
	1	2	3	4	5	6
15 m, s	6.5	6.5	6.4	6.4	6.4	6.3
45 m, s	36.5	33.9	31.8	30.7	29.1	27.1
50 m, s	40.6	37.8	35.7	33.9	32.6	30.2
15-45 m, s	30.0	27.4	25.4	24.3	22.7	20.8
Velocity, m · s^{-1}	1.00	1.09	1.18	1.23	1.32	1.44
Stroke rate, strokes · min^{-1}	32.8	37.1	42.9	45.1	50.5	54.7
Stroke length, m	1.83	1.77	1.65	1.64	1.57	1.58
Stroke efficiency, index	1.83	1.94	1.95	2.03	2.08	2.28
Stroke count, strokes / 50 m	22	21	23	22	24	25

References

Anderson, M.E., Hopkins, W.G., Roberts AD, and Pyne DB. 2006. Monitoring seasonal and long-term changes in test performance in elite swimmers. *European Journal of Sports Science.* 6(3):145-154.

Anderson, M.E., Hopkins, W.G., Roberts, A.D., and Pyne, D.B. 2008. Ability of test measures to predict competitive performance in elite swimmers. *Journal of Sports Science.* 26(2):123-130.

Baguet, A., Bourgois, J., Vanhee, L., Achten, E., and Derave, W. 2010. Important role of muscle carnosine in rowing performance. *Journal of Applied Physiology.* 109(4):1096-101.].

Borresen, J., and Lambert, M.I. 2009. The quantification of training load, the training response and effect on performance. *Sports Medicine.* 39(9):779-795.

Bosquet, L., Montpetit, J., Arvisais, D., and Mujika, I. 2007. Effects of tapering on performance: A meta-analysis. *Medicine and Science in Sports and Exercise.* 39(8):1358-1365.

Carter, J.E.L., and Ackland, T.R. 1994. *Kinanthropometry in Aquatic Sports—A Study of World Class Athletes.* Champaign, IL: Human Kinetics.

Counsilman, B.E., and Counsilman, J.E. 1991. The residual effects of training. *Journal of Swimming Research.* 7(1):5-12.

Dekerle, J., Brickley, G., Alberty, M., and Pelayo, P. 2009. Characterising the slope of the distance-time relationship in swimming. *Journal of Science and Medicine in Sport.* 13(3):365-370.

Edelmann-Nusser, J., Hohmann, A., and Henneberg, B. 2002. Modeling and prediction of competitive performance in swimming upon neural networks. *European Journal of Sports Science.* 2(2):1-10.

Foster, C., Cohen, J., Donovan, K., Gastrau, P., Killian P.J., Schrager, M., and Snyder, A.C. 1993. Fixed time versus fixed distance protocols for the blood lactate profile in athletes. *International Journal of Sports Medicine.* 14(5):264-268.

Foster, C., Fitzgerald, D.J., and Spatz, P. 1999. Stability of the blood lactate—Heart rate relationship in competitive athletes. *Medicine and Science in Sports and Exercise.* 31(4):578-582.

Greenwood, J.D., Moses, G.E., Bernardino, F.M., Gaesser, G.A., and Weltman, A. 2008. Intensity of exercise recovery, blood lactate disappearance, and subsequent swimming performance. *Journal of Sports Science.* 26(1):29-34.

Hein, M.S., Kelly, J.M., and Zeballos, R.J. 1989. A fixed velocity swimming protocol for determination of individual anaerobic threshold. *Journal of Swimming Research.* 5(3):15-19.

Kilding, A.E., Brown, S., and McConnell, A.K. 2010. Inspiratory muscle training improves 100 and 200 m swimming performance. *European Journal of Applied Physiology.* 108(3):505-511.

Maw ,G. and Volkers, S. 1996. Measurement and application of stroke dynamics during training in your own pool. *Australian Swimming Coach.* 12(3):34-38.

Mujika, I., Busso, T., Lacoste, L., Borale, F., Geyssant, A., and Chatard, J-C. 1996. Modeled responses to training and taper in competitive swimmers. *Medicine and Science in Sports and Exercise.* 28(2):251-258.

Mujika, I., Padilla, S., Pyne, D., and Busso, T. 2004. Physiological changes associated with the pre-event taper in athletes. *Sports Medicine.* 34(13):891-927.

Parouty, J., Haddad, H.A., Quod, M., Lepetre, P-M., Ahmaidi, S., and Buchheit, M. 2010. Effect of cold water immersion on 100-m sprint performance in well-trained swimmers. *European Journal of Applied Physiology.* 109(3):483-490.

Pruscino, C.L., Ross, M.L., Gregory, J.R., Savage, B., and Flanagan, T.R. 2008. Effects of sodium bicarbonate, caffeine, and their combination on repeated 200-m freestyle performance. *International Journal of Sport Nutrition and Exercise Metabolism.* 18(2):116-130.

Pyne, D.B., and Telford, R.D. 1988. Classification of swimming training sessions by blood lactate and heart rate responses. *Excel.* 5(2):9-12.

Pyne, D.B., Lee, H., and, Swanwick, K.M. 2001. Monitoring the lactate threshold in world-ranked swimmers. *Medicine and Science in Sports and Exercise.* 33(2):291-297.

Pyne, D.B., Anderson, M.E., and Hopkins, W.G. 2006. Monitoring changes in lean mass of elite male and female swimmers. *International Journal of Sports Physiology and Performance.* 1:14-26.

Ribeiro, L.F., Lima, M.C., and Gobatto, C.A. 2010. Changes in physiological and stroking parameters during interval swims at the slope of the d-t relationship. *Journal of Science and Medicine in Sport.* 13(1):141-145.

Robertson, E.Y., Aughey, R.J., Anson, J.M., Hopkins, W.G., and Pyne, D.B. 2010. Effects of simulated and real altitude exposure in elite swimmers. *Journal of Strength and Conditioning Research.* 24(2):487-493.

Smith, D.J., Norris, S.R., and Hogg, J.M. 2002. Performance valuation of swimmers: Scientific tools. *Sports Medicine.* 32(9):539-554.

Tegtbur, U., Busse, M.W., and Braumann, K.M. 1993. Estimation of an individual equilibrium between lactate production and catabolism during exercise. *Medicine and Science in Sports and Exercise.* 25(5):620-627.

Toubekis, A.G., Peyrebrune, M.C., Lakomy, H.K., and Nevill, M.E. 2008. Effects of active and passive recovery on performance during repeated-sprint swimming. *Journal of Sports Science.* 26(14):1497-1505.

Trujens, M.J., Rodriguez, F.A., Townsend, N.E., Stray-Gundersen, J., Gore, C.J., and Levine, B.D. 2008. The effect of intermittent hypobaric hypoxic exposure and sea level training on submaximal economy in well-trained swimmers and runners. *Journal of Applied Physiology.* 104(2):328-337.

Wallace, L.K., Slattery, K.M., and Coutts, A.J. 2009. The ecological validity and application of the session-RPE method for quantifying training loads in swimming. *Journal of Strength and Conditioning Research.* 23(1):33-38.

Tennis Players

Machar Reid, Narelle Sibte, Sally Clark, and David Whiteside

The time–motion characteristics of tennis help determine the most relevant physiological parameters that influence tennis performance and, thus, the development of related testing protocols for the physiological assessment of the game's players. In the upper echelons of the international junior or professional game, players compete for almost 11 months of the year, rendering most conventional forms of periodization difficult. These extensive competition calendars are further complicated by uncertain weekly schedules, in which playing times and the number of matches to be played are unpredictable (Reid et al. 2009), and by the continuous travel and need to adapt to different climatic conditions and court surfaces.

The typical match lasts for between 1 and 4 h. Wimbledon 2010, however, saw two players compete for more than 11 h. Throughout, players need to repeat short-duration, explosive bursts of activity hundreds of times. Efforts to classify tennis as anaerobic or aerobic, although common (Bergeron et al. 1991), seem misplaced. The sport would appear best understood as a predominantly anaerobic activity requiring high levels of aerobic conditioning to aid recovery between points and avoid fatigue (Kovacs 2006).

Within an average point or rally, players hit three shots and cover 3 m to each ball contact (Weber 2001). Generating high ball speeds is another key determinant of elite tennis success (Pugh et al. 2003; Reid and Schnieker 2008). Players therefore need to generate and coordinate forces through varying ranges of joint motion in both the lower and upper body (Verstegen 2003). Although an absolute level of strength is required, alone it will not produce optimal stroke speed or movement. Rather, this is the corollary of player physiological characteristics and technique.

Given the open nature of the game, elite tennis players require technical expertise, tactical sense, and psychological skill (Reid et al. 2007). However, in the absence of well-developed physiological fitness, a player's potential is unlikely to be fully realized (Konig et al. 2001).

Testing may be used by coaches to monitor and assess a player's physiological fitness and the effectiveness of training interventions or programs. In an attempt to standardize test procedures in tennis, the following protocols have been documented. The relevance and specificity of the tests, in addition to the ease of administration and practical application, have been considered in the selection process.

Athlete Preparation

To ensure test reliability, the following specific pretest conditions should be observed in addition to those outlined in chapter 2, Pretest Environment and Athlete Preparation.

- *Diet.* The dietary and hydration status of the individuals should be standardized, when possible.

- *Training.* As mentioned, the competitive calendar of tennis is extensive. Efforts to plan players' tournament schedules in advance are often complicated by uncertain draws and tournament cutoffs. As a consequence, identifying key dates of the year when training load can be controlled to the extent preferred by sport scientists can be challenging. Nevertheless, when testing dates are identified, the preference is for players not to undertake heavy training in the 24 h prior to testing and, more specifically, to only undertake light exercise on the day preceding testing.

- *Testing.* Athletes should be familiar with all test procedures. Test order and time of day should be standardized. The preference is for testing to take place in the morning.

Test Environment

Environmental conditions often influence field testing, so documentation of ambient temperature, humidity, and wind is essential. Field-based testing involving locomotion should be performed on a rubberized track.

Recommended Test Order

The tests should be completed in the same order on each testing occasion to allow for reliable comparison. Where practically possible, the order should be as follows:

Day	Tests
1	Anthropometry
	20 m sprint
	5-0-5 agility
	Vertical jump
	Repeat sprint ability
	Multistage shuttle run
2	Muscular strength

Equipment Checklist

Anthropometry

- [ ] Stadiometer (wall mounted)
- [ ] Balance scales (accurate to ±0.05 kg)
- [ ] Anthropometry box
- [ ] Skinfold calipers (Harpenden skinfold caliper)
- [ ] Marker pen
- [ ] Anthropometric measuring tape
- [ ] Recording sheet (refer to the anthropometry data sheet template)
- [ ] Pen

20 m Sprint Test

- [ ] Electronic light gate equipment (e.g., dual-beam light gates)
- [ ] Measuring tape
- [ ] Field marking tape
- [ ] Witches hats
- [ ] Recording sheet (refer to the sprint test data sheet template)
- [ ] Pen

5-0-5 Agility Test

- [ ] Electronic light gate equipment (e.g., dual-beam light gates)
- [ ] Measuring tape
- [ ] Field marking tape
- [ ] Witches hats
- [ ] Recording sheet (refer to the agility test data sheet template)
- [ ] Pen

Vertical Jump Test

- [ ] Yardstick jumping device (e.g., Swift Performance Yardstick)
- [ ] Measuring tape
- [ ] Recording sheet (refer to the vertical jump data sheet template)
- [ ] Pen

10 × 20 m Repeat Sprint Ability Test

- [ ] Electronic light gate equipment (e.g., dual-beam light gates)
- [ ] Measuring tape
- [ ] Field marking tape
- [ ] Witches hats
- [ ] Chalk
- [ ] Stopwatch
- [ ] Recording sheet (refer to the repeat sprint ability test data sheet template)
- [ ] Pen

Multistage Shuttle Run Test

- [ ] Measuring tape
- [ ] Witches hats
- [ ] Sound box or CD player
- [ ] Multistage shuttle run test CD (or MP3 file)
- [ ] Stopwatch
- [ ] Recording sheet (refer to the multistage shuttle run test data sheet template)
- [ ] Pen

Muscular Strength Tests

- [ ] Bench press bench
- [ ] Squat rack or power cage
- [ ] Chin-up bar
- [ ] Barbell (Olympic 20 kg)
- [ ] Weight plates (2.5-25 kg increments)
- [ ] Recording sheet (refer to the strength and power test data sheet templates)
- [ ] Pen

ANTHROPOMETRY

RATIONALE

For players to have good capacity for thermoregulation as well as power–weight ratios that contribute positively to on-court movement, injury prevention, and racket speed, they should carry as little adipose tissue as possible. However, adipose levels should not be so low so as to compromise player health and performance.

TEST PROCEDURE

Skinfolds are recorded over seven sites (triceps, biceps, subscapular, supraspinale, abdominal, front thigh, and medial calf). The individual skinfold measures as well as the sum of the seven sites should be reported. Refer to anthropometry protocols outlined in chapter 11, Assessment of Physique, for a detailed description of all anthropometric test procedures. More advanced anthropometric assessment including muscle girths, bone breadths, and limb lengths can be conducted if required.

Although the description of skinfold measurement procedures seems simple, a high degree of technical skill is essential for consistent results. It is therefore important that these measurements be taken by an experienced tester who has been trained in these techniques. It is also important that the same tester conduct each retest to ensure reliability.

DATA ANALYSIS

An anthropometric assessment gives a detailed overview of a player's body composition. Changes in body composition targeted through specific training can be tracked through repeated anthropometric measures. This allows an evaluation of the training effect on the athlete's body composition. Depending on the training goal, the focus should be on changes in lean muscle mass or adiposity rather than purely on weight gains or losses. In tennis, the constant travel can challenge the consistency of players' diets and training schedules, so anthropometric measurement can be important. These measurements are taken four to eight times per year for older athletes, generally coinciding with the start and end of training blocks or before and after tours. Importantly, these data are captured and interpreted as part of a comprehensive physical profile.

NORMATIVE DATA

Table 30.1 presents anthropometric normative data for female and male tennis players.

TABLE 30.1 Anthropometric Data for National Academy Female and Male Tennis Players: Mean ± SD (Range)

Athlete group	Height, cm	Body mass, kg*	Σ7 skinfolds, mm[†]
Female			
16+ years	170.6 ± 6.4 (161.4-183.0)	62.7 ± 7.3 (47.1-77.0)	94.1 ± 20.0 (53.5-122.9)
15-16 years	167.5 ± 5.7 (160.0-179.0)	57.4 ± 7.6 (46.7-73.0)	105.5 ± 28.9 (73.1-151.8)
13-14 years	165.0 ± 7.0 (148.1-180.6)	54.8 ± 7.7 (43.7-73.0)	88.5 ± 23.6 (55.9-148.8)
11-12 years	155.7 ± 7.7 (146.4-170.7)	43.1 ± 6.5 (34.8-52.3)	n/a
Male			
16+ years	183.6 ± 6.7 (174.6-199.2)	76.5 ± 8.8 (62.8-97.4)	46.5 ± 9.8 (35.9-67.5)
15-16 years	180.4 ± 5.1 (174.0-189.8)	71.0 ± 5.4 (61.4-82.0)	59.5 ± 9.8 (52.1-74.0)
13-14 years	171.7 ± 6.8 (152.4-183.2)	59.4 ± 9.0 (41.1-79.4)	55.1 ± 9.6 (38.9-73.9)
11-12 years	159.8 ± 10.7 (149.0-191.3)	49.2 ± 10.4 (36.5-81.0)	n/a

Typical error: height = 1 cm; body mass = 1.5 kg; skinfolds = ~1.5 mm. n/a = not available. Data source: AIS & National academy squads (16+ years); female, n = 16*/11[†]; male, n = 33*/14[†]; 2008-2010. National academy (15-16 years); female, n = 20*/8[†]; male, n = 14a/5[†]; 2008-2010. National academy (13-14 years); female, n = 33*/17[†]; male, n = 39*/22[†]; 2008-2010. National academy (11-12 years); female, n = 13*; male, n = 16*; 2008-2010.

20 M SPRINT TEST

RATIONALE

Being able to successfully retrieve a drop shot or a lob or scramble to make a pass can be the difference between winning and losing in tennis. These are skills that challenge the qualities of linear acceleration and speed. So although the development of agility (discussed subsequently) is often considered more fundamental to the tennis player, speed is important.

The 20 m sprint test, with splits at 5 and 10 m, can be used as a measure of an athlete's linear acceleration and speed. Although the distances of 5 and 10 m are most specific to those covered in any one effort during a match, the evaluation of speed over 20 m (or intervals to 20 m) can also be informative.

TEST PROCEDURE

For tennis, the sprint test is conducted over 20 m, with intermediate distances of 5 and 10 m. Athletes should be encouraged to start the sprint with their body mass over their front foot, shoulders and hips square in a crouch "ready" position, heel up on back foot, and no rocking allowed. Refer to straight line sprint testing protocol in chapter 14, Field Testing Principles and Protocols, for a description of the procedures involved in sprint testing.

DATA ANALYSIS

In analyzing the results of individual athletes, the strength and conditioning coach often pays close attention to the comparison between the 5 and 20 m split times. The 5 m time is more specific to tennis performance and points to the player's capacity to generate and coordinate forces for horizontal or forward progression. Although related, the 20 m time seems to better complement a coach's observations regarding a player's running technique or movement efficiency.

Analyses of the results for age group and gender data are insightful. On the surface, comparison of the split times across age groups and between genders suggests that the relationship between age and sprint performance is more pronounced among male tennis players. There is a trend for the sprint times of the male players to decrease with age, but that same trend is not as readily observed among the female players. Working hypotheses for these observed differences include task constraint (i.e., straight-line running features less in the female game than in the male game) as well as physiological constraint (i.e., developmental differences in hip flexor strength are common).

NORMATIVE DATA

Table 30.2 presents 20 m sprint normative data for female and male tennis players.

TABLE 30.2 Sprint Data for National Academy Tennis Players: Mean ± SD (Range)

Athlete group	5 m time, s	10 m time, s	20 m time, s
Female			
16+ years	1.19 ± 0.05 (1.10-1.28)	2.01 ± 0.07 (1.84-2.14)	3.49 ± 0.14 (3.27-3.88)
15-16 years	1.21 ± 0.07 (1.10-1.34)	2.03 ± 0.10 (1.86- 2.23)	3.52 ± 0.16 (3.24-3.83)
13-14 years	1.20 ± 0.08 (1.06- 1.36)	2.05 ± 0.11 (1.84-2.26)	3.54 ± 0.18 (3.16-3.92)
11-12 years	1.24 ± 0.08 (1.12- 1.38)	2.11 ± 0.13 (1.93-2.93)	3.69 ± 0.25 (3.30-4.24)
Male			
16+ years	1.09 ± 0.06 (0.97-1.25)	1.81 ± 0.08 (1.67-2.03)	3.09 ± 0.10 (2.80-3.36)
15-16 years	1.12 ± 0.07 (0.91-1.24)	1.88 ± 0.10 (1.64-2.09)	3.21 ± 0.15 (2.85-3.69)
13-14 years	1.19 ± 0.08 (1.01-1.42)	2.00 ± 0.12 (1.75-2.29)	3.44 ± 0.20 (3.00-3.93)
11-12 years	1.21 ± 0.09 (1.05-1.45)	2.06 ± 0.12 (1.81-2.35)	3.57 ± 0.19 (3.14-3.98)

Typical error: 20 m time = 0.03 s. Data source: AIS & National academy squads (16+ years); female, $n = 21$; male: $n = 57$; 2008-2010. National academy (15-16 years); female, $n = 25$; male, $n = 42$; 2008-2010. National academy (13-14 years); female, $n = 53$; male, $n = 56$; 2008-2010. National academy (11-12 years); female, $n = 17$; male, $n = 22$; 2008-2010.

5-0-5 AGILITY TEST

RATIONALE

Players change direction regularly during point play, moving both laterally and forward and backward. Being agile generally affords players more time to make appropriate decisions as well as the capacity to execute shots on balance.

The 5-0-5 agility test helps coaches evaluate the qualities associated with a player's capacity to perform a single, rapid 180° change of direction. Because tennis players often traverse distances less than 15 m before changing direction, a modified 5-0-5 is also administered.

TEST PROCEDURE

Refer to the 5-0-5 agility testing protocol in chapter 14 for standard setup details and test procedure.

Modified 5-0-5 Agility Test

Instead of including an initial 10 m lead-in to the timing gates, as in the standard 5-0-5 agility test, the modified test requires athletes to begin from the position of the timing gates (5 m from the pivot line). When comfortable, athletes sprint 5 m to the pivot line, turn 180°, and sprint 5 m back to the finish line. The time it takes athletes to return to the 5 m line on which they began is the recorded value. Six repeats are performed, with three pivots on each foot. The best times from right and left feet pivots are recorded.

DATA ANALYSIS

Previous efforts to evaluate agility in tennis have attempted to simulate specific footwork actions (as in the first edition of this book). The reintroduction of the 5-0-5 agility test to the tennis testing protocols was motivated by the desire of strength and conditioning coaches to delimit the number of interacting physiological and mechanical influences within the one test and therefore to simplify the interpretation of agility test data.

Performance of the 5-0-5 test in its standard form as well as in its modified version has been active in Australia since July 2009. Younger athletes (≤15 years old) perform only the modified 5-0-5 (which is thought to present less of a strength and coordination challenge to players), whereas older athletes complete both versions of the test. Data exploring the strength of the relationship between the two tests (and potential test redundancy) are expected to become available.

As with the 20 m sprint age group and gender comparison, the 5-0-5 times suggest that the effect of age (or maturation) is more pronounced among the male players.

At the individual level, comparison of 5-0-5 times between right and left sides is generally interpreted alongside the one-leg vertical jump data to highlight any differences in movement laterality. As with the one-leg vertical jump, the capability to generate unilateral lower-limb force (in this case, changing horizontal direction of both feet) appears amenable to learning and thus potentially affects the early test results of any one individual.

NORMATIVE DATA

Normative data for the standard 5-0-5 agility test and the tennis modified 5-0-5 agility test for female and male tennis players are presented in table 30.3 and table 30.4, respectively.

TABLE 30.3 Standard 5-0-5 Agility Test Data for National Academy Female and Male Tennis Players: Mean ± *SD* (Range)

Athlete group	5-0-5 time right, s	5-0-5 time left, s	5-0-5 time difference, s
Female 16+ years	2.46 ± 0.07	2.48 ± 0.10	0.05 ± 0.04
	(2.34-2.58)	(2.29-2.65)	(0.01-0.14)
Male 16+ years	2.30 ± 0.13	2.30 ± 0.13	0.05 ± 0.04
	(2.08-2.65)	(2.08-2.63)	(0.00-0.16)

Data source: AIS & National academy squads (16+ years); female, *n* = 15; male, *n* = 55; 2008-2011.

TABLE 30.4 Modified 5-0-5 Agility Test Data for National Academy Female and Male Tennis Players: Mean ± *SD* (Range)

Athlete group	5-0-5 time right, s	5-0-5 time left, s	5-0-5 time difference, s
Female			
16+ years	2.80 ± 0.10 (2.54-3.00)	2.81 ± 0.10 (2.59-3.01)	0.05 ± 0.06 (0.00-0.32)
15-16 years	2.82 ± 0.11 (2.56-3.11)	2.82 ± 0.11 (2.61-3.13)	0.06 ± 0.07 (0.01-0.32)
13-14 years	2.84 ± 0.12 (2.60-3.25)	2.85 ± 0.14 (2.63-3.39)	0.06 ± 0.05 (0.00-0.29)
11-12 years	2.86 ± 0.13 (2.64-3.18)	2.85 ± 0.13 (2.64-3.23)	0.06 ± 0.05 (0.00-0.13)
Male			
16+ years	2.53 ± 0.09 (2.33-2.77)	2.54 ± 0.10 (2.35-2.77)	0.06 ± 0.05 (0.00-0.23)
15-16 years	2.59 ± 0.15 (2.18-2.90)	2.60 ± 0.15 (2.20-2.99)	0.05 ± 0.04 (0.00-0.25)
13-14 years	2.73 ± 0.18 (2.33-3.21)	2.74 ± 0.17 (2.38-3.09)	0.06 ± 0.05 (0.00-0.22)
11-12 years	2.82 ± 0.11 (2.56-3.03)	2.83 ± 0.10 (2.66-3.02)	0.06 ± 0.06 (0.00-0.23)

Data source: AIS & National academy squads (16+ years); female, *n* = 46; male, *n* = 60; 2008-2011. National academy (15-16 years); female, *n* = 41; male, *n* = 60; 2008-2011. National academy (13-14 years); female, *n* = 53; male, *n* = 65; 2008-2011. National academy (11-12 years); female, *n* = 23; male, *n* = 28; 2008-2011.

VERTICAL JUMP TEST

RATIONALE

Explosive unilateral and bilateral lower limb movements are central to successful stroke and movement production (Reid and Schneiker 2008). Although tests like the 20 m sprint inform coaches as to a player's capacity to develop force rapidly in a predominantly horizontal direction, the assessment of sport-specific one- and two-leg jumping ability provides insight into a player's capacity to develop force vertically.

The vertical jump test is an established measure of explosive and anaerobic power of the lower limbs and hips that is easy to perform, requires limited equipment, and is common to many power-related sports, allowing easy comparison.

TEST PROCEDURE

Refer to vertical jump protocol in chapter 14 for setup details and test procedures.

Two-Leg Vertical Jump

The two-leg vertical jump test is performed as a standard vertical jump (i.e., two feet, shoulder-width apart).

One-Leg Vertical Jump

The one-leg vertical jump test is also performed as per the standard vertical jump protocol; however, the takeoff must be from one foot with no preliminary steps or shuffling.

DATA ANALYSIS

While past tennis testing batteries have assessed both countermovement jump (CMJ) vertical jump performance, the current protocols focus their attention on the latter. The shortcomings of this test are well noted (Luhtanen and Komi 1978; Narita and Anderson 1992; Young 1994), yet comparison of previously collected data with the CMJ suggested test redundancy, and the sport-specificity of the vertical jump test was also preferred.

As with the speed and agility measures, analysis of the group results for jumping ability points to an increase in vertical jump performance with age. Interestingly, in players between 13 and 16 years of age, the magnitude of increases appears more pronounced, which is consistent with links of peak adolescent growth to improvements in explosive leg strength as measured by vertical jump ability (Philippaerts et al. 2006). Critique of the normative data for one-leg jumping infers minimal side-to-side differences and can be misleading. In this regard, for the individual athlete, interpretation and measurement of jump height difference are key. Developing male players appear to be characterized by slightly higher side to side differences in jumping ability than female players.

NORMATIVE DATA

Vertical jump normative data for female and male tennis players are presented in table 30.5.

TABLE 30.5 Vertical Jump Data for National Academy Female and Male Tennis Players: Mean ± *SD* (Range)

Athlete group	CMJ jump height, cm*	Single leg jump height right, cm[†]	Single leg jump height left, cm[†]	R/L jump height difference, cm[†]
Female				
16+ years	40 ± 6 (32-52)	30 ± 3 (24-38)	29 ± 4 (24-40)	2 ± 1 (0-5)
15-16 years	39 ± 6 (31-53)	30 ± 5 (23-40)	30 ± 5 (20-41)	2 ± 3 (0-12)
13-14 years	37 ± 5 (27-49)	27 ± 5 (16-38)	28 ± 5 (19-38)	2 ± 2 (0-8)
11-12 years	36 ± 6 (28-49)	26 ± 5 (18-36)	25 ± 5 (16-35)	2 ± 2 (0-7)
Male				
16+ years	54 ± 6 (40-68)	38 ± 5 (25-49)	38 ± 7 (22-54)	4 ± 3 (0-10)
15-16 years	51 ± 7 (40-72)	37 ± 6 (28-51)	38 ± 7 (27-52)	3 ± 2 (0-10)
13-14 years	43 ± 7 (31-63)	30 ± 6 (18-47)	30 ± 6 (19-47)	3 ± 2 (0-12)
11-12 years	37 ± 6 (29-50)	26 ± 4 (21-35)	25 ± 4 (17-34)	3 ± 3 (0-14)

Typical error: jump height = 1.5 cm. Data source: AIS & National academy squads (16+ years); female, *n* = 22*/18[†]; male, *n* = 56*/54[†]; 2008-2010. National academy (15-16 years); female, *n* = 28*/21[†]; male, *n* = 46*/37[†]; 2008-2010. National academy (13-14 years); female, *n* = 52*/44[†]; male, *n* = 62*/58[†]; 2008-2010. National academy (11-12 years); female, *n* = 23*/18[†]; male, *n* = 26*/19[†]; 2008-2010.

10 × 20 M REPEAT SPRINT ABILITY TEST

RATIONALE

The inclusion of a repeat sprint ability (RSA) test, which assesses a player's capacity to sustain high-intensity movement efforts with limited rest, fits with specific aspects of the sport's time–motion profile. The interval distance of 20 m was selected for practical reasons, and the prescription of 10 repetitions is consistent with published recommendations (Chaouachi et al. 2010). Similar tests have been shown to assess a different physiological parameter than the multistage shuttle run test (Castagna et al. 2007).

TEST PROCEDURE

- The RSA test requires the athletes to perform 10 maximal 20 m sprints departing every 20 s for both females and males.

- The tester measures 20 m along the court or track, clearly setting a start and finish mark and placing the electronic light gates at these marks.

- The electronic light gate equipment must be correctly set up prior to the commencement of the test. The tester should select lap mode and enter 10 splits. It is probably easier to record the trial number from the timing unit and recall the data after the testing session than to write down data during the test.

- Pretest preparation should include an active warm-up of jogging to striding intensity for 5 min in addition to approximately 2 sprints of 15 to 20 m. A recovery period of approximately 5 min should be given before commencing the test.

- The start position for each sprint is the same as for the 20 m sprint test. Each sprint must be performed maximally.

- A 5 s warning is given, and the athlete assumes the ready position and awaits the "go" command to commence the next sprint (i.e., tester calls "3, 2, 1, go").

- The tester starts a stopwatch in line with the commencement of the first sprint or split. Players start each subsequent sprint every 20 s (i.e., 20 s, 40 s, 60 s). They need to resume their start position in preparation for the tester's countdown and "go" command. Players do not need to return to the start line of the first sprint after each sprint; rather every subsequent sprint starts where the preceding sprint finished (i.e., at the 20 m mark).

- Once the final sprint has been completed, the tester presses the "recall" button to retrieve the data. The individual sprint or spilt times should be displayed as well as the cumulative time for the whole test.

- The test produces two scores, the absolute (total time in seconds) and relative (% decrement) scores. The calculation for percentage decrement is explained subsequently.

$$\% \; Decrement = 100 - \left(100 \times \frac{BST \times n}{TST} \right)$$

where BST is the best 20 m sprint time, TST is the total sprint time, and n is the number of repetitions (i.e., 10 in this protocol).

- The total sprint time has been shown to be very reliable, with a typical error (TE) of 0.7%. However, the percentage decrement was less reliable, with a TE of 14.9% (Spencer et al. 2004). Therefore, the major measure of performance should be the total sprint time.

DATA ANALYSIS

The effect of age on the total time of the repeat sprint ability (RSA) test is evident in our male but not female tennis playing cohort. In other words, RSA total time decreased with age among male players, which would appear to indicate improvement in certain anaerobic qualities. Interestingly, independent of gender, the percentage decrement seems to plateau or remain constant throughout the average player's development from 11 years of age. As reported elsewhere in the literature, it could be that the meaningfulness of this measure is constrained by the player's knowledge of the number of sprints and therefore their potential to use a pacing strategy.

NORMATIVE DATA

Table 30.6 presents repeat sprint ability test normative data for female and male tennis players.

TABLE 30.6 Repeat Sprint Ability Data for National Academy Female and Male Tennis Players: Mean ± *SD* (Range)

Athlete group	Total time, s	% Decrement
Female		
16+ years	38.5 ± 2.4	4.0 ± 1.7
	(35.0-42.2)	(1.3-6.1)
15-16 years	38.2 ± 2.4	5.6 ± 2.0
	(35.1-42.3)	(2.3-8.1)
13-14 years	37.9 ± 2	4.1 ± 1.9
	(33.7-41.3)	(1.1-8.5)
11-12 years	39.6 ± 2.6	4.4 ± 2.2
	(35.3-44.7)	(1.0-9.5)
Male		
16+ years	32.9 ± 1.2	3.5 ± 1.3
	(31.0-35.1)	(1.5-6.4)
15-16 years	34.4 ± 1.7	3.3 ± 1.5
	(32.2-38.4)	(0.9-6.3)
13-14 years	36.5 ± 2.6	3.8 ± 1.6
	(32.1-45.5)	(1.4-8.0)
11-12 years	38.9 ± 1.6	4.0 ± 1.4
	(35.3-41.4)	(1.9-6.3)

Data source: AIS & National academy squads (16+ years); female, $n = 6$; male, $n = 15$; 2008-2010. National academy (15-16 years); female, $n = 13$; male, $n = 22$; 2008-2010. National academy (13-14 years); female, $n = 25$; male, $n = 34$; 2008-2010. National academy (11-12 years); female, $n = 14$; male, $n = 15$; 2008-2010.

MULTISTAGE SHUTTLE RUN TEST

RATIONALE

The multistage shuttle run test (MSRT) has been selected as the test of aerobic power for tennis as it has been found to be a sufficiently accurate estimate of aerobic power (Leger and Lambert 1982; Ramsbottom et al. 1988); is easily administered both in the laboratory and on the road; is inexpensive; and is similar to tennis in respect to the stop, start, and change of direction movement patterns.

TEST PROCEDURE

Refer to the multistage shuttle run test protocol in chapter 14 for detailed description of this test.

DATA ANALYSIS

Perhaps this test more than any of the others mentioned is affected by player motivation, technique, and environmental conditions. Among adolescent males, a notable increase in test performance and, by inference, aerobic capacity is observed between the ages of 13 to 16. Similar trends have punctuated the MSRT performance of other adolescent male populations (Philippaerts et al. 2006). Of some surprise is that a like increase in the MSRT performance of the female tennis cohort was not evident until players (as a group) reached their 16th year. The open nature of tennis coupled with the different game styles and court surfaces would appear to underpin the observed variety in the baseline level of aerobic fitness for each individual. This variety in MSRT performance is arguably more pronounced in tennis than in many other sports and can present a challenge to the implementation of minimum MSRT standards.

NORMATIVE DATA

Table 30.7 presents multistage shuttle run test normative data for female and male tennis players.

TABLE 30.7 Multistage Shuttle Run Test Data for National Academy Female and Male Tennis Players: Mean ± *SD* (Range)

Athlete group	Level/shuttle
Female	
16+ years	11/2 ± 1/2
	(8/9-12/8)
15-16 years	10/1 ± 1/5
	(7/2-12/8)
13-14 years	10/2 ± 1/4
	(6/7-13/2)
11-12 years	10/1 ± 1/4
	(6/2-12/2)
Male	
16+ years	13/4 ± 1/0
	(11/1-15/4)
15-16 years	13/2 ± 1/1
	(8/9-15/5)
13-14 years	11/9 ± 1/1
	(9/1-14/5)
11-12 years	10/6 ± 1/8
	(6/9-13/9)

Typical error: shuttle = 4. Data source: AIS & National academy squads (16+ years); female, $n = 18$; male, $n = 55$; 2008-2010. National academy (15-16 years); female, $n = 26$; male, $n = 41$; 2008-2010. National academy (13-14 years); female, $n = 49$; male, $n = 55$; 2008-2010. National academy (11-12 years); female, $n = 25$; male, $n = 30$; 2008-2010.

STRENGTH TESTING

RATIONALE

Modern tennis is a showcase of strong and powerful athletes plying their trade over the course of three to five sets. The goals of any physical preparation program for a tennis player focus on two main areas: injury prevention and performance enhancement. Strength and power development is aimed at improving the explosive power of the upper body and trunk (in various planes of motion) to aid the development of racket speed, postural strength to enhance movement economy, and eccentric leg strength to assist speed and agility around the court.

A large proportion of injuries to tennis players are due to insufficient eccentric strength both in the upper body during the deceleration of the racket after serves, ground strokes, and volleys and in the lower body during the deceleration of the body to establish a stable base for stroke production (Kovacs et al. 2008).

Although field-based measures (e.g., push-up or modified pull-up) are appropriate for testing a large group of children (e.g., a physical education class), the use of RM strength testing procedures may provide more useful information for strength and conditioning coaches who need to assess strength performance in trained youth. Current research findings indicate that the maximal force-producing capabilities of healthy children and adolescents can be safely evaluated by 1-3RM testing procedures, provided that (a) youth participate in a habituation period before testing to learn proper exercise technique and (b) qualified strength and conditioning coaches closely supervise and administer each test. No injuries have been reported in prospective studies that used adequate warm-up periods, appropriate progression of loads, close and qualified supervision, and critically chosen maximal strength tests (1RM performance lifts, maximal isometric tests, and maximal isokinetic tests) to evaluate resistance training–induced changes in children (Faigenbaum et al. 2009).

Tennis Australia currently performs 3RM testing on athletes entering its Pro Tour Program, usually at 16 years of age. Players aged approximately 15+ in National academies and with a resistance training age of at least 12 months can be strength tested if the qualified coach believes that they are technically competent to perform the exercises correctly and that the information obtained will be useful in assisting ongoing program development.

TEST PROCEDURE

The strength tests used in tennis include 3RM squat, bench press, bench pull, and dead lift. Refer to chapter 14, Strength and Power Assessment Protocols, for a detailed description of the squat, bench press, and bench pull strength tests. Test procedures for the dead lift are detailed below. Warm-up and submaximal attempts of the squat and bench press should be performed in an appropriate rack with spotters present.

Dead Lift

- The use of a weight belt is optional but should be consistent between tests.

- Recommended assessor position is side on to the athlete to facilitate observation of posture.

- In the start position, feet should be placed under the bar so as to have the bar directly above the balls of the feet.

- Grip should be slightly wider than shoulders. The grip can be of a reverse nature, meaning one hand can be over-grip and the other under-grip.

- Shoulders should be forward of the bar, and the back should be held in a straight or concave set position.

- Initial movement of the bar should be generated from the legs by knee extension, with no change to the back angle at the pelvis.

- During the movement little or no change in back position should be noted until the final extension of the hips occurs, bringing the back into line with the rest of the fully upright body.

- The back must not bend or show excessive flexion during the lift.

- In the finish position, the body should be upright with fully extended arms.

- Repetitions are to be completed as soon as possible after each other; a single rest of no more than 5 seconds is allowed to re-position hands or feet.

- The bar may be dropped to the floor between repetitions.

- A valid repetition is one in which the bar is lifted in one continuous motion to a point at which the body reaches a fully upright position with the knee and hip joints extended to a neutral position (i.e., the back in line with the rest of the fully upright body).

Technical Violations

If any of the following technical violations occur, the trial will be considered invalid and the athlete will perform a second trial at the same weight:

- Failing to have the correct set-up, and lifting techniques

- Failing to lift the bar in one continuous motion

- Collapsing the back position during the lift

- Having greater than 5 seconds rest between repetitions

DATA ANALYSIS

Data have been collected and reported on athletes aged at least 16 years of age. Given the relatively recent addition of strength testing to Tennis Australia's physiological testing protocols, the data are limited, particularly in the female game. Likewise, although all players undertaking strength testing have demonstrated technical competence, bias introduced by the effect of learning cannot be discounted. This point would appear to find support in the relatively large standard deviations presented in table 30.8. The upper body 3RM measures indicate that players are comparably strong in both the push and the pull. These data are similar to the ratios commonly promoted as desirable for safe shoulder function using other upper-body strength tests (Ellenbecker and Roetert 2003).

NORMATIVE DATA

Table 30.8 presents strength testing normative data for female and male tennis players.

TABLE 30.8 Strength Testing Normative Data for Female and Male Tennis Players: Mean ± SD (Range)

Athlete group	3RM squat* (relative to mass)	3RM push[†] (relative to mass)	3RM pull[‡] (relative to mass)	3RM dead lift[#] (relative l to mass)
Female 16+ years	1.47 ± 0.28 (1.12-1.78)	0.69 ± 0.11 (0.54-0.84)	0.72 ± 0.07 (0.61-0.85)	1.06 ± 0.36 (0.71-1.43)
Male 16+ years	1.38 ± 0.26 (0.77-1.95)	0.87 ± 0.14 (0.66-1.28)	0.87 ± 0.09 (0.70-1.11)	1.48 ± 0.26 (1.11-2.13)

Data source: AIS & National academy squads (16+ years); female, n = 5*/7[†]/9[‡]/3[#]; male, n = 29*/24[†]/34[‡]/17[#]; 2008-2010.

References

Bergeron, M.F., Maresh, C.M., Kraemer, W.J., Abraham, A., Conroy, B., and Gabaree, C. 1991. Tennis: A physiological profile during match play. *International Journal of Sports Medicine* 12(5):474-479.

Castagna, C., Manzi, V., D'Ottavio, S., Annino, G., Padua, E., and Bishop, D. 2007. Relation between maximal aerobic power and the ability to repeat sprints in young basketball players. *Journal of Strength and Conditioning Research* 21(4):1172-1176.

Chaouachi, A., Manzi, V., del Wong, P., Chaalali, A., Laurencelle, L., Chamari, K., and Castagna C. 2010. Intermittent endurance and repeated sprint ability in soccer players. *Journal of Strength and Conditioning Research* 24(10):2663-2669.

Ellenbecker, T., and Roetert, E.P. 2003. Age specific isokinetic glenohumeral internal and external rotation strength in elite junior tennis players. *Journal of Science and Medicine in Sport* 6(1):63-70.

Faigenbaum, A., Kraemer, W., Blimkie, C., Jeffreys, I., Micheli, L., Nitka, M., and Rowland, T. 2009. Youth resistance training: Updated position statement from the National Strength and Conditioning Association. *Journal of Strength and Conditioning Research* 23(5):S60-S79.

Konig, D., Huonker, M., Schmid, A., Halle, M., Berg, A., and Keul, J. 2001. Cardiovascular, metabolic, and hormonal parameters in professional tennis players. *Medicine and Science in Sports and Exercise* 33(4):654-658.

Kovacs, M. 2006. Applied physiology of tennis performance. *British Journal of Sports Medicine* 40, 381-386.

Kovacs, M., Roetert, P., and Ellenbecker, T. 2008. Fixing the brakes. Deceleration: The forgotten factor in tennis specific training. *ITF Coaching and Sports Science Review* 15(46):6-8.

Leger, L.A. and Lambert, J. 1982. A maximal multistage 20 m shuttle run test to predict VO$_{2max}$. *European Journal of Applied Physiology* 49(1):1-5.

Luhtanen, P., Komi, P.V., 1978. Segmental contribution to forces in vertical jump. *European Journal of Applied Physiology* 38:181–188.

Narita, S., & Anderson, T. 1992. Effects of upper body strength training on vertical jumping ability of high school volleyball players (abstract). *Sports Medicine Training and Rehabilitation,* 3:34.

Philippaerts, R,M, Vaeyens, R., Janssens, M., Van Renterghem, B., Matthys, D., Craen, R., Bourgois, J., Vrijens, J., Beunen, G., and Malina, R.M. 2006. The relationship between peak height velocity and physical performance in youth soccer players. *Journal of Sports Sciences* 24(3):221-230.

Pugh, S.F., Kovaleski, J.E., Heitman, R.J., and Gilley, W.F. 2003. Upper and lower body strength in relation to ball speed during a serve by male collegiate tennis players. *Perceptual and Motor Skills* 97(3, Pt 1):867-872.

Ramsbottom, R., Brewer, J., and Williams, C. 1988. A progressive shuttle run test to estimate maximal oxygen uptake. *British Journal of Sports Medicine* 22:141-5.

Reid, M., Crespo, M., Lay, B., and Berry, J. 2007. Skill acquisition in tennis: Current research and practice. *Journal of Science and Medicine in Sport* 10(1):1-10.

Reid, M., and Schneiker, K. 2008. Strength and conditioning in tennis: Current research and practice. *Journal of Science and Medicine in Sport* 11(3):248-256.

Reid, M., Quinlan, G., Kearney, S., and Jones, D. 2009. Planning and periodisation for the elite junior tennis player. *Strength and Conditioning Journal* 31(4):69-76.

Spencer, M., Lawrence, S., Rechichi, C., Bishop, D., Dawson, B., and Goodman, C. 2004. Time-motion analysis of elite field hockey, with special reference to repeat sprint activity. *Journal of Sports Sciences,* 22(9):843-850.

Verstegen, M. 2003. Developing strength. In: M. Reid, A. Quinn, A., and M. Crespo (eds), *ITF Strength and Conditioning for Tennis.* London: ITF Ltd, pp 113-135.

Weber, K. 2001. Demand profile and training of running speed in elite tennis. In: M. Crespo, M. Reid, and D. Miley (eds), *Applied Sport Science for High Performance Tennis.* London: ITF Ltd: London, pp 41-48.

Young, W. 1994. Specificity of jumping ability and implications for training and testing athletes. *Proceedings of the National Coaching Conference, Canberra Australian Coaching Council.* 217-221.

Triathletes

Shaun D'Auria, Nicola Bullock, Katie Slattery,
Danielle Stefano, and Sally A. Clark

The physiological testing of triathletes has the aim of objectively assessing variables believed to be directly related to elite performance, establishing training intensities, and indicating whether athletes are adapting to training loads.

The battery of tests includes anthropometry, performance, and physiologically based tests for swimming, cycling, and running. The contents of this chapter have been heavily influenced by the testing protocols prepared for the individual sports of cycling and middle-distance running in an attempt to promote uniform sport science testing of athletes throughout Australia. The chapter's contents are also influenced by the coaching philosophies and practices of Australia's elite coaches. One such philosophy is that for success on the world stage, an Australian triathlete should have physiological capabilities similar to those of Australian national team swimmers, cyclists, and middle-distance runners.

Establishing Guidelines for Quantifying Training and Racing

In triathlon, as with most other endurance-based sports, there is a need to describe exercise intensity when prescribing training and analyzing race performance. Triathlon, in keeping with other endurance-based sports in Australia, has adopted standardized nomenclature for training zones. These zones are based on discernible points on the blood lactate concentration–workload curve, namely, the first and second lactate thresholds. The first lactate threshold (LT1) is the first intensity at which a sustained increase in blood lactate concentration above resting is discernible, and the second lactate threshold (LT2) is the intensity indicating the upper limit of equilibrium between lactate production and clearance (refer to chapter 6, Blood Lactate Thresholds: Concepts and Application). Triathlon also uses a supplementary, more performance-based framework for analyzing and describing cycle race data and training (see the section describing the cycle power profile test).

Athlete Preparation and Test Environment

Standardized pretest preparation is recommended to enable reliable and valid physiological data to be obtained. The key principal is that athletes present for testing in as close to a standard state of preparedness as possible (i.e., injury free, rested, hydrated, and with adequate carbohydrate stores). From a training perspective this requires an adjustment to training load, so that a minimum of 24 h of easy to moderate training precedes any test session. Refer to chapter 2, Pretest Environment and Athlete Preparation, for specific information relating to athlete preparation and test environment.

Equipment Checklist

Anthropometry

- [ ] Stadiometer (wall mounted)
- [ ] Balance scales (accurate to ±0.05 kg)
- [ ] Skinfold calipers (Harpenden skinfold caliper)
- [ ] Marker pen
- [ ] Anthropometric measuring tape
- [ ] Recording sheet (refer to the anthropometry data sheet template)
- [ ] Pen

1 km Swim Time Trial

- [ ] 50 m swimming pool
- [ ] Stopwatch (with rate function)
- [ ] Heart rate monitoring equipment (optional)
- [ ] Video camera (optional)
- [ ] Recording sheet (refer to the triathlon 1 km swim time trial data sheet template)
- [ ] Pen

Cycle Power Profile Test

- [ ] Modified air-braked cycle ergometer with power measuring device (e.g., SRM power meter, science version) or wind trainer bike and power meter
- [ ] Metabolic cart (and associated equipment)
- [ ] Heart rate monitoring equipment
- [ ] Stopwatch
- [ ] RPE scale
- [ ] Blood lactate analyzer and capillary blood sampling equipment
- [ ] Recording sheet (refer to the cycling power profile data sheet template)
- [ ] Pen

Cycle Incremental Test

- [ ] Calibrated, electromagnetically braked ergometer (e.g., Lode Excalibur Sport)
- [ ] Metabolic cart (and associated equipment)
- [ ] Heart rate monitoring equipment
- [ ] RPE scale
- [ ] Electric fan
- [ ] Blood lactate analyzer and capillary blood sampling equipment
- [ ] Recording sheet (refer to the cycling long graded exercise test data sheet templates)
- [ ] Pen

Two-Phase Treadmill Test

- [ ] Treadmill (calibrated for velocity and elevation)
- [ ] Metabolic cart (and associated equipment)
- [ ] Blood lactate analyzer and capillary blood sampling equipment
- [ ] Heart rate monitoring equipment
- [ ] Stopwatch (with rate function)
- [ ] RPE scale
- [ ] Video camera/Optojump device (optional)
- [ ] Recording sheet (refer to the treadmill test data sheet template)
- [ ] Pen

ANTHROPOMETRY

RATIONALE

In addition to being useful in terms of describing an athlete's physical size, the measurement of basic anthropometric characteristics (height, body mass, and skinfolds) can be used to estimate body composition (i.e., adiposity, muscularity) (Norton 1996). Landers and colleagues (2000) reported that low adiposity positively relates to performance in Olympic distance world championship triathlon. Our records also support this and highlight that relatively large differences in adiposity can be expected throughout a 12-month period (off-season vs. primary performance peak) in some triathletes. Thus, basic anthropometric measures when used in conjunction with other physiological and performance-related measures may add to the coach's and sport scientist's understanding of athletes in terms of their performance cycle. An athlete's body composition may also give valuable insight into how he or she ia adapting to training and the appropriateness of his or her total energy intake with respect to body composition goals.

TEST PROCEDURE

To assist with monitoring body composition, measurement of height, body mass, and skinfolds should be completed in conjunction with key testing blocks at regular periods throughout the year. Skinfolds are recorded over seven sites (triceps, biceps, subscapular, supraspinale, abdominal, front thigh, and medial calf) for both males and females using techniques endorsed by the International Society for the Advancement of Kinanthropometry. The individual skinfold measures as well as the sum of the seven sites should be reported. Refer to anthropometry protocols outlined in chapter 11, Assessment of Physique, for a detailed description of anthropometric test procedures.

NORMATIVE DATA

Table 31.1 presents anthropometric normative data for female and male triathletes.

TABLE 31.1 Anthropometric Data for Female and Male Triathletes: Mean ± *SD* (Range)

Athlete group	Height, cm	Body mass, kg	Σ7 skinfolds, mm
Female			
Junior	166.2 ± 4.9 (159.3-170.1)	52 ± 3 (49-58)	55 ± 15 (34-72)
Under 23/Elite	165.7 ± 3.6 (159.2-170.5)	55 ± 4 (49-64)	52 ± 10 (40-70)
Male			
Junior	179.9 ± 11.1 (172.0-187.7)	67 ± 8 (62-73)	46 ± 11 (38-54)
Under 23/Elite	178.0 ± 6.4 (169.3-193.8)	66 ± 7 (53-78)	38 ± 7 (27-50)

Typical error: height = 0.13%; body mass = 0.2%; Σ7 skinfolds = 1.7%. Data source: female, Junior n = 5, Under 23/Elite n = 10; male, Junior n = 2, Under 23/Elite n = 18; all measures in racing phase of season.

1 KM SWIM TIME TRIAL

RATIONALE

The aim of the 1 km swim time trial is to provide objective physiological information on the swimming performance of triathletes.

TEST PROCEDURE

- Athletes should present for testing in a standard state of preparedness (i.e., injury free, rested, hydrated, and carbohydrate fuelled). Refer to chapter 2 for specific information relating to pretest environment and athlete preparation.

- The athlete's personal best 1 km swim time from the previous season or recent performances should be determined to provide an estimate of split times and pacing for the time trial.

- The athlete should use a push start to commence the test. The first movement is used as the starting time, the foot touch for turns, and the hand touch for the finish.

- Time for each 200 m split (i.e., end of laps 4, 8, 12, 16, and 20) is recorded.

- Swimming stroke rate should be measured on the third lap of each 200 m (i.e., laps 3, 7, 11, 15, and 19) and stroke count on the fourth lap of each 200 m (i.e., laps 4, 8, 12, 16, and 20).

- Emphasis should be on maintaining an even pace for the time trial. A common mistake is to go too fast on the first 200 m.

- If available, a waterproof heart rate monitor can be used to collect heart rate data during the time trial.

- It is recommended that a section of the first and last 100 m of the 1 km time trial be videoed for review of technique when the athlete is fresh and when he or she is tired (above- and below-water views are recommended).

- A sample data table and results are provided in table 31.2.

DATA ANALYSIS

Consideration of the following may assist with the interpretation of testing results:

- Was the overall time faster or slower than previous tests?

- Was the final 200 m time faster or slower than before and how does it relate to the athlete's personal best time for the 200 m?

- Was the swim completed with even or appropriate 200 m splits?

- Were any stroke measurements made (e.g., stroke rate or stroke count) and, if so, are they consistent with the physiological data?

- What is the athlete's recent training history?

Apart from physiological factors, the primary cause of changes in swim performance is likely related to a change in swimming efficiency or technique. For example, at a standard velocity, when a swimmer's technique is poor, the heart rate may be elevated even though the swimmer has not lost fitness; this simply reflects the extra effort required to compensate for a loss in efficiency or technique. Wakayoshi and colleagues (1995) showed that a change in the slope of stroke rate–velocity relationship compared with a parallel displacement distinguished between technical and physiological swimming changes. It is recommended that coaches and sport scientists analyze and compare the stroke rate needed to achieve a given velocity. For example, if an increase in heart rate is accompanied by a significant increase in stroke rate between corresponding tests for the same velocity, it might be deduced that the swimmer's technique has suffered instead of his or her underlying physiological characteristics. Of course, this does not discount the possibility that the technical changes are themselves caused by changes in fitness, and further investigation may be necessary.

NORMATIVE DATA

Normative data for female and male triathletes for the 1 km swim time trial are presented in table 31.3.

TABLE 31.2 Sample 1 km Swimming Time Trial Data

Measure	0-200 m	200-400 m	400-600 m	600-800 m	800-1,000 m
Time, m:ss.00	2:30.40	2:29.15	2:31.14	2:29.22	2:29.01
Heart rate, b/min	170	173	175	177	181
Stroke rate, strokes/min	35.5	38.9	39	39.1	39.2
Stroke count, strokes/50 m	38	38	38	38	39

TABLE 31.3 Data for 1 km Swim Time Trial for Female and Male Triathletes: Mean ± *SD* (Range)

Athlete group	Time trial time, mm:ss
Female	
Junior	13:54 ± 00:42
	(12:25-14:53)
Under 23/Elite	13:19 ± 00:20
	(12:12-15:34)
Male	
Junior	13:28 ± 01:05
	(11:46-16:16)
Under 23/Elite	12:24 ± 00:36
	(11:15-14:01)

Typical error: 1 km time = 13 s or 1.5%. Data source: female, Junior n = 23, Under 23/Elite n = 5; male, Junior n = 37, Under 23/Elite n = 22; all measures preseason or early phase of season.

CYCLE POWER PROFILE TEST

RATIONALE

The cycle leg in International Triathlon Union, Junior Elite, and Elite competition is draft legal. This likely relates to varying physiological, performance, and skill requirements depending on the course and the specific race environment. Its importance to overall race outcome will also likely vary depending on particular circumstances (e.g., course profile, wetsuit or non-wetsuit swim, and field composition). Success in the cycling leg of the triathlon race, particularly the ability to maintain contact with the bunch and distribute energy output appropriately, requires a balanced performance profile and skill level. Awareness of these considerations and the availability of on-bicycle power-measuring devices (e.g., SRM power meter) have led to greater monitoring of performance in training and competition. In turn this has warranted the development of a more performance-based analysis framework (i.e., time-based power continuum) to complement the traditional blood lactate response framework for describing training intensity and performance.

The cycle power profile (time-based power continuum), shown to predict actual road cycling performance (Quod et al. 2010), has been adopted in triathlon. The test aims to determine the highest power output (maximal mean power, MMP) an athlete can hold for a particular duration of effort. This time–power continuum provides insight into athletes' performance strengths and weaknesses and provides useful prescriptive training data and race analysis framework (additional to blood lactate response).

TEST PROCEDURE

- The cycle power profile test should be performed on a calibrated air-braked cycle ergometer or using a calibrated power meter (e.g., SRM) mounted to the cyclist's bicycle, which is then placed in a wind trainer or rear wheel ergometer.

- When choosing the ergometer on which the test is to be performed, it is important to consider the kinetic energy associated with the system so as to replicate, as closely as possible, the feel of riding on the road. The ergometer's gearing should be adaptable to allow the cyclist to self-select a cadence that will allow him or her to produce a maximal effort over the duration of each effort.

- The power meter should be set to a sampling rate of 1 s (or less) and appropriately calibrated.

- The athlete's height and body mass (without shoes and minimal clothing) are measured and recorded.

- The metabolic cart must be adequately warmed up and calibrated for ventilation and gas composition. It is recommended that three primary gravimetric-standard reference gases (±0.02%), spanning the physiological range, be used for oxygen and carbon dioxide analyzer calibration.

- The tester attaches the heart rate monitoring equipment to the athlete and ensures that it is functioning correctly.

- A piece of tape is placed across the bridge of the athlete's nose. This is used to help prevent the nose clip from slipping during the test.

- A 5 min warm-up at 75 to 100 W is conducted to familiarize the cyclist with the bike setup and should include at least two 3 to 5 s sprints to ensure that the correct gearing is selected for the test (this gearing should be recorded when possible). The gear selected should enable the cyclist to self-select a cadence that is appropriate to the duration of the effort and similar to that used when road riding.

- Following the warm-up, the power meter should be zeroed and calibrated to the manufacturer's standard.

- The athlete is instructed to produce as much power as possible for the duration of each effort. For the shorter efforts (5-30 s), the cyclist will typically engage in an all-out sprint. However, for the longer efforts (1-10 min), an element of pacing is required.

- The test protocol is outlined in table 31.4. Note that the first two sprints are conducted from a standing start, with the remaining efforts completed from a rolling start between 70 and 80 revolutions per minute (rpm). An indication of an athlete's $\dot{V}O_2$peak can be obtained from metabolic analysis performed in the 4 min effort (dependent on maximal effort and relatively even pacing).

DATA COLLECTION

- Record power output and cadence data throughout the test using a sampling rate of 1 s or less.

- Record heart rate throughout the test using a sampling rate of 1 s. Record the heart rate at the end of each effort.

- Record $\dot{V}O_2$, $\dot{V}CO_2$, ventilation ($\dot{V}_E$), and respiratory exchange ratio (RER) during the 4 and 10 min efforts.

- Collect a capillary blood sample from the fingertip or earlobe 1 min prior to and 1 min after the 1, 4, and 10 min efforts. Analyze the blood sample for lactate concentration ([La$^-$]) and (if required) pH, blood gases, and bicarbonate concentration.

DATA ANALYSIS

In addition to calculating and presenting all maximal mean powers (MMPs) in absolute and relative terms (W and W · kg^{-1}), the coach and sport scientist may determine the percentage decrement between effort durations relative to peak (5 s MMP). Results can be plotted comparing time and power output in W and W · kg^{-1}.

NORMATIVE DATA

Table 31.5 presents normative data for female and male triathletes for the cycle power profile test.

TABLE 31.4 Cycle Power Profile Test Protocol

Time, min:s	Effort	Power output	Heart rate	Metabolic analysis	Blood sample	Notes
0:00-0:05	5 s, small gear	*	*			Standing start
0:05-1:00	Active recovery (55 s)	*	*			50-100 W
1:00-1:05	5 s, big gear	*	*			Standing start
1:05-4:00	Active recovery (175 s)	*	*			50-100 W
4:00-4:15	15 s	*	*			Rolling start (70-80 rpm)
4:15-8:00	Active recovery (225 s)	*	*			50-100 W
8:00-8:30	30 s	*	*			Rolling start (70-80 rpm)
8:30-14:00	Active recovery (330 s)	*	*			50-100 W
14:00-15:00	1 min	*	*		†	Rolling start (70-80 rpm)
15:00-23:00	Active recovery (480 s)	*	*			50-100 W
23:00-27:00	4 min	*	*	*	†	Rolling start (70-80 rpm)
27:00-37:00	Active recovery (600 s)	*	*			50-100 W
37:00-47:00	10 min	*	*	*	†	Rolling start (70-80 rpm)

*Sampling during effort.

†Sampling before and after effort.

TABLE 31.5   Cycle Power Profile Data for Female and Male Triathletes: Mean ± *SD* (Range)

Athlete group	MMP$_{5s}$	MMP$_{15s}$	MMP$_{30s}$	MMP$_{60s}$	MMP$_{4min}$	MMP$_{10min}$
Female						
Junior	11.2 ± 1.0 (9.8-12.4)	8.1 ± 0.6 (7.5-8.9)	6.5 ± 0.9 (5.0-7.1)	5.7 ± 0.5 (4.9-6.2)	4.7 ± 0.2 (4.2-4.9)	4.3 ± 0.3 (3.9-4.6)
Under 23/Elite	10.5 ± 1.5 (8.5-12.8)	8.5 ± 0.9 (6.8-9.6)	7.0 ± 0.6 (5.8-7.6)	5.9 ± 0.5 (5.3-6.7)	4.6 ± 0.4 (4.0-5.1)	4.1 ± 0.4 (3.6-4.7)
Male						
Junior	13.5 ± 0.4 (13.1-14.0)	11.0 ± 1.0 (9.9-12.4)	9.3 ± 1.0 (7.9-10.3)	7.3 ± 0.8 (6.1-8.1)	5.2 ± 0.6 (4.4-5.8)	4.7 ± 0.6 (3.9-5.3)
Under 23/Elite	13.4 ± 0.7 (12.6-15.0)	11.2 ± 0.8 (10.2-12.7)	9.0 ± 0.6 (7.8-10.0)	7.5 ± 0.6 (6.1-8.6)	5.6 ± 0.5 (4.7-6.4)	4.9 ± 0.5 (4.0-5.7)

All data given as W · kg^{-1}. MMP = maximal mean power. Typical error: MMP$_{5s}$ = 2.3%; MMP$_{15s}$ = 4.5%; MMP$_{30s}$ = 4.2%; MMP$_{60s}$ = 3.8%; MMP$_{4min}$ = 3.4%; MMP$_{10m}$ = 3.2% (Quod et al. 2010). Data source: female, Junior n = 5, Under 23/Elite n = 6; male, Junior n = 4, Under 23/Elite n = 14; all measures preseason or early phase of season.

CYCLE INCREMENTAL TEST

RATIONALE

The cycle incremental test is designed to allow for the determination of maximal oxygen consumption ($\dot{V}O_2$max) and peak power output (PPO) and for the development of training zones that are directly related to progressive blood lactate concentration.

Although the cycle power profile test is routinely used as the primary cycling test and the benchmark cycling performance test in triathlon, many athletes and coaches also use the cycle incremental test. This test has the benefit of a significant data history and direct measurement of various physiological variables, including progressive blood lactate response, which can be used to determine training zones.

TEST PROCEDURE

- The cycle incremental test should be performed using an electronically braked cycle ergometer (e.g., Lode or SRM). The ergometer should be well serviced and calibrated prior to testing.
- Set up the ergometer using the athlete's pedals and adjust the ergometer so that the seat height and handlebar position match the athlete's bicycle. Record these details and positions for use in future tests.
- Measure and record the athlete's height and body mass (without shoes and with minimal clothing).

- Ensure that the metabolic cart has been adequately warmed up and has been calibrated for ventilation and gas composition. It is recommended that three primary gravimetric-standard reference gases (±0.02%), spanning the physiological range, be used for oxygen and carbon dioxide analyzer calibration.
- Place a piece of tape across the bridge of the athlete's nose. This is used to help prevent the nose clip from slipping during the test.
- Attach heart rate monitoring equipment to the athlete and ensure that it is functioning correctly.
- Athletes should select a cadence to suit their riding style in order to optimize work output (cadence generally should be between 85 and 115 rpm).
- Refer to table 31.6 for information regarding starting workloads, step increments, step durations, and the recommended cadence range. If the ergometer does not operate in hyperbolic mode (whereby power output is independent of cadence), it may be necessary to adjust the gearing and change the resistance to ensure that the cyclist's pedaling cadence stays within the required range.
- The athlete is required to pedal at a constant workload for 3 min (female) or 5 min (male) stages, with the workload increasing

TABLE 31.6 Workloads for Cycle Incremental Test

Gender	Starting workload, W	Step increment, W	Step duration, min
Female	125	25	3
Male	100	50	5

following the completion of each stage. The test is continuous; the athlete stops only when he or she can no longer maintain the required power output. Failure occurs when the desired cadence can no longer be maintained or the cyclist opts to terminate the test. The cyclist should be strongly encouraged to complete a full minute or 30 s interval of the final workload, as this duration typically coincides with metabolic measurements.

- The test requires the mouthpiece and nose clip to be worn for the last half of each stage to ensure a stable reading for metabolic analysis (e.g., 1.5 min for 3 min stages, 2.5 min for 5 min stages). The cyclist is permitted to drink during the first half of each stage if required. However, towards the end of the test, the mouthpiece and nose clip should be in place continuously.

- The test is typically completed within 18 to 30 min for females and 25 to 40 min for males.

- Allow the athlete to perform a brief warm-up at power outputs below the applicable starting workload. During this period, outline the test procedures to the athlete (i.e., step increments, step duration, timing of blood sampling).

- When the athlete is ready to commence the test, connect the athlete to the calibrated metabolic cart.

DATA COLLECTION

- Instruct the athlete to commence cycling; simultaneously start the metabolic cart and timing devices. During each workload, provide verbal feedback to the athlete regarding the time remaining and check with the athlete about his ability to continue.

- Blood samples should be collected from the fingertip or earlobe in the final 30 s of each workload as well at 3 to 4 min posttest. The posttest sample should be a rested sample and should be collected with the athlete seated (i.e., *not* in an active recovery).

- With a few seconds remaining in each 30 s period throughout the expected final workload, ask the athlete whether he or she is capable of cycling for at least a further 30 s. The athlete should indicate his or her decision with hand signals (i.e., thumbs-up is yes; thumbs-down is no).

- The test ends when the athlete stops voluntarily or indicates that he or she cannot complete a further 30 s.

- Data typically collected during the cycle incremental test include $\dot{V}_E$, $\dot{V}O_2$, $\dot{V}CO_2$, RER, heart rate (HR), and power output (PO) for each 30 s of the test plus blood lactate concentration for each step.

- For steady-state values, average the values recorded for the last 1 min of each 3 min stage and the last 2 min of each 5 min stage (L · min^{-1} and ml · kg^{-1} · min^{-1}).

- Peak values are reported as the highest value attained over a period of one full minute (L · min^{-1} and ml · kg^{-1} · min^{-1}) collected during the test.

DATA ANALYSIS

Lactate thresholds (LT1 and LT2) are calculated using the modified Dmax method, chapter 6, Blood Lactate Thresholds. The following indices should also be reported for each lactate threshold: $\dot{V}O_2$ as L · min^{-1} and ml · kg^{-1} · min^{-1}, PO in W as well as W · kg^{-1}, %$\dot{V}O_2$peak, HR, %HRmax. Test data and calculated lactate thresholds should be used to determine endurance training intensities (T1-T5, see chapter 6 for description of T1-T5 zones).

NORMATIVE DATA

Normative data for female and male triathletes for the cycle incremental test are presented in table 31.7.

TABLE 31.7 Cycle Incremental Test Data for Female and Male Triathletes: Mean ± *SD* (Range)

Athlete group	$\dot{V}O_2$peak, ml · kg⁻¹ · min⁻¹	Maximal aerobic power, W · kg⁻¹	Power at T1, W · kg⁻¹	Power at T2, W · kg⁻¹
Female				
Junior	58.1 ± 7.9 (49.6-76.5)	4.6 ± 0.4 (3.7-5.0)	2.5 ± 0.3 (1.9-2.9)	3.7 ± 0.3 (3.3-4.3)
Under 23/Elite	60.7 ± 5.5 (49.8-69.3)	4.5 ± 1.0 (1.8-5.1)	2.9 ± 0.3 (2.4-3.6)	4.0 ± 0.2 (3.5-4.4)
Male				
Junior	69.5 ± 4.5 (61.8-75.8)	5.4 ± 0.4 (4.6-6.0)	3.0 ± 0.3 (2.5-3.5)	4.4 ± 0.5 (3.3-5.4)
Under 23/Elite	70.0 ± 5.6 (59.6-80.3)	5.3 ± 0.4 (4.7-6.2)	3.1 ± 0.3 (2.4-3.7)	4.3 ± 0.3 (3.7-4.9)

Typical error: $\dot{V}O_2$peak (ml · kg⁻¹ · min⁻¹) = 2.88%, maximal aerobic power (W · kg⁻¹) = 1.33%, power at T1 (W · kg⁻¹) = 7.03%, power at T2 (W · kg⁻¹) = 1.39%. Data source: female, Junior $n = 9$, Under 23/Elite $n = 9$; male, Junior $n = 12$, Under 23/Elite $n = 20$; all measures preseason or early phase of season.

TWO-PHASE TREADMILL TEST

RATIONALE

The two-phase treadmill test is designed to allow for the determination of key physiological variables known to be of importance to running performance: peak oxygen consumption (Foster 1983), running economy (Lucia et al. 2006), and lactate threshold (Bird et al. 2003) with an emphasis on simplifying and minimizing testing procedures.

To streamline data collection and minimize testing time, a number of assumptions must be made in the data analysis phase in order to get maximum output from one test. It is acknowledged that this approach potentially weakens the physiological basis (potential validity and reliability) for some calculated variables. Test–retest reliability, however, demonstrates relatively strong reliability, so it can be argued that test practicality and reliability outweigh any potential validity issues in this setting.

TEST PROCEDURE

The two-phase treadmill test involves four to six 4 min submaximal efforts separated by 1 min rest periods. Efforts are performed at 0% gradient and increase by 1 km · h⁻¹ each effort. The submaximal efforts are followed by a progressive run to exhaustion with velocity or gradient increasing every 30 s. The progressive run is commenced 4 min after completion of the final 4 min submaximal effort.

The starting velocity for the submaximal efforts should be determined based on previous tests or on coaches' and sport scientists' recommendations. For athletes who have not completed the test previously and those with whom the scientist and coach are not familiar, suggested starting velocities are junior males 14 to 15 km · h⁻¹, junior females 12 to 13 km · h⁻¹, senior males 15 to 16 km · h⁻¹, and senior females 13 km · h⁻¹.

The starting velocity should be slow enough to allow a minimum of four submaximal stages to be completed. The termination of the submaximal phase of the test should be made at the discretion of the sport scientist when taking into account HR, blood lactate concentration, rating of perceived exertion (RPE), RER, and $\dot{V}O_2$. The intent is that this phase terminates around or just above the second lactate threshold (e.g., 3-4 mmol · L⁻¹). Guidelines for the progressive run to exhaustion are presented in table 31.8.

- Ensure that the treadmill being used is clean and has been recently calibrated for velocity and gradient. Run the treadmill through the range of velocities required in the test before the athlete arrives to verify proper operation.

- Measure and record the athlete's height and body mass (without shoes and minimal clothing).

- Ensure that the metabolic cart has been adequately warmed up and has been calibrated for ventilation and gas composition. It is

TABLE 31.8 Guidelines for Two-Phase Treadmill Test Progressive Run to Exhaustion

Time	Velocity
0:00-0:30	Final submaximal effort velocity minus 3.0 km · h⁻¹, 0% gradient
0:30-1:00	Final submaximal effort velocity minus 2.5 km · h⁻¹, 0% gradient
1:00-1:30	Final submaximal effort velocity minus 2.0 km · h⁻¹, 0% gradient
1:30-2:00	Final submaximal effort velocity minus 1.5 km · h⁻¹, 0% gradient
2:00-2:30	Final submaximal effort velocity minus 1.0 km · h⁻¹, 0% gradient
2:30-3:00	Final submaximal effort velocity minus 0.5 km · h⁻¹, 0% gradient
3:00-3:30	Final submaximal effort velocity 0% gradient
3:30-4:00	0.5 km · h⁻¹ > Final submaximal effort velocity, 0% gradient
4:00-4:30	1.0 km · h⁻¹ > Final submaximal effort velocity, 0% gradient
4:30-5:00	1.0 km · h⁻¹ > Final submaximal effort velocity, +0.5% gradient
5:00-5:30	1.0 km · h⁻¹ > Final submaximal effort velocity, +1% gradient
5:30-6:00	1.0 km · h⁻¹ > Final submaximal effort velocity, +1.5% gradient
6:00-6:30	1.0 km · h⁻¹ > Final submaximal effort velocity, +2% gradient
6:30-7:00	1.0 km · h⁻¹ > Final submaximal effort velocity, +2.5% gradient

recommended that three primary gravimetric-standard reference gases (±0.02%), spanning the physiological range, be used for oxygen and carbon dioxide analyzer calibration.

- Attach HR monitoring equipment to the athlete and ensure that it is functioning correctly.
- Place a piece of tape across the bridge of the athlete's nose. This is used to help prevent the nose clip from slipping during the test.
- Instruct the athlete to take position at the front of the treadmill, and when he or she is ready, allow the athlete a brief warm-up and familiarization period on the treadmill. Warm-up speed should be less than the starting velocity of the test. During this period, outline the test procedures to the athlete (i.e., step increments, step duration, timing of blood sampling).
- Position the headset and mouthpiece on the athlete, ensuring maximum comfort. Instruct the athlete to take hold of the nose clip and position it comfortably to block nasal air flow.
- When the athlete is ready to commence the test, connect the athlete to the calibrated metabolic cart.
- Prior to commencement of the test, have the athlete hold the handrails and straddle the treadmill. Start the treadmill, and when it is operating at the desired velocity, have the athlete start running while you simultaneously start the metabolic cart and time device.

- During each workload, provide verbal feedback to the athlete regarding the time remaining and check with the athlete about his or her ability to continue.
- Stop the treadmill at the end of each 4 min effort, giving the athlete prior warning with a countdown over the last 5 s.
- Blood samples should be collected from the fingertip or earlobe on completion of each submaximal effort. Record rating of perceived exertion (RPE) for each submaximal effort.
- Approximately 10 to 15 s prior to the start of the subsequent submaximal effort, instruct the athlete to hold the handrails and straddle the treadmill. Start the treadmill, and when it is operating at the desired velocity, have the athlete start running as the 1 min rest period concludes.
- Data typically collected during the two-phase treadmill test include $\dot{V}O_2$, $\dot{V}CO_2$, $\dot{V}_E$, RER, and HR for the final minute of each submaximal effort and posteffort blood lactate concentration ([La⁻]) and RPE.
- A video camera and tripod can be used to film the last 60 s of each submaximal effort to measure stride rate, frequency, flight time, and contact time. Alternatively, an Optojump light screen (or similar) may be used if available.
- Encourage athletes to run to exhaustion but to finish on a 30 s increment where possible.

With a few seconds remaining in each 30 s period, ask the athlete whether he or she is capable of completing another 30 s effort. The athlete should indicate his or her decision with hand signals (i.e., thumbs-up is yes; thumbs-down is no).

- The test ends when the athlete voluntarily presses the emergency stop button on the treadmill or indicates to the tester that she cannot complete the next 30 s effort.

- Immediately on completion of the test, record maximal HR and collect and analyze a blood sample. Ask the athlete to indicate his or her RPE. Collect a posttest lactate sample after 3 to 4 min to ensure that the final value recorded was indeed maximum.

- Submaximal oxygen uptakes should be reported as the average value for the final 1 min of each 4 min submaximal stage.

- Peak values are reported as the highest value attained over a period of one full minute collected during the test.

DATA ANALYSIS

To determine the theoretical velocity at $\dot{V}O_2$peak, graph the submaximal speed and $\dot{V}O_2$ data. Using a linear regression equation, calculate the velocity for the known peak oxygen uptake. This value should be used in conjunction with the maximal [La⁻], HR, and peak $\dot{V}O_2$ when calculating subsequent variables. Lactate thresholds (LT1 and LT2) are calculated using the modified Dmax method (see chapter 6). The following indices should also be reported for each lactate threshold: $\dot{V}O_2$, %$\dot{V}O_2$peak, HR, and % HRmax. Velocity should be presented in km · h⁻¹ as well as min · km⁻¹ and s · km⁻¹. $\dot{V}O_2$ should be presented as L · min⁻¹ and ml · kg⁻¹ · min⁻¹. Test data and calculated lactate thresholds should be used to determine endurance training intensities (T1-T5, see chapter 6 for a description of the T1-T5 zones).

NORMATIVE DATA

Table 31.9 presents normative data for the two-phase treadmill test for female and male triathletes.

TABLE 31.9 Two-Phase Treadmill Test Data for Female and Male Triathletes: Mean ± *SD* (Range)

Athlete group	$\dot{V}O_2$peak, ml · kg⁻¹ · min⁻¹	Velocity at T1, km · h⁻¹	$\dot{V}O_2$ at T1, ml · kg⁻¹ · min⁻¹	Velocity at T2, km · h⁻¹	$\dot{V}O_2$ at T2, ml · kg⁻¹ · min⁻¹
Female					
Junior	63.3 ± 1.7 (60.6-65.3)	14.2 ± 1.3 (13.0-16.0)	48.0 ± 3.7 (43.6-53.4)	15.7 ± 1.3 (13.8-17.6)	54.5 ± 4.1 (47.8-58.8)
Under 23/Elite	63.3 ± 3.6 (59.9-69.0)	15.0 ± 1.1 (13.0-16.0)	51.6 ± 3.9 (45.6-57.5)	16.7 ± 0.8 (15.5-17.6)	59.0 ± 3.2 (55.3-63.5)
Male					
Junior	69.5 ± 1.8 (67.6-72.3)	15.6 ± 0.5 (15.0-16.0)	54.1 ± 2.4 (50.9-56.4)	17.9 ± 0.6 (17.1-18.5)	62.5 ± 2.6 (60.2-66.6)
Under 23/Elite	72.9 ± 4.9 (63.3-80.8)	15.7 ± 0.8 (14.0-17.0)	56.1 ± 2.4 (47.6-65.3)	18.0 ± 0.6 (16.8-19.0)	66.1 ± 4.0 (58.2-72.2)

Typical error: $\dot{V}O_2$peak (ml · kg⁻¹ · min⁻¹) = 2.47%, velocity at T1 (km · h⁻¹) = 0.17%, $\dot{V}O_2$ at LT1 (ml · kg⁻¹ · min⁻¹) = 1.50%, velocity at T2 (km · h⁻¹) = 2.63%, $\dot{V}O_2$ at T2 (ml · kg⁻¹ · min⁻¹) = 2.69%. Data source: female, Junior *n* = 5, Under 23/Elite *n* = 7; male, Junior *n* = 5, Under 23/Elite *n* = 20; all measures preseason or early phase of season.

References

Bird, S.R., Theakston, S.C., Owen A., and Nevill, M. 2003. Characteristics associated with 10 km running performance among a group of highly trained male endurance runners age 21-63 years. *Journal of Aging and Physical Activity* 11(3):333-350.

Borg, G. 1962. Physical performance and perceived exertion. *Studia Psychologica et Paedagogica Series altera Investigationes XI. Lund Sweden: Gleerup*, 1-64

Foster, C. 1983. VO2 max and training indices as determinants of competitive running performance. *Journal of Sport Sciences* 1(1):13-22.

Landers, G.J., Blanksby, B.A., Ackland, T.R., and Smith, D. 2000. Morphology and performance of world championship triathletes. *Annals of Human Biology* 27(4):387-400.

Lucia, A., Esteve-Lanao, J., Olivan, F., Gomez-Gallego, F., SanJuan, A.F., Santiago, C., Perez, M., Chamorro-Vina, C., and Foster, C. 2006. Physiological characteristics of the best Eritrean runners—Exceptional running economy. *Applied Physiology in Nutrition and Metabolism* 31(5):530-540.

Norton, K.I. 1996. Anthropometric estimation of body fat. In: K.I. Norton and T. Olds (eds), *Anthropometrica*. Sydney: University of New South Wales Press, pp 171-198.

Wakayoshi, K., D'Acquisto, L.J., Cappaert, J.M., and Troup, J.P. 1995. Relationship between oxygen uptake, stroke rate and swimming velocity in competitive swimming. *International Journal of Sports Medicine* 16:19-23.

Quod, M.J., Martin, D.T., Martin, J.C., and Laursen, P.B. 2010. The power profile predicts road cycling MMP. *International Journal of Sports Medicine* 31(6):397-401.

Indoor and Beach Volleyball Players

Jeremy M. Sheppard, Tim J. Gabbett, and Michael P. Riggs

Volleyball is a team sport played by two teams on a court divided by a net. The object of the game is to send the ball over the net in order to ground it on the opponent's side. Teams are allowed three hits to return the ball over the net. Two competitive forms of volleyball are played around the world: indoor volleyball and beach volleyball.

Indoor Volleyball

Competitive indoor volleyball uses a player rotation system; the positional duties are a *setter,* two *middles,* three *outsides,* and a backcourt defensive specialist, the *libero.* There are six athletes on the 9 m × 9 m court at any one time; the libero interchanges into the backcourt for either of the two middles. The demands of a match change with the demands of offensive and defensive duties and an individual's placement on court (e.g., frontcourt or backcourt in the rotation).

The vast majority of jumping in volleyball occurs in the frontcourt, consisting of attacking spike jumps (SPJ) and defensive block jumps (BLJ), with setters also performing jump sets (JS). However, backcourt offensive attacks (i.e., SPJ) are a significant component of play, particularly with men (Polglaze and Dawson 1992; Sheppard and Gabbett 2007; Sheppard et al. 2007; Viitasalo 1991; Viitasalo et al. 1987). Furthermore, defensive diving efforts are a physiological stress that must be considered (Sheppard et al. 2007).

Several published volleyball time–motion analysis studies were conducted on competition prior to several rule changes, including the player substitution rules and the change from service scoring to rally-point scoring in 1999 (Dyba 1982; Polgaze and Dawson 1992). These rule changes appear to have had some minor influences on the specific physiological demands of volleyball matches (Sheppard et al. 2007b). It was determined that 77% of rallies were 12 s or less, whereas the average rally time was approximately 11 s. However, the range of durations included rallies as short as 3 s (ace service) and as long as 40 s. In addition, 44% of rest periods between rallies were 12 s or less, with the average rest time being 14 s in duration. Rest periods were observed to be as short as 4 s and as long as 38 s (contested call or time-out).

Observations of match conditions (Polgaze and Dawson 1992; Sheppard et al. 2007; Viitasalo et al. 1987) reveal that volleyball is characterized by frequent short bouts of high-intensity exercise, followed by periods of low-intensity exercise and brief rest periods. The high-intensity bouts of exercise with relatively short recovery periods, coupled with the total duration of the match (~90 min), suggest that volleyball players require well-developed anaerobic alactic (phosphocreatine) and anaerobic lactic (anaerobic glycolytic) energy systems as well as reasonably well-developed aerobic capabilities (Dyba 1982; Polgaze and Dawson 1992; Viitasalo et al. 1987). Testing results and observation of match conditions also indicate that considerable demands are placed on the neuromuscular system during the various sprints, dives, jumps, and multidirectional court movements that occur repeatedly during competition (Sheppard and Borgeaud 2008; Sheppard et al. 2008b, 2008c, 2009d). As a result, it can be assumed that volleyball players require well-developed speed and muscular power (Sheppard et al. 2007, 2008b; Stanganelli et al. 2008) and the ability to perform these repeated maximal efforts with limited recovery for the duration of the match (Gabbett et al. 2007; Polgaze and Dawson 1992; Sheppard et al. 2007).

Vertical jumping ability is likely the single most important performance indicator in volleyball. Indeed, testing measures such as spike jump (with three- or four-step approach), block jump (countermovement jump with two-arm reach), and countermovement vertical jump (countermovement jump with one-arm reach) are typical of most programs (Gladden and Colacino 1978; Heimer et al. 1988; Marques et al. 2008; Newton et al. 1999; Smith et al. 1987; Spence et al. 1980). The use of these measures is supported by their ability to discriminate between higher and lower performers within the sport (Heimer et al. 1988; Smith et al. 1992; Spence et al. 1980; Viitasalo 1991). Considering the tactical nature and importance of jumping activities and the frequency with which they occur in a typical match, countermovement jump ability (i.e., jump and reach height) and approach jump ability (i.e., spike jump height) are considered critical performance indicators for volleyball and are a feature of the physiological profile of a volleyball player (Sheppard et al. 2008b, 2009a, 2009b; Spence et al. 1980; Viitasalo 1991). However, we have observed large variability in the performance of the block jump, likely due to athletes' inability to complete the two-arm reach with their hands at the same height consistently. The typical error of this measure, in our experience, is too large to be useful to the practitioner, making it difficult to detect real changes versus changes attributable to alterations in execution of the test by the athlete.

Beach Volleyball

Beach volleyball has been played since the early 1900s and is primarily based on the game of indoor volleyball. Beach volleyball became an Olympic sport at the 1996 Atlanta summer games. The sport has evolved through a variety of rule and regulation changes made by the international governing body, the FIVB (Fédération Internationale de Volleyball 2009). The sport is played on a court measuring 8 m × 16 m with an 8 m wide net splitting the court into two 8 m × 8 m halves. The net height is different for male and female athletes, with the top of the net reaching 2.43 m and 2.24 m, respectively.

Two opposing teams compete against each other on opposite sides of the net, with each team consisting of two competitors, traditionally one taking on the role of backcourt play and the other frontcourt. A single match follows a best of three set format (first to win two sets). A set is complete when a team scores 21 points or greater and is 2 points clear of the opposition; the third tiebreaker set is won by the first team that reaches 15 points or greater and is 2 points clear of the opposition. In the first and second set, teams will alternate playing from each end, swapping ends every time the total combined score is a multiple of 7. For example, if team A has scored 6 points and team B has scored 8 points, the total combined points equals 14, so the teams will swap ends before commencing play for the next (15th) point. During the third set, teams change sides of the court at every combined multiple of 5.

After winning a point, the winning team has 12 s to return to the baseline and serve the ball to start play for the next point. During the first and second sets (but not the third), there is a 30 s "technical time-out" when the combined scores equal 21. In association with this, each team is allowed to call a 30 s time-out at any stage during the set; this applies in a tiebreaker set also.

Data from the 2009 FIVB World Tour indicate that the mean duration of a female two-set match was 38.3 ± 5.3 min and the combined mean total points scored was 74.5 ± 6.1. In a three-set match, mean duration and scores were 55.0 ± 7.0 min and 104.6 ± 6.9 points, respectively. Male two-set matches had a mean duration of 41.5 ± 6.2 min and combined mean total points scored of 75.3 ± 6.7. A three-set match had a mean duration and combined points score of 59.8 ± 7.6 and 105.3 ± 7.9, respectively (www.fivb.org/EN/BeachVolleyball/Competitions/WorldTour/2009x/).

Beach volleyball consists of multiple high-intensity maximal efforts lasting approximately 6 to 8 s broken up by short, low-intensity recovery periods (Homberg and Papageorgiou 1994; Turpin et al. 2008). The ability to reproduce and maintain the quality of these efforts is a key component of performance. Therefore it is suggested that when planning a training session, coaches consider the repeated-effort nature of the sport and ensure that there is a substantial anaerobic bias. To maintain training specificity and develop a conditioning program that challenges the athlete's ability to perform maximal efforts repeatedly, coaches should vary the duration of effort and recovery periods to stimulate the desired energy system. Like indoor volleyball players, beach volleyball players require well-developed anaerobic alactic (ATP-CP) and anaerobic lactic (anaerobic glycolytic) energy systems as well as reasonably well-developed aerobic capabilities.

At the time of publication the authors were in the process of developing a relevant, reliable, and valid repeated-effort test specific to beach volleyball. The test and related data are not included in this chapter; however, we believed it necessary to suggest this as an important research area. If a valid and reliable test can be developed, it should be included in the test battery for beach volleyball.

In association with the repeated-effort nature of the sport, consideration must be given to the influence of jumping. A major goal for any beach volleyball conditioning program should be to maximize vertical jump height and minimize excessive stress in both takeoff and landing. Two studies highlight the importance of jumping: During the course of play an elite male German beach volleyball athlete averaged 85 maximal jumps per game (Homberg and Papageorgiou 1994), whereas Turpin and colleagues (2008) found that the average number of jumps per team per match was 219 ± 7.4. This is supported by research with the Australian Institute of Sport (AIS) beach volleyball program showing that within an FIVB World Tour game, an Australian men's team performed on average 145 ± 22.1 maximal jumps during the course of play across seven games.

Key performance skills (such as jump serving, spiking, and blocking) in beach volleyball depend on an athlete's vertical jump ability (Batista et al. 2008; Giatsis and Tzetzis 2003; Grgantov et al. 2005). The higher an athlete can jump when performing these skills allows for a higher ball contact point and as a result improved hitting angles, a greater range of attacking options, and, in the case of blocking, improved defending options attributable to the greater potential for a reduction in effectiveness of the attacking opponent. From an attacking perspective, Koch and Tilp (2009) demonstrated that despite a high error rate, the jump serve was the most successful serving technique in relation to direct points won and that the crosscourt spike action or hit was the most used attacking technique by both genders (male 38.9%, female 30.6%). From a defensive perspective, these authors assessed the variations in blocking technique and showed that an active block (block in which the defender reaches over the net and penetrates the opponent's side) was the most common style of block for both male and female athletes. Giatsis (2001) concluded that blocking accounted for 27% of the total number of jumps throughout a game of beach volleyball.

Because of the sizeable impact of the athlete's vertical jump ability on performance and some associated major factors influencing beach volleyball players' jump heights (e.g., somatotype, body composition, ground reaction force characteristics), an important component of athletes' conditioning is to maximize their power to weight ratio, which requires the athlete to minimize nonessential (fat) mass (Riggs and Sheppard 2009). This can vary between athletes and is dependent on athlete maturation, gender, and training history.

Accordingly, key training outcomes and goals for beach volleyball include maximizing vertical jump reach height, multidirectional acceleration, and repeated maximal efforts.

Athlete Preparation

Standardized pretest preparation is recommended to enable reliable and valid physiological data to be obtained. Refer to chapter 2, Pretest Environment and Athlete Preparation, for specific information relating to athlete and environment preparation.

It is generally unrealistic to expect that elite athletes can be completely recovered for testing occasions. However, performing tests after a complete rest day, and several days after a player's last match and international travel, is considered good practice.

Test Environment

For indoor volleyball, vertical jump testing, as well as any repeated-effort or endurance testing, should be performed in the volleyball indoor sports hall, if possible, or a suitable indoor facility with a nonslip surface. A laboratory setting is most suitable for anthropometry testing as well as the assessment of strength qualities (i.e., using a strength laboratory).

For beach volleyball, vertical jump and reach testing, as well as any repeated-effort or endurance testing, should be performed on a beach volleyball court. Environmental conditions can significantly affect test results when beach volleyball athletes are tested in their competition and training environment. Issues like altered compressive characteristics of wet sand versus hot dry sand, air temperature, and wind speed can all alter performance outcome. After rain, the sand on a beach volleyball court will tend to become harder and more compliant, resulting in a surface that provides better energy return compared with hot and dry sand. Consequently, it is important that when conducting physiological testing on beach volleyball athletes, testers record environmental data (temperature, humidity, air pressure, wind speed) and time of testing. It is also relevant to make note of the sand characteristics (e.g., hot, dry, and fluffy or cool, wet, and hard).

Equipment Checklist

Anthropometry

- ☐ Stadiometer (wall mounted)
- ☐ Balance scales (accurate to ±0.05 kg)
- ☐ Anthropometry box
- ☐ Skinfold calipers (Harpenden skinfold caliper)

- ☐ Marker pen
- ☐ Anthropometric measuring tape
- ☐ Recording sheet (refer to anthropometry data sheet template)
- ☐ Pen

Vertical Jump Test

- ☐ Yardstick jumping device (e.g., Swift Performance Yardstick)
- ☐ Measuring tape
- ☐ Field marking tape

- ☐ Recording sheet (refer to vertical jump data sheet template)
- ☐ Pen

Lower-Body Speed-Strength and 1RM Strength Testing

- ☐ Lifting platform
- ☐ Squat rack with safety bars
- ☐ Olympic bench press
- ☐ Chin-up bar
- ☐ Barbell (Olympic 20 kg)

- ☐ Weight plates (2.5-25 kg increments)
- ☐ Recording sheet (refer to strength and power data sheet template)
- ☐ Pen

Indoor Volleyball Repeated-Effort Test

- ☐ Electronic light gate equipment (e.g., dual-beam light gates)
- ☐ Yardstick jumping device (e.g., Swift Performance Yardstick)
- ☐ 2 mounted volleyballs on apparatus
- ☐ Stopwatch

- ☐ Measuring tape
- ☐ Field marking tape
- ☐ Recording sheet (refer to indoor volleyball repeated-effort test data sheet template)
- ☐ Pen

Recommended Test Order

It is important that field and strength tests are completed in the same order to control the interference between tests. This order also allows valid comparison of different test occasions. The order is as follows:

Day	Tests
1	Anthropometry
	Vertical jump
2	Lower-body speed-strength and strength
3	Indoor volleyball repeated-effort

ANTHROPOMETRY

RATIONALE

The height of volleyball athletes is a performance consideration and may also influence court position (Sheppard and Borgeaud 2008; Sheppard et al. 2009a, 2009b). Low fat mass is imperative in elite volleyball players and is a major consideration in improving relative strength qualities and jump heights in elite players (Riggs and Sheppard 2009; Sheppard et al. 2009b).

TEST PROCEDURE

Measurement of height, body mass, and skinfolds should be carried out prior to field testing protocols. Skinfolds are recorded over seven sites (triceps, biceps, subscapular, supraspinale, abdominal, front thigh, and medial calf). The individual skinfold measures as well as the sum of the seven sites should be reported. Refer to anthropometry protocols outlined in chapter 11, Assessment of Physique, for a detailed description of anthropometric test procedures.

Although the description of skinfold measurement procedures seems simple, a high degree of technical skill is essential for consistent results. It is therefore important that these measurements be taken by an experienced tester who has been trained in these techniques. It is also important that the same tester conduct each retest to ensure reliability.

DATA ANALYSIS

Elite volleyball athletes in general appear to be taller than previously reported (Sheppard et al. 2009b). However, height alone does not determine the suitability of a player, and although athletes who play certain positions (e.g., middles) tend to be quite tall, height does not exclude an athlete from consideration for any one position. Generally, height can be used as a talent potential indicator in that tall players who are taught to play well may achieve elite status, and it can also be used to track growth in preelite players.

A useful ratio comparing body mass and sum of seven skinfolds ($\Sigma 7$ skinfolds) can be used to reflect lean mass. The formula of body mass/$\Sigma 7$ skinfolds provides a simple ratio that allows comparison of athletes who are vastly different in height and body mass, in regard to their lean mass. This ratio (e.g., 100 kg/50 mm = 2.0) also provides a useful, straightforward context for educating athletes on the role of lean muscle mass increases while reducing or maintaining skinfold levels.

NORMATIVE DATA

Table 32.1 presents anthropometric normative data for Australian female and male indoor volleyball athletes. Table 32.2 presents anthropometric normative data for Australian female and male beach volleyball athletes.

TABLE 32.1 Anthropometric Data for Male and Female Indoor Volleyball Players: Mean ± *SD* (Range)

Gender	Height, cm	Body mass, kg	$\Sigma 7$ skinfolds, mm
Female	183.0 ± 6.4 (173.7-194.2)	73 ± 5 (65-84)	101 ± 21 (81-136) Desired <60 mm
Male	200.4 ± 5.0 (187.3-207.5)	95 ± 6 (82-109)	57 ± 9 (40-72) Desired <50 mm

Data source: female, Australian junior (*n* = 9) and senior national (*n* = 2); male, Australian junior (*n* = 8) and senior national (*n* = 14).

TABLE 32.2 Anthropometric Data for Male and Female Beach Volleyball Players: Mean ± *SD* (Range)

Gender	Height, cm	Body mass, kg	$\Sigma 7$ skinfolds, mm
Female	183.7 ± 4.3 (175.9-191.1)	71 ± 5 (63-82)	84 ± 22 (53-108) Desired <55 mm
Male	195.2 ± 3.76 (190.7-199.7)	90 ± 6 (84-99)	48 ± 13 (33-71) Desired <50 mm

Data source: female, Australian junior (*n* = 5) and senior national (*n* = 4); male, Australian junior (*n* = 2) and senior national team (*n* = 4).

VERTICAL JUMP TEST

RATIONALE

Vertical jump height is likely the single most important measured physical quality in elite volleyball players (Sheppard et al. 2009a, 2009b; Stanganelli et al. 2008). Improving vertical jump height in volleyball players greatly increases their options for angle of attack in spiking as well as their defensive abilities in blocking, two major components of volleyball (Sheppard et al. 2007).

TEST PROCEDURE

Refer to the vertical jump test (countermovement with arm swing) protocol in chapter 14, Field Testing Principles and Protocols, and drop jump and squat jump protocols in chapter 13, Strength and Power Assessment Protocols.

DATA ANALYSIS

Indoor Volleyball

Jump test results can be used not only to identify potential elite talent but also to track progress and evaluate the effectiveness of training methods in improving volleyball performance (Sheppard et al. 2008c, 2008d, 2008e, 2008f). In volleyball, jump height is a quality that is not optimized but rather needs to be maximized to each athlete's potential and is likely the single largest discriminator between junior and senior elite players (Sheppard et al. 2009b).

Useful comparisons can be made between the vertical jump (CMJ), drop jump (DJ), and spike jump (SPJ; vertical jump with approach) to gain insight into the training needs of a volleyball player. For example, the jump height difference between SPJ and CMJ provides an indication of the athlete's ability to use the approach (run-up) to gain greater jump heights (Sheppard et al. 2007). As seen from table 32.3, on average the SPJ is 8 cm higher than CMJ for females and 14 cm for males. It is suggested that practitioners consider 6 cm and 10 cm a desired minimum for females and males, respectively. If the difference between SPJ and CMJ is lower than these values, more attention should be paid to approach and jump technique, horizontal to vertical transition, or specific physical qualities (e.g., stretch–shorten cycle [SSC] function).

Comparing DJ and CMJ heights is useful to determine training needs. In general, DJ height should be superior to CMJ height (Bobbert 1990; Bobbert et al. 1986). However, this is not always the case with each athlete; the relationship between DJ height and CMJ height is quite sensitive to training activities (Sheppard et al. 2008f), and the DJ height of volleyball players tends to increase with increased training (Sheppard et al. 2009a). It is proposed that if an athlete's DJ is lower than their CMJ, then training with an increased stretch load (e.g., drop jump training, accentuated eccentric loads) is warranted, because it presents a potential to increase DJ height (Sheppard et al. 2008a, 2008e, 2008f). This approach stands to reason with volleyball athletes, because increases in DJ height are strongly associated with increases in the jumps that occur in volleyball (CMJ and SPJ) (Sheppard et al. 2008d, 2009a).

Beach Volleyball

As discussed for indoor volleyball, to gain insight into the training needs of a beach volleyball player and identify possible weaknesses in their ability to maximize their jump height reach, we must understand the technical and skill demands of the sport. In beach volleyball the skill to perform an effective block has a lot to do with the athlete's ability to read and assess the opponent early, position themselves early, drop into a deep squat out of the hitting opponent's field of vision, hold this deep squat, and then jump maximally from this static position without using an arm swing due to hand positioning. Therefore, the need to assess block jump height and the ability to execute a maximal jump from a static deep hold are most relevant. Batista and colleagues (2008) assessed the block jump (BJ) height as well as spike jump height of 38 male beach volleyball athletes, comparing the elite and the subelite players. The investigators identified a significant difference between the two group mean jump heights, and the greatest difference was seen in the block jump height (elite group 8% higher). Well-trained beach volleyball athletes often obtain similar CMJ and BJ heights on both a hard and a sand surface (see tables 32.4 and 32.5). These findings suggest that beach volleyball athletes need to generate force and power from active muscle recruitment more than from heavy reliance on the SSC (Riggs and Sheppard 2009).

In the indoor volleyball section we suggested that the jump height difference between SPJ and CMJ provides an indication of the athlete's ability to use their approach (run-up) to gain greater jump heights (Sheppard et al. 2008d). For beach volleyball athletes being tested on sand, this appears not to be the case. Results presented in normative data tables demonstrate that scores obtained on sand are less than those obtained on a hard surface and that the difference between the two jump tests is greater on the hard surface. This reinforces the notion that a beach volleyball athlete's vertical jump ability on sand may not reflect SSC performance but rather reflects his or her ability to generate force and power through active force recruitment.

NORMATIVE DATA

Table 32.3 presents vertical jump normative data for female and male Australian indoor volleyball athletes. Tables 32.4 and 32.5 present vertical jump normative data for female and male Australian beach volleyball athletes on sand and hard surfaces, respectively.

TABLE 32.3 Vertical Jump Data for Male and Female Indoor Volleyball Players: Mean ± *SD* (Range)

Gender	Countermovement jump, cm	Depth jump (from 30 cm), cm	Spike jump, cm
Female	291 ± 11	292 ± 11	299 ± 12
	(276-308)	(279-311)	(284-315)
Male	330 ± 5	330 ± 6	344 ± 7
	(318-338)	(318-340)	(332-366)

Typical error: countermovement jump = 2.1 cm; depth jump = 2.2 cm; spike jump = 2.1 cm. Data source: female, Australian junior (*n* = 9) and senior national (*n* = 2); male, Australian junior (*n* = 14) and senior national team (*n* = 12).

TABLE 32.4 Vertical Jump Data for Male and Female Beach Volleyball Players on Sand: Mean ± *SD* (Range)

Gender	Countermovement jump, cm	Block jump, cm	Spike jump, cm
Female	293 ± 11	285 ± 12	296 ± 13
	(277-310)	(268-302)	(278-312)
Male	332 ± 6	317 ± 4	335 ± 4
	(323-338)	(310-321)	(330-342)

Typical error: countermovement jump = 1 cm; block jump = 2 cm; spike jump = 1 cm. Data source: female, Australian junior (*n* = 4) and senior national (*n* = 2); male, Australian junior (*n* = 2) and senior national team (*n* = 4).

TABLE 32.5 Vertical Jump Data for Male and Female Beach Volleyball Players on Hard Surface: Mean ± *SD* (Range)

Gender	Countermovement jump, cm	Block jump, cm	Spike jump, cm
Female	297 ± 13	286 ± 11	303 ± 12
	(278-316)	(270-303)	(287-316)
Male	331 ± 6	319 ± 5	341 ± 5
	(321-336)	(310-323)	(333-349)

Typical error: countermovement jump = 2 cm; block jump = 1 cm; spike jump = 2 cm. Data source: female, Australian junior (*n* = 4) and senior national (*n* = 2); male, Australian junior (*n* = 2) and senior national team (*n* = 4).

LOWER BODY SPEED-STRENGTH AND STRENGTH TESTING ──

RATIONALE

Speed-strength measures, maximal strength (e.g., 1RM squat), and heavy load power (e.g., 1RM clean) are important measures for a volleyball player, because they are associated with direct performance measures such as CMJ and SPJ (Sheppard et al. 2008d).

TEST PROCEDURE

Specific guidelines for the conduct of upper- and lower-body maximal strength tests are contained in chapter 13. Typical strength movements tested for volleyball are 1RM squat, clean, and bench press. Speed-strength assessment includes the incremental power load profile for jump squats with loads beginning with body mass, body mass + 25%, and body mass + 50%.

A large range of tests are available for coaches to use in assessing strength. Given that muscle imbalances are linked to injury, it is important to assess muscular strength of both the upper and lower body and to assess strength in the anterior and posterior planes bilaterally. Because fatigue accumulates quickly during maximal efforts, the number of tests should be kept to a minimum. Therefore, the squat, bench press, and bench pull are the lifts to be tested.

Although the single repetition maximum (1RM) test is generally considered the most representative of strength, there are concerns about safety, reliability, and the maintenance of technique when using 1RM tests for individuals with low training age (Baechle and Earle 2008). Therefore, 3RM testing is recommended for athletes not specifically trained for strength.

The following general guidelines must be adhered to for all tests:

- Strength testing should be performed on a separate day from the field tests. Strength and field tests should be separated by 48 h.
- The athlete must perform an appropriate warm-up. As a minimum, all athletes are required to perform a trial at approximately 90% of specified repetition maximum for each test. If this is the first test, the athlete should perform an initial trial at approximately 90% of weight lifted in training.
- Lowering and lifting actions must be performed in a continuous manner. A single rest of no more than 2 s is allowed between repetitions.
- A maximum of 5 min recovery between trials is allowed.
- Minimum weight increments of 2.5 kg should be used between trials. However, increments should be guided by ease of each trial.
- The specified RM test should be completed within four trials (not including the warm-up).
- If the athlete is unable to complete tests as per protocol, this should be noted on testing results information, and values should not be included in any mathematical calculations (e.g., average, TE).
- A spotter other than the supervising coach should be used when spotting is required.

NORMATIVE DATA

Table 32.6 presents strength norms for male Australian indoor volleyball athletes.

TABLE 32.6 Strength Data for Male Indoor Volleyball Players: Mean ± *SD* (Range)

Player group	3RM squat, kg	3RM bench press, kg	1RM power clean, kg
National senior	139 ± 30 (110-210)	91 ± 13 (68-110)	104 ± 13 (85-130)
National junior	115 ± 25 (80-155)	71 ± 15 (55-100)	85 ± 13 (70-105)

Data source: Australian junior (*n* = 2) and senior national team (*n* = 4).

INDOOR VOLLEYBALL REPEATED-EFFORT TEST ─────────

RATIONALE

The indoor volleyball repeated-effort test (RET) involves jumping and movement activity that is specific to indoor volleyball, with duration and rest periods that replicate a portion of the most extreme demands of a match (Sheppard and Gabbett 2007). Research results have demonstrated that the RET is a reliable method of assessing repeated-effort ability in volleyball players, is able to discriminate between higher and lower performers, and is sensitive to specific training interventions (Sheppard et al. 2007, 2008a). As such, the RET is a useful test for practitioners to evaluate volleyball players' repeated jump and speed movements.

TEST PROCEDURE

The RET encompasses spiking, blocking, and lateral movements and was developed to mimic the physical demands of frontcourt play in volleyball. Because frontcourt play generally involves the most activities (jumping and lateral movement), the test was designed to reflect these demands. It was decided that to assess the most extreme demands of volleyball play, the test should include repetitions that reflected the typical rally time and involve rest periods that reflected the most extreme demands of international play (Sheppard et al. 2007). Four repetitions of the RET are performed to reflect these extreme demands, with the repetitions commencing at 20 s intervals to allow approximately 4 to 8 s of rest between each repetition depending on the speed at which the athlete executes each repetition.

As figure 32.1 illustrates, two spike jumps are measured using a vertical jump apparatus, and the speed of movements is measured using a timing light system. The timing of the specific movement is separated from the spike jumping task. In other words, the timing as measured by the timing lights is the movement speed performance measure and represents a portion of the total repetition time (which is fixed for all subjects as a 20 s interval). The block jumps are performed on a specially designed apparatus that is adjustable for height and position, and these movements are included in the time measure on each repetition. The apparatus, which is 2 m wide, is placed on the opposite side of the net from the player being tested. Two volleyballs are secured to 2 separate supports that are mounted onto an aluminum beam and supported by a tripod or other suitable apparatus. The bottom of the ball is 15 cm above the top of the net (258 cm high for males, 239 cm for females), and the side of the ball nearest the athlete is placed 15

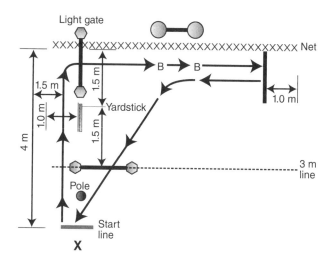

FIGURE 32.1 Illustration of the repeated-effort test for volleyball.

cm into the opposing court. To correctly execute a block, the athlete is required to perform a block jump and place both hands on the ball without contacting the net. Any attempts during the execution of the RET in which the athlete does not correctly execute a block is recorded as an error, and this is included in his or her RET results.

Players should be well familiarized with all testing procedures prior to beginning data collection. Practitioners are encouraged to perform the test as a training drill two or more times before collecting data. After a thorough explanation of the test procedures, the data collection session should begin with the athlete's typical volleyball match-specific warm-up. Following this, one or two trials of one repetition are performed at a submaximal intensity, after which feedback can be provided to the athlete to ensure clarity of instructions and adherence to the test protocol. The athlete then performs one maximal repetition of the test at full intensity. Following 3 to 5 min of rest, the athlete performs the actual test battery, which includes four repetitions of the test. The measured performance outcomes of *each* of the four repetitions include 2 spike jump scores (cm), lateral movement time (s), and any errors that may have occurred. The first spike jump and the movement time in the first repetition are used as the "ideal jump" and "ideal time" scores. The mean of all spike jumps recorded (total of 8) and the mean of all movement times recorded (total of 4) across the four repetitions are used as the actual jump and actual time scores. Errors are recorded if the blocking task, as outlined previously, is not performed correctly (e.g., one hand touching or hand on net).

DATA ANALYSIS

Reliability data from a group of junior national team players are presented in table 32.7. Reliability of the percentage decrement of time and jump variables calculated from the volleyball RET has resulted in very large TE scores (Sheppard et al. 2007), which is in agreement with previous findings using this calculation (Spencer et al. 2006). With such large TE scores observed for jump and time decrements, it is unlikely that practitioners can find utility in these calculations because extremely large changes would need to occur for practitioners to confidently interpret these changes as being real versus being due to normal variation.

It has been observed that within each group, the fastest athletes on the movement time test and the highest jumpers demonstrated the largest percentage decrement when expressed in relative terms. When expressed in absolute terms, these superior athletes' scores can be interpreted differently. In other words, when absolute ideal and actual scores are viewed, the superiority of higher-performing athletes is clearly demonstrated despite a higher percentage decrement (table 32.7), which is also evidenced in comparisons of higher and lower performers (figures 32.2 and 32.3). Practitioners are encouraged to forego calculating a percentage decrement for the volleyball RET and instead plot absolute scores across repetitions. This will allow the practitioner to take into consideration the best scores as well as the average scores achieved as a reflection of fatigue. Five major variables are used to evaluate performance in the volleyball RET: ideal jump, actual jump, ideal time, actual time, and errors. Because the actual score for both spike jumps and movement time takes into consideration all of the efforts performed, this score reflects the fatigue resistance of the athlete. This allows the practitioner to evaluate the athlete based on their ideal and actual

TABLE 32.7 Reliability and Normative Data for the Indoor Volleyball Repeated-Effort Test (RET) for Junior National Team Athletes: Mean ± SD

	Day 1 scores	Day 2 scores	TE	%TE	ICC
Actual time, s	7.43 ± 0.6	7.27 ± 0.6	0.16	2.24	.93
Ideal time, s	7.15 ± 0.6	6.97 ± 0.7	0.23	3.54	.87
Actual jump, cm	327.76 ± 8.6	328.48 ± 9.0	1.81	0.55	.96
Ideal jump, cm	334.25 ± 10.3	335.25 ± 9.8	3.15	0.95	.90
Errors	2.82 ± 2.3	2.91 ± 2.9	1.49	22.08	.57
% Jump decrement	1.97 ± 0.9	2.06 ± 1.0	1.03	88.01	.15
% Time decrement	3.94 ± 2.3	4.51 ± 3.5	2.49	82.38	.31

Results are presented as typical error (TE), relative typical error (%TE), and intraclass correlations (ICC) ($n = 12$).

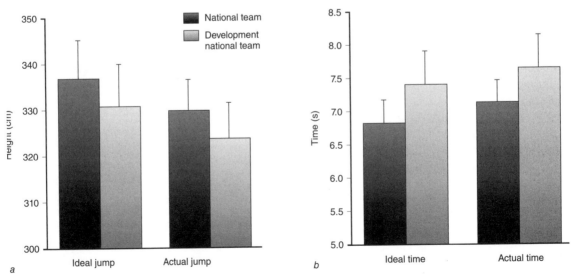

FIGURE 32.2 Mean ± standard deviation of (a) ideal jump height and actual jump height and (b) ideal time and actual time for the national team ($n = 8$) and the development national team ($n = 8$). All measures depicted were observed to have large (>0.50) effect size differences between groups.

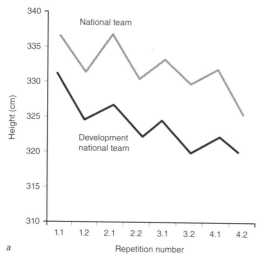

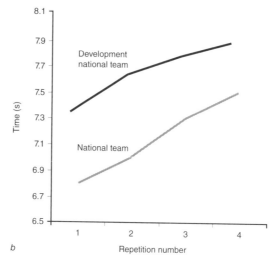

FIGURE 32.3 Mean movement time (a) and mean jump height (b) for the national team (n = 8) and the development national team (n = 8) across the four repetitions of the repeated-effort test.

score as a measure of fatigue resistance. Examining the ideal time and ideal jump results of the repeated-effort test provides the practitioner insight into the movement speed and jumping ability of volleyball athletes. By comparing the actual time and ideal time, the practitioner can determine the athlete's fatigue resistance. Finally, an examination of the frequency, type, and progression (over each repetition) of errors allows the practitioner insight into the movement

and jump technique of the athlete as well as the influence of her progressing fatigue on technique.

NORMATIVE DATA

Table 32.8 provides reliability and normative data for the indoor volleyball RET for junior national team athletes, whereas figures 32.2 and 32.3 provide normative data comparing junior and senior national team athletes on the RET.

TABLE 32.8 Comparison of Absolute Scores and Percentage Decrement Scores Between Two Athletes

	ATHLETE A			ATHLETE B		
	Ideal	Actual	% Decrement	Ideal	Actual	% Decrement
Jumps, cm	347	334	3.66	320	316	1.11
Time, s	6.88	7.12	3.60	7.92	8.09	2.08

References

Baechle, T.R., and Earle, R.W. 2008. Essentials of strength training and conditioning. Champaign, IL: Human Kinetics.

Batista, G.R., De Araujo, R.F., and Guerra, R.O. 2008. Comparison between vertical jumps of high performance athletes on the Brazilian men's beach volleyball team. *Journal of Sports Medicine and Physical Fitness* 48:172-176. Bobbert, M.F. 1990. Drop jumping as a training method for jumping ability. *Sports Medicine* 9(1):7-22.

Bobbert, M.F., Mackay, M., Schinkelshoek, D., Huijing, P.A., and van Ingen Schenau, G.J. 1986. Biomechanical analysis of drop and countermovement jumps. *European Journal of Applied Physiology and Occupational Physiology* 54(6):566-573.

Dyba, W. 1982. Physiological and activity characteristics of volleyball. *Volleyball Technical Journal* 6(3):33-51.

Fédération Internationale de Volleyball 2009, Fédération Internationale de Volleyball Château Les Tourelles, Edouard-Sandoz 2-4, 1006 Lausanne, Switzerland. www.fivb.org/EN/BeachVolleyball/Competitions/WorldTour/2009x/.

Gabbett, T., Georgieff, B., and Domrow, N. 2007. The use of physiological, anthropometric, and skill data to predict selection in a talent-identified junior volleyball squad. *Journal of Sports Sciences* 25(12):1337-1344.

Giatsis, G. 2001. Jumping quality and quantitative analysis of beach volleyball game. In: S. Tokmakidis (ed), 9th International Congress on Physical Education and Sport. p 95.

Giatsis, G., and Tzetzis, G. 2003. Comparison of performance for winning and losing beach volleyball teams on different court dimensions. *International Journal of Performance Analysis in Sport* 3(1):65-74.

Gladden, B.L., and Colacino, D. 1978. Characteristics of volleyball players and success at a national tournament. *Journal of Sports Medicine and Physical Fitness* 18:57-64.

Grgantov, Z., Katie, R., and Marelic, N. 2005. Effect of new rules on the correlation between situation parameters and performance in beach volleyball. *Collegium Antropologicum* 29(2):717-722.

Heimer, S., Misogoj, M., and Medved, V. 1988. Some anthropological characteristics of top volleyball players in FSR Yugoslavia. *Journal of Sports Medicine and Physical Fitness* 28(2):200-208.

Homberg S., and Papageorgiou, A. *Handbook for Beach Volleyball*. Aachen: Meyer & Meyer Verlag, 1994.

Koch, C., and Tilp, M. 2009. Analysis of beach volleyball action sequences of female top athletes. *Journal of Human Sport and Exercise* 4(3):272-283.

Marques, M.C., Tillaar, R.V.D., Vescovi, J.D., and Gonzalez-Badillo, J.J. 2008. Changes in strength and power performance in elite senior female professional volleyball players during the in-season: A case study. *Journal of Strength and Conditioning Research* 22(4):1147-1155.

Newton, R.U., Kraemer, W.J., and Hakkinen, K. 1999. Effects of ballistic training on preseason preparation of elite volleyball players. *Medicine and Science in Sports and Exercise* 31(2):323-330.

Polglaze, T., and Dawson, B. 1992. The physiological requirements of the positions in state league volleyball. *Sports Coach* 15(1):32-37.

Riggs, M.P., and Sheppard, J.M. 2009. The relative importance of strength and power qualities to vertical jump height of elite beach volleyball players during the counter-movement and squat jump. *Journal of Human Sport and& Exercise* 4(3):221-236.

Sheppard, J.M., and Gabbett, T. 2007. The development and evaluation of a repeated effort test for volleyball. Paper presented at the National Strength and Conditioning Association Annual Conference, Atlanta, Georgia, July 10-12.

Sheppard, J.M., Gabbett, T.J., Taylor, K.L. Dorman, J., Lebedew, A.J., and Borgeaud, R. 2007. Development of a repeated-effort test for elite men's volleyball. *International Journal of Sports Physiology and Performance* 2(3):292-304.

Sheppard, J.M., and Borgeaud, R. 2008. Influence of stature on movement speed and repeated efforts in elite volleyball players. *Journal of Australian Strength and Conditioning* 16(3):12-14.

Sheppard, J.M., Gabbett, T., and Borgeaud, R. 2008a. Training repeated effort ability in national team male volleyball players. *International Journal of Sports Physiology and Performance* 3(3):397-400.

Sheppard, J.M., Chapman, D., Gough, C., McGuigan, M., and Newton, R.U. 2008b. The association between changes in vertical jump and changes in strength and power qualities in elite volleyball players over 1 year. *National Strength and Conditioning Association Annual Conference Abstracts - Journal of Strength and Conditioning Research* 22(6):1-115.

Sheppard, J.M., Cormack, S., Taylor, K.L., McGuigan, M, and Newton, R.U. 2008c. Assessing the force-velocity characteristics of well trained athletes: The incremental load power profile. *Journal of Strength and Conditioning Research* 22(4):1320-1326.

Sheppard, J.M., Cronin, J., Gabbett, T.J., McGuigan, M., Etxebarria, N., and Newton, R.U. 2008d. Relative importance of strength and power qualities to jump performance in elite male volleyball players. *Journal of Strength and Conditioning Research* 22(3):758-765.

Sheppard, J.M., Hobson, S., Barker, M., Taylor, K.L., Chapman, D., McGuigan, M., and Newton, R.U. 2008e. The effect of training with accentuated eccentric load countermovement jumps on strength and power characteristics of high-performance volleyball players. *International Journal of Sports Science and Coaching* 3(3):355-363.

Sheppard, J.M., McGuigan, M.R., and Newton, R.U. 2008f. The effects of depth-jumping on vertical jump performance of elite volleyball players: An examination of the transfer of increased stretch-load tolerance to spike jump performance. *Journal of Australian Strength and Conditioning* 16(4):3-10.

Sheppard, J. M., Chapman, D., Gough, C., McGuigan, M., and Newton, R.U. 2009a. Twelve month training induced changes in elite international volleyball players. *Journal of Strength and Conditioning Research* 23(7):2096-2101.

Sheppard, J.M., Gabbett, T., and Stanganelli, L. 2009b. An analysis of playing positions in elite men's volleyball: Considerations for competition demands and physiological characteristics. *Journal of Strength and Conditioning Research* 23(6):1858-1866.

Smith, D.J., Roberts, D., and Watson, B. 1992. Physical, physiological and performance differences between Canadian national team and university volleyball players. *Journal of Sports Sciences* 10(2):131-138.

Smith, D.J., Stokes, S., and Kilb, B. 1987. Effects of resistance training on isokinetic and volleyball performance measures. *Journal of Applied Sports Science Research* 1(3):42-44.

Spence, D.W., Disch, J.G., Fred, H.L., and Coleman, A.E. 1980. Descriptive profiles of highly skilled women volleyball players. *Medicine and Science in Sports and Exercise* 12(4):299-302.

Spencer, M., Fitzsimmons, M., Dawson. B., Bishop, D., and Goodman, C. 2006. Reliability of a repeated-sprint test for field-hockey. *Journal of Science and Medicine in Sport* 9(1):181-184.

Stanganelli, L.C., Dourado, A.C., Oncken, P., Mancan, S., and da Costa, S.C. 2008. Adaptations on jump capacity of Brazilian volleyball players prior to the under 19 World championships. *Journal of Strength and Conditioning Research* 22(3):741-749.

Turpin, J.P.r.A., Cortell, J.M., Chinchilla, J.J., Cejuela, R., and Suarez, C. 2008. Analysis of jump patterns in competition for elite male beach volleyball players. *International Journal of Performance Analysis in Sport* 8(2):94-101.

Viitasalo, J.T. 1991. Evaluation of physical performance characteristics in volleyball. *International Volleyball Technique* 3:4-8.

Viitasalo, J.T., Rusko, H., Pajala., O and Rahkila, P. 1987. Endurance requirements in volleyball. *Canadian Journal of Applied Sports Science* 12(4):194-201.

Water Polo Players

Ted Polglaze, Frankie Tan, Shaun D'Auria, and Sally A. Clark

Water polo is played by two opposing teams in a pool measuring 25 m × 20 m (women) and 30 m × 20 m (men). Teams are made up of seven players—six field players and one goalkeeper. There are six substitutes on the bench who interchange regularly with those in the pool. A match is divided into four quarters, each comprising 8 min of actual playing time. All field players are involved at both ends of the pool, in both attack and defense. Field players are divided into two main positional groups—center and perimeter.

There have been significant rule changes in recent years that have affected the demands of the game and consequently the fitness requirements of the players. The current battery of tests has been developed following extensive analysis of international-level women's match play under prevailing rules. The tests have been selected because they provide a profile of the physiological characteristics necessary for elite-level performance in water polo.

Athlete Preparation

To ensure test reliability, the following specific pretest conditions should be observed in addition to those outlined in chapter 2, Pretest Environment and Athlete Preparation.

- *Diet.* Athletes should be encouraged to consume a series of high-carbohydrate meals in the 36 h before testing and to arrive on test day well hydrated. No food, cigarettes, or beverages containing alcohol or caffeine are to be consumed in the 2 h prior to testing.

- *Training.* No training that might induce severe physiological or neural fatigue should be undertaken in the 24 h prior to testing. This includes any high-intensity skills or physical training.

- *Testing.* All test sessions should be scheduled at the same time of day to avoid fluctuations in physiological response to the set protocols due to circadian rhythm (Winget et al. 1985). All tests must be completed in the recommended order, with adequate and consistent recovery time between tests. This will promote optimal performance and allow for a valid comparison of results on different test occasions.

Test Environment

All testing should be carried out in a pool of sufficient depth that players cannot touch the bottom while in a vertical body position with head out of water (i.e., not less than 1.8 m). Pool temperature should comply with FINA guidelines that require the water temperature to be no less that 26 ± 1 °C. Side lane ropes or wave arresting edges are advisable to minimize turbulence.

Recommended Test Order

It is important that tests are completed in the same order to control the interference between tests. This order also allows valid comparison of different test occasions. Anthropometry can be completed either at the start of the test session or at a separate session, provided it is done prior to training. The recommended order is as follows:

Day	Tests
1	Anthropometry
	10 m sprint
	In-water vertical jump
	Multistage shuttle swim
2	Repeat sprint

Equipment Checklist

Anthropometry

- ☐ Stadiometer (wall mounted)
- ☐ Balance scales (accurate to ±0.05 kg)
- ☐ Anthropometry box
- ☐ Skinfold calipers (Harpenden skinfold caliper)
- ☐ Marker pen
- ☐ Anthropometric measuring tape
- ☐ Recording sheet (refer to the anthropometry data sheet template)
- ☐ Pen

10 m Sprint Test and Repeat Sprint Test

- ☐ Digital video camera, tripod, and tape
- ☐ Measuring tape
- ☐ Starting rope (two for repeat sprint test)
- ☐ Calibration rope
- ☐ Marking cones
- ☐ Stopwatch
- ☐ Recording sheet (refer to the water polo testing data sheet template)
- ☐ Pen

In-Water Vertical Jump Test

- ☐ Yardstick jumping device (e.g., Swift Performance Yardstick) including specific attachments for water jump
- ☐ Extra weight for base
- ☐ Measuring tape
- ☐ Spirit level
- ☐ Recording sheet (refer to the water polo testing data sheet template)
- ☐ Pen

Multistage Shuttle Swim Test

- ☐ Sound box or CD player and extension cord
- ☐ Multistage swim test CD (or MP3 file)
- ☐ Measuring tape
- ☐ Two swimming lane ropes
- ☐ Tightening tool
- ☐ Recording sheet (refer to the water polo testing data sheet template)
- ☐ Pen

ANTHROPOMETRY

RATIONALE

Water polo involves direct physical contact between opponents, so physical size and body mass are both important factors in performance. Tan and colleagues (2009a) demonstrated that International standard players were taller and heavier than their National standard counterparts and that perimeter players were lighter and leaner than center players.

TEST PROCEDURE

Height, body mass, and sum of skinfolds are measured. For senior athletes, height need only be measured at the start of the season. Skinfolds are recorded over seven sites (triceps, biceps, subscapular, supraspinale, abdominal, front thigh, and medial calf). The individual skinfold measures as well as the sum of the seven sites should be reported. Refer to anthropometry protocols outlined in chapter 11, Assessment of Physique, for a detailed description of all anthropometric test procedures.

Trochanter height and reach height need to be assessed to determine the hip-to-tip component for the in-water jump. Trochanter height is measured according to the procedures outlined by Norton and colleagues (1996), using an anthropometric box and segmometer. Reach height is measured using a Yardstick jumping device in the same manner as for a vertical jump; the only difference is that the athlete must be barefoot. Ensure that the athlete has feet together and heels flat, reaches as high as possible directly overhead, elevates the scapula, and has straight arms and fingers. The trochanter and reach height—and hence hip-to-tip component—need only be measured at the start of each season.

DATA ANALYSIS

Although the description of skinfold measurement procedures seem simple, a high degree of technical skill is essential for consistent results. It is therefore important that these measurements be taken by an experienced tester who has been trained in these techniques. It is also important that the same tester conduct each retest to ensure reliability.

NORMATIVE DATA

Table 33.1 presents anthropometric normative data for female water polo players.

TABLE 33.1 Anthropometric Data for National Female Water Polo Players: Mean ± *SD* (Range)

Position	Height, m	Body mass, kg	Σ7 skinfolds, mm
Center (n = 7)	1.77 ± 0.06 (1.70-1.83)	80.5 ± 6.9 (71.7-91.7)	102.0 ± 21.5 (65.6-126.6)
Perimeter (n = 12)	1.74 ± 0.06 (1.67-1.82)	70.2 ± 5.6 (61.7-83.6)	90.7 ± 20.3 (65.9-142.7)
Goalkeeper (n = 4)	1.74 ± 0.05 (1.69-1.83)	76.8 ± 5.3 (69.5-82.1)	98.9 ± 18.3 (75.7-114.0)

Typical error: body mass = 0.02 kg; Σ7 skinfolds = 2.6 mm. Data source: National senior women's squad, 2007-2008.

10 M SPRINT TEST

RATIONALE

The requirement to accelerate and sprint at maximal or near-maximal intensity is an important aspect of water polo. The acceleration to full speed can occur either from a stationary position or from a lower-paced swim and sometimes involves a change of stroke. Therefore, both off-the-mark acceleration and top-end speed are necessary. Analyses of both international competition (Rechichi et al. 2005) and individual time trials (Polglaze, unpublished data 2005) indicate that elite water polo players reach peak velocity in approximately 3 m (from a stationary start) and maintain that velocity to approximately 10 m. Furthermore, nearly 80% of all swims are less than 10 m in distance, particularly those at peak velocity (Rechichi et al. 2005). The reliability of the 10 m sprint test was established by Tan and colleagues (2009a), who demonstrated that this test can discriminate between players of different competition standards and playing positions.

TEST PROCEDURE

The positioning of all equipment for the 10 m sprint test is outlined in figure 33.1.

- A 0 m start line is set across the pool, coinciding with the anchor points for a lane rope. From this point, markers are placed at 3, 7, and 10 m from the start line on either side of the pool. The start line should be at least 3 m from the side of the pool to ensure that athletes cannot push off from the wall.

- A digital video camera, filming at 50 or 100 Hz, is positioned on the opposite side of the pool, preferably in the stands, with the viewing width set to ensure that both start and finish markers are clearly visible. The camera is placed approximately halfway between the start and finish lines.

- If it is dark outside or there are no windows, spotlights over the pool should be turned on. These are not required if testing is done during daylight hours. A polarizing filter should be fitted to the camera to minimize glare.

- Once the camera is set in position, it must not be moved until all trials are completed and the calibration rope has been filmed.

- A start rope is attached to the anchor point at the 0 m mark on the opposite side of the pool. On the main side, the rope is held by an assistant so that it is level with the water surface and goes directly over the 0 m marking on the side of the pool.

- Either before or after the testing, the calibration rope needs to be filmed so that virtual lines can be overlaid onto the video analysis program. A brightly colored, thick rope should be used to ensure visibility. While the camera is recording, the tester holds the rope taut between the 0 m points on either side of the pool. The tester waves to the camera to signify the rope is in position and keeps it there for approximately 3 s to ensure it can be seen clearly when the tape is played back. This procedure is repeated for 3, 7, and 10 m lines.

- Athletes should wear either a black or brightly colored fluorescent swim cap to ensure adequate contrast with splashing water.

- The start list of the athletes is recorded on a datasheet, and athletes are told their starting order.

- The "record" button on the camera is pressed just prior to the first athlete commencing the first trial. The camera is left running until all trials have been completed and the calibration rope has been filmed (if being completed after testing).

- The athlete starts in front of the start rope, with some part of her head touching it. The rope must be held tightly so that the athlete does not push it behind the 0 m line. The athlete can start from either a front-on or side-on position, as long as she maintains contact with the rope.

- In her own time, the athlete starts swimming as quickly as possible toward and past the finish marker. As soon as she starts moving, the rope should be quickly pulled up to avoid interfering with the athlete's kick. The athlete should continue swimming as fast as possible until she has passed the 10 m marker. (Ensure that the athlete swims parallel to the side of the pool rather than at an angle.)

- Once the first athlete has taken off, the next athlete is able to position herself on the rope and commence the trial.

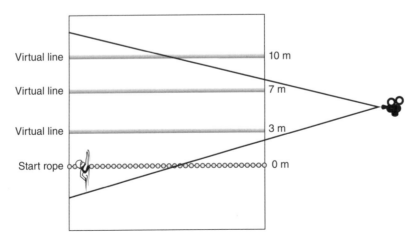

FIGURE 33.1   Setup for 10 m sprint test.

- Athletes complete at least three attempts, with at least 2 min of recovery between trials.
- Each athlete completes the first trial before the group moves to the second and then third trial. Athletes must complete their trials in the order on the datasheet so that they can be identified in subsequent video analysis.

DATA ANALYSIS

Timing starts from one frame before the head leaves the start rope.

- Split times are recorded when the head (swim cap) intercepts the virtual line at each position.

- Fastest measured 0 to 3 m time is recorded as the acceleration score.
- Fastest measured 7 to 10 m time is recorded as the speed score.
- Fastest measured 0 to 10 m time is recorded as a combined acceleration and speed score.

NORMATIVE DATA

Normative data for female water polo players for the 10 m sprint test are presented in table 33.2.

TABLE 33.2   10-m Sprint Test Data for National Female Water Polo Players: Mean ± *SD* (Range)

Position	0-3 m split, s	0-10 m time, s	7-10 m split, s
Center (*n* = 6)	1.85 ± 0.05 (1.80-1.92)	6.05 ± 0.21 (5.80-6.34)	1.77 ± 0.07 (1.68-1.86)
Perimeter (*n* = 10)	1.77 ± 0.05 (1.70-1.86)	5.91 ± 0.24 (5.66-6.38)	1.78 ± 0.10 (1.62-1.94)
Goalkeeper (*n* = 3)	1.84 ± 0.05 (1.80-1.90)	6.15 ± 0.34 (5.80-6.48)	1.82 ± 0.11 (1.70-1.92)

Typical error: 0-3 m = 0.06 s, coefficient of variation = 3.0%; 0-10 m = 0.11 s, coefficient of variation = 1.9%; 7-10 m = 0.05s, coefficient of variation = 3.0%. Data source: National senior women's squad, 2007-2008.

IN-WATER VERTICAL JUMP

RATIONALE

The ability to raise the body as high as possible out of the water is an important component of fitness for water polo. Not only does it allow the athlete to pass, receive, and intercept more effectively (Platanou 2005), it is also significantly related to shooting speed (Sanders 1999). The reliability of the in-water vertical jump test has been established (Platanou 2005; Tan et al. 2009a). Absolute jump height can discriminate players of different competition standards, and relative jump height varies according to playing position (Tan et al. 2009a).

The in-water vertical jump test is not a test of pure leg power, because factors such as segment speed, pitch, and range of motion are major contributors to final jump height (Sanders 1999). Assessment of these technical factors via video analysis at the time of testing will give a more complete indication of individual jumping ability and progress.

TEST PROCEDURE

The setup and position of the Yardstick for water testing are illustrated in figure 33.2.

FIGURE 33.2 The Yardstick jumping device can be fitted with customized connections that allow testing to be done in a pool (i.e., to measure jump height for water polo players).

Photo courtesy of Frankie Tan.

Setup

- Connect the base stands, upright pole, and extension pole for each of the two Yardsticks.

- Clip the connecting sleeve attachment into the extension pole of each of the two Yardsticks. The upright poles must be the same width apart as the clips on the connecting sleeve.

- Clip the elbow joint onto the extension arm and then slide the arm into the connecting sleeve.

- Invert the vane stack from one of the Yardsticks and connect it to the elbow joint, ensuring that the spring-loaded button is firmly in place.

- Align all the vanes so that they are parallel to the connecting sleeve, and then tighten the plastic sleeve. (If vanes are not aligned, the stack will hang down at an angle rather than vertically.)

- Check from behind to ensure that the two uprights and the vane stack are parallel and aligned in the vertical plane.

- When the device is assembled for water testing, the vane stack is hanging upside down, and hence an additional set of number stickers are required that run in the direction opposite to those used for land testing.

- Adjust the height by raising the extension pole in both uprights, ensuring that the connecting sleeve remains horizontal. (With both poles set at 160 cm, the bottom vane is 35 cm above pool deck.)

- Adjust distance from edge by either moving the front clip on the connecting sleeve back (thus bringing the two uprights closer) or sliding the extending arm farther out. (The first option is preferable, as the connecting sleeve is a more rigid structure and less likely to bend.)

- Position the Yardstick so that vane stack is at least 1 m in from the edge of the pool.

- Set the two uprights to the same offset value (normally third hole for women, fifth hole for men). Use a spirit level to check that the connecting sleeve is horizontal. If not, adjust the height of the back pole as required. The front pole must stay at the set height; otherwise, it will affect the offset value for the first vane. Record the offset value on the datasheet.

- Place weights on the base stands to hold them in place.

- For venues where the pool deck and water surface level are at different heights, record the difference in height between the two.

- Ensure that all vanes are aligned and parallel with the central crossbar. For athletes with previous results, rotate lower vanes 180° (i.e., back toward the edge of the pool) so that only those in likely jumping range are facing out (e.g., ~10 cm below previous result).

Procedure

- Instruct the athlete to assume a self-selected starting position directly underneath the device, and then propel herself as high as possible out of the water, reaching directly upward at the peak of the jump to displace as many vanes as possible.

- Allow at least three trials but continue until the athlete fails to improve on two consecutive occasions.

- Record the height of the highest displaced vane.

DATA ANALYSIS

- Calculate the height of the bottom vane by summing the offset height and the difference in height between pool deck and water level.

- Determine maximal vertical jump height based on the distance from the water surface to the bottom vane plus the number of vanes displaced. Report this score as the athlete's absolute jump height.

- To allow for comparison between players of varying height and arm lengths, report jump height relative to the athlete's "hip-to-tip" length, using the following equations:

$$\text{Hip-to-Tip} = \text{Standing Reach Height} - \text{Trochanter Height}$$

$$\text{Relative Jump Height} = (\text{Jump Height} / \text{Hip-to-Tip}) \times 100.$$

Thus, hip-to-tip measurement is reported as a percentage score—100% indicates that the athlete raised her hip to surface level in the water. Scores above 100% indicate that the hip went above surface level, whereas scores below 100% indicate that the hip did not reach surface level.

Relative jump can also be calculated as a distance:

$$\text{Relative Jump Height} = \text{Jump Height} - \text{Hip-to-Tip}.$$

with scores reflecting the height of the hip either above (positive) or below (negative) water surface level.

NORMATIVE DATA

In-water vertical jump normative data for female water polo players are presented in table 33.3.

Table 33.3 In-Water Vertical Jump Data for National Female Water Polo Players: Mean ± *SD* (Range)

Position	Absolute jump height, cm	Relative jump height, %
Center (n = 6)	141 ± 8 (128-150)	101 ± 3 (97-104)
Perimeter (n = 10)	141 ± 5 (136-153)	104 ± 3 (100-109)
Goalkeeper (n = 3)	142 ± 2 (140-144)	106 ± 5 (100-111)

Typical error: absolute = 1.5 cm; coefficient of variation = 1.1%. Data source: National senior women's squad, 2007-2008.

MULTISTAGE SHUTTLE SWIM TEST

RATIONALE

Endurance, and more specifically aerobic power, is an important fitness component for team sports, allowing the athlete to maintain her work rate throughout a match (Helgerud et al. 2001; McMahon and Wenger 2000). This is particularly important for water polo players, who are continually involved in the play and must move between both ends of the pool as the game unfolds rather than rest when play moves to the other end.

The reliability and validity of the multistage shuttle swim test (MSST) were established by Rechichi and colleagues (2000), who showed that aerobic power accounted for 78% of the variance for MSST score; speed and agility also contributed to the result. The movement patterns required for the test (stop, start, and change of direction) replicate those observed in games, meaning the test has good logical validity compared with constant-pace, straight-line swimming protocols. Tan and colleagues demonstrated that the MSST can discriminate players of different competition standards and playing positions (2009a) and that it is a valid tool to determine peak heart rate (2009b). Goalkeepers are not required to complete the MSST.

TEST PROCEDURE

Setup

- Verify the CD player using the timing verification (track 24).
- Set up the test in an area so that the depth of the pool prevents the player from touching or pushing off the bottom.
- Measure 10 m and place lane ropes across the pool at either end of the measured distance. (Lanes are 2.5 m in width, so using four lanes should give you the correct distance.) Lane ropes need to be extremely tight to prevent movement during the test.

- Place each lane rope at least 2 m from the pool wall to prevent the player from pushing off the wall when turning.
- If more than one player is being tested, there should be approximately 2 m between players. It is advisable to test no more than 5 athletes at one time.

A description of tracks on the MSST CD is detailed in table 33.4.

Procedure

- The athlete swims the 10 m distance between lane ropes (shuttle) within a progressively decreasing cued time frame. The timing is based on an audio cue, where the interval between signals is reduced approximately every minute (level). Level 1 requires a swimming velocity of 0.9 m · s⁻¹, which is increased by 0.05 m · s⁻¹ with each subsequent level.
- At the end of each shuttle, athletes are required to touch and immediately release the lane rope. If they reach the rope before the signal, they should then remain stationary while awaiting the cue for the commencement of the next shuttle.
- Individual athletes should be given a warning if they fail to be within one arm stroke of the lane rope as the audio cue is sounded.
- An athlete should be removed from the test if she fails to be within one stroke of the lane rope on two consecutive occasions.
- The final level and shuttle completed immediately before the athlete was eliminated is recorded as her score.
- Given the varying number of shuttles within each level, a look-up table is required to convert scores to decimal values. This conversion should be used when calculating squad averages and other descriptive statistics.

TABLE 33.4 Description of Multistage Shuttle Swim Test

Track	Content	
1	Introduction	
2	Subject instructions	
3	Countdown to the start of the test	

Track	Level	Shuttles
4	1	1 2 3 4 5
5	2	1 2 3 4 5 6
6	3	1 2 3 4 5 6
7	4	1 2 3 4 5 6
8	5	1 2 3 4 5 6 7
9	6	1 2 3 4 5 6 7
10	7	1 2 3 4 5 6 7
11	8	1 2 3 4 5 6 7 8
12	9	1 2 3 4 5 6 7 8
13	10	1 2 3 4 5 6 7 8
14	11	1 2 3 4 5 6 7 8
15	12	1 2 3 4 5 6 7 8 9
16	13	1 2 3 4 5 6 7 8 9
17	14	1 2 3 4 5 6 7 8 9
18	15	1 2 3 4 5 6 7 8 9 10
19	16	1 2 3 4 5 6 7 8 9 10
20	17	1 2 3 4 5 6 7 8 9 10
21	18	1 2 3 4 5 6 7 8 9 10 11
22	19	1 2 3 4 5 6 7 8 9 10 11
23	20	1 2 3 4 5 6 7 8 9 10 11
24	Timing verification	

Instructions for Players

- Before performing the MSST, complete a suitable warm-up including stretching, swimming, egg-beater kick, and turning.

- To obtain the best result in the test, keep to the pace of the beeps of the CD by gradually increasing your speed as the interval is reduced.

- "Racing" or attempting to get ahead of the beeps may be detrimental to your test result.

- This test requires a maximal effort from you and will be over when you fall behind the pace set by the beeps.

- The test supervisor will warn you if you do not make it to the end of a shuttle before the beep sounds and will remove you from the test if this occurs on two consecutive occasions.

- Your result is the number of the last level and shuttle called out before you stop or are removed from the test.

- The test starts with you treading water and touching the starting lane rope with your hand.

- A 5 s countdown will be given prior to a double beep signal to begin swimming.

- The aim is to swim for as long as possible back and forth across the 10 m distance, pacing each swim so that touching the lane rope at each end coincides with the beep.

- Once you touch the lane rope, immediately remove your hand so as not to move the lane rope away or down.

- If you arrive before the signal, turn around and remain stationary until you hear the beep and then recommence swimming.

- The starting speed will be quite slow. A triple beep will signal the beginning of a new level and a faster pace.

- You are not permitted to touch the boundaries of the pool or hang on to the lane ropes at any time during the test.

DATA ANALYSIS

The score reported to the coach and athlete should be in the format level/shuttle (e.g., 9/4 or 11/2).

These are not decimal scores, given the varying (and increasing) number of shuttles within each level. To convert scores to decimal values and enable statistical analysis, see table 33.5.

NORMATIVE DATA

Table 33.6 presents multistage shuttle swim test normative data for female water polo players.

TABLE 33.5 Decimal Conversion of Multistage Shuttle Swim Test Scores

Shuttle	LEVEL					
	1	2, 3, 4	5, 6, 7	8, 9, 10, 11	12, 13, 14	15, 16, 17
1	0.00	0.00	0.00	0.00	0.00	0.00
2	0.20	0.16	0.14	0.12	0.11	0.10
3	0.40	0.33	0.28	0.25	0.22	0.20
4	0.60	0.50	0.43	0.37	0.33	0.30
5	0.80	0.66	0.57	0.50	0.44	0.40
6		0.83	0.72	0.62	0.55	0.50
7			0.86	0.75	0.66	0.60
8				0.87	0.77	0.70
9					0.88	0.80
10						0.90

TABLE 33.6 Multistage Shuttle Swim Test Data for National Female Water Polo Players: Mean ± *SD* (Range)

Position	Level/shuttle	Distance, m
Center (*n* = 6)	10/1 ± 4 (8/3-11/2)	606 ± 85 (470-700)
Perimeter (*n* = 10)	10/6 ± 10 (8/3-12/5)	664 ± 105 (470-810)

Typical error: shuttles = 2.5; coefficient of variation = 4.5%. Data source: National senior women's squad, 2007-2008.

REPEAT SPRINT TEST

RATIONALE

The ability to repeat short, high-intensity efforts with minimal recovery has been identified as an important fitness component in intermittent team sports (Spencer et al. 2005). Analysis of international women's water polo competition revealed that a typical bout of repeated high-intensity activity (RHIA) consisted of approximately 4 efforts, ranging up to 10, with mean effort and recovery times of 5.1 and 10.2 s respectively (Polglaze et al. 2008). There were, on average, 8 RHIA bouts per player per game. Therefore, it is suggested that an assessment of repeat sprint ability specific to water polo is important to include in the routine field testing of elite water polo players.

TEST PROCEDURE

Setup for the repeat sprint test is illustrated in figure 33.3. The setup for the repeat sprint test is identical to the 10 m sprint test with the exceptions that only 0 m and 10 m markers are required and start ropes and operators are needed at each end.

- The in-water repeat sprint test requires the players to perform six maximal 10 m sprints departing every 17 s. Each sprint must be performed maximally from start to finish.

- Players should be fully warmed up, performing at least two maximal 10 m sprints, before commencing the test.

- The start position for each sprint is the same as for the 10 m maximal sprint test (i.e., head in contact with taut start rope).

- Two players can perform the test simultaneously, each starting at opposite ends. They should be in separate lanes to avoid colliding.

- With each player in position and the video recording, the test operator counts down "3, 2, 1, go" and starts a manual stopwatch to keep track of departure time.

- On "go," each athlete sprints maximally to, and past, the 10 m mark.

- Once the athlete departs, the start rope operators should hold their ropes up high enough for the incoming athlete to swim underneath.

- Once the athlete has passed the 10 m line, she can stop and recover. The start rope operator repositions the rope so that it is ready for the next sprint, and the athlete moves in front of the rope.

- The test operator should give a verbal warning with 5 s remaining, at which time the athlete assumes the ready position and awaits the "go" command to commence the next sprint. The test operator again counts down "3, 2, 1, go" for the commencement of the next sprint.

- This procedure continues until six repetitions have been completed. Departure times for the sprints are 0, 17, 34, 51, 68, and 85 s.

It is possible to prerecord the commands and countdowns in a digital format and play them back on an audio system to ensure consistent timing and clearly audible signals for the athletes.

DATA ANALYSIS

Timing procedures are the same as for the 10 m maximal sprint test. The test produces three scores: total time (s), deficit time (s), and relative decrement (%). The following provides an example of the methods used to calculate these parameters.

Repetition	Total time (s)
1	5.68
2	5.88
3	6.00
4	6.04
5	6.12
6	6.18

Total Time (s) = 35.90.

Ideal Time (s) = Best Time $\times$ 6

$$= 5.68 \times 6$$

$$= 34.08.$$

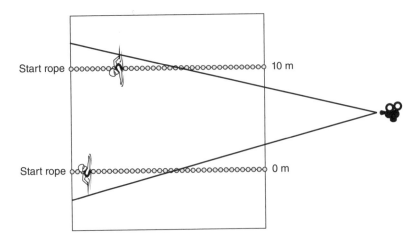

FIGURE 33.3 Setup for repeat sprint test.

$$\text{Deficit Time (s)} = \text{Total Time} - \text{Ideal Time}$$

$$= 35.90 - 34.08$$

$$= 1.82.$$

$$\text{Decrement (\%)} = (\text{Total / Ideal} \times 100) - 100$$

$$= (35.90 / 34.08 \times 100) - 100$$

$$= 5.3\%.$$

The criterion score for this test is total time (Tan et al. 2010). Calculations for decrement can help with the interpretation of the result, but are less reliable.

NORMATIVE DATA

Table 33.7 presents normative data for the repeat sprint test for female water polo players.

TABLE 33.7 Repeat Sprint Test Data for National Female Water Polo Players: Mean ± *SD* (Range)

Position	Total time, s	Deficit time, s	Decrement, %
Center (*n* = 6)	38.10 ± 1.23 (36.56-39.80)	1.54 ± 0.26 (1.16-1.92)	4.2 ± 0.7 (3.2-5.3)
Perimeter (*n* = 9)	36.95 ± 1.51 (35.66-40.28)	2.02 ± 0.64 (1.54-3.68)	5.8 ± 1.7 (4.3-10.0)

Typical error: total time = 0.43 s, coefficient of variation = 1.2%; deficit time = 0.53 s, coefficient of variation = 26.0%; decrement % = 1.5%; coefficient of variation = 27.2%. Data source: National senior women's squad, 2007-2008.

References

Helgerud, J., Engen, L.C., Wisloff, U., and Hoff, J. 2001. Aerobic endurance training improves soccer performance. *Medicine and Science in Sports and Exercise* 33(11):1925-1931.

McMahon, S., and Wenger, H. 2000. The relationship between aerobic fitness and both power output and subsequent recovery during maximal intermittent exercise. *Journal of Science & Medicine in Sport* 1(4):219-227.

Norton, K., Whittingham, N., Carter, L., Kerr, D., Gore, C., and Marfell-Jones, M. 1996. Measurement techniques in anthropometry. In: K. Norton and T. Olds (eds), *Anthropometrica*. Sydney: University of New South Wales Press, pp 25-73.

Platanou, T. 2005. On-water and dry land vertical jump in water polo players. *Journal of Sports Medicine and Physical Fitness* 45(1):26-31.

Polglaze, T. Rechichi, C., Tan, F., Hankin, S., and McFadden, G. 2008. The repeat high intensity activity characteristics of elite women's water polo. *Coaching and Sport Science Journal* 3(2):58.

Rechichi, C., Dawson, B., and Lawrence, S.R. 2000. A multistage shuttle swim test to assess aerobic fitness in competitive water polo players. *Journal of Science and Medicine in Sport* 3(1):55-64.

Rechichi, C., Lyttle, A., Doyle, M., and Polglaze, T. 2005. Swimming velocity patterns in elite women's water polo: a case study. *International Journal of Performance Analysis in Sport* 5(3):139-148.

Sanders, R. 1999. A model of kinematic variables determining height achieved in water polo boosts. *Journal of Applied Biomechanics* 15(3):270-283.

Spencer, M., Bishop, D., Dawson, B., and Goodman, C. 2005. Physiological and metabolic responses of repeated-sprint activities: specific to field-based team sports. *Sports Medicine* 35(12):1025-1044.

Tan, F., Polglaze, T., Dawson, B., and Cox, G. 2009a. Anthropometric and fitness characteristics of elite Australian female water polo players. *Journal of Strength and Conditioning Research* 23(5):1530-1536.

Tan, F., Polglaze, T., and Dawson, B. 2009b. Comparison of progressive maximal swimming tests in elite female water polo players. *International Journal of Sports Physiology and Performance* 4(2):206-217.

Tan, F., Polglaze, T., and Dawson, B. 2010. Reliability of an in-water repeated sprint test for water polo. *International Journal of Sports Physiology and Performance* 5(1):117-120.

Winget, C.M., DeRoshia, C.W., and Holley, D.C. 1985. Circadian rhythms and athletic performance. *Medicine and Science in Sports and Exercise* 17(5):498-516.

APPENDIX

GENERAL DATA **Sheets**

SPORT-SPECIFIC DATA **Sheets**

ANTHROPOMETRY DATA SHEET

Tester: _____   Date: _____

Sport: _____   Squad: _____

Skinfold caliper ID: _____

| Athlete | DOB | Height, cm | Body mass, kg | SKINFOLDS (MM) | | | | | | | |
				Triceps	Biceps	Subscap.	Supra.	Abdom.	Front thigh	Medial calf	Σ7

DOB = date of birth.

From Australian Institute of Sport, 2013, *Physiological tests for elite athletes*, 2nd ed. (Champaign, IL: Human Kinetics).

VERTICAL JUMP DATA SHEET

Tester: _____ Date: _____

Sport: _____ Squad: _____

Testing time: _____ Testing location: _____

Testing surface: _____ Temperature: _____

Humidity: _____

Jump device: _____

Jump protocol: _____

Athlete	DOB	Pole setting height, cm	Reach height, cm	Trial 1, cm	Trial 2, cm	Trial 3, cm

DOB = date of birth.

From Australian Institute of Sport, 2013, *Physiological tests for elite athletes*, 2nd ed. (Champaign, IL: Human Kinetics).

SPRINT TEST DATA SHEET

Tester: _____   Date: _____

Sport: _____   Squad: _____

Testing time: _____   Testing location: _____

Testing surface: _____   Temperature: _____

Humidity: _____ Wind speed (m · s⁻¹): _____ Wind direction: _____

Light gates: _____

Training phase

☐ Transition ☐ General preparation ☐ Specific preparation ☐ Competition ☐ Peak ☐ Rehabilitation

Athlete	DOB	TRIAL 1			TRIAL 2			TRIAL 3		
		__ m*	__ m	__ m	__ m	__ m	__ m	__ m	__ m	__ m

DOB = date of birth.

*Insert relevant sprint test split distances.

From Australian Institute of Sport, 2013, *Physiological tests for elite athletes*, 2nd ed. (Champaign, IL: Human Kinetics).

AGILITY TEST DATA SHEET

Tester: _____ Date: _____

Sport: _____ Squad: _____

Testing time: _____ Testing location: _____

Testing surface: _____ Temperature: _____

Humidity: _____ Wind speed (m · s⁻¹): _____ Wind direction: _____

Light gates: _____

Athlete	DOB	LEFT			RIGHT		
		Trial 1, s	Trial 2, s	Trial 3, s	Trial 1, s	Trial 2, s	Trial 3, s

DOB = date of birth.

From Australian Institute of Sport, 2013, *Physiological tests for elite athletes*, 2nd ed. (Champaign, IL: Human Kinetics).

REPEAT SPRINT ABILITY TEST DATA SHEET

Tester: _____   Date: _____

Sport: _____   Squad: _____

Testing time: _____   Testing location: _____

Testing surface: _____   Temperature: _____

Humidity: _____   Wind speed (m · s⁻¹): _____   Wind direction: _____

Light gates: _____

Athlete	DOB	Trial 1*, s	Trial 2, s	Trial 3, s	Trial 4, s	Trial 5, s	Trial 6, s	Trial 7, s	Trial 8, s	Trial 9, s	Trial 10, s	Trial 11, s	Trial 12, s	Total, s

DOB = date of birth.

*Complete form for number of relevant trials (AFL, hockey, rugby union = 6 × 30 m; football/soccer = 6 × 20 m; rugby league = 12 × 20 m; tennis = 10 × 20 m).

From Australian Institute of Sport, 2013, *Physiological tests for elite athletes*, 2nd ed. (Champaign, IL: Human Kinetics).

MULTISTAGE SHUTTLE RUN TEST DATA SHEET

Tester: _____ Date: _____

Sport: _____ Squad: _____

Testing time: _____ Testing location: _____

Testing surface: _____ Temperature: _____

Humidity: _____ Wind speed (m · s⁻¹): _____ Wind direction: _____

Training phase

☐ Transition ☐ General preparation ☐ Specific preparation ☐ Competition ☐ Peak ☐ Rehabilitation

Athlete	DOB	Level	Shuttle	Athlete	DOB	Level	Shuttle

DOB = date of birth.

From Australian Institute of Sport, 2013, *Physiological tests for elite athletes*, 2nd ed. (Champaign, IL: Human Kinetics).

YO-YO INTERMITTENT RECOVERY TEST DATA SHEET

Tester: _____ Date: _____

Sport: _____ Squad: _____

Testing time: _____ Testing location: _____

Testing surface: _____ Temperature: _____

Humidity: _____ Wind speed (m · s⁻¹): _____ Wind direction: _____

Training phase

☐ Transition ☐ General preparation ☐ Specific preparation ☐ Competition ☐ Peak ☐ Rehabilitation

Athlete	DOB	Level per shuttle	Total distance, m	Comments

DOB = date of birth.

From Australian Institute of Sport, 2013, *Physiological tests for elite athletes*, 2nd ed. (Champaign, IL: Human Kinetics).

1RM STRENGTH AND POWER TEST DATA SHEET

Tester: _____ Date: _____

Sport: _____ Squad: _____

Testing time: _____ Testing location: _____

Temperature: _____ Humidity: _____

| Athlete | DOB | Body mass, kg | 1RM, KG | | |
			Test: _____*	Test: _____	Test: _____

DOB = date of birth.

*Insert relevant test title.

From Australian Institute of Sport, 2013, *Physiological tests for elite athletes*, 2nd ed. (Champaign, IL: Human Kinetics).

3RM STRENGTH AND POWER TEST DATA SHEET

Tester: _____ Date: _____

Sport: _____ Squad: _____

Testing time: _____ Testing location: _____

Temperature: _____ Humidity: _____

| Athlete | DOB | TEST: _____ * | | | TEST: _____ | | | TEST: _____ | | |
		Attempt 1, kg	Attempt 2, kg	Attempt 3, kg	Attempt 1, kg	Attempt 2, kg	Attempt 3, kg	Attempt 1, kg	Attempt 2, kg	Attempt 3, kg
____										
____										
____										
____										
____										
____										
____										
____										
____										
____										
____										
____										
____										
____										
____										
____										
____										

DOB = date of birth.

*Insert relevant test title.

From Australian Institute of Sport, 2013, *Physiological tests for elite athletes*, 2nd ed. (Champaign, IL: Human Kinetics).

BASIC DATA SHEET

Tester: _____ Date: _____

Sport: _____ Squad: _____

Testing time: _____ Testing location: _____

Testing surface: _____ Temperature: _____

Humidity: _____

Equipment: _____

Athlete	DOB	Trial 1	Trial 2	Trial 3	Best score

DOB = date of birth.

From Australian Institute of Sport, 2013, *Physiological tests for elite athletes*, 2nd ed. (Champaign, IL: Human Kinetics).

TREADMILL TEST DATA SHEET

Athlete: _____

Date of birth: _____

Body mass: _____

Ergometer: _____

Tester: _____ Date: _____

Sport: _____ Squad: _____

Testing time: _____ Testing location: _____

Temperature: _____ Humidity: _____

Stage	Speed, km · h⁻¹	Duration, min	Stage time, min	Stride rate, steps · min⁻¹	HR, beats · min⁻¹	BLa, mmol · L⁻¹	RPE	V̇O₂, L · min⁻¹	RER	VE, L · min⁻¹	Respiratory rate, breaths · min⁻¹
1		0-1 min	1			x	x				
		1-2 min	2			x	x				
		2-3 min	3			x	x				
		3-4 min	4								
2		5-6 min	1			x	x				
		6-7 min	2			x	x				
		7-8 min	3			x	x				
		8-9 min	4								
3		10-11 min	1			x	x				
		11-12 min	2			x	x				
		12-13 min	3			x	x				
		13-14 min	4								
4		15-16 min	1			x	x				
		16-17 min	2			x	x				
		17-18 min	3			x	x				
		18-19 min	4								
5		20-21 min	1			x	x				
		21-22 min	2			x	x				
		22-23 min	3			x	x				
		23-24 min	4								
6		25-26 min	1			x	x				
		26-27 min	2			x	x				
		27-28 min	3			x	x				
		28-29 min	4								

Stage	Speed, km · h⁻¹	Duration, min	Stage time, min	Stride rate, steps · min⁻¹	HR, beats · min⁻¹	BLa, mmol · L⁻¹	RPE	$\dot{V}O_2$, L · min⁻¹	RER	VE, L · min⁻¹	Respiratory rate, breaths · min⁻¹
Start time/final submaximal 3 km											
0%		x	30 s								
0%		x	30 s								
0%		x	30 s								
0%		x	30 s								
0%		x	30 s								
0%		x	30 s								
0%		x	30 s								
0%		x	30 s								
0%		x	30 s								
0.50%		x	30 s								
1.00%		x	30 s								
1.50%		x	30 s								
2.00%		x	30 s								
2.50%		x	30 s								
3.00%		x	30 s								
3.50%		x	30 s								
x	x	x	x	x	3 min post		x	x	x	x	x

DOB = date of birth; x = data not collected.

From Australian Institute of Sport, 2013, *Physiological tests for elite athletes*, 2nd ed. (Champaign, IL: Human Kinetics).

MAXIMAL ACCUMULATED OXYGEN DEFICIT TEST DATA SHEET

Athlete: _____

Date of birth: _____

Body mass: _____

Ergometer: _____

Tester: _____ Date: _____

Sport: _____ Squad: _____

Testing time: _____ Testing location: _____

Temperature: _____ Humidity: _____

Submaximal stage	Speed, km · h⁻¹	Stage time, min	HR, beats · min⁻¹	BLa, mmol · L⁻¹	RPE
1		1		x	x
		2		x	x
		3		x	x
		4			
2		1		x	x
		2		x	x
		3		x	x
		4			
3		1		x	x
		2		x	x
		3		x	x
		4			
4		1		x	x
		2		x	x
		3		x	x
		4			
5		1		x	x
		2		x	x
		3		x	x
		4			

Maximal stage	Speed, km · h⁻¹	Gradient, %	Test time, min	HR, beats · min⁻¹	3 min post BLa, mmol · L⁻¹

DOB = date of birth; x = data not collected.

From Australian Institute of Sport, 2013, *Physiological tests for elite athletes*, 2nd ed. (Champaign, IL: Human Kinetics).

AFL
VERTICAL JUMP DATA SHEET

Tester: _____ Date: _____

Squad: _____ Testing time: _____

Testing location: _____ Testing surface: _____

Temperature: _____ Humidity: _____

Jump device: _____

Athlete	Pole setting height, cm	Reach height, cm	VERTICAL JUMP (CMJ WITH ARM SWING)			RUNNING VERTICAL JUMP, LEFT			RUNNING VERTICAL JUMP, RIGHT		
			Trial 1	Trial 2	Trial 3	Trial 1	Trial 2	Trial 3	Trial 1	Trial 2	Trial 3

CMJ = countermovement jump.

From Australian Institute of Sport, 2013, *Physiological tests for elite athletes,* 2nd ed. (Champaign, IL: Human Kinetics).

AFL
MULTISTAGE SHUTTLE RUN YO-YO IRT
AND 3 KM TIME TRIAL DATA SHEET

Tester: _____ Date: _____

Squad: _____ Testing time: _____

Testing location: _____ Testing surface: _____

Temperature: _____ Humidity: _____

Wind speed and direction: _____

Athlete	DOB	MSRT/Yo-Yo IRT score, level per shuttle	3 km time, min:s

DOB = date of birth; MSRT = multistage shuttle run test; Yo-Yo- IRT = Yo-Yo Intermittent Recovery Test.

From Australian Institute of Sport, 2013, *Physiological tests for elite athletes*, 2nd ed. (Champaign, IL: Human Kinetics).

AFL
STRENGTH AND POWER TEST DATA SHEET

Tester: _____ Date: _____

Squad: _____ Testing time: _____

Testing location: _____ Testing surface: _____

Humidity: _____

| Athlete | DOB | Body mass, kg | BENCH PRESS | | CHIN-UP | | SQUAT | SQUAT JUMP | | CMJ | |
			1RM, kg	3RM, kg	1RM, kg	3RM, kg	1RM, kg	BM	BM + 40	BM	BM + 40

BM = body mass (kg); CMJ = countermovement jump; DOB = date of birth; RM = repetition maximum.

From Australian Institute of Sport, 2013, *Physiological tests for elite athletes*, 2nd ed. (Champaign, IL: Human Kinetics).

BASKETBALL
FUNCTIONAL MOVEMENT SCREENING DATA SHEET

Athlete: _____

Tester: _____ Date: _____

Squad: _____ Testing time: _____

Testing location: _____

Functional movement screen	0	1	2	3
Overhead squat				
Hurdle step				
In-line lunge				
Shoulder mobility				
Active straight leg raise				
Trunk push-up				
Rotary stability test				
Total score				

From Australian Institute of Sport, 2013, *Physiological tests for elite athletes*, 2nd ed. (Champaign, IL: Human Kinetics).

CYCLING
SHORT GRADED EXERCISE TEST DATA SHEET: FEMALE SENIOR

Athlete: _____ Date of birth: _____

Body mass: _____ Ergometer: _____

Bike setup

Seat height: _____ Seat nose-BB: _____

Pedals: _____ Blood analyzer: _____

Operator: _____ Comments: _____

Tester: _____ Date: _____

Squad: _____ Testing time: _____

Testing location: _____ Temperature: _____ Humidity: _____

Time, min	Power, W	RPE (6-20)	Cadence, rpm	HR, beats · min⁻¹
0:00-0:30	125			
0:30-1:00				
1:00-1:30	150			
1:30-2:00				
2:00-2:30	175			
2:30-3:00				
3:00-3:30	200			
3:30-4:00				
4:00-4:30	225			
4:30-5:00				
5:00-5:30	250			
5:30-6:00				
6:00-6:30	275			
6:30-7:00				
7:00-7:30	300			
7:30-8:00				
8:00-8:30	325			
8:30-9:00				
9:00-9:30	350			
9:30-10:00				
10:00-10:30	375			
10:30-11:00				
11:00-11:30	400			
11:30-12:00				
12:00-12:30	425			
12:30-13:00				
13:00-13:30	450			
13:30-14:00				
14:00-14:30	475			
14:30-15:00				
Final values				
Post values	BLa, mmol · L⁻¹	pH	HCO₃⁻, mmol · L⁻¹	Glucose, mmol · L⁻¹
1 min				
2 min				
5 min				
7 min				

BLa = blood lactate; HR = heart rate; RPE = rating of perceived exertion.

From Australian Institute of Sport, 2013, *Physiological tests for elite athletes*, 2nd ed. (Champaign, IL: Human Kinetics).

CYCLING
SHORT GRADED EXERCISE TEST DATA SHEET: FEMALE JUNIOR

Athlete: _____ Date of birth: _____

Body mass: _____ Ergometer: _____

Bike setup

Seat height: _____ Seat nose-BB: _____

Pedals: _____ Blood analyzer: _____

Operator: _____ Comments: _____

Tester: _____ Date: _____

Squad: _____ Testing time: _____

Testing location: _____ Temperature: _____ Humidity: _____

Time, min	Power, W	RPE (6-20)	Cadence, rpm	HR, beats · min^{-1}
0:00-0:30	75			
0:30-1:00				
1:00-1:30	100			
1:30-2:00				
2:00-2:30	125			
2:30-3:00				
3:00-3:30	150			
3:30-4:00				
4:00-4:30	175			
4:30-5:00				
5:00-5:30	200			
5:30-6:00				
6:00-6:30	225			
6:30-7:00				
7:00-7:30	250			
7:30-8:00				
8:00-8:30	275			
8:30-9:00				
9:00-9:30	300			
9:30-10:00				
10:00-10:30	325			
10:30-11:00				
11:00-11:30	350			
11:30-12:00				
12:00-12:30	375			
12:30-13:00				
13:00-13:30	400			
13:30-14:00				
14:00-14:30	425			
14:30-15:00				
Final values				
Post values	BLa, mmol · L^{-1}	pH	HCO$_3^-$, mmol · L^{-1}	Glucose, mmol · L^{-1}
1 min				
2 min				
5 min				
7 min				

BLa = blood lactate; HR = heart rate; RPE = rating of perceived exertion.

From Australian Institute of Sport, 2013, *Physiological tests for elite athletes,* 2nd ed. (Champaign, IL: Human Kinetics).

CYCLING SHORT GRADED EXERCISE TEST DATA SHEET: MALE SENIOR

Athlete: _____ Date of birth: _____

Body mass: _____ Ergometer: _____

Bike setup

Seat height: _____ Seat nose-BB: _____

Pedals: _____ Blood analyzer: _____

Operator: _____ Comments: _____

Tester: _____ Date: _____

Squad: _____ Testing time: _____

Testing location: _____ Temperature: _____ Humidity: _____

Time, min	Power, W	RPE (6-20)	Cadence, rpm	HR, beats · min⁻¹
0:00-0:30	175			
0:30-1:00				
1:00-1:30	200			
1:30-2:00				
2:00-2:30	225			
2:30-3:00				
3:00-3:30	250			
3:30-4:00				
4:00-4:30	275			
4:30-5:00				
5:00-5:30	300			
5:30-6:00				
6:00-6:30	325			
6:30-7:00				
7:00-7:30	350			
7:30-8:00				
8:00-8:30	375			
8:30-9:00				
9:00-9:30	400			
9:30-10:00				
10:00-10:30	425			
10:30-11:00				
11:00-11:30	450			
11:30-12:00				
12:00-12:30	475			
12:30-13:00				
13:00-13:30	500			
13:30-14:00				
14:00-14:30	525			
14:30-15:00				
Final values				
Post values	BLa, mmol · L⁻¹	pH	HCO₃⁻, mmol · L⁻¹	Glucose, mmol · L⁻¹
1 min				
2 min				
5 min				
7 min				

BLa = blood lactate; HR = heart rate; RPE = rating of perceived exertion.

From Australian Institute of Sport, 2013, *Physiological tests for elite athletes*, 2nd ed. (Champaign, IL: Human Kinetics).

CYCLING
SHORT GRADED EXERCISE TEST DATA SHEET: MALE JUNIOR

Athlete: _____ Date of birth: _____

Body mass: _____ Ergometer: _____

Bike setup

Seat height: _____ Seat nose-BB: _____

Pedals: _____ Blood analyzer: _____

Operator: _____ Comments: _____

Tester: _____ Date: _____

Squad: _____ Testing time: _____

Testing location: _____ Temperature: _____ Humidity: _____

Time, min	Power, W	RPE (6-20)	Cadence, rpm	HR, beats · min⁻¹
0:00-0:30	150			
0:30-1:00				
1:00-1:30	175			
1:30-2:00				
2:00-2:30	200			
2:30-3:00				
3:00-3:30	225			
3:30-4:00				
4:00-4:30	250			
4:30-5:00				
5:00-5:30	275			
5:30-6:00				
6:00-6:30	300			
6:30-7:00				
7:00-7:30	325			
7:30-8:00				
8:00-8:30	350			
8:30-9:00				
9:00-9:30	375			
9:30-10:00				
10:00-10:30	400			
10:30-11:00				
11:00-11:30	425			
11:30-12:00				
12:00-12:30	450			
12:30-13:00				
13:00-13:30	475			
13:30-14:00				
14:00-14:30	500			
14:30-15:00				
Final values				
Post values	BLa, mmol · L⁻¹	pH	HCO₃⁻, mmol · L⁻¹	Glucose, mmol · L⁻¹
1 min				
2 min				
5 min				
7 min				

BLa = blood lactate; HR = heart rate; RPE = rating of perceived exertion.

From Australian Institute of Sport, 2013, *Physiological tests for elite athletes*, 2nd ed. (Champaign, IL: Human Kinetics).

CYCLING
LONG GRADED EXERCISE TEST DATA SHEET: FEMALE SENIOR

Athlete: _____   Date of birth: _____

Body mass: _____   Ergometer: _____

Bike setup

Seat height: _____   Seat nose-BB: _____

Pedals: _____   Blood analyzer: _____

Operator: _____   Comments: _____

Tester: _____   Date: _____

Squad: _____   Testing time: _____

Testing location: _____   Temperature: _____ Humidity: _____

Time	Power, W	RPE (6-20)	Cadence, rpm	HR, beats · min^{-1}
0:00-3:00	125			
3:00-6:00	150			
6:00-9:00	175			
9:00-12:00	200			
12:00-15:00	225			
15:00-18:00	250			
18:00-21:00	275			
21:00-24:00	300			
24:00-27:00	325			
27:00-30:00	350			
30:00-33:00	375			
33:00-36:00	400			
36:00-39:00	425			
Final values				
Post values	BLa, mmol · L^{-1}	pH	HCO$_3^-$, mmol · L^{-1}	Glucose, mmol · L^{-1}
1 min				
3 min				
6 min				
10 min				

BLa = blood lactate; HR = heart rate; RPE = rating of perceived exertion.

From Australian Institute of Sport, 2013, *Physiological tests for elite athletes*, 2nd ed. (Champaign, IL: Human Kinetics).

CYCLING
LONG GRADED EXERCISE TEST DATA SHEET:
FEMALE JUNIOR

Athlete: _____ Date of birth: _____

Body mass: _____ Ergometer: _____

Bike setup

Seat height: _____ Seat nose-BB: _____

Pedals: _____ Blood analyzer: _____

Operator: _____ Comments: _____

Tester: _____ Date: _____

Squad: _____ Testing time: _____

Testing location: _____ Temperature: _____ Humidity: _____

Time	Power, W	RPE (6-20)	Cadence, rpm	HR, beats · min⁻¹
0:00-3:00	75			
3:00-6:00	100			
6:00-9:00	125			
9:00-12:00	150			
12:00-15:00	175			
15:00-18:00	200			
18:00-21:00	225			
21:00-24:00	250			
24:00-27:00	275			
27:00-30:00	300			
30:00-33:00	325			
33:00-36:00	350			
36:00-39:00	375			
Final values				
Post values	BLa, mmol · L⁻¹	pH	HCO₃⁻, mmol · L⁻¹	Glucose, mmol · L⁻¹
1 min				
3 min				
6 min				
10 min				

BLa = blood lactate; HR = heart rate; RPE = rating of perceived exertion.

From Australian Institute of Sport, 2013, *Physiological tests for elite athletes*, 2nd ed. (Champaign, IL: Human Kinetics).

CYCLING
LONG GRADED EXERCISE TEST DATA SHEET: MALE SENIOR

Athlete: _____ Date of birth: _____

Body mass: _____ Ergometer: _____

Bike setup

Seat height: _____ Seat nose-BB: _____

Pedals: _____ Blood analyzer: _____

Operator: _____ Comments: _____

Tester: _____ Date: _____

Squad: _____ Testing time: _____

Testing location: _____ Temperature: _____ Humidity: _____

Time	Power, W	RPE (6-20)	Cadence, rpm	HR, beats · min^{-1}
0:00-5:00	150			
5:00-10:00	200			
10:00-15:00	250			
15:00-20:00	300			
20:00-25:00	350			
25:00-30:00	400			
30:00-35:00	450			
35:00-40:00	500			
40:00-45:00	550			
45:00-50:00	600			
50:00-55:00	650 W			
55:00-60:00	700 W			
Final values				
Post values	BLa, mmol · L^{-1}	pH	HCO$_3^-$, mmol · L^{-1}	Glucose, mmol · L^{-1}
1 min				
3 min				
6 min				
10 min				

BLa = blood lactate; HR = heart rate; RPE = rating of perceived exertion.

From Australian Institute of Sport, 2013, *Physiological tests for elite athletes*, 2nd ed. (Champaign, IL: Human Kinetics).

CYCLING
LONG GRADED EXERCISE TEST DATA SHEET: MALE JUNIOR

Athlete: _____ Date of birth: _____

Body mass: _____ Ergometer: _____

Bike setup

Seat height: _____ Seat nose-BB: _____

Pedals: _____ Blood analyzer: _____

Operator: _____ Comments: _____

Tester: _____ Date: _____

Squad: _____ Testing time: _____

Testing location: _____ Temperature: _____ Humidity: _____

Time	Power, W	RPE (6-20)	Cadence, rpm	HR, beats · min⁻¹
0:00-5:00	100			
5:00-10:00	150			
10:00-15:00	200			
15:00-20:00	250			
20:00-25:00	300			
25:00-30:00	350			
30:00-35:00	400			
35:00-40:00	450			
40:00-45:00	500			
45:00-50:00	550			
Final values				
Post values	BLa, mmol · L⁻¹	pH	HCO₃⁻, mmol · L⁻¹	Glucose, mmol · L⁻¹
1 min				
3 min				
6 min				
10 min				

BLa = blood lactate; HR = heart rate; RPE = rating of perceived exertion.

From Australian Institute of Sport, 2013, *Physiological tests for elite athletes*, 2nd ed. (Champaign, IL: Human Kinetics).

CYCLING
30 MIN TIME TRIAL DATA SHEET

Athlete: _____ Date of birth: _____

Body mass: _____ Ergometer: _____

Linear factor: _____

Bike setup

Seat height: _____ Seat nose-BB: _____

Pedals: _____ Blood analyzer: _____

Operator: _____ Comments: _____

Tester: _____ Date: _____

Squad: _____ Testing time: _____

Testing location: _____ Temperature: _____ Humidity: _____

Time, min	Power, W	HR, beats · min⁻¹	Cadence, rpm	RPE (6-20)	BLa, mmol · L⁻¹	pH	HCO₃, mmol · L⁻¹	Glucose, mmol · L⁻¹	V̇O₂, L · min⁻¹	RER
1			x	x	x	x	x	x		
2			x	x	x	x	x	x		
3			x	x	x	x	x	x		
4			x	x	x	x	x	x		
5										
6			x	x	x	x	x	x	x	x
7			x	x	x	x	x	x	x	x
8			x	x	x	x	x	x	x	x
9			x	x	x	x	x	x	x	x
10									x	x
11			x	x	x	x	x	x		
12			x	x	x	x	x	x		
13			x	x	x	x	x	x		
14			x	x	x	x	x	x		
15										
16			x	x	x	x	x	x	x	x
17			x	x	x	x	x	x	x	x
18			x	x	x	x	x	x	x	x
19			x	x	x	x	x	x	x	x
20									x	x
21			x	x	x	x	x	x		
22			x	x	x	x	x	x		
23			x	x	x	x	x	x		
24			x	x	x	x	x	x		
25										
26			x	x	x	x	x	x	x	x
27			x	x	x	x	x	x	x	x
28			x	x	x	x	x	x	x	x
29			x	x	x	x	x	x	x	x
30									x	x

BLa = blood lactate; HR = heart rate; RER = respiratory exchange ratio; RM = repetition maximum; RPE = rating of perceived exertion; x = data not collected.

From Australian Institute of Sport, 2013, *Physiological tests for elite athletes*, 2nd ed. (Champaign, IL: Human Kinetics).

CYCLING
POWER PROFILE DATA SHEET

Athlete: _____ Date of birth: _____

Body mass: _____ Ergometer: _____

#Vanes: _6 or 18_ _____

Tester: _____ Date: _____

Squad: _____ Testing time: _____

Testing location: _____ Temperature: _____ Humidity: _____

Time, min:s	Effort	Gear	Power, W	Cadence, rpm	HR, beats · min⁻¹	RPE (6-20)	BLa, mmol · L⁻¹	$\dot{V}O_2$, L · min⁻¹	Notes
0:00-5:00	Warm-up ~100 W	x	x	x	x	x	x	x	
5:00-5:03	Sprint (70%)	x	x	x	x	x	x	x	
5:03-5:30	Recovery	x	x	x	x	x	x	x	
5:30-5:33	Sprint (90%)	x	x	x	x	x	x	x	
2 min	Recovery (50-100 W)	x	x	x	x	x	x	x	
0:00-0:06	MMP$_{6s}$ small (5 Hz)					x	x	x	Standing start
1:00-1:06	MMP$_{6s}$ bigger (5 Hz)						x	x	Standing start
2:00									
3:00									
4:00-4:15	MMP$_{15s}$ (5 Hz)						x	x	Start from 70-80 rpm
5:00									
6:00	3:45 recovery (501-100 W)	x	x	x	x	x	x	x	
7:00									
8:00-8:30	MMP$_{30s}$ (5 Hz)						x	x	Start from 70-80 rpm
9:00									
10:00									
11:00	5:30 recovery (50-100 W)	x	x	x	x	x	x	x	
12:00									
13:00								x	
14:00	MMP$_{1min}$ (5 Hz)						x	x	Start from 70-80 rpm
15:00								x	

Time, min:s	Effort	Gear	Power, W	Cadence, rpm	HR, beats · min⁻¹	RPE (6-20)	BLa, mmol · L⁻¹	$\dot{V}O_2$, L · min⁻¹	Notes
16:00									
17:00	Download Red Box	x	x	x	x	x	x	x	
18:00	8:00 recovery (50-100 W)	x	x	x	x	x	x	x	
19:00									
20:00									
21:00	Fan on (450 Hz; 10 km · h⁻¹)	x	x	x	x	x	x	x	
22:00		x	x	x	x	x			
23:00	MMP$_{4min}$ (5 Hz)		x	x	x	x	x		Start from 70-80 rpm
24:00		x	x	x	x	x	x		
25:00		x	x	x	x	x	x		
26:00		x	x	x	x	x	x		
27:00		x							
28:00									
29:00									
30:00									
31:00									
32:00	10:00 recovery (50-100 W)	x	x	x	x	x	x	x	
33:00									
34:00									
35:00									
36:00									
37:00	MMP$_{10min}$ (5 Hz)		x	x	x	x	x		Start from 70-80 rpm
38:00		x	x	x	x	x	x		
39:00		x	x	x	x	x	x		
40:00		x	x	x	x	x	x		
41:00		x	x	x	x	x	x		
42:00		x	x	x	x	x	x		
43:00		x	x	x	x	x	x		
44:00		x	x	x	x	x	x		
45:00		x	x	x	x	x	x		
46:00		x	x	x	x	x	x		
47:00		x							
48:00		x	x	x	x	x		x	

BLa = blood lactate; HR = heart rate; MMP = maximal mean power; RPE = rating of perceived exertion; x = data not collected.

From Australian Institute of Sport, 2013, *Physiological tests for elite athletes*, 2nd ed. (Champaign, IL: Human Kinetics).

CYCLING
6 S POWER TEST DATA SHEET

Athlete: _____ Date of birth: _____

Body mass: _____ Ergometer: _____

Tester: _____ Date: _____

Squad: _____ Testing time: _____

Testing location: _____ Temperature: _____ Humidity: _____

Gear	Peak power, W
21	
19	

Comments:

BMX
6 × 3 S REPEAT SPRINT TEST DATA SHEET

Athlete: _____ Date of birth: _____

Body mass: _____ Ergometer: _____

Tester: _____ Date: _____

Squad: _____ Testing time: _____

Testing location: _____ Temperature: _____ Humidity: _____

Gear	Peak power, W
21	
19	
Stage	
1	
2	
3	
4	
5	
6	

From Australian Institute of Sport, 2013, *Physiological tests for elite athletes*, 2nd ed. (Champaign, IL: Human Kinetics).

ROWING
POWER PROFILE DATA SHEET

Tester: _____ Date: _____

Squad: _____ Testing time: _____

Testing location: _____ Temperature: _____ Humidity: _____

Barometric pressure: _____

Athlete	Gender, M/F	Category, L/H	Body mass, kg	Ergometer model ID	Drag factor	Distance, m	Time, min:s.0	Average 500 m pace, min:s	Power output, W	Average stroke rate, strokes · min⁻¹
						100				
						500				
						2,000				
						6,000				
						100				
						500				
						2,000				
						6,000				
						100				
						500				
						2,000				
						6,000				
						100				
						500				
						2,000				
						6,000				
						100				
						500				
						2,000				
						6,000				
						100				
						500				
						2,000				
						6,000				

From Australian Institute of Sport, 2013, *Physiological tests for elite athletes*, 2nd ed. (Champaign, IL: Human Kinetics).

ROWING
INCREMENTAL TEST DATA SHEET

Athlete: _____ Date of birth: _____

Body mass: _____ Ergometer: _____

Drug factor: _____

Tester: _____ Date: _____

Squad: _____ Testing time: _____

Testing location: _____ Temperature: _____ Humidity: _____

Barometric pressure: _____

Stage	Target power, W	Pace/500 m, min.s.0	Time, min	HR, beats · min⁻¹	Stroke rate, strokes · min⁻¹	Average power, W	BLa, mmol · L⁻¹	pH	RPE	V̇O₂, L · min⁻¹	RER
1			1				x	x	x	x	x
			2				x	x	x	x	x
			3				x	x	x	x	x
			4								
2			1				x	x	x	x	x
			2				x	x	x	x	x
			3				x	x	x	x	x
			4								
3			1				x	x	x	x	x
			2				x	x	x	x	x
			3				x	x	x	x	x
			4								
4			1				x	x	x	x	x
			2				x	x	x	x	x
			3				x	x	x	x	x
			4								
5			1				x	x	x	x	x
			2				x	x	x	x	x
			3				x	x	x	x	x
			4								
6			1				x	x	x	x	x
			2				x	x	x	x	x
			3				x	x	x	x	x
			4								
7			1				x	x	x	x	x
			2				x	x	x	x	x
			3				x	x	x	x	x
			4								
Post values	x	x	x	x	x	x	x	x	x	x	x
4 min	x	x	x	x	x	x			x	x	x

BLa = blood lactate; HR = heart rate; RER = respiratory exchange ratio; RPE = rating of perceived exertion; x = data not collected.

From Australian Institute of Sport, 2013, *Physiological tests for elite athletes*, 2nd ed. (Champaign, IL: Human Kinetics).

RUGBY UNION
PHYSICAL COMPETENCY SCREENING DATA SHEET

Tester: _____   Date: _____

Squad: _____   Testing time: _____

Testing location: _____   Temperature: _____ Humidity: _____

Athlete	DOB	BB OH SQUAT		A-LUNGE		BB RDL		SL SQUAT TO BENCH		PUSH-UP		PULL-UP		SL CALF RAISE		TRUNK BRIDGE	
		Pass	Fail	Pass	Fail	Pass	Fail	Pass	Fail	Pass	Fail	Pass	Fail	Pass	Fail	Pass	Fail

DOB = date of birth; BB OH = barbell overhead; BB RDL = barbell Romanian deadlift; SL = single leg.

From Australian Institute of Sport, 2013, *Physiological tests for elite athletes*, 2nd ed. (Champaign, IL: Human Kinetics).

SAILING
INDIVIDUAL ATHLETE TEST
DATA SHEET

Athlete: _____ Date of birth: _____

Body mass: _____ Boat class: _____

Tester: _____ Date: _____

Squad: _____ Testing time: _____

Testing location: _____ Temperature: _____ Humidity: _____

Strength Tests

Chin-ups, number	3RM squat, kg	3RM bench press, kg	3RM bench pull, kg	Best

Rowing 30 s Sprint Test

Distance, m	Average power, W	Average 500 m pace	Notes

Rowing O'Neill 4 min Test

Ergometer model		Drag factor setting*		Notes
Distance, m		Average 500 m pace		
RPE (6-20)	HR (final)	HR (1 min post)	HR (2 min post)	

HR = heart rate; RM = repetition maximum; RPE = rating of perceived exertion.

*Drag factor settings: Women ≤59 kg = 110. Women >59 kg = 120. Men ≤72 kg = 120. Men >72 kg = 130.

From Australian Institute of Sport, 2013, *Physiological tests for elite athletes*, 2nd ed. (Champaign, IL: Human Kinetics).

SPRINT KAYAK
7 × 4 MIN DATA SHEET

Athlete: _____ Date of birth: _____

Height: _____ Body mass: _____

Ergometer: _____ Drug factor: _30/40_____

Seat: _____ Footbar: _____

BLa analyzer #: _____

Tester: _____ Date: _____

Squad: _____ Testing time: _____

Testing location: _____ Temperature: _____ Humidity: _____

Stage	Desired workload, W	Screen workload, W	Power achieved, W	Distance, m	HR, beats · min⁻¹	Average stroke rate, strokes · min⁻¹	RPE, (6-20)	BLa, mmol · L⁻¹	Notes
1									
2									
3									
4									
5									
6									
7									
Max									

BLa = blood lactate; HR = heart rate; RPE = rating of perceived exertion.

From Australian Institute of Sport, 2013, *Physiological tests for elite athletes*, 2nd ed. (Champaign, IL: Human Kinetics).

SPRINT KAYAK
ON-WATER INCREMENTAL TEST
DATA SHEET

Water temperature: _____ Water depth: _____

Water currents: _____ Wind speed and direction: _____

Athlete: _____ Date of birth: _____

Height: _____ Body mass: _____

Kayak: _____ BLa analyzer #: _____

Tester: _____ Date: _____

Squad: _____ Testing time: _____

Testing location: _____ Temperature: _____ Humidity: _____

Stage	Desired speed, m · s⁻¹	Finish time, min:s	HR, beats · min⁻¹	Average stroke rate, strokes · min⁻¹	RPE (6-20)	BLa, mmol · L⁻¹	Notes
1							
2							
3							
4							
5							
6							
7							
Max							

BLa = blood lactate; HR = heart rate; RPE = rating of perceived exertion.

From Australian Institute of Sport, 2013, *Physiological tests for elite athletes*, 2nd ed. (Champaign, IL: Human Kinetics).

SWIMMING
5 × 200 M STEP TEST
DATA SHEET

Athlete: _____

Tester: _____   Date: _____

Squad: _____   Coach: _____

Testing time: _____   Testing location: _____

Stroke: _____

Swim	1	2	3	4	5
100 m, s					
200 m, min:s					
HR, beats · min^{-1}					
BLa, mmol · L^{-1}					
Stroke rate, strokes · min^{-1}					
Stroke count, strokes · 50 m^{-1}					
RPE					

BLa = blood lactate; HR = heart rate; RPE = rating of perceived exertion.

From Australian Institute of Sport, 2013, *Physiological tests for elite athletes*, 2nd ed. (Champaign, IL: Human Kinetics).

SWIMMING
8 × 100 M STEP TEST
DATA SHEET

Athlete: _____

Tester: _____ Date: _____

Squad: _____ Coach: _____

Testing time: _____ Testing location: _____

Stroke: _____

Swim	3 × 100 M			2 × 100 M		1 × 100 M	1 × 100 M	1 × 100 M
	1	2	3	4	5	6	7	8
100 m, min:s								
50 m, s								
HR, beats · min⁻¹								
BLa, mmol · L⁻¹								
Stroke rate, strokes · min⁻¹								
Stroke count, strokes · 50 m⁻¹								
RPE								

BLa = blood lactate; HR = heart rate; RPE = rating of perceived exertion.

From Australian Institute of Sport, 2013, *Physiological tests for elite athletes*, 2nd ed. (Champaign, IL: Human Kinetics).

SWIMMING
6 × 50 M STROKE MECHANICS
DATA SHEET

Athlete: _____

Tester: _____ Date: _____

Squad: _____ Coach: _____

Testing time: _____ Testing location: _____

Stroke: _____

Swim	1	2	3	4	5	
50 m, s						
15 m, s						
45 m, s						
Stroke rate, strokes · min^{-1}						
Stroke count, strokes · 50 m^{-1}						
Notes						

From Australian Institute of Sport, 2013, *Physiological tests for elite athletes*, 2nd ed. (Champaign, IL: Human Kinetics).

TRIATHLON
1 KM SWIM TIME TRIAL
DATA SHEET

Athlete: _____

Tester: _____ Date: _____

Squad: _____ Testing time: _____

Testing location: _____ Temperature: _____ Humidity: _____

Light gates: _____

	0-200 m	200-400 m	400-600 m	600-800 m	800-1,000 m
Time, m:ss.00					
HR, beats · min⁻¹					
Stroke rate, strokes · min⁻¹					
Stroke count, strokes · 50 m⁻¹					

HR = heart rate.

From Australian Institute of Sport, 2013, *Physiological tests for elite athletes*, 2nd ed. (Champaign, IL: Human Kinetics).

VOLLEYBALL
REPEATED-EFFORT TEST DATA SHEET

Tester: _____ Date: _____

Squad: _____ Testing time: _____

Testing location: _____ Testing surface: _____

Temperature: _____ Humidity: _____

Light gates: _____

Athlete	HR monitor #	Repetition	SPJ 1	Time	Errors	SPJ 2
		1				
		2				
		3				
		4				
		1				
		2				
		3				
		4				
		1				
		2				
		3				
		4				
		1				
		2				
		3				
		4				
		1				
		2				
		3				
		4				
		1				
		2				
		3				
		4				
		1				
		2				
		3				
		4				
		1				
		2				
		3				
		4				

HR = heart rate; SPJ = spike jump.

From Australian Institute of Sport, 2013, *Physiological tests for elite athletes*, 2nd ed. (Champaign, IL: Human Kinetics).

WATER POLO
TESTING DATA SHEET

Tester: _____ Date: _____

Squad: _____ Testing time: _____

Testing location: _____ Temperature: _____

Humidity: _____

Pole setting: _____

Water–deck difference: _____

Training phase

☐ Transition ☐ General preparation ☐ Specific preparation ☐ Competition ☐ Peak ☐ Rehabilitation

REPEAT SPRINT TEST				
Athlete	DOB	Test order	Partner	Cap color

IN-WATER VERTICAL JUMP AND MULTISTAGE SHUTTLE SWIM TESTS				
Athlete	DOB	Sprint order	Jump height	Level/shuttle

DOB = date of birth.

From Australian Institute of Sport, 2013, *Physiological tests for elite athletes*, 2nd ed. (Champaign, IL: Human Kinetics).

INDEX

PLEASE NOTE: Page numbers followed by an italicized *f*, or *t*, indicate that a figure or table will be found on those pages, respectively. Page numbers that are italicized indicate that the page is a form.

ABOUT THE EDITORS

Rebecca K. Tanner, BS, has been working as the Manager of the National Sport Science Quality Assurance (NSSQA) program at the Australian Institute of Sport (AIS) since January 2001. The NSSQA program delivers nationally a best practice program to support world class athlete services underpinning athlete performance, and operates quality assurance accreditation programs across sport science disciplines. Rebecca has significant experience in the development and implementation of strategies that address quality assurance challenges in high performance sport and processes to ensure the accuracy, reliability and comparability of athlete performance evaluations. She also has extensive experience in the validation and calibration of a range of physiology testing equipment, especially metabolic carts, and portable blood lactate analysers.

Prior to her position with the NSSQA program, Rebecca was employed as the National Sport Science Coordinator for Rowing. Her major responsibility in this position was to oversee the development of a national approach to the use of sport science by rowers, and to ensure that scientific input was clearly directed toward achieving performance gains. She also contributed substantially to several large applied research projects focused on rowing.

Rebecca has also been closely associated with the Department of Physiology at the AIS. In 1993, she held a one-year postgraduate scholarship awarded by the Department to enable young scientists to gain experience in working directly with elite sporting squads. Subsequent to this position she was appointed as a permanent member of the Physiology staff and worked as a physiologist supporting the AIS rowing program. Rebecca was also employed for a time as laboratory manager and played a major role in the development and implementation of quality control procedures for laboratory and field testing. She was primarily responsible for producing the extensive documentation needed for the physiology laboratory to achieve national accreditation.

Christopher J. Gore, PhD, FACSM, obtained his doctoral degree from Flinders University of South Australia in 1989. Since graduating, he has worked mostly for the Australian Institute of Sport (AIS), from 1992 until now. For the first 10 years at the AIS he established an Australia-wide, world-class quality assurance scheme for accreditation of laboratories that conduct the fitness and performance testing of Australia's elite athletes. Between 2001 and 2005 he was the manager for the AIS's collaboration with four universities and a micro-technology research organisation, the objective of which was to develop novel methods to monitor athletes in the field instead of the laboratory. For the last seven years he has been Head of Physiology, which has approximately 20 staff and 10 PhD scholars.

Christopher has more than 130 publications in peer-reviewed journals with about one-third of these is the area of hypoxia / altitude training.

In 2000, Christopher was awarded a Sport Australia Medal for his contribution to the team that developed a blood-based test for EPO that was implemented at the Sydney Olympic Games. In 2006, he received one of the 40 inaugural Distinguished Alumni awards from Flinders University (those 40 were selected from approximately 50,000 graduates). In 2007, he was awarded a Professorial fellowship at Flinders University, and an in 2009 was granted an adjunct Professorial position at the University of Canberra.

ABOUT THE AUSTRALIAN INSTITUTE OF SPORT

The **Australian Institute of Sport (AIS)** is the High Performance Division of the Australian Sports Commission and Australia's premier sport training institute. The AIS has gained an international reputation as a best-practice model for high-performance athlete development delivering outstanding results through a combination of skilled coaches, world-class facilities, and cutting-edge sport science and sports medicine services. The institute offers 36 sport programs in 26 sports, with hundreds of scholarships offered annually to Australia's finest athletes.

The headquarters of the AIS is located in Canberra, in the Australian Capital Territory. AIS programs are also located throughout Australia in Adelaide, Brisbane, Gold Coast, Melbourne, Perth, and Sydney as well as in Spain, Italy, and the United Kingdom.